Lecture Notes in Computer Science

Lecture Notes in Artificial Intelligence 15350

Founding Editor

Jörg Siekmann

The series Lecture Notes in Artificial Intelligence (LNAI) was established in 1988 as a topical subseries of LNCS devoted to artificial intelligence.

The series publishes state-of-the-art research results at a high level. As with the LNCS mother series, the mission of the series is to serve the international R & D community by providing an invaluable service, mainly focused on the publication of conference and workshop proceedings and postproceedings.

Sébastien Destercke · Maria Vanina Martinez ·
Giuseppe Sanfilippo
Editors

Scalable Uncertainty Management

16th International Conference, SUM 2024
Palermo, Italy, November 27–29, 2024
Proceedings

Editors
Sébastien Destercke
Université de Technologie de Compiègne
Compiègne, France

Maria Vanina Martinez
Artificial Intelligence Research Institute
Barcelona, Spain

Giuseppe Sanfilippo
University of Palermo
Palermo, Italy

ISSN 0302-9743 ISSN 1611-3349 (electronic)
Lecture Notes in Artificial Intelligence
ISBN 978-3-031-76234-5 ISBN 978-3-031-76235-2 (eBook)
https://doi.org/10.1007/978-3-031-76235-2

LNCS Sublibrary: SL7 – Artificial Intelligence

This Springer imprint is published by the registered company Springer Nature Switzerland AG
The registered company address is: Gewerbestrasse 11, 6330 Cham, Switzerland

Preface

These are the proceedings of the 16th International Conference on Scalable Uncertainty Management (SUM 2024), held during November 27–29, 2024, at University of Palermo, Italy. The SUM conferences were originally held annually. However, starting from 2020, they became biennial events, occurring every two years.

The SUM conferences aim to gather together researchers with a common interest in managing and analyzing imperfect information from a wide range of fields, such as artificial intelligence and machine learning, uncertain reasoning, databases, information retrieval and data mining, the semantic web, and risk analysis, with the goal of fostering collaboration and cross-fertilization of ideas from these different communities.

The first SUM conference was held in Washington DC in 2007. Since then, the SUM conferences have successively taken place in Naples in 2008, Washington DC in 2009, Toulouse in 2010, Dayton in 2011, Marburg in 2012, Washington DC in 2013, Oxford in 2014, Québec in 2015, Nice in 2016, Granada in 2017, Milan in 2018, Compiègne in 2019, Bolzano in 2020 (fully virtual), and Paris in 2022.

SUM 2024 solicited three types of paper submissions: Long papers reporting on original research or providing surveys that synthesize current research trends, short papers describing promising work in progress, systems, or positions on controversial issues, and extended abstracts (which were not included in the proceedings) in which participants present their recently published work or preliminary ideas to discuss with the community.

We received a reasonable number of submissions (43 in total), mostly from reputable research groups. Three to five reviewers from the Program Committee were assigned per paper, and only those papers with an overall positive evaluation were accepted, and papers with an overall positive evaluation having received at least one negative review were revised again before acceptance. As a result, 35 papers (28 as long and 7 as short papers) were included in the proceedings.

The conference also featured three invited talks by Meghyn Bienvenu (CNRS - University of Bordeaux), Lluis Godo (Artificial Intelligence Research Institute - IIIA, CSIC), and Barbara Vantaggi (University of Rome "La Sapienza"), as well as four tutorials by Andrea Capotorti (University of Perugia), Cassio de Campos (Eindhoven University of Technology), Niki Pfeifer (University of Regensburg), and Michele Tumminello (University of Palermo).

We would like to extend our gratitude to all authors, invited speakers, and tutorial speakers for their valuable contributions to the conference program, as well as to the members of the Program Committee and external reviewers for their constructive feedback. Special thanks go to the Local Organization and the SUM Steering Committee for their support in organizing the event. We also thank the head of the Department of Mathematics and Computer Science, University of Palermo, Roberto Livrea, and the administrative secretary, Cesare Ferrante, for their assistance in organizing the event.

We gratefully acknowledge the financial support provided by the University of Palermo, as well as the Department of Mathematics and Computer Science, for their contribution to both the organization and the venue of this event. We would like to express our gratitude for its financial support to the EurAI (European Association for Artificial Intelligence) and to the GNAMPA-INdAM Research group (Italian National Group for Mathematical Analysis, Probability and their Applications). We also acknowledge the support of the MUR PRIN 2022 project "Models for dynamic reasoning under partial knowledge to make interpretable decisions" (Project number: 2022AP3B3B, CUP: J53D23004340006) funded by the European Union – Next Generation EU.

We are grateful to Springer Nature for providing a grant of 500 euros for the best young researcher award of the conference and for the smooth collaboration when preparing the proceedings. Finally, EasyChair proved to be a convenient platform for handling submissions, reviewing, and receiving final papers for the proceedings of SUM 2024, which was greatly appreciated.

September 2024

Sébastien Destercke
Maria Vanina Martinez
Giuseppe Sanfilippo

Organization

Conference Chair

Giuseppe Sanfilippo	University of Palermo, Italy

Program Committee Chairs

Sébastien Destercke	Université de technologie de Compiègne, France
Maria Vanina Martinez	IIIA-CSIC, Spain

Steering Committee

Salem Benferhat	Artois University, France
Didier Dubois	IRIT-CNRS, France
Lluis Godo	IIIA-CSIC, Spain
Eyke Hüllermeier	Universität Paderborn, Germany
Anthony Hunter	University College London, UK
Henri Prade	IRIT-CNRS, France
Steven Schockaert	Cardiff University, UK
V. S. Subrahmanian	Northwestern University, USA

Program Committee

Alessandro Antonucci	Dalle Molle Institute for Artificial Intelligence, Switzerland
Vaishak Belle	University of Edinburgh, UK
Nahla Ben Amor	LARODEC, Institut Supérieur de Gestion de Tunis, Tunisia
Salem Benferhat	CRIL, CNRS UMR 8188, Université d'Artois, France
Leopoldo Bertossi	SKEMA Business School Canada Inc., Canada
Meghyn Bienvenu	LaBRI research lab, University of Bordeaux, France
Fernando Bobillo	University of Zaragoza, Spain

Imen Boukhris	LARODEC, Institut Supérieur de Gestion de Tunis, Tunisia
Andrea Campagner	University of Milano-Bicocca, Italy
Lydia Castronovo	University of Messina, Italy
Davide Ciucci	University of Milano-Bicocca, Italy
Fabio Cozman	University of São Paulo, Brazil
Cassio de Campos	Eindhoven University of Technology, Netherlands
Thierry Denoeux	HeuDiaSyc, Université de Technologie de Compiègne, France
Sylvie Doutre	IRIT, CNRS UMR 5505, Université Toulouse 1 Capitole, France
Florence Dupin de Saint-Cyr	IRIT, CNRS UMR 5505, Université Paul Sabatier, France
Saint-Cyr	IRIT, CNRS UMR 5505, Université Paul Sabatier, France
Zied Elouedi	LARODEC, Institut Supérieur de Gestion de Tunis, Tunisia
Tommaso Flaminio	Artificial Intelligence Research Institute, IIIA - CSIC, Spain
Angelo Gilio	University of Rome "La Sapienza", Italy
Lluis Godo	Artificial Intelligence Research Institute, IIIA - CSIC, Spain
Christophe Gonzales	LIS, Aix-Marseille Université, France
John Grant	Towson University, USA
Manuel Gómez-Olmedo	Universidad de Granada, Spain
Anthony Hunter	University College London, UK
Gabriele Kern-Isberner	Technische Universität Dortmund, Germany
Sébastien Konieczny	CRIL, CNRS UMR 8188, Université d'Artois, France
Eric Lefevre	LGI2A, Université d'Artois, France
Thomas Lukasiewicz	University of Oxford, UK
Francesca Mangili	Dalle Molle Institute for Artificial Intelligence, Switzerland
Silviu Maniu	Université Grenoble Alpes, France
Nicolas Maudet	LIP6, Sorbonne Université, France
Tommie Meyer	University of Cape Town and CAIR, South Africa
Enrique Miranda	University of Oviedo, Spain
Meltem Ozturk	LAMSADE, Université Paris Dauphine, France
Francesco Parisi	University of Calabria, Italy
Davide Petturiti	University of Perugia, Italy
Rafael Peñaloza	University of Milano-Bicocca, Italy
Frédéric Pichon	LGI2A, Université d'Artois, France
Nico Potyka	Cardiff University, UK

Henri Prade	IRIT, CNRS UMR 5505, Université Paul Sabatier, France
Kamal Premaratne	University of Miami, USA
Andrea Pugliese	University of Calabria, Italy
Benjamin Quost	HeuDiaSyc, Université de Technologie de Compiègne, France
Miguel Romero Orth	Universidad Adolfo Ibáñez, Chile
Kai Sauerwald	University of Hagen, Germany
Steven Schockaert	Cardiff University, UK
Pierre Senellart	ENS, PSL University, France
Umberto Straccia	ISTI-CNR, Italy
V. S. Subrahmanian	Northwestern University, USA
Marco Elio Tabacchi	University of Palermo, Italy
Karim Tabia	CRIL, CNRS UMR 8188, Université d'Artois, France
Andrea Tettamanzi	Université Côte d'Azur, France
Martin Theobald	University of Luxembourg, Luxembourg
Barbara Vantaggi	University of Rome "La Sapienza", Italy

Additional Reviewers

Filippone, Giuseppe
La Rosa, Gianmarco
Liu, Xiaolong
Mancini, Manuel

Local Committee

Giuseppina Castiglione	University of Palermo, Italy
Lydia Castronovo	University of Messina, Italy
Giuseppe Filippone	Univeristy of Palermo, Italy
Davide Petturiti	University of Perugia, Italy
Giuseppe Sanfilippo	Univeristy of Palermo, Italy
Marco Elio Tabacchi	University of Palermo, Italy

Contents

MIP Outer Belief Approximations of Lower Conditional Joint CDFs in Statistical Matching Problems

Marco Baioletti[1], Andrea Capotorti[1], Davide Petturiti[1(✉)], and Barbara Vantaggi[2]

[1] University of Perugia, Perugia, Italy
{marco.baioletti,andrea.capotorti,davide.petturiti}@unipg.it
[2] Sapienza University of Rome, Rome, Italy
barbara.vantaggi@uniroma1.it

Abstract. We propose a mixed integer programming (MIP) procedure to find an outer belief approximation of a lower conditional joint cumulative distribution function (lower conditional joint CDF) obtained by the statistical matching of several sources of information, given a common variable. We assume that the variables have finite supports and we provide a procedure based on the MIP technique that produces a sparse solution with at most a given finite number of focal elements, permitting to obtain an outer approximation with a conditional belief function. In turn, the family of sparse solutions given the common variable, allows us to efficiently perform coherent inferences on new items, relying on the generalized Bayesian conditioning rule. We finally show the effectiveness of the proposed approach in the domain of company fraud detection.

Keywords: Outer belief approximation · Mixed integer programming · Lower conditional joint CDF · Fraud detection

1 Introduction

The statistical matching problem (see, e.g., [12]) refers to a finite number $k \geq 2$ of sources giving rise to random vectors $(X, Y_1), \ldots, (X, Y_k)$, where X denotes a common variable. The input of the problem is a probabilistic assessment on the conditional random variables $Y_i|X$'s and the common random variable X. The conditional probability distributions of $Y_i|X$'s are estimated from each single source, while the marginal probability distribution of X is estimated simultaneously from all sources. The issue is to find a joint probability distribution for the random vector $(X, Y_1, \ldots, Y_k)$, which is compatible with the estimated distributions.

In general, the statistical matching problem gives rise to a class of compatible joint distributions for $(X, Y_1, \ldots, Y_k)$ or, more generally, of compatible full conditional probabilities on the related Boolean algebra (see [24]). Choosing a particular element of the class gives rise to a loss of information, so the

S. Destercke et al. (Eds.): SUM 2024, LNAI 15350, pp. 1–13, 2025.
https://doi.org/10.1007/978-3-031-76235-2_1

issue of non-uniqueness can be faced by working with either the whole class or its envelopes. The drawback of this approach is that inferential procedures become computationally hard, since the resulting lower envelope may fail 2-monotonicity (see [11]).

In this paper, we deal with statistical matching by addressing the computation of an outer approximation of the lower conditional joint CDFs through conditional belief functions. We follow the path already detailed in the recent work [20] and, in particular, we deal with the operational goal of finding the aforementioned outer belief approximation through a Mixed Integer Programming (MIP in the sequel) procedure. Such outer belief approximation will minimize the L_1 distance.

Other authors have recently tackled the problem of outer approximating a lower probability on a finite Boolean algebra (see e.g. [19]), by pointing out its complexity, even restricting to belief functions. In fact, even if the Dempster-Shafer theory [9,23] has been recognized as one of the best frameworks for handling uncertain and partial knowledge, one of the arguments often raised against this theory is its high computational complexity: a belief function requires an exponential size representation, in general.

To bypass such a problem, different restrictions, e.g. to 2-additive belief functions, have been proposed like in [20], or approximations through particular structures, e.g. hierarchical clustering as proposed in [10].

Here, by profiting from the particular structure of lattice of events involved in the evaluation of the lower conditional joint CDF and by adopting the L_1 distance to measure the "closeness" between evaluations, we are able to profit from MIP potentialities to let the focal sets being part of the solution. In this way, we skip the complex step of generating all possible focal sets. In fact the proposed procedure generates a polynomial number of variables and constraints with respect to the number N of the input values of the lower conditional joint CDF.

The motivation for the choice of the L_1 distance (that is also adopted in [19]), apart from the immediate benefit of giving a linear objective function, is in its direct interpretability since it obeys to the basic principle of minimal change in a numerical uncertainty setting. Moreover, L_1 distance is commonly used in Machine Learning and in Stochastic Programming to search for ambiguous sets of distributions close to a given nominal assessment not fully trusted (see e.g. [5,14,21]). Finally, in a finite setting like ours, L_1 distance is proportional to (it is actually the double) the widely adopted total variation measure (see e.g. Proposition 4.2 in [17]).

The rest of the paper is organized in the following way. In Sect. 2 we recall the necessary material on belief functions and conditioning. Section 3 describes the MIP procedure to find outer belief approximations minimizing the L_1 distance from lower conditional joint CDFs of variables expressed by different sources of information conditioned to a common one, in line with the statistical matching problem. In Sect. 4 we report a numerical application of our procedure to an example of matching information about fraudulent or not companies. A short conclusive Sect. 5 closes the contribution.

2 Preliminaries

Let $\mathcal{A}$ be a finite Boolean algebra of events, where $\bot$ and $\top$ stand for the impossible and sure events. Denote by $\vee$, $\wedge$, $\neg$, the disjunction, conjunction, and negation operations on $\mathcal{A}$. Furthermore, let $\leq$ be the partial order relation on $\mathcal{A}$ induced by $\vee$ and $\wedge$, which we interpret as implication, whose symmetric part is $=$. We also denote $\mathcal{A}^0 = \mathcal{A} \setminus \{\bot\}$. Recall that, since $\mathcal{A}$ is finite, it is completely determined by the set of **atoms**$(\mathcal{A}) = \{A_1, \ldots, A_d\}$ which is the largest subset of $\mathcal{A}^0$ such that $A_i \wedge A_j = \bot$, for all distinct i, j, and $\bigvee_{i=1}^d A_i = \top$.

A function $\varphi : \mathcal{A} \to [0,1]$ such that $\varphi(\bot) = 0$ and $\varphi(\top) = 1$ is said to be a:

probability measure: if $\varphi(A \vee B) = \varphi(A) + \varphi(B)$, for all $A, B \in \mathcal{A}$ with $A \wedge B = \bot$;
belief function: if, for every $k \geq 2$ and every $E_1, \ldots, E_k \in \mathcal{A}$,

$$\varphi\left(\bigvee_{i=1}^k E_i\right) \geq \sum_{\emptyset \neq I \subseteq \{1,\ldots,k\}} (-1)^{|I|+1} \varphi\left(\bigwedge_{i \in I} E_i\right);$$

2-monotone capacity: if $\varphi(A \vee B) \geq \varphi(A) + \varphi(B) - \varphi(A \wedge B)$, for all $A, B \in \mathcal{A}$;
(coherent) lower probability: if there exists a closed (in the product topology) set $\mathcal{P}$ of probability measures on $\mathcal{A}$ such that, for all $A \in \mathcal{A}$,

$$\varphi(A) = \min_{P \in \mathcal{P}} P(A).$$

In particular, a lower probability [25,26] is normally denoted as $\underline{P}$ and $\mathcal{P}$ is taken equal to the set

$$\mathcal{P} = \{P : P \text{ is a probability measure on } \mathcal{A},\ P \geq \underline{P}\},$$

called *core* [15] or *credal set* [16]. Furthermore, every function φ has an associated *dual* function ψ defined, for every $A \in \mathcal{A}$, as $\psi(A) = 1 - \varphi(\neg A)$.

In the rest of the paper, we will be mainly concerned with belief functions, whose theory has been essentially developed by Dempster and Shafer in [9,23]. A belief function is usually denoted as Bel and its dual, called *plausibility function*, as Pl. Every belief function is completely characterized by its *Möbius inverse* $m : \mathcal{A} \to [0,1]$, also called *basic mass assignment (bma)*, which is a non-negative function with $m(\bot) = 0$ and summing up to 1 such that, for all $A \in \mathcal{A}$,

$$Bel(A) = \sum_{\substack{B \leq A \\ B \in \mathcal{A}}} m(B).$$

The elements of $\mathcal{A}$ where m is positive are called *focal sets*.

In the rest of the paper, we refer to the notion of conditional probability due to Dubins [13]. A function $P : \mathcal{A} \times \mathcal{A}^0 \to [0,1]$ is a *full conditional probability on* $\mathcal{A}$ if it satisfies:

(i) $P(E|H) = P(E \wedge H|H)$, for every $E \in \mathcal{A}$ and $H \in \mathcal{A}^0$;
(ii) $P(\cdot|H)$ is a probability measure on $\mathcal{A}$, for every $H \in \mathcal{A}^0$;
(iii) $P(E \wedge F|H) = P(E|H) \cdot P(F|E \wedge H)$, for every $H, E \wedge H \in \mathcal{A}^0$ and $E, F \in \mathcal{A}$.

More generally, given a non-empty $\mathcal{G} \subseteq \mathcal{A} \times \mathcal{A}^0$, a function $P : \mathcal{G} \to [0,1]$ is a *coherent conditional probability* (see [8]) if there exists a full conditional probability Q on $\mathcal{A}$ such that $P(E|H) = Q(E|H)$, for all $E|H \in \mathcal{G}$.

This allows one to introduce the notion of *coherent lower conditional probability* due to Williams [27] as a function $\underline{P} : \mathcal{G} \to [0,1]$ such that there exists a closed (in the product topology) set $\mathcal{Q}$ of full conditional probabilities on $\mathcal{A}$ such that $\underline{P}(E|H) = \min_{Q \in \mathcal{Q}} Q(E|H)$, for all $E|H \in \mathcal{G}$.

As usual, we identify $P(E|\top) = P(E)$ and $\underline{P}(E|\top) = \underline{P}(E)$, for all $E \in \mathcal{A}$, and in this case we automatically obtain coherence for unconditional assessments.

Remark 1. All notions introduced so far adapt in a straightforward way if we take a field of sets in place of $\mathcal{A}$, switching to set-theoretic operations. Moreover, given a belief function Bel on $\mathcal{A}$ with dual plausibility function Pl, setting $\underline{P}(\cdot|\top) = Bel(\cdot)$, we get a coherent lower conditional probability. Next, taking as $\mathcal{Q}$ the largest set of full conditional probabilities on $\mathcal{A}$ dominating $\underline{P}$, we get that for all $E|H \in \mathcal{A} \times \mathcal{A}^0$ with $Bel(E \wedge H) + Pl(\neg E \wedge H) > 0$ we can compute the least committal coherent extension (which is a conditional belief function) applying a *generalized Bayesian conditioning rule* (see [7])

$$Bel(E|H) = \min_{Q \in \mathcal{Q}} Q(E|H) = \frac{Bel(E \wedge H)}{Bel(E \wedge H) + Pl(\neg E \wedge H)}. \tag{1}$$

3 MIP Implementation for Statistical Matching

We directly report the details of the MIP procedure adopted in this paper. For basic notions about statistical matching and outer belief approximation we refer to [20] and the related bibliography. For the MIP formulation, we follow the same approach already adopted for probability, conditional or not, and statistical matching corrections as detailed in [1–4].

Let $(Y_1, \ldots, Y_k)$ be a vector of k random variables with finite supports. For each $i = 1, \ldots, k$, let the support of Y_i be $D_{Y_i} = \{y_{i.1}, \ldots, y_{i.n_i}\}$ where $n_i = |D_{Y_i}|$. These values are assumed to be ordered increasingly: $y_{i.1} < y_{i.2} < \cdots < y_{i.n_i}$.

The set of all possible combinations of values for $(Y_1, \ldots, Y_k)$ is denoted by

$$\mathbf{C} = \bigtimes_{i=1}^{k} D_{Y_i} \tag{2}$$

and is composed by $N = \prod_{i=1}^{k} n_i$ elements $c_1, \ldots, c_N$. The generic element c_h of $\mathbf{C}$ can be interpreted as a value assignment for the variables in $(Y_1, \ldots, Y_k)$ by setting $Y_i = c_h(i) \in D_{Y_i}$, for $i = 1, \ldots, k$ and it is in one-to-one correspondence with the mapping

$$c_h^* : \mathbf{C} \to \bigtimes_{i=1}^{k} \{1, \ldots, n_i\} \tag{3}$$

defined by

$$c_h^*(i) = l \Longleftrightarrow c_h(i) = y_{i.l}, \quad i = 1, \ldots, k. \tag{4}$$

Let X be another random variable, whose support is $D_X = \{x_1, \ldots, x_m\}$.

We assume that the variables in the global vector $(X, Y_1, \ldots, Y_k)$ are not linked by logical relations. This means that

$$\mathbf{atoms}(\mathcal{A}) = \left\{ (X = x_t) \wedge \bigwedge_{i=1}^{k} (Y_i = c_h(i)) \ : \ x_t \in D_X,\, c_h \in \mathbf{C} \right\}$$

is the set of atoms of a finite Boolean algebra $\mathcal{A}$, which is isomorphic to the field of sets $2^{D_X \times \mathbf{C}}$. We also notice that

$$\mathbf{atoms}(\mathcal{B}) = \left\{ \bigwedge_{i=1}^{k} (Y_i = c_h(i)) \ : \ c_h \in \mathbf{C} \right\},$$

is the set of atoms of a Boolean sub-algebra $\mathcal{B}$ of $\mathcal{A}$, which is isomorphic to the field of sets $2^{\mathbf{C}}$. Instead, the matching problem under logical relations between the involved variables has been considered in [6,12,20,24].

For a given value $x_t \in D_X$, where $t \in \{1, \ldots, m\}$, for each $i = 1, \ldots, k$ and $l = 1, \ldots, n_i$ let $\pi_{i.l}^{x_t}$ denote the conditional probability value $P(Y_i = y_{i.l} | X = x_t)$.

In the same context, let $F_i^{x_t}(y_{i.l})$ denote the marginal cumulative conditional probability value

$$F_i^{x_t}(y_{i.l}) = \sum_{s=1}^{l} \pi_{i,s}^{x_t} = P(Y_i \leq y_{i.l} | X = x_t). \tag{5}$$

Given a value assignment $c_h \in \mathbf{C}$, let

$$\underline{\mathrm{F}}^{x_t}(c_h) = T_L(F_1^{x_t}(c_h(1)), \ldots, F_k^{x_t}(c_h(k))) \tag{6}$$

where T_L is the Lukasiewicz's t-norm in its k-ary definition

$$T_L(x_1, \ldots, x_k) = \max\left\{ 0, \sum_{i=1}^{k} x_i - (k-1) \right\}. \tag{7}$$

For each $x_t \in D_X$, the function $c_h \mapsto \underline{\mathrm{F}}^{x_t}(c_h)$ determines an assessment on the family of conditional events $\{E_{c_h} | X = x_y \ : \ c_h \in \mathbf{C}\}$, where

$$E_{c_h} = \bigwedge_{i=1}^{k} (Y_i \leq c_h(i)). \tag{8}$$

As is well-known (see, e.g., [22]), when $k > 2$, the quoted assessment is generally not a coherent conditional probability but is only guaranteed to be a coherent lower conditional probability. For this reason, $\underline{\mathrm{F}}^{x_t}$ is dubbed a *lower conditional joint cumulative distribution function (lower conditional joint CDF).*

The problem we face here is to find a conditional belief function Bel on $\mathcal{B}$ given $X = x_t$ such that for all $t = 1, \ldots, m$ and for all $h = 1, \ldots, N$

$$Bel(E_{c_h}|X = x_t) \leq \underline{\mathrm{F}}^{x_t}(c_h). \tag{9}$$

We point out that in computing outer approximations (see, e.g., [19,20]) we should require the above inequality to hold for all $E \in \mathcal{B}$. In turn, to impose such a set of constraints, one needs explicitly to build the lower envelope of the set of coherent probabilities given $X = x_t$, compatible with the marginal information coming from the different sources. This problem is difficult per se, especially when we have more than 2 sources. For this reason, in this paper we require dominance to hold, only on the events of type (8).

We also notice that the events of type (8) form a lattice with respect to the partial order $\leq$ on $\mathcal{B}$. This last fact is particularly relevant for the MIP formulation presented below since it is used when constructing the basic mass assignment of an outer approximation.

Fixing a value $x_t \in D_X$, we solve the following MIP problem to find Bel, using the fact that it is possible to find a sparse solution with "only" at most $N+1$ focal elements $\Phi_1^{x_t}, \ldots, \Phi_{N+1}^{x_t}$. Due to the isomorphism between $\mathcal{B}$ and $2^{\mathbf{C}}$, each focal element $\Phi_i^{x_t}$ can be identified with a subset of $\mathbf{C}$, hence with a union of some vectors $c_h \in \mathbf{C}$.

We employ two sets of binary variables a_{ijl} and e_{ijl}, for $i = 1, \ldots, N+1$ (i.e., i varies in the index set of the focal elements $\Phi_i^{x_t}$), $j = 1, \ldots, k$ (i.e., j varies in the index set of the variables Y_j) and $l = 1, \ldots, n_j$ (i.e., l varies in the index set of each j-th variable modality $y_{j.l}$). The definition of a_{ijl} is that $a_{ijl} = 1$ if and only if there is at least one $c_h \in \Phi_i^{x_t}$, such that $c_h(j) = y_{j.l}$. The definition of e_{ijl} is that $e_{ijl} = 1$ if and only if $\sum_{s=l+1}^{n_j} a_{ijs} = 0$, which is equivalent to the condition that for all $c_h \in \Phi_i^{x_t}$, $c_h(j) \leq y_{j.l}$.

The constraints for a_{ijl} are

$$\sum_{l=1}^{n_i} a_{ijl} \geq 1 \tag{10}$$

for $i = 1, \ldots, N+1$ and $j = 1, \ldots, k$, which corresponds to the fact that each focal element $\Phi_i^{x_t}$ contains at least one combination of values for all the variables in the random vector $(Y_1, \ldots, Y_k)$.

The constraints for e_{ijl} are

$$(n_j - l)e_{ijl} \leq (n_j - l) - \sum_{s=l+1}^{n_j} a_{ijs} \tag{11}$$

and

$$e_{ijl} \geq 1 - \sum_{s=l+1}^{n_j} a_{ijs} \tag{12}$$

for $i = 1, \dots, N+1$, $j = 1, \dots, k$ and $l = 1, \dots, n_j$.

Let $q_1, \dots, q_{N+1}$ be a set of non-negative real variables, with the constraint

$$\sum_{i=1}^{N+1} q_i = 1. \tag{13}$$

The value q_i corresponds to the basic mass assignment of $\Phi_i^{x_t}$, for $i = 1, \dots, N+1$.

Let $b_{i,h}$, for $i = 1, \dots, N+1$ and $h = 1, \dots, N$, be a second set of non-negative real variables, with the following constraints

$$b_{ih} \leq e_{ijc_h^*(j)} \text{ for all } j = 1, \dots, k \tag{14}$$

and

$$\sum_{j=1}^{k} e_{ijc_h^*(j)} - k + q_i \leq b_{ih} \leq q_i \tag{15}$$

for $i = 1, \dots, N+1$ and $h = 1, \dots, N$.

It is possible to prove that

$$b_{ih} = q_i \cdot (e_{i1c_h^*(1)} \wedge e_{i2c_h^*(2)} \wedge \cdots \wedge e_{ikc_h^*(k)}).$$

In fact, if $e_{i1c_h^*(1)} = e_{i2c_h^*(2)} = \cdots = e_{ikc_h^*(k)} = 1$, all the assignments c_h's in $\Phi_i^{x_t}$ imply the event E_h, and then $b_{ih} = q_i$. Otherwise, $b_{ih} = 0$.

The constraint on the belief function values is therefore

$$\sum_{i=1}^{N+1} b_{ih} + r_h = \underline{\mathrm{F}}^{x_t}(c_h) \tag{16}$$

for all $h = 1, \dots, N$, so that the objective function to minimize is

$$\sum_{h=1}^{N} r_h \tag{17}$$

which corresponds to minimizing the L_1 distance.

Proceeding as in [20], once we get the entire family of outer approximating conditional beliefs $\{Bel(\cdot|X = x_t) : x_t \in D_X\}$ we can compute a joint belief function for the whole vector $(X, Y_1, \dots, Y_k)$ relying on the marginal probability distribution of X. For every $A \in 2^{D_X \times \mathbf{C}}$, denoting by $[A]_{x_t} = \{(x_t, c_h) \in A : c_h \in \mathbf{C}\}$, we set

$$Bel((X, Y_1, \dots, Y_k) \in A) = \sum_{t=1}^{m} Bel((Y_1, \dots, Y_k) \in [A]_{x_t} | X = x_t) P(X = x_t). \tag{18}$$

In turn, inferences can be computed relying on the generalized Bayesian conditioning rule, recalled in Eq. (1).

We can synthesize our method as follows:

- Acquisition of the marginal conditional cumulative distributions from the different sources of information (Eq. (5));
- Computation of joint lower conditional distribution via the Lukasiewicz's t-norm (Eq. (3));
- Implementation and solution of the MIP problem with constraints (10)–(16) and objective function (17);
- Extrapolation from the solution of the MIP problem of the focal elements $\Phi_1^{x_t}, \ldots, \Phi_{N+1}^{x_t}$, obtained through the binary values a_{ijl}, and their associate basic mass assignments q_i;
- Inference on any conditional event of interest through the generalized Bayes formula (Eq. (1)).

4 Application to Fraud Detection

We consider the dataset described in [18] about 286 Chinese companies related to financial anomalies or not during the fiscal year 2020. The different sources of information are identified with the different financial and operational disclosures, one for each feature reported in the dataset. In order to explicitly show all the steps of the methodology, we limit to variables whose support has two elements.

We take as common variable X the indicator of fraudulent activity, hence X is a binary variable with $D_X = \{0, 1\}$. As explanatory variables, we choose the first three variables Y_1, Y_2, Y_3 with highest absolute correlation with X, specifically $Y_1 =$ "Finanomalies142", $Y_2 =$ "Leverage9", both belonging to the "Operation" category, while $Y_3 =$ "Finanomalies122", that belongs to the "Profitability" category. These three features result also among those with the highest Information Values IV (see [18] for more details). Note that the dataset is perfectly balanced with respect to the classification label X since each fraudulent company has been paired with a healthy one from the same industry class of a similar scale and business pattern. Hence we have $P(X = 0) = P(X = 1) = 0.5$.

Fixing one specific value $X = x_t$, we discretize Y_1, Y_2, Y_3 into the classes $y_{i.1}^{x_t} = [0, Q3_i^{x_t}]$, $y_{i2}^{x_t} = (Q3_i^{x_t}, Max_i^{x_t}]$, with $Q3_i^{x_t}$ and $Max_i^{x_t}$ the third quartile and maximum, respectively, of each $Y_i|X = x_t$, $i = 1, 2, 3$, in the original dataset. We chose these values on the one hand to limit the zero values stemming from the application of Lukasiewicz's t-norm (6), and, on the other hand, to refer to some significant threshold.

By a bootstrap simulation, with 100 resamplings, we obtained for $X = 0$ the conditional marginal CDFs reported in Table 1: the problem has been studied through a simple technique in order to focus on statistical matching.

Hence, conditioned to each event $X = x_t$, we have the 8-value lower conditional joint CDFs in Table 2 stemming from the Lukasiewicz's t-norm (6).

By applying the MIP correction procedure described in Sect. 3, we obtained two outer approximations of type (9), that is $Bel(E_h|X = 0)$ and $Bel(E_h|X = 1)$. The first conditional belief function has 6 focal elements (see Table 3) while the second with 5 focal elements (see Table 4). Such focal elements permit on the one hand to reconstruct the outer approximations (see Table 5 for the numeric

Table 1. Conditional marginal CDFs for $X = 0$ and $X = 1$.

	$y^0_{i.1}$	$y^0_{i.2}$
F^0_1	0.6363636	1.0
F^0_2	0.6713287	1.0
F^0_3	0.6713287	1.0

(a) $X = 0$

	$y^1_{i.1}$	$y^1_{i.2}$
F^1_1	0.6643357	1.0
F^1_2	0.6363636	1.0
F^1_3	0.6643357	1.0

(b) $X = 1$

Table 2. Lower conditional joint CDFs for the Fraud Detection example obtained through Lukasiewicz's t-norm (6) of values in Table 1.

$X = x_t$	$X = 0$	$X = 1$
$\underline{F}^{x_t}(y^{x_t}_{1.1}, y^{x_t}_{2.1}, y^{x_t}_{3.1})$	0.0	0.0
$\underline{F}^{x_t}(y^{x_t}_{1.1}, y^{x_t}_{2.1}, y^{x_t}_{3.2})$	0.3076923	0.3006993
$\underline{F}^{x_t}(y^{x_t}_{1.1}, y^{x_t}_{2.2}, y^{x_t}_{3.1})$	0.3076923	0.3286714
$\underline{F}^{x_t}(y^{x_t}_{1.1}, y^{x_t}_{2.2}, y^{x_t}_{3.2})$	0.6363636	0.6643357
$\underline{F}^{x_t}(y^{x_t}_{1.2}, y^{x_t}_{2.1}, y^{x_t}_{3.1})$	0.3426574	0.3006993
$\underline{F}^{x_t}(y^{x_t}_{1.2}, y^{x_t}_{2.1}, y^{x_t}_{3.2})$	0.6713287	0.6363636
$\underline{F}^{x_t}(y^{x_t}_{1.2}, y^{x_t}_{2.2}, y^{x_t}_{3.1})$	0.6713287	0.6643357
$\underline{F}^{x_t}(y^{x_t}_{1.2}, y^{x_t}_{2.2}, y^{x_t}_{3.2})$	1.0	1.0

values and the associated corrections), and on the other to perform any inference step as described in Sect. 2 (see also [20]).

About the approximation, note that $Bel(E_h|X = 0)$ has a total L_1 distance of 0.020979 from $\underline{F}^0$ and changes values on the two events $E_{y_{1.1}y_{2.2}y_{3.2}}|X = 0$ and $E_{y_{1.2}y_{2.2}y_{3.1}}|X = 0$, while $Bel(E_h|X = 1)$ has a total L_1 distance of 0.034965 from $\underline{F}^1$ and changes only the value on $E_{y_{1.1}y_{2.2}y_{3.2}}|X = 1$.

About inference, if, e.g., we observe a new factory with a value of Y_1 and Y_3 below both their empirical third-quartiles (i.e., $Y_1 \leq \min(y^0_{1.1}, y^1_{1.1})$ and $Y_3 \leq \min(y^0_{3.1}, y^1_{3.1})$), while we don't have any information about Y_2 (hence $Y_2 \leq \max(y^0_{2.2}, y^1_{2.2})$) we can infer

$$Bel(X = 0|Y_1 \leq \min(y^0_{1.1}, y^1_{1.1}), Y_2 \leq \max(y^0_{2.2}, y^1_{2.2}), Y_3 \leq \min(y^0_{3.1}, y^1_{3.1}))$$

$$= \frac{Bel(E_{y^0_{1.1}y^0_{2.2}y^0_{3.1}}|X=0)P(X=0)}{Bel(E_{y^0_{1.1}y^0_{2.2}y^0_{3.1}}|X=0)P(X=0)+Pl(E_{y^1_{1.1}y^1_{2.2}y^1_{3.1}}|X=1)P(X=1)}$$

$$= \frac{0.3076923 \cdot 0.5}{0.3076923 \cdot 0.5 + 1.0 \cdot 0.5} = 0.2352941,$$

Table 3. Focal elements and their associated basic mass assignments (bma) q_i for $Bel(E_h|X = 0)$.

Φ_i^0	bma q_i
$(y_{1.1}^0, y_{2.1}^0, y_{3.1}^0) \cup (y_{1.2}^0, y_{2.1}^0, y_{3.1}^0)$	0.3426574
$(y_{1.1}^0, y_{2.1}^0, y_{3.1}^0) \cup (y_{1.1}^0, y_{2.1}^0, y_{3.2}^0)$	0.3076923
$(y_{1.1}^0, y_{2.1}^0, y_{3.1}^0) \cup (y_{1.1}^0, y_{2.2}^0, y_{3.1}^0)$	0.3076923
$(y_{1.1}^0, y_{21}^0, y_{3.1}^0) \cup (y_{1.1}^0, y_{2.1}^0, y_{3.2}^0) \cup (y_{1.2}^0, y_{2.1}^0, y_{3.1}^0) \cup (y_{1.2}^0, y_{2.1}^0, y_{3.2}^0)$	0.020979
$(y_{1.1}^0, y_{2.1}^0, y_{3.1}^0) \cup (y_{1.1}^0, y_{2.1}^0, y_{3.2}^0) \cup (y_{1.1}^0, y_{2.2}^0, y_{3.1}^0) \cup (y_{1.1}^0, y_{2.2}^0, y_{3.2}^0)$	0.013986
$(y_{1.1}^0, y_{2.1}^0, y_{3.1}^0) \cup (y_{1.1}^0, y_{2.2}^0, y_{3.1}^0) \cup (y_{1.2}^0, y_{2.1}^0, y_{3.1}^0) \cup (y_{1.2}^0, y_{2.2}^0, y_{3.1}^0)$	0.006993

Table 4. Focal elements and their associated basic mass assignments (bma) q_i for $Bel(E_h|X = 1)$.

Φ_i^1	bma q_i
$(y_{1.1}^1, y_{2.1}^1, y_{3.1}^1) \cup (y_{1.2}^1, y_{2.1}^1, y_{3.1}^1)$	0.3006993
$(y_{1.1}^1, y_{2.1}^1, y_{3.1}^1) \cup (y_{1.1}^1, y_{2.1}^1, y_{3.2}^1)$	0.3006993
$(y_{11}^1, y_{2.1}^1, y_{3.1}^1) \cup (y_{1.1}^1, y_{2.2}^1, y_{3.1}^1)$	0.3286714
$(y_{1.1}^1, y_{2.1}^1, y_{3.1}^1) \cup (y_{1.1}^1, y_{2.1}^1, y_{3.2}^1) \cup (y_{1.2}^1, y_{2.1}^1, y_{3.1}^1) \cup (y_{1.2}^1, y_{2.1}^1, y_{3.2}^1)$	0.03496498
$(y_{1.1}^1, y_{2.1}^1, y_{3.1}^1) \cup (y_{1.1}^1, y_{2.2}^1, y_{3.1}^1) \cup (y_{1.2}^1, y_{2.1}^1, y_{3.1}^1) \cup (y_{1.2}^1, y_{2.2}^1, y_{3.1}^1)$	0.03496498

Table 5. Outer approximations $Bel(E_h|X = 0)$ and $Bel(E_h|X = 1)$ and their associated corrections r_i's with respect to $\underline{F}^0(c_h)$ and $\underline{F}^1(c_h)$.

$X = x_t$	$X = 0$	$X = 1$
$Bel(E_{y_{1.1}^{x_t} y_{2.1}^{x_t} y_{3.1}^{x_t}}|X = x_t)$	0.0	0.0
r_1	0.0	0.0
$Bel(E_{y_{1.1}^{x_t} y_{2.1}^{x_t} y_{3.2}^{x_t}}|X = x_t)$	0.3076923	0.3006993
r_2	0.0	0.0
$Bel(E_{y_{1.1}^{x_t} y_{2.2}^{x_t} y_{3.1}^{x_t}}|X = x_t)$	0.3076923	0.3286714
r_3	0.0	0.0
$Bel(E_{y_{1.1}^{x_t} y_{2.2}^{x_t} y_{3.2}^{x_t}}|X = x_t)$	0.6293706	0.6293707
r_4	0.006993	0.034965
$Bel(E_{y_{1.2}^{x_t} y_{2.1}^{x_t} y_{3.1}^{x_t}}|X = x_t)$	0.3426574	0.3006993
r_5	0.0	0.0
$Bel(E_{y_{1.2}^{x_t} y_{2.1}^{x_t} y_{3.2}^{x_t}}|X = x_t)$	0.6713287	0.6363636
r_6	0.0	0.0
$Bel(E_{y_{1.2}^{x_t} y_{2.2}^{x_t} y_{3.1}^{x_t}}|X = x_t)$	0.6573427	0.6643357
r_7	0.013986	0.0
$Bel(E_{y_{1.2}^{x_t} y_{2.2}^{x_t} y_{3.2}^{x_t}}|X = x_t)$	1.0	1.0
r_8	0.0	0.0

while we have

$$Bel(X = 1 | Y_1 \leq \min(y^0_{1.1}, y^1_{1.1}), Y_2 \leq \max(y^0_{2.2}, y^1_{2.2}), Y_3 \leq \min(y^0_{3.1}, y^1_{3.1}))$$

$$= \frac{Bel(E_{y^1_{1.1} y^1_{2.2} y^1_{3.1}} | X=1) P(X=1)}{Bel(E_{y^1_{1.1} y^1_{2.2} y^1_{3.1}} | X=1) P(X=1) + Pl(E_{y^0_{1.1} y^0_{2.2} y^0_{3.1}} | X=0) P(X=0)}$$

$$= \frac{0.3286714 \cdot 0.5}{0.3286714 \cdot 0.5 + 1.0 \cdot 0.5} = 0.2473685,$$

showing a slightly greater belief in having to deal with a fraudulent firm.

5 Conclusion

We have shown how to profit from the lattice structure of the events in the domain of a lower conditional joint CDF to implement MIP programs. This permits to effectively find outer belief approximations at minimal L_1 distance with at most $N + 1$ focal elements, one approximation for each conditioning value x_t of the common variable X, in a general statistical matching problem.

Through a suitable introduction of several binary variables and relative constraints, it was possible to let the focal elements of the solution appears as part of the MIP solution, avoiding an a priori construction of all the potential ones with the computational benefits it gives.

Note that the L_1 objective function (17) can be easily modified with the introduction of further slack variables to an L_∞ objective.

We have shown, through an example of real fraud detection data, the potentiality and effectiveness of our approach. Being MIP linear and having very efficient solvers for it, the only real computational limits to our approach are those imposed by the number of variables and constraints a solver can deal with.

In the next future we plan to perform systematic simulation studies with both real and simulated data to measure the trade-off between complexity reduction and belief approximation, possibly in comparison with other approaches present in literature.

Acknowledgments. We acknowledge the support of the PRIN 2022 project "Models for dynamic reasoning under partial knowledge to make interpretable decisions" (Project number: 2022AP3B3B, CUP Master: J53D23004340006, CUP: B53D23009860006) funded by the European Union - Next Generation EU (Missione 4 Componente 2).

References

1. Baioletti, M., Capotorti, A.: Efficient L1-based probability assessments correction: algorithms and applications to belief merging and revision. In: ISIPTA'15 Proceedings of the 9th International Symposium on Imprecise Probability: Theories and Applications, Pescara (IT), pp. 37–46. ARACNE (2015)
2. Baioletti, M., Capotorti, A.: A $L1$ based probabilistic merging algorithm and its application to statistical matching. Appl. Intell. **49**, 1–13 (2018)

3. Baioletti, M., Capotorti, A.: A L1 minimization optimal corrective explanation procedure for probabilistic databases. In: Information Processing and Management of Uncertainty in Knowledge-Based Systems, pp. 80–92 (2020)
4. Baioletti, M., Capotorti, A.: A further step for efficient corrections of inconsistent probabilistic data sets. Int. J. Approx. Reason. **159**, 108942 (2023)
5. Bayraksan, G., Love, D.: Data-driven stochastic programming using phi-divergences, pp. 1–19 (2015)
6. Brozzi, A., Capotorti, A., Vantaggi, B.: Incoherence correction strategies in statistical matching. Int. J. Approx. Reason. **53**(8), 1124–1136 (2012)
7. Coletti, G., Petturiti, D., Vantaggi, B.: Conditional belief functions as lower envelopes of conditional probabilities in a finite setting. Inf. Sci. **339**, 64–84 (2016)
8. Coletti, G., Scozzafava, R.: Probabilistic Logic in a Coherent Setting, Trends in Logic, vol. 15. Kluwer Academic Publisher, Dordrecht/Boston/London (2002)
9. Dempster, A.: Upper and lower probabilities induced by a multivalued mapping. Ann. Math. Stat. **38**(2), 325–339 (1967)
10. Denœux, T.: Inner and outer approximation of belief structures using a hierarchical clustering approach. Int. J. Uncertainty Fuzz. Knowl.-Based Syst. **9**, 437–460 (2001)
11. Di Zio, M., Vantaggi, B.: Partial identification in statistical matching with misclassification. Int. J. Approx. Reason. **82**, 227–241 (2017)
12. D'Orazio, M., Di Zio, M., Scanu, M.: Statistical Matching: Theory and Practice. Wiley, Hoboken (2006)
13. Dubins, L.: Finitely additive conditional probabilities, conglomerability and disintegrations. Ann. Probab. **3**(1), 89–99 (1975)
14. Duchi, J., Shalev-Shwartz, S., Singer, Y., Chandra, T.: Efficient projections onto the L1-ball for learning in high dimensions. In: Proceedings of the 25th International Conference on Machine Learning, ICML 2008, pp. 272–279. Association for Computing Machinery, New York (2008)
15. Grabisch, M.: Set Functions, Games and Capacities in Decision Making. Theory and Decision Library C. Springer, Heidelberg (2016). https://doi.org/10.1007/978-3-319-30690-2
16. Levi, I.: The Enterprise of Knowledge. MIT Press, Cambridge (1980)
17. Levin, D., Peres, Y.: Markov Chains and Mixing Times, 2nd edn. AMS American Mathematical Society (2017)
18. Lin, K., Gao, Y.: Model interpretability of financial fraud detection by group shap. Expert Syst. Appl. **210**, 118354 (2022)
19. Montes, I., Miranda, E., Vicig, P.: Outer approximating coherent lower probabilities with belief functions. Int. J. Approx. Reason. **110**, 1–30 (2019)
20. Petturiti, D., Vantaggi, B.: Probability envelopes and their Dempster-Shafer approximations in statistical matching. Int. J. Approx. Reason. **150**, 199–222 (2022)
21. Rahimian, H., Bayraksan, G., de Mello, T.H.: Identifying effective scenarios in distributionally robust stochastic programs with total variation distance. Math. Program. **173**, 393–430 (2019)
22. Rüschendorf, L.: Mathematical Risk Analysis: Dependence, Risk Bounds, Optimal Allocations and Portfolios. Springer, Heidelberg (2012). https://doi.org/10.1007/978-3-642-33590-7
23. Shafer, G.: A Mathematical Theory of Evidence. Princeton University Press, Princeton (1976)
24. Vantaggi, B.: Statistical matching of multiple sources: a look through coherence. Int. J. Approx. Reason. **49**, 701–711 (2008)

25. Walley, P.: Coherent lower (and upper) probabilities. Technical report. Department of Statistics, University of Warwick (1981)
26. Walley, P.: Statistical Reasoning with Imprecise Probabilities. Chapman and Hall, London (1991)
27. Williams, P.: Notes on conditional previsions. Int. J. Approx. Reason. **44**, 366–383 (2007)

Integrating User Preferences into Gradual Bipolar Argumentation for Personalised Decision Support

Elisa Battaglia[1], Pietro Baroni[1(✉)], Antonio Rago[2], and Francesca Toni[2]

[1] DII, University of Brescia, Brescia, Italy
e.battaglia001@studenti.unibs.it, pietro.baroni@unibs.it
[2] Department of Computing, Imperial College London, London, UK
{a.rago,ft}@imperial.ac.uk

Abstract. Gradual bipolar argumentation has been shown to be an effective means for supporting decisions across a number of domains. Individual user preferences can be integrated into the domain knowledge represented by such argumentation frameworks and should be taken into account in order to provide personalised decision support. This however requires the definition of a suitable method to handle user-provided preferences in gradual bipolar argumentation, which has not been considered in previous literature. Towards filling this gap, we develop a conceptual analysis on the role of preferences in argumentation and investigate some basic principles concerning the effects they should have on the evaluation of strength in gradual argumentation semantics. We illustrate an application of our approach in the context of a review aggregation system, which has been enhanced with the ability to produce personalised outcomes based on user preferences.

Keywords: Gradual argumentation · Preferences · Decision support

1 Introduction

Bipolar Argumentation Frameworks (BAFs) [9] endowed with a gradual argumentation semantics have been shown to provide a suitable formal basis for the development of applications for decision support in a variety of contexts, such as the evaluation of design alternatives [6], multiparty cooperative work [4], forecasting [13] and review aggregation [10]. In a nutshell, BAFs provide an argumentative representation of the network of reasons underlying the uncertain assessment of a given issue, which are related by attack and support relations. A gradual argumentation semantics provides a numerical assessment of the *strength* of the arguments belonging to a BAF. Strength values may then be used as the basis for informed decisions. When decision support concerns some personal choice (e.g. the selection of a product to purchase or a movie to watch) the issue of providing personalised outcomes, taking into account different user

S. Destercke et al. (Eds.): SUM 2024, LNAI 15350, pp. 14–28, 2025.
https://doi.org/10.1007/978-3-031-76235-2_2

preferences, emerges. In the formal context sketched above, this requires user preferences in gradual argumentation semantics for BAFs to be considered, a problem which has not been addressed in previous literature.

This work contributes to fill this gap by investigating some general principles concerning the effects that user preferences should have on the evaluation of argument strength in gradual argumentation semantics and illustrating their application in an enhanced version of the ADA review aggregation system [10].

The paper is organised as follows. After recalling some background notions in Sect. 2, we carry out in Sect. 3 a conceptual analysis on the role of preferences in formal argumentation, pointing out the different natures and uses that can be found in the literature. We then investigate in Sect. 4 general principles concerning the role of user preferences in the evaluation of argument strength and illustrate an application of the proposed approach in Sect. 5, while Sect. 6 concludes.

2 Background

Our work lies in the family of abstract argumentation formalisms, which are focused on the evaluation of the acceptability of arguments based on the relations among them. Dung's argumentation framework [11], considering only an attack relation between arguments, is the simplest model in this area.

Definition 1 [11]. *An argumentation framework (AF) is a pair $\mathcal{X}, \mathcal{A}$ where $\mathcal{X}$ is a finite set of arguments and $\mathcal{A} \subseteq \mathcal{X} \times \mathcal{X}$ is a binary (attack) relation.*

In the context of AFs, an *extension-based semantics* is a criterion specifying which sets of arguments, called *extensions*, are collectively acceptable.

We will use the more expressive BAFs, which also encompass a relation of support. Further, we will consider *gradual* argumentation semantics, where argument evaluation is expressed by a *strength* value on a given scale and arguments are equipped with an initial *base score*. These notions are formalised by the following definition [5,9].

Definition 2 [5,9]. *A Quantitative Bipolar Argumentation Framework (QBAF) is a quadruple $\langle \mathcal{X}, \mathcal{A}, \mathcal{S}, \tau \rangle$ where $\mathcal{X}$ is a finite set of arguments, $\mathcal{A} \subseteq \mathcal{X} \times \mathcal{X}$ is a binary (attack) relation, $\mathcal{S} \subseteq \mathcal{X} \times \mathcal{X}$ is a binary (support) relation and $\tau : \mathcal{X} \to \mathbb{I}$ is a total function, where $\mathbb{I}$ is a set equipped with a preorder $\leq$. For any $\alpha \in \mathcal{X}$, we call $\tau(\alpha)$ the base score of α. A* gradual semantics *σ is a criterion that, given a QBAF $\mathcal{Q} = \langle \mathcal{X}, \mathcal{A}, \mathcal{S}, \tau \rangle$, returns a strength function $\sigma_{\mathcal{Q}} : \mathcal{X} \to \mathbb{I}$ representing the strength evaluation of the arguments in $\mathcal{Q}$ according to the semantics σ.*

With a minor abuse of notation, for $S \subseteq \mathcal{X}$ we will denote as $\sigma(S)$ the multiset of the strengths of the elements of S, i.e., given $S = \{x_1, \dots x_n\}$, $\sigma(S) = \{\sigma(x_1), \dots \sigma(x_n)\}$. Given $x \in \mathcal{X}$ the set of attackers of x is denoted as $\mathcal{A}(x) \triangleq \{y \mid (y, x) \in \mathcal{A}\}$ and the set of supporters of x is denoted as $\mathcal{S}(x) \triangleq \{y \mid (y, x) \in \mathcal{S}\}$, the set of influencers of x is denoted as $\mathcal{I}(x) \triangleq \mathcal{A}(x) \cup \mathcal{S}(x)$.

We focus on applications of QBAFs for decision support: in this context some of the arguments have a distinguished role, since they represent the possible *answers* (or options) of the decision process, while the reasons in favour or against each option are represented by other arguments (called pro and con arguments), which in turn can be supported or attacked by other reasons corresponding to other pro and con arguments, and so on. QBAFs featuring this structure provide a formal counterpart to the IBIS method for decision making [8,12] as illustrated in [6]. As it emerges from the description sketched above, QBAFs for decision support can be represented as sets of trees, with the root of each tree corresponding to an answer argument, other vertices corresponding to pro and con arguments and the edges corresponding to the attack and support relations. Considering more general topologies, e.g. encompassing cycles of attacks, is left to future work. For brevity, we will consider the treatment of preferences in QBAFs consisting of a single tree, the extension to the case of a set of trees being straightforward. Definition 3 [18] captures the structure of the QBAFs we focus on, namely QBAFs for decision support about an answer r.

Definition 3 [18]. *Let $\mathcal{Q}$ be a QBAF $\langle \mathcal{X}, \mathcal{A}, \mathcal{S}, \tau \rangle$. For any arguments $a, b \in \mathcal{X}$, let a* path *from a to b be defined as $(c_0, c_1), ..., (c_{n-1}, c_n)$ for some $n > 0$ (referred to as the length of the path) where $c_0 = a, c_n = b$ and, for any $1 \leq i \leq n, (c_{i-1}, c_i) \in \mathcal{A} \cup \mathcal{S}$. Then, for $r \in \mathcal{X}$, $\mathcal{Q}$ is a* QBAF for r *iff i)$\nexists a \in \mathcal{X} \setminus \{r\}$ such that $(r, a) \in \mathcal{A} \cup \mathcal{S}$ ii)$\forall a \in \mathcal{X} \setminus \{r\}$, there is a unique path from a to r; and iii)$\nexists a \in \mathcal{X}$ with a path from a to a. Given a* QBAF for r $\mathcal{Q}$, *the* pro *arguments and* con *arguments in $\mathcal{Q}$ are defined respectively as: $pro(\mathcal{Q}) = \{a \in \mathcal{X} | \exists p \in paths(a, r)$ where $|p \cap \mathcal{A}|$ is even$\}$; $con(\mathcal{Q}) = \{a \in \mathcal{X} | \exists p \in paths(a, r)$ where $|p \cap \mathcal{A}|$ is odd$\}$.*

In the following, if not otherwise specified, we will assume that all QBAFs are for a root argument r, denoted as $root(\mathcal{Q})$. We will investigate the use of preferences in these QBAFs. In general a preference is a preorder over a set.

Definition 4. *Given a set S, a preference $\preceq_p$ over S is a reflexive and transitive relation on S. As usual, given $a, b \in S$, $a \preceq_p b$ and $a \not\succeq_p b$ will be denoted as $a \prec_p b$, while $a \preceq_p b$ and $b \preceq_p a$ will be denoted as $a \simeq_p b$.*

A preference over a set S can induce a preference over the powerset of S based on some criterion (examples are the `Elitist` and `Democratic` criteria in [16]). When assuming the existence of such a preference-inducing criterion C, we will denote as $\preceq_p^C$ the preference relation between subsets of S induced by $\preceq_p$ according to C. In argumentation frameworks preferences are over arguments, with gradual semantics defined analogously as for QBAFs.

Definition 5. *A preference-based argumentation framework (PAF) [3,14] is a 3-tuple $\langle \mathcal{X}, \mathcal{A}_p, \preceq_p \rangle$ where $\langle \mathcal{X}, \mathcal{A}_p \rangle$ is an AF and $\preceq_p$ is a preference over $\mathcal{X}$. A preference-based QBAF (PQBAF) is a pair $\langle \mathcal{Q}, \preceq_p \rangle$ where $\mathcal{Q} = \langle \mathcal{X}, \mathcal{A}, \mathcal{S}, \tau \rangle$ is a QBAF and $\preceq_p$ is a preference over $\mathcal{X}$. A gradual semantics σ is a criterion that, given a PQBAF $\mathcal{P}$, returns a strength function $\sigma_{\mathcal{P}} : \mathcal{X} \rightarrow \mathbb{I}$ representing the strength evaluation of the arguments in $\mathcal{P}$ according to the semantics σ.*

3 Preferences in Formal Argumentation

Several approaches to the treatment of preferences have been considered in formal argumentation literature. In the context of abstract argumentation a major line of investigation has concerned the treatment of so called *critical attacks* in PAFs. An attack is said to be *critical* [3] if the attacked element is strictly preferred to the attacker. The question then arises on whether and how the preference for the attacked element influences the attack relation. Several approaches use preferences to reduce a PAF $\mathcal{P}$ to an AF, that we will denote as $r(\mathcal{P})$, with the same set of arguments but a different attack relation. The acceptability of arguments is then evaluated by applying a semantics to $r(\mathcal{P})$.

Notation 1. *Given a PAF* $\mathcal{P} = \langle \mathcal{X}, \mathcal{A}_p, \succ_p \rangle$, *the reduced argumentation framework corresponding to* $\mathcal{P}$ *is denoted as* $r(\mathcal{P}) = \langle \mathcal{X}, \mathcal{A} \rangle$.

Four main reduction methods have been proposed in the literature (see the relevant references for details):

- [2] $\forall a, b, \in \mathcal{X}$, $(a, b) \in \mathcal{A}$ iff $(a, b) \in \mathcal{A}_p, b \not\succ_p a$.
- [3] $\forall a, b, \in \mathcal{X}$, $(a, b) \in \mathcal{A}$ iff $((a, b) \in \mathcal{A}_p, b \not\succ_p a)$ or $((b, a) \in \mathcal{A}_p, (a, b) \notin \mathcal{A}_p, a \succ_p b)$.
- [14] $\forall a, b, \in \mathcal{X}$, $(a, b) \in \mathcal{A}$ iff $((a, b) \in \mathcal{A}_p, b \not\succ_p a)$ or $((a, b) \in \mathcal{A}_p, (b, a) \notin \mathcal{A}_p)$.
- [14] $\forall a, b, \in \mathcal{X}$, $(a, b) \in \mathcal{A}$ iff $((a, b) \in \mathcal{A}_p, b \not\succ_p a)$ or $((b, a) \in \mathcal{A}_p, (a, b) \notin \mathcal{A}_p, a \succ_p b)$ or $((a, b) \in \mathcal{A}_p, (b, a) \notin \mathcal{A}_p)$.

The first reduction suppresses the critical attack; this technique has been criticized in [3] because it can lead to extensions which are not conflict-free with respect to the original PAF. For this reason, the second reduction, aims to "repair" the AF and avoids that drawback by reversing the direction of the critical attack. In [14], Kaci et al. argued that the second reduction implies a strong constraint since a preferred argument can never be successfully attacked, hence they proposed the third reduction, which deletes a critical attack only if the opposite attack belongs to $\mathcal{A}_p$ too. As a further alternative, the fourth reduction comprises both the second and the third reduction. Each choice corresponds to a different intuition and is subject to potential criticisms.

In the context of the ASPIC+ formalism [16], a rule-based approach to argument construction is proposed, leading to the identification of different forms of attack between arguments, which are classified as preference-dependent or preference-independent. Only preference-dependent attacks are affected by preferences: they are ignored when the attacked argument is strictly preferred to the attacker. Since in ASPIC+ preference-dependent attacks are always symmetrical, this bears some similarity with the third reduction mentioned above.

While the above approaches concern extension-based semantics, the use of preferences in the context of *gradual* argumentation has received lesser attention, as remarked in [15]. The authors propose a property called *Preference Precedence* (PP) stating that the preference relation should have a direct impact on the strength evaluation. In a nutshell, if an argument x_1 is preferred to x_2, then the strength of x_1 should be not lesser than the strength of x_2.

It can be remarked that this principle imposes a rather strong requirement on the evaluation of arguments, independently of the relations holding between them. Moreover, the role of preferences is quite different, as they are not used to modify the attack relation but rather are meant to affect the final evaluation.

The potential twofold role of preferences is also evidenced in [3], where, in addition to handling critical attacks in PAFs, preferences are used to induce an ordering on the extensions prescribed by a given semantics.

With respect to the goals of the present paper, two main limitations emerge from the above surveyed approaches: i) none of them concerns BAFs, i.e. they do not consider the support relation, which is needed in our context; ii) a conceptual analysis about the motivations underlying the different proposals is lacking.

As to the latter point, we remark in particular that different uses of preferences may be required by different application contexts. In this respect, we propose here a simple taxonomy on the uses of preferences in argumentation based on two classification dimensions: i) the origin of preferences, which can be *endogenous* or *exogenous* with respect to the argument construction process; ii) the purpose of the formalization, which can be *normative* or *descriptive*.

Concerning the first point, we call *endogenous* preferences those which are induced on arguments from preferences concerning their constitutive elements, e.g. premises and rules as in [16], while *exogenous* preferences are ascribed to arguments based on elements which are not involved in their construction, like for instance the values they promote, as in [7]. Concerning the second point, a *normative* approach aims to define a standard behavior on the basis of certain rationality principles, while a *descriptive* approach aims to represent how people actually behave, possibly in an unprincipled manner. We suggest that some links can be drawn between these notions and the uses of preferences in the literature.

For instance the PP principle in [15], where, in a sense, preferences determine the evaluation outcomes overriding any relations between arguments, can be justified in a descriptive approach with exogenous preferences. For instance, if some people have a preference for an information source they trust, they may accept all arguments from that source, no matter what their content is. This behavior would be in contrast with a normative approach, where arguments' contents and relations should play a role also in presence of preferences.

Concerning the treatment of critical attacks, suppressing all of them independently of any other condition [2] appears in line with a descriptive approach, where, as above, preferences have a sort of absolute priority over other factors, with the possible production of outcomes which are not conflict-free. On the other hand, the treatment proposed in [16] concerns endogenous preferences with a normative approach, where they have the role of converting mutual attacks into unidirectional ones when appropriate.

While an extended discussion of these aspects is beyond the scope of this paper, the observations above indicate that a proper characterization of the application context is necessary to lay the foundations of the approach we aim to propose. In particular, the preferences we are interested in are exogenous, since they can be provided by users as an additional element with respect to a QBAF representing domain knowledge in a given decision support context. Moreover, we

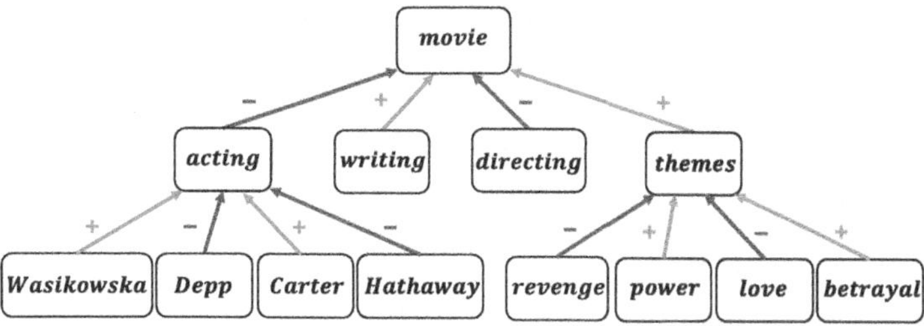

Fig. 1. A simple BAF in the movie domain. Arguments are represented by vertices, attacks by red edges labelled '−' and supports by green edges labelled '+'. Arguments correspond to features and subfeatures of a given movie. An attack (support) relation indicates that a feature has a negative (positive) effect on the assessment of its parent. For more details on the domain modelling, see Sect. 5. (Color figure online)

aim for a normative approach where preferences are used to provide personalised recommendations in a principled manner, as discussed in the next section.

4 Adding Preferences to QBAFs for Decision Support

In this section we illustrate the basic ideas of our approach to encompass user preferences in gradual bipolar argumentation for decision support. To support our presentation, we will use as a running example the simple framework presented in Fig. 1, taken from the movie recommendation domain.

The first question we consider concerns the pairs of arguments on which preferences are given. In this respect, some differences with the approaches reviewed in the previous section have to be underlined. In particular, while, in principle, endogenous preferences can refer to any pair of arguments (since they involve constitutive elements common to all arguments), user-defined exogenous preferences can only be given on arguments whose comparison is meaningful to the user. In the family of frameworks we are considering, it is then natural to consider preferences between sibling nodes (i.e. between influencers of the same node) since they contribute together to the evaluation of the influenced node. For instance, in the example of Fig. 1, it is reasonable to imagine that a user may give more importance to the *themes* of the movie than to the quality of *acting*, or may prefer one actor to another, while it does not seem meaningful to express preferences between an influencer and an influenced node (e.g. between *love* and *themes*) and more generally preferences across different levels of the tree. This represents a significant difference with respect to approaches whose main focus is the treatment of critical attacks.

Towards defining general principles for the treatment of preferences, a further question then concerns identifying the cases where their effect on argument evaluation can be univocally determined. In this respect, we distinguish the cases of preferences concerning arguments of different polarity (e.g. an attacker vs. a supporter) with respect to arguments with the same polarity (e.g. a supporter vs. a supporter). In the first case, the expected effect of preferences can be clearly identified. For instance, if an attacker is preferred to a supporter of a node x, it can be expected that the strength of x should be lower with respect to the case where this preference is the opposite or does not hold. In the example of Fig. 1, if a user gives more importance to *acting* (a negative feature of the considered movie) than to *themes* (a positive feature of the movie) it is reasonable to consider the movie less appropriate for this user with respect to a user who has the opposite preference (or no preference at all). In the second case, the effect of preferences is undetermined: if a user prefers an attacker a_1 to another attacker a_2 and then you consider another user who prefers a_2 to a_1, you have two structurally indistinguishable situations for which a different behavior cannot be prescribed. The same holds for a preference involving two supporters.

While the examples above concern a single preference between a pair of arguments, they can be extended to the case where multiple preferences are given, from which a preference relation on sets of arguments is derived. On this basis, we introduce a property of *local coherence* specifying the effects of preferences between the set of attackers and the set of supporters of a given node. The property refers to the comparison between two PQBAFs which differ only in the preference relation concerning the influencers of a given argument x.

Definition 6. *Given a QBAF, $\mathcal{Q} = \langle \mathcal{X}, \mathcal{A}, \mathcal{S}, \tau \rangle$, let $\mathcal{P} = \langle \mathcal{Q}, \succeq_p \rangle$ and $\mathcal{P}' = \langle \mathcal{Q}, \succeq_{p'} \rangle$ be two PQBAFs. Given $x \in \mathcal{X}$ we say that $\mathcal{P}'$ is a* x-local modification *of $\mathcal{P}$ if $\succeq_p \cap ((\mathcal{X} \times \mathcal{X}) \setminus (\mathcal{I}(x) \times \mathcal{I}(x))) = \succeq_{p'} \cap ((\mathcal{X} \times \mathcal{X}) \setminus (\mathcal{I}(x) \times \mathcal{I}(x)))$.*

Definition 7. *Given a QBAF, $\mathcal{Q} = \langle \mathcal{X}, \mathcal{A}, \mathcal{S}, \tau \rangle$, let $\mathcal{P} = \langle \mathcal{Q}, \succeq_p \rangle$ and $\mathcal{P}' = \langle \mathcal{Q}, \succeq_{p'} \rangle$ be two PQBAFs, such that $\mathcal{P}'$ is a x-local modification of $\mathcal{P}$ for some $x \in \mathcal{X}$. Let C be a preference inducing criterion and σ a gradual semantics for PQBAFs. The property of* local coherence with preferences *is satisfied iff the following conditions hold:*

- *if $\mathcal{A}(x) \preceq_p^C \mathcal{S}(x)$ and $\mathcal{A}(x) \succeq_{p'}^C \mathcal{S}(x)$ then $\sigma_{\mathcal{P}}(x) \geq \sigma_{\mathcal{P}'}(x)$;*
- *if $\mathcal{A}(x) \succeq_p^C \mathcal{S}(x)$ and $\mathcal{A}(x) \preceq_{p'}^C \mathcal{S}(x)$ then $\sigma_{\mathcal{P}}(x) \leq \sigma_{\mathcal{P}'}(x)$.*

The property of strict local coherence with preferences *is satisfied iff the following conditions hold:*

- *if $\mathcal{A}(x) \preceq_p^C \mathcal{S}(x)$ and $\mathcal{A}(x) \succ_{p'}^C \mathcal{S}(x)$ then $\sigma_{\mathcal{P}}(x) > \sigma_{\mathcal{P}'}(x)$;*
- *if $\mathcal{A}(x) \succeq_p^C \mathcal{S}(x)$ and $\mathcal{A}(x) \prec_{p'}^C \mathcal{S}(x)$ then $\sigma_{\mathcal{P}}(x) < \sigma_{\mathcal{P}'}(x)$.*

Let us illustrate Definitions 6 and 7, with reference to the example of Fig. 1. Consider a relation $\succeq_p = \emptyset$ expressing the absence of any preference, and a relation $\succeq_{p'}$ expressing the view of a user who strictly prefers Depp to Wasikowska and Hathaway to Carter. Then, $\succeq_{p'}$ gives rise to a x-local modification with x=*acting*. Since the user prefers two attackers to two supporters, we may assume that, given any criterion C, $\mathcal{A}(acting) \succ_{p'}^C \mathcal{S}(acting)$. As a consequence, (strict) local coherence requires that $\sigma_{\mathcal{P}'}(acting)$ is (strictly) lesser that $\sigma_{\mathcal{P}}(acting)$.

Under mild requirements on the considered semantics, it is possible to show that local coherence ensures that the effects of preferences are coherent with the roles of pro and con arguments along the structure of the tree. In particular, we show in Proposition 1 that a preference for pros over cons among the influencers of an argument x has the effect of increasing the strength of pros and decreasing the strength of cons in the path from x to the root of the PQBAF, and vice versa in the case of a preference for cons over pros.

Towards this result, we require first of all that the strength of an argument is determined by its base score, the strengths of its attackers and supporters, and the preferences between them. This is the extension to the case of presence of preferences of a property which is common to most gradual argumentation semantics for BAFs in the literature (see e.g. [9]).

Definition 8. *A gradual semantics σ for PQBAFs is* based on local evaluation *if for every PQBAF $\mathcal{P} = \langle \mathcal{Q}, \preceq_p \rangle$, for every $x \in \mathcal{X}$,*
$\sigma_{\mathcal{P}}(x) = f(\tau(x), \sigma_{\mathcal{P}}(\mathcal{A}(x)), \sigma_{\mathcal{P}}(\mathcal{S}(x)), \preceq_p \cap (\mathcal{I}(x) \times \mathcal{I}(x))$ *for some function f.*

Moreover, we assume a semantics based on a *monotonic* strength function σ. The relevant definitions from [5] are adapted below, in a simplified form which requires a notion of similarity of arguments with respect to their influencers.

Definition 9. *Given two PQBAFs $\mathcal{P} = \langle \mathcal{Q}, \preceq_p \rangle$, $\mathcal{P}' = \langle \mathcal{Q}', \preceq_{p'} \rangle$ two arguments $x \in \mathcal{X}$, $x' \in \mathcal{X}'$ are $\mathcal{I}$-similar iff there is a bijective function $h : \mathcal{I}(x) \to \mathcal{I}(x')$ such that: i) $\forall y \in \mathcal{I}(x)$, $y \in \mathcal{A}(x)$ iff $h(y) \in \mathcal{A}'(x')$ and $y \in \mathcal{S}(x)$ iff $h(y) \in \mathcal{S}'(x')$; ii) $\forall x_1, x_2 \in \mathcal{I}(x)$, $x_1 \preceq_p x_2$ iff $h(x_1) \preceq_{p'} h(x_2)$.*

In words, two arguments are $\mathcal{I}$-similar if they have the "same" (modulo a bijection) attackers and supporters with the "same" preference relation among them.

Definition 10. *Given two PQBAFs $\mathcal{P} = \langle \mathcal{Q}, \preceq_p \rangle$, $\mathcal{P}' = \langle \mathcal{Q}', \preceq_{p'} \rangle$, for $A \subseteq \mathcal{X}$, $B \subseteq \mathcal{X}'$ A is* strength equivalent *to B, denoted $A = B$ iff $\sigma_{\mathcal{P}}(A) = \sigma_{\mathcal{P}'}(B)$; A is* at least as strong as *B, denoted $A \geq B$ iff there exists an injective mapping f from B to A such that $\forall \alpha \in B, \sigma_{\mathcal{P}}(f(\alpha)) \geq \sigma_{\mathcal{P}'}(\alpha)$; A is* stronger than *B, denoted $A > B$ iff $A \geq B$ and $B \ngeq A$.*

Intuitively, two sets of arguments are strength equivalent if the multisets of their strength values are the same (i.e. the two sets have the same cardinality and one can establish a bijection linking arguments with the same strength). A set of arguments A is at least as strong as a set B if A has a greater or equal cardinality than B and for each element of B one can identify a distinct element of A which has a greater or equal strength.

Definition 11. *For any argument x in a PQBAF $\mathcal{P}$, the* shaping triple *of x is $(\tau(x), \mathcal{S}(x), \mathcal{A}(x))$, denoted $ST_{\mathcal{P}}(x)$. Given two PQBAFs $\mathcal{P} = \langle \mathcal{Q}, \preceq_p \rangle$, $\mathcal{P}' = \langle \mathcal{Q}', \preceq_{p'} \rangle$, let $x_1 \in \mathcal{X}$, $x_2 \in \mathcal{X}'$ be $\mathcal{I}$-similar. $ST_{\mathcal{P}'}(x_2)$ is said to be:* as boosting as *$ST(x_1)$, denoted $ST_{\mathcal{P}}(x_1) \simeq ST_{\mathcal{P}'}(x_2)$ iff $\tau(x_1) = \tau'(x_2), \mathcal{S}(x_1) = \mathcal{S}'(x_2)$ and $\mathcal{A}(x_1) = \mathcal{A}'(x_2)$;* at least as boosting as *$ST_{\mathcal{P}}(x_1)$, denoted $ST_{\mathcal{P}}(x_1) \preceq ST_{\mathcal{P}'}(x_2)$, iff $\tau(x_1) \leq \tau'(x_2), \mathcal{S}(x_1) \leq \mathcal{S}'(x_2), \mathcal{A}(x_1) \leq \mathcal{A}'(x_2)$;* strictly more boosting than *$ST_{\mathcal{P}}(x_1)$, denoted $ST_{\mathcal{P}}(x_1) \prec ST_{\mathcal{P}'}(x_2)$, iff $ST_{\mathcal{P}}(x_1) \preceq ST_{\mathcal{P}'}(x_2)$ and*

$ST_{\mathcal{P}'}(x_2) \not\preceq ST_{\mathcal{P}}(x_1)$. A strength function σ is monotonic *iff the following conditions hold: i) if $ST_{\mathcal{P}}(x_1) \simeq ST_{\mathcal{P}'}(x_2)$, then $\sigma_{\mathcal{P}}(x_1) = \sigma_{\mathcal{P}'}(x_2)$; ii) if $ST_{\mathcal{P}}(x_1) \preceq ST_{\mathcal{P}'}(x_2)$, then $\sigma_{\mathcal{P}}(x_1) \leq \sigma_{\mathcal{P}'}(x_2)$. A strength function σ is* strictly monotonic *iff σ is monotonic and if $ST_{\mathcal{P}}(x_1) \prec ST_{\mathcal{P}'}(x_2)$, then $\sigma_{\mathcal{P}}(x_1) < \sigma_{\mathcal{P}'}(x_2)$.*

The shaping triple collects the elements affecting the strength evaluation of an argument: its base score and its supporters and attackers. The boosting relations are based on an element-wise comparison between shaping triples and essentially check whether two shaping triples are equal or one (strictly) dominates the other with respect to the strength values. A strength function is (strictly) monotonic if its outcomes on arguments (strictly) follow the (in)equalities between the relevant shaping triples.

On this basis, Proposition 1 shows that if a local modification corresponds to a preference for pros over cons then it can only induce an increase of the strength of other pros and a decrease of the strength of other cons, and vice versa in the case of a preference for cons over pros. This can be regarded as a globally coherent behavior induced by the local coherence property.

Proposition 1. *Given a QBAF, $\mathcal{Q} = \langle \mathcal{X}, \mathcal{A}, \mathcal{S}, \tau \rangle$, let $\mathcal{P} = \langle \mathcal{Q}, \succeq_p \rangle$ and $\mathcal{P}' = \langle \mathcal{Q}, \succeq_{p'} \rangle$ be two PQBAFs, such that $\mathcal{P}'$ is a x-local modification of $\mathcal{P}$ for some $x \in \mathcal{X}$. Let C be a preference inducing criterion and σ a monotonic semantics based on local evaluation. If the property of local coherence holds, for any $y \in \{x\} \cup \{z \mid z$ is in the path from x to $root(\mathcal{Q})\}$ it holds that:*

(a) if $\mathcal{I}(x) \cap pro(\mathcal{Q}) \preceq_p^C \mathcal{I}(x) \cap con(\mathcal{Q})$ and $\mathcal{I}(x) \cap pro(\mathcal{Q}) \succeq_{p'}^C \mathcal{I}(x) \cap con(\mathcal{Q})$ then:
- *if $y \in pro(\mathcal{Q})$, then $\sigma_{\mathcal{P}}(y) \leq \sigma_{\mathcal{P}'}(y)$;*
- *if $y \in con(\mathcal{Q})$, then $\sigma_{\mathcal{P}}(y) \geq \sigma_{\mathcal{P}'}(y)$*

(b) if $\mathcal{I}(x) \cap con(\mathcal{Q}) \preceq_p^C \mathcal{I}(x) \cap pro(\mathcal{Q})$ and $\mathcal{I}(x) \cap con(\mathcal{Q}) \succeq_{p'}^C \mathcal{I}(x) \cap pro(\mathcal{Q})$ then:
- *if $y \in pro(\mathcal{Q})$, then $\sigma_{\mathcal{P}}(y) \geq \sigma_{\mathcal{P}'}(y)$;*
- *if $y \in con(\mathcal{Q})$, then $\sigma_{\mathcal{P}}(y) \leq \sigma_{\mathcal{P}'}(y)$*

Proof. Proof is by induction on the length of the path from any element of $\mathcal{I}(x)$ to y. *Induction base.* Suppose that $y = x$ and let $w \in \mathcal{I}(x) \cap pro(\mathcal{Q})$. Then, by Definition 3, the number of attacks in the path from w to $root(\mathcal{Q})$ is even, while for $u \in \mathcal{I}(x) \cap con(\mathcal{Q})$ the number of attacks in the path from u to $root(\mathcal{Q})$ is odd. Therefore, if $y = x \in pro(\mathcal{Q})$ then $w \in \mathcal{S}(x)$ and $u \in \mathcal{A}(x)$. As a consequence $\mathcal{I}(x) \cap pro(\mathcal{Q}) = \mathcal{S}(x)$ and $\mathcal{I}(x) \cap con(\mathcal{Q}) = \mathcal{A}(x)$. In the case (a), this entails $\mathcal{S}(x) \preceq_p^C \mathcal{A}(x)$ and $\mathcal{S}(x) \succeq_{p'}^C \mathcal{A}(x)$. It then follows from the local coherence property that $\sigma_{\mathcal{P}}(x) \leq \sigma_{\mathcal{P}'}(x)$. If instead $y = x \in con(\mathcal{Q})$ we get $\mathcal{I}(x) \cap pro(\mathcal{Q}) = \mathcal{A}(x)$ and $\mathcal{I}(x) \cap con(\mathcal{Q}) = \mathcal{S}(x)$ and we analogously derive $\sigma_{\mathcal{P}}(y) \geq \sigma_{\mathcal{P}'}(y)$. The proof for case (b) is similar.

Induction step. We inductively suppose that the statement above is valid for every y in the path from any child w of x to $root(\mathcal{Q})$ such that the path has length n, and we want to show that it holds for length $n + 1$.

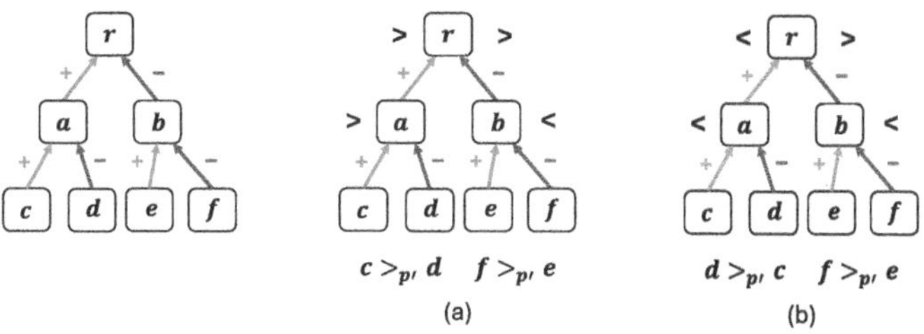

Fig. 2. Two simple examples of the effects of multiple preferences. The non-leaf nodes are labeled by $>$ and $<$ to indicate, respectively, an increase and a decrease in their strengths, assuming the strict version of Proposition 1. The effects on the root strength of the two preferences added in case (a) are concordant, while they are discordant in case (b), hence, the overall effect on the strength of r is not determined in the second case.

Considering the case (a), for every z such that the path from w to z has length n, we know from the induction hypothesis that if $z \in pro(\mathcal{Q})$ then $\sigma_{\mathcal{P}}(z) \leq \sigma_{\mathcal{P}'}(z)$.

Let now y be the argument attacked or supported by z. We get that: (i) if $y \in pro(\mathcal{Q})$, then $z \in \mathcal{S}(y)$, hence $ST_{\mathcal{P}'}(y) \geq ST_{\mathcal{P}}(y)$ (y has a stronger supporter in $\mathcal{P}'$ while all other elements of its shaping triple are the same) and, by the monotonicity property, $\sigma_{\mathcal{P}'}(y) \geq \sigma_{\mathcal{P}}(y)$; (ii) if $y \in con(\mathcal{Q})$, then $z \in \mathcal{A}(y)$, hence $ST_{\mathcal{P}'}(y) \leq ST_{\mathcal{P}}(y)$ and, by the monotonicity property $\sigma_{\mathcal{P}'}(y) \leq \sigma_{\mathcal{P}}(y)$. The treatment of the case (b) is analogous. □

Assuming a stricly monotonic semantics and the property of strict local coherence, a strict version of Proposition 1 where the inequalities between $\sigma_{\mathcal{P}}(y)$ and $\sigma_{\mathcal{P}'}(y)$ are strict can be derived. We omit the obvious statement and the analogous proof due to space limitations.

While the above results concern only preferences over the influencers of a specific argument they can be used as a basis to reason about the addition of an arbitrary set of preferences to a QBAF. In particular a QBAF without preferences can be regarded as a special case of *preference uniform* PQBAF where for every pair x, y of sibling nodes it holds that $x \sim_p y$. A preference uniform PQBAF can be transformed into a generic PQBAF by a series of local modifications, each concerning the preferences over the children of a single node.

By iterated application of Proposition 1 it is possible to characterize the variations induced by the preferences on the argument strengths, when they are determined. In particular, when the hypotheses of Proposition 1 are satisfied, a x-local modification determines effects on the arguments in the path from x to the root. If there are multiple modifications, they affect concurrently the arguments shared in the relevant paths: these include at least the root and may involve also other arguments (see Fig. 2). For the arguments shared among multiple paths, multiple variations are induced by multiple preferences. If the variations are all concordant the overall effect on the arguments is determined, otherwise it is not.

Figure 2 presents a simple case where the propagation of the effects of preferences on the strength of the root node is determined, and a case where it is not.

5 Integrating Preferences in the ADA System

ADA is a review aggregation system introduced in [10] within the movie domain The ADA pipeline is organised as follows (see [10] for details):

- ADA employs natural language processing to break down the reviews into a *feature-based review aggregation* where each feature corresponds to an argument and is assigned a polarity, based on sentiment analysis;
- ADA generates a tree-structured QBAF to represent the feature-based review aggregation and the identified polarities (Fig. 1 shows a simple example);
- the base score of each feature is derived from the review aggregation, in a nutshell the base score reflects how much the reviews about that feature are coherent: if a feature has consistently positive or negative reviews its base score will be higher, if reviews are more mixed its base score will be lower;
- A gradual semantics is applied to the QBAF to compute argument strengths;
- The extracted QBAF provides the underlying structure for generating dialogical explanations for users, taking into account the computed strengths.

As a preliminary application of the concepts we previously introduced, we investigated a method to integrate user preferences in ADA in compliance with the properties introduced in Sect. 4. We assume that, for a given user u, preferences are given as a set $\mathcal{P}_u$ of pairs of features, where each pair (x, y) is such that x and y are sibling nodes in the QBAF, and indicates that x is strictly preferred over y. As a simple example, that we will use also later, we consider a user who, with reference to the framework shown in Fig. 1, prefers the actor *Johnny Depp* over the actress *Mia Wasikowska*, and the movie's *writing* over its *directing*.

Our main goal was to carry out an initial analysis of the issues to be faced when putting our general notions into practice, before developing further investigations. To this purpose, the idea was to adopt a simple parametric approach to combine preferences with existing gradual argumentation semantics and then carry out a preliminary assessment of the impact of the use of preferences on strength evaluation. The approach we adopted consists in decreasing the base score of the arguments which are less preferred, by multiplying them by a given *discount factor* $\delta < 1$. Formally, given a set of preferences $\mathcal{P}_u$, for each $(x, y) \in \mathcal{P}_u$, we let $\tau_p(y) = \delta \cdot \tau(y)$. Then a gradual semantics is applied to the framework with these modified base scores. Simple as it is, this method is coherent with the properties introduced in Sect. 4 whenever the preferences over the influencers of a given argument are uniform in terms of polarity (i.e. all indicate the superiority of an attacker over a supporter or vice versa) and the adopted semantics is monotonic. In this context, in order to carry out a preliminary evaluation of the effects on the argument strengths, we considered two main choices affecting the final outcome: the factor δ used to decrease the base scores of the less preferred arguments and the chosen semantics.

As to the first choice, we considered the following values for δ: 0.25, 0.50, 0.75. As to the second choice we considered four gradual semantics, namely, QuAD [6], DF-QuAD [19], REB [1] and Quadratic Energy Model (QEM) [17]. All these semantics produce strength values belonging to the $[0, 1]$ real interval.

We draw some general considerations on the roles of the elements involved in our approach, before presenting some examples of quantitative assessments.

Role of the Base Scores. The idea of using preferences to adjust the value of the base scores has the advantage of simplicity and of enabling the use of existing gradual semantics as interchangeable alternatives, while ensuring the satisfaction of the properties introduced in Sect. 4. However, it makes the effect of preferences dependent on the way the base score of arguments is determined. In the case of ADA, the base score of an argument corresponding to a feature is derived from a normalised count of the positive and negative judgments about that feature found in the set of reviews under consideration (see [10] for details). The normalisation method used in the original version of ADA, gave rise typically to rather small base scores (around 0.05 with the dataset in the movie domain). While these small base scores were in line with the original purposes of ADA, they turned out to be somehow problematic, since the quantitative effects of preferences were in fact dampened since they were conveyed through the adjustment of small values. To avoid this problem we experimented with a different normalisation method which led to base scores' with values on average ten times bigger, leading in turn to a greater quantitative impact of the preferences.

Role of the Gradual Semantics. As mentioned above, we experimented our approach with four gradual semantics, which, while sharing the property of being monotonic, are rather different by design (a comparison is beyond the scope of the present paper). This entails that a given change in base scores, as determined by preferences in our approach, may have significantly different impacts on strength evaluation outcomes, depending on the semantics adopted. In a sense, we can say that a gradual semantics can be more or less sensitive to preferences in the context of our approach. While the study of a formal notion of sensitivity to preferences is an interesting issue for future work, we will draw some preliminary considerations concerning the semantics we experimented with.

Definition of the Discount Factor. The value of the discount factor δ modulates the entity of the modification of base scores due to preferences, the bigger this modification, the bigger the expected impact on argument strengths, though it depends also on the sensitivity of semantics, as mentioned previously. Indeed, one can regard as an open question, whose answer is context dependent, how heavily preferences should affect strength evaluation with respect to the rest of the framework. In the specific case of ADA, the question concerns balancing the individual inclinations of a user and the indications emerging from the reviews produced by the community. It can be imagined that a proper balance is in turn user dependent, with some users being more radical in their preferences and others who are more open to take into account the opinions of the crowd. While a

Table 1. Strength values and percent variations with *Depp* $\succ_p$ *Wasikowska* and *writing* $\succ_p$ *directing*, $\delta = 0.75$.

	old base score normalisation				new base score normalisation			
	QuAD	DF-QuAD	REB	QEM	QuAD	DF-QuAD	REB	QEM
Depp	0.0111	0.0111	0.0111	0.0111	0.2	0.2	0.2	0.2
Wasikowska	0.0167	0.0167	0.0167	0.0167	0.375	0.375	0.375	0.375
directing	0.0250	0.0250	0.0250	0.0250	0.45	0.45	0.45	0.45
writing	0.0444	0.0444	0.0444	0.0444	0.5	0.5	0.5	0.5
acting	0.0275	0.0407	0.0223	0.0223	0.0894	0.3288	0.0915	0.0910
movie	0.5429	0.5279	0.5219	0.5171	0.6194	0.5336	0.5465	0.5700
Δ% acting	−16,17	−6.13	−0.53	−0.41	−41.00	−10.33	−9.89	−14.81
Δ% movie	1.19	0.52	0.27	0.04	14.22	6.01	5.14	6.48

Table 2. Focus on *acting* and *movie* for $\delta = 0.25$ and $\delta = 0.5$.

		old base score normalisation				new base score normalisation			
		QuAD	DF-QuAD	REB	QEM	QuAD	DF-QuAD	REB	QEM
$\delta = 0.25$	acting	0.0021	0.0354	0.0221	0.0222	0.0742	0.2530	0.0743	0.0859
	movie	0.5533	0.5334	0.5247	0.5178	0.7560	0.6030	0.6023	0.6605
	Δ% acting	−32.71	−18.38	−1.58	−0.54	−51.00	−30.99	−26.81	−19.57
	Δ% movie	3.14	1.57	0.81	0.17	39.41	19.80	15.89	23.38
$\delta = 0.5$	acting	0.0222	0.0380	0.0222	0.0222	0.0818	0.2909	0.0825	0.0897
	movie	0.5493	0.5306	0.5222	0.5174	0.6872	0.5668	0.5741	0.6131
	Δ% acting	−32.34	−12.25	−1.06	−0.54	−46.00	−20.66	−18.80	−16.01
	Δ% movie	2.39	1.04	0.54	0.10	26.72	12.61	10.46	14.52

proper modelling of these different attitudes represents another interesting subject of future investigation, the use of different discount factors can be regarded as a first crude method to give them a counterpart in our approach.

As a specific but informative illustration of the issue discussed above, Tables 1 and 2 show the results of applying our approach to the example of Fig. 1 with different choices of base score normalisation, discount factor and semantics. Table 1 shows strength values of the arguments affected by preferences and evidences the variations of *acting* and *movie* with respect to the case of no preferences, for $\delta = 0.75$. Table 2 focuses on *acting* and *movie* for other values of δ.

We discuss the effects on the strength of the *movie* representing the root of the framework, and of the feature *acting*, which are both affected by the preferences between the underlying elements. The main comments are as follows.

With the old normalisation method the dampening effect of small base scores values is evident on the variations of the *movie* strength which, in all cases are below 4%. A more significant effect is visible on the *acting* strength in the case of QuAD and DF-QuaD semantics, while the other semantics are less sensitive.

With the new normalisation method, as expected, the impact of preferences on the strength of both *movie* and *acting* is more significant and its amplitude is modulated by the choice of the discount factor, thus achieving the goal of taking preferences into account more effectively. In all cases QuAD is definitely more sensitive to preferences than the other semantics. Depending on the discount factor, the variation of the *movie* strength with QuAD ranges from 14.22% to 39.41%, while REB is the least sensitive semantics in all cases, with a range from 5.14% to 15.89%. As a side remark, we note that DF-QuAD is more sensitive than QEM with small base scores, while the converse holds with the new normalisation method.

6 Conclusions

With the aim of enabling personalised recommendations, we explored a normative approach for the use of exogenous preferences in QBAFs for decision support. In particular, we introduced a property of local coherence concerning the expected effects of preferences on argument strength and proved that, under the assumption of a monotonic semantics, it ensures that these effects are in line with the roles of pro and con arguments along the structure of the framework. Based on this approach, we extended a review aggregation system with the ability to deal with user preferences and carried out a preliminary experiment, showing how the quantitative effects of preferences are significantly affected by alternative design choices. Among future research directions we mention the study of further methods to deal with preferences in gradual argumentation semantics and a more extensive experimentation in the domain of review aggregation. We also plan to investigate the relationships of our argumentative approach with methods adopted in other fields, like Multi-Criteria Decision Analysis and Bayesian decision theory.

Acknowledgements. Baroni was partially supported by MUR project PRIN 2022 EPICA "Enhancing Public Interest Communication with Argumentation" (CUP D53D23008860006) funded by the European Union - Next Generation EU. Rago and Toni were partially funded by J.P. Morgan and by the Royal Academy of Engineering under the Research Chairs and Senior Research Fellowships scheme. Rago and Toni were partially funded by the European Research Council (ERC) under the European Union's Horizon 2020 research and innovation programme (grant agreement No. 101020934). Any views or opinions expressed herein are solely those of the authors listed.

References

1. Amgoud, L., Ben-Naim, J.: Evaluation of arguments in weighted bipolar graphs. Int. J. Approx. Reason. **99**, 39–55 (2018)
2. Amgoud, L., Cayrol, C., Lagasquie-Schiex, M.C.: On the bipolarity in argumentation frameworks. In: Proceedings of the 10th International Workshop on Non-Monotonic Reasoning (NMR), pp. 1–9 (2004)

3. Amgoud, L., Vesic, S.: Rich preference-based argumentation frameworks. Int. J. Approx. Reason. **55**(2), 585–606 (2014)
4. Aurisicchio, M., Baroni, P., Pellegrini, D., Toni, F.: Comparing and integrating argumentation-based with matrix-based decision support in *Arg&Dec*. In: Black, E., Modgil, S., Oren, N. (eds.) TAFA 2015. LNCS (LNAI), vol. 9524, pp. 1–20. Springer, Cham (2015). https://doi.org/10.1007/978-3-319-28460-6_1
5. Baroni, P., Rago, A., Toni, F.: From fine-grained properties to broad principles for gradual argumentation: a principled spectrum. Int. J. Approx. Reason. **105**, 252–286 (2019)
6. Baroni, P., Romano, M., Toni, F., Aurisicchio, M., Bertanza, G.: Automatic evaluation of design alternatives with quantitative argumentation. Argument Comput. **6**(1), 24–49 (2015)
7. Bench-Capon, T.: Persuasion in practical argument using value-based argumentation frameworks. J. Log. Comput. **13**(3), 429–448 (2003)
8. Buckingham Shum, S., Hammond, N.: Argumentation-based design rationale: what use at what cost? Int. J. Hum.-Comput. Stud. **40**(4), 603–652 (1994)
9. Cayrol, C., Lagasquie-Schiex, M.C.: Gradual valuation for bipolar argumentation frameworks. In: Godo, L. (ed.) ECSQARU 2005. LNCS (LNAI), vol. 3571, pp. 366–377. Springer, Heidelberg (2005). https://doi.org/10.1007/11518655_32
10. Cocarascu, O., Rago, A., Toni, F.: Extracting dialogical explanations for review aggregations with argumentative dialogical agents. In: Proceedings of the 18th International Conference on Autonomous Agents and MultiAgent Systems, (AAMAS), pp. 1261–1269 (2019)
11. Dung, P.M.: On the acceptability of arguments and its fundamental role in nonmonotonic reasoning, logic programming and n-person games. Artif. Intell. **77**(2), 321–358 (1995)
12. Fischer, G., Lemke, A.C., McCall, R., Morch, A.I.: Making argumentation serve design. Hum.-Comput. Interact. **6**(3), 393–419 (1991)
13. Irwin, B., Rago, A., Toni, F.: Forecasting argumentation frameworks. In: Proceedings of the 19th International Conference on Principles of Knowledge Representation and Reasoning (KR), pp. 533–543 (2022)
14. Kaci, S., van der Torre, L.W.N., Villata, S.: Preference in abstract argumentation. In: Proceedings of the 7th International Conference on Computational Models of Argument (COMMA), pp. 405–412 (2018)
15. Mailly, J., Rossit, J.: Argument, I choose you! preferences and ranking semantics in abstract argumentation. In: Proceedings of the 17th International Conference on Principles of Knowledge Representation and Reasoning, (KR), pp. 647–651 (2020)
16. Modgil, S., Prakken, H.: A general account of argumentation with preferences. Artif. Intell. **195**, 361–397 (2013)
17. Potyka, N.: Continuous dynamical systems for weighted bipolar argumentation. In: Proceedings of the 16th International Conference on Principles of Knowledge Representation and Reasoning (KR), pp. 148–157 (2018)
18. Rago, A., Li, H., Toni, F.: Interactive explanations by conflict resolution via argumentative exchanges. In: Proceedings of the 20th International Conference on Principles of Knowledge Representation and Reasoning, (KR), pp. 582–592 (2023)
19. Rago, A., Toni, F., Aurisicchio, M., Baroni, P.: Discontinuity-free decision support with quantitative argumentation debates. In: Proceedings of the 15th International Conference on Principles of Knowledge Representation and Reasoning (KR), pp. 63–73 (2016)

Scaling Up Reasoning from Conditional Belief Bases

Christoph Beierle[1(✉)], Jonas Haldimann[1,2,3], Arthur Sanin[1], Leon Schwarzer[1], Aron Spang[1], Lars-Phillip Spiegel[1], and Martin von Berg[1]

[1] Knowledge-Based Systems, Faculty of Mathematics and Computer Science, FernUniversität in Hagen, 58084 Hagen, Germany
christoph.beierle@fernuni-hagen.de

[2] Institute of Logic and Computation, TU Wien, 1040 Vienna, Austria

[3] University of Cape Town and CAIR, Cape Town 7700, South Africa

Abstract. This paper introduces the online reasoning platform InfOCF-Web 2.0 that provides easy access to implementations of various inference methods for conditional belief bases. We present an overview of the realization of the inductive inference operators p-entailment, system Z, c-inference, and system W. In order to address the fact that the possible worlds to be taken into account grow exponentially with the propositional signature over which the conditionals in the belief base are defined, the implementations employ SAT and Partial MaxSAT concepts and use the power of current SAT and SMT solvers. Our evaluation shows that each of the four inference operators can handle belief bases over signatures containing more than 100 variables and with more than 100 conditionals. Thus, InfOCF-Web 2.0 scales up nonmonotonic reasoning from conditionals to a new dimension because apart from the implementations now available in InfOCF-Web 2.0, there is no other implementation of an inference operator for conditional belief bases for which such problem sizes are feasible.

1 Introduction

Conditionals of the form "If A, then usually B" encode a plausible, yet uncertain relation between the antecedent A and the consequent B. The question what a belief base consisting of a set of such conditionals should entail has been studied extensively in the area of logic-based knowledge representation and reasoning, and different semantic approaches have been proposed for dealing with conditionals (e.g. [1,11,29,33]). Besides the well known inference methods of p-entailment that coincides with reasoning according to the axioms of system P [1,29] and system Z that coincides with rational closure [19,36], newer approaches include c-inference [5] that takes all c-representations [25] of a belief base into account, and the more recently introduced system W [28]. Notable properties of c-inference and system W include capturing and going beyond p-entailment, and unlike system P and system Z, avoiding the drowning problem [12,36] and fully complying with syntax splitting [22,26]; furthermore, system W captures and properly extends not only c-inference, but also rational closure.

In contrast to these proposals and investigations on the theoretical side, the actual implementation of the resulting nonmonotonic inference methods have attracted much

S. Destercke et al. (Eds.): SUM 2024, LNAI 15350, pp. 29–44, 2025.
https://doi.org/10.1007/978-3-031-76235-2_3

less attention, e.g., [6,7,9,30,34,39,40], and only three of these approaches [7,30, 39] also address the question of how to implement reasoning when the size of belief bases increases. However, all implementations of inference operators provided by [7, 30] are limited in the sense that they can deal with belief bases involving only up to approximately 25 conditionals over about 25 propositional variables, and the approach to reasoning with system P in [39] shows that it can deal with belief bases over up to 50 propositional variables, but instead of a sound implementation it only provides an approximation because an incomplete heuristic is used.

The main reason underlying these restrictions is that the number of propositional interpretations, called possible worlds, grows exponentially with the number of propositional variables. For dealing with this, recently, new approaches for implementing c-inference [4] and system W [28] have been proposed that employ the power of modern SAT and SMT solvers. In this paper, we build upon these approaches, develop new implementations for system P and system Z, and integrate them in the new online reasoning platform InfOCF-Web 2.0.[1] Our evaluation shows that the implementations in InfOCF-Web 2.0 outperform all previous implementations of system P, system Z, c-inference, and system W, and scale up reasoning from conditionals to a new dimension, supporting belief bases of 100+ conditionals over more than 2^{100} possible worlds.

After briefly recalling the necessary basics in Sec. 2, we describe the inference operators and their implementations in Sect. 3. In Sect. 4, we present the web interface of InfOCF-Web 2.0 and our evaluation of its underlying implementations, before concluding and pointing our future work in Sect. 5.

2 Background on Conditional Logic

A *(propositional) signature* is a finite set Σ of propositional variables, and the usual propositional language over Σ is here denoted by $\mathcal{L}_\Sigma$. Usually, we denote elements of the signatures with lowercase letters $a, b, c, \ldots$ and formulas with uppercase letters $A, B, C, \ldots$. We may denote a conjunction $A \wedge B$ by AB and a negation $\neg A$ by $\overline{A}$ for brevity of notation. The set of interpretations over Σ, also called *worlds*, is denoted as Ω_Σ. An interpretation $\omega \in \Omega_\Sigma$ is a *model* of a formula $A \in \mathcal{L}_\Sigma$ if A holds in ω, denoted as $\omega \models A$, and the set of models of A is $\Omega_A = \{\omega \in \Omega_\Sigma \mid \omega \models A\}$. A formula A *entails* a formula B, denoted by $A \models B$, if $\Omega_A \subseteq \Omega_B$.

A *conditional* $(B|A)$ connects two formulas $A, B \in \mathcal{L}_\Sigma$ and represents the rule "If A then usually B". The conditional language over a signature Σ is denoted as $(\mathcal{L}|\mathcal{L})_\Sigma = \{(B|A) \mid A, B \in \mathcal{L}_\Sigma\}$. A finite set Δ of conditionals is called a *(conditional) belief base*. In this paper, we use a three-valued semantics of conditionals [18]. For a world ω, a conditional $(B|A)$ is either *verified* by ω if $\omega \models AB$, *falsified* by ω if $\omega \models A\overline{B}$, or *not applicable* to ω if $\omega \models \overline{A}$. A common formalism for modelling conditionals and conditional belief bases are ranking functions, introduced in a more general form in [38]. A ranking function, also called *ordinal conditional function (OCF)*, is a function $\kappa : \Omega_\Sigma \to \mathbb{N}_0$ such that $\kappa^{-1}(0) \neq \emptyset$, the intuition being that more plausible worlds are assigned lower ranks by κ. Ranking functions are lifted to formulas

[1] InfOCF-Web 2.0 is available at https://www.fernuni-hagen.de/wbs/research/infocf-web/.

by letting $\kappa(A) = \min_{\omega \in \Omega_A} \kappa(\omega)$, and a ranking function κ models or *accepts* a conditional $(B|A)$, denoted by $\kappa \models (B|A)$, if $\kappa(AB) < \kappa(A\overline{B})$. A belief base Δ is called *consistent* if there is a ranking function κ accepting every conditional in Δ, denoted by $\kappa \models \Delta$ [19].

3 Inductive Inference Operators Available in InfOCF-Web 2.0

An inductive inference operator completes an explicitly given belief base to the inference relation representing all conditional beliefs an agent can derive [26]. In this section, we present an overview of the inductive inference operators available in InfOCF-Web 2.0 and how they are implemented.

3.1 Implementation of p-Entailment

Nonmonotonic inference relations are usually evaluated by means of properties. In particular, the axiom system P [1] provides an important standard for plausible, nonmonotonic inferences. The most conservative inference operator satisfying system P is *p-entailment* [1]. While originally introduced in a probabilistic framework [1] and later characterized via preferential models [32], it can also be defined in terms of ranking functions. Every OCF κ induces an inference relation $\mathrel{|\!\sim}_\kappa$ that satisfies system P by

$$A \mathrel{|\!\sim}_\kappa B \quad \text{iff} \quad A \equiv \bot \quad \text{or} \quad \kappa(AB) < \kappa(A\overline{B}), \tag{1}$$

and p-entailment takes all ranking models of a belief base into account.

Definition 1 (p-entailment [19]**).** *Let Δ be a belief base and let A, B be formulas. A* p-entails *B in the context of Δ, written $A \mathrel{|\!\sim}^p_\Delta B$, if $A \mathrel{|\!\sim}_\kappa B$ for all $\kappa \models \Delta$.*

p-Entailment is sound and complete for inferences from belief bases via system P, i.e., given a belief base Δ, inference with the axioms of system P coincides with p-entailment [1,19,35]. p-Entailment can be reduced to testing for consistency.

Proposition 1 ([19]). *$A \mathrel{|\!\sim}^p_\Delta B$ iff $\Delta \cup \{(\overline{B}|A)\}$ is inconsistent.*

The consistency of a belief base Δ can be determined via the ordered partition of Δ induced by the notion of tolerance.

Definition 2 (tolerance, $OP(\Delta)$ [19]**).** *A conditional $(B|A)$ is* tolerated *by a set of conditionals $\Delta = \{(B_i|A_i) \mid i = 1, \ldots, n\}$ if there is a world $\omega \in \Omega$ such that ω verifies $(B|A)$ and ω does not falsify any conditional in Δ, i.e., $\omega \models AB$ and $\omega \models \bigwedge_{i=1}^n (\overline{A_i} \vee B_i)$. The* ordered partition *of Δ is $OP(\Delta) = (\Delta^0, \ldots, \Delta^k)$ where each Δ^i is the inclusion maximal subset of $\bigcup_{j=i}^k \Delta^j$ that is tolerated by $\bigcup_{j=i}^k \Delta^j$.*

It is well-known that $OP(\Delta)$ exists iff Δ is consistent, and because the Δ^i are inclusion-maximal, $OP(\Delta)$ is unique [36].

Algorithm 1. isConsistent(Δ)

Input: belief base Δ
Output: *Yes* if Δ is consistent and *No* otherwise.

let $i = 0$
while $\Delta \neq \emptyset$ **do**
  $\Delta^i \leftarrow \{(B|A) \mid (B|A) \text{ is tolerated by } \Delta\}$
  if $\Delta^i = \emptyset$ **then**
    return *No* ▷ $OP(\Delta)$ does not exist
  else
    $\Delta \leftarrow \Delta \setminus \Delta^i$
    $i \leftarrow i + 1$
return *Yes* ▷ $OP(\Delta) = (\Delta^0, \ldots, \Delta^k)$

Algorithm 2. Pinf(Δ, A, B)

Input: consistent belief base Δ and formulas A, B
Output: *Yes* if $A \mathrel{|\!\!\sim}^p_\Delta B$ and *No* otherwise

let $\Delta' = \Delta \cup (\overline{B}|A)$
return if isConsistent(Δ') **then** *No* **else** *Yes*

Example 1 (Δ_{bird}). Let $\Sigma = \{b, p, f, w\}$ representing birds, penguins, flying things and winged things, and let Δ_{bird} contain $r_1 = (B_1|A_1) = (f|b)$, $r_2 = (B_2|A_2) = (\overline{f}|p)$, $r_3 = (B_3|A_3) = (b|p)$, and $r_4 = (B_4|A_4) = (w|b)$. For instance, r_1 expresses *"birds usually fly"*. To check whether b entails $\overline{p}$, we check the consistency of $\Delta^{q1} = \Delta_{bird} \cup \{(p|b)\}$ by trying to construct $OP(\Delta^{q1})$. Because Δ^{q1} is inconsistent, we know that $b \mathrel{|\!\!\sim}^p_{\Delta_{bird}} \overline{p}$. For checking whether $\overline{b}$ entails $\overline{p}$, we check the consistency of $\Delta^{q2} = \Delta_{bird} \cup \{(p|\overline{b})\}$. We have that Δ^{q2} is consistent and therefore $\overline{b} \mathrel{|\!\!\not\sim}^p_{\Delta_{bird}} \overline{p}$.

The realization of p-entailment in InfOCF-Web 2.0 uses the characterization in Proposition 1 and implements the consistency test as given in [19] (Algorithms 1 and 2). The complexity of computing $\mathrel{|\!\!\sim}^p_\Delta$ is thus determined by the complexity of the consistency test. The computation of $OP(\Delta)$ needed for the consistency test requires $O(n^2)$ propositional satisfiability tests where $n = |\Delta|$ [19, Collorary 6]. These satisfiability tests involve 2^m possible worlds where $m = |\Sigma|$. In InfOCF-Web 2.0, these tests are implemented using the power of the SAT solver Z3 [16], supporting entailments with belief bases containing 100+ conditionals over 100+ signature elements.

3.2 Implementation of System Z

The ordered partition $OP(\Delta)$ groups the conditionals in Δ according to how "exceptional" they are: Δ^0 contains the most general conditionals and the later sets in the partition contain increasingly more specific conditionals. Pearl's *system Z* [36] (that coincides with rational closure [19]) is an inductive inference operator that uses $OP(\Delta)$, also called *Z-partition* of Δ, to rank worlds according to their plausibility: worlds that

Algorithm 3. $\mathsf{SZinf}(\Delta, A, B)$

Input: consistent belief base Δ and formulas A, B
Output: *Yes* if $A \mathrel{|\!\sim}^{z}_{\Delta} B$, and *No* otherwise

let $OP(\Delta) = (\Delta^0, \dots, \Delta^k)$
function *recZinf*(j)
  $V \leftarrow SAT(\bigcup_{i=j}^{k} nf(\Delta^i) \cup \{AB\})$
  if $V = UNSAT$ **then**
    return *No*
  $F \leftarrow SAT(\bigcup_{i=j}^{k} nf(\Delta^i) \cup \{A\overline{B}\})$
  if $F = SAT$ **then**
    if $j = 0$ **then**
      return *No*
    if *recZinf*($j - 1$) $= No$ **then**
      return *No*
  return *Yes*
end function
return if $A \equiv \bot$ **then** *Yes* **else** *recZinf*(k)

falsify a specific conditional are considered less likely than worlds falsifying no conditional or only a more general conditional.

Definition 3 (system Z, $\mathrel{|\!\sim}^{z}_{\Delta}$[36]). *Let Δ be a belief base with $OP(\Delta) = (\Delta^0, \dots, \Delta^k)$. The Z-ranking function κ^z_Δ is defined as follows: For each world $\omega \in \Omega_\Sigma$, let Δ^j be the last part in $OP(\Delta)$ that contains a conditional falsified by ω. Then, let $\kappa^z_\Delta(\omega) := j + 1$. If ω does not falsify any conditional in Δ, then let $\kappa^z_\Delta(\omega) := 0$. The inductive inference operator* system Z *maps Δ to the relation $\mathrel{|\!\sim}^{z}_{\Delta}$ induced by κ^z_Δ via* (1).

Example 2 (Example 1 *cont'd).* For Δ_{bird} from Example 1 we have $OP(\Delta_{bird}) = ({\Delta_{bird}}^0, {\Delta_{bird}}^1)$ with ${\Delta_{bird}}^0 = \{r_1, r_4\}$ and ${\Delta_{bird}}^1 = \{r_2, r_3\}$. For the Z-ranking function induced by Δ_{bird} we have $\kappa^z_{\Delta_{bird}}(\overline{p}b) = 0$ and $\kappa^z_{\Delta_{bird}}(p\overline{b}) = 2$, thus $\overline{b} \mathrel{|\!\sim}^{z}_{\Delta_{bird}} \overline{p}$. Analogously, we can check that $\kappa^z_{\Delta_{bird}}(wp) = \kappa^z_{\Delta_{bird}}(\overline{w}p) = 1$, yielding $p \not\mathrel{|\!\sim}^{z}_{\Delta_{bird}} w$.

A straightforward way to implement system Z would be to determine the ranking function κ^z_Δ and then compare the ranks $\kappa^z_\Delta(AB)$ and $\kappa^z_\Delta(A\overline{B})$ as in (1) to determine whether A entails B. This approach may soon become infeasible for larger examples because the number of possible worlds in the ranking function grows exponentially with the size of the signature; for an approach avoiding to work explicitly with possible worlds in the related area of iterated belief revision see [37]. Instead, InfOCF-Web 2.0implements the algorithm SZinf (Algorithm 3) using the SAT solver Z3 and exploiting the following observation.

Assume we have a consistent belief base Δ with $OP(\Delta) = (\Delta^0, \ldots, \Delta^k)$. Observe that if every model of a formula F falsifies a conditional in Δ^j, then $\kappa^z_\Delta(F) \geqslant j+1$, and that if a model of F does not falsify any conditional in $\Delta^j \cup \cdots \cup \Delta^k$, then $\kappa^z_\Delta(F) < j+1$. For a conditional $(D|C)$, the formula $\overline{C} \vee D$ expressing its *non-falsification* is denoted by $nf(D|C)$, and nf is extended canonically to a set Δ of condtionals. Thus,

$$nf(\Delta) = \{\overline{C} \vee D \mid (D|C) \in \Delta\} \tag{2}$$

is the set of non-falsifying formulas for the conditionals in Δ. For testing whether A entails B, $\mathsf{SZinf}(\Delta, A, B)$ considers the parts in the Z-partition beginning with the latest Δ^k. The main part of SZinf is a recursive function $recZinf(j)$ whose argument is the index of the part of $OP(\Delta)$ considered next. For Δ^j, the algorithm determines $V^j = SAT(\bigcup_{i=j}^k nf(\Delta^i) \cup \{AB\})$ and $F^j = SAT(\bigcup_{i=j}^k nf(\Delta^i) \cup \{A\overline{B}\})$. Note that for $j < k$, $recZinf(j)$ is only called if $F^{j+1} = SAT$ (cf. Line 7), implying that $\kappa^z_\Delta(A\overline{B}) < j+2$. If $V = UNSAT$, then $\kappa^z_\Delta(AB) \geqslant j+1$ and thus $\kappa^z_\Delta(AB) \not< \kappa^z_\Delta(A\overline{B})$, implying $A \not\mid\!\sim^z_\Delta B$. If $V = SAT$ and $F = UNSAT$, then we can conclude that $\kappa^z_\Delta(AB) < j+1$ and $\kappa^z_\Delta(A\overline{B}) \geqslant j+1$ and thus $\kappa^z_\Delta(AB) < \kappa^z_\Delta(A\overline{B})$, implying $A \mid\!\sim^z_\Delta B$. If $V = SAT$ and $F = SAT$, we continue by recursively calling $recZinf(j-1)$ to check the parts of $OP(\Delta)$ with a lower index. Line 14 contains the initial call to $recZinf$ and the handling of the border case that the antecedent is an inconsistent formula.

3.3 Implementation of c-Inference

Other than system Z, the approach of c-representations does not use the most severe falsification of a conditional, but assigns an individual impact to each conditional and generates the world ranks as a sum of impacts of falsified conditionals.

Definition 4 (c-representation [24,25]**).** *A* c-representation *of a belief base* $\Delta = \{(B_1|A_1), \ldots, (B_n|A_n)\}$ *is a ranking function* κ *constructed from integer impacts* $\eta_i \in \mathbb{N}_0$ *assigned to each* $(B_i|A_i) \in \Delta$ *such that* κ *accepts* Δ *and is given by:*

$$\kappa(\omega) = \sum_{1 \leqslant i \leqslant n,\ \omega \models A_i\overline{B}_i} \eta_i \tag{3}$$

Example 3. Verification (v), falsification (f), and non-applicability (–) of the conditionals in Δ_{bird} are given in Table 1, along with the three impact vectors $\vec{\eta}_1, \vec{\eta}_2, \vec{\eta}_3$, each representing a possible assignment of impacts to conditionals, and their induced ranking functions $\kappa_{\vec{\eta}_1}, \kappa_{\vec{\eta}_2}, \kappa_{\vec{\eta}_3}$. Using Table 1, one can check that $\kappa_{\vec{\eta}_i} \models \Delta_{bird}$ holds for $i = 1, 2, 3$, ensuring that $\kappa_{\vec{\eta}_i}$ is a c-representaton of Δ_{bird}.

Table 1. c-Representations for Δ_{bird} in Example 3.

ω	r_1: $(f\|b)$	r_2: $(\overline{f}\|p)$	r_3: $(b\|p)$	r_4: $(w\|b)$	impact on ω	$\kappa_{\vec{\eta}_1}(\omega)$	$\kappa_{\vec{\eta}_2}(\omega)$	$\kappa_{\vec{\eta}_3}(\omega)$
$b\,p\,f\,w$	v	f	v	v	η_2	2	4	5
$b\,p\,f\,\overline{w}$	v	f	v	f	$\eta_2 + \eta_4$	3	7	12
$b\,p\,\overline{f}\,w$	f	v	v	v	η_1	1	3	4
$b\,p\,\overline{f}\,\overline{w}$	f	v	v	f	$\eta_1 + \eta_4$	2	6	11
$b\,\overline{p}\,f\,w$	v	–	–	v	0	0	0	0
$b\,\overline{p}\,f\,\overline{w}$	v	–	–	f	η_4	1	3	7
$b\,\overline{p}\,\overline{f}\,w$	f	–	–	v	η_1	1	3	4
$b\,\overline{p}\,\overline{f}\,\overline{w}$	f	–	–	f	$\eta_1 + \eta_4$	2	6	11
$\overline{b}\,p\,f\,w$	–	f	f	–	$\eta_2 + \eta_3$	4	8	11
$\overline{b}\,p\,f\,\overline{w}$	–	f	f	–	$\eta_2 + \eta_3$	4	8	11
$\overline{b}\,p\,\overline{f}\,w$	–	v	f	–	η_3	2	4	6
$\overline{b}\,p\,\overline{f}\,\overline{w}$	–	v	f	–	η_3	2	4	6
$\overline{b}\,\overline{p}\,f\,w$	–	–	–	–	0	0	0	0
$\overline{b}\,\overline{p}\,f\,\overline{w}$	–	–	–	–	0	0	0	0
$\overline{b}\,\overline{p}\,\overline{f}\,w$	–	–	–	–	0	0	0	0
$\overline{b}\,\overline{p}\,\overline{f}\,\overline{w}$	–	–	–	–	0	0	0	0
impacts:	η_1	η_2	η_3	η_4				
$\vec{\eta}_1$	1	2	2	1				
$\vec{\eta}_2$	3	4	4	3				
$\vec{\eta}_3$	4	5	6	7				

Every c-representation κ induces an inference relation according to (1), exhibiting desirable inference properties [25]. The inductive inference operator *c-inference* takes all c-representations of Δ into account [2].

Definition 5 (c-inference, $\mid\!\sim^{c}_{\Delta}$[2]). *Let Δ be a belief base and $A, B \in \mathcal{L}_\Sigma$. B is* a c-inference *from A in the context of Δ, denoted by $A \mid\!\sim^{c}_{\Delta} B$, iff $A \mid\!\sim^{\kappa} B$ holds for all c-representations κ for Δ.*

E.g., $p \mid\!\sim^{c}_{\Delta_{bird}} w$ holds because $p \mid\!\sim^{\kappa} w$ holds not only for the three c-representations given in Table 1, but for all c-representations of Δ_{bird}; for detailed explanations of c-representations and of c-inference we refer to [4,25]. While p-entailment can be reduced to ckecking the inconsistency of a set of conditionals, c-inference can be reduced to checking the unsolvability of a constraint satisfaction problem (CSP).

Proposition 2. ($CR(\Delta, A, B)$[2]). *For $\Delta = \{(B_1|A_1), \ldots, (B_n|A_n)\}$, $A, B \in \mathcal{L}$, let*

$$CR(\Delta, A, B) = CR(\Delta) \cup \{\neg CR_\Delta(B|A)\} \tag{4}$$

with the constraint satisfaction problem $CR(\Delta)$ on the constraint variables $\{\eta_1, \ldots, \eta_n\}$ ranging over $\mathbb{N}_0$ given by the constraints, for all $i \in \{1, \ldots, n\}$:

$$\eta_i \geqslant 0 \tag{5}$$

$$\eta_i > \min_{\omega \models A_i B_i} \sum_{\substack{j \neq i \\ \omega \models A_j \overline{B}_j}} \eta_j - \min_{\omega \models A_i \overline{B}_i} \sum_{\substack{j \neq i \\ \omega \models A_j \overline{B}_j}} \eta_j \tag{6}$$

and with $\neg CR_{\Delta}(B|A)$ *being the constraint:*

$$\min_{\omega \models AB} \sum_{\substack{1 \leqslant i \leqslant n \\ \omega \models A_i \overline{B}_i}} \eta_i \geqslant \min_{\omega \models A\overline{B}} \sum_{\substack{1 \leqslant i \leqslant n \\ \omega \models A_i \overline{B}_i}} \eta_i \tag{7}$$

Then, $A \mid\!\sim^{c}_{\Delta} B$ *iff* $CR(\Delta, A, B)$ *is not solvable.*

The implementation of c-inference in InfOCF-Web 2.0 is based on the characterization given in Proposition 2. Note that the constraints in $CR(\Delta)$ are the constraints that ensure that $\kappa_{\vec{\eta}}$ accepts all conditionals in Δ [24], while $\neg CR_{\Delta}(B|A)$ expresses that $(B|A)$ is not accepted. A major challenge for realizing the check for unsatisfiability of $CR(\Delta, A, B)$ are the multiple min-expressions that are defined over the set of all possible worlds. To address this problem, a Partial MaxSAT approach has been developed that reduces both the size and the number of the min-expressions [15] and that is implememented using the solver RC2 [23]. For the details of transforming $CR(\Delta, A, B)$ into the thereby obtained satisfiability equivalent $CR_{opt}(\Delta, A, B)$, we refer to [15] and present an example for illustration here.

Example 4. In addition to $\eta_i \geqslant 0$, $CR(\Delta_{bird}, p, w)$ contains the constraints

$$\eta_1 > \min_{\substack{\omega \in \Omega_\Sigma \\ \omega \models bf}} \sum_{\substack{j \neq 1 \\ \omega \models A_j \overline{B}_j}} \eta_j - \min_{\substack{\omega \in \Omega_\Sigma \\ \omega \models b\overline{f}}} \sum_{\substack{j \neq 1 \\ \omega \models A_j \overline{B}_j}} \eta_j$$

$$\eta_2 > \min_{\substack{\omega \in \Omega_\Sigma \\ \omega \models p\overline{f}}} \sum_{\substack{j \neq 2 \\ \omega \models A_j \overline{B}_j}} \eta_j - \min_{\substack{\omega \in \Omega_\Sigma \\ \omega \models pf}} \sum_{\substack{j \neq 2 \\ \omega \models A_j \overline{B}_j}} \eta_j$$

$$\eta_3 > \min_{\substack{\omega \in \Omega_\Sigma \\ \omega \models pb}} \sum_{\substack{j \neq 3 \\ \omega \models A_j \overline{B}_j}} \eta_j - \min_{\substack{\omega \in \Omega_\Sigma \\ \omega \models p\overline{b}}} \sum_{\substack{j \neq 3 \\ \omega \models A_j \overline{B}_j}} \eta_j$$

$$\eta_4 > \min_{\substack{\omega \in \Omega_\Sigma \\ \omega \models bw}} \sum_{\substack{j \neq 4 \\ \omega \models A_j \overline{B}_j}} \eta_j - \min_{\substack{\omega \in \Omega_\Sigma \\ \omega \models b\overline{w}}} \sum_{\substack{j \neq 4 \\ \omega \models A_j \overline{B}_j}} \eta_j$$

$$\min_{\substack{\omega \in \Omega_\Sigma \\ \omega \models pw}} \sum_{\substack{1 \leqslant i \leqslant 4 \\ \omega \models A_i \overline{B}_i}} \eta_i \geqslant \min_{\substack{\omega \in \Omega_\Sigma \\ \omega \models p\overline{w}}} \sum_{\substack{1 \leqslant i \leqslant 4 \\ \omega \models A_i \overline{B}_i}} \eta_i$$

which in $CR_{opt}(\Delta_{bird}, p, w)$ are reduced and simplified to:

$$\eta_2 > \min\{\eta_1, \eta_3\},\ \eta_3 > \min\{\eta_2, \eta_1\},\ \min\{\eta_1, \eta_2, \eta_3\} \geqslant \min\{\eta_3, \eta_1 + \eta_4, \eta_2 + \eta_4\}$$

$CR_{opt}(\Delta, A, B)$ is then further transformed into an SMT problem $SMT_{opt}(\Delta, A, B)$ with linear arithmetic [17] following the approach developed in [3,13]. One of the key ideas of the transformation from $CR_{opt}(\Delta, A, B)$ to $SMT_{opt}(\Delta, A, B)$ is the replacement of minimum expressions according to the following principle: For integers m, a, b, c, the equation $m = min\{a, b, c\}$ is equivalent to $(m \leqslant a) \wedge (m \leqslant b) \wedge (m \leqslant c) \wedge \neg((m < a) \wedge (m < b) \wedge (m < c))$. The (un-)

Algorithm 4. $\mathsf{Cinf}(\Delta, A, B)$

Input: consistent belief base Δ and formulas A, B
Output: *Yes* if $A \mathrel{|\!\!\sim}^{c}_{\Delta} B$, and *No* otherwise

1: $\mathcal{S}_1 \leftarrow CR_{opt}(\Delta, A, B)$ ▷ reduce size and number of min-expressions in $CR(\Delta, A, B)$
2: $\mathcal{S}_2 \leftarrow SMT_{opt}(\Delta, A, B)$ ▷ transform into SMT_{LIA}-problem
3: **return if** $\mathcal{S}_2 = $ *UNSAT* **then** *Yes* **else** *No*

satisfiability of $SMT_{opt}(\Delta, A, B)$, which is satisfiability equivalent to $CR(\Delta, A, B)$, is then checked using the SMT solver Z3 [16]. A high-level description of the implementation of c-inference in InfOCF-Web 2.0 is given in Algorithm 4. For instance, because $CR(\Delta_{bird}, p, w)$ is unsatisfiable, Algorithm 4 determines that $p \mathrel{|\!\!\sim}^{c}_{\Delta_{bird}} w$ holds.

3.4 Implementation of System W

While system Z only considers which parts of $OP(\Delta)$ contain a falsified conditional for determining the rank of a world ω, system W [27,28] also takes into account the structural information which conditionals are falsified by ω. The definition of system W is based on a relation called *preferred structure on worlds* $<^{\mathsf{w}}_{\Delta}$ over Ω_{Σ} induced by Δ.

Definition 6 **(ξ^j, preferred structure $<^{\mathsf{w}}_{\Delta}$ on worlds** [28]**).** *Let* $\Delta = \{(B_1|A_1), \ldots, (B_n|A_n)\}$ *with* $OP(\Delta) = (\Delta^0, \ldots, \Delta^k)$*. For* $j = 0, \ldots, k$*, the function* ξ^j *maps each world* ω *to the set of conditionals from* Δ^j *that are falsified by* ω*:*

$$\xi^j(\omega) := \{(B_i|A_i) \in \Delta^j \mid \omega \models A_i\overline{B_i}\}. \tag{8}$$

The preferred structure on worlds *induced by* Δ *is given by the binary relation* $<^{\mathsf{w}}_{\Delta} \subseteq \Omega \times \Omega$ *defined by, for any* $\omega, \omega' \in \Omega$,

$$\begin{aligned} \omega <^{\mathsf{w}}_{\Delta} \omega' \quad \text{iff} \quad & \text{there exists an } m \in \{0, \ldots, k\} \text{ such that} \\ & \xi^i(\omega) = \xi^i(\omega') \quad \text{for all } i \in \{m+1, \ldots, k\} \text{ and} \\ & \xi^m(\omega) \subsetneqq \xi^m(\omega'). \end{aligned} \tag{9}$$

Thus, $\omega <^{\mathsf{w}}_{\Delta} \omega'$ if and only if ω falsifies strictly fewer conditionals than ω' in the last part of $OP(\Delta)$ where the conditionals falsified by ω and ω' differ. Note that the relation $<^{\mathsf{w}}_{\Delta}$is a strict partial order [28, Lemma 3].

Definition 7 **(system W,** $\mathrel{|\!\!\sim}^{\mathsf{w}}_{\Delta}$[28]**).** *Let* Δ *be a consistent belief base and* $A, B \in \mathcal{L}$*. Then,* B *is a* system W inference *from* A *(in the context of* Δ*), denoted* $A \mathrel{|\!\!\sim}^{\mathsf{w}}_{\Delta} B$*, iff for every* $\omega' \in Mod_{\Sigma}(A\overline{B})$ *there is an* $\omega \in Mod_{\Sigma}(AB)$ *such that* $\omega <^{\mathsf{w}}_{\Delta} \omega'$*.*

Example 5 (Example 2 cont'd). Let Δ_{bird} and $OP(\Delta_{bird}) = (\Delta^0, \Delta^1)$ as in Example 2. We have $Mod\,(pw) = \{bpfw, bp\overline{f}w, \overline{b}pfw, \overline{b}p\overline{f}w\}$ and $Mod\,(p\overline{w}) = \{bpf\overline{w}, bp\overline{f}\overline{w}, \overline{b}pf\overline{w}, \overline{b}p\overline{f}\overline{w}\}$. We check that $bpfw <^{\mathsf{w}}_{\Delta_{bird}} bpf\overline{w}$ because $\xi^1(bpfw) = \xi^1(bpf\overline{w})$ and $\xi^0(bpfw) \subsetneqq \xi^0(bpf\overline{w})$. Similarly, we can check that for every $\omega' \in Mod_{\Sigma}(p\overline{w})$ there is an $\omega \in Mod_{\Sigma}(pw)$ with $\omega <^{\mathsf{w}}_{\Delta_{bird}} \omega'$, implying $p \mathrel{|\!\!\sim}^{\mathsf{w}}_{\Delta_{bird}} w$.

Algorithm 5. SWinf($\varDelta, A, B$)

Input: consistent belief base $\varDelta$ and formulas A, B
Output: *Yes* if $A \mid\!\sim^{\mathsf{w}}_{\varDelta} B$, and *No* otherwise

let $OP(\varDelta) = (\varDelta^0, \dots, \varDelta^k)$
function *recWinf*(j, H)
  $\mathcal{V} \leftarrow MCS(nf(\varDelta^j), H \cup \{AB\})$
  $\mathcal{F} \leftarrow MCS(nf(\varDelta^j), H \cup \{A\overline{B}\})$
  if $\neg(\forall N' \in \mathcal{F}\, \exists N \in \mathcal{V} \,.\, N \subseteq N')$ **then**
    return *No*
  for all $N \in \mathcal{V} \cap \mathcal{F}$ **do**
    if $j = 0$ **then**
      return *No*
    $H_{new} \leftarrow (nf(\varDelta^j) \setminus N) \cup \overline{N}$
    if *recWinf*$(j-1, H \cup H_{new}) = No$ **then**
      return *No*
  return *Yes*
end function
return *recWinf*$(k, \emptyset)$

The implementation of system W in InfOCF-Web 2.0 uses the algorithm SWinf (Algorithm 5) introduced in [10] that does not compute the full preferred structure on worlds but answers queries to system W utilizing Partial MaxSAT concepts [31]. Given a set of formulas S of *soft constraints* and a set of formulas H of *hard constraints*, $M \subseteq S$ is a *maximal satisfiable subset (MSS)* w.r.t. (S, H) if $M \cup H$ is consistent and M is maximal with this property. Every MSS M corresponds to a *minimal correction set (MCS)* $N = S \setminus M$. The set of all MCS w.r.t. (S, H) is denoted by $MCS(S, H)$.

Example 6. Let $\varDelta_{bird}$ and $OP(\varDelta_{bird}) = (\varDelta^0, \varDelta^1)$ as in Example 2. For $S = nf(\varDelta^1)$ and $H = \{pw\}$ we get $MSS(nf(\varDelta^1), \{pw\}) = MSS(\{\overline{p} \vee \overline{f}, \overline{p} \vee b\}, \{pw\}) = \{\{\overline{p} \vee \overline{f}, \overline{p} \vee b\}\}$, and thus $MCS(nf(\varDelta^1), \{pw\}) = \{\emptyset\}$. For $S = nf(\varDelta^0)$ and $H = nf(\varDelta^1) \cup \{pw\}$ we get $MSS(nf(\varDelta^0), nf(\varDelta^1) \cup \{pw\}) = MSS(\{\overline{b} \vee f, \overline{b} \vee w\}, \{\overline{p} \vee \overline{f}, \overline{p} \vee b, pw\}) = \{\{\overline{b} \vee w\}\}$ and thus $MCS(nf(\varDelta^0), nf(\varDelta^1) \cup \{pw\}) = \{\{\overline{b} \vee f\}\}$.

Because $<^{\mathsf{w}}_{\varDelta}$ is a strict partial order, Definition 7 directly implies that is suffices to consider only the $<^{\mathsf{w}}_{\varDelta}$-minimal worlds for checking whether $\mid\!\sim^{\mathsf{w}}_{\varDelta}$ holds:

$$A \mid\!\sim^{\mathsf{w}}_{\varDelta} B \text{ iff } \begin{array}{l}\text{for every } \omega' \in \min(Mod_{\Sigma}(A\overline{B}), <^{\mathsf{w}}_{\varDelta}) \\ \text{there is an } \omega \in \min(Mod_{\Sigma}(AB), <^{\mathsf{w}}_{\varDelta}) \text{ such that } \omega <^{\mathsf{w}}_{\varDelta} \omega'.\end{array}$$

In addition to only using the minimal worlds verifying (or falsifying, respectively) the query for computing $\mid\!\sim^{\mathsf{w}}_{\varDelta}$, SWinf exploits the fact that the underlying relation $\omega <^{\mathsf{w}}_{\varDelta} \omega'$ (cf. Equation (9)) can be determined by going from the the highest partition element $\varDelta^k$ in $OP(\varDelta) = (\varDelta^0, \dots, \varDelta^k)$ down to a lower element $\varDelta^j$ only in case that $\xi^l(\omega) = \xi^l(\omega')$ for all $l \in \{j+1, \dots, k\}$.

Assume we have Δ with $OP(\Delta) = (\Delta^0, \dots, \Delta^k)$. Central part of SWinf is the recursive function *recWinf* that takes the index for one of the sets in the tolerance partition and a set of formulas as arguments. It is initially called for the last part Δ^k of the Z-partition. Each set in $\mathcal{V} = MCS(nf(\Delta^k), \{AB\})$ corresponds to a maximal selection of conditionals in Δ^k such that there is a model of AB not falsifying them, and thus to a selection of worlds that falsify a minimal set of conditionals in Δ^k. Analogously, this holds for $\mathcal{F} = MCS(nf(\Delta^k), \{A\overline{B}\})$ and $A\overline{B}$. If there is an $N \in \mathcal{F}$ for which there is no $N' \in \mathcal{V}$ with $N \subseteq N'$, there is a world $\omega' \in Mod_{\Sigma}(A\overline{B})$ for which there is no world $\omega \in Mod_{\Sigma}(AB)$ with $\xi^k(\omega) \subseteq \xi^k(\omega')$. Thus, if the condition in line 5 holds, we have that $A \not\mid\!\sim^{\mathsf{w}}_{\Delta} B$. Otherwise, we continue to consider the intersection $\mathcal{V} \cap \mathcal{F}$. Each set $N \in \mathcal{V} \cap \mathcal{F}$ corresponds to a selection of conditionals that is a minimal set of falsified conditionals in Δ^k both for some models of AB and $A\overline{B}$. If there is no such N, then all inclusions we considered in Line 5 are strict inclusions and we have $A \mid\!\sim^{\mathsf{w}}_{\Delta} B$. Otherwise, for each such N we need to consider the parts of the Z-partition with lower indices to check whether for each world $\omega' \in Mod_{\Sigma}(AB)$ that falsifies the conditionals in N there is a world $\omega \in Mod_{\Sigma}(A\overline{B})$ that falsifies the conditionals in N. To do this, we add $(nf\Delta^k \setminus N) \cup \overline{N}$ as hard constraints to fix the falsification behaviour on Δ^k and call the function *recWinf* recursively.

Example 7. Executing $\mathsf{SWinf}(\Delta_{bird}, p, w)$ results in two succesive calls of *recWinf* involving the following values and conditions, cf. Example 6:
(1) *recWinf*($j = 1, H = \emptyset$), $\mathcal{V} \leftarrow \{\emptyset\}$, $\mathcal{F} \leftarrow \{\emptyset\}$, $(\forall N' \in \mathcal{F}\, \exists N \in \mathcal{V} \,.\, N \subseteq N') =$ *true*, $\mathcal{V} \cap \mathcal{F} = \{\emptyset\}$, $j = 1 > 0$, return *Yes* iff *recWinf*($0, \{\overline{p} \vee \overline{f}, \overline{p} \vee b\}$) = *Yes*
(2) *recWinf*($j = 0, H = \{\overline{p} \vee \overline{f}, \overline{p} \vee b\}$), $\mathcal{V} \leftarrow \{\{\overline{b} \vee f\}\}$, $\mathcal{F} \leftarrow \{\{\overline{b} \vee f, \overline{b} \vee w\}\}$, $(\forall N' \in \mathcal{F}\, \exists N \in \mathcal{V} \,.\, N \subseteq N') =$ *true*, $\mathcal{V} \cap \mathcal{F} = \emptyset$ $\Rightarrow$ return *Yes*

Thus, $\mathsf{SWinf}(\Delta_{bird}, p, w)$ returns *Yes*, and $p \mid\!\sim^{\mathsf{w}}_{\Delta_{bird}} w$.

4 System Walkthrough and Evaluation

InfOCF-Web 2.0 provides implementations for all inference systems presented in Sect. 3; its graphical user interface is shown in Fig. 1. The user can enter a belief base directly, load it from a file, or load one of several example belief bases of different sizes. For a query "Does A entail B?", the user can select which inference systems are used to answer it. Figure 1 shows the results for the query "Does p entail w?", corresponding to the question "Do penguins (usually) have wings" in our running example Δ_{bird}.

For evaluating InfOCF-Web 2.0, belief bases and queries were constructed by a randomized scheme which takes as input a signature Σ. In the following, we give a brief overview of this generation scheme; the algorithms realizing it are given in [14]. For each conditional for a belief base or a query to be constructed, the number of literals to be included is randomly chosen in the integer interval $[4, 11]$. Next, an ordered list of randomly chosen literals of this number is created and split after a randomly chosen element. The elements of the two resulting lists are connected to formulas by randomly applying conjunctive and disjunctive connections of the first two list elements or negation of the first list element until only one element in each list remains; these are taken as antecedent and as consequent of the conditional. Only consistent belief bases build

Table 2. Evaluation of InfOCF-Web 2.0 implementations of system **P**, system **Z**, c-inference (**C**) and system **W**; belief bases are over signatures with size $|\Sigma|$ and contain $|\Delta|$ conditionals. Time is in milliseconds; timeout (t.o.) was set at 5 min for preprocessing, and additionally 5 min for inference time. Times are calculated as a mean over 10 different queries for each of 100 different belief bases of the 16 $(|\Sigma|, |\Delta|)$-combinations; belief bases for which at least one query lead to a timeout were not considered in the mean. **solved**: Percentage of belief bases where all queries were answered without a timeout. **pp-time**: Time for preprocessing the belief base without taking the query into account, e.g. determining $OP(\Delta)$ for system Z and for system W. **inf-time**: Time required for answering a single query after preprocessing is completed. **time**: Sum of pp-time and inf-time. Evaluation environment: The evaluation was performed on a mobile workstation equipped with an i9-11950H OctaCore processor, 128 DDR4-3200 working memory and a PCIe-x4 SSD hard drive. The operating system consists of Windows 11 running Ubuntu 22.04.04 in a Windows Subsystem for Linux 2 environment.

$\|\Sigma\|$	$\|\Delta\|$	$\mid\!\sim$	solved %	pp-time	inf-time	total time
60	60	P	100	–	54	54
		Z	100	37	20	57
		C	98	689	145	834
		W	100	38	77	115
60	80	P	100	–	40	40
		Z	100	51	23	74
		C	96	33754	7809	41563
		W	100	50	337	388
60	100	P	100	–	54	54
		Z	100	69	27	96
		C	3	70485	4767	75252
		W	92	62	2666	2728
60	120	P	100	–	60	60
		Z	100	74	27	101
		C	0	t.o.	t.o.	t.o.
		W	86	71	5866	5937
80	60	P	100	–	29	29
		Z	100	38	21	59
		C	100	275	65	340
		W	100	37	60	97
80	80	P	100	–	36	36
		Z	100	47	24	72
		C	98	11444	1433	12877
		W	100	48	145	193
80	100	P	100	–	48	48
		Z	100	63	28	91
		C	54	55850	7023	62873
		W	99	57	1953	2010
80	120	P	100	–	59	59
		Z	100	74	28	102
		C	7	73706	4267	77972
		W	88	67	5788	5856

$\|\Sigma\|$	$\|\Delta\|$	$\mid\!\sim$	solved %	pp-time	inf-time	total time
100	60	P	100	–	26	26
		Z	100	37	21	58
		C	100	214	65	280
		W	100	36	57	93
100	80	P	100	–	37	37
		Z	100	50	26	76
		C	100	616	84	700
		W	100	47	105	152
100	100	P	100	–	44	44
		Z	100	58	27	84
		C	94	17817	2043	19860
		W	100	54	275	329
100	120	P	100	–	53	53
		Z	100	69	30	99
		C	58	52114	3991	56105
		W	94	62	2187	2248
120	60	P	100	–	26	26
		Z	100	36	21	57
		C	100	215	64	279
		W	100	38	59	97
120	80	P	100	–	37	37
		Z	100	52	26	78
		C	100	449	82	531
		W	100	46	98	144
120	100	P	100	–	43	43
		Z	100	55	25	80
		C	100	1702	117	1819
		W	100	53	181	233
120	120	P	100	–	51	51
		Z	100	65	29	94
		C	89	22712	1869	24581
		W	100	60	750	810

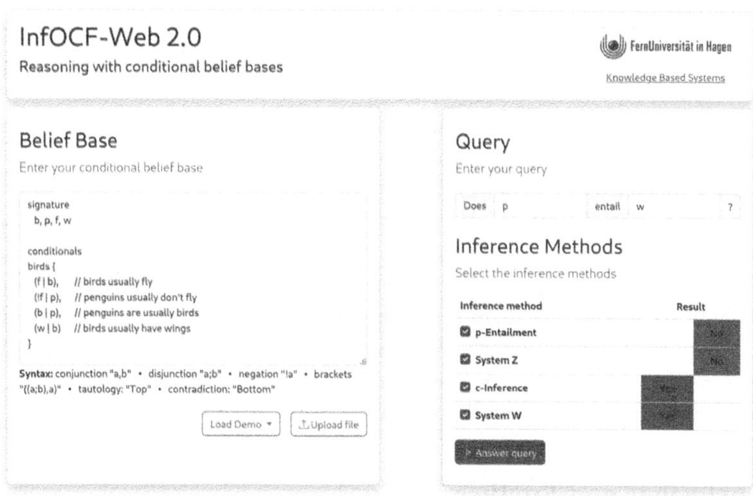

Fig. 1. InfOCF-Web 2.0 at https://www.fernuni-hagen.de/wbs/research/infocf-web/

during this process were used for benchmarking because reasoning with c-inference and system W with respect to weakly consistent belief bases has been introduced only very recently [20,21]. In the evaluation, belief bases with signature sizes $|\Sigma|$ and number of conditionals $|\Delta|$ both ranging from 60 to 120 were considered, incremented in steps of 20. For each of the 16 $(|\Sigma|, |\Delta|)$-combinations in the evaluation, 100 belief bases were created along with 10 queries for each belief base, resulting in 1600 belief bases and 16000 queries; the problem set CLKR-PS003 obtained thereby is available at the CLKR repository [8]. The evaluation times for each $(|\Sigma|, |\Delta|)$-combination are calculated as means over the corresponding 1000 queries. Table 2 shows the results of our evaluation.

Of the 16 different $(|\Sigma|, |\Delta|)$-combinations, both p-entailment and system Z solved 100% of the queries. c-Inference solved 100% of 6 combinations, between 89% and 98% of another 5 combinations, and only for one of the remaining 5 combinations, no solution was obtained due to timeout for all belief bases. System W solved 100% of 11 combinations and between 86% and 99% of the remaining 5 combinations. Thus, in the given range of 60 to 120 conditionals over 2^{60} to 2^{120} possible worlds, the results clearly demonstrate that for the realized inductive inference operators, InfOCF-Web 2.0 indeed scales up nonmonotonic reasoning from conditional belief bases to a new dimension.

In the following, we look at some of the results in more detail. System P and system Z run considerably faster than c-inference and system W and almost independently of $|\Sigma|$ and $|\Delta|$ (26 to 60 ms and 57 to 102 ms, respectively). For a fixed signature size and increasing number of conditionals, we observe an increasing preprocessing as well as and an increasing solving time with c-inference and system W. Somewhat surprisingly, for c-inference and system W, solving times decrease considerably when increasing $|\Sigma|$ from to $|\Sigma| = 60, |\Delta| = 120$ to $|\Sigma| = 120, |\Delta| = 120$. For c-inference, there is a timeout for all belief bases at $(60, 120)$, while at $(120, 120)$, 89% of belief bases could be fully queried successfully. This phenomenon is very likely caused by the fact that in the randomly generated belief basesover a fixed signature size, recurrences of literals become more frequent with decreasing belief base size. This in turn leads to more interactions among the conditionals and thus to more shared effects on possible worlds, and especially the implementation of c-inference has to take these interactions into account, resulting in higher computational demands.

The implementation of system P in [39] achieves only approximate results within hours for a signature size of at most 50 and 100 conditionals whereas our approach delivers exact results. The smallest problem size in our evaluation involves belief bases with $|\Sigma| = 60$, and to the best of our knowledge, apart from the implementations now available in InfOCF-Web 2.0, there is no other implementation of an inference operator for conditional belief bases for which such problem sizes are feasible. Therefore, we have refrained from including other approaches in our evaluation because they would run into timeout consistently for all problem sizes that we have examined here.

5 Conclusions and Further Work

The online reasoning platform InfOCF-Web 2.0 employs SAT, Partial MaxSAT and SMT concepts for implementing p-entailment, system Z, c-inference, and system W. Our evaluation demonstrates that belief base sizes can be handled that were fully infeasible before, thus scaling up nonmonotonic inference from conditional belief bases to a new dimension. Our future work includes investigating the algorithmic complexity of the described algorithms, and to evaluate them with respect to benchmarks obtained from, e.g., other randomized schemes or from translations of real-world problems. Furthermore, we will extend the implementations in InfOCF-Web 2.0 to cover also only weakly consistent belief bases and we will add further inference operators.

Acknowledgments. This work was supported by the Deutsche Forschungsgemeinschaft (DFG, German Research Foundation) - 512363537, grant BE 1700/12-1 awarded to Christoph Beierle.

References

1. Adams, E.: The logic of conditionals. Inquiry **8**(1–4), 166–197 (1965)
2. Beierle, C., Eichhorn, C., Kern-Isberner, G.: Skeptical inference based on C-representations and its characterization as a constraint satisfaction problem. In: Gyssens, M., Simari, G. (eds.) FoIKS 2016. LNCS, vol. 9616, pp. 65–82. Springer, Cham (2016). https://doi.org/10.1007/978-3-319-30024-5_4
3. Beierle, C., von Berg, M., Sanin, A.: Realization of c-inference as a SAT problem. In: The International FLAIRS Conference Proceedings, vol. 35 (2022).https://doi.org/10.32473/flairs.v35i.130663
4. Beierle, C., Eichhorn, C., Kern-Isberner, G., Kutsch, S.: Properties of skeptical c-inference for conditional knowledge bases and its realization as a constraint satisfaction problem. Ann. Math. Artif. Intell., 247–275 (2018). https://doi.org/10.1007/s10472-017-9571-9
5. Beierle, C., Eichhorn, C., Kern-Isberner, G., Kutsch, S.: Properties and interrelationships of skeptical, weakly skeptical, and credulous inference induced by classes of minimal models. Artif. Intell. **297**, 103489 (2021). https://doi.org/10.1016/j.artint.2021.103489

6. Beierle, C., Eichhorn, C., Kutsch, S.: A practical comparison of qualitative inferences with preferred ranking models. KI - Künstliche Intelligenz **31**(1), 41–52 (2016). https://doi.org/10.1007/s13218-016-0453-9
7. Beierle, C., Haldimann, J., Kollar, D., Sauerwald, K., Schwarzer, L.: An implementation of nonmonotonic reasoning with system W. In: KI 2022. LNCS, vol. 13404, pp. 1–8. Springer, Heidelberg (2022). https://doi.org/10.1007/978-3-031-15791-2_1
8. Beierle, C., Haldimann, J., Schwarzer, L.: CLKR - conditional logic and knowledge representation. KI – Künstliche Intelligenz (2024). https://doi.org/10.1007/s13218-024-00842-z
9. Beierle, C., Kutsch, S., Sauerwald, K.: Compilation of static and evolving conditional knowledge bases for computing induced nonmonotonic inference relations. Ann. Math. Artif. Intell. **1**, 5–41 (2019). https://doi.org/10.1007/s10472-019-09653-7
10. Beierle, C., Spang, A., Haldimann, J.: A partial maxsat approach to nonmonotonic reasoning with system w. In: The International FLAIRS Conference Proceedings, vol. 37, no. 1 (2024). https://doi.org/10.32473/flairs.37.1.135330
11. Benferhat, S., Dubois, D., Prade, H.: Possibilistic and standard probabilistic semantics of conditional knowledge bases. J. Logic Comput. **9**(6), 873–895 (1999)
12. Benferhat, S., Cayrol, C., Dubois, D., Lang, J., Prade, H.: Inconsistency management and prioritized syntax-based entailment. In: Proceedings of IJCAI 1993, vol. 1, pp. 640–647. Morgan Kaufmann Publishers, San Francisco (1993)
13. von Berg, M., Sanin, A., Beierle, C.: Representing nonmonotonic inference based on c-representations as an SMT problem. In: ECSQARU-2023. LNCS, vol. 14249, pp. 210–223. Springer, Heidelberg (2023). https://doi.org/10.1007/978-3-031-45608-4_17
14. von Berg, M., Sanin, A., Beierle, C.: An implementation of nonmonotonic reasoning with c-representations using an SMT solver. Int. J. Approx. Reason. **175**, 109285 (2024). https://doi.org/10.1016/j.ijar.2024.109285
15. von Berg, M., Sanin, A., Beierle, C.: Scaling up nonmonotonic c-inference via partial MaxSAT problems. In: FoIKS-2024. LNCS, vol. 14589, pp. 182–200. Springer, Heidelberg (2024). https://doi.org/10.1007/978-3-031-56940-1_10
16. Bjørner, N., de Moura, L., Nachmanson, L., Wintersteiger, C.M.: Programming Z3. In: Engineering Trustworthy Software Systems: 4th International School, SETSS 2018, Chongqing, China, 7–12 April 2018, Tutorial Lectures, vol. 4 pp. 148–201 (2019)
17. Dutertre, B., de Moura, L.: A fast linear-arithmetic solver for DPLL(T). In: Ball, T., Jones, R.B. (eds.) CAV 2006. LNCS, vol. 4144, pp. 81–94. Springer, Heidelberg (2006). https://doi.org/10.1007/11817963_11
18. de Finetti, B.: La prévision, ses lois logiques et ses sources subjectives. Ann. Inst. H. Poincaré **7**(1), 1–68 (1937). Engl. transl. *Theory of Probability*, J. Wiley & Sons, 1974
19. Goldszmidt, M., Pearl, J.: Qualitative probabilities for default reasoning, belief revision, and causal modeling. Artif. Intell. **84**(1–2), 57–112 (1996)
20. Haldimann, J., Beierle, C., Kern-Isberner, G.: Syntax splitting and reasoning from weakly consistent conditional belief bases with c-inference. In: FoIKS 2024. LNCS, vol. 14589, pp. 85–103. Springer, Heidelberg (2024). https://doi.org/10.1007/978-3-031-56940-1_5
21. Haldimann, J., Beierle, C., Kern-Isberner, G., Meyer, T.: Conditionals, infeasible worlds, and reasoning with system W. In: The International FLAIRS Conference Proceedings, vol. 36, no. 1 (2023). https://doi.org/10.32473/flairs.36.133268
22. Heyninck, J., Kern-Isberner, G., Meyer, T., Haldimann, J.P., Beierle, C.: Conditional syntax splitting for non-monotonic inference operators. In: AAAI-2023, vol. 37, pp. 6416–6424 (2023). https://doi.org/10.1609/aaai.v37i5.25789
23. Ignatiev, A., Morgado, A., Marques-Silva, J.: RC2: an efficient MaxSAT solver. J. Satisf. Boolean Model. Comput. **11**(1), 53–64 (2019). https://doi.org/10.3233/SAT190116
24. Kern-Isberner, G.: Conditionals in nonmonotonic reasoning and belief revision, LNAI, vol. 2087. Springer, Heidelberg (2001). https://doi.org/10.1007/3-540-44600-1

25. Kern-Isberner, G.: A thorough axiomatization of a principle of conditional preservation in belief revision. Ann. Math. Artif. Intell. **40**(1–2), 127–164 (2004)
26. Kern-Isberner, G., Beierle, C., Brewka, G.: Syntax splitting = relevance + independence: new postulates for nonmonotonic reasoning from conditional belief bases. In: KR-2020, pp. 560–571 (2020). https://doi.org/10.24963/kr.2020/56
27. Komo, C., Beierle, C.: Nonmonotonic inferences with qualitative conditionals based on preferred structures on worlds. In: Schmid, U., Klügl, F., Wolter, D. (eds.) KI 2020. LNCS (LNAI), vol. 12325, pp. 102–115. Springer, Cham (2020). https://doi.org/10.1007/978-3-030-58285-2_8
28. Komo, C., Beierle, C.: Nonmonotonic reasoning from conditional knowledge bases with system W. Ann. Math. Artif. Intell. **90**(1), 107–144 (2021). https://doi.org/10.1007/s10472-021-09777-9
29. Kraus, S., Lehmann, D.J., Magidor, M.: Nonmonotonic reasoning, preferential models and cumulative logics. Artif. Intell. **44**(1–2), 167–207 (1990)
30. Kutsch, S., Beierle, C.: InfOCF-web: an online tool for nonmonotonic reasoning with conditionals and ranking functions. In: IJCAI 2021, pp. 4996–4999. ijcai.org (2021). https://doi.org/10.24963/ijcai.2021/711
31. Larrosa, J., Rollon, E.: Towards a better understanding of (partial weighted) MaxSAT proof systems. In: Pulina, L., Seidl, M. (eds.) SAT 2020. LNCS, vol. 12178, pp. 218–232. Springer, Cham (2020). https://doi.org/10.1007/978-3-030-51825-7_16
32. Lehmann, D., Magidor, M.: What does a conditional knowledge base entail? Artif. Intell. **55**, 1–60 (1992)
33. Lewis, D.: Counterfactuals. Harvard University Press, Cambridge (1973)
34. Olivetti, N., Pozzato, G.L.: NESCOND: an implementation of nested sequent calculi for conditional logics. In: Demri, S., Kapur, D., Weidenbach, C. (eds.) IJCAR 2014. LNCS (LNAI), vol. 8562, pp. 511–518. Springer, Cham (2014). https://doi.org/10.1007/978-3-319-08587-6_39
35. Pearl, J.: Probabilistic Reasoning in Intelligent Systems. Morgan Kaufmann, San Mateo (1988)
36. Pearl, J.: System Z: a natural ordering of defaults with tractable applications to nonmonotonic reasoning. In: Proceedings of TARK'1990, pp. 121–135. Morgan Kaufmann (1990)
37. Schwind, N., Konieczny, S., Lagniez, J., Marquis, P.: On computational aspects of iterated belief change. In: Bessiere, C. (ed.) Proceedings of the Twenty-Ninth International Joint Conference on Artificial Intelligence, IJCAI 2020. pp. 1770–1776. ijcai.org (2020). https://doi.org/10.24963/ijcai.2020/245
38. Spohn, W.: Ordinal conditional functions: a dynamic theory of epistemic states. In: Harper, W., Skyrms, B. (eds.) Causation in Decision, Belief Change, and Statistics, II, pp. 105–134. Kluwer Academic Publishers (1988)
39. Stojanovic, T., Ikodinovic, N., Davidovic, T., Ognjanovic, Z.: Automated non-monotonic reasoning in system P. Ann. Math. Artif. Intell. **89**(5–6), 471–509 (2021)
40. Thimm, M.: Tweety: A comprehensive collection of java libraries for logical aspects of artificial intelligence and knowledge representation. In: KR 2014. AAAI Press (2014)

On Decisive Revision Operators and Totally Ordered Information

Khaled Belahcène[1], Jérome Gaigne[2], and Sylvain Lagrue[2](✉)

[1] MICS, CentraleSupélec, Université Paris-Saclay, Gif-sur-Yvette, France
khaled.belahcene@centralesupelec.fr
[2] Université de technologie de Compiègne, CNRS, Heudiasyc (Heuristics and Diagnosis of Complex Systems), CS 60 319 – 60 203, Compiègne Cedex, France
{gaigne,lagrue}@hds.utc.fr

Abstract. This paper focuses on decisive belief revision operators, i.e. operators leading to totally informed situations. Such situations can be represented in the Katsuno-Mendelzon (KM) revision framework by a complete propositional formula, from which either any formula or its negation can be entailed. From a semantic point of view, this kind of operator leads to a single, most plausible interpretation. Despite their prevalence in decision theory, this class of operators has not been previously studied in the context of the KM revision framework.

We propose in this paper to characterize decisive operators by a set of postulates. We also provide a representation theorem leading to total orders on interpretations. Finally, we exhibit a concrete operator family satisfying this new set of postulates by combining KM revision operators with tie-breaking functions.

Keywords: KM Belief Revision · Totally ordered information · Decisive Revision Operators

1 Introduction

The process of belief revision is defined as the issue that arises when an agent encounters new information about the world that may be in conflict with their existing beliefs. This article focuses on framework by Katsuno Mendelzon (KM), wherein the agent's beliefs, or its representation of the world (the situation), are represented by a propositional formula, designated as φ, and the new information is represented by another propositional formula, designated as α. In this framework, when the new information is consistent with the current beliefs the result has to be the conjunction $\varphi \wedge \alpha$. In this case, the number of interpretations compatible to the agent's beliefs decreases and the agent has a more precise vision of the situation.

This article takes this reasoning even further and focuses on *decisive revision operators* leading to *totally informed situations*. This kind of situation can be represented by a complete propositional formula from which either any formula or its negation can be entailed. In other words, a formula satisfied by a

S. Destercke et al. (Eds.): SUM 2024, LNAI 15350, pp. 45–52, 2025.
https://doi.org/10.1007/978-3-031-76235-2_4

single, most plausible interpretation. Such unequivocal outcomes are particularly important in decision situations. For instance, real-life voting procedures are *resolute*, in the sense they output a single winner, often by complementing an idealized anonymous rule (such as plurality voting or using the Borda score) with a tie-breaking procedure, e.g. by giving precedence to some specific voter or referring to an exogenous variable like "electing the oldest candidate". In decision under complete uncertainty (see for instance [3,6]), the authors of [20] coined the word *pignistic* ("suitable for betting") to extend the approach consisting in reasoning as if the latent truth were uniformly distributed among possible worlds, and therefore selecting the option supported by a majority of these. In multiple criteria decision aiding, it often happens that the parameterized procedures used to aggregate viewpoints are insufficiently elicited. While the use of cautious aggregation based on the version space of the aggregator has been advocated, the most common trend, by far, is to use soft computing techniques to select a "representative" value function that guarantees to obtain clear-cut decisions (see e.g. [4] for a discussion of this issue in the scope of sorting problems).

To capture this behavior for KM revision operators. For this purpose, we propose a set of postulates, capturing the family of *decisive revision operators*. We also provide a representation theorem leading to a *total order* on the set of interpretations. Finally, we define an operator based on the one proposed by Dalal, equipped with a *tie-breaking* function [7,10], which satisfies all our postulates.

Outline. Section 2 gives first some necessary background on ordered structures and on belief revision. Section 3 focuses on the main contribution: the characterization of decisive belief revision operators. It provides an original postulate and a representation theorem. Finally, Sect. 4 presents a concrete operator, based on Dalal's supplemented with a tie-breaking function.

2 Preliminaries

We provide in this section the needed material used in the sequel of this paper. The first part is a reminder on different notions of orders. The second part treats propositional logic, belief revision and its links with orders.

Binary Relations. Given $\mathcal{R}$ a binary relation over a set X, it is said to be a *preorder*, denoted here by $\precsim$, when it is a reflexive ($\forall x \in X$, x$\mathcal{R}x$) and transitive relation ($\forall x, y, z \in X$, if $x\mathcal{R}y$ and $y\mathcal{R}z$, then $x\mathcal{R}z$). An *order*, denoted by $\preceq$, is a preorder which is antisymmetric ($\forall x, y \in X$, if $x\mathcal{R}y$ and $y\mathcal{R}x$, then $x = y$). A *total order* (aka a linear order), is an order which is total ($\forall x, y \in X$, $x\mathcal{R}y$ or $y\mathcal{R}x$).

For a given (pre)order $\precsim$, we denote by $\prec$ its asymmetric part, i.e. $x \prec y$ when $x \precsim y$ and not $y \precsim x$; by $\approx$ its symmetric part, i.e. $x \approx y$ when both $x \precsim y$ and $y \precsim x$; and by $\sim$ its incomparability relation, i.e. $x \sim y$ when neither $x \precsim y$ nor $y \precsim x$.

Posets. A set X associated with a (pre)order $\mathcal{R}$ is a (pre)ordered set denoted $P = (X, \mathcal{R})$. A partially ordered set is often called a *poset.*

Given a preordered set $P = (X, \precsim)$ and $Y \subseteq X$, we use the following definition to denote the set of the *minimal elements* of Y :

$$\mathrm{Min}\,(Y, \precsim) = \{x \in Y : \forall y \in Y, y \prec x \text{ does not hold}\}$$

Propositional Logic. We denote by $\mathcal{P}$ a **finite** set of propositional variables and by $\mathcal{L}$ the set of propositional formulas built over $\mathcal{P}$. We will use the usual connectors $\vee$, $\wedge$, $\neg$, $\rightarrow$, $\leftrightarrow$ and the symbols $\top$ and $\bot$ to respectively represent the always-true formula and the always false formula. We use Greek letters, except Ω and ω, to represent an element of $\mathcal{L}$. Ω is the set of all interpretations of $\mathcal{L}$ and we denote by $[\![\alpha]\!]$ the models of α, that is the set $\{\omega \in \Omega : \omega \models \alpha\}$. The symbols $\vdash$ and $\models$ have their usual meanings. A formula φ is *complete* if $\forall \mu \in \mathcal{L}$, we have either $\varphi \vdash \mu$ or $\varphi \vdash \neg\mu$ but not both. In other words, $[\![\varphi]\!]$ is a singleton.

Belief Revision. The set of rationality postulates proposed in [1] constitutes the base of the AGM framework for belief revision. A translation of the AGM postulates into propositional logic, introduced in [14] by Katsuno Mendelzon, leads to the following ones:

($\boldsymbol{R_1}$) $\varphi \circ \alpha \vdash \alpha$
($\boldsymbol{R_2}$) if $\varphi \wedge \alpha \nvdash \bot$, then $\varphi \circ \alpha \equiv \varphi \wedge \alpha$
($\boldsymbol{R_3}$) if $\alpha \nvdash \bot$, then $\varphi \circ \alpha$ is consistent
($\boldsymbol{R_4}$) if $\varphi_1 \equiv \varphi_2$ and $\alpha \equiv \beta$ then $\varphi_1 \circ \alpha \equiv \varphi_2 \circ \beta$
($\boldsymbol{R_5}$) $(\varphi \circ \alpha) \wedge \beta \vdash \varphi \circ (\alpha \wedge \beta)$
($\boldsymbol{R_6}$) if $(\varphi \circ \alpha) \wedge \beta \nvdash \bot$, then $\varphi \circ (\alpha \wedge \beta) \vdash (\varphi \circ \alpha) \wedge \beta$

A revision operator $\circ$ satisfying all these postulates is equivalent to searching the minimum in a specific totally *preordered* set of interpretations provided by a faithful assignment of $\circ$.

Definition 1 (Faithful assignment, [14]**).** *A function assigning, to a formula* φ*, a total preorder* $\precsim_\varphi$ *over* Ω *is a faithful assignment iff:*

(1) if $\omega_1, \omega_2 \in [\![\varphi]\!]$ *then* $\omega_1 \prec_\varphi \omega_2$ *does not hold*
(2) if $\omega_1 \in [\![\varphi]\!]$ *and* $\omega_2 \notin [\![\varphi]\!]$ *then* $\omega_1 \prec_\varphi \omega_2$ *holds*
(3) if $\varphi \equiv \psi$*, then* $\precsim_\varphi = \precsim_\psi$

The first condition states that the models of the current beliefs are equivalent. The second one requires that all models are preferred to all countermodels. The third condition expresses independence from the syntax.

Theorem 1 (Representation, [14]**).** *A revision operator* $\circ$ *satisfies postulates* ($\boldsymbol{R_1}$)-($\boldsymbol{R_6}$) *iff there exists a faithful assignment that maps* φ *to a total preorder* $\precsim_\varphi$ *such that:*

$$[\![\varphi \circ \mu]\!] = \mathrm{Min}\,([\![\mu]\!], \precsim_\varphi)\,.$$

It can be noticed that this set can potentially contain more than one element.

3 Decisive Revision Operators

If a new piece of information α is consistent with the initial agent's belief, by satisfying R_2, the result of the revision always leads to a refinement of the initial belief $\varphi \circ \alpha \equiv \varphi \wedge \alpha$. The situation represented by $\varphi \circ \alpha$ is more specific than the initial beliefs φ of the agent, in the sense that it can decide more formulas (i.e. deduce it or its negation). Equivalently, if we focus on models, when $\varphi \wedge \alpha \not\vdash \bot$, the possible worlds of the agent represented by $[\![\varphi \circ \alpha]\!]$ form a subset of the models of the initial beliefs $[\![\varphi]\!]$.

The objective of this section is to go even further by focusing on operators that lead to a totally informed situation, i.e. a complete formula from where the agent can decide any formula. Equivalently, in terms of models, this leads to a propositional formula that admits only one model.

Example 1. Let us consider the expert beliefs of a doctor on flu (f) and coughing patients (c). The two situations that are the most plausible to them is $\neg c \wedge \neg f$ where a patient does not cough and is not suffering from the flu, and $c \wedge f$ where a patient coughs and is suffering from the flu. In other words, $[\![\varphi]\!] = \{\{\neg c, \neg f\}, \{c, f\}\}$.

Given the new information that the patient is coughing, modelled by $\alpha = c$, the result of the revision process is $\varphi \circ \alpha \equiv c \wedge f$. In this case the situation is said to be totally informed, since the new beliefs of the agent have exactly one model, $\{c, f\}$. The doctor can make a decision about their beliefs by being fully informed about all the variables of the language.

We formalize the concept of decisive revision operators in the following subsection, even if the new information is contradictory with the current agent's beliefs, and capture them by means of a set of rationality postulates. A revision operator $\circ$ is a *decisive operator* (DC-operator) if it satisfies the following property:

($\boldsymbol{DC}$) $\forall \alpha \in \mathcal{L}$, if $\alpha \nvdash \bot$, then $\varphi \circ \alpha$ is complete.

In other words, if the new information is not a contradiction, a DC-operator leads to a totally informed situation, i.e. a formula having only one model. As stated by the following proposition, an operator satisfying ($\boldsymbol{R_2}$) cannot be a DC revision operator.

Proposition 1. *Postulates* ($\boldsymbol{R_2}$) *and* ($\boldsymbol{DC}$) *are mutually inconsistent.*

For instance, let us take $\omega_1, \omega_2 \in \Omega$ such that $\omega_1, \omega_2 \in [\![\varphi]\!]$ and α such that $[\![\alpha]\!] = \{\omega_1, \omega_2\}$. From ($\boldsymbol{R_2}$), $\varphi \circ \alpha \equiv \varphi \wedge \alpha \equiv \alpha$. From ($\boldsymbol{DC}$), $\forall \beta \in \mathcal{L}$, either $\varphi \circ \alpha \vdash \beta$ or $\varphi \circ \alpha \vdash \neg\beta$. Thus $\forall \beta \in \mathcal{L}$, either $\alpha \vdash \beta$ or $\alpha \vdash \neg\beta$. If we consider β such that $[\![\beta]\!] = \{\omega_1\}$, we obtain a contradiction.
This proposition leads us to consider the following weakening of the postulate ($\boldsymbol{R_2}$):

($\boldsymbol{R_{2w}}$) if $\varphi \wedge \alpha \nvdash \bot$ then $\varphi \circ \alpha \vdash \varphi \wedge \alpha$

Such a weakening of $(\boldsymbol{R_2})$ can be found in the literature for instance in [5, 17], or in [11] where $(\boldsymbol{R_{2w}})$ appears under the name of $(\boldsymbol{R_7})$. They present it as the *Vacuity* postulate inside the AGM formulation and argue it allows the agent to drop some information of its beliefs even if it is consistent with the new information.

Now we propose an original postulate, named $(\boldsymbol{R_L})$, to characterize decisive revision operators. It states that if α and β are consistent but together inconsistent, then the disjunction between $\varphi \circ \alpha$ and $\varphi \circ \beta$ does not entail the revision of the disjunction between α and β.

$(\boldsymbol{R_L})$ if $\alpha \nvdash \bot$, $\beta \nvdash \bot$, and $\alpha \wedge \beta \vdash \bot$ then $(\varphi \circ \alpha) \vee (\varphi \circ \beta) \nvdash \varphi \circ (\alpha \vee \beta)$

Proposition 2. *Postulates* $(\boldsymbol{R_2})$ *and* $(\boldsymbol{R_L})$ *are mutually inconsistent.*

For instance, let us consider $\omega_1, \omega_2 \in \Omega$ such that $\omega_1, \omega_2 \in [\![\varphi]\!]$, α such that $[\![\alpha]\!] = \{\omega_1\}$, and β such that $[\![\beta]\!] = \{\omega_2\}$. From R_2, $\varphi \circ \alpha \equiv \varphi \wedge \alpha \equiv \alpha$. From R_2, $\varphi \circ \beta \equiv \varphi \wedge \beta \equiv \beta$. From R_2, $\varphi \circ (\alpha \vee \beta) \equiv \varphi \wedge (\alpha \vee \beta) \equiv \alpha \vee \beta$. From R_L, $(\varphi \circ \alpha) \vee (\varphi \circ \beta) \nvdash \varphi \circ (\alpha \vee \beta)$. Thus, $\alpha \vee \beta \nvdash \alpha \vee \beta$, a contradiction.

The following proposition states that in the presence of $(\boldsymbol{R_1})$, $(\boldsymbol{R_{2w}})$, $(\boldsymbol{R_3})$–$(\boldsymbol{R_6})$, the postulates $(\boldsymbol{R_L})$ and $(\boldsymbol{DC})$ are equivalent.

Proposition 3. *Let us consider a revision operator* $\circ$ *satisfying* $(\boldsymbol{R_1})$, $(\boldsymbol{R_{2w}})$,$(\boldsymbol{R_3})$–$(\boldsymbol{R_6})$, *then* $\circ$ *satisfies* $(\boldsymbol{R_L})$ *iff* $\circ$ *satisfies* $(\boldsymbol{DC})$.

Finally, we propose a representation theorem for DC-revision operators. This theorem refers to the notion of DC-faithful assignment which maps a formula to a total order where all counter-models of the formula φ are dominated by its models (1). It is also independent of the syntax (2). The main difference with standard faithful assignment, if we exclude the total order, is that the models of the formula can be *strictly* compared.

Definition 2 (DC-faithful assignment). *A function assigning to a formula* φ *a* total order $\preceq_\varphi$ *over* Ω *is a* decisive faithful assignment *iff:*

(1) if $\omega_1 \in [\![\varphi]\!]$ *and* $\omega_2 \notin [\![\varphi]\!]$ *then* $\omega_1 \prec_\varphi \omega_2$ *holds*
(2) if $\varphi \equiv \psi$, *then* $\preceq_\varphi = \preceq_\psi$

Now we are able to present the following representation theorem. Roughly speaking, this theorem states that, given a DC-revision operator, it is possible to associate to the beliefs of the agent a corresponding total order on the set of interpretations.

Theorem 2. *A revision operator* $\circ$ *satisfies* $(\boldsymbol{R_1})$, $(\boldsymbol{R_{2w}})$, $(\boldsymbol{R_3})$–$(\boldsymbol{R_6})$ *and* $(\boldsymbol{R_L})$ *iff there exists a DC-faithful assignment that maps* φ *to a total order* $\preceq_\varphi$ *such that:*

$$[\![\varphi \circ \alpha]\!] = \mathrm{Min}\left([\![\alpha]\!], \preceq_\varphi\right).$$

4 A Concrete Decisive Revision Operator

We present in this section a concrete operator showing the consistency of our set of postulates. In practice, a decisive revision operator can be obtained by coupling any KM revision operator [14] with a tie-breaking function [7,10], denoted $\mathcal{T}$ hereafter. In the sequel of this paper, we consider that all tie-breaking functions are decisive choice functions satisfying the following independence property (denoted α in [19]):

$$\forall E, E' \in 2^{\Omega} \text{ s.t. } \mathcal{T}(E) \in E' \subseteq E, \text{ we have } \mathcal{T}(E) = \mathcal{T}(E').$$

Example 2. Let us take $\mathcal{P} = \{a, b\}$, and $\Omega = \{\omega_0, \omega_1, \omega_2, \omega_3\}$ such that $\omega_0 = \{\neg a, \neg b\}$, $\omega_1 = \{\neg a, b\}$, $\omega_2 = \{a, \neg b\}$, $\omega_3 = \{a, b\}$. We consider $\varphi = \neg b$ and thus $[\![\varphi]\!] = \{\omega_0, \omega_2\}$. As an example, we can couple Dalal revision operator [8] with a tie-breaking function. The result of Dalal revision is equivalent to the set of minimal elements for this order:

$$[\![\varphi \circ_{d_H} \alpha]\!] = Min([\![\alpha]\!], \precsim_{\varphi}^{d_H}),$$

where d_H is the Hamming distance and $\precsim^{d_H}$ the preorder induced by the Hamming distance between interpretations and the agent's initial beliefs' models.

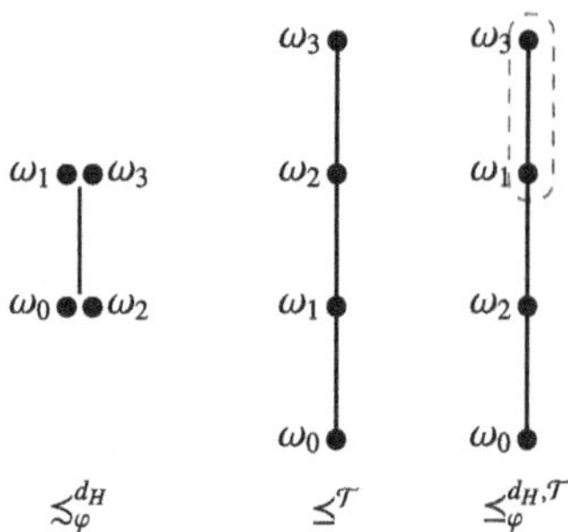

Fig. 1. Hasse diagrams representing φ from Example 2 with $\precsim_{\varphi}^{d_H}$ being the preorder obtained with the Hamming distance, $\precsim_{\mathcal{T}}$ the total order derived from $\mathcal{T}$, and $\preceq_{\varphi}^{d_H,\mathcal{T}}$ the order obtained by applying the tie-breaking function

We can then decide on ties using a tie-breaking function. As an example, we can use the lexicographic order over the interpretations $\prec^{\mathcal{T}}$, induced by the following order on the falsified propositional variables $a \prec b$ as our tie-breaking function. Figure 1 represents, using Hasse diagrams[1], φ firstly as the Dalal preorder $\precsim_{\varphi}^{d_H}$ corresponding solely to Hamming distance and then as the order $\preceq_{\varphi}^{d_H,\mathcal{T}}$ obtained by applying the tie-breaking function to it. Let us now say that φ is revised by $\alpha = b$. Thus $[\![\alpha]\!] = \{\omega_1, \omega_3\}$, and we have $[\![\varphi \circ \alpha]\!] = \{\omega_1\}$.

[1] Hasse diagrams are a graphical representation of preorders where neither reflexivity nor transitivity are represented and where the minimal elements are at the bottom of the figure.

Proposition 4. *The revision operators obtained from a KM revision operator coupled with a tie-breaking function satisfies* $(\boldsymbol{R_1})$, $(\boldsymbol{R_{2w}})$, $(\boldsymbol{R_3})$–$(\boldsymbol{R_6})$, $(\boldsymbol{R_L})$.

It can be noticed that *maxichoice* contraction operators [2] are closely related to our decisive revision operators. Firstly, we recall that contraction and revision operators can define each other through the identities of Levi [16] and Harper [12]. Using Levi identity, a maxichoice contraction can generate a maxichoice revision operator, which generate a complete theory if the negation of the revising information is entailed by the initial beliefs. So in this particular case, a maxichoice revision operator behaves like a decisive revision operator. In every other cases, the maxichoice revision operator can generate an incomplete theory, which is impossible with a decisive revision operator. On the contrary, if we use Harper identity on a decisive revision operator to define a contraction operator, then the obtained operator is necessarily a maxichoice contraction operator.

Maxichoice revision operators have also been shown to be the only possible operators when adding iteration postulates into the original AGM postulates [13]. The difference between decisive and maxichoice revision operators can be seen in their respective representation theorem. A maxichoice revision operator is compatible with a preorder where the countermodels of the new information are linearly ordered. While decisive revision operator leads to a linear order over *all* interpretations.

5 Conclusion and Perspectives

We define in this article the notion of decisive operators $(\boldsymbol{DC})$ and provide a new postulate $(\boldsymbol{R_L})$. We provide a set of postulates based on the KM ones (except for a weakening of $(\boldsymbol{R_2})$) and a corresponding representation theorem leading to total orders on interpretations. In this context, $(\boldsymbol{R_L})$ turns out to be equivalent to $(\boldsymbol{DC})$. We provide an example of a decisive operator, based on Dalal operator coupled with a tie-breaking function.

These works open some perspectives. The first that comes to mind is applying them to iterated belief revision [9]. In particular, an interesting question could be how to stay in the same ordered structures throughout the iteration process. Moreover some other theoretical links can be made with other frameworks, such as dynamic-limited revision operators [18]. It should also be interesting to further investigate our works with other faces of belief change, as well as the links with Nonmonotonic Reasoning [15].

Another question is to find some operators useful in practice. Those proposed in this article have been made for illustration purpose and to prove the consistency of our sets of postulates. Finally, the theoretical complexity of our revision operators remains an open question.

References

1. Alchourrón, C.E., Gärdenfors, P., Makinson, D.: On the logic of theory change: partial meet functions for contraction and revision. J. Symb. Log. **50**, 510–530 (1985)

2. Alchourron, C.E., Makinson, D.: On the logic of theory change: contraction functions and their associated revision functions. Theoria **48**(1), 14–37 (1982)
3. Arrow, K., Hurwicz, L.: An optimality criterion for decision-making under ignorance. In: Uncertainty and Expectations in Economics: Essays in Honour of G.L.S. Shackle, pp. 1–11 (1972)
4. Belahcène, K., Mousseau, V., Ouerdane, W., Pirlot, M., Sobrie, O.: Multiple criteria sorting models and methods-Part I: survey of the literature. 4OR 1–46 (2023)
5. Benferhat, S., Lagrue, S., Papini, O.: Revision of partially ordered information: axiomatization, semantics and iteration. In: Proceedings of the 19th International Joint Conference on Artificial Intelligence (IJCAI 2005), pp. 376–381 (2005)
6. Bossert, W., Pattanaik, P.K., Xu, Y.: Choice under complete uncertainty: axiomatic characterizations of some decision rules. Econ. Theor. **16**(2), 295–312 (2000)
7. Bubboloni, D., Gori, M.: Breaking ties in collective decision-making. Decis. Econ. Finan. **44**, 411–457 (2021)
8. Dalal, M.: Investigations into a theory of knowledge base revision. In: Proceedings of the 7th National Conference on Artificial Intelligence (AAAI 1988), pp. 475–479 (1988)
9. Darwiche, A., Pearl, J.: On the logic of iterated belief revision. Artif. Intell. **89**, 1–29 (1997)
10. Freeman, R., Brill, M., Conitzer, V.: General tiebreaking schemes for computational social choice. In: Proceedings of the 2015 International Conference on Autonomous Agents and Multiagent Systems, AAMAS 2015, pp. 1401–1409 (2015)
11. Haret, A., Woltran, S.: Belief revision operators with varying attitudes towards initial beliefs. In: Proceedings of the 28th International Joint Conference on Artificial Intelligence (IJCAI 2019), pp. 1726–1733 (2019)
12. Harper, W.L.: Rational conceptual change. PSA: Proc. Biennial Meet. Phil. Sci. Assoc. **1976**(2), 462-494 (1976)
13. Herzig, A., Konieczny, S., Perrussel, L.: On iterated revision in the AGM framework. In: Nielsen, T.D., Zhang, N.L. (eds.) ECSQARU 2003. LNCS (LNAI), vol. 2711, pp. 477–488. Springer, Heidelberg (2003). https://doi.org/10.1007/978-3-540-45062-7_39
14. Katsuno, H., Mendelzon, A.O.: Propositional knowledge base revision and minimal change. Artif. Intell. **52**(3), 263–294 (1991)
15. Kraus, S., Lehmann, D., Magidor, M.: Nonmonotonic reasoning, preferential models and cumulative logics. Artif. Intell. **44**, 167–207 (1990)
16. Levi, I.: Subjunctives, dispositions and chances. Synthese **34**, 423–455 (1977)
17. Ryan, M.: Belief revision and ordered theory presentations. In: Logic, Action, and Information, pp. 129–151 (1996)
18. Sauerwald, K., Kern-Isberner, G., Beierle, C.: On limited non-prioritised belief revision operators with dynamic scope. CoRR (2021). arXiv:2108.07769
19. Sen, A.K.: Choice functions and revealed preference. Rev. Econ. Stud. **38**(3), 307–317 (1971)
20. Smets, P.: Decision making in the TBM: the necessity of the pignistic transformation. Int. J. Approx. Reason. **38**(2), 133–147 (2005)

Judicial Support Tool: Finding the k Most Likely Judicial Worlds

Maksim Bolonkin[1], Sayak Chakrabarty[2], Cristian Molinaro[3(✉)], and V. S. Subrahmanian[2]

[1] Dartmouth College, Hanover, NH, USA
mbolonkin@cs.dartmouth.edu
[2] Northwestern University, Evanston, IL, USA
{sayak.chakrabarty,vss}@northwestern.edu
[3] University of Calabria, Rende, CS, Italy
cmolinaro@dimes.unical.it

Abstract. Judges sometimes make mistakes. We propose JUST, a logical framework within which judges can record propositions about a case and witness statements where a witness says that certain propositions are true or false. JUST allows the judge (or a jury) to assign a rating of credibility to witness statements. A world is an assignment of true/false to each proposition, which is required to satisfy case-specific integrity constraints. We first develop JUST's *explicit* algorithm, which calculates the k most likely worlds without using independence assumptions between propositions. The judge may use these calculated top-k most likely worlds to make her final decision. For this computation, JUST uses a suite of "combination" functions. We also develop JUST's *implicit* algorithm, which is far more efficient. We test JUST on 5 real-world court cases and 19 TV court cases, showing that JUST works well in practice.

1 Introduction

According to a 2017 study [26], 11.7% of convictions in the US were wrong. In cases where DNA evidence was present, that number rose to 12.6%. Moreover, according to a report[1] submitted to the UN, "African-American adults are 5.9 times as likely to be incarcerated than whites and Hispanics are 3.1 times as likely" (in the US). These findings portray fundamental flaws in the judicial process in the US.

We propose JUST, a logical framework that can potentially assist judges/juries to consider the evidence more objectively. JUST proceeds in four broad steps (cf. Fig. 1). In the first step, JUST encodes witness statements as propositions in logic. This can be done by any one of many companies (e.g. CourtScribes) that today transcribe witness statements and depositions in US courtroom proceedings. Such companies may not do all of this automatically. Some automatically transcribe the spoken language using commercial off-the-shelf tools to generate natural language transcription and then have

[1] https://www.sentencingproject.org/publications/un-report-on-racial-disparities/.

S. Destercke et al. (Eds.): SUM 2024, LNAI 15350, pp. 53–69, 2025.
https://doi.org/10.1007/978-3-031-76235-2_5

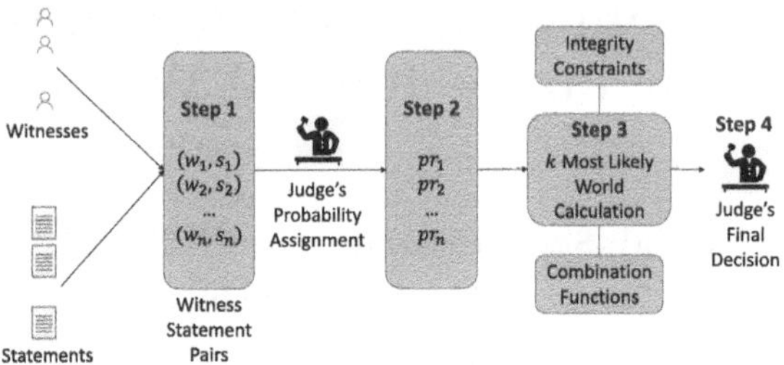

Fig. 1. Architecture of the JUST framework

workers manually clean that automatically generated text. This task can be enhanced (with enthusiasm by such companies that could charge an additional fee) by adding the logical formulas that capture the content of the text using an appropriate graphical user interface. In a second step, judges/juries can then use their own judgement to assign probabilities (or more qualitative credibility ratings) to whether those statements are true. To reduce dependence on a biased judge, one could instead have the judge and jury independently assign probabilities and then take the median value. In the third step, JUST would come up with a space of k most likely possible worlds, along with their probabilities. These possible worlds will need to satisfy various integrity constraints. By showing these k possible worlds to a judge, who will make a final decision in the last step, JUST makes judges aware of different ways in which the evidence in the case can be viewed, along with the probabilities of those worlds based on the judge's own assessment. At the end of the day, we hope this means that the judge can make a final better informed decision, keeping in mind the alternative worlds presented to them by the JUST system.

We provide more details on the problem solved by JUST. There is a universe of propositions. In Step 1, a witness may say some propositions are true, others are false, and offer no opinion about some propositions. Witness testimonies can conflict. Possible worlds are interpretations in logic that satisfy case-specific integrity constraints. In Step 2, judges/juries can assign a probability (or more qualitative ratings) to each witness statement denoting how much they believe that witness statement. For instance, in the famous Casey Anthony murder trial[2] in the US, Casey's father said "Casey sedated her daughter". A judge may assign a 75% probability that the witness was telling the truth, while another witness might say the opposite and the judge may assign a different probability to that witness's statement. We use a class of "combination functions" (CFs) to combine these probabilities and show that they induce probability distributions on the space of possible worlds in different ways. We then present two approximation algorithms to find the k most likely possible worlds (Step 3), which can be presented to a judge (or jury) as options to consider before rendering a final decision (Step 4). Possible worlds can be explained to judges simply as possible

[2] https://en.wikipedia.org/wiki/Death_of_Caylee_Anthony.

ways of interpreting the evidence, e.g., via simple statements of the form "Scenario #1 is true with probability p_1, Scenario #2 is true with probability p_2, and so forth. We developed a prototype implementation of our framework and evaluated its performance over 19 TV legal cases and 5 real-world court cases. JUST is able to perform very well, achieving high prediction performance when comparing the ground truth and the top-k possible worlds.

2 The JUST Framework

We assume the existence of a set $\mathcal{S}$ of "sources" and a set $\mathcal{P}$ of (Boolean) propositions. For example, sources might be those who provided testimony about the Casey Anthony case. The propositions refer to the specific statements that they made, such as "Casey sedated her daughter Caylee". W.l.o.g. we assume that $\mathcal{P}$ does not contain propositions with "complementary" meaning, e.g., $p =$ "*Casey sedated her daughter Caylee*" and $p' =$ "*Casey did not sedate her daughter Caylee*".

Witness functions capture statements made by witnesses.

Definition 1. *A* witness *function is a partial mapping* $\omega : \mathcal{S} \times \mathcal{P} \rightarrow \{0, 1\}$.

Intuitively, $\omega(s,p) = 1$ when source s said proposition p is true, while $\omega(s,p) = 0$ when s said p is false. Moreover, ω is a partial mapping, and $\omega(s,p)$ is not defined in cases when s did not say anything about p.

We now introduce the notion of an *assessment function*, which provides a "confidence" for each statement by some source s about some proposition p. We can think of this assessment function reflecting the judge's assessment of the veracity of a specific statement by a witness. It can be implemented via a GUI in which the judge can log her degree of belief in witness's remarks on a specific statement.

Definition 2. *An* assessment *function for a witness function* ω *is a partial mapping* $\alpha : \mathcal{S} \times \mathcal{P} \rightarrow [0, 1]$ *such that for every source* $s \in \mathcal{S}$ *and every proposition* $p \in \mathcal{P}$, $\alpha(s,p)$ *is defined iff* $\omega(s,p)$ *is defined.*

Intuitively, $\alpha(s,p)$ returns a real value in $[0, 1]$ assessing the confidence that the judge has in s's statement about proposition p. For example, $\alpha(s,p) = 0.7$ says that the assessment function believes the *statement* by source s about proposition p is true with a confidence of 70%. Such assessment functions are provided by the judge or jury.

Example 1. Figure 2 shows a sample case (extracted from the Casey Anthony case) with 4 sources $\mathcal{S} = \{s_1, \ldots, s_4\}$ and 4 propositions $\mathcal{P} = \{p_1, \ldots, p_4\}$, depicted as a bipartite graph. The witness function is captured by the edges connecting the sources to the propositions. A solid (resp., dashed) edge denotes $\omega(s,p) = 1$ (resp., $\omega(s,p) = 0$). The assessment function is captured via the edge weights. For example, the dashed edge from s_1 to p_3 with weight of 0.6 specifies that s_1 said that p_3 (namely, "Caylee was with the nanny") is false, and the assessment function assesses s_1's testimony to be true with 60% confidence.

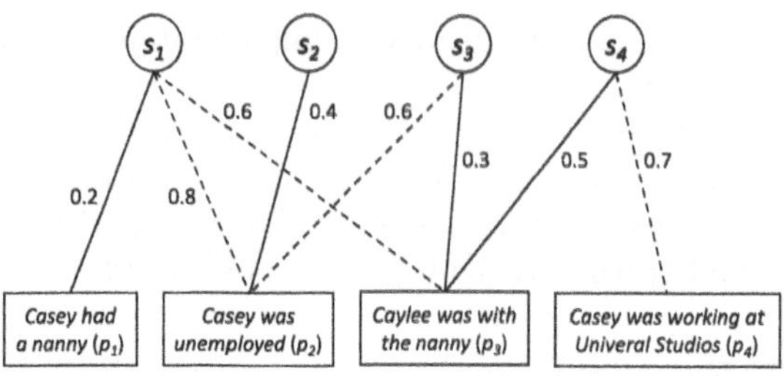

Fig. 2. Sample witness and assessment functions. Solid (resp., dashed) lines represent that a source said the corresponding proposition was true (resp., false). Numbers represent the confidence in veracity of the statements made by sources

Definition 3. *A* possible world *is a total mapping* $\gamma : \mathcal{P} \to \{0, 1\}$.

Intuitively, a possible world labels every single proposition as being either true (1) or false (0). We use PW to denote the set of all possible worlds over $\mathcal{P}$.

Example 2. Our running example has 16 possible worlds, depending on whether the function assigns 0 or 1 to each of the four propositions. Table 1 (the column ρ_{IC} will be discussed later on) shows these worlds. As an example, according to possible world γ_{12}, which assigns 1 (resp., 0) to p_1 and p_2 (resp., p_3 and p_4), Casey had a nanny, was unemployed, did not work at Universal Studios, and Caylee was not with the nanny.

Table 1. Possible worlds for $\mathcal{P} = \{p_1, p_2, p_3, p_4\}$ with a probability distribution ρ_{IC}

γ	$p_1p_2p_3p_4$	ρ_{IC}	γ	$p_1p_2p_3p_4$	ρ_{IC}	γ	$p_1p_2p_3p_4$	ρ_{IC}	γ	$p_1p_2p_3p_4$	ρ_{IC}
γ_0	0 0 0 0	0.2	γ_4	0 1 0 0	0.1	γ_8	1 0 0 0	0.2	γ_{12}	1 1 0 0	0
γ_1	0 0 0 1	0.1	γ_5	0 1 0 1	0	γ_9	1 0 0 1	0.1	γ_{13}	1 1 0 1	0
γ_2	0 0 1 0	0	γ_6	0 1 1 0	0	γ_{10}	1 0 1 0	0.15	γ_{14}	1 1 1 0	0.1
γ_3	0 0 1 1	0	γ_7	0 1 1 1	0	γ_{11}	1 0 1 1	0.05	γ_{15}	1 1 1 1	0

An important point to note from the above example is that the set of all possible worlds is exponential in the set of propositions. This will prove to be a major challenge in computing the k most likely worlds later in this paper.

In some real-world situations, we may have partial or total ground truth, but in situations like judicial cases, and in general in the presence of uncertainty, we do *not* necessarily have ground truth about what occurred and what did not occur. A possible world captures one possible scenario that *might* have occurred. To find the k most likely worlds, we must in principle consider all possible worlds. However, some of them may be inconsistent with the semantics of the propositions considered. As usual, we capture semantics of cases via two common types of integrity constraints, introduced below.

Definition 4. *A* denial constraint *is of the form* $\neg(p_1 \wedge \cdots \wedge p_n)$, *with* $n \geq 1$ *and each* $p_i \in \mathcal{P}$.

Intuitively, a denial constraint says that the propositions $p_1, \ldots, p_n$ cannot all be simultaneously true.

Definition 5. *A* definite constraint *is of the form* $p_1 \wedge \cdots \wedge p_n \rightarrow p_0$, *with* $n \geq 0$ *and each* $p_i \in \mathcal{P}$.

Intuitively, a definite constraint says that if the propositions $p_1, \ldots, p_n$ are all simultaneously true, then p_0 must be true. When $n = 0$, then p_0 must be true.

Example 3. In our running example, we may have the following two constraints (referring to some fixed time point):

$\neg$("Casey was unemployed" (p_2) $\wedge$ "Casey was working at Universal Studios" (p_4)),
"Caylee was with the nanny" (p_3) $\rightarrow$ "Casey had a nanny" (p_1)

The first is a denial constraint saying "Casey was unemployed" and "Casey was working at Universal Studios" cannot both be true. The second is a definite constraint saying that if "Caylee was with the nanny" is true, then 'Casey had a nanny" must be true.

We now define what it means for a possible world to satisfy an integrity constraint.

Definition 6. *A possible world* γ satisfies *a denial constraint* denc *of the form* $\neg(p_1 \wedge \cdots \wedge p_n)$ *iff there exists* $1 \leq j \leq n$ *such that* $\gamma(p_j) = 0$. *A possible world* γ satisfies *a definite constraint* defc *of the form* $p_1 \wedge \cdots \wedge p_n \rightarrow p_0$ *iff* $\gamma(p_0) = 1$ *or there exists* $1 \leq j \leq n$ *such that* $\gamma(p_j) = 0$. *We write* $\gamma \models \mathsf{denc}$ *and* $\gamma \models \mathsf{defc}$ *to denote satisfaction in these two cases.*

Example 4. The possible worlds $\gamma_5, \gamma_7, \gamma_{13}, \gamma_{15}$ (cf. Table 1) do not satisfy the denial constraint of Example 3, while the other possible worlds satisfy it. Also, $\gamma_2, \gamma_3, \gamma_6, \gamma_7$ violate the definite constraint of Example 3, while the other possible worlds satisfy it.

Suppose IC is a set of integrity constraints. A possible world γ *satisfies* IC, denoted $\gamma \models \mathsf{IC}$, iff $\gamma \models ic$ for all $ic \in \mathsf{IC}$, otherwise γ does not satisfy IC, denoted $\gamma \not\models \mathsf{IC}$. In our running example, the possible worlds that satisfy the integrity constraints IC of Example 3 are $\gamma_0, \gamma_1, \gamma_4, \gamma_8, \gamma_9, \gamma_{10}, \gamma_{11}, \gamma_{12}, \gamma_{14}$. We now associate a probability that a possible world is the actual world. We first define a *possible world distribution*.

Definition 7. *A* possible world distribution *w.r.t. a set* IC *of integrity constraints is a probability distribution* ρ_{IC} *over* PW *such that* $\rho_{\mathsf{IC}}(\gamma) = 0$ *for all possible worlds* $\gamma \in \mathsf{PW}$ *with* $\gamma \not\models \mathsf{IC}$.[3]

[3] A probability distribution over PW is a function $\rho_{\mathsf{IC}} : \mathsf{PW} \rightarrow [0, 1]$ s.t. $\Sigma_{\gamma \in \mathsf{PW}}\, \rho_{\mathsf{IC}}(\gamma) = 1$.

The last column of Table 1 shows a sample possible world distribution ρ_{IC} w.r.t. the set IC of integrity constraints of our running example. For instance, this possible world distribution assigns a 10% probability of being correct to each of $\gamma_1, \gamma_4, \gamma_9, \gamma_{14}$, a 15% probability of being correct to γ_{10}, a 20% probability of being correct to each of γ_0 and γ_8, and a 5% probability to γ_{11}.

Example 5. Let us consider proposition p_3 in Fig. 2. Both sources that say p_3 is true have low credibility—s_3 is assessed as lying with 70% probability, while s_4 is assessed as lying with 50% probability. This places us in a *dilemma*. What should we infer from this? It is clear that this situation can be resolved in many different ways.

To address this issue, we introduce *combination functions*, which return a confidence about a proposition being true, based on a witness and an assessment function.

Definition 8. *A* combination function *(CF) is a function* cf *that takes as input a proposition* p*, a witness function* ω*, and an assessment function* α *for* ω*, and returns a value in* $[0, 1]$ *as output.*

Rather than committing to a specific combination function, the definition above is general and allows different concrete instances that can accommodate different needs, depending on the application. Special cases can satisfy different probabilistic intuitions such as independence, ignorance, positive or negative correlation [12], but also can be non-probabilistic (e.g., capture fuzzy or possibilistic intuitions [6]). Below we propose some concrete combination functions. Of course, this library of combination functions can be extended if future research suggests better ones.

We introduce CFs using aggregate operators Φ (we consider the 5 aggregates min, max, avg, median, mode in our experiments) that take a multiset of real numbers in $[0, 1]$ and return a number in the $[0, 1]$ interval. Specifically, the basic idea to determine the confidence of a proposition p is to combine all statements made about p as follows: take the value $\alpha(s, p)$ for each statement (made by some source s) claiming that p is true, take the value $1 - \alpha(s', p)$ for each statement (made by some source s') claiming that p is false, and then apply an aggregate operator Φ over all such collected values.

Definition 9. *For an aggregate operator* Φ *and a proposition* p*, we define the combination function:* $\mathsf{cf}_\Phi(p, \omega, \alpha) = \Phi\big(\{\alpha(s, p) \mid \omega(s, p) = 1\} \cup \{1 - \alpha(s, p) \mid \omega(s, p) = 0\}\big)$.

Given a witness function ω, an assessment function α for ω, a set of integrity constraints IC, and a combination function cf, we can find possible world distributions by solving the following set LC of linear constraints:

$$\begin{cases} 0 \leq \sum\limits_{\gamma_j \in \mathsf{PW}, \gamma_j \models \mathsf{IC}, \gamma_j(p_i) = 1} X_j \leq \mathsf{cf}(p_i, \omega, \alpha), & \text{for } p_i \in \mathcal{P}. \\ 0 \leq X_j \leq 1, \quad \text{for } \gamma_j \in \mathsf{PW} \text{ s.t. } \gamma_j \models \mathsf{IC}. \\ \sum\limits_{\gamma_j \in \mathsf{PW}, \gamma_j \models \mathsf{IC}} X_j = 1. \end{cases}$$

In the set of linear constraints above, each variable X_j stands for the (unknown) probability of a possible world γ_j satisfying the integrity constraints. In the first kind of linear constraint, we introduce, for each proposition p_i, one linear constraint imposing that the sum of the probabilities of all possible worlds satisfying the integrity constraints and assigning true to p_i is at most the value returned by the combination function for proposition p_i. Notice that the assessment function, as well as the witness function, are taken into account by the combination function, which in turn provides the upper bound for the first kind of linear constraints The second kind of linear constraint simply requires each possible world satisfying the integrity constraints to have a probability between 0 and 1. The third kind of linear constraint imposes a probability distribution over the possible worlds satisfying the integrity constraints.

Example 6. For our running example, the set LC of linear constraints is:

$$\begin{cases} 0 \leq X_8 + X_9 + X_{10} + X_{11} + X_{12} + X_{14} \leq \mathsf{cf}(p_1, \omega, \alpha). \\ 0 \leq X_4 + X_{12} + X_{14} \leq \mathsf{cf}(p_2, \omega, \alpha). \\ 0 \leq X_{10} + X_{11} + X_{14} + \leq \mathsf{cf}(p_3, \omega, \alpha). \\ 0 \leq X_1 + X_9 + X_{11} + \leq \mathsf{cf}(p_4, \omega, \alpha). \\ 0 \leq X_j \leq 1, \quad \text{for } X_j \in \{X_0, X_1, X_4, X_8, X_9, X_{10}, X_{11}, X_{12}, X_{14}\}. \\ X_0 + X_1 + X_4 + X_8 + X_9 + X_{10} + X_{11} + X_{12} + X_{14} = 1. \end{cases}$$

Each solution of LC corresponds to a possible world distribution. In general, LC can have multiple solutions because we do not assume independence between propositions. Thus, a fixed possible world γ can have different probabilities w.r.t. different possible world distributions. One way to get a single probability for γ is to define it as its average probability across all possible world distributions. Then, a set of top-k possible worlds is a set of k possible worlds with highest average probability.

Qualitative Assessments. Judges or juries may be uncomfortable assigning probabilities to witness statements. Suppose they prefer to assign a "low", "medium", "high" rating to the credibility of a witness statement. These can be converted to the intervals $[0, 1/3]$, $[1/3, 2/3]$, $[2/3, 1]$, respectively. This can be easily handled via a small modification to our framework: (i) the assessment function would return a probability interval rather than a point probability; (ii) the definition of combination function would be replaced by a function cf' that behaves just like cf except that $\mathsf{cf}'(p_i, \omega, \alpha)$ would return a probability interval rather than a point probability; (iii) the first constraint shown above would be replaced by: $\sum_{\gamma_j \in \mathsf{PW}, \gamma_j \models \mathsf{IC}, \gamma_j(p_i)=1} X_j \in \mathsf{cf}'(p_i, \omega, \alpha), \quad \text{for } p_i \in \mathcal{P}$.

3 JUST Algorithms

We propose two approximation algorithms to compute most likely possible worlds. The *explicit* approach ($\mathsf{JUST}^{\mathsf{exp}}$) makes no independence assumptions; it

Algorithm 1. $\mathsf{JUST}^{\mathsf{exp}}$

Input: A witness function ω, an assessment function α for ω, a set IC of integrity constraints, a combination function cf, and positive integers $k \leq 2^{|\mathcal{P}|}$ and s.
Output: A set of k possible worlds.
1: Randomly sample a set S of solutions of LC with $|S| = s$.
2: **for each** $\gamma \in \mathsf{PW}$ **do**
3: $\quad EP(\gamma) = \frac{\sum_{\rho \in S} \rho(\gamma)}{|S|}$.
4: **end for**
5: Let Top be top-k (w.r.t. EP) possible worlds.
6: **return** Top.

computes the average probability of each possible world over a sample of solutions of LC and then returns the set of k most likely ones (w.r.t. the average probability). The *implicit* approach ($\mathsf{JUST}^{\mathsf{imp}}$) assumes only that propositions appearing in no integrity constraints are independent, reducing runtime.

$\mathsf{JUST}^{\mathsf{exp}}$ (cf. Algorithm 1) works as follows. First, a set S of solutions of LC (i.e., probability distributions over possible worlds) is randomly sampled (line 1) using the well-known *Hit-and-Run* walk [16,23]. Then, for each possible world γ, its *expected probability* $EP(\gamma)$ (w.r.t. S) is computed (lines 2–3). Finally, a set of top-k (w.r.t. EP) possible worlds is returned (lines 4–5).

Before presenting $\mathsf{JUST}^{\mathsf{imp}}$, we introduce some notation. We will need to consider possible worlds defined w.r.t. subsets $\mathcal{P}'$ of $\mathcal{P}$, i.e., functions of the form $\gamma' : \mathcal{P}' \rightarrow \{0, 1\}$. For notational convenience, γ' is represented also as the set $\{p_i \mid p_i \in \mathcal{P}' \text{ and } \gamma'(p_i) = 1\} \cup \{\overline{p_i} \mid p_i \in \mathcal{P}' \text{ and } \gamma'(p_i) = 0\}$. We use $\mathsf{PW}[\mathcal{P}']$ (resp., $\mathsf{LC}[\mathcal{P}']$) to denote the set of possible worlds (resp., linear constraints) w.r.t. only the propositions in $\mathcal{P}'$. Under the independence assumption, the *probability* $\mathbf{P}(\gamma')$ of a possible world γ' in $\mathsf{PW}[\mathcal{P}']$ is: $\left(\prod_{p \in \mathcal{P}', \gamma'(p)=1} \mathsf{cf}(p, \omega, \alpha)\right) \cdot \left(\prod_{p \in \mathcal{P}', \gamma'(p)=0} (1 - \mathsf{cf}(p, \omega, \alpha))\right)$

When $\gamma' = \emptyset$, $\mathbf{P}(\gamma')$ is defined to be 0. For a set IC of integrity constraints, $\mathcal{P}_{\mathsf{IC}}$ is the set of propositions in $\mathcal{P}$ that appear in IC, while $\overline{\mathcal{P}_{\mathsf{IC}}} = \mathcal{P} \setminus \mathcal{P}_{\mathsf{IC}}$.

$\mathsf{JUST}^{\mathsf{imp}}$ (cf. Algorithm 2) works as follows. First, a set S of solutions of $\mathsf{LC}[\mathcal{P}_{\mathsf{IC}}]$ is randomly sampled (line 1). For each possible world γ in $\mathsf{PW}[\mathcal{P}_{\mathsf{IC}}]$, its *expected probability* $EP(\gamma)$ (w.r.t. S) is computed (lines 2–3). Then, a set of top-k' possible worlds from $\mathsf{PW}[\overline{\mathcal{P}_{\mathsf{IC}}}]$ is computed assuming independence (line 4)—we will discuss this further shortly. Then, for each pair of possible worlds γ in $\mathsf{PW}[\mathcal{P}_{\mathsf{IC}}]$ and γ' from the previous step, γ and γ' are combined into a possible world γ^* with probability $EP(\gamma) \times \mathbf{P}(\gamma')$ (lines 5–7). Finally, a set of top-k possible worlds from those computed at the previous step is returned (lines 8–9).

We still need to show how to compute a set of top-k' possible worlds under the independence assumption (line 4 of $\mathsf{JUST}^{\mathsf{imp}}$). We address this problem in the rest of this section. We provide a dynamic programming algorithm, called $\mathsf{JUST}^{\mathsf{ind}}$ (cf. Algorithm 3), whose worst-case time complexity is $O(m \cdot k')$, where m is the number of propositions in $\overline{\mathcal{P}_{\mathsf{IC}}}$, and thus it runs in pseudo-polynomial time.

Algorithm 2. $\mathsf{JUST}^{\mathsf{imp}}$

Input: A witness function ω, an assessment function α for ω, a set IC of integrity constraints, a combination function cf, and positive integers $k \leq 2^{|\mathcal{P}|}$, $k' \leq 2^{|\overline{\mathcal{P}_{\mathsf{IC}}}|}$, and s.

Output: A set of k possible worlds.

Randomly sample a set S of solutions of $\mathsf{LC}[\mathcal{P}_{\mathsf{IC}}]$ with $|S| = s$.
for each $\gamma \in \mathsf{PW}[\mathcal{P}_{\mathsf{IC}}]$ **do**
    $EP(\gamma) = \frac{\sum_{\rho \in S} \rho(\gamma)}{|S|}$.
end for
Compute a set Top' of top-k' possible worlds from $\mathsf{PW}[\overline{\mathcal{P}_{\mathsf{IC}}}]$ under the independence assumption.
for each $\gamma \in \mathsf{PW}[\mathcal{P}_{\mathsf{IC}}]$ and $\gamma' \in Top'$ **do**
    $\gamma^* = \gamma \cup \gamma'$.
    $EP(\gamma^*) = EP(\gamma) \times \mathbf{P}(\gamma')$.
end for
Let Top be top-k (w.r.t. EP) possible worlds from PW.
return Top.

One key idea of Algorithm 3 is to build possible worlds bottom-up. It starts by considering only one proposition and then iteratively considers the remaining ones, one at a time. We now go into the details of $\mathsf{JUST}^{\mathsf{ind}}$. On line 1, m is set to the number of propositions. On lines 2–4, three arrays Top, L_t, and L_f are introduced, each with k' entries initialized to the empty set. The array Top stores the most likely possible world for the propositions currently considered, while L_t and L_f are auxiliary arrays whose role will be explained shortly. Line 5 introduces the integer n, which is initialized to 1 and then updated on line 11, whose meaning is as follows. The value of n after being updated on line 11 is the number of most likely possible worlds to be computed: n is strictly lower than k' when i propositions are being considered and their number of possible worlds is $2^i < k'$—indeed, in such a case, $n = 2^i$; when i becomes high enough, i.e., it is such that $2^i \geq k'$, n becomes equal to k'. The **for** loop on lines 6–13 performs m iterations. At any iteration i, the top-n most likely possible worlds are computed on lines 7–13 *by considering only the first* i *propositions* (an arbitrary order over the propositions can be used). Specifically, on lines 7–9, each possible world in Top is augmented with proposition p_i being true (resp., false) and the resulting world is stored in L_t (resp., L_f)—see line 8 (resp., line 9). Next, L_t and L_f are merged into an array L sorted by descending $\mathbf{P}$ value (line 10). Then, the top-n possible worlds in L are copied into Top (lines 12–13). Finally, after m iterations of the **for** loop on lines 6–13, the array Top, containing the top-k' most likely possible worlds (sorted by descending probability) *w.r.t. all propositions in* $\overline{\mathcal{P}_{\mathsf{IC}}}$, is returned (line 14). The following two theorems state the correctness and the worst-case time complexity of Algorithm 3.

Theorem 1. *Algorithm 3 computes a set of top-k' possible worlds under the independence assumption.*

Theorem 2. *The worst-case time complexity of Algorithm 3 is $O(m \cdot k')$.*

Algorithm 3. $\mathsf{JUST}^{\mathsf{ind}}$

Input: $\mathsf{cf}(p_i, \omega, \alpha)$ for each $p_i \in \overline{\mathcal{P}_{\mathsf{IC}}}$, a positive integer k' s.t. $k' \leq 2^{|\overline{\mathcal{P}_{\mathsf{IC}}}|}$.
Output: A set of top-k' possible worlds.

$m = |\overline{\mathcal{P}_{\mathsf{IC}}}|$.
$Top[1..k']$; $L_t[1..k']$; $L_f[1..k']$.
for $i = 1$ **to** k' **do**
  $Top[i] = \emptyset$; $L_t[i] = \emptyset$; $L_f[i] = \emptyset$.
end for
$n = 1$.
for $i = 1$ **to** m **do**
  for $j = 1$ **to** n **do**
    $L_t[j] = Top[j] \cup \{p_i\}$.
    $L_f[j] = Top[j] \cup \{\overline{p_i}\}$.
  end for
  Merge $L_t[1..n]$ and $L_f[1..n]$ into $L[1..2n]$ sorted by descending $\mathbf{P}$.
  $n = min\{2 \times n, k'\}$.
  for $j = 1$ **to** n **do**
    $Top[j] = L[j]$.
  end for
end for
return Top.

4 Experimental Evaluation

In this section, we report on experiments we conducted with 24 datasets. We implemented $\mathsf{JUST}^{\mathsf{exp}}$, $\mathsf{JUST}^{\mathsf{imp}}$, and $\mathsf{JUST}^{\mathsf{ind}}$ in Python. We ran experiments on a 3.10GHz Intel Core i9-9960X CPU with 131GB of RAM, running Ubuntu 18.04.5. The number of samples (i.e., the s input in Algorithms 1 and 2) was set to 20. For $\mathsf{JUST}^{\mathsf{imp}}$, the number k' of possible worlds considered under the independence assumption was $k' = |\mathsf{PW}[\mathcal{P}_{\mathsf{IC}}]|$, that is, the same number of possible worlds induced by propositions appearing in the integrity constraints. While in general $\mathsf{JUST}^{\mathsf{imp}}$ can generate the same number of possible worlds as $\mathsf{JUST}^{\mathsf{exp}}$, our choice of k' limits the maximum number of possible worlds that can be generated by $\mathsf{JUST}^{\mathsf{imp}}$ to $2^{2k'}$. Nonetheless, as shown in the rest of this section, under the setting we chose, $\mathsf{JUST}^{\mathsf{imp}}$ is able to converge to high-quality results with few possible worlds saving computation time. We used min, max, avg, mode, and median in our CFs as per Definition 9.

Datasets. We considered 5 real-word trial cases: Casey Anthony (**CA**) (30.5 h of video), Ashley McArthur (**AMA**) (14 h of video), Jacob Cayer (**JC**) (22 h of video), Nathaniel Rowland (**NR**) (16 h of video), Joshua Aide (**JA**) (17.5 h of video). We also considered 19 additional cases taken from the US TV show "Judge Judy" (**JJ**), which are 10 min long on average.

We manually annotated the sources, propositions and witness functions. Though we have discussed JUST with a couple of judges and received their advice, we were unable to convince them to spend the time necessary to provide the assessment function. We then used an online "deception detection algorithm" [27] that was trained on a real world courtroom dataset [20]. Of course, this is just a proxy for a judge's assessments and we do not suggest that deception detection algorithms be used for assessment.

For the CA, AMA, JC, NR, JA datasets we defined 7, 3, 7, 5, 6 integrity constraints, respectively (these were the natural number of ICs for the propositions in the case). The number of ICs for the JJ dataset ranged from 0 to 3 as these cases were short (10 min compared to, e.g., 30.5 h for CA) and had a smaller number of propositions. We will publicly release our annotated data and our code upon publication of the paper.

Prediction Performance. For each dataset, we manually watched its video(s) and captured the real ground truth (which is a possible world) based on what the judge decided to be true vs. false, that is, a proposition is true (resp., false) in the ground truth iff the proposition is true (resp., false) according to the judge's final decision. In the following, we define two metrics to measure prediction performance, $PP[i]$ and $PP^*[i]$, whose goal is to assess the quality of the first i possible worlds, for $1 \leq i \leq k$.

Let us start with the definition of $PP[i]$. For a given dataset, let γ^* be the ground truth possible world and $\gamma_1, \ldots, \gamma_k$ be the possible worlds we computed sorted by descending probability. For $1 \leq i \leq k$ and for each proposition $p_j \in \mathcal{P}$, we define $F[i][j] = 1$ if there exists a possible world γ_h with $h \leq i$ s.t. $\gamma^*(p_j) = \gamma_h(p_j)$, that is, there is a possible world among the first i ones that agrees with the ground truth on p_j's truth value; otherwise, $F[i][j] = 0$. Then, for each $1 \leq i \leq k$, we define the *prediction performance of the top-i possible worlds*, denoted $PP[i]$, as the average of $F[i][j]$ across all propositions p_j's.

The second metric we propose is more demanding and is based on the following idea. In an "ideal" set of top-k possible worlds, the first one (i.e., the most likely one) is the ground truth, then we have all possible worlds that are the same as the ground truth except for the truth value of one proposition, then we have all possible worlds that are the same as the ground truth except for the truth value of two propositions, and so on.

In general, if m is the number of propositions, then the number of possible words that differ from the ground truth by j proposition truth values is $\binom{m}{j}$, with $0 \leq j \leq m$. Thus, there is $\binom{m}{0} = 1$ possible world differing from the ground truth by 0 propositions (this is the ground truth itself), there are $\binom{m}{1} = m$ possible worlds differing from the ground truth by 1 proposition, and so on. In an ideal set of top-k possible worlds, the i-th possible world $(1 \leq i \leq k)$ is the same as the ground truth except for the truth value of x_i propositions, where $0 \leq x_i \leq m$ is the smallest integer s.t. $i \leq \sum_{h=0}^{x_i} \binom{m}{h}$. Equivalently, the i-th possible world agrees with the ground truth function on $m - x_i$ proposition truth values—we define $n_i = m - x_i$. Thus, n_i must be read as the number of propositions the ground truth and the i-th possible world agree on in an ideal result.

Now let $\gamma_1, \ldots, \gamma_k$ be the possible worlds we computed sorted by descending probability. We want to measure how much it differs from the ideal result discussed above by assigning a score to each possible world γ_i measuring how

Table 2. $\mathsf{JUST}^{\mathsf{imp}}$ and $\mathsf{JUST}^{\mathsf{exp}}$ runtime (secs) with cf_{avg}. Here – means $\mathsf{JUST}^{\mathsf{exp}}$ did not terminate in 30 min

	imp	exp		imp	exp		imp	exp		imp	exp		imp	exp
CA	123	–	JJ1	2.41	–	JJ6	0.1	5.42	JJ11	0.064	33.2	JJ16	0.064	4.37
AMA	0.015	–	JJ2	0.64	–	JJ7	0.77	–	JJ12	0.15	–	JJ17	0.15	–
JC	0.92	–	JJ3	0.9	–	JJ8	0.514	–	JJ13	0.053	–	JJ18	0.17	–
NR	0.063	–	JJ4	0.52	–	JJ9	0.25	–	JJ14	0.047	2.87	JJ19	0.058	0.25
JA	0.13	6.3	JJ5	0.58	–	JJ10	0.16	–	JJ15	0.17	–			

much γ_i differs from a possible world in the same position in an ideal result. We then compute the average score across all possible worlds. For a possible world γ_i, let n_{γ_i} be the number of propositions on which γ_i and the ground truth agree. Then, we define $score(\gamma_i) = \min\{1, \frac{n_{\gamma_i}}{n_i}\}$, and, for $1 \leq i \leq k$, define $PP^*[i] = avg\{score(\gamma_h) \mid 1 \leq h \leq i\}$.

Results. We start by reporting on running times. While $\mathsf{JUST}^{\mathsf{imp}}$ terminated in all 24 cases, $\mathsf{JUST}^{\mathsf{exp}}$ terminated within half an hour only on the JA, JJ6, JJ11, JJ14, JJ16, and JJ19 datasets. As results did not vary substantially across different CFs, we discuss a representative case, which is cf_{avg}. Running times (in seconds) for all datasets are reported in Table 2. We see that $\mathsf{JUST}^{\mathsf{imp}}$ took less than 1 s in all cases but two, one of which is the CA dataset, which has a much higher number of possible worlds, hence higher run time. $\mathsf{JUST}^{\mathsf{imp}}$ is always 1–3 orders of magnitude faster than $\mathsf{JUST}^{\mathsf{exp}}$.

We now analyze prediction performances. Figures 3 and 4 show the prediction performances PP and PP^*, respectively, of the $\mathsf{JUST}^{\mathsf{imp}}$ and $\mathsf{JUST}^{\mathsf{exp}}$ algorithms as the number considered possible worlds increases, considering all datasets for which $\mathsf{JUST}^{\mathsf{exp}}$ was able to terminate within 30 min (cf. Table 2). More precisely, for each figure, we computed the top-k possible worlds with k being the highest value on the x-axis; the figure then reports the prediction performance as the number i of possible worlds (one might want to inspect) varies from 1 to k.

As for PP (Fig. 3), $\mathsf{JUST}^{\mathsf{exp}}$ always reaches a PP value of 1, even though this might require considering a high number of possible worlds (up to 524 and 297 on average across the 6 datasets). $\mathsf{JUST}^{\mathsf{imp}}$ does not always reach a PP value of 1, but still gets to high PP values with few possible worlds: besides JJ14 where the highest $PP[i]$ is 0.55 but with only 3 possible worlds, on the remaining 5 cases $\mathsf{JUST}^{\mathsf{imp}}$ reaches the PP values 0.81, 0.91, 0.91, 1, 1 after considering only 7, 5, 6, 5, 13 possible worlds.

Regarding PP^* (Fig. 4), we can state again that with few possible worlds $\mathsf{JUST}^{\mathsf{imp}}$ reaches PP^* values similar to those reached by $\mathsf{JUST}^{\mathsf{exp}}$ with more possible worlds—indeed, for the JA and JJ11 datasets, $\mathsf{JUST}^{\mathsf{imp}}$ reaches higher values, while $\mathsf{JUST}^{\mathsf{imp}}$ values are lower for the JJ14 dataset. Notice that PP^* values are lower than those of the PP metric: as already mentioned before, PP^* is a demanding metric (e.g., it can give a low value to a result where the most

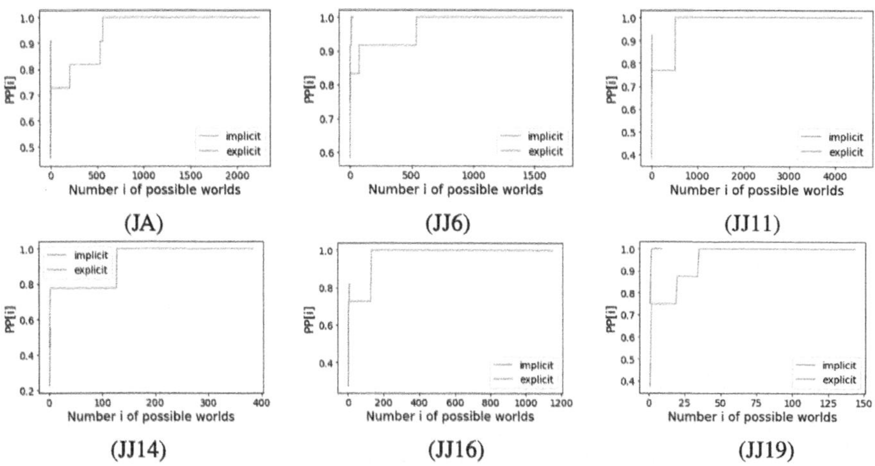

Fig. 3. $PP[i]$ of JUST^{imp}/JUST^{exp} on all datasets for which JUST^{exp} terminated in 30 min

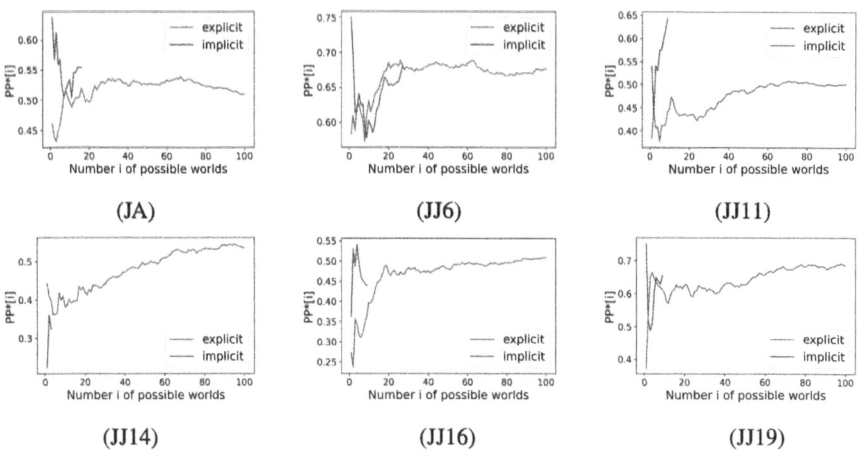

Fig. 4. $PP^*[i]$ of JUST^{imp}/JUST^{exp} on all datasets for which JUST^{exp} terminated in 30 min

likely possible world is the ground truth, but the remaining possible worlds are not the expected ones in their respective positions).

Recall that JUST^{imp} was able to provide results on the remaining 18 datasets (in little time) while the JUST^{exp} did not. Thus, JUST^{imp} achieves a good trade-off between result quality and runtime. We next analyze the behavior of JUST^{imp} over all datasets.

Figure 5 shows the average prediction performance (across all datasets) for both PP (left) and PP^* (right) as the number of considered possible worlds

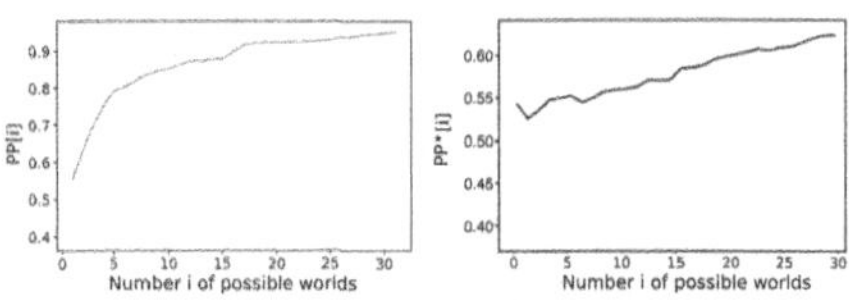

Fig. 5. $PP[i]$ and $PP^*[i]$ of $\mathsf{JUST}^{\mathsf{imp}}$ across all datasets

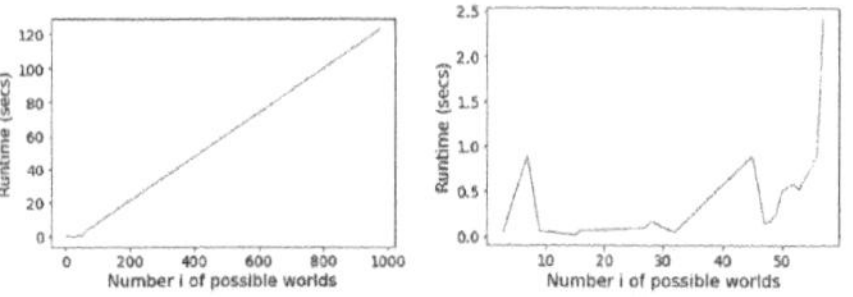

Fig. 6. Average runtime of $\mathsf{JUST}^{\mathsf{imp}}$ w/ (left) and w/o (right) the CA dataset

increases. The results show that we reach good prediction performance with only 30 possible worlds.

Figure 6 shows the average run time of $\mathsf{JUST}^{\mathsf{imp}}$ across all datasets as the number of possible world increases, w/ (left) and w/o (right) the CA dataset (the run time for this dataset is higher because of the much higher number of possible worlds). $\mathsf{JUST}^{\mathsf{imp}}$ scales well even when considering the CA dataset, with very low running time when the latter is not considered.

To summarize, our evaluation shows that $\mathsf{JUST}^{\mathsf{imp}}$ provides results of high quality, has low running times, with no significant difference across the different CF aggregates.

5 Related Work

There is much work on legal systems using logical methods [10,11]. [3] introduced evidence logic as a way to model epistemic agents facing contradictory evidence from different sources, where each agent has a collection of possible worlds, one of which the agent believes to be true. [15] introduced a logic where an agent accumulates evidence in support of a given fact from other agents and the body of evidence in support of that fact can become a reason to believe it. [22] proposed a case-based sentencing support system. [14] created an automated decision support system for probation officers. [9] developed an argumentation framework to reason about debate outcomes. [25] designed a case-based decision support system allowing plaintiff and defendant to solve their legal case without involvement of judges. [18] developed a model for matching a set of facts to relevant legal sources. [8] develop a logic for dealing with uncertainties in courtroom settings. [13] analyzes and compares Bayesian networks with embedded scenarios and formal analysis of argument validity. There is also work on detecting deception in trials [7,20]. A few efforts try to predict court decisions [1,5,17].

We also mention that the problem of assessing plausible truth values to a set of propositions is quite general, and is not limited to justice only. For instance, we may have to understand what is going on in a scene form partial observations, e.g. in a battle field. An example of such a problem can be found in [2], where a nonmonotonic reasoning, more qualitative approach is used.

[24] provides a connectionist approach to resolving semantically indeterminate terms like "suitable employment" in a legal context—such an approach might complement JUST in encoding language via propositions. [21] combines connectionist and symbolic reasoning for legal information retrieval—such methods may be useful in automatically extracting semantic knowledge (e.g., propositions) from legal transcripts.

6 Limitations and Discussion

Changing centuries of bias cannot be done via one paper—so our paper is a small step. Many questions remain about how JUST can be deployed. (Q1) Won't a biased judge just assign biased probabilities to witness statements? There are at least two approaches to address this. If the judge's assessments are part of the record, lawyers for the aggrieved party can appeal based on the judge's assessments. Another option would have each jury member assign probabilities and take say the median probability. (Q2) Isn't it challenging to encode natural language statements in logic? We did not find this hard to do *manually* when we coded our data. In the real world, a court reporting company could write down the propositions (along with the sentences from which the propositions are derived). A judge or jury can modify these statements to suit their understanding using an easy to use GUI, along with their understanding of the relevant integrity constraints applicable to the scenario leveraging GUIs to express integrity constraints, e.g., see [4]. Recently, research in the natural language processing community has developed methods to code sentences in natural language in logic [19] and such techniques can also be useful, especially if they are incorporated within a useful GUI. The problem of developing the right combination of GUIs, manual and automated effort for this purpose is an interesting topic for future work. (Q3) What about inconsistencies between statements made by a witness? The integrity constraints within JUST address this need. If a given set of integrity constraints does not flag some inconsistencies, the judge/jury can add more integrity constraints using a GUI of the kind mentioned above. (Q4) Are there alternatives to the proposed approach for the computation of most likely possible worlds? This is a direction for future work to be explored, e.g., one could impose all propositions to be independent by decomposing dependencies into independent factors and then either assess these independent variables or have assessments on propositional formulas over these independent variables.

Acknowledgement. Cristian Molinaro acknowledges the support of the PNRR project FAIR - Future AI Research (PE00000013), Spoke 9 - Green-aware AI, under the NRRP MUR program funded by the NextGenerationEU.

Contributions. The authors are listed in alphabetical order, not by magnitude of contribution.

References

1. Bagherian-Marandi, N., Ravanshadnia, M., Akbarzadeh-T, M.R.: Two-layered fuzzy logic-based model for predicting court decisions in construction contract disputes. Artif. Intell. Law **29**(4), 453–484 (2021)
2. Benferhat, S., Chehire, T., Monai, F.: Possibilistic ATMS in a data fusion problem. In: Wiley, J., Sons (eds.) Publié In Fuzzy Set Methods in Information Engineering: A Guided Tour of Applications, pp. 417–435 (1998)
3. van Benthem, J., Fernández-Duque, D., Pacuit, E.: Evidence logic: a new look at neighborhood structures. In: Proceedings Advances in Modal Logic, pp. 97–118 (2012)
4. Brägger, R.P., Dudler, A., Rebsamen, J., Zehnder, C.A.: Gambit: an interactive database design tool for data structures, integrity constraints, and transactions. IEEE Trans. Software Eng. **11**(7), 574–583 (1985)
5. Deeks, A.: The judicial demand for explainable artificial intelligence. Columbia Law Rev. **119**(7), 1829–1850 (2019)
6. Dubois, D., Prade, H.: A class of fuzzy measures based on triangular norms a general framework for the combination of uncertain information. Int. J. Gen. Syst. **8**(1), 43–61 (1982)
7. Fornaciari, T., Poesio, M.: Automatic deception detection in Italian court cases. Artif. Intell. Law **21**(3), 303–340 (2013)
8. Jøsang, A., Bondi, V.A.: Legal reasoning with subjective logic. Artif. Intell. Law **8**(4), 289–315 (2000)
9. Karamlou, A., Cyras, K., Toni, F.: Deciding the winner of a debate using bipolar argumentation. In: Proceedings of AAMAS, pp. 2366–2368 (2019)
10. Kowalski, R., Datoo, A.: Logical english meets legal english for swaps and derivatives. Artif. Intell. Law **30**(2), 163–197 (2021)
11. Kowalski, R.A.: Legislation as logic programs. In: Informatics and the Foundations of Legal Reasoning, pp. 325–356 (1995)
12. Lakshmanan, L.V.S., Leone, N., Ross, R.B., Subrahmanian, V.S.: Probview: a flexible probabilistic database system. ACM Trans. Database Syst. **22**(3), 419–469 (1997)
13. van Leeuwen, L., Verheij, B.: A comparison of two hybrid methods for analyzing evidential reasoning. Front. Artif. Intell. Appl. **322**, 53–62 (2019)
14. Leith, P.: The judge and the computer: how best 'decision support'? Artif. Intell. Law **6**(2), 289–309 (1998)
15. Liu, F., Lorini, E.: Reasoning about belief, evidence and trust in a multi-agent setting. In: Proceedings of PRIMA, vol. 10621, pp. 71–89 (2017)
16. Lovász, L., Vempala, S.S.: Hit-and-run from a corner. SIAM J. Comput. **35**(4), 985–1005 (2006)
17. Medvedeva, M., Vols, M., Wieling, M.: Using machine learning to predict decisions of the European court of human rights. Artif. Intell. Law **28**(2), 237–266 (2020)
18. Mokanov, I., Shane, D., Cerat, B.: Facts2law: using deep learning to provide a legal qualification to a set of facts. In: Proceedings of ICAIL, pp. 268–269 (2019)
19. Parmar, M., et al.: Logicbench: towards systematic evaluation of logical reasoning ability of large language models. In: Proceedings of ACL, pp. 13679–13707 (2024)
20. Pérez-Rosas, V., Abouelenien, M., Mihalcea, R., Burzo, M.: Deception detection using real-life trial data. In: Proceedings of ICMI, pp. 59–66 (2015)
21. Rose, D.E.: A Symbolic and Connectionist Approach to Legal Information Retrieval. Psychology Press, London (2013)

22. Schild, U.J.: Criminal sentencing and intelligent decision support. Artif. Intell. Law **6**(2), 151–202 (1998)
23. Smith, R.L.: Efficient Monte Carlo procedures for generating points uniformly distributed over bounded regions. Oper. Res. **32**(6), 1296–1308 (1984)
24. Van Opdorp, G., Walker, R., Schrickx, J., Groendijk, C., Van den Berg, P.: Networks at work: a connectionist approach to non-deductive legal reasoning. In: Proceedings of the 3rd International Conference on Artificial Intelligence and Law, pp. 278–287 (1991)
25. Wah, T.K., Muniandy, M.: Courtroom decision support system using case based reasoning. Procedia. Soc. Behav. Sci. **129**, 489–495 (2014)
26. Walsh, K., Hussemann, J., Flynn, A., Yahner, J., Golian, L.: Estimating the prevalence of wrongful convictions (2017). https://www.ojp.gov/pdffiles1/nij/grants/251115.pdf
27. Wu, Z., Singh, B., Davis, L.S., Subrahmanian, V.S.: Deception detection in videos. In: Proceedings of AAAI, pp. 1695–1702 (2018)

Flexible Risk Aware Sequential Decision Making

Nadjet Bourdache(✉)

Université de Caen Normandie, ENSICAEN, CNRS, Normandie Univ, GREYC UMR 6072, 14000 Caen, France
nadjet.bourdache@unicaen.fr

Abstract. In this work, we study risk aware sequential decision making in a Markov Decision Process (MDP). Unlike many works in the literature, where MDPs are solved by optimizing expected rewards (ER), and thus assuming neutrality w.r.t. risk, we use a more sophisticated operator: the Weighted Ordered Weighted Average (WOWA), a parameterized operator that allows to model a wide range of behaviors, from extreme risk seeking to extreme risk aversion (as well as compromises between both behaviors). This operator has thus a high descriptive capacity, but is rather difficult to optimize in an MDP because of its non-linearity that makes standard solving algorithms sub-optimal. In this paper, we introduce and justify a ranking algorithm that allows to determine an optimal (or nearly optimal) policy for a wide range of attitudes w.r.t. risk (averse, seeking, neutral, intermediate) using WOWA. Empirical results are given to illustrate the relevance and the efficiency of the approach.

Keywords: Sequential decision making · Markov decision processes · decision theory under risk · preference modeling

1 Introduction

Designing a decision support system or an automated decision making system for sequential decision making under risk is a widely studied task [1,2,21]. On the one hand, the need for such systems is considerably important as sequential decision making under risk has applications in several fields (business, finance, health, navigation, ...). On the other hand, dealing with risk in decision making is a hard task, since it requires to both know and consider all the probable consequences of a decision (or action) at short, medium and/or long term.

An agent faces risk when the outcomes of her decisions (or actions) are uncertain. Uncertainty can be found in different forms, we consider here situations where several outcomes can result from an action, and where the agents knows these outcomes, as well as their probabilities, but cannot predict with certainty which one will occur. For example, playing the roulette game enables to win 35 times the bet played with a probability 1/37, and to lose the bet otherwise. In such decision processes, the uncertainty makes the decision complex, and

S. Destercke et al. (Eds.): SUM 2024, LNAI 15350, pp. 70–84, 2025.
https://doi.org/10.1007/978-3-031-76235-2_6

when decisions are taken sequentially, uncertainty can greatly increase, as the succession of stochastic events can greatly increase the number of possible consequences, making the decision much harder.

Markov Decision Processes (MDPs) [19] provide natural tools to formalize sequential decision making under risk processes. Thus, solving the decision problem consists in computing an optimal policy for the corresponding MDP. In the literature, the most commonly used operator to solve an MDP is the expected reward (ER) criterion, suggesting the (very strong) assumption that any agent, human or artificial, is perfectly risk-neutral. Yet, this assumption is often unrealistic. Let us consider a simple example:

Example 1. An agent has to chose between two actions: action c that gives $15k\$$ with probability $2/3$ and 0 otherwise; and action d that gives $10k\$$ for sure.

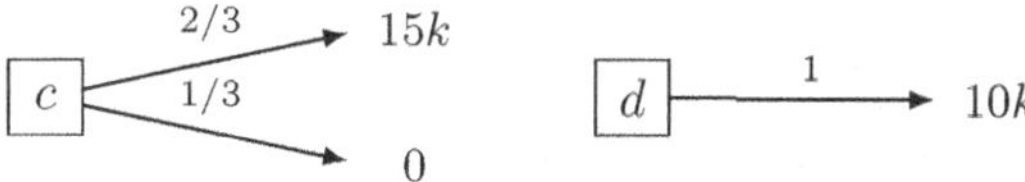

In this example, both actions have the same ER value, yet, it seems obvious that any rational agent would not choose indifferently one of the two actions. The first option offers the opportunity of a greater gain in the best case, but it is riskier as it involves having nothing at all in the worst case. It is clear that, in such situations, optimality is subjective, and we can easily admit that neutrality does not coincide with any agent behavior. Some, being strongly risk-averse, would always prefer the certainty over the risk (choosing action d in the example), while others, being strongly risk-seeking, would always be attracted by the possibility of a bigger payoff (choosing action c in the example). And finally, some decision makers would have intermediate behaviors being risk-averse in some situations and risk-seeking in others. Using ER fails to express such basic behaviors, simply because it does not allow to measure risk, even in the simple case where a random variable X is opposed to the certain gain $ER(X)$. Whereas in real-life situations, agents are often faced with more complex situations where they have to chose between actions representing different levels of risk.

As we will see in Sect. 2, several works in the literature proposed to replace ER by other operators to model non-neutral attitudes w.r.t. risk. Most of them focused on risk-averse decision making and by proposing pessimistic operators that evaluate the worst possible outcome(s): the min-max (optimizing the worst possible outcome), the min-max regret (minimizing the worst regret over a choice), the expected shortfall (optimizing the average reward of a given proportion of the worst possible outcomes). However, it is important to note that risk-averse decision making is not only about avoiding risk or danger, it is also about the compromise between the risk and the potential benefit from this risk, which can lead to intermediate behaviors that are not model by the cited operators. For example, many operators (including many of those discussed later) fail to express the widely observed Allais paradox [3] that states that the independence axiom (which is often considered as verified) is not always relevant. We illustrate this paradox through the following example:

Example 2. Before the agent of Example 1 makes her choice, we flip a coin. If the coin lands on head (with probability α), the agent chooses her action and gets her reward as described in Example 1. If it lands on tail (with probability $1 - \alpha$), she gets nothing. When $\alpha = 0.9$, this game is equivalent to a choice between actions c' and d' pictured below:

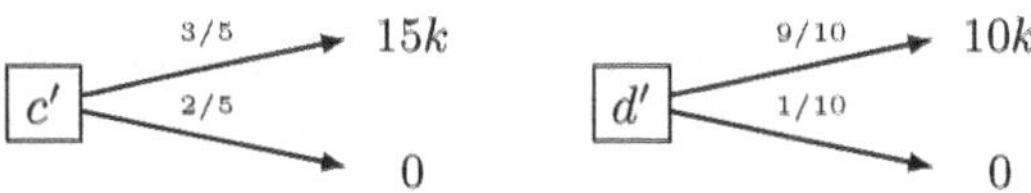

The independence axiom states that action c' is preferred to d' ($c' \succ d'$) *if and only if* $c \succ d$ (whatever the value of α). Nevertheless, the experimental study of Allais showed that the majority of surveyed agents violates this axiom.

The aim of this work is to propose a generic approach to solve sequential decision making problems under risk, considering as many behaviors as possible w.r.t. risk, including the one highlighted by the Allais paradox. For this, we will focus on the Weighted Ordered Weighted Average operator (WOWA for short). An operator that offers a rich descriptive power thanks to its parameter which, when well chosen, allows to model a wide range of attitudes w.r.t. risk. We will see that the main drawback of WOWA is its non-linearity and its lack of time-consistency, which prevents the use of standard algorithms like dynamic and linear programming. We propose in this paper a ranking algorithm that allows to determine a WOWA-optimal or nearly WOWA-optimal policy in an MDP.

2 Related Works

There are many works in the literature addressing risk-aware optimization in MDPs. The most popular operator used to substitute ER is the expected utility (EU) criterion [25], that consists in replacing the reward values in ER by their subjective utility values. Thus, in order to determine an optimal policy following a risk-averse (resp. risk-seeking) attitude, EU is optimized using a concave (resp. convex) utility function [15,23]. However, this operator has some descriptive limits, including the inability to explain the Allais paradox described above. Other criticisms of this operator can be found in [7,12,16].

Prospect theory [16], that generalizes EU, has been proposed to overcome the limits of ER and EU. The main idea is to use a transform function on probability values in order to express the subjective way agents perceive them. Nevertheless, this model has not been widely used in the literature because it can promote dominated solutions.

Another popular risk measure is the Conditional Value at Risk (CVaR), a.k.a. expected shortfall. It consists in optimizing the total mean reward in the worst α-fraction of runs, α being a parameter to be priorly fixed. This operator has been widely used to solve MDPs [5,8,9,22], but it has two major drawbacks: first, it only allows to model more or less pessimistic attitudes w.r.t. risk, and second, it can promote dominated solutions since the evaluation is focused on the worst possible consequences. Authors of [20] indeed show that there could exist several policies that optimize the CVaR value, and thus, an algorithm optimizing CVaR, without considering all the consequences of the policies, would indifferently return one of them, whether dominated or not. Authors of [20] introduced in their paper a lexicographic method that returns the best ER policy among the CVAR-optimal ones, thus responding to the criticism of the dominated solutions. But the lack of flexibility in terms of descriptive capacity remains.

To sum up, many papers studied the question of modeling an agent's attitude w.r.t. risk, but these papers have two main limits that we want to overcome in this paper. First, they are focused on the representation of risk-averse attitudes while other types of behaviors can be observed (risk-seeking and intermediate behaviors). Secondly, they suffer from a lack of flexibility regarding the modeling of different levels of attitudes. An agent can in fact be more or less risk-averse than another, or even have a behavior mixing the two attitudes (as in the Allais paradox). There are therefore an infinite number of possible behaviors that are not modeled by many operators used in the literature.

In this paper, we will focus on the Weighted Ordered Weighted Average operator (WOWA for short) [26]. A parameterized operator that offers a rich descriptive power. It allows to model different levels of risk-aversion, risk-seeking, and intermediate behaviors. However, optimizing WOWA in an MDP is a challenging issue because of its non-linearity and time inconsistency. Thus, the use of standard algorithms as dynamic and linear programming is not possible (or at least not obvious). Note that there exists an LP formulation for optimizing WOWA [18], but this solution does not apply here. The linear program indeed only apply to cases were the number of scenarios/consequences is fixed and priorly known. In an MDP, as we will see in the next section, policies have a variable (see Example 1) and unpredictable number of outcomes. We propose in this paper a ranking algorithm that allows to determine a WOWA-optimal or nearly WOWA-optimal policy. The idea of a ranking algorithm is not new. Outside the MDP framework, the authors of [6] proposed a ranking algorithm to determine a robust solution for the assignment and shortest path problems. They assume that the result of a decision is certain but the satisfaction of the agent is not. This uncertainty is related to the existence of a fixed and known number of scenarios. The adaptation of their method to MDPs is not easy as the context is different. The number of consequences is neither known or fixed in our case. In addition, their approach focused on the case of risk-averse behaviors, while we extend the method to deal with a larger range of behaviors, we will see next that this generalization implies additional difficulties.

To conclude this section, it is important to note that WOWA has already been used in the MDP framework but for multi-criteria decision making [17]. The main difference with our work lies in the policy evaluation, which implies both a descriptive difference and different algorithmic issues. They associate a reward vector to each pair (state, action) to describe the immediate reward of the pair according to every considered criterion. A policy is then associated to a vector giving, for every criteria, the expected discounted reward of the policy.

Thus assuming neutrality to obtain every criterion value. WOWA is only used to evaluate the compromise between criteria values. While in our paper, we consider only one reward function, and evaluate a policy with a lottery that summarizes all its possible consequences (thus not assuming neutrality). The lottery is then evaluated using WOWA. Thus, the difference between both works is similar to the difference with [6] except that one considers different criteria while the other considers different scenarios.

3 Backround and Notations

We define a Risk aware MDP (R-MDP) by a tuple $\mathcal{M} = \langle \mathcal{S}, \mathcal{A}, \mathcal{P}, \mathcal{R}, H, f \rangle$ where $\mathcal{S}$ is a finite set of fully observable states; $\mathcal{A}$ is a finite set of actions; $\mathcal{P} : \mathcal{S} \times \mathcal{A} \times \mathcal{S} \rightarrow [0,1]$ is the transition function where $\mathcal{P}(s, a, s')$ gives the probability of reaching state s' after performing action a in state s; $\mathcal{R} : \mathcal{S} \times \mathcal{A} \times \mathcal{S} \rightarrow \mathbb{R}^+$ is the reward function where $\mathcal{R}(s, a, s')$ is the reward value obtained when performing action a in state s and reaching state s'; H is the maximum time step; and finally f is a function that model the attitude of the agent w.r.t. risk. Depending on the considered problem, an R-MDP can be defined with two additional elements: an initial state $s_0 \in \mathcal{S}$ and a set of final states $\mathcal{S}_f \subset \mathcal{S}$ if applicable.

A solution of an MDP is a policy $\pi : H \times \mathcal{S} \rightarrow \mathcal{A}$ that gives the action to perform in state $s \in \mathcal{S}$ at time step $h < H$. Due to the stochasticity of the process, the result of an action is uncertain, and thus, a policy π induces a lottery $\langle r_1^\pi : p_1^\pi, \ldots, r_{|\mathcal{T}_\pi|}^\pi : p_{|\mathcal{T}_\pi|}^\pi \rangle$ where $\mathcal{T}_\pi = \{t_1^\pi, \ldots, t_{|\mathcal{T}_\pi|}^\pi\}$ is the set of trajectories induced by π, each trajectory t_i^π occurring with probability p_i^π (defined by $\mathcal{P}$) and leading to a cumulative reward r_i^π (defined by $\mathcal{R}$). Note that we can obtain the lottery using a dynamic programming method similar to value iteration [4].

In order to evaluate policies, function f will generally be defined on its induced lottery. Thus, solving an R-MDP means determining the best policy according to f. We call such a policy an f-optimal policy. In the literature, the most commonly used operator to evaluate policies is the Expected Reward (ER):

$$\mathrm{ER}(\pi) = \sum_{i=1}^{|\mathcal{T}_\pi|} p_i^\pi r_i^\pi \tag{1}$$

As ER is linear, Eq. (1) can be formulated dynamically thanks to Bellman's equations. Thus, the value of each state $s \in \mathcal{S}$ at each time step $h \in \{0, \ldots, H-1\}$ is given by:

$$V_h^\pi(s) = \sum_{s' \in \mathcal{S}} \mathcal{P}(s, \pi(h, s), s') \times \left[\mathcal{R}(s, \pi(h, s), s') + V_{h+1}^\pi(s')\right] \tag{2}$$

with $V_H^\pi(s') = 0$. Using Eq. (2), an ER-optimal policy can be obtained with linear programming [11] or using the well known value or policy iteration [4,14].

As illustrated in the introduction, ER's has descriptive limits that represent a serious drawback when it comes to model realistic behaviors in risky situations.

Thus, we will not use it to model the agent preferences w.r.t. risk, but it is still useful as we will see in the remaining of the paper. The following section gives a definition of the operator we use to better model risk-aware agents as well as the motivation behind this choice.

4 WOWA Model for Risk-Aware Optimization

The Weighted Ordered Weighted Average (WOWA) model is also known as the Yaari's model [24,26] because it has been introduced and justified by Yaari in [26] in the context of decision making under risk. It is a parameterized model that generalizes ER and offers a much richer descriptive power. It is define as follows.

Definition 1. *For any policy π and transform probability function $\varphi : [0,1] \to [0,1]$ that is continuous, increasing and such that $\varphi(0) = 0$ and $\varphi(1) = 1$, we have:*

$$\mathrm{W}_\varphi(\pi) = \sum_{i=1}^{|\mathcal{T}_\pi|} \left[(r^\pi_{\sigma(i)} - r^\pi_{\sigma(i-1)}) \right] \varphi \left(\sum_{k=i}^{|\mathcal{T}_\pi|} p^\pi_{\sigma(k)} \right) \tag{3}$$

where $r^\pi_{\sigma(0)} = 0$, and σ is a permutation of $\{1, \ldots, |\mathcal{T}_\pi|\}$ that reorders the elements of r^π in the increasing order of reward values, i.e., $r^\pi_{\sigma(1)} \leq \cdots \leq r^\pi_{\sigma(|\mathcal{T}_\pi|)}$.

Example 3. Let us consider a policy π that induces the lottery: $\langle 0 : \frac{1}{3}, 10 : \frac{1}{2}, 15 : \frac{1}{6} \rangle$, and a function $\varphi(p) = p^2, \forall p \in [0,1]$. We have:

$$\mathrm{W}(\pi) = 0 + (10-0)\varphi(\frac{1}{2} + \frac{1}{6}) + (15-10)\varphi(\frac{1}{6}) \approx 4.58$$

The parameter function φ allows to model a subjective perception of the probability values. Note that, as long as φ is increasing on $[0,1]$, the preferences induced by W_φ are monotonic with respect to the first order stochastic dominance (FSD). FSD expresses the rational behavior of preferring ℓ to ℓ' as long as for all $x \in \mathbb{R}$, the probability of getting a reward higher than x with ℓ is greater than with ℓ'. In addition to this objective and rational preference, φ allows to control the type of decision behavior we want to model, depending on its specific shape: a convex function (resp. concave) allows to model risk-aversion (resp. risk-seeking) [13,26]; and a linear function allows to model neutrality as $\mathrm{W}_\varphi(\pi) = ER(\pi)$ when $\varphi(p) = p, \forall p \in [0,1]$. We can also use S-shaped or inverse S-shaped functions in order to model more sophisticated behaviors. In particular, the function proposed by *Kahneman and Tversky* [16], defined by $\varphi(p) = \exp^{-\sqrt{-ln(p)}}, \forall p \in [0,1]$, allows to express the behavior highlighted by the Allais paradox. This is illustrated by the following examples.

Example 1 (continued). *Table 1 gives WOWA values of actions c and d for three different φ functions: p^2, $\sqrt{p}$* and $\exp^{-\sqrt{-ln(p)}}$ *(noted kt[1]) for short:*

[1] After Kahneman and Tversky.

Here, unlike ER, WOWA gives different values for c and d. We can see that the convex function $\varphi(p) = p^2$ (as well as the Kahneman and Tversky function) favors d which is less risky, while the concave function $\varphi(p) = \sqrt{p}$ gives the opposite preference order.

Table 1. Wowa values for actions c and d

action	W_{p^2}	$W_{\sqrt{p}}$	W_{kt}
c	6666.6667	**12247.4487**	7935.0431
d	**10000**	10000	**10000**

The following example shows that the precise definition of φ allows to model different level of a certain type of preferences, as well as the behavior highlighted by the Allais paradox.

Example 4. Let us consider a simple sequential decision problem modeled by the decision tree pictured in Fig. 1, where circle nodes (s_0 to s_6) are states and rectangle nodes (a to d) are actions. For any transition (s, a, s'), the probability $\mathcal{P}(s, a, s')$ is given above the edge (a, s'), and the reward $\mathcal{R}(s, a, s')$ is given on the right of node state s'.

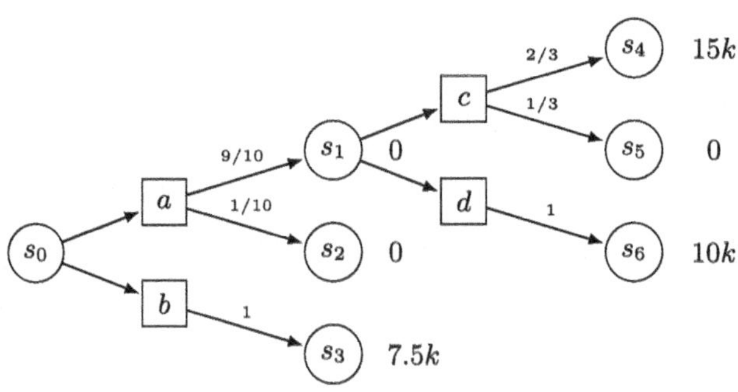

Fig. 1. Decision tree of Example 4

There are three possible policies: choosing a in s_0 and c in s_1 noted π_{ac}; choosing a in s_0 and d in s_1 noted π_{ad}; choosing b in s_0 noted π_b. The lotteries induced by these policies and their evaluations with ER and WOWA with 4 different φ functions are given in Table 2.

The table illustrates several aspect of the descriptive power of WOWA: 1. it allows to discriminate between policies having equal ER values (π_{ac} and π_{ad} for example), but also to favor a policy that is strictly dominated w.r.t. ER (π_b and π_{ac} for example); 2. different shapes for φ can give different preference orders. We

Table 2. Lotteries of Example 1 and their WOWA values

Policy	Induced lotteries	ER	W_{p^2}	W_{p^5}	$W_{\sqrt{x}}$	W_{kt}
π_{ac}	$\langle 15k : \frac{3}{5}, 0 : \frac{2}{5}\rangle$	**9000**	5400	1166.4	**11618.95**	7339.93
π_{ad}	$\langle 10k : \frac{9}{10}, 0 : \frac{1}{10}\rangle$	**9000**	**8100**	5904.9	9486.83	7228.21
π_b	$\langle 7.5k : 1\rangle$	7500	7500	**7500**	7500	**7500**

can also see that two functions modeling the same kind of behavior can also give different preference orders as they represent different levels of the behavior. For example, the two convex functions $\varphi(p) = p^2$ and $\varphi(p) = p^5$ do not give the same preference order for π_{ad} and π_b; 3. WOWA explains the Allais paradox with the Kahneman and Tversky function. We can indeed see that π_{ac} is preferred to π_{ad} while Table 1 shows that c is preferred to d (note that the lotteries induced by π_{ac} and π_{ad} are the same as those induced by c' and d' in Example 2); 4. the preference orders in the previous point also show that WOWA is not time consistent. Note that this is a consequence of the non-linearity of the operator, which comes from the sorting operation in WOWA.

The absence of time consistency has a dual effect. On the one hand it reflects a very realistic behavior, expressing (among other things) the Allais paradox. On the other hand, it prevents the use of practical tools as linear programming and value/policy iteration that are based on Bellman's equations. In the remaining of the paper, we introduce a ranking algorithm allowing to determine a WOWA-optimal or nearly WOWA-optimal policy.

5 Computing a WOWA-Optimal Policy

The main idea of the algorithm is to enumerate policies using a linear operator, that is easier to optimize, until a satisfactory policy has been found. The algorithm has the following steps:

1. determine a close linear bound for WOWA noted B (see Subsect. 5.1),
2. enumerate policies by decreasing order of B values (see Subsect. 5.2),
3. stop enumeration when we can prove that a (nearly) WOWA-optimal policy has been enumerated (see Subsect. 5.3).

The steps are detailed in the remaining of the section.

5.1 Bounding WOWA

The next proposition allows to define a bounding linear (on $\mathrm{ER}(\pi)$) function $B : \Pi \rightarrow \mathbb{R}^+$, where Π is the set of all possible policies.

Proposition 1. *Let $\pi \in \Pi$ be a feasible policy, T_π be the set of induced trajectories, and $\langle r_1^\pi : p_1^\pi, \ldots, r_{|T_\pi|}^\pi : p_{|T_\pi|}^\pi\rangle$ be the associated lottery. For any linear* *function* $g : [0,1] \rightarrow \mathbb{R}^+$ *of the form* $g(p) = ap + b$ *where* $a, b \in \mathbb{R}^+$ *chosen such that* $\varphi(p) \leq g(p), \forall p \in [0,1]$, *we have:*

$$\mathrm{W}_\varphi(\pi) \leq B(\pi) = aER(\pi) + b \max_{i \in \{1,\ldots,|\mathcal{T}_\pi|\}} r_i^\pi \tag{4}$$

Proof. As $\varphi(p) \leq g(p) = ap + b, \forall p \in [0,1]$, and using Eq. (3) we have:

$$\begin{aligned}
\mathrm{W}_\varphi(\pi) &\leq \sum_{i=1}^{|\mathcal{T}_\pi|} \left[r_{(i)}^\pi - r_{(i-1)}^\pi \right] \left[a \left(\sum_{k=i}^{|\mathcal{T}_\pi|} p_{(k)}^\pi \right) + b \right] \\
&\leq a \sum_{i=1}^{|\mathcal{T}_\pi|} r_{(i)}^\pi \left[\sum_{k=i}^{|\mathcal{T}_\pi|} p_{(k)}^\pi - \sum_{k=i+1}^{|\mathcal{T}_\pi|} p_{(k)}^\pi \right] + b\, r_{(|\mathcal{T}_\pi|)}^\pi \leq a \sum_{i=1}^{|\mathcal{T}_\pi|} r_{(i)}^\pi p_{(i)}^\pi + b\, r_{(|\mathcal{T}_\pi|)}^\pi \\
&\leq a \sum_{i=1}^{|\mathcal{T}_\pi|} r_i^\pi p_i^\pi + b\, r_{(|\mathcal{T}_\pi|)}^\pi \leq aER(\pi) + b \max_{i \in \{1,\ldots,|\mathcal{T}_\pi|\}} r_t^\pi
\end{aligned}$$

Thus, to define B we only need to find values for a and b such that $\varphi(p) \leq ap + b$ in $[0,1]$. Note that there is an infinite number of values verifying this inequality. In order to obtain the most efficient algorithm possible, we will take values that makes g as close as possible to φ in $[0,1]$. Thus, for convex functions, we define g by $g(p) = p$, for concave and (inverse) S-shaped functions we will take a tangent line having the minimum distance to φ[2].

5.2 Enumerating Policies

The idea of the enumeration method is simple, it consists in exploring and partitioning Π in a specific way to enumerate policies by decreasing order of B values, and this without missing any policy. For this, we will need to: 1. determine a procedure to efficiently partition Π and to explore its subsets. 2. find an optimal policy in a specific subset of Π.

Partitioning Π. The procedure described below is very similar to the one proposed in [10]. The idea is simple: we first optimize B for the initial MDP (see the MILP formulation paragraph below), and we obtain a policy π_1^*. Using π_1^* we partition Π as follows. Let us note E the support of π_1^*[3] and $\{e_1, ..., e_{|E|}\}$ the different elements of E. We define the set $\mathcal{X}^i, \forall i \in \{0, \ldots, |E|\}$, as

$$\{\pi \in \Pi | \pi(e_i) \neq \pi_1^*(e_i) \wedge \pi(e_k) = \pi_1^*(e_k), \forall k < i\}$$

It is easy to see that $\{\pi_1^*\} \cup \mathcal{X}^0 \cup \cdots \cup \mathcal{X}^{|E|}$ is a partition of Π. Thus, the second best policy of Π (according to B), noted π_2^*, will be the B-optimal policy

---

[2] The distance of a function f to a function g on $[0,1]$ is defined by $\int_0^1 |f(x) - g(x)| dx$. Note that in our experiments, we used a sum to approximate this integral.

[3] The set of couples $(h, s) \in \{1, \ldots, H\} \times \mathcal{S}$ such that s is reachable at time h when we apply π_1^*.

of one of the sets $\mathcal{X}^i, i \in \{0, \ldots, |E|\}$. Thus, to find it we optimize B in every subset and take the best one. Let us note j the index of the set $\mathcal{X}^j$ containing π_2^*. The next step is to partition $\mathcal{X}^j$ using π_2^* as we did for Π using π_1^*. The total partition of Π is then obtained by replacing $\mathcal{X}^j$ (in the initial partition) by the union of $\{\pi_2^*\}$ and the obtained partition of $\mathcal{X}^j$. These operations are repeated as many times as necessary to find a satisfying policy (see Subsect. 5.3).

Optimizing B. To optimize the B value in each subset of the partition, we propose in the following a MILP formulation that is composed of the main MILP (given below), which solves the initial problem (considering Π as the set of possible policies), and additional constraints to determine the optimal policy for a subset $\mathcal{X}^i$ of Π. The main MILP is:

$$\max \sum_{h=0}^{H-1} \sum_{(s,a,s')\in\mathcal{T}} \mathcal{R}(s,a,s') \left(x_{sa}^h \mathcal{P}(s,a,s') + y_{sas'}^h\right)$$

$$\sum_a x_{sa}^h - \sum_{s'} \sum_a \mathcal{P}(s',a,s) x_{s'a}^{h-1} \leq \mathbb{1}_{[s=s_0,h=0]} \quad \forall s \in \mathcal{S}, h < H \tag{5}$$

$$y_{sas'}^h \leq x_{sa}^h + 1 - \epsilon \quad \forall s \in \mathcal{S}, a \in \mathcal{A}, h < H \tag{6}$$

$$\sum_{s'a} y_{s'as}^h = \sum_{s'a} y_{sas'}^{h+1} \quad \forall s \in \mathcal{S}, h < H \tag{7}$$

$$\sum_{sas'} y_{sas'}^h = 1 \quad \forall h < H \tag{8}$$

$$x_{sa}^h \geq 0 \quad \forall s \in \mathcal{S} \setminus \mathcal{S}_f, a \in \mathcal{A}, h < H-1 \tag{9}$$

$$y_{sas'}^h \in \{0,1\} \quad \forall s, s' \in \mathcal{S} \setminus \mathcal{S}_f, a \in \mathcal{A}, h < H-1 \tag{10}$$

Two type of variables are used:

Continuous variables x_{sa}^h that are the standard decision variables of the LP formulation of an MDP [11]. Their values can be interpreted as follows:

$$x_{sa}^h = \begin{cases} p(s,h \mid s_0) & \text{if } a \text{ is performed in state } s \text{ at time } h \\ 0 & \text{otherwise} \end{cases}$$

where $p(s,h \mid s_0)$ is the probability of reaching s at time h when s_0 is the initial state. Note that these variables will be implicitly constrained to be less than 1 (by definition of the MILP).

Binary variables $y_{sas'}^h$ that indicates whether the transition (s,a,s') is in the maximum reward trajectory induced by the policy (the max term in Proposition 1) or not.

The objective function of the MILP expresses the linear function B defined in (4). In this expression, $\mathcal{T}$ is the set of all probable transitions, i.e., all $(s,a,s') \in$

$\mathcal{S}\times\mathcal{A}\times\mathcal{S}$ such that $\mathcal{P}(s,a,s') > 0$. The first constraint is the standard constraint of the MDP's LP formulation [11]. The second constraint expresses the fact that a transition (s,a,s') (at time h) can be in the maximum reward trajectory only if a is performed in s at time h in the policy. Here we consider that a probability is significant only if it is greater than a given ϵ. The third constraint expresses the fact that a the trajectory gets out of a state $s \in \mathcal{S}$ at time $h+1$ if and only if it enters in at time h. The forth constraint expresses the fact that only one transition is crossed at time h.

An optimal solution to the defined MILP gives a B-optimal solution for the initial MDP. In order to determine the optimal policy for a subset $\mathcal{X}^i$ of Π, we need to define some additional constraints:

- for any tuple (h,s,a) for which there is a constraint $\pi(h,s) \neq a$:

$$x^h_{sa} = 0$$

- for any tuple (h,s,a) for which there is a constraint $\pi(h,s) = a$, we add the constraint $\pi(h,s) \neq a', \forall a' \neq a$ to prevent the choice of another action[4].

5.3 Stopping Condition

Proposition 2 gives a condition that, when fulfilled at a given step of the ranking algorithm, guaranties that a WOWA-optimal policy has been found.

Proposition 2. *Let Π be the set of all possible policies, and let $\Pi^k = (\pi^1,\ldots,\pi^k)$ be the list of the k best elements of Π according to the linear bound B. We have:*

$$\max_{\pi\in\Pi^k} \mathrm{W}_\varphi(\pi) > B(\pi^k) \Rightarrow \max_{\pi\in\Pi^k} \mathrm{W}_\varphi(\pi) = \max_{\pi\in\Pi} \mathrm{W}_\varphi(\pi)$$

Proof. We will prove that, at enumeration k, if there exists a policy $\pi' \in \arg\max_{\pi\in\Pi^k} \mathrm{W}_\varphi(\pi)$ satisfying $\mathrm{W}_\varphi(\pi') > B(\pi^k)$, then any policy in $\Pi\setminus\Pi^k$ cannot be WOWA-optimal: a policy $\pi'' \in \Pi\setminus\Pi^k$ is such that $B(\pi'') \leq B(\pi^k)$ (otherwise it would have been enumerated before π^k). From Proposition 1 we have $\mathrm{W}_\varphi(\pi'') \leq B(\pi'')$, we then deduce that $\mathrm{W}_\varphi(\pi'') \leq \mathrm{W}_\varphi(\pi') = \max_{\pi\in\Pi^k} \mathrm{W}_\varphi(\pi)$, which concludes the proof.

The enumeration can then be stopped as soon as this condition is fulfilled. Note that this algorithm is anytime, we can, at any step k, stop the enumeration and return the best found policy (a policy in $\arg\max_{\pi\in\Pi^k} \mathrm{W}_\varphi(\pi)$) with $B(\pi^k) - \max_{\pi\in\Pi^k} \mathrm{W}_\varphi(\pi)$ as a bound on the distance to the optimal value.

Finally, we could relax the stopping condition and stop enumerations as soon as $\max_{\pi\in\Pi^k} \mathrm{W}_\varphi(\pi) > B(\pi^k) - \delta$ where δ is a prefixed threshold in order to save some computation time and determine a δ-optimal policy.

[4] Note that we could instead define a constraint $x^h_{sa} \geq \epsilon$ but our experiments showed that this option is slower.

Proposition 3. *The ranking algorithm ends and returns an (δ-)optimal solution.*

The optimality results from Proposition 2, while the termination results from the definition of the algorithm: either the stopping condition is verified and the algorithm stops, or the condition is never fulfilled and all the policies of Π are enumerated exactly once. Since Π is a finite set, the algorithm necessarily stops.

6 Experimental Results

We have implemented[5] and tested[6] the ranking algorithm introduced in the previous section, and we give in this section a part of the obtained results.

We tested the algorithm on 100 randomly generated[7] MDPs with 10 states, 3 actions, and $H = 5$. In order to simulate the attitude of an agent w.r.t. risk, we used a WOWA operator with multiple parameter functions φ: $p^5, p^{0.25}$ and $\exp -\sqrt{-ln(p)}$. The histograms on Fig. 2 give the rank (in the enumeration) of the WOWA-optimal policy. These histograms focus on the first 1000 enumerations, but we can see that this was not much constraining as, most of the time, the optimal policy is found before 1000 enumerations. Finally, the figure in the bottom right of Fig. 2 shows the evolution of ER (or more accurately $aER + b$), WOWA, and CVaR values throughout the running of the ranking algorithm on a specific instance[8].

We can see in the histograms that, for any considered φ function, the WOWA-optimal policy is generally different from the ER-optimal policy. Note that there is a proportion of instances for which the first enumerated policy is optimal. This is due to the fact that random generation often produces uninteresting instances w.r.t. decision under risk, as it is the case when, for example, there exists a policy that (strongly) dominates all the other ones, regardless of the considered type of preferences. Besides these instances, the rank of the optimal policy is relatively well distributed between rank 2 and rank 1000. Thus, the WOWA-optimal policy is often far from the ER-optimal one in term of ranks. Considering the figure on the bottom right, it clearly shows that WOWA can discriminate between solutions considered equivalent either by ER or by the CVaR operator. It also shows that the stopping condition can take time to be verified (more than 1000 enumerations here), but the approach has the advantage of allowing the determination of a policy in a more accurate (and may be less hazardous) way than by simply optimizing ER and/or CVaR with value iteration or lexicographic algorithms such as the one in [20].

---

[5] The implementation was performed in Python, and the linear programs were solved using the `gurobi` Python library.

[6] Tests were performed on an Intel(R) Core(TM) i7-1165G7 CPU with 15 GB of RAM.

[7] Using the `mdptoolbox` library: https://pymdptoolbox.readthedocs.io.

[8] A betting game instance, a definition can be found in [20].

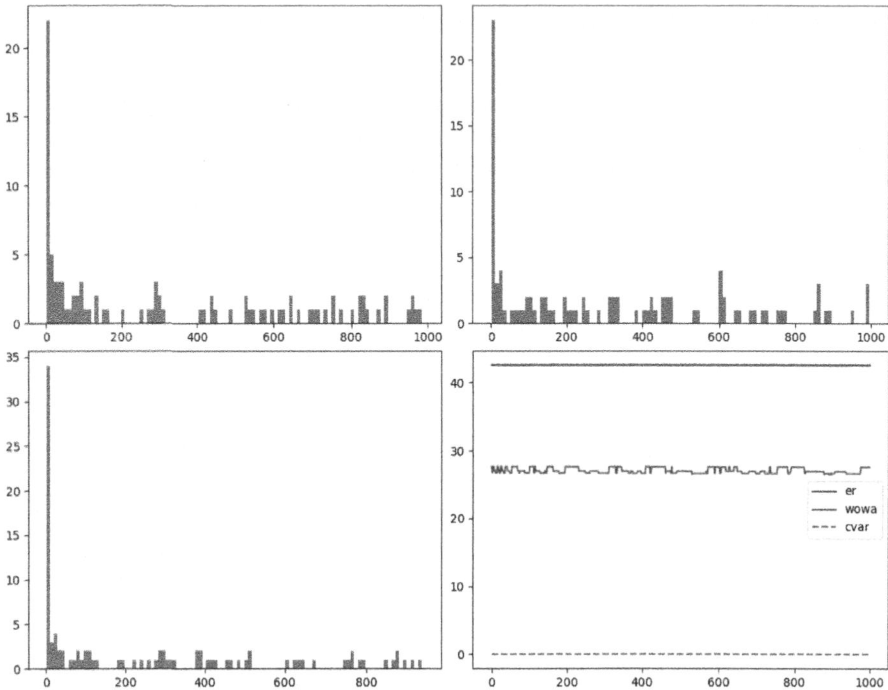

Fig. 2. Rank of the optimal policies for $\varphi(p) = p^5$ (top left), $\varphi(p) = p^{0.25}$ (top right), $\varphi(p) = e^{-\sqrt{-ln(p)}}$ (bottom left) and an execution trace of the algorithm (bottom right)

7 Conclusion

We have introduced an algorithm to solve risk-aware MDPs. This work differentiates from previous works in the MDP literature in proposing a more flexible risk measure that has never been used (as far as we know) in MDPs as a measure of risk. In comparison with the decision making under risk literature, this work provides an extension to sequential decision problems and to a more general case in term of preference modeling. Similar works indeed focus on the case of risk-averse agents.

In this work, we make the assumption that the attitude of the agent w.r.t. risk is precisely known, and that the parameter function φ can be fixed accordingly. However, this assumption is restrictive and may be unrealistic in some situations. Thus, our next step is to extend this approach to the case of imprecisely known preferences w.r.t. risk. This is a challenging task because if φ is not fixed, its bound cannot be defined with the same precision, and consequently, it is harder to determine an efficient bound on WOWA.

References

1. Ahiska, S.S., Appaji, S.R., King, R.E., Warsing, D.P., Jr.: A Markov decision process-based policy characterization approach for a stochastic inventory control problem with unreliable sourcing. Int. J. Prod. Econ. **144**(2), 485–496 (2013)
2. Alexander, G.J., Baptista, A.M.: A comparison of VAR and CVAR constraints on portfolio selection with the mean-variance model. Manage. Sci. **50**(9), 1261–1273 (2004)
3. Allais, M.: Le comportement de l'homme rationnel devant le risque: critique des postulats et axiomes de l'école américaine. Econometrica: journal of the Econometric Society, pp. 503–546 (1953)
4. Bellman, R.: A Markovian decision process. J. Math. Mech. **6**(5), 679–684 (1957)
5. Borkar, V., Jain, R.: Risk-constrained Markov decision processes, pp. 2664–2669 (2010)
6. Bourdache, N., Perny, P.: Anytime algorithms for adaptive robust optimization with OWA and WOWA, pp. 93–107 (2017)
7. Chateauneuf, A., Cohen, M., Meilijson, I.: Four notions of mean-preserving increase in risk, risk attitudes and applications to the rank-dependent expected utility model. J. Math. Econ. **40**(5), 547–571 (2004)
8. Chow, Y., Ghavamzadeh, M.: Algorithms for CVAR optimization in MDPs. In: Advances in Neural Information Processing Systems, vol. 27 (2014)
9. Chow, Y., Tamar, A., Mannor, S., Pavone, M.: Risk-sensitive and robust decision-making: a CVAR optimization approach. In: Advances in Neural Information Processing Systems, vol. 28 (2015)
10. Dai, P., Goldsmith, J.: Finding best k policies, pp. 144–155 (2009)
11. Denardo, E.V.: On linear programming in a Markov decision problem. Manage. Sci. **16**(5), 281–288 (1970)
12. Gonzales, C., Perny, P.: Decision under uncertainty. In: A Guided Tour of Artificial Intelligence Research: Volume I: Knowledge Representation, Reasoning and Learning, pp. 549–586 (2020)
13. Hong, C.S., Karni, E., Safra, Z.: Risk aversion in the theory of expected utility with rank dependent probabilities. J. Econ. Theory **42**(2), 370–381 (1987)
14. Howard, R.A.: Dynamic programming and Markov processes (1960)
15. Howard, R.A., Matheson, J.E.: Risk-sensitive Markov decision processes. Manage. Sci. **18**(7), 356–369 (1972)
16. Kahneman, D., Tversky, A.: Prospect theory: an analysis of decision under risk. Econometrica **47**(2), 263–291 (1979)
17. Ogryczak, W., Perny, P., Weng, P.: A compromise programming approach to multiobjective Markov decision processes. Int. J. Inf. Technol. Decis. Making **12**(05), 1021–1053 (2013)
18. Ogryczak, W., Śliwiński, T.: On efficient WOWA optimization for decision support under risk. Int. J. Approx. Reason. **50**(6), 915–928 (2009)
19. Puterman, M.L.: Markov Decision Processes: Discrete Stochastic Dynamic Programming, 1st edn. Wiley, Hoboken (1994)
20. Rigter, M., Duckworth, P., Lacerda, B., Hawes, N.: Planning for risk-aversion and expected value in MDPs, vol. 32, pp. 307–315 (2022)
21. Rigter, M., Lacerda, B., Hawes, N.: Risk-averse Bayes-adaptive reinforcement learning. Adv. Neural. Inf. Process. Syst. **34**, 1142–1154 (2021)
22. Rockafellar, R.T., Uryasev, S., et al.: Optimization of conditional value-at-risk. J. Risk **2**, 21–42 (2000)

23. Rothschild, M., Stiglitz, J.E.: Increasing Risk: I. A Definition. Elsevier (1978)
24. Torra, V.: The weighted OWA operator. Int. J. Intell. Syst. **12**(2), 153–166 (1997)
25. Von Neumann, J., Morgenstern, O.: Theory of Games and Economic Behavior, 2nd rev (1947)
26. Yaari, M.E.: The dual theory of choice under risk. Econometrica J. Econometric Soc. 95–115 (1987)

Entropy and Extropy for Partial Probability Assessments on Arbitrary Families of Events

Lydia Castronovo[1(✉)] and Giuseppe Sanfilippo[2]

[1] Department of Mathematics and Computer Sciences, Physical Sciences and Earth Sciences, University of Messina, Messina, Italy
lydia.castronovo@studenti.unime.it

[2] Department of Mathematics and Computer Science, University of Palermo, Palermo, Italy
giuseppe.sanfilippo@unipa.it

Abstract. In subjective probability theory, a coherent probability assertion represents an honest expression of its promoter's uncertain knowledge about the value of an unknown quantity. The theory of proper scoring rules was central to de Finetti's ideas about assessing the relative values of different subjective probability assessments. In this paper, we consider an asymmetric proper scoring rule for the probability of an event, which belongs to the 2-parameter Beta family of scoring rules. Then we consider the associated loss function of a probability assessment on an arbitrary family of n events. We observe that, in the particular case of a probability mass distribution of a random quantity, the expected loss function associated with the asymmetric score coincides with the Shannon entropy. Likewise, we show similar properties between the notion of extropy and the expected loss function associated with the complement of the asymmetric score. Then, we suitably extend the notion of entropy and extropy from partitions of events to arbitrary families of events. We also introduce Bregman divergences associated with these measures of information. Finally, we introduce a symmetric proper scoring rule for an event, showing that the associated expected loss function for an arbitrary family of events coincides with the sum of entropy and extropy.

Keywords: Bregman divergence · Entropy · Extropy · Kullback-Leibler divergence · Measures of information · Penalty criterion · Proper scoring rules · Subjective probability

1 Introduction

Scoring rules are a quite popular tool in statistical and probabilistic research and they have been widely investigated and extended to various fields of application (see, e.g., [13–16,19,23,24,26,27,29]). In subjective probability theory, a coherent probability assertion represents an honest expression of its promoter's uncertain knowledge about the value of an unknown quantity. The theory of proper

S. Destercke et al. (Eds.): SUM 2024, LNAI 15350, pp. 85–99, 2025.
https://doi.org/10.1007/978-3-031-76235-2_7

scoring rules was central to de Finetti's ideas about assessing the relative values of different subjective probability assessments. In his theory, de Finetti discussed the significance of proper scoring rules and Dutch Book in evaluating subjective probability assessments. In particular, he studied the connections between the admissibility of a probability assessment and the concept of coherence by means of the penalty criterion and the Brier quadratic score [8,9]. Then, this context is a perfect setting for the notion of scoring rule that measures the "quality" of a probability assessment and encourages sincerity since an individual interest lies in announcing probabilities that conforms to her own beliefs [25]. The relationship between the concepts of coherence and non-dominance, in the context of generic strictly proper scoring rules, has been in particular examined in [25] (for unconditonal events) and in [12] (for conditional events). These papers have also investigated the notion of scoring rule for a probability assessment on an event in the context of Bregman divergence. Scoring rules for probability distributions, particularly for probability assessments on finite partitions of events, can also be related to the notions of Shannon entropy [18] and its dual complement, extropy [22]. In this paper, by exploiting (asymmetric) scoring rules for the probability of an event, we suitably extend the definitions of entropy and extropy from a partition of events to the case of arbitrary families of events. Consequently, we also extend the well-known Kullback-Leibler divergence (and its dual) from a simplex of $\mathbb{R}^n$ to the larger domain $[0,1]^n$.

The paper is organized as follows. In Sect. 2 we recall some basic notions on events, scoring rules, Bregman divergence, entropy, extropy, and the penalty criterion for defining coherence. In Sect. 3 we introduce the asymmetric logarithmic scoring rule, by showing that it is strictly proper. We also illustrate its relation to the 2-parameter Beta family of scoring rules.

In Sect. 4 we extend the notion of entropy from partitions to an arbitrary finite family of events and we notice that the entropy can be seen as the expected value of the random penalty associated to the asymmetric logarithmic score. We also introduce a Bregman divergence which generalizes the Kullback-Leibler divergence. In Sect. 5, we introduce the complement of the asymmetric logarithmic score, which allows us to define the notion of extropy for an arbitrary family of events and the associated Bregman divergence. In Sect. 6, we propose a suitable symmetric strictly proper scoring rule, showing that the associated expected loss coincides, not only for the probability mass distribution of a random quantity but also for probability assessments on arbitrary families of events, with the sum of entropy and extropy. Finally in Sect. 7 we give some conclusions and possible future developments.

2 Preliminary Notions

An event E is a two-valued logical entity which can be true, or false. The indicator of an event E, denoted by the same symbol, takes the values 1, or 0, according to whether E is true, or false. The sure event and impossible event are denoted by Ω and $\emptyset$, respectively. Given two events E_1 and E_2, we denote by $E_1 \wedge E_2$, or simply by E_1E_2, the logical conjunction and by $E_1 \vee E_2$ the logical disjunction. The negation of E is denoted $\bar{E}$. We simply write $E_1 \subseteq E_2$ to denote that E_1 logically implies E_2, that is $E_1\bar{E}_2 = \emptyset$. We recall that n events $E_1, \ldots, E_n$ are said to be logically independent when the number m of constituents, or possible worlds, generated by them is 2^n, that is when the all 2^n conjunctions $E_1^* \cdots E_n^*$, with $E_i^* \in \{E_i, \bar{E}_i\}$, $i = 1, 2, \ldots, n$, are not impossible. In the subjective approach to probability of de Finetti, the probability $P(E)$ that an individual attributes to an event E, in his current state of knowledge, is a measure of its degree of belief on E being true. In order to operatively assess probabilities, de Finetti proposed two (equivalent) criteria: the betting scheme and the penalty criterion, both based on the concept of coherence.

2.1 Strictly Proper Scoring Rules and Coherence

We recall the notion of strictly proper scoring rule for an event E. In the literature, a score can be interpreted as a reward or as a penalty. In the following, we look at scores as penalties and hence improving the score means reducing it.

Definition 1. *[25] A function $s : \{0,1\} \times [0,1] \to [0,+\infty]$ is said to be a* strictly proper scoring rule *if the following conditions are satisfied:*
(a) for every $x, p \in [0,1]$, with $x \neq p$, it is

$$p\, s(1,p) + (1-p)\, s(0,p) < p\, s(1,x) + (1-p)\, s(0,x)\,; \tag{1}$$

(b) the functions $s(1,x)$ and $s(0,x)$ are continuous.

We observe that x is your announced probability for the event E while p is your degree of belief on E and hence the quantity $p\, s(1,x) + (1-p)\, s(0,x)$ represents your expected score.

We consider for any strictly proper scoring rule $s : \{0,1\} \times [0,1] \to [0,\infty]$, the extension S on the set $[0,1] \times [0,1]$, defined as (see [12])

$$S(p,x) = ps(1,x) + (1-p)s(0,x).$$

We observe that if x is your announced probability for the event E, and p represents your degree of belief in E, then the quantity $S(p,x)$ represents your expected score. Using S, condition (a) in Definition 1 can be rewritten as

$$S(p,p) < S(p,x),\ \forall x, p \in [0,1],\ x \neq p.$$

That is, given a strictly proper scoring rule s, your expected score is minimum only when your announced probability x coincides with your degree of belief p. Furthermore, if $s(1,x)$ and $s(0,x)$ are continuous functions, then $S(p,x)$, for any given $p \in [0,1]$ is a continuous function in the variable x.

Remark 1. Condition (b) in Definition 1 is consistent with s assuming value $+\infty$. As noticed in [25], the only cases where a strictly proper scoring rule s can assume value $+\infty$ are arguments $(1,0)$ and $(0,1)$, that is for categorically mistaken judgements. Therefore, when we consider the function S we have that $S(p,0) = ps(1,0) + (1-p)s(0,0) = +\infty$ for $p > 0$, if $s(1,0) = +\infty$; $S(p,1) = ps(1,1) + (1-p)s(0,1) = +\infty$ for $p < 1$, if $s(0,1) = +\infty$. In particular $S(p,p) < \infty$ for all $p \in [0,1]$.

The notion of coherence is crucial in the subjective approach to probability and it can be defined through a penalty criterion. Given a strictly proper scoring rule s, with the assessment $\mathcal{P} = (p_1, \ldots, p_n)$ on $\mathcal{F} = \{E_1, \ldots, E_n\}$ we associate a random loss $\mathcal{L}(\mathcal{F}, \mathcal{P}) = \sum_1^n s(E_i, p_i)$. Then, based on [9,10], coherence can be characterized by the following definition [11,12].

Definition 2. *Given a strictly proper scoring rule s and an arbitrary family of events $\mathcal{K}$, a probability assessment P on the events in $\mathcal{K}$ is said to be* coherent *if and only if, for every integer n, for every finite subfamily $\mathcal{F} = \{E_1, \ldots, E_n\} \subseteq \mathcal{K}$, denoting by $\mathcal{P} = (p_1, \ldots, p_n)$ the restriction of P to $\mathcal{F}$, it does not exist another assessment $\mathcal{P}^* = (p_1^*, \ldots, p_n^*)$ on $\mathcal{F}$ such that $\mathcal{L}(\mathcal{F}, \mathcal{P}^*) \leq \mathcal{L}(\mathcal{F}, \mathcal{P})$, with $\mathcal{L}(\mathcal{F}, \mathcal{P}^*) < \mathcal{L}(\mathcal{F}, \mathcal{P})$ in at least one case.*

Definition 2 defines coherence in terms of the loss $\mathcal{L}$, which in turn depends on the chosen scoring rule s. However, as shown in [12, Theorem 4] (see also [25]), the set of all coherent assessments on $\mathcal{K}$ does not depend on the choice of the strictly proper scoring rule s. Then, Definition 2 coincides with the definition of coherence originally based on the Brier score, $s = (E - p)^2$ (see, [9,10]).

Remark 2. We observe (see the proof of [12, Theorem 3]) that if the assessment $\mathcal{P} = (p_1, \ldots, p_n)$ on $\mathcal{F} = \{E_1, \ldots, E_n\}$ is coherent, then for any coherent extension $(\lambda_1, \ldots, \lambda_m)$ (that is, with $\sum_{h:C_h \subseteq E_i} \lambda_h = p_i$, $i = 1, \ldots, n$, $\sum_{h=1}^m \lambda_h = 1$ and $\lambda_h \geq 0$, $h = 1, \ldots, m$) on the set of constituents $\{C_1, \ldots, C_m\}$ associated with $\mathcal{F}$, and for any $\mathcal{P}^* \neq \mathcal{P}$, it holds that

$$\sum_{h=1}^{m} \lambda_h L_h = \sum_{i=1}^{n} S(p_i, p_i) < \sum_{i=1}^{n} S(p_i, p_i^*) = \sum_{h=1}^{m} \lambda_h L_h^*, \tag{2}$$

where L_h (resp., L_h^*) represents the value of $\mathcal{L}(\mathcal{F}, \mathcal{P})$ (resp., of $\mathcal{L}(\mathcal{F}, \mathcal{P}^*)$) when the constituent C_h is true. In other words, for any coherent extension $(\lambda_1, \ldots, \lambda_m)$ of the coherent assessment $\mathcal{P}$, formula (2) implies that

$$\mathbb{E}(\mathcal{L}(\mathcal{F}, \mathcal{P})) < \mathbb{E}(\mathcal{L}(\mathcal{F}, \mathcal{P}^*)), \;\; \forall \mathcal{P}^* \neq \mathcal{P}.$$

2.2 Bregman Divergence, Entropy and Extropy

Let C be a convex subset of $\mathbb{R}^n$ with non empty interior and let $\phi : C \to \mathbb{R}$ be a strictly convex function, differentiable in the interior of C whose gradient $\nabla\phi$ extends to a bounded, continuous function on C. In the definition below we recall the notion of Bregman divergence [7].

Definition 3. *For $x, y \in C$ the Bregman divergence $D_\phi : C \times C \to \mathbb{R}$ corresponding to ϕ is given by $D_\phi(x, y) = \phi(x) - \phi(y) - \nabla\phi(y)(x - y)$.*

We can construct a Bregman divergence starting from a given scoring rule s. We recall that it can be proved that $S(p, x) = S(x, x) + S'(x, x)(p - x)$ and, by setting $\phi(p) = -S(p, p)$, it holds that

$$D_\phi(p, x) = \phi(p) - \phi(x) - \nabla\phi(x)(p - x) = S(p, x) - S(p, p). \tag{3}$$

We recall now two important objects strongly related to the notion of scoring rule. Given a finite random quantity X with possible values $x_1, \ldots, x_n$, $n \geq 1$, we set $E_i = (X = x_i)$, $i = 1, \ldots, n$. Then, the family $\mathcal{F} = \{E_1, E_2, \ldots, E_n\}$ is a partition of Ω. Let $\mathcal{P} = (p_1, \ldots, p_n)$, with $p_i = P(E_i)$, be a coherent probability assessment on $\mathcal{F}$. If $\mathcal{P}$ is coherent, then it is a probability mass function on X.

Definition 4. *The* entropy *associated with the pair* $(X, \mathcal{P})$ *is defined as*

$$H(X, \mathcal{P}) = -\sum_{i=1}^{n} p_i \log(p_i). \tag{4}$$

We recall that the entropy is related with the expected value of the logarithmic strictly proper scoring rule of a distribution [18, pag. 343], that is with the expected value of

$$s_{\log}(\mathcal{F}, \mathcal{P}) = -\sum_{i=1}^{n} E_i \log(p_i). \tag{5}$$

Definition 5. *The* extropy *associated with the pair* $(X, \mathcal{P})$ *is defined as*

$$J(X, \mathcal{P}) = -\sum_{i=1}^{n} (1 - p_i) \log(1 - p_i). \tag{6}$$

The extropy can be seen as dual and complement of entropy [22]. We also recall that the extropy can be seen as the expected value of the following strictly proper scoring rule of a distribution [1,21,22]

$$s^{c}_{\log}(\mathcal{F}, \mathcal{P}) = -\sum_{i=1}^{n} (1 - E_i) \log(1 - p_i). \tag{7}$$

3 Asymmetric Logarithmic Score

In this section we analyze the asymmetric logarithmic score and we show that it is strictly proper. First, we recall the definition of symmetric score [30].

Definition 6. *A scoring rule s is symmetric if and only if*

$$s(i, x) = s(1 - i, 1 - x)\,, i = 0, 1, \textit{for all } x \in [0, 1].$$

The well known Brier score $s(E, x) = (E - x)^2$ is an example of symmetric (strictly proper) scoring rule [4]. Indeed $s(1 - E, 1 - x) = (1 - E - 1 + x)^2 = (E - x)^2$.

By Definition 6, a score s is said to be *asymmetric* if $s(i, x) \neq s(1 - i, 1 - x)\,, i = 0, 1$ for some $x \in [0, 1]$. An example of asymmetric score is given by the following logarithmic score.

Definition 7. *Given an event E, the* asymmetric logarithmic *scoring rule is defined as*

$$s(E, x) = x - E\log(x) - E = \begin{cases} x - \log(x) - 1, & \text{if } E = 1, \\ x, & \text{if } E = 0, \end{cases}$$

where

$$s(1, 0) = \lim_{x \to 0^+} s(1, x) = +\infty.$$

Of course $s(E, x)$ is not symmetric, because $s(1, x) = x - \log(x) - 1 \neq s(0, 1-x) = 1 - x$. We observe that the extension $S(p, x)$ is given by

$$S(p, x) = p\,(x - \log(x) - 1) + (1 - p)\,x = -p\log(x) - p + x, \tag{8}$$

and in particular $S(1, x) = s(1, x) = x - \log(x) - 1$ and $S(0, x) = s(0, x) = x$. As a consequence of the asymmetry of the score, it follows that $S(p, x) \neq S(1-p, 1-x)$ as illustrated in Fig. 1. We show now that the previous score is a strictly proper scoring rule according to Definition 1.

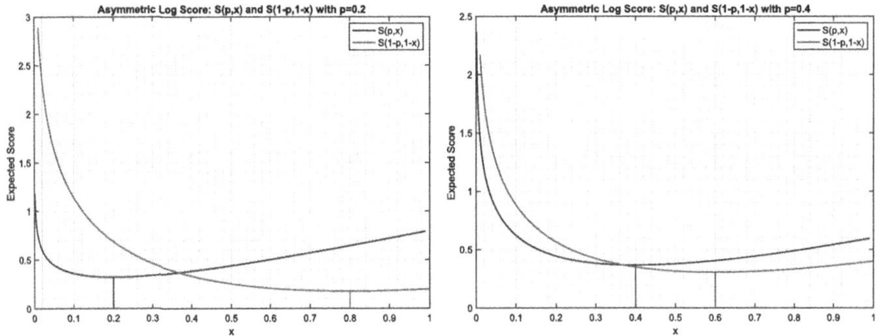

Fig. 1. The functions $S(p, x)$ and $S(1 - p, 1 - x)$ for the asymmetric log score with $p = 0.2$ (left) and $p = 0.4$ (right) illustrate the asymmetry of the score.

Proposition 1. *The asymmetric scoring rule $s(E, x) = x - E\log(x) - E$ is a strictly proper score.*

Proof. We begin by observing that $s(1, x) = x - \log(x) - 1$ and $s(0, x) = x$ are continuous functions in $[0, 1]$ and hence condition (b) of Definition 1 is satisfied. By Definition 1, to prove that $s(E, x)$ is strictly proper we have to show that $S(p, x) = ps(1, x) + (1 - p)s(0, x)$ is uniquely minimized at $x = p$. We know that $p = P(E)$ so that we can consider it fixed and looking at $S(p, x)$ as a function of x only. We distinguish two cases: $(i)\, p \in (0, 1]$ and $(ii)\, p = 0$.

Case (i). Let $p \in (0, 1]$ and $x \in [0, 1]$. We first observe that for $x = 0$,

$$\lim_{x \to 0+} S(p, x) = \lim_{x \to 0+} (x - p\log(x) - 1) = +\infty$$

and hence $S(p,p) < S(p,0)$. Then, for $x \in (0,1]$, $x \neq p$, it holds that

$$\begin{aligned} S(p,p) < S(p,x) &\iff p - p\log(p) - p < x - p\log(x) - p \\ &\iff x - p - p\log(x) + p\log(p) > 0 \iff \tfrac{x}{p} - \log(\tfrac{x}{p}) - 1 > 0. \end{aligned}$$

For $x = p$ we have of course that $S(p,p) = S(p,x)$. Therefore for every $p \in (0,1]$ and for every $x \in [0,1]$ it holds that $S(p,p) < S(p,x)$ if $x \neq p$ and $S(p,p) = S(p,x)$ if and only if $x = p$, that is $S(p,x)$ is uniquely minimized at $x = p$.

Case (ii). It is immediate to see that $S(0,x) = s(0,x) = x$ takes its minimum in $x = p = 0$. □

The property of being strictly proper for the asymmetric logarithmic score can also be also observed in Fig. 1, indeed for $p = 0.2$ (left) and $p = 0.4$ (right) the expected score $S(p,x)$ reaches its minimum only when $x = p$.

Remark 3 (Beta family). We note that this score belongs to the 2-parameter Beta family of scoring rules (see [5], see also [13, pag. 364]), that is the two-parameter family

$$s(1,x) = \int_x^1 t^{\alpha-1}(1-t)^{\beta}dt,$$

$$s(0,x) = \int_0^x t^{\alpha}(1-t)^{\beta-1}dt,$$

with $\alpha, \beta > -1$. Indeed, we obtain the asymmetric logarithmic scoring rule by setting the parameters $\alpha = 0, \beta = 1$ and we have that

$$s(1,x) = x - \log(x) - 1 = \int_x^1 t^{\alpha-1}(1-t)^{\beta}dt, \quad \alpha = 0, \beta = 1\,.$$

$$s(0,x) = x = \int_0^x t^{\alpha}(1-t)^{\beta-1}dt, \quad \alpha = 0, \beta = 1\,.$$

We also recall that this family provides symmetric scoring rules when $\alpha = \beta$, indeed by setting $c = 1 - t$ it holds that

$$\begin{aligned} s(1,x) &= \textstyle\int_x^1 t^{\alpha-1}(1-t)^{\beta}dt = \int_{1-x}^0 (1-c)^{\alpha-1}c^{\alpha}(-dc) = \\ &= \textstyle\int_0^{1-x} c^{\alpha}(1-c)^{\alpha-1}dc = s(0,1-x). \end{aligned}$$

Given the asymmetric logarithmic score $s(E,x) = x - E\log(x) - E$, using Eq. (3), we can find its associated Bregman divergence. Therefore, by Eq. (8), setting $\phi(p) = -S(p,p) = p\log(p)$, it holds that

$$\begin{aligned} D_{\phi}(x,y) &= S(p,x) - S(p,p) \\ &= x - p\log(x) - p - p + p\log(p) + p = p\log(\tfrac{p}{x}) - (p - x). \end{aligned} \tag{9}$$

The divergence in Eq. (9) will be used in Sect. 4 to generalize the Kullback-Leibler divergence (KLD).

4 Entropy for a Probability Assessment on an Arbitrary Family of Events

Looking at Eq. (4), we observe that the quantities $x_1, \ldots, x_n$, are not involved in the definition of the entropy and hence we can consider the entropy for any pair $(\mathcal{F}, \mathcal{P})$, where $\mathcal{F}$ is a partition of Ω, even if $\mathcal{F}$ is not associated with a random quantity X. Then, given a partition $\mathcal{F}$ of Ω and a coherent probability assessment $\mathcal{P}$ on $\mathcal{F}$ we set the entropy associated with the pair $(\mathcal{F}, \mathcal{P})$ as

$$H(\mathcal{F}, \mathcal{P}) = -\sum_{i=1}^{n} p_i \log(p_i) \,.$$

Given a coherent probability assessment $\mathcal{P} = (p_1, \ldots, p_n)$ on an arbitrary family of event $\mathcal{F} = \{E_1, \ldots, E_n\}$, we consider the penalty $\mathcal{L}$ w.r.t. the asymmetric strictly proper scoring rule $s(E, x) = x - E\log(x) - E$,

$$\begin{aligned}\mathcal{L}(\mathcal{F}, \mathcal{P}) &= \textstyle\sum_{i=1}^{n} s(E_i, p_i) = \sum_{i=1}^{n} (p_i - E_i \log(p_i) - E_i) \\ &= \textstyle\sum_{i=1}^{n} p_i - \sum_{i=1}^{n} E_i \log(p_i) - \sum_{i=1}^{n} E_i \,.\end{aligned}$$

If $\mathcal{F}$ is a partition of Ω, as $\sum_{i=1}^{n} p_i = 1$, we have that the penalty

$$\mathcal{L}(\mathcal{F}, \mathcal{P}) = \sum_{i=1}^{n} p_i - \sum_{i=1}^{n} E_i \log(p_i) - 1 = -\sum_{i=1}^{n} E_i \log(p_i) = s_{\log}(\mathcal{F}, \mathcal{P}).$$

In other words, the logarithmic score of a distribution coincides with the sum of the asymmetric logarithmic scoring rule applied to each event in the partition $\mathcal{F}$.

Given a coherent probability assessment $\mathcal{P} = (p_1, \ldots, p_n)$ and a probability assessment $\mathcal{Q} = (q_1, \ldots, q_n)$ on an arbitrary family of events $\mathcal{F} = \{E_1, \ldots, E_n\}$, we set

$$S(\mathcal{P}, \mathcal{Q}) = \sum_{i=1}^{n} S(p_i, q_i) = \sum_{i=1}^{n} q_i - \sum_{i=1}^{n} p_i \log(q_i) - \sum_{i=1}^{n} p_i,$$

which, by the linearity property of the expectation (see Remark 2), coincides with the expected value of your penalty $\mathbb{E}(\mathcal{L}(\mathcal{F}, \mathcal{Q}))$ when your announced probability assessment on $\mathcal{F}$ is $\mathcal{Q}$, while your degrees of belief on $\mathcal{F}$ are given by the assessment $\mathcal{P}$. In particular, $S(\mathcal{P}, \mathcal{P})$ is the expected value of $\mathcal{L}(\mathcal{F}, \mathcal{P})$, that is

$$S(\mathcal{P},\mathcal{P}) = \mathbb{E}(\mathcal{L}(\mathcal{F},\mathcal{P})) = -\sum_{i=1}^{n} p_i \log(p_i)\,. \tag{10}$$

By setting $\Phi(p_1,\ldots,p_n) = \sum_{i=1}^{n} \phi(p_i)$, where $\phi(p_i) = p_i \log(p_i)$, it follows from Eq. (9) that the Bregman divergence $D_\Phi(\mathcal{P},\mathcal{Q})$ is given by

$$D_\Phi(\mathcal{P},\mathcal{Q}) = S(\mathcal{P},\mathcal{Q}) - S(\mathcal{P},\mathcal{P}) = \textstyle\sum_{i=1}^{n}[p_i \log(\frac{p_i}{q_i}) - (p_i - q_i)]. \tag{11}$$

The divergence in Eq. (11) is the Generalized I-divergence [3] which generalizes the Kullback-Leibler divergence (KLD) to the larger domain $[0,1]^n$.

If $\mathcal{F}$ is a partition of Ω, then Eq. (10) shows that the expected value of the penalty $\mathcal{L}(\mathcal{F},\mathcal{P})$, that is $S(\mathcal{P},\mathcal{P})$, when your announced probability assessment coincides with your degree of belief, is the entropy associated with the pair $(\mathcal{F},\mathcal{P})$. Moreover, in this case, as $\sum_{i=1}^{n} p_i = \sum_{i=1}^{n} q_i = 1$, it follows that

$$S(\mathcal{P},\mathcal{Q}) = -\sum_{i=1}^{n} p_i \log(q_i),$$

which coincides with the cross entropy. The divergence $D_\Phi(\mathcal{P},\mathcal{Q})$ given in Eq. (11) reduces to KLD, that is

$$D_\Phi(\mathcal{P},\mathcal{Q}) = \sum_{i=1}^{n} p_i \log\left(\frac{p_i}{q_i}\right).$$

In the following definition, by exploiting the asymmetric logarithmic score, we extend the notion of entropy to an arbitrary finite family of events $\mathcal{F}$.

Definition 8. *Given a family of n events $\mathcal{F}$ and a coherent probability assessment $\mathcal{P}$ on the $\mathcal{F}$, let $s(E,x) = x - E\log(x) - E$ be the asymmetric logarithmic strictly proper scoring rule on an event E and $\mathcal{L}$ be the corresponding penalty associated with $(\mathcal{F},\mathcal{P})$. We define the entropy associated with the pair $(\mathcal{F},\mathcal{P})$, denoted by $H(\mathcal{F},\mathcal{P})$, as the expected value w.r.t. $\mathcal{P}$ of the penalty $\mathcal{L}$, that is*

$$H(\mathcal{F},\mathcal{P}) = \mathbb{E}(\mathcal{L}(\mathcal{F},\mathcal{P})) = S(\mathcal{P},\mathcal{P}) = -\sum_{i=1}^{n} p_i \log(p_i).$$

Remark 4. In particular, if $n = 1$, that is $\mathcal{F} = \{E\}$, $\mathcal{P} = \{p\}$, we have

$$H(E,p) = S(p,p) = -p\log(p),$$

which coincides with the well-known expected surprise for an event with probability p. Then, as the entropy of a random quantity X coincides with the expected surprise of a probability mass function on a partition of events, the entropy given in Definition 8 can be interpreted as the expected surprise of an assessment $\mathcal{P}$ on an arbitrary family of events $\mathcal{F}$.

5 Extropy for a Probability Assessment on an Arbitrary Family of Events

Similarly to the previous section, we define the notion of extropy for an arbitrary family of events by considering the complement of the asymmetric logarithmic score $s(E,x) = x - E\log(x) - E$. We define the complement asymmetric logarithmic score of $s(E,x)$ as

$$s^c(E,x) = s(1-E, 1-x) = E - x - (1-E)\log(1-x).$$

The score $s^c(E,x)$ is asymmetric (see Fig. 1), strictly proper and its expected score is given by

$$S^c(p,x) = p - x - (1-p)\log(1-x).$$

Given a coherent probability assessment $\mathcal{P} = (p_1, \dots, p_n)$ on an arbitrary family of n event $\mathcal{F} = \{E_1, \dots, E_n\}$, the penalty $\mathcal{L}^c$ w.r.t. $s^c(E,x)$ is given by

$$\begin{aligned}\mathcal{L}^c(\mathcal{F},\mathcal{P}) &= \textstyle\sum_{i=1}^n s^c(E_i, p_i)\\ &= \textstyle\sum_{i=1}^n E_i - \sum_{i=1}^n p_i - \sum_{i=1}^n (1-E_i)\log(1-p_i)\,.\end{aligned}$$

If $\mathcal{F}$ is a partition of Ω, as $\sum_{i=1}^n p_i = 1$, we have that

$$\mathcal{L}^c(\mathcal{F},\mathcal{P}) = 1 - 1 - \sum_{i=1}^n (1-E_i)\log(1-p_i) = -\sum_{i=1}^n (1-E_i)\log(1-p_i) = s^c_{\log}(\mathcal{F},\mathcal{P}).$$

By recalling that $\mathcal{Q} = (q_1, \dots, q_n)$ is your announced probability assessment on $\mathcal{F}$ is $\mathcal{Q}$, and $\mathcal{P} = (p_1, \dots, p_n)$ the vector of your degrees of belief on $\mathcal{F}$, we set

$$S^c(\mathcal{P},\mathcal{Q}) = \sum_{i=1}^n S^c(p_i, q_i) = \sum_{i=1}^n p_i - \sum_{i=1}^n (1-p_i)\log(1-q_i) - \sum_{i=1}^n q_i,$$

which coincides with $\mathbb{E}(\mathcal{L}^c(\mathcal{F},\mathcal{Q}))$. In particular, $S^c(\mathcal{P},\mathcal{P})$, that is the expected value of the penalty when $\mathcal{Q} = \mathcal{P}$, is

$$S^c(\mathcal{P},\mathcal{P}) = \mathbb{E}(\mathcal{L}^c(\mathcal{F},\mathcal{P})) = -\sum_{i=1}^n (1-p_i)\log(1-p_i)\,. \tag{12}$$

As done in Sect. 4, we set $\Phi^c(p_1, \dots, p_n) = \sum_{i=1}^n \phi^c(p_i)$, where $\phi^c(p_i) = (1-p_i)\log(1-p_i)$. Then, if $\mathcal{P}$ is your degree of belief on $\mathcal{F}$ and $\mathcal{Q}$ your announced probability, it follows from Eq. (3) that the Bregman divergence $D_{\Phi^c}(\mathcal{P},\mathcal{Q})$ is given by

$$D_{\Phi^c}(\mathcal{P},\mathcal{Q}) = S^c(\mathcal{P},\mathcal{Q}) - S^c(P,P) = \textstyle\sum_{i=1}^n [(1-p_i)\log(\frac{1-p_i}{1-q_i}) - (q_i - p_i)]. \tag{13}$$

If $\mathcal{F}$ is a partition of Ω, then Eq. (12) shows that the expected value of the penalty $\mathcal{L}^c(\mathcal{F},\mathcal{P})$, that is $S^c(\mathcal{P},\mathcal{P})$, when your announced probability assessment coincides with your degree of belief, is the extropy associated with the pair $(\mathcal{F}, \mathcal{P})$. Moreover, in this case it holds that

$$S^c(\mathcal{P}, Q) = \sum_{i=1}^{n} S^c(p_i, q_i) = -\sum_{i=1}^{n}(1 - p_i)\log(1 - q_i),$$

coincides with the cross extropy [20].

Now, by exploiting the complement asymmetric logarithmic score, we extend the notion of extropy to an arbitrary finite family of events $\mathcal{F}$.

Definition 9. *Given a family of n events $\mathcal{F}$ and a coherent probability assessment $\mathcal{P} = (p_1, \ldots, p_n)$ on $\mathcal{F} = \{E_1, \ldots, E_n\}$, let $s^c(E, x)$ be the complement asymmetric logarithmic score on an event E and $\mathcal{L}^c$ be the corresponding penalty associated with $(\mathcal{F}, \mathcal{P})$. We define the extropy associated with the pair $(\mathcal{F}, \mathcal{P})$, denoted by $J(\mathcal{F}, \mathcal{P})$, as the expected value w.r.t. $\mathcal{P}$ of the penalty $\mathcal{L}^c$, that is*

$$J(\mathcal{F}, \mathcal{P}) = \mathbb{E}(\mathcal{L}^c(\mathcal{F}, \mathcal{P})) = S^c(\mathcal{P}, \mathcal{P}) = -\sum_{i=1}^{n}(1 - p_i)\log(1 - p_i).$$

6 Symmetrized Logarithmic Score of the Probability of an Event

The notions of entropy and extropy for a random quantity, as well as our extensions to a probability assessment on an arbitrary family of events, can be interpreted as the expected value of the penalties associated with asymmetric scores of the probability of an event. We consider now the symmetrized logarithmic score obtained as the sum of $s(E, x)$ and $s^c(E, x)$, that is

$$\begin{aligned} s_{\text{sym}}(E, p) &= s(E, p) + s^c(E, p) = s(E, p) + s(1 - E, 1 - p) \\ &= (p - E\log(p) - E) + (1 - p - (1 - E)\log(1 - p) - 1 + E) \\ &= -E\log(p) - (1 - E)\log(1 - p). \end{aligned} \tag{14}$$

Of course, the symmetric score s_{sym} is strictly proper because it is the sum of two strictly proper scoring rules. Denoting by p you degree of belief on E, and by x your announced probability, the expected value of s_{sym} is

$$S_{\text{sym}}(p, x) = -p\log(x) - (1 - p)\log(1 - x),$$

with

$$S_{\text{sym}}(p, p) = -p\log(p) - (1 - p)\log(1 - p) < S_{\text{sym}}(p, x), \quad \text{if } x \neq p.$$

As a consequence of the symmetry of the score, it follows that $S_{\text{sym}}(p, x) = S_{\text{sym}}(1 - p, 1 - x)$ as illustrated in Fig. 2.

Given a coherent probability assessment $\mathcal{P} = (p_1, \ldots, p_n)$ on an arbitrary family of n events $\mathcal{F} = \{E_1, \ldots, E_n\}$, the penalty $\mathcal{L}_{\text{sym}}$ w.r.t. $s_{\text{sym}}(E, x)$ is given by

$$\begin{aligned} \mathcal{L}_{\text{sym}}(\mathcal{F}, \mathcal{P}) &= \textstyle\sum_{i=1}^{n} s_{\text{sym}}(E_i, p_i) = \sum_{i=1}^{n}(s(E_i, p_i) + s^c(E_i, p_i)) \\ &= -\textstyle\sum_{i=1}^{n} E_i\log(p_i) - \sum_{i=1}^{n}(1 - E_i)\log(1 - p_i). \end{aligned}$$

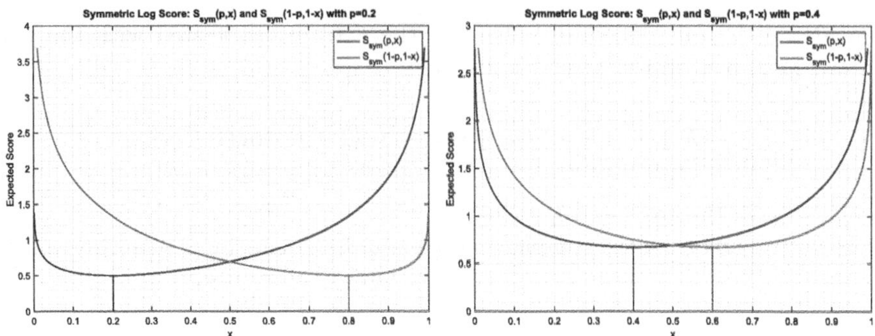

Fig. 2. The functions $S_{\rm sym}(p,x)$ and $S_{\rm sym}(1-p,1-x)$ for the symmetric log score with $p=0.2$ (left) and $p=0.4$ (right) illustrate the symmetry of the score.

Still denoting by $\mathcal{Q}=(q_1,\dots,q_n)$ your announced probability assessment on $\mathcal{F}$, and by $\mathcal{P}=(p_1,\dots,p_n)$ the vector of your degrees of belief on $\mathcal{F}$, we set

$$S_{\rm sym}(\mathcal{P},\mathcal{Q}) = S(\mathcal{P},\mathcal{Q})+S^c(\mathcal{P},\mathcal{Q}) = -\sum_{i=1}^n p_i\log(q_i)-\sum_{i=1}^n(1-p_i)\log(1-q_i). \quad (15)$$

Then,

$$S_{\rm sym}(\mathcal{P},\mathcal{Q}) = \mathbb{E}(\mathcal{L}_{\rm sym}(\mathcal{F},\mathcal{Q})) = \mathbb{E}(\mathcal{L}(\mathcal{F},\mathcal{Q})) + \mathbb{E}(\mathcal{L}^c(\mathcal{F},\mathcal{Q})).$$

Considering the Definitions 8 and 9 of Entropy and Extropy on an arbitrary family of events $\mathcal{F}$, given a coherent assessment $\mathcal{P}$ on $\mathcal{F}$, the expected value of the penalty associated w.r.t. $s_{\rm sym}$ given in Eq. (15) coincides with the sum of entropy and extropy, that is

$$S_{\rm sym}(\mathcal{P},\mathcal{P}) = H(\mathcal{F},\mathcal{P}) + J(\mathcal{F},\mathcal{P}). \quad (16)$$

Since both entropy and extropy represent the expected value of the penalty with respect to asymmetric scores, a useful measure of information can be obtained by considering the expected value of the corresponding symmetrized score. This measure is given by Eq. (16), that is by $H(\mathcal{F},\mathcal{P})+J(\mathcal{F},\mathcal{P})$, which take into account both entropy and extropy for arbitrary family of events.

To look at the Bregman divergence that we can obtain using the expected score in Eq. (16), from Eqs. (11) and (13), by setting $\Phi_{\rm sym}(p_1,\dots,p_n) = \sum_{i=1}^n \phi_{\rm sym}(p_i)$, where $\phi_{\rm sym}(p_i)=\phi(p_i)+\phi^c(p_i) = p_i\log(p_i)+(1-p_i)\log(1-p_i)$, it holds that

$$\begin{aligned} D_{\Phi_{\rm sym}}(\mathcal{P},\mathcal{Q}) &= D_{\Phi}(\mathcal{P},\mathcal{Q}) + D_{\Phi^c}(\mathcal{P},\mathcal{Q}) \\ &= \textstyle\sum_{i=1}^n [p_i\log(\frac{p_i}{q_i}) - (p_i-q_i)] + \sum_{i=1}^n[(1-p_i)\log(\frac{1-p_i}{1-q_i}) - (q_i-p_i)] \\ &= \textstyle\sum_{i=1}^n \left[p_i\log\left(\frac{p_i}{q_i}\right) + (1-p_i)\log\left(\frac{1-p_i}{1-q_i}\right)\right]. \end{aligned} \quad (17)$$

Then, the Bregman divergence of the symmetrized score is nothing but the sum of the divergence obtained in Eqs. (11) and (13) using the expected value of the asymmetric logarithmic score (Eq. (5)) and its "complementary" asymmetric score (Eq. (7)). Notice that, since we are considering that $\mathcal{P}$ and $\mathcal{Q}$ are probability assessments on an arbitrary family of n events $\mathcal{F}$, unlike the KLD, in Eq. (17) it is not required that $\mathcal{P}$ and $\mathcal{Q}$ belong to the simplex of $\mathbb{R}^n$. Finally, if $\mathcal{F}$ is a partition, coherence requires that $\mathcal{P}$ belongs to the simplex and hence Eq. (16) coincides with the expected value of the total score associated with the following score for a distribution [22]

$$s_{\text{totallog}}(\mathcal{F}, \mathcal{P}) = s_{\text{log}}(\mathcal{F}, \mathcal{P}) + s^c_{\text{log}}(\mathcal{F}, \mathcal{P}) = \sum_{i=1}^{n} [-E_i \log(p_i) - (1 - E_i) \log(1 - p_i)],$$

which in our case is obtained as the penalty $\mathcal{L}_{\text{sym}}$ w.r.t. the symmetric logarithmic score s_{sym} for the probability of an event.

7 Conclusion

We introduced and analyzed the asymmetric logarithmic score for an event $s(E, x) = x - E\log(x) - E$, showing that it is a strictly proper scoring rule belonging to the 2-parameter Beta family of scores. Then, we suitably extended the definitions of entropy from a partition of events to the case of arbitrary families of events. Consequently, we also extended the well-known Kullback-Leibler divergence from a simplex of $\mathbb{R}^n$ to the larger domain $[0, 1]^n$. We noticed that entropy can be seen as the expected value of the random penalty associated with the asymmetric logarithmic score. Furthermore, we introduced the complement of the asymmetric logarithmic score, $s^c(E, x) = s(1 - E, 1 - x)$. This score allows us to define the notion of extropy for an arbitrary family of events and the associated Bregman divergence. Then, we considered a symmetrization of the asymmetric score $s_{\text{sym}}(E, x) = s(E, x) + s^c(E, x)$ showing that the associated expected loss coincides, not only for the probability mass distribution of a random quantity but also for probability assessments on arbitrary families of events, with the sum of entropy and extropy. In order to extend the notions of entropy, extropy, and the associated Bregman divergences to an arbitrary family of events, a key role is played by the fact that the involved scoring rules for a distribution, s_{log} and s^c_{log}, can be seen as the random penalty associated with the asymmetric scores $s(E, x)$ and $s^c(E, x)$ on the probability of an event E. Further developments could concern the study of the properties of entropy, extropy, cross-entropy, and cross-extropy in the case of an arbitrary family of events (or conditional events), by also exploiting the connections between scoring rules and Bregman divergence. Moreover, the structure free approach used in this paper may also be of interest when considering partial assessments, as it allows for a connection to work on scoring rules for imprecise models [17,24,28]. Possible applications include correcting incoherent probability assessments [2,6], eliciting subjective probabilities [26], and evaluating the predictive quality of alternative sequential forecast partial assessments [20].

Acknowledgement. We thank the three anonymous reviewers for their useful comments and suggestions. We also thank Angelo Gilio and Frank Lad for past discussions on this topic. G. Sanfilippo is supported by the FFR2024 project of University of Palermo and by the MIUR-PRIN project 2022AP3B3B funded by Next Generation EU. Both authors are affiliated with the INdAM-GNAMPA research group and contributed equally to the paper.

Competing Interests. The authors have no competing interests to declare that are relevant to the content of this article.

References

1. Agró, G., Lad, F., Sanfilippo, G.: Sequentially forecasting economic indices using mixture linear combinations of EP distributions. J. Data Sci. **8**(1), 101–126 (2010). http://www.unipa.it/giuseppe.sanfilippo/pdf/2010/jds/JDS-473.pdf. ISSN 1680-743X
2. Baioletti, M., Capotorti, A.: A L1 based probabilistic merging algorithm and its application to statistical matching. Appl. Intell. **49**(1), 112–124 (2019). https://doi.org/10.1007/s10489-018-1233-z
3. Banerjee, A., Merugu, S., Dhillon, I.S., Ghosh, J.: Clustering with Bregman divergences. J. Mach. Learn. Res. **6**(58), 1705–1749 (2005). http://jmlr.org/papers/v6/banerjee05b.html
4. Brier, G.W.: Verification of Forecasts expressed in terms of probability. Mon. Weather Rev. **78**(1), 1–3 (1950)
5. Buja, A., Stuetzle, W., Shen, Y.: Loss functions for binary class probability estimation and classification: structure and applications. http://www-stat.wharton.upenn.edu/~buja/
6. Capotorti, A., Regoli, G., Vattari, F.: Correction of incoherent conditional probability assessments. Int. J. Approx. Reasoning **51**(6), 718–727 (2010). https://doi.org/10.1016/j.ijar.2010.02.002
7. Censor, Y.A., Zenios, S.A.: Parallel Optimization: Theory, Algorithms and Applications. Oxford University Press, Oxford (1997)
8. de Finetti, B.: Does it make sense to speak of 'good probability appraisers'? In: Good, I.J. (ed.) The Scientist Speculates: An Anthology of Partly-Baked Ideas, pp. 357–364. Heinemann, London (1962)
9. de Finetti, B.: Teoria delle probabilitá. Ed. Einaudi, 2 voll., Torino (1970), English version: Theory of Probability 1 (2), Chichester, Wiley, 1974 (1975)
10. Gilio, A.: Criterio di penalizzazione e condizioni di coerenza nella valutazione soggettiva della probabilità. Boll. Un. Mat. Ital., [7a] 4-B(3): 645-660 (1990)
11. Gilio, A., Sanfilippo, G.: Subjective probability, trivalent logics and compound conditionals. In: Égré, P., Rossi, L. (eds.) Handbook of Three-Valued Logics. MIT Press, Cambridge (forthcoming). http://arxiv.org/abs/2301.09327
12. Gilio, A., Sanfilippo, G.: Coherent conditional probabilities and proper scoring rules. In: Coolen, F., de Cooman, G., Fetz, T., Oberguggenberger, M. (eds.) ISIPTA 2011: Proceedings of the Seventh International Symposium on Imprecise Probability: Theories and Applications, pp. 189–198. SIPTA, Innsbruck (2011)
13. Gneiting, T., Raftery, A.E.: Strictly proper scoring rules, prediction, and estimation. J. Am. Stat. Assoc. **102**, 359–378 (2007). http://EconPapers.repec.org/RePEc:bes:jnlasa:v:102:y:2007:p:359-378

14. Good, I.J.: Rational decisions. J. Roy. Stat. Soc. Ser. B (Methodol.) **14**(1), 107–114 (1952). http://www.jstor.org/stable/2984087
15. Hendrickson, A., Buehler, R.: Proper scores for probability forecasters. Ann. Math. Statist. **42**(6), 1916–1921 (1971). https://doi.org/10.1214/aoms/1177693057
16. Jose, V., Nau, R., Winkler, R.L.: Scoring rules, generalized entropy, and utility maximization. Oper. Res. **56**(5), 1146–1157 (2008)
17. Konek, J.: Evaluating imprecise forecasts. In: Miranda, E., Montes, I., Quaeghebeur, E., Vantaggi, B. (eds.) Proceedings of the Thirteenth International Symposium on Imprecise Probability: Theories and Applications. Proceedings of Machine Learning Research, vol. 215, pp. 270–279. PMLR (2023). https://proceedings.mlr.press/v215/konek23a.html
18. Lad, F.: Coherent prevision as a linear functional without an underlying measure space: the purely arithmetic structure of conditional quantities. In: Coletti, G., et al. (eds.) Mathematical Models for Handling Partial Knowledge in Artificial Intelligence, pp. 101–112. Plenum Press, New York (1995)
19. Lad, F.: Operational Subjective Statistical Methods. Wiley, Hoboken (1996)
20. Lad, F., Sanfilippo, G.: Scoring alternative forecast distributions: completing the Kullback symmetric distance complex. Global Local Econ. Rev. **22**(02), 63–90 (2018). http://www.gler.it/archivio/ISSUE/gler_22_2.pdf
21. Lad, F., Sanfilippo, G., Agró, G.: Completing the logarithmic scoring rule for assessing probability distributions. AIP Conf. Proc. **1490**(1), 13–30 (2012). https://doi.org/10.1063/1.4759585. http://link.aip.org/link/?APC/1490/13/1. ISBN: 978-0-7354-1102-9, ISSN 0094-243X
22. Lad, F., Sanfilippo, G., Agró, G.: Extropy: complementary dual of entropy. Stat. Sci. (J. Inst. Math. Stat.) **30**(1), 40–58 (2015). https://doi.org/10.1214/14-STS430
23. Lindley, D.V.: Scoring rules and the inevitability of probability. Int. Stat. Rev./Revue Internationale de Statistique **50**(1), 1–11 (1982)
24. Petturiti, D., Vantaggi, B.: How to assess coherent beliefs: a comparison of different notions of coherence in dempster-shafer theory of evidence. In: Augustin, T., Cozman, F.G., Wheeler, G. (eds.) Reflections on the Foundations of Probability and Statistics, pp. 161–185. Springer, Cham (2022). https://doi.org/10.1007/978-3-031-15436-2_8
25. Predd, J., Seiringer, R., Lieb, E., Osherson, D., Poor, H., Kulkarni, S.: Probabilistic coherence and proper scoring rules. IEEE Trans. Inf. Theory **55**(10), 4786–4792 (2009). https://doi.org/10.1109/TIT.2009.2027573
26. Savage, L.S.: Elicitation of personal probabilities and expectations. J. Am. Stat. Assoc. **66**(336), 783–801 (1971)
27. Schervish, M.J., Seidenfeld, T., Kadane, J.B.: Proper scoring rules, dominated forecasts, and coherence. Decis. Anal. **6**(4), 202–221 (2009). https://doi.org/10.1287/deca.1090.0153
28. Seidenfeld, T., Schervish, M.J., Kadane, J.B.: Forecasting with imprecise probabilities. Int. J. Approximate Reasoning **53**(8), 1248–1261 (2012). https://doi.org/10.1016/j.ijar.2012.06.018
29. Winkler, R.L.: Scoring rules and the evaluation of probability assessors. J. Am. Stat. Assoc. **64**(327), 1073–1078 (1969)
30. Winkler, R.L.: Evaluating probabilities: asymmetric scoring rules. Manag. Sci. **40**(11), 1395–1405 (1994). http://www.jstor.org/stable/2632926

Robust Discrete Bayesian Classifier Under Covariate and Label Noise

Wenlong Chen, Cyprien Gilet(✉), Benjamin Quost, and Sébastien Destercke

Université de technologie de Compiègne, CNRS, Heudiasyc (Heuristics and Diagnosis of Complex Systems), CS 60 319, 60 203 Compiègne Cedex, France
{wenlong.chen,cyprien.gilet,benjamin.quost,sebastien.destercke}@hds.utc.fr

Abstract. In this paper, we focus on the Discrete Bayesian Classifier (DBC), which discretizes the input space into regions where class probabilities are estimated. We investigate fuzzy partitioning as an alternative to the hard partitioning classically used to discretize the space. We show that our approach not only boosts the DBC's performance and resilience to noise, but also mitigates the loss of information due to discretization. The benefits of soft partitioning are demonstrated experimentally on several synthetic and real datasets.

Keywords: Bayesian classifier · Discretization · Robustness

1 Introduction

In this paper, we address the problem of learning a classifier from data. Our goal is to train a model to accurately classify any new instance. We are particularly interested in the case where the data are corrupted by noise. More precisely, we study how softening/randomizing a discretized classifier can improve robustness to label or feature noise. We distinguish label noise, which corresponds to mistakes in the training labels, from feature (or attribute) noise [10], referred in the title as covariate noise, which might corrupt training as well as test features.

A wide range of techniques have been proposed to determine a classifier with theoretical performance guarantees. In order to well approximate the Bayes decision rule when processing numeric or mixed features in a high dimensional feature space, a relevant approach consists in partitioning the feature space so as to determine a discrete, nonparametric version of the Bayes classifier (DBC) [4,6,8,11,12]. In a nutshell, this approach discretizes the input space into regions (also called discrete profiles), into which the Bayes classifier is analytically determined by estimating the class frequencies; test instances are classified based on the estimates in the region to which they belong. Discrete profiles can correspond to the regions associated with the leaves of a tree when using supervised decision trees [5,13,21], or to the Voronoi cells derived from a K-means partitioning [18].

Most discretization methods are based on hard partitioning [24]: a test instance is mapped to a unique discrete profile, all the instances assigned to this profile are assumed to share similar features and are associated to the same

S. Destercke et al. (Eds.): SUM 2024, LNAI 15350, pp. 100–114, 2025.
https://doi.org/10.1007/978-3-031-76235-2_8

predicted class. Discretization has been shown to mitigate the impact of noise or outliers [15]. Yet, the hard partitioning may result in very similar instances in different profiles being assigned different outputs by the subsequent "hard DBC", and in the estimates for a given profile being unaffected by instances close but outside of this profile, thus potentially missing important information. The hard DBC may therefore still suffer from label and covariate noise.

Randomization, as well as softening strategies, have been shown to regularize classifiers while retaining their main features. For instance, some research [22] investigated the use of soft partitioning together with the naive Bayesian classifier with promising results, for a specific classifier with strong assumptions, together with a specific (0/1) loss function. Label smoothing, i.e., replacing hard labels with probabilistic ones, is commonly used in deep learning to reduce overfitting: introducing a small amount of uncertainty during training prevents the model from becoming overly confident, thus enhancing its generalization performance. Label smoothing can also improve robustness to label noise [17].

The goal of this paper is to investigate the use of soft clustering with the DBC, so as to benefit from the theoretical guarantees of this latter while improving classification performances by computing smoother decision boundaries. We first recall the basics of the DBC and discuss its limitations. Through detailed mathematical modeling and algorithms, we explain how soft partitioning can enhance its performances. Building on this, we propose the soft probabilistic DBC (SPDBC), which allows each data point to belong to different classes with a certain probability, thus enabling the model to handle overlapping or noisy data more flexibly. In contrast with [22], our present approach allows to consider any kind of loss/cost function.

Experiments on synthetic and standard real datasets, and in particular with a controlled level of noise, demonstrate how the SPDBC improves classification accuracy in noisy environments while maintaining stability. The experimental results confirm the effectiveness of our proposal in dealing with complex data structures and showcase its potential for practical applications.

The paper is organized as follows. Section 2 provides reminders on the discrete Bayesian classifier, talks about how to make DBC probabilistic(PDBC) and presents some criticisms of this classifier. Section 3 proceeds with our approach to deal with the limitation of DBC, resulting in the so-called soft probabilistic discrete Bayesian classifier (SPDBC). Section 4 reports some experiments, which notably stress the robustness of the SPDBC compared to the DBC when facing noisy data. Section 5 provides some conclusions to this preliminary work, mentioning several possible future directions.

2 Discrete Bayesian Classifier

2.1 Setting

We aim to compute a function $\delta : \mathcal{X} \to \mathcal{Y}$ able to provide, for any instance in the input space $x \in \mathcal{X}$, an estimate (or "guess") $\delta(x)$ of its (unknown) actual class $Y \in \mathcal{Y} = \{1, \ldots, K\}$. For this purpose, we leverage a training set $\{(x_1, y_1), \ldots, (x_n, y_n)\}$ composed of instances $x_i \in \mathcal{X}$ associated with labels $y_i \in \mathcal{Y}$, assumed to be observations of the actual classes of the instances.

We consider here asymmetrical decision problems, i.e. where the decision costs may not be the same (notably across errors). We assume to have access to a matrix $L = (L_{kl})$, whose general term $L_{kl} \geq 0$ quantifies the cost incurred from predicting $\delta(x) = l$ when the ground truth (i.e., actual class) is k. While it is reasonable to assume $L_{kk} = 0$ for all $k \in \mathcal{Y}$, we may have $L_{kl} \neq L_{lk}$ for some $k \neq l$: for instance, erroneously raising an alarm regarding the condition of a patient may be considered as less harmful (and therefore more acceptable) than failing to detect an actual condition of this same patient.

Bayes' Decision Strategy. Bayesian decision theory [2,9,20] provides a theoretical solution to this learning problem. Under the assumption that data are generated according to a joint probability $\mathbb{P}$ over $\mathcal{X} \times \mathcal{Y}$, it establishes that the decision strategy minimizing the *expected risk* (or misclassification loss) should be based on the posterior probabilities of the classes and on the misclassification costs:

$$\delta^B(x) = \arg \min_{k=1,\ldots,K} R_k(\delta|x), \quad \text{with} \quad R_l(\delta|x) = \sum_{k=1}^{K} L_{kl} \, \mathbb{P}\left(Y = k|X = x\right). \tag{1}$$

Generative models typically derive the posterior probabilities $\mathbb{P}(Y = k|X = x)$ from the class-conditional distributions $\mathbb{P}(X = x|Y = k)$ and the prior probabilities $\pi_k = \mathbb{P}(Y = k)$ using Bayes' rule:

$$\mathbb{P}(Y = k|X = x) = \frac{\pi_k \mathbb{P}(X = x|Y = k)}{\mathbb{P}(X = x)}, \quad \mathbb{P}(X = x) = \sum_{l} \pi_l \mathbb{P}(X = x|Y = l).$$

Therefore, a wide range of approaches aim at estimating the prior probabilities π_k and the conditional distributions $\mathbb{P}(X = x|Y = k)$ so as to implement the Bayes classifier [1,14,15], for instance via maximum likelihood (ML). While π_k can be estimated using the class frequencies in the training set, $\mathbb{P}(X = x|Y = k)$ usually requires additional assumptions. Thus, many strategies postulate a (semi-)parametric model for $\mathbb{P}(X = x|Y = k)$ and focus on estimating the parameters of the distribution, for instance as in discriminant analysis [16].

Discrete Bayesian Classifier. When the distributional assumption is not satisfied, the resulting classifier may be a biased estimate of the actual Bayes classifier [15], and thus be far from optimal even if the training sample is large. An alternative then consists in using a nonparametric approach. First, the input space $\mathcal{X}$ is partitioned into T regions or *profiles* $\{\phi_1, \ldots, \phi_T\} = \mathcal{P}$. We introduce the mapping $\Phi : \mathcal{X} \to \mathcal{P}$, which maps any instance x to a profile $\Phi(x)$; we may interchangeably write $\Phi(x) = \phi_t$ or $x \in \phi_t$ whenever the instance x falls into the region ϕ_t (or, put another way, when x corresponds to profile ϕ_t).

Then, the discrete Bayesian classifier (DBC) amounts to estimate the class-conditional distributions $\mathbb{P}(X = x|Y = k)$ by the fractions of input samples from a given class falling into the various profiles:

$$\widehat{p}_{kt} = \widehat{\mathbb{P}}\left(\Phi(X) = \phi_t \mid Y = k\right) := \frac{1}{n_k} \sum_{i \in \mathcal{I}_k} \mathbb{1}_{\{\Phi(x_i) = \phi_t\}}, \tag{2}$$

with $\mathcal{I}_k = \{i \in \{1, \ldots, n\} : Y_i = k\}$ the set of indexes corresponding to instances from class k, $n_k = |\mathcal{I}_k|$ the cardinal of this set and $\mathbb{1}_{\{\}}$ is the indicator function:

$$\mathbb{1}_{\{\Phi(x) = \phi_t\}} = \begin{cases} 1 & \text{if } x \text{ is assigned to cluster } \phi_t \\ 0 & \text{otherwise.} \end{cases} \tag{3}$$

Two approximations are made here: one is due to the input space being discretized into profiles, and thus to the class-conditional density $\mathbb{P}(X = x|Y = k)$ being replaced with the probability of the profile $\mathbb{P}(\Phi(X) = \phi_t \mid Y = k)$; the second one results from this latter probability being estimated with the relative frequency of the profile ϕ_t in the instances of class k in the training set. Figure 1 displays a 2D example, where the input space is discretized using K-means.

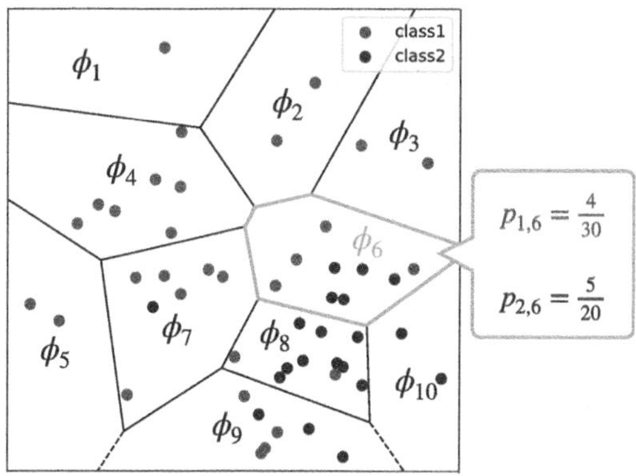

Fig. 1. Bivariate classification problem addressed using the DBC; class 1 (blue) counts $n_1 = 30$ training instances, class 2 (red) $n_2 = 20$ training instances. (Color figure online)

The deterministic DBC consists in classifying any instance x into the class which minimizes the expected risk:

$$\delta_\pi^B(x) = \arg\min_{l \in \mathcal{Y}} f_l(\pi, x), \tag{4a}$$

$$f_l(\pi, x) := \sum_{k \in \mathcal{Y}} L_{kl} \sum_{t=1}^{T} \frac{\pi_k \hat{p}_{kt}}{\sum_j \pi_j \hat{p}_{jt}} \mathbb{1}_{\{\Phi(x) = \phi_t\}}, \quad \forall l \in \mathcal{Y}. \tag{4b}$$

Notice the similarity of Eq. (4b) with Eq. (1), where the posterior probabilities $\mathbb{P}(Y = k|x)$ have been replaced with their profile-based estimates

$$\mathbb{P}(Y = k|\Phi(x)) := \sum_{t=1}^{T} \frac{\pi_k \hat{p}_{kt}}{\sum_j \pi_j \hat{p}_{jt}} \mathbb{1}_{\{\Phi(x) = \phi_t\}}. \tag{5}$$

We note that since f_l is a risk associated with a specific class, theoretically the class with smaller f_l is more likely to be less risky as a prediction. We propose to map f_l into a probability of predicting class l, written $\mathbb{P}\left(\delta_\pi^B(x) = l\right)$, so as to derive a probabilistic counterpart to the deterministic DBC presented above. For this purpose, Definition 1 proposes a relative compensation probability assignment technique, that allows each point to belong to different classes with a certain estimated probability reflecting the associated risk. We first calculate a compensatory score λ_l for each class, such that if a class's score f_l is relatively lower, λ_l will be comparatively higher, and vice versa. We then normalize these λ_l values across all classes to calculate the final probabilities, so as to ensure that these probabilities sum up to 1.

Definition 1. *For any initial feature point $x \in \mathcal{X}$, the estimated probability for the discrete Bayes classifier δ_π^B to assign a class $l \in \mathcal{Y}$ is given by*

$$\begin{aligned} \mathbb{P}\left(\delta_\pi^B(x) = l\right) &= \frac{\lambda_l(\pi, x)}{\sum_{k \in \mathcal{Y}} \lambda_k(\pi, x)} \\ \textit{where} \quad \lambda_l(\pi, x) &= \sum_{k \in \mathcal{Y}} f_k(\pi, x) - f_l(\pi, x). \end{aligned} \tag{6}$$

This method naturally balances differences in scores between classes, reducing the undue influence of any single class due to scale discrepancies in scoring. Note that whenever the 0/1 loss function is considered, these probabilities boil down to the profile-based posterior probability estimates defined by Eq. (5).

Example 1. Assume a test instance x in Fig. 1 such that $\Phi(x) = \phi_6$, together with a loss matrix L satisfying $L_{11} = L_{22} = 0$, $L_{21} = 3$, $L_{12} = 2$; we can calculate

$$\begin{aligned} f_1(\pi, x) &= L_{21} \frac{\pi_2 \hat{p}_{2,6}}{\pi_1 \hat{p}_{1,6} + \pi_2 \hat{p}_{2,6}} = 3 \times \frac{\frac{20}{50} \times \frac{5}{20}}{\frac{30}{50} \times \frac{4}{30} + \frac{20}{50} \times \frac{5}{20}} = \frac{5}{3}, \\ f_2(\pi, x) &= L_{12} \frac{\pi_1 \hat{p}_{1,6}}{\pi_1 \hat{p}_{1,6} + \pi_2 \hat{p}_{2,6}} = 2 \times \frac{\frac{30}{50} \times \frac{4}{30}}{\frac{30}{50} \times \frac{4}{30} + \frac{20}{50} \times \frac{5}{20}} = \frac{8}{9}. \end{aligned}$$

Thus, λ_1 and λ_2 are

$$\begin{aligned} \lambda_1(\pi, x) &= (f_1(\pi, x) + f_2(\pi, x)) - f_1(\pi, x) = \frac{5}{3} + \frac{8}{9} - \frac{5}{3} = \frac{8}{9}, \\ \lambda_2(\pi, x) &= (f_1(\pi, x) + f_2(\pi, x)) - f_2(\pi, x) = \frac{5}{3} + \frac{8}{9} - \frac{8}{9} = \frac{5}{3}. \end{aligned}$$

Finally, we obtain

$$\mathbb{P}\left(\delta_\pi^B(x) = 1\right) = \frac{\lambda_1(\pi, x)}{\lambda_1(\pi, x) + \lambda_2(\pi, x)} = \frac{\frac{8}{9}}{\frac{8}{9} + \frac{5}{3}} = \frac{8}{23}, \quad \mathbb{P}\left(\delta_\pi^B(x) = 2\right) = \frac{15}{23}.$$

Proposition 1. *The decision rule of the DBC in Eq. (4a) is equivalent to picking the class with highest probability $\mathbb{P}\left(\delta_\pi^B(x) = l\right)$ defined in Eq. (6):*

$$\delta_\pi^B : x \mapsto \underset{l \in \mathcal{Y}}{\operatorname{argmax}}\, \mathbb{P}\left(\delta_\pi^B(x) = l\right). \tag{7}$$

Proof. Let $f_l(\pi, x)$ be the minimum value among the set $\{f_k(\pi, x) : k \in \mathcal{Y}\}$; then,

$$\begin{aligned} & \forall k \in \mathcal{Y}, \quad f_l(\pi, x) \leq f_k(\pi, x) \\ \Leftrightarrow \quad & \forall k \in \mathcal{Y}, \quad \forall j \neq l, \quad f_k(\pi, x) - f_j(\pi, x) \leq f_k(\pi, x) - f_l(\pi, x) \\ \Leftrightarrow \quad & \forall j \neq l, \quad \underbrace{\sum_{k \in \mathcal{Y}} f_k(\pi, x) - f_j(\pi, x)}_{\lambda_j(\pi, x)} \leq \underbrace{\sum_{k \in \mathcal{Y}} f_k(\pi, x) - f_l(\pi, x)}_{\lambda_l(\pi, x)}. \end{aligned}$$

In other words,

$$f_l(\pi, x) = \min_{k \in \mathcal{Y}} f_k(\pi, x) \quad \Leftrightarrow \quad \lambda_l(\pi, x) = \max_{k \in \mathcal{Y}} \lambda_k(\pi, x).$$

Moreover, given the definition of $\mathbb{P}\left(\delta_\pi^B(x) = l\right)$ in Eq. (6),

$$\lambda_l(\pi, x) = \max_{k \in \mathcal{Y}} \lambda_k(\pi, x) \quad \Leftrightarrow \quad \mathbb{P}\left(\delta_\pi^B(x) = l\right) = \max_{k \in \mathcal{Y}} \mathbb{P}\left(\delta_\pi^B(x) = k\right).$$

To conclude, we have

$$\underset{l \in \mathcal{Y}}{\operatorname{argmin}}\, f_l(\pi, x) = \underset{l \in \mathcal{Y}}{\operatorname{argmax}}\, \mathbb{P}\left(\delta_\pi^B(x) = l\right),$$

and thus $\delta_\pi^B(x) = \arg\min_{l \in \mathcal{Y}} f_l(\pi, x)$ and $\delta_\pi^B = \arg\max_{l \in \mathcal{Y}} \mathbb{P}\left(\delta_\pi^B(x) = l\right)$ are equivalent. □

This stresses that our probabilized classifier is consistent with Bayes' decision strategy, as it will assign the highest probability to Bayes' optimal decision.

2.2 Impact of Discretization

Advantages. As discussed in the Introduction, discretization makes the resulting classifier less sensitive to noise or outliers [15], two issues which may significantly affect parameter estimates of continuous distributions, and therefore impede approximating the actual Bayes' classifier. Last, for large-scale data sets, discretization can significantly reduce the overall computational complexity.

Limitations. As can be seen from Eq. (4b), the value of $f_l(\pi, x)$ is the same for all instances in the same profile. This results in the decision boundary of DBC being solely determined by the boundaries of the discrete profiles. Therefore, the partitioning algorithm may have a significant impact on the performances of the classifier. In particular, two instances close to each other in the input space may nevertheless be associated with different decisions. Figure 2 displays the profiles obtained for a synthetic 2D dataset: as can be seen, some of them (like the leftmost profile in red) are associated with mixed subsets of training instances, in which case associating the profile with a single class will result in errors.

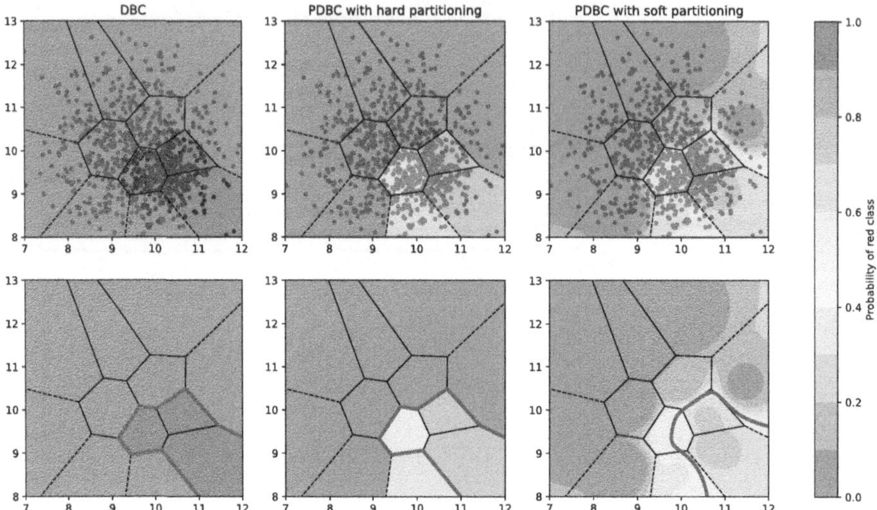

Fig. 2. An example for three different decision regions of DBC, the purple line is the decision boundary (Color figure online)

Besides, linking the instances to the profiles using a "hard" mapping (determined by the indicator function) results in instances close to x in the input space being associated to different profiles, hence having no influence on the class-conditional risks computed via Eq. (4b). This results in the decision boundary of DBC being a combination of multiple local optimal boundaries between profiles, determined entirely by the instances in these profiles.

These two considerations motivate determining the profile using soft clustering, so that instances can belong to several profiles with different degrees, resulting in DBC decisions being based on more global information.

3 PDBC with Soft Partitioning

We discuss here how soft profile memberships can be leveraged to improve the performances of the DBC. The main idea is to replace the indicator function $\mathbb{1}_{\{\Phi(x)=\phi_t\}}$ in Eq. (4a) by a probability $\mathbb{P}\left(\Phi(x)=\phi_t\right)$, so as to regularise/soften the boundaries of the clusters. This means that any clustering methods producing probabilities such as Gaussian mixture [23] or other approaches producing probabilities [25] can be relevant.

In our case, we use the fuzzy C-means (FCM) algorithm [3] for this purpose. The choice of FCM is motivated by several reasons. This robust and versatile approach comes with a geometrical interpretation; it provides cluster representatives (the cluster centers), to which profiles can be associated, and based on which cluster memberships can be derived (thus, even among the points that would be associated with a given profile, cluster memberships can vary).

We first introduce the FCM algorithm, and then discuss how it can be combined with PDBC to improve the robustness of the decision boundaries.

3.1 Reminder on the Fuzzy C-Means Partitioning

Fuzzy C-means (FCM) is a typical fuzzy partitioning method [3] akin to the K-means algorithm. In this algorithm, any instance x is associated with a set of coefficients modelling its degrees of membership to each of the C clusters: in the following, $u_t(x)$ stands for the membership of instance x to the t-th cluster.

Centroids. In FCM, a cluster is characterized by its centroid c_t, defined as the average of all instances weighted by their membership degrees to the cluster:

$$c_t = \frac{\sum_x u_t(x)^m x}{\sum_x u_t(x)^m},$$

with m the hyper-parameter that controls the level of fuzziness of the cluster (the higher m, the fuzzier the cluster).

Algorithm. The FCM algorithm attempts to partition the set of n instances at hand $\{x_1, \ldots, x_n\}$ into a collection of T fuzzy clusters $\{\phi_1, \ldots, \phi_T\}$ as follows. Given a finite set of data, the algorithm returns a list of T cluster centers $C = \{c_1, ...c_T\}$ and a partition matrix $U = u_{it} \in [0,1], i = 1, ..., n, t = 1, ..., T$, where each element u_{it} tells the degree to which element x_i belongs to cluster ϕ_t. The FCM aims to minimize an objective function:

$$J(U, C) = \sum_{i=1}^{n} \sum_{t=1}^{T} u_{it}^m ||x_i - c_t||^2, \tag{8a}$$

$$\text{with } u_{it} = \frac{1}{\sum_{j=1}^{T} \left(\frac{||x_i - c_t||}{||x_i - c_j||}\right)^{\frac{2}{m-1}}}. \tag{8b}$$

When m is low (close to 1), the cluster memberships become close to binary (0 or 1): FCM then behaves like the K-means algorithm, producing (almost) hard partitions. When m is high, the cluster memberships become fuzzier. The membership degrees of a data point tend to be distributed across different clusters. In this scenario, the cluster boundaries become softer, better capturing the fuzziness and uncertainty in the data.

3.2 Soft PDBC

According to [19], the membership degree $u_t(x)$ of x to the cluster ϕ_t, defined in Eq. (8b), can be interpreted as a posterior probability of x belonging to a ϕ_t given some assumptions, as it satisfies

$$0 \leq u_t(x) \leq 1, \quad \sum_{t=1}^{K} u_t(x) = 1. \tag{9}$$

In the following, given a collection of T fuzzy clusters $\{\phi_1, \ldots, \phi_T\}$ and their associated cluster centers $\{c_1, \ldots c_T\}$, we define, for all $x \in \mathcal{X}$ and $t \in \{1, \ldots, T\}$,

$$\mathbb{P}(\Phi(x) = \phi_t) = u_t(x) = \frac{1}{\sum_{j=1}^{T} \left(\frac{||x-c_t||}{||x-c_j||} \right)^{\frac{2}{m-1}}}. \tag{10}$$

We propose to substitute these estimated posterior probabilities to the indicator functions in our PDMC model.

Definition 2. *The estimated probability that an instance has the discrete feature profile ϕ_t given that its actual class label is $y = k$, is given by*

$$\hat{p}_{kt} = \frac{1}{n_k} \sum_{i \in \mathcal{I}_k} \mathbb{P}(\Phi(x_i) = \phi_t) \tag{11}$$

with $\mathcal{I}_k$ and n_k defined as in Sect. 2.1 and $\mathbb{P}(\Phi(x_i) = \phi_t)$ as in Eq. (10).

Definition 3. *Under Definition 2, the risk f_l of predicting class $l \in \mathcal{Y}$ given a feature instance $x \in \mathcal{X}$ initially defined by Eq.* (4b) *becomes*

$$f_l(\pi, x) := \sum_{k \in \mathcal{Y}} L_{kl} \sum_{t=1}^{T} \frac{\pi_k \, \hat{p}_{kt}}{\sum_j \pi_j \, \hat{p}_{jt}} \mathbb{P}(\Phi(x) = \phi_t). \tag{12}$$

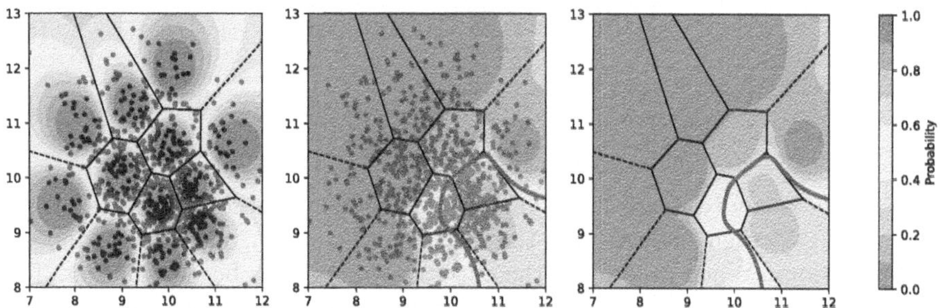

Fig. 3. Soft probabilistic DBC obtained using FCM: (left) probability that each position belongs to the current profile, (center and right) probability that each position in the feature space belongs to the red class. (Color figure online)

Each data point may thus belong to various profiles with specific probabilities, allowing the model to consider the neighboring information in different profiles, and providing a softer decision boundary which is not constrained by the profile boundaries, as can be seen in Fig. 3. This will mitigate the impact of a crude partitioning, and arguably result in better generalization performances.

4 Experiments

The purpose of our experimental study is to validate the robustness and accuracy of our proposed model when dealing with noise. We use eight datasets from the UCI Machine Learning Repository: Iris, Breast Cancer, Diabetes, Heart Disease, Raisin, Zoo, Glass, Energy (see Table 1). For each dataset, we conduct two experiments. The code is available in our Github repository: Menamot/SUM-experiments.

Label Noise. First, we add label noise to the training set on Y with levels $s \in \{0, 0.05, 0.1, 0.15, 0.2, 0.25\}$, meaning that a fraction s of training labels are switched at random according to a uniform distribution on $\mathcal{Y} \setminus y_i$. We then compute the accuracy on the test set for each noise level, via 20 times 5-fold cross validation. The purpose of this experiment is to study whether the model can learn accurate decision boundaries from imperfectly-labeled training instances, which may occur whenever the data are manually labeled.

Covariate Noise. In a second step, we add noise to the features X_j in the test set according to a Gaussian distribution:

$$X_j^{\text{noise}} = X_j + \epsilon_j \quad \text{with} \quad \epsilon_j \sim \mathcal{N}\left(0, \sigma_j^2\right), \quad \sigma_j = \text{std}(\text{X}_\text{j}) \times s, \tag{13}$$

and with $s \in \{0, 0.1, 0.2, 0.3, 0.4, 0.5\}$. This experiment aims at testing the robustness of the model to feature noise; whereas training data can be collected carefully, using high-quality sensors or by skilled operators, it is often not the case for test data, which are generally processed without additional treatment.

Table 1. Datasets used in the experiments.

Dataset	nb. samples	nb. classes	class frequencies	nb. features
Iris	150	3	[0.33, 0.33, 0.33]	4
Breast Cancer	569	2	[0.37, 0.62]	30
Diabetes	768	2	[0.65, 0.35]	8
Heart Disease	303	2	[0.54, 0.46]	13
Raisin	900	2	[0.5, 0.5]	7
Zoo	101	7	[0.41, 0.20, 0.05, 0.13, 0.04, 0.08, 0.10]	16
Glass	214	6	[0.33, 0.36, 0.08, 0.06, 0.04, 0.14]	9
Energy	358	6	[0.31, 0.17, 0.20, 0.13, 0.13, 0.05]	34

Partitioning Methods. We use K-means as hard clustering approach, and Fuzzy C-means for soft partitioning. In order to remove the influence of the two partitioning methods on the profile position, we set the cluster centers in FCM to be the same as for K-means, which means that the profiles of the two clusters will be exactly the same, except that C-means allows data to belong to different

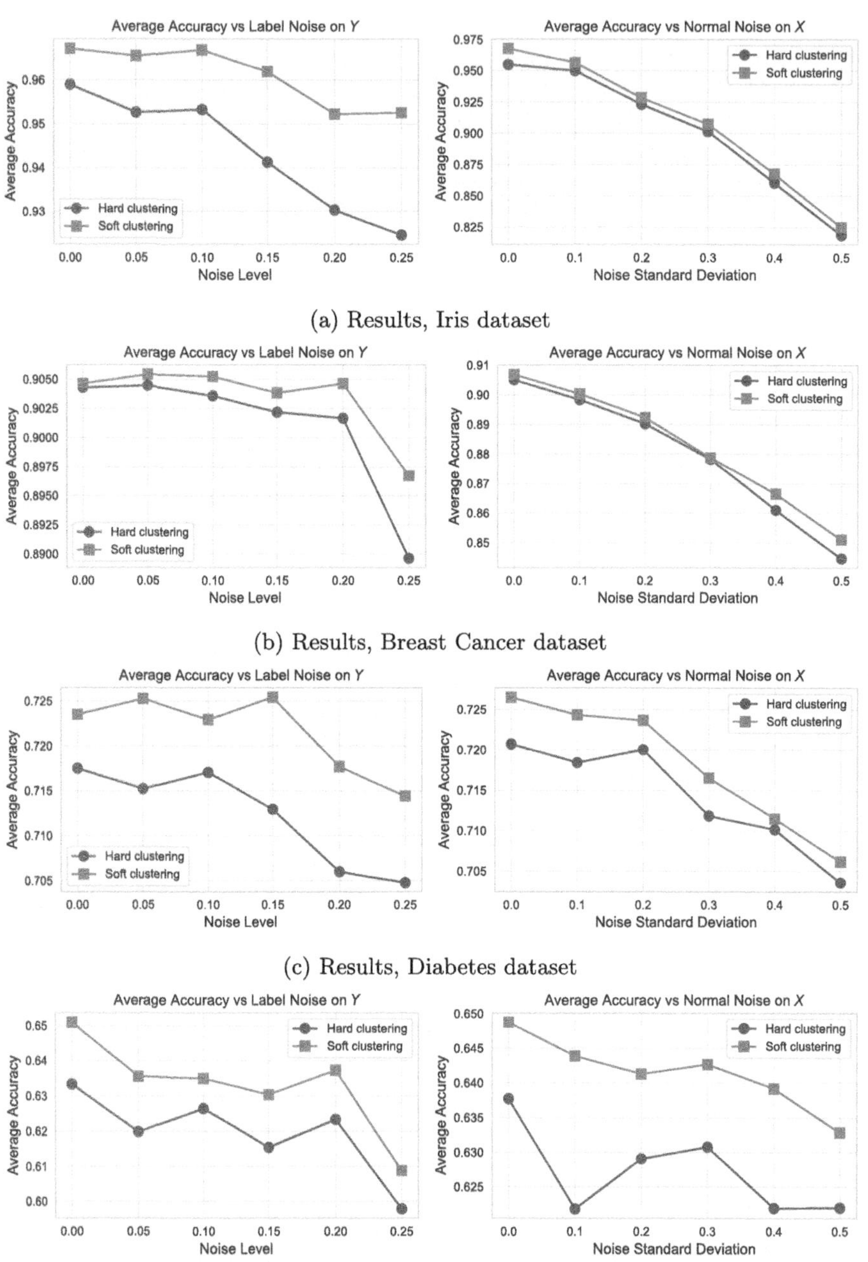

(a) Results, Iris dataset

(b) Results, Breast Cancer dataset

(c) Results, Diabetes dataset

(d) Results, Heart Disease dataset

Fig. 4. Experimental results on different data sets: from top to bottom, Iris, Breast Cancer, Diabetes and Heart Disease.

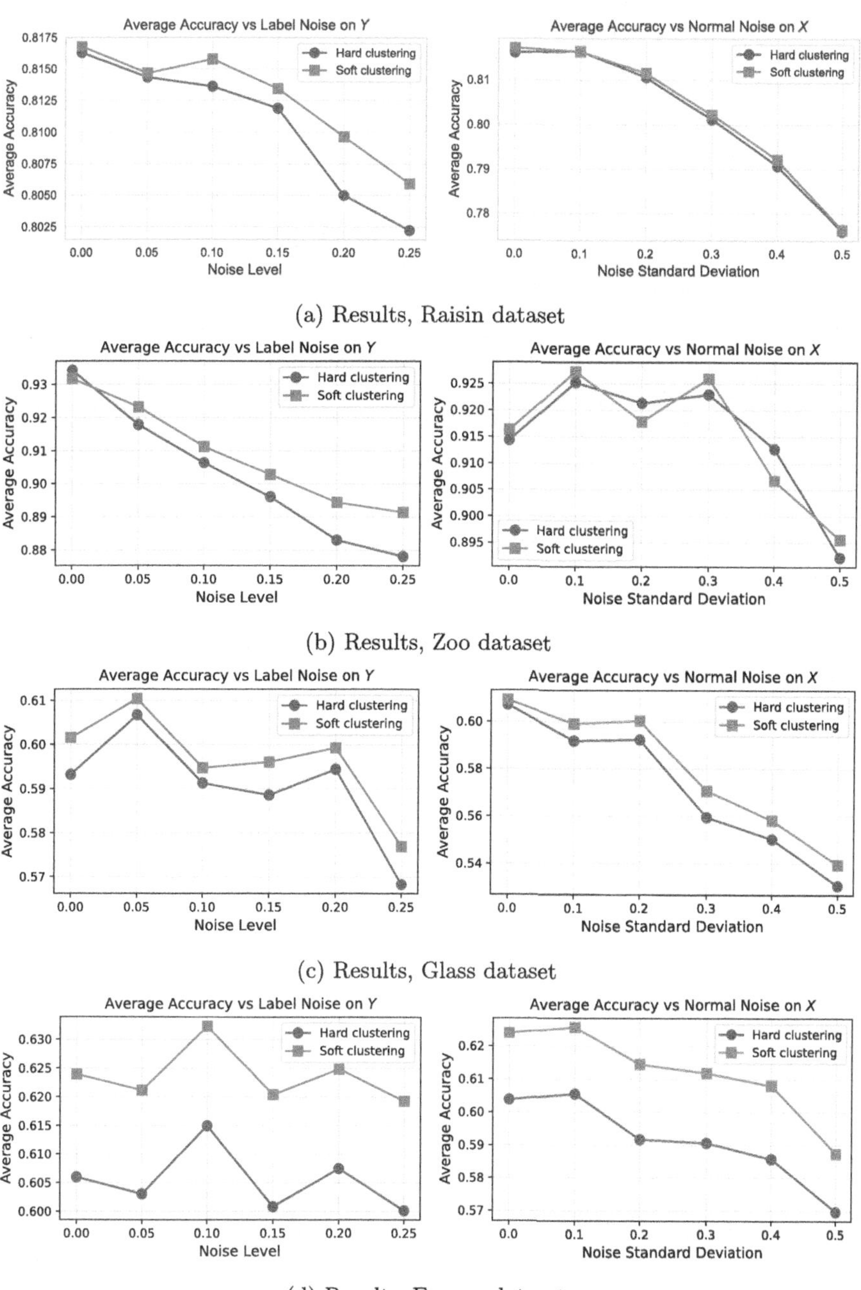

(a) Results, Raisin dataset

(b) Results, Zoo dataset

(c) Results, Glass dataset

(d) Results, Energy dataset

Fig. 5. Experimental results on different data sets: from top to bottom, Raisin, Zoo, Glass, and Energy.

profiles with specific degrees of membership [3], calculated using Eq. (10). In this way, we can determine how replacing hard profile memberships with soft ones affects the results of the DBC. Note that since the cluster centers are defined as the K-means centers, the FCM algorithm only needs to calculate these membership degrees: the runtime of the two models is therefore almost identical to that of K-means. We select hyper-parameters using 10×10 cross validation.

Results. Figures 4 and 5 show that our model is indeed more robust than the original DBC. As discussed in Sects. 2.2 and 3.2, we can see that our model exhibits better performances than the original DBC, even when the data are not corrupted by noise. This clearly demonstrates how our soft clustering-based approach improves the robustness and accuracy compared to the traditional hard clustering-based DBC, especially in presence of both label and covariate noise.

As *label noise* increases, soft clustering consistently outperforms hard clustering for all datasets. For instance, in the Iris and Breast Cancer datasets, the performance of soft clustering remains more stable, with accuracy decreasing less significantly compared to hard clustering when noise levels reach 0.25. This highlights the ability of soft clustering to mitigate the effects of noisy labels, resulting in smoother decision boundaries and improved robustness in classification.

Similarly, when *covariate noise* is introduced (with varying levels of standard deviation), soft clustering maintains a higher accuracy across the datasets. Even in the worst-case scenarios, such as with the Raisin and Zoo datasets, the performance of SPDBC is on par with the standard DBC—we speculate that this is due to the boundaries between the classes being already clear enough, thereby allowing K-means to effectively separate instances from different classes. In such cases, refining the decision boundaries such as with the SPDBC no longer results in a significant increase in accuracy.

Overall, these experimental results validate the effectiveness of soft probabilistic discrete Bayesian classifiers in noisy environments. Soft clustering improves the robustness of the model, resulting in a higher classification accuracy even in challenging conditions, such as when data are noisy or uncertain.

5 Conclusion

In this paper, we explored the capabilities of discrete Bayesian classification and its adaptation using soft partitioning techniques to address challenges posed by imperfect datasets. Our proposed approach, called soft probabilistic discrete Bayes classifier, allies the efficiency of discrete Bayesian classification with the flexibility of probabilistic profiles, thereby enhancing the classifier's effectiveness in practical scenarios.

Our experimental results confirm that our approach not only improves the classification accuracy in noisy environments, but also maintains stability across different data distributions, making it a valuable tool for applications where robustness to noise is crucial. Additionally, the probabilistic approach of soft partitioning within the discrete Bayesian framework helps achieving more regular decision boundaries, which are essential for complex class structures.

Future work may explore further enhancements to the SPDBC model. We may for instance use advanced machine learning algorithms that can dynamically adjust the partitioning granularity based on data complexity and distribution shifts. We may also use neural networks to apply our soft probabilistic minimax approach to advanced image classification. Another interesting line of research would be to robustify this softening strategy, for instance by considering clustering methods delivering not a single probability for each instance but a set of probabilities, such as evidential clustering approaches [7].

References

1. Ahmad, A., Quegan, S.: Analysis of maximum likelihood classification on multispectral data. Appl. Math. Sci. **6**(129), 6425–6436 (2012)
2. Berger, J.O.: Statistical Decision Theory and Bayesian Analysis. Springer Series in Statistics, 2nd edn. Springer, New York (1985). https://doi.org/10.1007/978-1-4757-4286-2
3. Bezdek, J.C., Ehrlich, R., Full, W.: FCM: the fuzzy c-means clustering algorithm. Comput. Geosci. **10**(2–3), 191–203 (1984)
4. Braga-Neto, U., Dougherty, E.R.: Exact performance of error estimators for discrete classifiers. Pattern Recogn. **38**(11), 1799–1814 (2005)
5. Breiman, L., Friedman, J.H., Olshen, R.A., Stone, C.J.: Classification and Regression Trees, 1st edn. Chapman and Hall/CRC, Boca Raton (1984)
6. Dalton, L.A., Dougherty, E.R.: Bayesian minimum mean-square error estimation for classification error - part i: definition and the Bayesian MMSE error estimator for discrete classification. IEEE Trans. Signal Process. **59**, 115–129 (2011)
7. Denœux, T., Masson, M.H.: Evclus: evidential clustering of proximity data. IEEE Trans. Syst. Man Cybern. Part B (Cybern.) **34**(1), 95–109 (2004)
8. Devroye, L., Györfi, L., Lugosi, G.: A Probabilistic Theory of Pattern Recognition, 2nd edn. Springer, New York (1996). https://doi.org/10.1007/978-1-4612-0711-5
9. Ferguson, T.: Mathematical Statistics: A Decision Theoretic Approach. Academic Press, Cambridge (1967)
10. Frénay, B., Verleysen, M.: Classification in the presence of label noise: a survey. IEEE Trans. Neural Netw. Learn. Syst. **25**(5), 845–869 (2013)
11. Gilet, C.: Discrete minimax classifier for personalized diagnosis in medicine. Ph.D. thesis, Université Côte d'Azur (2021). https://tel.archives-ouvertes.fr/tel-03553934
12. Gilet, C., Barbosa, S., Fillatre, L.: Discrete box-constrained minimax classifier for uncertain and imbalanced class proportions. IEEE Trans. Pattern Anal. Mach. Intell. **44**(6), 2923–2937 (2020)
13. Gilet, C., Guyomard, M., Barbosa, S., Fillatre, L.: Adjusting decision trees for uncertain class proportions. In: Workshop on Uncertainty in Machine Learning at ECML/PKDD 2020 (2020). https://sites.google.com/view/wuml-2020/program
14. Grossman, D., Domingos, P.: Learning Bayesian network classifiers by maximizing conditional likelihood. In: Proceedings of the Twenty-First International Conference on Machine Learning, p. 46 (2004)
15. John, G.H., Langley, P.: Estimating continuous distributions in Bayesian classifiers. arXiv preprint arXiv:1302.4964 (2013)
16. Klecka, W.R.: Discriminant Analysis. Sage, Thousand Oaks (1980)

17. Lukasik, M., Bhojanapalli, S., Menon, A., Kumar, S.: Does label smoothing mitigate label noise? In: International Conference on Machine Learning, pp. 6448–6458. PMLR (2020)
18. MacQueen, J.: Some methods for classification and analysis of multivariate observations. In: Proceedings of the 5th Berkeley Symposium on Mathematical Statistics and Probability, pp. 281–297 (1967)
19. Mencar, C., Castiello, C.: A Bayesian interpretation of fuzzy C-means. In: Massanet, S., Montes, S., Ruiz-Aguilera, D., González-Hidalgo, M. (eds.) EUSFLAT AGOP 2023. LNCS, vol. 14069, pp. 443–454. Springer, Cham (2023). https://doi.org/10.1007/978-3-031-39965-7_37
20. Rish, I., et al.: An empirical study of the Naive Bayes classifier. In: IJCAI 2001 Workshop on Empirical Methods in Artificial Intelligence, Seattle, WA, USA, vol. 3, pp. 41–46 (2001)
21. Scott, C., Nowak, R.D.: Minimax-optimal classification with dyadic decision trees. IEEE Trans. Inf. Theory **52**(4) (2006)
22. Tang, Y., Pan, W., Li, H., Xu, Y.: Fuzzy Naive Bayes classifier based on fuzzy clustering. In: IEEE International Conference on Systems, Man and Cybernetics, vol. 5, pp. 6–pp. IEEE (2002)
23. Yang, M.S., Lai, C.Y., Lin, C.Y.: A robust EM clustering algorithm for gaussian mixture models. Pattern Recogn. **45**(11), 3950–3961 (2012)
24. Yang, Y., Webb, G.I.: A comparative study of discretization methods for Naive-Bayes classifiers. In: Proceedings of PKAW, vol. 2002 (2002)
25. Zass, R., Shashua, A.: A unifying approach to hard and probabilistic clustering. In: Tenth IEEE International Conference on Computer Vision (ICCV 2005), vol. 1, pp. 294–301. IEEE (2005)

Imprecise Dynamic Value-at-Risk Induced by a DS-Bivariate Random Walk

Andrea Cinfrignini[1], Davide Petturiti[1(✉)], and Barbara Vantaggi[2]

[1] Department of Economics, University of Perugia, Perugia, Italy
{andrea.cinfrignini,davide.petturiti}@unipg.it
[2] Department of MEMOTEF, "La Sapienza" University of Rome, Rome, Italy
barbara.vantaggi@uniroma1.it

Abstract. Referring to Dempster-Shafer theory, we introduce a bivariate random walk enforcing Markovianity and time-homogeneity under a pessimistic view towards ambiguity. This is done through a suitable family of joint t-step transition belief functions, generalizing the product of two independent binomial transition probabilities, where ambiguity is expressed by a parameter. Given a real-valued function of the pair at a fixed time horizon, we define the dynamic lower and upper Value-at-Risk (VaR), generated by the corresponding dynamic p-box.

Keywords: DS-bivariate random walk · Lower and upper dynamic Value-at-Risk · Dynamic p-box

1 Introduction

In decision theory, finance and artificial intelligence, usually uncertainty is modeled through a probability measure, but deviations from the classical theory are justified by empirically documented phenomena. In this stream of research stochastic processes under imprecise probabilities have been recently investigated (see, e.g., [6,14,17,21,23]).

For instance, in [21] probability distributions are replaced by interval probabilities that can derive from partial knowledge or missclassified probabilities. An alternative is to settle the problem in the framework of upper and lower expectations referring to Walley's theory [25]. On the other hand, a different way for introducing imprecision in a stochastic process is to abandon the probabilistic setting and refer to the purely non-additive framework of Choquet theory [4]. This is particularly meaningful in view of financial applications since Choquet expectations give rise to non-linear pricing rules [1,2] and coherent risk measures [9].

In [6] a univariate multiplicative binomial process under Dempster-Shafer uncertainty [10,20] has been introduced. The aim of the quoted model is to generalize a multiplicative binomial process by allowing for reasoning under partial knowledge. The model is completely determined by the choice of only two

S. Destercke et al. (Eds.): SUM 2024, LNAI 15350, pp. 115–129, 2025.
https://doi.org/10.1007/978-3-031-76235-2_9

parameters, b_u and b_d, summing up to a value less than or equal to one, that characterize the transition belief functions, while the excess to unity $\epsilon = 1-(b_u+b_d)$ represents a quantification of imprecision. Indeed, the quoted model fixes the structure of the entire set of t-step transition belief functions such that the process is Markovian and time-homogeneous (namely, a DS-multiplicative binomial process, where DS stands for Dempster-Shafer). In turn, the existence of a global belief function compatible with the t-step transitions is proved, by referring to the geometric conditioning rule [22]. We also notice that a logarithmic transformation leads us to a DS-random walk [5].

Inspired by [6], we consider a bivariate process whose evolution is jointly modeled as a bivariate random walk under partial knowledge. Thus, its uncertainty is characterized by a family of joint transition belief functions, parameterized by two pairs of strictly positive parameters. The resulting bivariate process can be used to model the time-varying total losses of investments in two independent markets faced by a company, complying with imprecision of belief evaluations.

In the proposed characterization we assume that the t-step joint transition belief function is the sum of two parts: the first part corresponds to a generalization of the product of two independent binomial probability distributions while the second part takes into account partial trajectories with independent binomial weighing deflated by an excessive weight to unity given by the two pairs of parameters. We prove that this choice for the family of transition belief functions makes the process Markovian and time-homogeneous (namely, a *DS-bivariate random walk*).

In this setting, for every real-valued function of the DS-bivariate random walk at a fixed time horizon T, we define a lower and an upper conditional cumulative distribution function, which give rise to a *dynamic* p-box [12,25]. In turn, the dynamic p-box allows us to define the corresponding lower and upper dynamic Value-at-Risk (VaR) measures. By recalling that the classical VaR measure is defined as the minimum loss that will occur at a fixed confidence level, it follows that, in case of imprecision and under a pessimistic/optimistic view, we are looking for the most pessimistic/optimistic minimum loss.

The paper is structured as follows. Section 2 collects some preliminaries while in Sect. 3 we characterize the DS-bivariate random walk with its family of transition belief functions. Section 4 defines the dynamic p-box and the corresponding lower and upper dynamic Value-at-Risk. Finally, Sect. 5 draws conclusions and future perspectives. Due to space limitations, proofs have been omitted and reserved for an extended version of the present paper.

2 Preliminaries

Let Ω be a finite non-empty set of states of the world. We denote by $\mathcal{F} = \mathcal{P}(\Omega)$ the power set of Ω, and by $\mathbb{R}^\Omega$ the set of all random variables on Ω.

The Dempster-Shafer theory of evidence [10,20] is based on a pair of non-additive set functions called, respectively, belief and plausibility functions.

A *belief function* is a mapping $\nu : \mathcal{F} \to [0,1]$ which is

(i) (*grounded and normalized*) $\nu(\emptyset) = 0$ and $\nu(\Omega) = 1$;
(ii) (*completely monotone*) for every $A_1, \ldots, A_k \in \mathcal{F}$ and $k \geq 2$, it holds that

$$\nu\left(\bigcup_{i=1}^{k} A_i\right) \geq \sum_{\emptyset \neq I \subseteq \{1,\ldots,k\}} (-1)^{|I|+1} \nu\left(\bigcap_{i \in I} A_i\right).$$

Probability measures are particular belief functions that turn out to satisfy *(ii)* with an equality sign.

Every belief function ν is associated with a dual set function, called *plausibility function*, and defined, for every $A \in \mathcal{F}$, as

$$\overline{\nu}(A) = 1 - \nu(A^c), \tag{1}$$

and satisfying $\nu \leq \overline{\nu}$ on $\mathcal{F}$. Belief and plausibility functions are both completely characterized by the Möbius inverse function of ν [13], denoted as $m : \mathcal{F} \to [0, 1]$, that satisfies the following properties:

(a) $m(\emptyset) = 0$ and $\sum_{B \in \mathcal{F}} m(B) = 1$;
(b) $m(B) \geq 0$ for every $B \in \mathcal{F}$;
(c) $\nu(A) = \sum_{B \subseteq A} m(B)$ and $\overline{\nu}(A) = \sum_{B \cap A \neq \emptyset} m(B)$.

Every belief function induces a non-empty, closed and convex set of probability measures on $\mathcal{F}$

$$\mathbf{core}(\nu) = \{\pi \;:\; \pi \text{ is a probability on } \mathcal{F},\ \pi \geq \nu\},$$

such that $\nu = \min \mathbf{core}(\nu)$ and $\overline{\nu} = \max \mathbf{core}(\nu)$ where minima and maxima are pointwise on $\mathcal{F}$.

In this paper we adopt the *product (or geometric)* conditioning rule for belief functions [22] that is, for every $A, B \in \mathcal{F}$ with $\nu(B) > 0$,

$$\nu(A|B) = \frac{\nu(A \cap B)}{\nu(B)}.$$

Fixing $B \in \mathcal{F}$ with $\nu(B) > 0$, it holds that $\nu(\cdot|B)$ is still a belief function on $\mathcal{F}$, therefore it is the lower envelope of $\mathbf{core}(\nu(\cdot|B))$, the latter being defined as in (2) by referring to $\nu(\cdot|B)$. This last fact highlights that $\nu(\cdot|B)$ can be given a lower probability interpretation, locally on B, nevertheless $\nu(\cdot|\cdot)$ is generally not a lower envelope of conditional probabilities in the sense of [8,25,26]. A decision-theoretic justification of this conditioning rule has been given in [19], while for a deeper discussion and justification in financial applications we refer to Remark 1 in [6]. In particular, we notice that the product conditioning rule consists in focusing on the evidence contained in the conditioning event [11].

For every belief function ν on $\mathcal{F}$ and $X \in \mathbb{R}^\Omega$ having range $\mathcal{X} = \{x_1, \ldots, x_n\}$ with $x_1 < \ldots < x_n$, we define the *lower* and *upper cumulative distribution functions* of X, as $\underline{F}, \overline{F} : \mathcal{X} \to [0, 1]$,

$$\underline{F}(x_i) = \nu(X \leq x_i) = \nu(X \in \{x_1, \ldots, x_i\}),$$
$$\overline{F}(x_i) = \overline{\nu}(X \leq x_i) = \overline{\nu}(X \in \{x_1, \ldots, x_i\}).$$

We notice that, for all $x \in \mathbb{R}$, since $\{X \leq x\} = \bigcup\{\{X = x_i\} : x_i \leq x, x_i \in \mathcal{X}\}$, both functions $\underline{F}, \overline{F}$ can be naturally extended as functions $\underline{F}, \overline{F} : \mathbb{R} \to [0,1]$.

It holds that $\underline{F} \leq \overline{F}$ and $\underline{F}(x_n) = \overline{F}(x_n) = 1$, and the pair $(\underline{F}, \overline{F})$ is called *probability box* (*p-box*) [12,25]. In turn, each p-box $(\underline{F}, \overline{F})$ induces a set of probability distributions on $\mathcal{H} = \mathcal{P}(\mathcal{X})$ for X, denoted as

$$\mathbf{core}(\underline{F}, \overline{F}) = \{P : P \text{ is a probability on } \mathcal{H},\ \underline{F} \leq F_P \leq \overline{F}\},$$

where F_P is the cumulative distribution function with respect to P, i.e., $F_P(x_i) = P(\{x_1, \dots, x_i\})$, with $x_i \in \mathcal{X}$. The lower and upper envelopes of $\mathbf{core}(\underline{F}, \overline{F})$ defined, for every $H \in \mathcal{H}$, as

$$\underline{P}_{(\underline{F},\overline{F})}(H) = \min_{P \in \mathbf{core}(\underline{F},\overline{F})} P(H) \quad \text{and} \quad \overline{P}_{(\underline{F},\overline{F})}(H) = \max_{P \in \mathbf{core}(\underline{F},\overline{F})} P(H),$$

are a belief function and a plausibility function, respectively [15,24], and the pair $(\underline{P}_{(\underline{F},\overline{F})}, \overline{P}_{(\underline{F},\overline{F})})$ is called *natural extension* of $(\underline{F}, \overline{F})$.

It holds that the natural extension coincides with ν and $\overline{\nu}$ on every $\{X \leq x_i\}$ with $x_i \in \mathcal{X}$, i.e., it holds that

$$\begin{aligned} \underline{P}_{(\underline{F},\overline{F})}(\{x_1, \dots, x_i\}) &= \underline{F}(x_i) = \nu(\{X \leq x_i\}), \\ \overline{P}_{(\underline{F},\overline{F})}(\{x_1, \dots, x_i\}) &= \overline{F}(x_i) = \overline{\nu}(\{X \leq x_i\}), \end{aligned}$$

but they do not coincide, generally, on all $\{X \in H\}$ with $H \in \mathcal{H}$.

The natural extension $(\underline{P}_{(\underline{F},\overline{F})}, \overline{P}_{(\underline{F},\overline{F})})$ can be explicitly computed through Proposition 4 in [24]. Referring to $X \in \mathbb{R}^\Omega$ with range $\mathcal{X} = \{x_1, \dots, x_n\}$ and $x_1 < x_2 < \dots < x_n$, each non-empty set $H \in \mathcal{H}$ can be written as

$$H = \bigcup_{j=0}^{k} (x_{i_{2j}}, x_{i_{2j+1}}],$$

with $x_{i_0}, x_{i_1}, \dots, x_{i_{2k+1}} \in \mathcal{X} \cup \{-\infty\}$, and $x_{i_0} < x_{i_1} < \dots < x_{i_{2k+1}}$ and $k \geq 0$, where $(x_{i_{2j}}, x_{i_{2j+1}}] := \{x \in \mathcal{X} : x_{i_{2j}} < x \leq x_{i_{2j+1}}\}$ and $\overline{F}(-\infty) := \underline{F}(-\infty) := 0$.

For every $H = \bigcup_{j=0}^{k}(x_{i_{2j}}, x_{i_{2j+1}}] \in \mathcal{H}$, the *lower* natural extension is given by

$$\underline{P}_{(\underline{F},\overline{F})}(H) = \sum_{j=0}^{k} \max\{0, \underline{F}(x_{i_{2j+1}}) - \overline{F}(x_{i_{2j}})\}, \tag{2}$$

while the *upper* natural extension $\overline{P}_{(\underline{F},\overline{F})}$ is found by duality (i.e., $\overline{P}_{(\underline{F},\overline{F})}(H) = 1 - \underline{P}_{(\underline{F},\overline{F})}(H^c)$). Moreover, the Möbius inverse of $\underline{P}_{(\underline{F},\overline{F})}$ can be directly computed as done in [16].

Finally, denoting by $\nu \circ X^{-1}$ the *pushforward belief function* induced by ν and X on $\mathcal{H}$, the following relation holds (see [24])

$$\mathbf{core}(\nu \circ X^{-1}) \subseteq \mathbf{core}(\underline{F}, \overline{F}) = \mathbf{core}(\underline{P}_{(\underline{F},\overline{F})}),$$

and $\underline{P}_{(\underline{F},\overline{F})} \leq \nu \circ X^{-1}$, where the inclusion above (and the related inequality between $\underline{P}_{(\underline{F},\overline{F})}$ and $\nu \circ X^{-1}$) holds since the probabilities in $\mathbf{core}(\underline{F},\overline{F}) = \mathbf{core}(\underline{P}_{(\underline{F},\overline{F})})$ are asked to satisfy less constraints than those in $\mathbf{core}(\nu \circ X^{-1})$. In other terms, this means that different belief functions can induce the same p-box, while each p-box has a unique natural extension.

3 Dempster-Shafer Bivariate Random Walks

We consider a discrete-time finite-horizon bivariate process $\{(X_n, Y_n)\}_{n=0}^T$, with $T \in \mathbb{N}$, where $(X_0, Y_0) = (x_0, y_0) \in \mathbb{R}^2$ and, for $n = 1, \dots, T$,

$$(X_n, Y_n) = \begin{cases} (X_{n-1} + u_X, Y_{n-1} + u_Y), \\ (X_{n-1} + u_X, Y_{n-1} + d_Y), \\ (X_{n-1} + d_X, Y_{n-1} + u_Y), \\ (X_{n-1} + d_X, Y_{n-1} + d_Y), \end{cases} \tag{3}$$

where $u_i > d_i$, for $i \in \{X, Y\}$, are constant parameters that denote marginal "up" and "down" additive terms in each time period. The process is called *bivariate random walk* and it can be represented on a 4-nomial tree. A typical choice is $(X_0, Y_0) = (0, 0)$ and $u_X = u_Y = 1$, $d_X = d_Y = -1$, in which case it is called bivariate *symmetric* random walk. Figure 1 shows a bivariate random walk for $T = 2$.

The bivariate random walk is defined on a filtered space $(\Omega, \mathcal{F}, \{\mathcal{F}_n\}_{n=0}^T)$, where $\Omega = \{1, \dots, 4^T\}$, $\mathcal{F}_n$ is the algebra generated by $\{(X_0, Y_0), \dots, (X_n, Y_n)\}$, for $n = 0, \dots, T$, with $\mathcal{F}_0 = \{\emptyset, \Omega\}$ and $\mathcal{F}_T = \mathcal{F} = \mathcal{P}(\Omega)$.

Let us define $\mathcal{A}_0 = \{(0, 0)\}$ and denote, for $n = 1, \dots, T$,

$$\mathcal{A}_n = \{(a_k^X, a_h^Y) = (ku_X + (n-k)d_X, hu_Y + (n-h)d_Y) \ : \ k, h = 0, \dots, n\}.$$

Each (X_n, Y_n) takes values in $\mathcal{U}_n = (x_0, y_0) + \mathcal{A}_n$, where the operation $(x, y) + \mathcal{A}_n$ is defined, for all $x, y \in \mathbb{R}$ and all $A \in \mathcal{P}(\mathcal{A}_n)$, as

$$(x, y) + A = \{(x + a_k^X, y + a_h^Y) \ : \ (a_k^X, a_h^Y) \in A\}.$$

We assume that the uncertainty on the evolution of the bivariate random walk is modeled by a strictly positive belief function $\nu : \mathcal{F} \to [0, 1]$.

In analogy with [5] we define the Markov and the time-homogeneity properties of a bivariate random walk under the Dempster-Shafer uncertainty.

Definition 1. *Let $\{(X_n, Y_n)\}_{n=0}^T$, with $T \in \mathbb{N}$, be a bivariate random walk as in (3), defined on the filtered belief space $(\Omega, \mathcal{F}, \{\mathcal{F}_n\}_{n=0}^T, \nu)$. Let us consider $0 \leq n \leq T$, $1 \leq t \leq T - n$, $A \in \mathcal{P}(\mathcal{A}_t)$ and a trajectory with positive belief $(x_0, y_0) \in \mathcal{U}_0, \dots, (x_n, y_n) \in \mathcal{U}_n$.*

We say that the bivariate random walk satisfies:

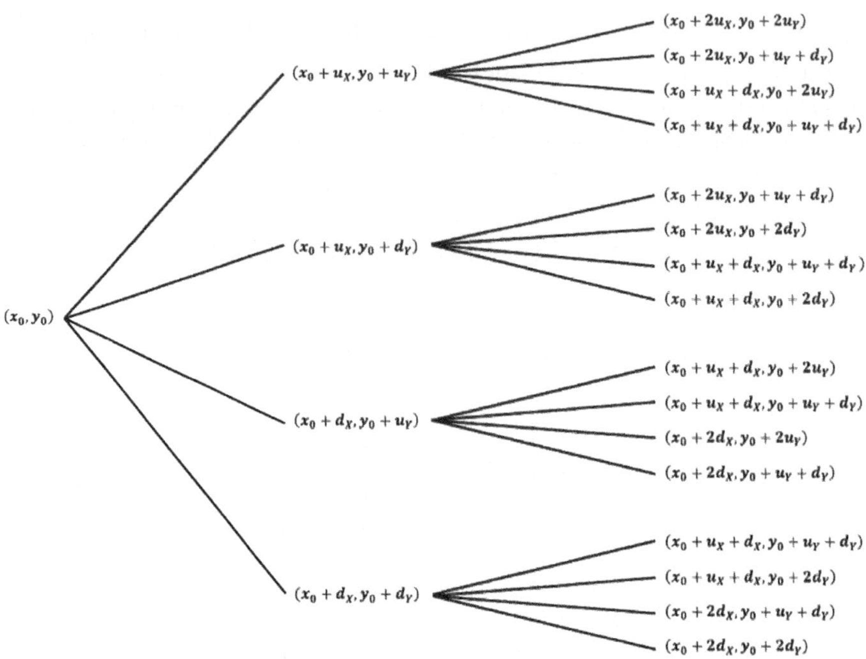

Fig. 1. Bivariate random walk for $T = 2$

the **Markov property,** *if it holds that*

$$\begin{aligned}\nu\left((X_{n+t}, Y_{n+t}) \in (x_n, y_n) + A|\{(X_i, Y_i) = (x_i, y_i)\}_{i=0}^{n}\right) \\ = \nu\left((X_{n+t}, Y_{n+t} \in (x_n, y_n) + A|(X_n, Y_n) = (x_n, y_n)\right);\end{aligned}$$

the **time-homogeneity property,** *if it holds that*

$$\nu\left((X_{n+t}, Y_{n+t}) \in (x_n, y_n) + A|\{(X_i, Y_i) = (x_i, y_i)\}_{i=0}^{n}\right) = \beta_t(A),$$

where $\beta_t : \mathcal{P}(\mathcal{A}_t) \to [0, 1]$ is a fixed belief function.

If the process $\{(X_n, Y_n)\}_{n=0}^{T}$ satisfies both properties, we call it ***DS-bivariate random walk*** *(where DS stands for Dempster-Shafer).*

Our aim is the characterization of a belief function ν such that the process is a DS-bivariate random walk. We have that $\{\beta_t\}_{t=1}^{T}$, defined on $\{\mathcal{P}(\mathcal{A}_t)\}_{t=1}^{T}$, is the family of *transition belief functions*. In the Dempster-Shafer theory it is well-known that the Chapman-Kolmogorov equations do not hold (see, e.g., [7]): this implies that we need to fix the entire family $\{\beta_t\}_{t=1}^{T}$ in order to characterize ν.

We assume that ν is parameterized by two pairs of strictly positive parameters b_u^X, b_d^X and b_u^Y, b_d^Y, with $b_u^X + b_d^X \leq 1$ and $b_u^Y + b_d^Y \leq 1$. We point out that b_u^i, b_d^i, for $i \in \{X, Y\}$, can be interpreted as the marginal one-step "up" and "down" conditional beliefs. Generalizing two independent binomial probability distributions, we require that the joint one-step "up" and "down" conditional beliefs assume the following form:

$$\begin{aligned}
\nu((X_{n+1}, Y_{n+1}) = (x_n + u_X, y_n + u_Y) | \{(X_i, Y_i) = (x_i, y_i)\}_{i=0}^n) = b_u^X b_u^Y, \\
\nu((X_{n+1}, Y_{n+1}) = (x_n + u_X, y_n + d_Y) | \{(X_i, Y_i) = (x_i, y_i)\}_{i=0}^n) = b_u^X b_d^Y, \\
\nu((X_{n+1}, Y_{n+1}) = (x_n + d_X, y_n + u_Y) | \{(X_i, Y_i) = (x_i, y_i)\}_{i=0}^n) = b_d^X b_u^Y, \\
\nu((X_{n+1}, Y_{n+1}) = (x_n + d_X, y_n + d_Y) | \{(X_i, Y_i) = (x_i, y_i)\}_{i=0}^n) = b_d^X b_d^Y.
\end{aligned}$$

Let us consider the function $\beta_t : \mathcal{P}(\mathcal{A}_t) \to [0,1]$, such that, for every $t = 1, \ldots, T$ and $A \in \mathcal{P}(\mathcal{A}_t)$,

$$\begin{aligned}
\beta_t(A) = & \sum_{(a_k^X, a_h^Y) \in A} \mu_t(\{(a_k^X, a_h^Y)\}) \\
& + \sum_{j=1}^{t} \sum_{\substack{\{(a_i^X, a_l^Y) + \mathcal{A}_j\} \subseteq A \\ (a_i^X, a_l^Y) \in \mathcal{A}_{t-j}}} \mu_t((a_i^X, a_l^Y) + \mathcal{A}_j),
\end{aligned} \tag{4}$$

where, for $k, h = 0, \ldots, t$

$$\mu_t(\{(a_k^X, a_h^Y)\}) = \binom{t}{k}\binom{t}{h} (b_u^X)^k (b_d^X)^{t-k} (b_u^Y)^h (b_d^Y)^{t-h}, \tag{5}$$

and for $j = 1, \ldots, t$, $(a_i^X, a_l^Y) \in \mathcal{A}_{t-j}$, $i, l = 0, \ldots, t-j$,

$$\mu_t((a_i^X, a_l^Y) + \mathcal{A}_j) = \binom{t-j}{i}\binom{t-j}{l} (b_u^X)^i (b_d^X)^{t-j-i} (b_u^Y)^l (b_d^Y)^{t-j-l} \delta, \tag{6}$$

and $\delta = 1 - (b_u^X + b_d^X)(b_u^Y + b_d^Y)$ is the *excessive weight to unity.*

Let us note that, whenever $j = t$, it follows that $(a_i^X, a_l^Y) \in \mathcal{A}_0$ are such that $i = 0, l = 0$ and $\mu_t((0,0) + \mathcal{A}_j) = \mu_t(\mathcal{A}_t) = \delta$.

In the following proposition we prove that $\{\beta_t\}_{t=1}^T$ is a belief function.

Proposition 1. *Let $\beta_t : \mathcal{P}(\mathcal{A}_t) \to [0,1]$ be, for $t = 1, \ldots, T$, the function defined in* (4). *Then, β_t is a belief function with Möbius inverse $\mu_t : \mathcal{P}(\mathcal{A}_t) \to [0,1]$ as defined in* (5)–(6).

In the next theorem, we prove the existence of a strictly positive global belief function on $\mathcal{F}$ that makes $\{(X_n, Y_n)\}_{n=0}^T$ a DS-bivariate random walk with transition belief functions $\{\beta_t\}_{t=1}^T$ defined as in (4).

Theorem 1. *Let $\{(X_n, Y_n)\}_{n=0}^T$ be a bivariate random walk as defined in* (3) *on the filtered space $(\Omega, \mathcal{F}, \{\mathcal{F}_n\}_{n=0}^T)$. Then, there exists a belief function $\nu : \mathcal{F} \to [0,1]$ such that:*

(i) $\nu(B) > 0$ for every $B \in \mathcal{F} \setminus \{\emptyset\}$;

(ii) $\{(X_n, Y_n)\}_{n=0}^T$ *is a DS-bivariate random walk, i.e., it satisfies the Markov and time-homogeneity properties, with family of transition belief functions* $\{\beta_t\}_{t=1}^T$ *in* (4).

We notice that the 1-step transition belief function β_1 is consistent with the joint one-step "up" and "down" conditional beliefs, since

$$\beta_1(\{(a_1^X, a_1^Y)\}) = b_u^X b_u^Y, \quad \beta_1(\{(a_1^X, a_0^Y)\}) = b_u^X b_d^Y,$$
$$\beta_1(\{(a_0^X, a_1^Y)\}) = b_d^X b_u^Y, \quad \beta_1(\{(a_0^X, a_0^Y)\}) = b_d^X b_d^Y.$$

The transition belief function β_t is the sum of two components: the first term (Eq. (5)) is the generalization of the product of two independent binomial probability distributions; the second term (Eq. (6)) represents the ambiguous component given by the partial trajectories of the process with independent binomial weighing, that are deflated by the excessive weight to unity, represented by δ.

In fact, the parameter δ quantifies the amount of ambiguity (i.e., the lack of information) in the model: if ν is additive, i.e., $b_u^X + b_d^X = b_u^Y + b_d^Y = 1$, then $\delta = 0$, meaning absence of ambiguity. Then β_t reduces to a bivariate binomial probability distribution with independent marginals.

4 Imprecise Dynamic Value-at-Risk

We assume that $\{(X_n, Y_n)\}_{n=0}^T$ is a DS-bivariate random walk with fixed transition belief functions $\{\beta_t\}_{t=1}^T$ defined as in (4). Let us consider a function $Z_T = \varphi(X_T, Y_T)$ at the fixed time horizon T that has range $\mathcal{Z}_T = \varphi(\mathcal{U}_T)$, where $\varphi : \mathcal{U}_T \to \mathbb{R}$.

In case the belief function ν reduces to a probability measure π, the distribution of Z_T, given the history up to time n, for $n = 0, \ldots, T-1$, is completely singled out by the conditional cumulative distribution function $F_n(z_T; (x_n, y_n)) = \pi(Z_T \leq z_T | (X_n, Y_n) = (x_n, y_n))$. Here, we extend this notion to the belief functions framework.

Definition 2. *Let* $\{(X_n, Y_n)\}_{n=0}^T$ *be a DS-bivariate random walk with fixed transition belief functions* $\{\beta_t\}_{t=1}^T$ *as in* (4). *Let* $Z_T = \varphi(X_T, Y_T)$ *with* $\varphi : \mathcal{U}_T \to \mathbb{R}$.

For every $0 \leq n \leq T-1$ *and* $t = T - n$, *the* **lower and upper dynamic cumulative distribution functions of Z_T given** $(x_n, y_n) \in \mathcal{U}_n$ *are*

$$\underline{F}_n(z_T; (x_n, y_n)) = \nu(Z_T \leq z_T | (X_n, Y_n) = (x_n, y_n)) = \beta_t(A_{z_T}^{\leq}(x_n, y_n)), \quad (7)$$

$$\begin{aligned}\overline{F}_n(z_T; (x_n, y_n)) &= \overline{\nu}(Z_T \leq z_T | (X_n, Y_n) = (x_n, y_n)) \\ &= 1 - \nu(Z_T > z_T | (X_n, Y_n) = (x_n, y_n)) \\ &= 1 - \beta_t(A_{z_T}^{>}(x_n, y_n)),\end{aligned} \quad (8)$$

where $z_T \in \mathcal{Z}_T$ *and*

$$A_{z_T}^{\leq}(x_n, y_n) = \{(x_t, y_t) \in \mathcal{A}_t : \varphi(x_n + x_t, y_n + y_t) \leq z_T\},$$
$$A_{z_T}^{>}(x_n, y_n) = \{(x_t, y_t) \in \mathcal{A}_t : \varphi(x_n + x_t, y_n + y_t) > z_T\}.$$

Remark 1. By the Markov and time-homogeneity properties of $\{(X_n, Y_n)\}_{n=0}^T$ we have that

$$\underline{\nu}(Z_T \leq z_T | \{(X_i, Y_i) = (x_i, y_i)\}_{i=0}^n) = \underline{\nu}(Z_T \leq z_T | (X_n, Y_n) = (x_n, y_n)),$$
$$\overline{\nu}(Z_T \leq z_T | \{(X_i, Y_i) = (x_i, y_i)\}_{i=0}^n) = \overline{\nu}(Z_T \leq z_T | (X_n, Y_n) = (x_n, y_n)).$$

Hence, for every $z_T \in \mathcal{Z}_T$, the family of functions $\{(\underline{F}_n, \overline{F}_n)\}_{n=0}^{T-1}$ gives rise to a pair of measurable stochastic processes, defined, for all $\omega \in \{(X_i, Y_i) = (x_i, y_i)\}_{i=0}^n$, as

$$\underline{F}_n(z_T)(\omega) = \underline{F}_n(z_T; (x_n, y_n)) \quad \text{and} \quad \overline{F}_n(z_T)(\omega) = \overline{F}_n(z_T; (x_n, y_n)),$$

that we simply call *lower and upper dynamic cumulative distribution functions of* Z_T.

The pairs in $\{(\underline{F}_n, \overline{F}_n)\}_{n=0}^{T-1}$ are, for every $(x_n, y_n) \in \mathcal{U}_n$, non-decreasing functions ranging in $[0, 1]$ satisfying $\underline{F}_n(z_T; (x_n, y_n)) \leq \overline{F}_n(z_T; (x_n, y_n))$ for every $z_T \in \mathcal{Z}_T$. Moreover, letting $z_T^* = \max \mathcal{Z}_T$, we have $\underline{F}_n(z_T^*; (x_n, y_n)) = \overline{F}_n(z_T^*; (x_n, y_n)) = 1$. For this reason, in what follows we call $\{(\underline{F}_n, \overline{F}_n)\}_{n=0}^{T-1}$ a *dynamic p-box*. Each pair $(\underline{F}_n, \overline{F}_n)$ can be interpreted as the bounds of a set of admissible probabilistic cumulative distribution functions of Z_T seen from time n.

Example 1. Let $T = 2$ and consider a DS-bivariate random walk $\{(X_n, Y_n)\}_{n=0}^2$ with $u_X = u_Y = 1$, $d_X = d_Y = -1$, whose family of transition belief functions is $\{\beta_t\}_{t=1}^2$ as in (4).

We keep the loss position of a bank that engages in two parallel sequences of two bets with two different bettors, each bet producing a unitary loss/gain under ambiguous information. Positive values are outflows for the bank, while negative values are inflows for the bank. Hence, we can interpret X_2 and Y_2 as the total losses at the end of the two sequences of lotteries. Both total losses X_2 and Y_2 range in $\{a_0^i, a_1^i, a_2^i\} = \{-\$2, \$0, \$2\}$ at time $T = 2$, with $i \in \{X, Y\}$.

Figure 2 shows the possible evolution of the bivariate symmetric random walk. We denote

$$\mathcal{A}_2 = \Bigg\{ \underbrace{(a_2^X, a_2^Y)}_{=a_1}, \underbrace{(a_2^X, a_1^Y)}_{=a_2}, \underbrace{(a_2^X, a_0^Y)}_{=a_3}, \underbrace{(a_1^X, a_2^Y)}_{=a_4}, \underbrace{(a_1^X, a_1^Y)}_{=a_5},$$
$$\underbrace{(a_1^X, a_0^Y)}_{=a_6}, \underbrace{(a_0^X, a_2^Y)}_{=a_7}, \underbrace{(a_0^X, a_1^Y)}_{=a_8}, \underbrace{(a_0^X, a_0^Y)}_{=a_9} \Bigg\}$$

and $A_{ijk\ldots} = \{a_i, a_j, a_k, \ldots\}$.

In Table 1 we report the focal elements of the Möbius inverse μ_2 of β_2, where $\delta = 1 - (b_u^X + b_d^X)(b_u^Y + b_d^Y)$ is the excessive weight to unity, quantifying ambiguity.

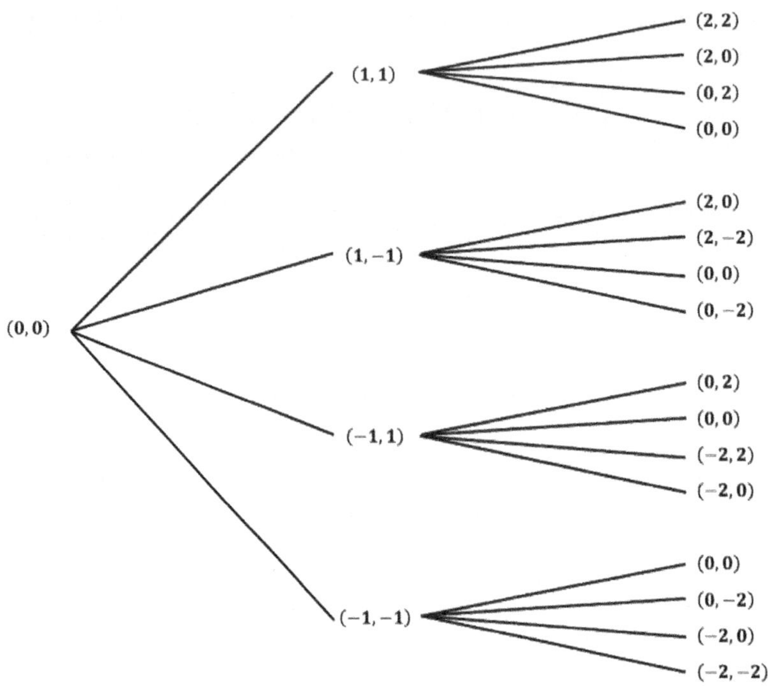

Fig. 2. Bivariate symmetric random walk for $T = 2$

Let us consider $Z_2 = \varphi(X_2, Y_2) = X_2 + Y_2$, that corresponds to the total loss of the bank at time $T = 2$, whose range is $\mathcal{Z}_2 = \{-4, -2, 0, 2, 4\}$, as reported in Table 2.

We compute the lower and upper dynamic cumulative distribution functions of Z_2 at time $n = 0$, i.e., conditional on $\{(X_0, Y_0) = (0, 0)\}$, as defined in (7)–(8) by taking $b_u^X = \frac{1}{4}$, $b_d^X = \frac{1}{2}$, $b_u^Y = \frac{1}{3}$, $b_d^Y = \frac{1}{5}$. With this choice of parameters, it follows that the quantification of the ambiguity is $\delta = 1 - \left(\frac{3}{4}\right)\left(\frac{8}{15}\right) = \frac{3}{5}$.

In Table 3 we report the computation of $\underline{F}_0(z_2; (0, 0))$, $\overline{F}_0(z_2; (0, 0))$ while in Fig. 3 we show the graph of the dynamic p-box $(\underline{F}_0, \overline{F}_0)$ at time $n = 0$.

As recalled in Sect. 2, for a random variable $X \in \mathbb{R}^\Omega$ and a belief function ν on $\mathcal{F}$, there is not a one-to-one correspondence between the pushforward belief function $\nu \circ X^{-1}$ and the induced p-box $(\underline{F}, \overline{F})$ since we can find more belief functions generating the same p-box (see, e.g., [24]). This fact continues to hold also in our dynamic setting, as the following example shows.

Example 2. Consider the DS-bivariate random walk, the total loss Z_2 and its corresponding dynamic p-box $(\underline{F}_0, \overline{F}_0)$ at time $n = 0$ of Example 1.

Referring to the field of sets $\mathcal{H}$ introduced in Sect. 2, let $B = (-2, 0] \in \mathcal{H}$, where $B = (-2, 0] = \{0\}$. We find that the natural extension induced by $(\underline{F}_0, \overline{F}_0)$

Table 1. Focal elements of the Möbius inverse μ_2 of β_2

Set	A_1	A_2	A_3	A_4	A_5	A_6
μ_2	$(b_u^X)^2(b_u^Y)^2$	$2(b_u^X)^2 b_u^Y b_d^Y$	$(b_u^X)^2(b_d^Y)^2$	$2b_u^X b_d^X (b_u^Y)^2$	$4b_u^X b_d^X b_u^Y b_d^Y$	$2b_u^X b_d^X (b_d^Y)^2$

Set	A_7	A_8	A_9	A_{1245}	A_{2356}	A_{4578}	A_{5689}	$\mathcal{A}_2$
μ_2	$(b_d^X)^2(b_u^Y)^2$	$2(b_d^X)^2 b_u^Y b_d^Y$	$(b_d^X)^2(b_d^Y)^2$	$b_u^X b_u^Y \delta$	$b_u^X b_d^Y \delta$	$b_d^X b_u^Y \delta$	$b_d^X b_d^Y \delta$	δ

Table 2. Values of Z_2 depending on $\mathcal{U}_2 = (0,0) + \mathcal{A}_2 = \mathcal{A}_2$

$\mathcal{U}_2$	a_1	a_2	a_3	a_4	a_5	a_6	a_7	a_8	a_9
Z_2	4	2	0	2	0	-2	0	-2	-4

on B conditional on $(0,0)$ is (see Eq. (2))

$$\underline{P}_{(\underline{F}_0,\overline{F}_0)}(B;(0,0)) = \max\{0, \underline{F}_2(0;(0,0)) - \overline{F}_2(-2;(0,0))\} = 0.$$

On the contrary, we have that

$$\begin{aligned}\nu(Z_2 = 0|(X_0,Y_0) = (0,0)) &= \beta_2(\{(x_2,y_2) \in \mathcal{A}_2 \,:\, x_2 + y_2 = 0\}) \\ &= \beta_2(A_{357}) = \frac{229}{3600},\end{aligned}$$

thus we can see that the natural extension $\underline{P}_{(\underline{F}_0,\overline{F}_0)}$ does not coincide with the conditional pushforward belief function $\nu(\cdot|(X_0,Y_0) = (0,0)) \circ Z_T^{-1}$.

Starting from the dynamic p-box $\{(\underline{F}_n,\overline{F}_n)\}_{n=0}^{T-1}$ of a random variable Z_T seen as a loss, we define the corresponding lower and upper dynamic Value-at-Risk (VaR) measures, that are inspired to a notion of imprecise VaR introduced in [18] for a reinsurance application.

Definition 3. *Let $Z_T = \varphi(X_T,Y_T)$ be a random loss, where $\{(X_n,Y_n)\}_{n=0}^T$ is a DS-bivariate random walk with transition belief functions $\{\beta_t\}_{t=1}^T$ as in (4). Let $\{(\underline{F}_n,\overline{F}_n)\}_{n=0}^{T-1}$ be the dynamic p-box of Z_T induced by $\{\beta_t\}_{t=1}^T$ as in (7)–(8).*

For every $\alpha \in (0,1)$, the **lower and upper dynamic Value-at-Risk (or VaR for short) measures of Z_T given** *(x_n,y_n) are, respectively,*

$$\underline{\mathrm{VaR}}_n^\alpha[Z_T;(x_n,y_n)] = \inf\{z_T \in \mathbb{R} \,:\, \underline{F}_n(z_T;(x_n,y_n)) \geq \alpha\}, \tag{9}$$

$$\overline{\mathrm{VaR}}_n^\alpha[Z_T;(x_n,y_n)] = \inf\{z_T \in \mathbb{R} \,:\, \overline{F}_n(z_T;(x_n,y_n)) \geq \alpha\}, \tag{10}$$

where we refer to the extensions of $(\underline{F}_n,\overline{F}_n)$ as functions on $\mathbb{R}$.

Note that the lower VaR is always greater than or equal to the upper VaR and this aspect has a natural interpretation: the amount required as a reserve for the potential loss Z_T is larger if the agent has a pessimistic attitude towards ambiguity (i.e., taking $\underline{\mathrm{VaR}}_n^\alpha$). We point out that lower and upper dynamic VaR measures constitute imprecise versions of the classical dynamic VaR (see, e.g., [3]).

Table 3. Dynamic p-box $(\underline{F}_0, \overline{F}_0)$ of the total loss $Z_2 = X_2 + Y_2$ at time $n = 0$

z_2	$A^{\leq}_{z_2}(0,0)$	$A^{>}_{z_2}(0,0)$	$\underline{F}_0(z_2;(0,0))$	$\overline{F}_0(z_2;(0,0))$
-4	A_9	$A_{12345678}$	$\frac{1}{100}$	$\frac{67}{100}$
-2	A_{689}	A_{123457}	$\frac{4}{75}$	$\frac{253}{300}$
0	A_{356789}	A_{124}	$\frac{637}{3600}$	$\frac{689}{720}$
2	$A_{23456789}$	A_1	$\frac{247}{720}$	$\frac{143}{144}$
4	A_2	$\emptyset$	1	1

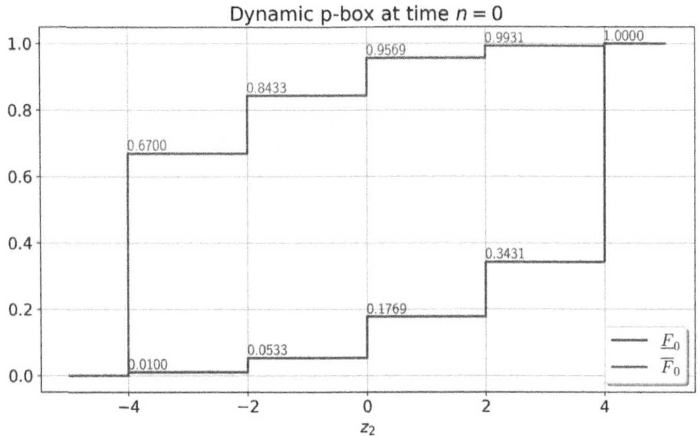

Fig. 3. Graph of the dynamic p-box $(\underline{F}_0, \overline{F}_0)$ of the total loss $Z_2 = X_2 + Y_2$ at time $n = 0$

Remark 2. Proceeding as in Remark 1 we can define two measurable stochastic processes by setting, for all $\omega \in \{(X_i, Y_i) = (x_i, y_i)\}_{i=0}^n$,

$$\underline{\text{VaR}}_n^\alpha[Z_T](\omega) = \underline{\text{VaR}}_n^\alpha[Z_T;(x_n, y_n)] \quad \text{and} \quad \overline{\text{VaR}}_n^\alpha[Z_T](\omega) = \overline{\text{VaR}}_n^\alpha[Z_T;(x_n, y_n)].$$

The processes $\{\underline{\text{VaR}}_n^\alpha[Z_T]\}_{n=0}^{T-1}$ and $\{\overline{\text{VaR}}_n^\alpha[Z_T]\}_{n=0}^{T-1}$ can be simply called *lower and upper dynamic VaR measures of* Z_T.

Example 3. Consider the DS-bivariate random walk, the total loss Z_2 and its corresponding dynamic p-box $(\underline{F}_0, \overline{F}_0)$ at time $n = 0$ of Examples 1–2.

Referring to Fig. 4, the lower and upper dynamic VaR of Z_2 at time $n = 0$ with $\alpha = 0.95$ are

$$\underline{\text{VaR}}_0^{0.95}[Z_2;(0,0)] = \inf\{z_2 \in \mathbb{R} : \underline{F}_0(z_2;(0,0)) \geq 0.95\} = 4,$$
$$\overline{\text{VaR}}_0^{0.95}[Z_2;(0,0)] = \inf\{z_2 \in \mathbb{R} : \overline{F}_0(z_2;(0,0)) \geq 0.95\} = 0.$$

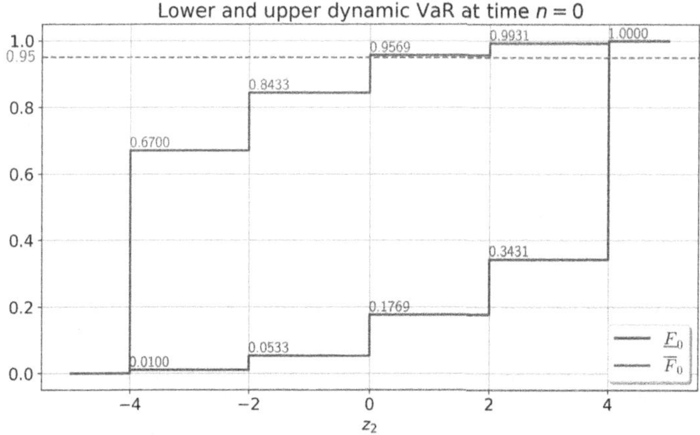

Fig. 4. Lower and upper dynamic Value-at-Risk of the total loss $Z_2 = X_2 + Y_2$ at time $n = 0$ and level $\alpha = 0.95$

5 Conclusion

We proposed a bivariate random walk whose evolution is ruled by a family of joint transition belief functions. We characterized the family of joint transition belief functions with two pairs of strictly positive parameters such that the process is Markovian and time-homogeneous (namely, a *DS-bivariate random walk*). Considering a real-valued function of the DS-bivariate random walk at a fixed time horizon T, we defined a dynamic p-box made by a lower and an upper conditional cumulative distribution function. In turn, the dynamic p-box allows us to define the corresponding lower and upper dynamic Value-at-Risk (VaR) measures.

A distinguished remark concerns the pessimistic and optimistic interpretation of lower and upper dynamic VaR measures, together with the nice parameterization of the introduced DS-bivariate random walk. Indeed, such features suggest a calibration procedure on a lower/upper dynamic VaR assessment as an imprecise dynamic quantile fitting problem.

As future research, we aim to characterize a bivariate multiplicative binomial process under Dempster-Shafer uncertainty. Assuming that we deal with a frictional market with bid-ask spreads, the bid price evolution of a pair of reference stocks can be modeled through the proposed bivariate multiplicative binomial process. In this market, we can price a basket option based on the two stocks proposing a conditional lower pricing rule in terms of conditional Choquet expectation.

Acknowledgement. We acknowledge the support of the PRIN 2022 project "Models for dynamic reasoning under partial knowledge to make interpretable decisions" (Project number: `2022AP3B3B`, CUP Master: `J53D23004340006`, CUP: `B53D23009860006`) funded by the European Union - Next Generation EU (Missione

4 Componente 2). The first and third authors have been supported by the Sapienza University of Rome research project "Ambiguity: its role in asset pricing and insurance" (Grant number: RM123188F744909D).

References

1. Bastianello, L., Chateauneuf, A., Cornet, B.: Put-call parities, absence of arbitrage opportunities, and nonlinear pricing rules. Math. Financ. (2024). https://doi.org/10.1111/mafi.12433
2. Cerreia-Vioglio, S., Maccheroni, F., Marinacci, M.: Put-Call parity and market frictions. J. Econ. Theory **157**, 730–762 (2015)
3. Cheridito, P., Stadje, M.: Time-inconsistency of VaR and time-consistent alternatives. Financ. Res. Lett. **6**(1), 40–46 (2009)
4. Choquet, G.: Theory of capacities. Annales de l'Institut Fourier **5**, 131–295 (1954)
5. Cinfrignini, A., Petturiti, D., Vantaggi, B.: Markov and time-homogeneity properties in Dempster-Shafer random walks. In: Ciucci, D., et al. (eds.) Information Processing and Management of Uncertainty in Knowledge-Based Systems. Communications in Computer and Information Science, vol. 1601, pp. 784–797. Springer, Cham (2022)
6. Cinfrignini, A., Petturiti, D., Vantaggi, B.: Dynamic bid-ask pricing under Dempster-Shafer uncertainty. J. Math. Econ. **107**, 102871 (2023)
7. Çinlar, E.: Introduction to Stochastic Processes. Prentice-Hall (1975)
8. Coletti, G., Petturiti, D., Vantaggi, B.: Conditional belief functions as lower envelopes of conditional probabilities in a finite setting. Inf. Sci. **339**, 64–84 (2016)
9. Delbaen, F.: Coherent risk measures on general probability spaces. In: Sandmann, K., Schönbucher, P.J. (eds.) Advances in Finance and Stochastics, pp. 1–37. Springer, Heidelberg (2002). https://doi.org/10.1007/978-3-662-04790-3_1
10. Dempster, A.: Upper and lower probabilities induced by a multivalued mapping. Ann. Math. Stat. **38**(2), 325–339 (1967)
11. Dubois, D., Prade, H.: Updating with belief functions, ordinal conditional functions and possibility measures. In: Proceedings of the Sixth Annual Conference on Uncertainty in Artificial Intelligence, UAI 1990. MIT, Cambridge, MA, USA, 27–29 July 1990, pp. 311–330. Elsevier (1991)
12. Ferson, S., Kreinovich, V., Ginzburg, L., Myers, D., Sentz, K.: Constructing probability boxes and Dempster-Shafer structures. Technical report. SAND2002-4015, Sandia National Laboratories (2003)
13. Grabisch, M.: Set Functions, Games and Capacities in Decision Making. Springer, Cham (2016)
14. Krak, T., T'Joens, N., de Bock, J.: Hitting times and probabilities for imprecise Markov chains. Proc. Mach. Learn. Res. **103**, 265–275 (2019)
15. Kriegler, E., Held, H.: Utilizing belief functions for the estimation of future climate change. Int. J. Approximate Reasoning **39**(2), 185–209 (2005)
16. Montes, I., Destercke, S.: On extreme points of p-boxes and belief functions. Ann. Math. Artif. Intell. **81**, 405–428 (2017)
17. Nendel, M.: On nonlinear expectations and Markov chains under model uncertainty. Int. J. Approximate Reasoning **130**, 226–245 (2021)
18. Petturiti, D., Stabile, G., Vantaggi, B.: Addressing ambiguity in randomized reinsurance stop-loss treaties using belief functions. Int. J. Approximate Reasoning **161**, 108986 (2023)

19. Petturiti, D., Vantaggi, B.: Conditional decisions under objective and subjective ambiguity in Dempster-Shafer theory. Fuzzy Sets Syst. **447**, 155–181 (2022)
20. Shafer, G.: A Mathematical Theory of Evidence. Princeton University Press, Princeton (1976)
21. Škulj, D.: Random walks on graphs with interval weights and precise marginals. Int. J. Approximate Reasoning **73**, 76–86 (2016)
22. Suppes, P., Zanotti, M.: On using random relations to generate upper and lower probabilities. Synthese **36**(4), 427–440 (1977)
23. T'Joens, N., De Bock, J., de Cooman, G.: A particular upper expectation as global belief model for discrete-time finite-state uncertain processes. Int. J. Approximate Reasoning **131**, 30–55 (2021)
24. Troffaes, M., Destercke, S.: Probability boxes on totally preordered spaces for multivariate modelling. Int. J. Approximate Reasoning **52**(6), 767–791 (2011)
25. Walley, P.: Statistical Reasoning with Imprecise Probabilities. Chapman and Hall, London (1991)
26. Williams, P.: Notes on conditional previsions. Int. J. Approximate Reasoning **44**(3), 366–383 (2007)

Boolean Weighting in Possibilistic Logic

Didier Dubois and Henri Prade[(✉)]

Institut de Recherche en Informatique de Toulouse (IRIT) - CNRS, 118, route de Narbonne, 31062 Toulouse Cedex 9, France
{dubois,prade}@irit.fr

Abstract. Propositional possibilistic logic handles pairs made of a proposition and a level expressing a degree of certainty; these levels belong to a totally ordered scale. This basic possibilistic logic only allows for the conjunction of possibilistic formulas, in agreement with the min decomposability of necessity measures for this connective. Generalized possibilistic logic extends this formalism to negation and disjunctions of weighted pairs of formulas. In this paper, we consider a class of possibilistic logics where propositions are labeled by elements of a Boolean algebra. We first consider the example where propositions are associated with groups of agents that believe in them. This multiagent logic is then extended by attaching degrees of necessity to pairs (proposition, set of agents), and a multiagent counterpart of generalized possibilistic logic is proposed as well. Other examples of Boolean-valued formulas are discussed, where the Boolean labels represent time intervals, or yet other propositional formulas representing reasons to believe propositions.

1 Introduction

Based on possibility theory, a framework for dealing with epistemic uncertainty, possibilistic logic has been developing for forty years and has been applied to many knowledge representation and reasoning problems in artificial intelligence ranging from the handling of uncertainty to nonmonotonic reasoning, belief revision, information fusion, possibilistic networks, or preference modeling; see [14,15] for introductory surveys.

Over time, it has become clear that logical formulas can be associated not only with certainty levels, but also with other modalities or labels such as time intervals, sources, or sets of agents, while keeping the capability of having these pieces of information pervaded with uncertainty; see [5] for a brief introduction in terms of the semi-rings involved in each case of logical formula and modal information pairs. This idea of formulas as pairs can be related to more general settings of labelled deduction systems [19]. Here we focus on possibilistic logic based on Boolean-valued versions of possibility and necessity set functions. After recalling standard and generalized possibilistic logic (allowing for negations and disjunctions of labelled propositions), we describe its Boolean-valued version, taking the case of the previously introduced multi-agent possibilistic

S. Destercke et al. (Eds.): SUM 2024, LNAI 15350, pp. 130–146, 2025.
https://doi.org/10.1007/978-3-031-76235-2_10

logic. Here we extend it to a Boolean-valued counterpart of generalized possibilistic logic. We describe structures of opposition [16] underlying Boolean valued possibility and necessity functions, to show the richness of their representational frameworks. Other examples of Boolean-labelled possibilistic logics are provided, like timed possibilistic logic, introduced quite early, and a new example: reason-based possibilistic logic.

More precisely, we deal with pairs of propositions made of a claim and a reason for it. The interest of such a representation is to keep separate some propositions that can serve as potential explanations, and also to have a link between claims and reasons that is a bit weaker than the one provided by material implication. More generally, these pairs claim-reason may be augmented with other modalities, such as certainty levels.

The paper is organized as follows. Section 2 provides a reminder on possibilistic logic, including generalized possibilistic logic (GPL) where negation or disjunction can also be applied to possibilistic logic formulas, in addition to conjunction. This is the occasion to present a new cube of opposition, both in the syntax and in the semantics, showing the interplay between possibilistic logic formulas possibly exhibiting inside or outside negations, or applying an order reversing function on the certainty weights. In Sect. 3, multiple agent possibilistic logic, where formulas are associated with subsets of agents (possibly also weighted with certainty levels), is recalled; a corresponding cube of opposition is presented, and the correctness and completeness of a generalized version of multiple agent possibilistic logic (in the sense of GPL) is established. The semantics is in terms of set-valued possibility and necessity functions. Another Boolean weighted possibilistic-like logic is finally presented in Sect. 4, where propositions expressing claims are associated with reasons. This logic is, formally speaking, isomorphic to multiple agent possibilistic logic, although with a different meaning.

2 Basics of Possibilistic Logic

Possibilistic logic [8,11] originates from the need to provide a refined version of classical logic, where certainty weights are attached to propositions, while keeping the capability of classical logic to handle incomplete information. Possibilistic logic handles weighted propositions we call possibilistic formulas. It expresses that it is certain at least at some level (a lower bound on a necessity value) that the proposition is true. It is syntactically denoted as a pair (p, α), where p is a proposition and α is a weight.

Contrary to probabilistic logic, possibilistic logic is a logic of accepted beliefs (one reasons as if beliefs were true). This logic, where the disjunction, or the negation of possibilistic formulas are not allowed is often referred as *basic* possibilistic logic.

2.1 Basic Possibilistic Logic

After providing a short background on possibility theory, we present the syntactic properties of basic possibilistic logic, before recalling its semantics.

A possibility distribution π is a mapping from a set (here of logical interpretations) W to a totally ordered scale L (typically the unit interval $[0, 1]$) upper bounded by 1 and lower bounded by 0. It represents an epistemic state by means of more or less plausible situations. The possibility distribution is said to be normalized if $\pi(w) = 1$ for some element $w \in W$. Let p be a formula in propositional logic. Two dual set functions are associated with a possibility distribution, namely a possibility measure Π, and a necessity measure N:

$$\forall p, \Pi(p) = \max_{w \models p} \pi(w); \quad N(p) = 1 - \Pi(\neg p) = \min_{w \not\models p} 1 - \pi(w),$$

where $1 - \cdot$ denotes the order-reversing map on L.

In this paper, for simplicity, we only consider possibilistic *propositional* logic, formulas (p, α) where p is a formula in a propositional language, and $\alpha > 0 \in L$. The pair (p, α) encodes the statement $N(p) \geq \alpha$. Formulas of the form $(p, 0)$, contain no information ($N(p) \geq 0$ always holds), and are not considered. A possibilistic formula can be viewed as a Boolean statement that is true (if $N(p) \geq \alpha$) or false (if $N(p) < \alpha$). However in basic possibilistic logic, it is only possible to express that statements $N(p) \geq \alpha$ are true. A possibilistic logic base can thus be viewed as a expressing a conjunction of such statements. Note that the definition of a necessity measure makes it clear that a possibilistic statement of the form $(p \wedge q, \alpha)$ will be equivalent to the base $\{(p, \alpha), (q, \alpha)\}$ since $N(p \wedge q) = \min(N(p), N(q))$. This minitivity property of necessity measures for conjunction thus implies that a possibilistic logic base, i.e., a set of possibilistic formulas, can be always put in clausal form.

The axioms of basic possibilistic logic (*PL*) [8] are those of propositional logic, where each axiom schema is now supposed to hold with certainty 1, together with the two inference rules:

- if $\beta \leq \alpha$ then $(p, \alpha) \vdash (p, \beta)$ (certainty weakening)
- $(\neg p \vee q, \alpha), (p, \alpha) \vdash (q, \alpha)$, $\forall \alpha \in (0, 1]$ (modus ponens).

Moreover, the following inference rule is also valid:

$(\neg p \vee q, \alpha), (p \vee r, \beta) \vdash (q \vee r, \min(\alpha, \beta))$ (resolution)

A formula weakening inference rule is a consequence of the previous ones: if $p \vdash q$ then $(p, \alpha) \vdash (q, \alpha)$, $\forall \alpha \in (0, 1]$ (formula weakening).

Let $K = \{(p_i, \alpha_i), i = 1, ..., m\}$ be a knowledge base made of a set of possibilistic formulas. Clearly, K can be viewed as the (classical) conjunction of its possibilistic formulas. In a way quite similar to propositional logic, proving (p, α) from K (denoted by $K \vdash (p, \alpha)$) amounts to proving $(\bot, \alpha)$ from $K \cup \{(\neg p, 1)\}$ by repeated application of the resolution rule. Possibilistic logic is close to classical logic. Indeed, let $K_\alpha = \{(p_i, \alpha_i) \in K, \alpha_i \geq \alpha\}$ the set of possibilistic formulas having certainty levels at least equal to α and $K^* = \{p_i \mid (p_i, \alpha_i) \in K\}$ the set of propositional formulas appearing in K. Then it can be checked that $K \vdash (p, \alpha)$

if and only if $K_\alpha \vdash (p, \alpha)$ if and only if $(K_\alpha)^* \vdash p$. Thus the certainty levels stratify the knowledge base K into nested level cuts K_α, i.e., $K_\alpha \subseteq K_\beta$ if $\beta \leq \alpha$. A consequence (p, α) from K can only be obtained by classical inference of p from formulas in the knowledge base $(K_\alpha)^*$, and we are interested in the maximal weight $\bar{\alpha} = \max\{\alpha | (K_\alpha)^* \vdash p\}$.

A distinctive feature of possibilistic logic is its behavior with respect to (in)consistency. The *inconsistency level* of K is defined by

$$inc(K) = \max\{\alpha \mid K \vdash (\bot, \alpha)\}.$$

The formulas in K whose level is strictly above $inc(K)$ are safe from inconsistency, namely $inc(\{(p_i, \alpha_i) | (p_i, \alpha_i) \in K \text{ and } \alpha_i > inc(K)\}) = 0$. Indeed, if $\alpha > inc(K)$, $(K_\alpha)^*$ is consistent. The classical consistency of K^* is equivalent to $inc(K) = 0$.

The semantics of basic possibilistic logic [8] is expressed in terms of possibility distributions, and necessity measures on the set W of interpretations of the propositional language used.

Since a pair (p, α) encodes the statement $N(p) \geq \alpha$, its semantics is given by the least informative possibility distribution $\pi_{(p,\alpha)}$ (i.e., maximizing possibility degrees):

$$\forall w \in W, \pi_{(p,\alpha)}(w) = \begin{cases} 1 & \text{if } w \models p \\ 1 - \alpha & \text{if } w \models \neg p \end{cases}$$

satisfying this constraint. It means that any interpretation that is a countermodel of p is all the less possible as p is more certain, i.e., as α is higher. It can be checked that the associated necessity measure is such that $N_{(p,\alpha)}(p) = \alpha$. A possibility distribution π satisfies (p, α) if and only if $\forall w, \pi(w) \leq \pi_{(p,\alpha)}(w)$, which is indeed equivalent to $N(p) \geq N_{(p,\alpha)}(p) = \alpha$.

A base K is semantically associated with the following possibility distribution that maximizes possibility degrees such that the constraints $N_{(p_i,\alpha_i)}(p_i) \geq \alpha_i$ hold:

$$\pi_K(w) = \min_{i=1}^{m} \pi_{(p_i,\alpha_i)}(w) = \min_{i=1}^{m} \max([p_i](w), 1 - \alpha_i)$$

where $[p_i]$ is the characteristic function of the set of models of p_i, i.e., $[p_i](w) = 1$ if $w \models p_i$ and $[p_i](w) = 0$ otherwise. Thus, the least informative induced possibility distribution π_K is obtained as the min-based conjunction of the possibility distributions representing each formula. It can be checked that $N_K(p_i) \geq \alpha_i$ for $i = 1, \ldots, m$, where N_K is the necessity measure defined from π_K. The semantic entailment is defined by

$$K \models (p, \alpha) \text{ if and only if } \forall w, \pi_K(w) \leq \pi_{\{(p,\alpha)\}}(w).$$

Possibilistic logic is sound and complete [8] with respect to this semantics: $K \vdash (p, \alpha)$ if and only if $K \models (p, \alpha)$. Moreover, we have $inc(K) = 1 - \max_{w \in W} \pi_K(w)$, which acknowledges the fact that the normalization of π_K (i.e., $\exists w \in W, \pi_K(w) = 1$) is equivalent to the classical consistency of K^*. So, it is clear that subnormalized possibility distributions express a form of inconsistency.

2.2 Generalized Possibilistic Logic

In basic possibilistic logic, only conjunctions of possibilistic logic formulas are allowed. But since (p, α) is semantically interpreted as $N(p) \geq \alpha$, a possibilistic formula can be manipulated as a propositional formula that is true (if $N(p) \geq \alpha$) or false (if $N(p) < \alpha$); it is then an atomic formula of a higher order propositional language. Then, viewed as atomic propositional formulae, possibilistic formulas can be combined with all propositional connectives, i.e., conjunction, but also negation, hence, disjunction and material implication, etc. This is *generalized possibilistic logic* (*GPL*) [13,18]. *GPL* is a two-tiered propositional logic, in which propositional formulas are encapsulated by weighted modal operators interpreted in terms of uncertainty measures from possibility theory. Let $\Lambda_k = \{0, \frac{1}{k}, \frac{2}{k}, \ldots, 1\}$ with $k \in \mathbb{N} \setminus \{0\}$ be a finite set of certainty degrees, and let $\Lambda_k^+ = \Lambda_k \setminus \{0\}$. The language of *GPL*, $\mathcal{L}_{\mathbf{N}}^k$, with $k+1$ certainty levels is built on top of the propositional language $\mathcal{L}$ as follows: i) If $p \in \mathcal{L}$, $\alpha \in \Lambda_k^+$, then $\mathbf{N}_\alpha(p) \in \mathcal{L}_{\mathbf{N}}^k$; ii) if $\varphi \in \mathcal{L}_{\mathbf{N}}^k, \psi \in \mathcal{L}_{\mathbf{N}}^k$, then $\neg\varphi$ and $\varphi \wedge \psi$ are also in $\mathcal{L}_{\mathbf{N}}^k$.

In [18], the notation $\mathbf{N}_\alpha(p)$ is used instead of (p, α), emphasizing the closeness with modal logic. So, an agent asserting $\mathbf{N}_\alpha(p)$ has an epistemic state π such that $N(p) \geq \alpha > 0$. Hence $\neg\mathbf{N}_\alpha(p)$ stands for $N(p) < \alpha$, which, given the finiteness of the set of considered certainty degrees, means $N(p) \leq \alpha - \frac{1}{k}$ and thus $\Pi(\neg p) \geq 1 - \alpha + \frac{1}{k}$. Let $\nu(\alpha) = 1 - \alpha + \frac{1}{k}$. Then, $\nu(\alpha) \in \Lambda_k^+$ iff $\alpha \in \Lambda_k^+$, and $\nu(\nu(\alpha)) = \alpha, \forall \alpha \in \Lambda_k^+$. Thus, we can write $\mathbf{\Pi}_\alpha(p) \equiv \neg\mathbf{N}_{\nu(\alpha)}(\neg p)$. In particular, $\mathbf{\Pi}_1(p) \equiv \neg\mathbf{N}_{\frac{1}{k}}(\neg p)$ if $k > 1$.[1] So, in GPL, one can distinguish between the absence of sufficient certainty that p is true ($\neg\mathbf{N}_\alpha(p)$) and the stronger statement that a is somewhat certainly false ($\mathbf{N}_\alpha(\neg p)$).

The semantics of *GPL* is as in basic possibilistic logic defined in terms of normalized possibility distributions over propositional interpretations, where possibility degrees are limited to Λ_k. A model of an atomic *GPL* formula $\mathbf{N}_\alpha(p)$ is any Λ_k-valued possibility distribution π such that $N(p) \geq \alpha$, where N is the necessity measure induced by π. Then, the standard definition for $\pi \models \varphi_1 \wedge \varphi_2$ and $\pi \models \neg\varphi$ applies (e.g., $\pi \models \neg\mathbf{N}_\alpha(p)$ if $N(p) < \alpha$). As usual, π is called a model of a set of *GPL* formulas Γ, written $\pi \models \Gamma$, if π is a model of each formula in Γ. We write $\Gamma \models \phi$, for Γ a set of *GPL* formulas and ϕ a *GPL* formula, iff every model of Γ is also a model of ϕ. Note that some formulas in *GPL* - for instance the disjunction '$\mathbf{N}_\alpha(p) \vee \mathbf{N}_\alpha(q)$'($= \neg(\neg\mathbf{N}_\alpha(p) \wedge \neg\mathbf{N}_\alpha(q))$) - will not always have a least specific possibility distribution that satisfies them, contrary to conjunctions in *PL*. The soundness and completeness of the following axiomatization of *GPL* (where $\phi \to \psi$ stands for $\neg\phi \vee \psi$) holds with respect to the above semantics [18]:

(PL) The Hilbert axioms of classical logic;
(K) $\mathbf{N}_\alpha(p \to q) \to (\mathbf{N}_\alpha(p) \to \mathbf{N}_\alpha(q))$;
(N) $\mathbf{N}_1(\top)$;

[1] If the set of necessity values were the unit interval, we should write that $\mathbf{\Pi}_1(p)$ stands for the infinitary formula $\wedge_{\beta>0}\neg\mathbf{N}_\beta(\neg p)$, which is very difficult to handle in practice.

(D) $\mathbf{N}_{\alpha}(p) \to \mathbf{\Pi}_1(p)$;
(W) $\mathbf{N}_{\alpha_1}(p) \to \mathbf{N}_{\alpha_2}(p)$, if $\alpha_1 \geq \alpha_2$.
with modus ponens as the only inference rule.

We can also express GPL as a propositional logic with atomic propositions of the form (p, α), rewriting the specialized axioms as follows:

(K) $(p \to q, \alpha) \to ((p, \alpha) \to (q, \alpha))$;
(N) $(\top, 1)$;
(D) $(p, \alpha) \to \neg(\neg p, 1/k)$;
(W) $(p, \alpha_1) \to (p, \alpha_2)$, if $\alpha_1 \geq \alpha_2$.

Note that axiom **(D)** can be written in a simple way because the necessity scale is finite.

Structures of opposition, such as square, hexagon, cubes, can be encountered in many knowledge representation settings [16]. For instance, the four set functions of possibility theory (possibility Π, necessity N, and two other set functions where max is replaced by min and vice-versa) applied to p and $\neg p$ can be organized in a graded cube of opposition [17]. The cube of Fig. 1 is a binary cube (the expression of each vertex, as inequality, can only be true or false) that only involves Π and N.

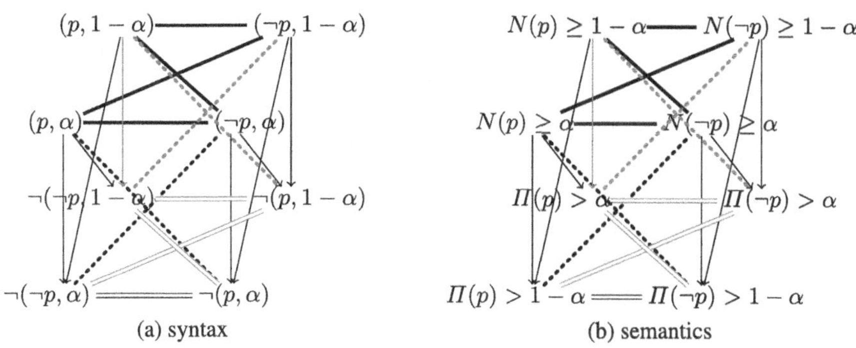

Fig. 1. The cube of opposition of generalized possibilistic logic $(1 > \alpha > 0)$

The diagonals of the front and back facets (thick dotted lines) join contradictories. Two vertices of the top facet linked by a thick black line cannot be true together; dually, two vertices of the bottom facet linked by a double thin line cannot be false together. The arrows from a vertex of the top facet to a vertex of the bottom facet express entailments.

3 Multiple Agent Possibilistic Logic

Multi-agent logic parallels possibilistic logic, but now the propositional formulas are associated with a subset of agents rather than a level in a linearly ordered scale; such a Boolean weighting introduces an important difference since the supremum of two proper subsets may be the whole universe (while the supremum of two non top levels is never the top level in a totally ordered scale). Like for possibilistic logic, we first recall the basic version, before studying the generalized case where disjunction and negation of multi-agent logic formulas are also allowed. We establish a soundness and completeness theorem with respect to a semantics in terms of set-valued possibility and necessity functions for this generalized case, before finally considering the situation where propositions are both associated with a certainty level and a set of agents.

3.1 A Basic Multi-agent Logic

In this section we consider a logic similar to basic possibilistic logic, where formulas are weighted by sets of agents. Namely, multiple agent Logic (*ma-L* for short) [3,12] handles pairs of the form (p, A), where p is a propositional formula, and $A \subseteq All$ denotes a subset of agents (All is the set of all agents). Each such atomic formula (p, A) means that *at least all* the agents in A believe that p is true. Only conjunctions of such atomic formulas are allowed. In spite of an obvious formal parallel with possibilistic logic, (p, A) is not only to be seen as a way of expressing the strength of the support in favor of p (the larger A, the stronger this support), but also as a piece of information linking a proposition with a particular subset of agents. One aim of this logic is to be able to infer what set of agents believe in some given proposition, given several belief bases, each attached to one agent.

The logic *ma-L* possesses all axioms of the form (p, A) for all A, whenever p is a tautology, and it has three inference rules:

- if $B \subseteq A$ then $(p, A) \vdash (p, B)$ (subset weakening);
- $(p, A), (p, B) \vdash (p, A \cup B)$ (strengthening);[2]
- $(\neg p \vee q, A), (p, A) \vdash (q, A)$, $\forall A \in 2^{All} \setminus \emptyset$ (modus ponens).

As a consequence, we also have the resolution rule

$$\text{if } A \cap B \neq \emptyset, \text{ then } (\neg p \vee q, A), (p \vee r, B) \vdash (q \vee r, A \cap B).$$

If $A \cap B = \emptyset$, the resulting information is trivial: $(p, \emptyset)$ is a non informative formula.

Given a multi-agent possibilistic knowledge base $\Gamma = \{(p_i, A_i), i = 1, \ldots, m\}$, the *subset of inconsistent agents* for Γ is defined as

$$inc\text{-}s(\Gamma) = \bigcup\{A \subseteq All \mid \Gamma \vdash (\bot, A)\}.$$

Namely each agent a in $inc\text{-}s(\Gamma)$ possesses inconsistent beliefs. Clearly, $inc\text{-}s(\Gamma) = \emptyset$ if $\nexists A$ s.t. $\Gamma \vdash (\bot, A)$, which does not imply the consistency of $\Gamma^\circ = \{p_i | (p_i, A_i) \in \Gamma,\ i = 1, \ldots, m\}$. It contrasts with possibilistic logic. Just consider the example $\Gamma = \{(p, A), (\neg p, A^c)\}$, where $A^c = All \setminus A$. We have

[2] This rule is used in [3] although not explicitly stated in the general presentation; in standard possibilistic logic, its counterpart is trivial and useless.

$\Gamma \vdash (\bot, A \cap A^c)$, then $inc\text{-}s(\Gamma) = \emptyset$ while Γ° is inconsistent. This is compatible with situations where consistent agents contradict each other. In contrast, the consistency of Γ° does entail $inc\text{-}s(\Gamma) = \emptyset$. What matters in *ma-L* is the collective consistency of *subsets* of agents (while the collection of beliefs held by the whole set of agents may be inconsistent).

The semantics of *ma-L* is expressed in terms of set-valued possibility distributions, and set-valued set functions standing for possibility and necessity measures. A set-valued possibility measure is a set function $\boldsymbol{\Pi}$ on W, valued on 2^{All}, i.e., by subsets of agents, and such that $\boldsymbol{\Pi}(p \vee q) = \boldsymbol{\Pi}(p) \cup \boldsymbol{\Pi}(q)$; moreover $\boldsymbol{\Pi}(\top) = All$ and $\boldsymbol{\Pi}(\bot) = \emptyset$. $\boldsymbol{\Pi}(p)$ represents the maximal set of agents who find p possible. The normalization condition $\boldsymbol{\Pi}(\top) = All$ means that all agents find the tautology possible, while the other limit condition says that all agents find the contradiction impossible. A multi-agent possibility distribution $\boldsymbol{\pi}$ is a mapping $W \to 2^{All}$ where $\boldsymbol{\pi}(w)$ is the *maximal* subset of agents that find the interpretation w possible. Then we have that $\boldsymbol{\Pi}(p) = \bigcup_{w \models p} \boldsymbol{\pi}(w)$.

The normalization condition $\boldsymbol{\Pi}(\top) = All$ is then equivalent to $\cup_{w \in W} \boldsymbol{\pi}(w) = All$, which means that each agent finds at least one w possible ($\forall a \in All, \exists w \in W$, $a \in \boldsymbol{\pi}(w)$), i.e. no agent is inconsistent (i-normalization). This condition is clearly weaker than the condition $\exists w, \boldsymbol{\pi}(w) = All$, which means that there is an interpretation that all agents believe possible, expressing a collective consistency condition we can call *ma*-normalization.

Given an i-normalized possibility distribution $\boldsymbol{\pi}$, we can define a multi-agent necessity function as

$$\boldsymbol{N}(p) = [\boldsymbol{\Pi}(\neg p)]^c = \bigcap_{w \models \neg p} [\boldsymbol{\pi}(w)]^c$$

It follows that and $\boldsymbol{N}(p \wedge q) = \mathrm{N(p)} \cap \boldsymbol{N}(q)$. $\boldsymbol{N}(p)$ is the minimal set of agents that certainly believe p. If $\boldsymbol{\pi}$ is i-normalized, then $\boldsymbol{N}(p) \subseteq \boldsymbol{\Pi}(p)$; in contrast, if $\boldsymbol{\pi}$ is *ma*-normalized, and $\boldsymbol{N}(p) \neq \emptyset$, then $\boldsymbol{\Pi}(p) = All$.

The semantics of a formula (p, A) is given by the set-valued distribution $\boldsymbol{\pi}_{\{(p,A)\}}$:

$$\forall w \in W, \boldsymbol{\pi}_{\{(p,A)\}}(w) = \begin{cases} All & \text{if} \quad w \models p, \\ A^c & \text{if} \quad w \models \neg p \end{cases}$$

Thus $\boldsymbol{\pi}_{\{(p,A)\}}(w)$ is the *maximal* subset of agents that find the interpretation w possible given the constraint (p, A). Namely, all agents may find any interpretation of p possible, while at most all agents in A^c can find any interpretation of $\neg p$ possible.

It can be checked that the necessity function associated to $\boldsymbol{\pi}_{\{(p,A)\}}$ is such that $\boldsymbol{N}_{(p,A)}(p) = A$, and $\boldsymbol{\pi}_{\{(p,A)\}}$ is the *least informative* possibility distribution (i.e., attaching the largest set of agents to each interpretation w) such that this constraint holds. In fact, any set-valued distribution $\boldsymbol{\pi}$ such that $\forall w, \boldsymbol{\pi}(w) \subseteq \boldsymbol{\pi}_{\{(p,A)\}}(\omega)$ is such that its associated necessity measure $\boldsymbol{N}$ satisfies $\boldsymbol{N}(p) \supseteq \boldsymbol{N}_{(p,A)}(p) = A$ (hence is more committed). Given the knowledge base $\Gamma = \{(p_i, A_i), i = 1, \ldots, m\}$ the formula $(p_j, A_j) \in \Gamma$ is understood as expressing the constraint $\mathbf{N}_\Gamma(p_i) \supseteq A_i$ where $\mathbf{N}_\Gamma$ is the set-valued necessity function associated with Γ, since $\forall w, \boldsymbol{\pi}_\Gamma(w) \subseteq \boldsymbol{\pi}_{\{(p_j,A_j)\}}(w)$. The least informative associated set-valued possibility distribution is

$$\boldsymbol{\pi}_\Gamma(w) = \begin{cases} All & \text{if } \forall (p_i, A_i) \in \Gamma, w \vDash p_i \\ \bigcap\{A_i^c, (p_i, A_i) \in \Gamma, w \vDash \neg p_i\} & \text{otherwise} \end{cases}$$

This distribution is *i*-normalized, but not always *ma*-normalized. Namely, consider the base $\Gamma = \{(p, A), (\neg p, A^c)\}$. It is easy to see that $\boldsymbol{\pi}_\Gamma(w) = A$ if $w \models p$, and $\boldsymbol{\pi}_\Gamma(w) = A^c$ otherwise. Note that if $inc\text{-}s(\Gamma) \neq \emptyset$, $\boldsymbol{\pi}_\Gamma$ would not be *i*-normalized. Soundness and completeness results for this logic have been established for this semantics [3].

Remark 1. A multiagent knowledge base can be viewed as a compact representation of knowledge bases of several agents, using a kind of factorisation. Moreover sets of agents assigned to formulas can be described in intension, using a specific vocabulary and propositional language, thus avoiding the enumeration of agents and facilitating multiagent reasoning in practice. Note that our multiagent logic is very different from multi-agent epistemic logics where agents can reason about the beliefs of other agents. Our aim is only to efficiently represent beliefs held by groups of agents, and to address questions of the form "which group of agents believe a given proposition?".

3.2 Generalized Multiagent Logic

Similarly to generalized possibilistic logic, the language of ma-L can be extended to allow for negation and disjunctions of atomic pairs (p, A), e.g., $\neg(p, A)$ and $(p, A) \vee (q, B)$. At the semantic level, $\neg(p, A)$ stands for $\boldsymbol{N}(p) \not\supseteq A$, i.e., $\exists a \in A, a \notin \boldsymbol{N}(p)$, so there is an agent in A that does not believe in p. Equivalently, we can write $\boldsymbol{\Pi}(\neg p) \not\subseteq A^c$ (or yet $\boldsymbol{\Pi}(\neg p) \cap A \neq \emptyset$). Note that the formula $\neg(p, A)$ also stands for $\exists a \in A \cap \boldsymbol{\Pi}(\neg p)$. Figure 2 shows the cubes of opposition corresponding to this generalized *ma-L* (G-ma-L). It resembles the cubes of Fig. 1 for GPL; the different types of lines have the same semantics.
Axioms of *G-ma-L* can thus be written as follows.

(CL) The Hilbert axioms of classical logic;
(K) $(p \to q, A) \to ((p, A) \to (q, A))$;
(N) $(\top, All)$;
(D) $(p, A) \to \wedge_{a \in A} \neg(\neg p, \{a\})$;
(W) $(p, A) \to (p, B)$, if $B \subseteq A$.
(S) $\{(p, A), (p, B)\} \vdash (p, A \cup B)$.

with modus ponens as the only inference rule. Axiom (D) says that (p, A) implies that no agent in A believes $\neg p$. The latter writes: $\wedge_{a \in A} \neg(\neg p, \{a\})$. The other axioms are easier to interpret.

The semantics of *G-ma-L* relies on *i*-normalized set-valued distributions $\boldsymbol{\pi}$ on the set of interpretations W: $\boldsymbol{\pi}$ is a model of the atomic formula (p, A) if and only if $\boldsymbol{N}(p) \supseteq A$ (where $\boldsymbol{N}$ is defined from $\boldsymbol{\pi}$). Then classically, $\boldsymbol{\pi}$ is a model of

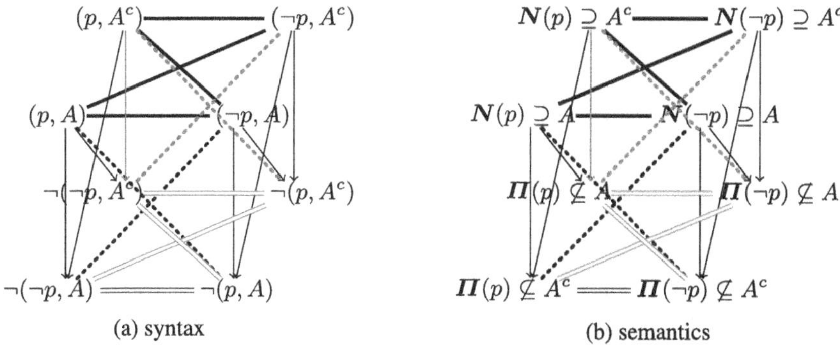

Fig. 2. The cube of opposition of multiple agent possibilistic logic ($A \neq \emptyset, A^c \neq \emptyset$)

a conjunction of atomic formulas if it is a model of each atomic formula in the conjunction. π is a model of $\neg(p, A)$ if it is not a model of (p, A). Finally, π is a model of a *G-ma-L* base Γ (written $\pi \models_{G\text{-}ma\text{-}L} \Gamma$ if it is a model of each formula in Γ).

We can now establish the proof of soundness and completeness of *G-ma-L*.

Theorem 1 (Soundness). *Let Γ be a set of formulas of G-ma-L. Suppose that the G-ma-L formula ϕ can be derived from Γ by modus ponens using the axioms (**CL**, **K**, **N**, **D**, **W**,**S**). Then it holds that $\Gamma \models_{G\text{-}ma\text{-}L} \phi$.*

Proof. The idea is that *G-ma-L* is a standard propositional logic (since **CL** holds), each interpretation of which assigns a standard truth value to each atomic formula (p, A). A set-valued distribution π can be viewed as such an interpretation which moreover makes each axiom formula **K**, **N**, **D**, **W**, **S** true. □

For the completeness we can use the fact that each standard valuation of *G-ma-L* that satisfies the axioms of this logic corresponds to a unique *i*-normalized set-valued distribution on W. A classical valuation v of *G-ma-L* assigns the value 0 or 1 to each atomic formula (p, A). Axiom (**W**) enforces a monotonicity property w.r.t sets of agents, i.e., if $v((p, A)) = 1$ and $B \subseteq A$, then $v((p, B)) = 1$. So, in agreement with axiom (**S**), we can define a set-function g_v on W valued on 2^{All}, in the form:

$$g_v([p]) = \bigcup\{A : v((p, A)) = 1\}.$$

Note that in *G-ma-L*, we can prove the *ma-L* modus ponens rule of the form $\{(p, A), (p \to q, A)\} \vdash (q, A)$. Indeed using axiom (**K**), and assuming (p, A) and $p \to q \equiv \top$ we derive (q, A) by modus ponens of classical logic. We thus establish the monotonicity of g_v, i.e., if $p \models q$ then $g_v([p]) \subseteq g_v([q])$. Other axioms enforce the following properties to g_v:

- Axiom (**N**) enforces $g_v(W) = 1$.
- For axiom (**D**), if $v((p, A)) = 1$, i.e., $g_v([p]) \supseteq A$, we have that for all $a \in A$, $(\neg p, \{a\})$ is false, so $v((\neg p, \{a\})) = 0$, i.e., for all $a \in A$, $a \notin g_v([\neg p])$. So

$g_v([\neg p]) \subseteq A^c$. In particular if p is a CL-tautology, then it yields $g_v(\emptyset) \subseteq All^c = \emptyset$.

- Since $p \to q$ is equivalent to $p \to (p \wedge q)$, axiom (**K**) can be written $(p \to q, A) \to ((p, A) \to (p \wedge q, A))$. Suppose $v((p, A)) = v((q, A)) = 1$. It implies $v((p \to q, A)) = 1$. By axiom (**K**), it follows that $v((p \wedge q, A)) = 1$. So, $g_v([p \wedge q]) \supseteq g_v([p]) \cap g_v([q])$. Since we know that by monotony of g_v, the converse inclusion $g_v([p \wedge q]) \subseteq g_v([p]) \cap g_v([q])$ holds, it follows that $g_v([p \wedge q]) = g_v([p]) \cap g_v([q])$.

It follows that there is a bijection between classical interpretations of *G-ma-L* that satisfy its axioms and set-valued necessity measures on W, or equivalently normalized set-valued distributions. Completeness of *G-ma-L* is then easily established:

Theorem 2 (Completeness). *Let Γ be a set of formulas of G-ma-L. Suppose that $\Gamma \models_{\text{G-ma-L}} \phi$. Then it holds that the G-ma-L formula ϕ can be syntactically derived from Γ by modus ponens using the axioms (***CL, K, N, D, W***).*

Proof. $\Gamma \vdash_{\text{G-ma-L}} \phi$ if and only if $\Gamma \cup \{\mathbf{K}, \mathbf{N}, \mathbf{D}, \mathbf{W}\} \vdash_{CL} \phi$ (in classical logic) if and only if $\Gamma \cup \{\mathbf{K}, \mathbf{N}, \mathbf{D}, \mathbf{W}\} \models_{CL} \phi$ (completeness of propositional logic), if and only if $\Gamma \models_{\text{G-ma-L}} \phi$ (using the bijection between set-valued distributions and classical valuations of *G-ma-L* satisfying its axioms). □

3.3 Generalized Possibilistic Multiagent Logic

Basic *PL* and *ma-L* may be combined to form a possibilistic multiple agent logic *P-ma-L*. Formulas are pairs $(p, \alpha/A)$ where A is a subset of All and $\alpha \in \Lambda_k$. They encode pieces of information of the form "at least all agents in A believe p at least at level α". The symbolic weight α/A represents a fuzzy set of agents with membership grades α if agent $a \in A$, and 0 otherwise. Note that (p, A) in *ma-L* is short for $(p, 1/A)$. More generally, one could consider formulas of the form (p, F) where F is a fuzzy set of agents, understood as "each agent $a \in All$ believes p at least at level $\mu_F(a)$", i.e. it is equivalent to define as many formulas as agents of the form $(p, \mu_F(a)/\{a\})$. A more concise decomposition is of the form $\{(p, \alpha_i/A_i) : i = 1, ..., k\}$ where $\mu_F(a) = \alpha_i, \forall a \in A_i$. These statements presuppose an inference rule of the form "$\{(p, F), (p, G)\}$ is equivalent to $(p, F \cup G)$", where $\cup$ is the max-based fuzzy set union.

Inference rules for *P-ma-L* extend the ones proposed for *PL* and *ma-L*, i.e., weakening reads $(p, \alpha/A) \vdash (p, \beta/B)$ whenever $B \subseteq A$ and $\beta \leq \alpha$. Strengthening can be expressed as $(p, \alpha/A), (p, \beta/B) \vdash \{(p, \alpha/A \setminus B), (p, \beta/B \setminus A), (p, \max(\alpha, \beta)/(A \cap B))\}$ Modus ponens holds replacing A by α/A. It gives the following inference rule:

$$(p, \alpha/A), (\neg p \vee q, \beta/B) \vdash (q, \min(\alpha, \beta)/A \cap B).$$

Given a multi-agent possibilistic knowledge base $\Gamma = \{(p_i, \alpha_i/A_i), i = 1, \ldots, m\}$, the *fuzzy subset of inconsistent agents* for Γ is defined as

$$inc\text{-}s(\Gamma) = \bigcup\{\alpha/A \mid \Gamma \vdash (\bot, \alpha/A)\}.$$

where $\bigcup$ is the max-based fuzzy set union.

At the semantic level we use fuzzy set-valued possibility distributions on W. Namely, $\boldsymbol{\pi}(w) = F$, means that for each agent $a \in All$, the interpretation w is possible at most at degree $\mu_F(a)$. The least specific possibility distribution compatible with a *P-ma-L-* formula $(p, \alpha/A)$ is defined by

$$\forall w \in W, \boldsymbol{\pi}_{\{(p,A)\}}(w) = \begin{cases} All & \text{if } \ w \models p, \\ ((1-\alpha)/A) \cup (1/A^c) & \text{if } \ w \models \neg p \end{cases}$$

Note that $((1-\alpha)/A) \cup (1/A^c)$ is nothing but the fuzzy set complement of α/A. Given the knowledge base $\Gamma = \{(p_i, \alpha_i/A_i), i = 1, \ldots, m\}$ the formula $(p_i, \alpha_i/A_i) \in \Gamma$ is understood as expressing the constraint $\mathbf{N}_\Gamma(p_i) \supseteq \alpha_i/A_i$ where $\mathbf{N}_\Gamma$ is the set-valued necessity function associated with Γ, since $\forall w, \boldsymbol{\pi}_\Gamma(w) \subseteq \boldsymbol{\pi}_{\{(p_i,\alpha_i/A_i)\}}(w)$. The least informative associated set-valued possibility distribution is

$$\boldsymbol{\pi}_\Gamma(w) = \begin{cases} All & \text{if } \forall (p_i, A_i) \in \Gamma, w \vDash p_i \\ \bigcap\{((1-\alpha_i)/A_i) \cup (1/A_i^c), (p_i, \alpha_i/A_i) \in \Gamma, w \vDash \neg p_i\} & \text{otherwise} \end{cases}$$

Soundness and completeness of *P-ma-L* has been established for basic possibilistic multiple agent logic [4].

We are then in a position to introduce the multi-agent extension of generalized possibilistic logic *GP-ma-L*. Similarly as before, the language of *P-ma-L* can be extended to allow for negation and disjunctions of atomic pairs $(p, \alpha/A)$. At the semantic level, $\neg(p, \alpha/A)$ stands for $\boldsymbol{N}(p) \not\supseteq \alpha/A$, i.e., there is an agent in A that does not believe p at level at least α. Axioms of *GP-ma-L* are similar to the crisp case:

(CL) The Hilbert axioms of classical logic;
(K) $(p \to q, \alpha/A) \to ((p, \alpha/A) \to (q, \alpha/A))$;
(N) $(\top, 1/All)$;
(D) $(p, \alpha/A) \to \wedge_{a \in A} \neg(\neg p, \frac{1}{k}/\{a\})$;
(W) $(p, \alpha/A) \to (p, \beta/B)$, if $\beta/B \subseteq \alpha/A$ (fuzzy set inclusion).
(S) $\{(p, \alpha/A), (p, \beta/B)\} \vdash \{(p, \alpha/A \setminus B), (p, \beta/B \setminus A), (p, \max(\alpha, \beta)/(A \cap B))\}$

with modus ponens as the only inference rule. The semantics of *GP-ma-L* relies on normalized fuzzy set-valued distributions $\boldsymbol{\pi}$ on the set of interpretations W: $\boldsymbol{\pi}$ is a model of the atomic formula $(p, \alpha/A)$ if and only if $\boldsymbol{N}(p) \supseteq \alpha/A$ (where $\boldsymbol{N}$ is defined from $\boldsymbol{\pi}$). Soundness and completeness of this logic can be proved similarly as for the all-or-nothing *G-ma-L*, proving that a standard interpretation v of the *GP-ma-L* language can be viewed as a fuzzy-set-valued set-function g_v on W (valued on Λ_k^{All}), in the form:

$$g_v([p]) = \bigcup\{\alpha/A : v((p, \alpha/A)) = 1\}.$$

where $\bigcup$ is the max-based fuzzy set union on *All*. Like before, the axioms of *GP-ma-L* ensure that g_v is a (fuzzy) set-valued necessity measure. Note that this logic handles fuzzy sets of agents using min and max connectives, so that the set of weights attached to formulas form a De Morgan lattice.

Remark 1. *Timed possibilistic logic [7] is a logic where each proposition is associated with a set of time instants when the proposition is believed to be true. It may be more generally a fuzzy set of time instants. Such a logic was the first example of a possibilistic-like logic where weights belong to a Boolean lattice or (more generally a De Morgan lattice); it is formally identical to* P-ma-L, *replacing 'agent' by 'time instant'.*

4 Reason-Based Possibilistic Logic

In the following we propose a logic for reasoning about reasons. This logic handles pairs of the form (p, x) where p and x are two propositional logical formulas expressed in two distinct languages (that use disjoint sets of propositional variables). In (p, x), p is called a claim, and x a reason. The formula (p, x) thus reads "x is a reason for p".

Such pairs of propositions differ from the use of basic possibilistic logic where the certainty weights may depend on the truth of other formulas, as previously proposed in [10]. In basic possibilistic logic, one may write equivalently $(\neg \mathrm{x} \vee p, 1)$ stating that x entails p for sure, or $(p, \tau(x))$ (where $\tau(x) \in \{0,1\}$ denotes the truth-value of proposition x) stating that p is fully certain provided x is true. More generally, the possibilistic formula $(\neg x \vee p, \alpha)$ (stating that it is α-certain that x entails p) is semantically equivalent to $(p, \min(\tau(x), \alpha))$ (stating that p is α-certain, provided x is true).

The idea here is that (p, x) is something weaker than $(\neg \mathrm{x} \vee p, 1)$ (but not in the sense of the strength of a certainty level). Namely, (p, x) does not entail $(\neg x \vee p, 1)$ because $\neg x \vee p$ cannot be expressed in the language of claims, which contrasts with logic-based argumentation literature [6], where p and x are expressed in the same language. The truth of (p, x) means that all the situations where x is true are reasons to believe p.

A reason-labelled base is a set of pairs $K = \{(p_i, x_i), p_i \in \mathcal{C}, x_i \in \mathcal{R}, i = 1, \ldots, n\}$ where $\mathcal{C}$ is the language of claims and $\mathcal{R}$ the language of reasons.

4.1 Syntax of Reason-Based Possibilistic Logic

It should be clear that this logic is one more example of Boolean-labelled logic, putting propositional labels in place of sets of interpretations thereof. In conformity with the previous descriptions of Boolean labelled logic, it is clear that reason-based logic has the same three inference rules as the multiagent logic *ma-L*, replacing sets A, B in the labels by propositional formulas x, y:

- if $x \vdash y$ then $(p, y) \vdash (p, x)$ (reason weakening[3])
- $(\neg p \vee q, x), (p, x) \vdash (q, x), \forall x \neq \bot$ (reason-based modus ponens)
- $(p, x), (p, y) \vdash (p, x \vee y)$ (reason strengthening)

The axioms of reason-based logic are again the axioms of propositional logic applied to pairs $(p, \top)$. Up to this change in the syntax of labels, this logic is exactly the same as *ma-L*.

Remark 2. *The pair (p, x) considered here should not be confused with a conditional object $p|x$ [9]. Conditional objects provide a semantics for nonmonotonic reasoning, and as such there is no entailment between $p|x$ and $p|(x \wedge x')$. On the contrary, if x is a reason for p, then $x \wedge x'$ is also a reason for p (provided that $x \wedge x' \neq \bot$). The formula x in $p|x$ describes a context rather than a reason.*

Again there are two types of consistency for a reason-based possibilistic base K. A strong version is the case where the classical base $K^* = \{p_i : (p_i, x_i) \in K\}$ is consistent. However contrary to basic possibilistic logic, another weaker form of consistency is defined by the property $\forall (p, x)$ such that $K \vdash (p, x)$, with $x \neq \bot$, then $p \neq \bot$. Note that the case when $K \vdash (p, x)$ implies $x = \bot$ does not question consistency, as it means that p is not supported. For instance (in agreement with *ma-L*), the base $K = \{(p, x), (\neg p, \neg x)\}$ is weakly consistent: K^* is inconsistent, but in this case we only have $K \vdash (\bot, \bot)$, i.e., inconsistency is not supported by K.

4.2 Semantics of Reason-Based Possibilistic Logic

We consider two sets of interpretations, the one for the logic of claims (W as previously) and the one for the logic of reasons, say $\mathcal{I}_{\mathcal{R}}$[4]. A reason-based possibility distribution is a function $\boldsymbol{\pi}$ from W to $2^{\mathcal{I}_{\mathcal{R}}}$. Namely $\boldsymbol{\pi}(w)$ represents the subset of situations where the state of facts w cannot be ruled out. The possibility distribution representing (p, x) is $\boldsymbol{\pi}_{\{(p,x)\}}(w) = \begin{cases} \mathcal{I}_{\mathcal{R}} & \text{if } w \models p \\ mod_{\mathcal{R}}(\neg x) & \text{if } w \models \neg p \end{cases}$ where $mod_{\mathcal{R}}(\neg x)$ is the subset of interpretations in $\mathcal{I}_{\mathcal{R}}$ that makes $\neg x$ true. For each $w \in W$, it is the maximal subset of situations in $\mathcal{I}_{\mathcal{R}}$ where w cannot be ruled out given the constraint $\boldsymbol{N}(p) \supseteq mod_{\mathcal{R}}(x)$, which captures the meaning of (p, x).

The semantics of the reason-based logic is isomorphic to the one of the multi-agent logic *ma-L*, and results of soundness and completeness of the latter can be adapted to the logic of reasons.

The latter can be extended to assign strengths to reasons (in the style of *P-ma-L*) and further extended to negations and disjunctions of pairs (p, x) in the style of *GP-ma-L*). In the latter case, axioms now read as follows:

[3] Even if one may argue that x is stronger than y as a reason, the name proposed is in agreement with what was done in the previous logic in Sect. 3.1 and acknowledges the fact that (p, x) covers less situations than (p, y) when $x \vdash y$.

[4] More generally, $\mathcal{I}_{\mathcal{R}}$ may be restricted to the subset of interpretations feasible according to a set of integrity constraints existing with respect to $\mathcal{R}$, expressed in the language of reasons.

(CL) The Hilbert axioms of classical logic;
(K) $(p \to q, \alpha/x) \to ((p, \alpha/x) \to (q, \alpha/x))$;
(N) $(\top, 1/\top)$;
(D) $(p, \alpha/x) \to \neg(\neg p, \frac{1}{k}/y)$, if $y \vdash x$, y consistent;
(W) $(p, \alpha/x) \to (p, \beta/y)$, if $y \vdash x$ and $\beta \leq \alpha$.
(S) $\{(p, \alpha/x), (p, \beta/y)\} \vdash \{(p, \alpha/(x \wedge \neg y)), (p, \beta/(y \wedge \neg x)), (p, \max(\alpha, \beta)/(x \wedge y))\}$

with modus ponens as the only inference rule.

Note that the above logical system can be used as well for the logic of agents, if, instead of using sets of agents in the labels, we use an intensional description of these subsets of agents, based on a specific vocabulary (e.g., old people, tall persons, etc.). The same logic could be used if the pairs of formulas are interpreted as (effect, cause) in a diagnosis situation.

Remark 3. *The idea of considering statements of the form "x is a reason for p" is due to Apothéloz [2] who views such statements as a building block for argumentation, and who observe that the statements can be transformed by applying negations to x, to p, and to the statement itself (outside negation). In [1] it was pointed out that the interplay of the negation on p and the outside negation led to the hexagon of opposition reproduced in Fig. 3(a)*[5]*; the interplay of the three negations yields the new cube of opposition of Fig. 3(b). Rudiments of the logic of reasons can also be found in the last two pages of [1].*

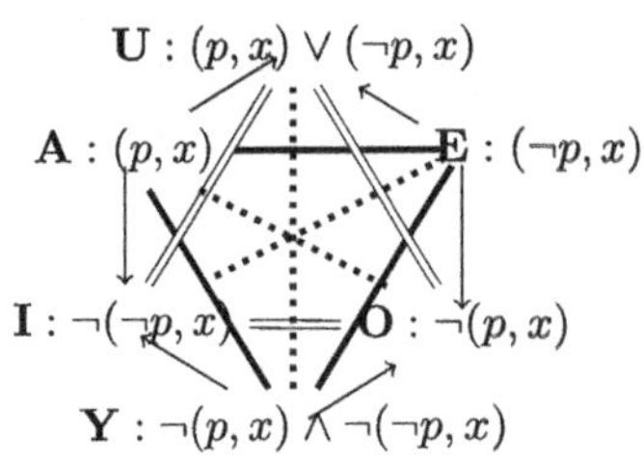

(a) (A, E, Y) are mutually exclusive [1]

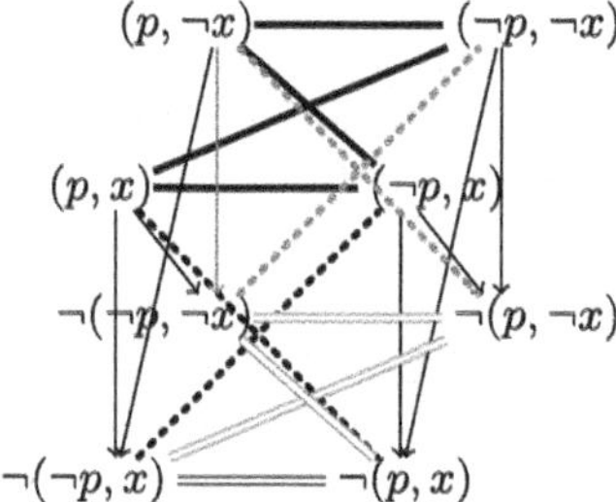

(b) cube induced by three negations

Fig. 3. The hexagon and the cube of opposition of reasons ($x \neq \bot, x \neq \top$)

Remark 4. *The logic of reasons is akin to the logic of supporters [20]. In the latter logic, each formula p is associated with a "supporter", modelled by a subset of subsets of "assumptions". It encodes a disjunction of conjunctions whose*

[5] In such a hexagon, **A**, **E**, and **Y** refer to pairwise mutually exclusive situations, while among the three situations **I**, **U**, and **O**, one is necessarily the case. The arrows forming the edges of the hexagon are entailments, and the diagonals link contradictories.

truth "supports" the truth of p. Supporters can thus be viewed as logical formulas without negation. The set of supporters forms a lattice. In their semantics, lattice-valued necessity measures are called support measures and are valued on specific subsets of the power set of the set of assumptions. our logic of reasons is more expressive and syntactically simpler.

5 Concluding Remarks

This paper has highlighted a variant of possibilistic logic where labels of propositional formulas are elements in a Boolean algebra. It has been shown that there is a Boolean-valued counterpart to generalised possibilistic logic handling negations and disjunctions of pairs (proposition, label). This setting has been exemplified in the case where propositions are labelled by subsets of agents that believe them, and when the labels are themselves propositions that stand as reasons for other propositions. The case of timed possibilistic logic, proposed a long time ago, also fits this framework. All such logics can be combined with the usual possibilistic logic, using certainty weights. The algebraic setting is then a pseudo-complemented distributive lattice of fuzzy sets, more precisely a De Morgan algebra based on max and min, less than a Boolean algebra.

Some perspectives can be envisaged. First, it would be interesting to consider the case of weights in non-distributive lattices; however at this point we miss a canonical example of this situation. Next it is possible to cumulate several kinds of labels of possibilistic pairs (e.g., time + multiple agents + uncertainty). Finally, in the case of reason-based possibilistic logic, we have assumed that p and x in (p, x) belong to two distinct languages. We may wonder if it would be possible to relax this hypothesis for at least allowing for formulas such as (x, x) (x is a reason for x!). Then it would hold that $(\neg x \vee p, 1), (x, x) \vdash (p, x)$, expressing that (p, x) is weaker than $(\neg x \vee p, 1)$ since from (p, x) we cannot go back to $(\neg x \vee p, 1)$ with the inference rules. Having claims and reasons as overlapping languages is a topic for further research.

References

1. Amgoud, L., Prade, H.: Towards a logic of argumentation. In: Hüllermeier, E., Link, S., Fober, T., Seeger, B. (eds.) SUM 2012. LNCS (LNAI), vol. 7520, pp. 558–565. Springer, Heidelberg (2012). https://doi.org/10.1007/978-3-642-33362-0_43
2. Apothéloz, D.: Esquisse d'un catalogue des formes de la contre-argumentation. Travaux du Centre de Recherches Sémiologiques **57**, 69–86 (1989)
3. Belhadi, A., Dubois, D., Khellaf-Haned, F., Prade, H.: Multiple agent possibilistic logic. J. Appl. Non Class. Logics **23**(4), 299–320 (2013)
4. Belhadi, A., Dubois, D., Khellaf-Haned, F., Prade, H.: Reasoning with multiple-agent possibilistic logic. In: Schockaert, S., Senellart, P. (eds.) SUM 2016. LNCS (LNAI), vol. 9858, pp. 67–80. Springer, Cham (2016). https://doi.org/10.1007/978-3-319-45856-4_5

5. Benferhat, S., Dubois, D., Prade, H.: Provenance calculus and possibilistic logic: a parallel and a discussion. In: Bouraoui, Z., Vesic, S. (eds.) ECSQARU 2023. LNCS, vol. 14294, pp. 427–441. Springer, Cham (2024). https://doi.org/10.1007/978-3-031-45608-4_32
6. Besnard, P., Hunter, A.: Elements of Argumentation. MIT Press, Cambridge (2008)
7. Dubois, D., Lang, J., Prade, H.: Timed possibilistic logic. Fundam. Informaticae **15**(3–4), 211–234 (1991)
8. Dubois, D., Lang, J., Prade, H.: Possibilistic logic. In: Gabbay, D., Hogger, C., Robinson, J., Nute, D. (eds.) Handbook of Logic in Artificial Intelligence and Logic Programming, vol. 3, pp. 439–513. Oxford University Press (1994)
9. Dubois, D., Prade, H.: Conditional objects as nonmonotonic consequence relationships. IEEE Trans. Syst. Man Cybern. **24**, 1724–1740 (1994)
10. Dubois, D., Prade, H.: Combining hypothetical reasoning and plausible inference in possibilistic logic. Int. J. Multiple-Valued Logic **1**, 219–239 (1996)
11. Dubois, D., Prade, H.: Possibilistic logic: a retrospective and prospective view. Fuzzy Sets Syst. **144**, 3–23 (2004)
12. Dubois, D., Prade, H.: Toward multiple-agent extensions of possibilistic logic. In: Proceedings IEEE International Conference on Fuzzy Systems (FUZZ-IEEE 2007), London, 23–26 July, pp. 1–6. IEEE (2007)
13. Dubois, D., Prade, H.: A crash course on generalized possibilistic logic. In: Ciucci, D., Pasi, G., Vantaggi, B. (eds.) SUM 2018. LNCS (LNAI), vol. 11142, pp. 3–17. Springer, Cham (2018). https://doi.org/10.1007/978-3-030-00461-3_1
14. Dubois, D., Prade, H.: Possibilistic logic: from certainty-qualified statements to two-tiered logics – a prospective survey. In: Calimeri, F., Leone, N., Manna, M. (eds.) JELIA 2019. LNCS (LNAI), vol. 11468, pp. 3–20. Springer, Cham (2019). https://doi.org/10.1007/978-3-030-19570-0_1
15. Dubois, D., Prade, H.: Possibility theory and possibilistic logic: tools for reasoning under and about incomplete information. In: Shi, Z., Chakraborty, M., Kar, S. (eds.) ICIS 2021. IAICT, vol. 623, pp. 79–89. Springer, Cham (2021). https://doi.org/10.1007/978-3-030-74826-5_7
16. Dubois, D., Prade, H., Rico, A.: The cube of opposition: a structure underlying many knowledge representation formalisms. In: Proceedings of the 24th International Joint Conference on Artificial Intelligence (IJCAI 2015), Buenos Aires, 25–31 July, pp. 2933–2939. AAAI Press (2015)
17. Dubois, D., Prade, H., Rico, A.: Structures of opposition and comparisons: Boolean and gradual cases. Log. Univers. **14**(1), 115–149 (2020)
18. Dubois, D., Prade, H., Schockaert, S.: Generalized possibilistic logic: foundations and applications to qualitative reasoning about uncertainty. Artif. Intell. **252**, 139–174 (2017)
19. Gabbay, D.: Labelled Deductive Systems, vol. 1. Oxford University Press, Oxford (1996)
20. Lafage, C., Lang, J., Sabbadin, R.: A logic of supporters. In: Bouchon-Meunier, B., Yager, R.R., Zadeh, L.A. (eds.) Information, Uncertainty and Fusion, pp. 381–392. Kluwer (1999)

Possibilistic Provenance

Didier Dubois and Henri Prade(✉)

Institut de Recherche en Informatique de Toulouse (IRIT) - CNRS,
118, route de Narbonne, 31062 Toulouse Cedex 9, France
{dubois,prade}@irit.fr

Abstract. Provenance calculus is based on two operations forming a semi-ring. Its aim is to combine and propagate annotations associated with data. This note emphasizes the point that this calculation, if based on max and min operations, exactly corresponds to query evaluation when data are labeled with levels of certainty (necessity) in the sense of possibility theory.

1 Introduction

The idea of provenance emerged in the early 2000s from some works on databases (e.g. [6]); it aimed at tracking tuples in relations through the operations of relational algebra. To this end, each tuple is associated with an annotation. The latter can also be used to encode uncertainty [15].

However, it was with the work of Green and Tannen [16] and others [13,14] that a calculation was proposed to combine these annotations. It is based on two operations forming a semi-ring. These two operations are assumed to be associative, with distinct neutral elements respectively denoted by 0 and 1; the first operation, considered as an "addition" (denoted by $+$), being commutative, and the second, considered as a "product" (denoted by $\cdot$) being distributive with respect to the first; moreover, the neutral element of the first (0) is an absorbing element for the second. When the product operation is commutative, we speak of a commutative semi-ring. Depending on the interpretation given to the annotations, the provenance calculation uses semi-rings involving integers, logical expressions, sets, and information access control levels in particular.

Recently, a parallel between provenance calculus and possibilistic logic [4] has been the subject of a study highlighting the use of different semi-rings in the two domains, depending on the interpretation given to data annotations, and to the weights associated with formulas in possibilistic logic or related logics.

But provenance calculation with possibilistic annotations has never been considered as such. Yet, it turns out to be similar to an approach developed in [18] for query evaluation in the face of an uncertain database where uncertainty takes the form of levels of certainty associated with attribute values or tuples. Highlighting this point is the purpose of this note.

The article first recalls the computation of provenance in Sect. 2 before applying it to possibilistic annotations and showing that it corresponds to the evaluation of queries to a database with certainty levels in Sect. 3.

S. Destercke et al. (Eds.): SUM 2024, LNAI 15350, pp. 147–153, 2025.
https://doi.org/10.1007/978-3-031-76235-2_11

2 Provenance Calculus

Provenance calculation aims to propagate tuple annotations when calculating query answers using relational algebra operations. We start with an example proposed by Green and Tannen, to illustrate the basic principles of provenance calculation [16]. See also [1].

Consider two relations R and S respectively defined on attributes (A, B, C) and (D, B, E). Let abc and dbe be two tuples of R and S respectively, with annotations p and r respectively. The product operation (denoted by $\cdot$) of the semi-ring associates the annotation $p \cdot r$ with the tuple $a\ b\ c\ d\ e$ of the relation $R \bowtie S$ obtained as the join of R and S on B and defined on the attributes (A, B, C, D, E). Given two relations R and S that both contain the tuple abc with annotations p and r respectively, an addition operation, denoted $+$, associates $p + r$ with the tuple abc in the union $R \cup S$ of R and S. This $+$ operation is also used in the case of projection, to keep track of the different tuples whose projection produces the same tuple.

Consider the query $Q = \sigma_{C=e}\pi_{AC}(\pi_{AC}R \bowtie \pi_{BC}R \cup \pi_{AB}R \bowtie \pi_{AC}R)$ addressed to a relation R defined on the attributes (A, B, C) (π denotes projection, σ selection, $\bowtie$ join). Suppose R contains the 3 tuples abc, dbe and fge, respectively annotated by p, r and s. That is, R corresponds to the Table 1.

Table 1. Relation R

A	B	C	
a	b	c	p
d	b	e	r
f	g	e	s

We get for $\pi_{AC}R \bowtie \pi_{BC}R$ (Table 2).

Table 2. Relation $\pi_{AC}R \bowtie \pi_{BC}R$

A	B	C	
a	b	c	p^2
d	b	e	r^2
d	g	e	$r \cdot s$
f	b	e	$s \cdot r$
f	g	e	s^2

Similarly for $\pi_{AB}R \bowtie \pi_{AC}R$; see Table 3.

Then we can check that the relationship obtained in the response is in Table 4, after applying the union, the projection π_{AC} and the selection $\sigma_{C=e}$ (which corresponds to a product by 1 if the selection condition is satisfied, and by 0 otherwise).

Table 3. Relation $\pi_{AB}R \bowtie \pi_{AC}R$

A	B	C	
a	b	c	p^2
a	b	e	$p \cdot r$
d	b	c	$r \cdot p$
d	b	e	r^2
f	g	e	s^2

Table 4. Result of the example

A	C	
a	c	$(p^2 + p^2) \cdot 0$
a	e	$p \cdot r \cdot 1$
d	c	$r \cdot p \cdot 0$
d	e	$(r^2 + r \cdot s + r^2) \cdot 1$
f	e	$(s^2 + s \cdot r + s^2) \cdot 1$

In these tables, p^2 means $p \cdot p$. One can replace $p + p$ by $2p$, and $p^2 + p^2$ by $2p^2$. In fact, the agreement of the calculus on formal annotations with the properties of union, join, projection and selection of relational algebra leads to the requirement that this calculus follows the properties of a commutative semi-ring [14].

As can be seen from the example, formal polynomials are used to encode the way in which tuples are obtained. For example, the tuple d e in Table 4 is annotated with the polynomial $2r^2 + r \cdot s$. It echoes the three ways of deriving d e from the relation R, two of which use r only (but twice), while the third uses r and s only once.

If we replace the formal annotations, p, r and s by integers (e.g., 2, 5 and 1 respectively) understood as the number of copies that exist in R for each tuple, we can count the number of ways of obtaining the tuple d e, for example, by applying the expression $2r^2+r \cdot s$ (i.e., $2 \times 5^2 + 5 \times 1 = 55$). This is bag semantics, where annotations are multiplicities. The semi-ring is then $(\mathbb{N}, +, \times, 0, 1)$.

Another example is provided by access control levels [2,13]. A database is assumed to be annotated with security information, where not only tuples as a whole, but also possible attribute values can be labeled with access levels belonging to the following fully ordered scale $\mathbb{A} = P < C < S < T < 0$, where P means "public", C "confidential", S "secret" and T "top-secret" (0 means something "completely inaccessible"). A database tuple can therefore already imply a "product" of access levels in the case of annotated attributes. Generally speaking, tuples responding to a query are associated with polynomials that must be interpreted using the semi-ring $(\mathbb{A}, \min, \max, 0, P)$, e.g. an output tuple annotated with $C \cdot S \cdot T + C^2 \cdot S = S$ corresponds to a "secret" piece of information, applying semi-ring operations.

3 Databases with Certainty Levels

Suppose now that annotations are levels of certainty associated with tuples. These levels of certainty are interpreted, as in possibilistic logic [9–11], i.e. as lower bounds of necessity measures. Since measures of necessity are min-decomposable for conjunction, the certainty attached to a conjunction is therefore the minimum of the certainties of the elements of the conjunction. If the same tuple can be obtained in several ways, we retain the maximum of the certainty levels, since these are lower bounds. We therefore work with a semi-ring $(\mathbb{S}, \max, \min, 1, 0)$, where $\mathbb{S}$ is the discrete or continuous scale used for degrees of necessity, e.g. $[0, 1]$, or a finite subset of ordered levels.

Using the example from the previous section, i.e. the Table 4, with $+ = \max$, $\cdot = \min$, $p = \alpha > r = \beta > s = \gamma$, we obtain that the level of certainty of the tuple ae is $\min(\alpha, \beta, 1) = \beta$, that of de is $\min(\max(\beta, \min(\beta, \gamma)), 1) = \beta$, and that of fe is $\min(\max(\gamma, \min(\beta, \gamma)), 1) = \gamma$. The answer to the Q query is therefore $\min(\max(\gamma, \min(\beta, \gamma)), 1) = \gamma$.

It is clear that this calculation is very similar to that of the access control levels, and that the semi-rings $(\mathbb{A}, \min, \max, 0, P)$ and $(\mathbb{S}, \max, \min, 1, 0)$ are isomorphic, for finite $\mathbb{S}$, up to a scale reversal. As already mentioned, the calculation of control levels allows us to attach levels not only to tuples, but also to attribute values. The similarity between the handling of access control levels and the treatment of uncertainty in description logics using possibility theory was first pointed out in [2].

The treatment of uncertainty in databases is considerably simplified if uncertainty takes the form of levels of certainty associated with attribute values or tuples for relational algebra queries (the complexity of query evaluation is then of the same order of magnitude as that of processing a standard query; see [18] for a detailed study. In addition, attribute values can be disjunctions of possible values.

In [18], we show that this treatment has an equivalent counterpart in possibilistic logic. Indeed, tuples are translated into a possibilistic logic base, applying the following principles: i) keys become variables; ii) attributes become predicates; iii) tuples become instantiated formulas. So, for example a tuple like:

$$t = \langle 113,\ \mathit{Jean}, (\mathit{Paris},\ 0.8),\ (\mathit{engineer},\ 0.7)\rangle$$

translates into the possibilistic logic formulas of the base K_1:

$$\begin{aligned} K_1 = \{&(\mathit{name}(113,\ \mathit{Jean}),\ 1), \\ &(\mathit{city}(113,\ \mathit{Paris}),\ 0.8), \\ &(\mathit{job}(113,\ \mathit{engineer}),\ 0.7)\}. \end{aligned}$$

The query "find x such that condition Q is satisfied", i.e., $\exists x\, Q(x)$? is processed by refutation, adding the formulas corresponding to $\neg Q(x) \vee \mathit{answer}(x)$. So the query "find employees who are engineers and live in Paris" is written as follows

$$q = \sigma_{city=\text{'}Paris\text{'}\ and\ job=\text{'}Engineer\text{'}}(emp)$$

which can be translated by:

$$\{((\neg ville(x,\ Paris)) \vee (\neg prof(x,\ engineer)) \vee answer(x),\ 1)\}.$$

And from K_1, we obtain by possibilistic refutation the answer: the tuple 113 with its level of certainty, i.e., $(answer(113),\ 0.7)$.

The compatibility of possibility theory with relational algebra operations was established in [18]. The product operation of the semi-ring, here min, corresponds to the join, which is a conjunction of attribute values, in accordance with the level of certainty of the attribute values concerned. The addition of the semi-ring, here max, maintains the highest level of certainty in the case of multiple ways of obtaining the same tuple through a union or projection. So the max − min specialization of the provenance calculation exactly corresponds to a query evaluation when data are labeled with levels of certainty in the sense of possibility theory.

There exist several extensions of possibilistic logic whose provenance counterpart is of interest for databases [4]. Let us give two examples: multi-source possibilistic logic [8] and timed possibilistic logic [7]. In the first logic, formulas are associated with fuzzy sets of sources, i.e., each source is associated with its level of certainty, and each formula with all the sources that support its truth (to the extent of the level of certainty attached to the source). This corresponds to working with the semi-ring $(\mathbb{S}^{\mathbf{S}}, \cup_{\max}, \cap_{\min}, \emptyset, \mathbf{S})$ where $\mathbf{S}$ is the set of all sources, and $\mathbb{S}$ the scale of the certainty level, $\cup_{\max}, \cap_{\min}$ being the fuzzy set union and intersection based on max and min respectively. This is a straightforward extension of the above case corresponding to one source. As above, tuples as a whole, but also attribute values can be associated with sources and their level of certainty.

As an illustration, going back to our running example, consider the tuple

$$t = \langle 113,\ Jean, (Paris,\ \{1/s_1, 0.7/s_2\}),\ (engineer,\ \{0.8/s_2\})\rangle$$

Here each piece of information is associated to a weight of the form $\{\alpha_1/s_1, \cdots, \alpha_n/s_n\}$, where s_i is a source and α_i is the certainty level associated to the information provided by this source [17]. Then t translates into the generalized possibilistic logic formulas of the base K_2:

$$\begin{aligned} K_2 = \{&(name(113,\ Jean),\ 1/\mathbf{S}),\\ &(city(113,\ Paris),\ \{1/s_1, 0.7/s_2\}),\\ &(job(113,\ engineer),\ \{0.8/s_2\})\}. \end{aligned}$$

As can be seen, the weights of this generalized possibilistic logic are no longer linearly ordered but belong to a De Morgan lattice. The details of the machinery of such a logic can be found in [12]. Roughly speaking, we deal with fuzzy-set valued necessity functions using the semi-ring $(\mathbb{S}^{\mathbf{S}}, \cup_{\max}, \cap_{\min}, \emptyset, \mathbf{S})$.

The query "find employees who are engineers and live in Paris" is (written as $\{((\neg ville(x,\ Paris)) \vee (\neg prof(x,\ engineer)) \vee answer(x),\ 1)\}$) returns

$$(answer(113),\ \{0.7/s_2\})$$

from K_2, i.e., the tuple 113 with a weight $\{0.7/s_2\}$ reflecting the source(s) that support it with the associated certainty level(s).

Similarly, in timed possibilistic logic, formulas are associated with *fuzzy* sets of time instants where the grade attached to a time instant is the certainty level with which the formula is true at that time. At the semantic level, it leads to an extension of necessity (and possibility) measures now valued in a distributive lattice structure where necessity functions are (fuzzy) set-valued. We are thus working with the commutative semi-ring $(\mathbb{S}^{\mathbf{T}}, \cup_{\max}, \cap_{\min}, \emptyset, \mathbf{T})$ where $\mathbf{T}$ is the set of time instants, $\mathbb{S}^{\mathbf{T}}$ the set of $\mathbb{S}$-graded fuzzy sets over $\mathbf{T}$. Thus, tuples as a whole, or attribute values can be associated with sets of time instants where the information is certain to some extent, using a machinery isomorphic to the one of the previous logic.

4 Conclusion

As already noticed in [4], DB and AI are fields that have developed separately, and there seems to have been no mutual exchange between the ideas underlying provenance calculus and epistemic uncertainty propagation, despite their formal proximity. The fact that the propagation, by means of the provenance calculus, of "possibilistic annotations" representing levels of certainty exactly corresponds to the possibilistic treatment of data uncertainty, confirms the close proximity between possibilistic logic and provenance calculus. In this sense, this note complements the essentially bibliographical parallel proposed in [4]. A similar treatment should also apply to possibilistic extensions of DL-Lite and other lightweight description logics [3,5], or possibilistic extensions of RDF language [19] in Web Semantics.

References

1. Abiteboul, S., Hull, R., Vianu, V.: Foundations of Databases. Addison-Wesley, Reading (1995)
2. Baader, F., Knechtel, M., Peñaloza, R.: Context-dependent views to axioms and consequences of Semantic Web ontologies. J. Web Semant. **12**, 22–40 (2012)
3. Benferhat, S., Bouraoui, Z.: Min-based possibilistic DL-Lite. J. Logic Comput. **27**(1), 261–297 (2017)
4. Benferhat, S., Dubois, D., Prade, H.: Provenance calculus and possibilistic logic: a parallel and a discussion. In: Bouraoui, Z., Vesic, S. (eds.) ECSQARU 2023. LNCS, vol. 14294, pp. 427–441. Springer, Cham (2024). https://doi.org/10.1007/978-3-031-45608-4_32
5. Bourgaux, C., Ozaki, A., Peñaloza, R.: Semiring provenance for lightweight description logics. arXiv, c s.LO, arXiv:2310.16472 (2023)
6. Cui, Y., Widom, J., Wiener, J.L.: Tracing the lineage of view data in a warehousing environment. ACM Trans. Database Syst. **25**(2), 179–227 (2000)
7. Dubois, D., Lang, J., Prade, H.: Timed possibilistic logic. Fundam. Informaticae **15**(3–4), 211–234 (1991)

8. Dubois, D., Lang, J., Prade, H.: Dealing with multi-source information in possibilistic logic. In: Neumann, B. (ed.) Proceedings of the 10th European Conference on Artificial Intelligence (ECAI 1992), Vienna, 3–7 August, pp. 38–42. Wiley (1992)
9. Dubois, D., Lang, J., Prade, H.: Possibilistic logic. In: Gabbay, D.M., Hogger, C.J., Robinson, J.A., Nute, D. (eds.) Handbook of Logic in Artificial Intelligence and Logic Programming, vol. 3, pp. 439–513. Oxford University Press (1994)
10. Dubois, D., Prade, H.: Possibilistic logic - an overview. In: Siekmann, J.H. (ed.) Computational Logic. Handbook of the History of Logic, vol. 9, pp. 283–342. Elsevier (2014)
11. Dubois, D., Prade, H.: Possibilistic logic: from certainty-qualified statements to two-tiered logics – a prospective survey. In: Calimeri, F., Leone, N., Manna, M. (eds.) JELIA 2019. LNCS (LNAI), vol. 11468, pp. 3–20. Springer, Cham (2019). https://doi.org/10.1007/978-3-030-19570-0_1
12. Kacprzyk, J., Nowacka, K., Zadrożny, S.: A possibilistic-logic-based information retrieval model with various term-weighting approaches. In: Rutkowski, L., Tadeusiewicz, R., Zadeh, L.A., Żurada, J.M. (eds.) ICAISC 2006. LNCS (LNAI), vol. 4029, pp. 1120–1129. Springer, Heidelberg (2006). https://doi.org/10.1007/11785231_117
13. Foster, J.N., Green, T.J., Tannen, V.: Annotated XML: queries and provenance. In: Lenzerini, M., Lembo, D. (eds.) Proceedings of the 27th ACM SIGMOD-SIGACT-SIGART Symposium on Principles of Database Systems (PODS 2008), Vancouver, 9–11 June, pp. 271–280 (2008)
14. Green, T., Karvounarakis, G., Tannen, V.: Provenance semirings. In: Libkin, L. (ed.) Proceedings of the 26th ACM SIGACT-SIGMOD-SIGART Symposium Principles of Database Systems (PODS 2007), Beijing, China, 11–13 June, pp. 31–40. ACM (2007)
15. Green, T.J., Tannen, V.: Models for incomplete and probabilistic information. IEEE Data Eng. Bull. **29**(1) (2006)
16. Green, T.J., Tannen, V.: The semiring framework for database provenance. In: Sallinger, E., Van den Bussche, J., Geerts, F. (eds.) Proceedings of the 36th ACM SIGMOD-SIGACT-SIGAI Symposium on Principles of Database Systems (PODS 2017), Chicago, 14–19 May, pp. 93–99. ACM (2017)
17. Pivert, O., Prade, H.: Querying uncertain multiple sources. In: Straccia, U., Calì, A. (eds.) SUM 2014. LNCS (LNAI), vol. 8720, pp. 286–291. Springer, Cham (2014). https://doi.org/10.1007/978-3-319-11508-5_24
18. Pivert, O., Prade, H.: A certainty-based model for uncertain databases. IEEE Trans. Fuzzy Syst. **23**(4), 1181–1196 (2015)
19. Zimmermann, A., Lopes, N., Polleres, A., Straccia, U.: A general framework for representing, reasoning and querying with annotated Semantic Web data. J. Web Semant. **11**, 72–95 (2012)

Towards a Logical Framework for Reasoning with Stereotypes

Florence Dupin de Saint-Cyr[1(✉)], Francis Faux[1(✉)], and Sabine Frittella[2(✉)]

[1] IRIT, Université Paul Sabatier, 118 route de Narbonne, 31062 Toulouse, France
Florence.Bannay@irit.fr, francis.faux@univ-jfc.fr

[2] INSA Centre Val de Loire, Univ. Orléans, LIFO EA, 4022 Orléans, France
sabine.frittella@centrale-marseille.fr

Abstract. Stereotypes are necessary for human cognition. Indeed our limited computational capabilities and our need for quick decision making require using shortcuts for reasoning. In this work, we discuss how to formalize reasoning with stereotypes using uncertain default rules with an anchorage degree.

Keywords: Uncertainty · Stereotypes · Belief revision

1 Introduction

Stereotypes are a widespread form of prejudice. A prejudice can be defined as "an a priori favourable or unfavourable opinion, adopted without examination, imposed by an environment, an education" (Montaigne, Essais, II, 12, ed. P. Villey and V.-L. Saulnier, p.506). Prejudices have the characteristic of being more or less entrenched: the more entrenched they are, the more difficult it is to refute them and the more they influence reasoning.[1] We can consider *dogmatism*, as the "extreme" prejudice, because it refers to closed mind, rigidity, inflexible system of thought not open to new information, new experiences or new environments [9].

Stereotypes are defined as "characterisations of social categories by which membership of a group is associated with the possession of certain attributes" (for example, scientists are intellectuals, Scots are greedy, men like the colour blue). Consequently, when faced with new information, prejudices can be seen as filters influenced by their polarities. In this paper, we consider that stereotypes are anchored beliefs of the form "if A, then generally B", some of them are objectively mostly true like "birds fly" and some are not, like "French are rude".

[1] In his book *The Nature of Prejudice*, the American psychologist Gordon [1] argues that "prejudice is essentially derived from the necessary mental shortcuts that the human brain uses to process the large amount of information it receives".

This research is part of the MOSAIC project financed by the European Union's Marie Skłodowska-Curie grant No. 101007627.

S. Destercke et al. (Eds.): SUM 2024, LNAI 15350, pp. 154–162, 2025.
https://doi.org/10.1007/978-3-031-76235-2_12

We use default rules enriched with a degree of certainty and anchorage to encode prejudices.

Within the framework of evidence theory, which is well suited to modeling uncertain or incomplete testimonies from different sources, the notion of prejudice was introduced and formalized by a state of belief [4,10]: prejudices can evolve, and their effect consists in weakening or fragmenting some focal sets of incoming information represented by a belief function. In this article, we choose to study stereotypes within the logical framework of uncertain default rules [7] based on possibility theory. It provides a natural way to encode a potentially unreliable, but still used, deduction rule without mentioning all its exceptions. Default rules are compact ways to express generic laws that are easy to apply even in the case of incomplete information, which is similar to what human beings expect from stereotypes: compactness means easy to memorize and easy to transmit to other people, genericity means easy to apply. In our approach, a stereotype is considered to be an information that is more or less reliable and more or less anchored, i.e., the prejudice is more or less likely to disappear or diminish. This is why we propose to represent stereotypes using two parameters, one for reliability and one for anchorage. The objective is to be able to simulate the behaviour of an agent with prejudices on the one hand and more or less certain information on the other.

In Sect. 2, we provide notations and present the notion of *belief base* used to encode the knowledge and stereotypes of an agent. We end this part by recalling a method proposed in [7] for transforming uncertain default rules into uncertain strict rules. In Sect. 3, we discuss the evolution of an agent's belief base given a new piece of information. The paper concludes with a look ahead to future work.

2 Definitions

2.1 Preliminaries

Let $\mathcal{X}, \mathcal{C}, \mathcal{P}$ be respectively sets of variables, constants, and predicates.
Let $\mathcal{L}$ be the first-order language s.t.:
$\mathcal{T} \ni t := x \mid a$,
$\mathcal{L} \ni \phi := P(t_1, \ldots, t_n) \mid \neg\phi \mid \phi \wedge \phi \mid \forall x, \phi$
with $x \in \mathcal{X}$, $a \in \mathcal{C}$, $P \in \mathcal{P}$ a predicate of arity n and $t_1, \ldots, t_n \in \mathcal{T}$. The other standard connectives $(\bot, \top, \vee, \rightarrow, \leftrightarrow, \exists)$ are defined as usual.

A *structure* for the language $\mathcal{L}$ is a tuple $\mathbb{S} = (\mathcal{D}, \mathcal{I}_\mathcal{C}, \mathcal{I}_\mathcal{P})$ such that $\mathcal{D}$ is a non-empty *domain*, $\mathcal{I}_\mathcal{C}$ is the *interpretation of constant symbols*, i.e., for every $a \in \mathcal{C}$, $\mathcal{I}_\mathcal{C}(c) \in \mathcal{D}$, $\mathcal{I}_\mathcal{P}$ is the *interpretation of predicate symbols*, i.e., for every $P \in \mathcal{P}$ of arity n, $\mathcal{I}_\mathcal{C}(P) \subseteq \mathcal{D}^n$. A *model* for the language $\mathcal{L}$ is a tuple $\mathbb{M} = (\mathbb{S}, \mathcal{I}_\mathcal{X})$ such that $\mathbb{S}$ is a structure for the language $\mathcal{L}$ and $\mathcal{I}_\mathcal{X} : \mathcal{X} \rightarrow \mathcal{D}$ is the *interpretation of variables*. The interpretation of formulas over a model $\mathbb{M}$ is defined in the usual manner. A *structure* $\mathbb{S}$ *satisfies a formula*, noted $\mathbb{S} \models \phi$, if $(\mathbb{S}, \mathcal{I}_\mathcal{X}) \models \phi$ for every interpretation $\mathcal{I}_\mathcal{X}$. In this article, we assume that the domain $\mathcal{D}$ is fixed and finite.

Let $\mathcal{L}_x$ be the restriction of $\mathcal{L}$ without the use of quantifiers and with one single variable x. Let $\mathcal{L}_C$ be the set of closed formulas in $\mathcal{L}$. We will only consider models with finite domains, therefore every formula can be translated into propositional logic. We use an additional symbol $\rightsquigarrow$ that connects two formulas ϕ and ψ from $\mathcal{L}_x$. Formulas of the form $\phi \rightsquigarrow \psi$ will be used to encode stereotypes via so-called *belief rules*. We will reason with those rules following the non-monotonic reasoning approaches of default rules [2].

Possibility theory enables us to associate two measures Π and N with a formula φ [6]. $\Pi(\varphi)$, called the *possibility* of φ, quantifies *how unsurprising* φ *is* ($\Pi(\varphi) = 0$ means that φ is necessarily false), and N, called the *necessity*, is the dual of Π defined by $N(\varphi) = 1 - \Pi(\neg\varphi)$ ($N(\varphi) = 1$ means that φ is necessarily true). In possibilistic logic, the resolution rule [5] is written $(a \vee b, \rho_1); (\neg a \vee c, \rho_2) \vdash_\pi (b \vee c, \min(\rho_1, \rho_2))$. A *possibilistic propositional belief base* is a set $B_\pi = \{(\varphi_i, \rho_i)\}_{i \in [1,m]}$ where each propositional formula φ_i is associated with a weight $\rho_i \in \,]0, 1]$ representing its certainty level and such that $N(\varphi) \geq \rho_i$. To compute the maximal certainty level to attach to a formula φ w.r.t constraints expressed in B_π, the user can add to B_π the clauses corresponding to the refutation of φ with a necessity level of 1 and then deduce $(\bot, \rho)$ with the possibilistic resolution rule $\vdash_\pi$. The weight ρ is a lower bound of the necessity of φ. In the following, we use $sk(B_\pi)$ to denote the skeleton of B_π, it is the set of formulas of B_π without their weights. Let $B_{\pi\rho} = \{\varphi_i \in \mathcal{L} \mid (\varphi_i, \rho_i) \in B_\pi \text{ and } \rho_i \geq \rho\}$ denote the "ρ-cut of the possibilistic base B_π", i.e., formulas of B_π (without their weights) that have a certainty level higher or equal to ρ. Now, $\text{Inc}(B_\pi) = \max\{\rho \in [0, 1] \mid B_{\pi\rho} \vdash \bot\}$ is called the inconsistency level of the possibilistic base B_π. When $sk(B_\pi)$ is inconsistent (then $\text{Inc}(B_\pi) > 0$), a refutation using only formulas strictly above the inconsistency level yields non-trivial conclusions (i.e., with a certainty strictly above the inconsistency level). We use $B_\pi \vdash_{>\text{Inc}} \varphi$ to denote a non-trivial inference, i.e., the existence of a value $\rho' \in [0, 1]$ and a refutation from $B_{\pi>\text{Inc}} = \{(\varphi_i, \rho_i) \in B_\pi \mid \rho_i > \text{Inc}(B_\pi)\}$ with $(\neg\varphi, 1)$ that leads to $(\bot, \rho')$. In the following, $\vdash_\pi$ is used to represent syntactic inference between possibilistic propositional formulas.

2.2 Stereotypes

The following definition proposes a formal structure, called a *belief base*, to encode both the *beliefs* and the *stereotypes* an agent uses to reason. Here the term *beliefs* refers to statements the agent believes to some degree of certainty. We choose to encode stereotypes by uncertain default rules [7] extended with an anchorage factor, below named *belief rules*.

Definition 1 (Belief Base). *A* belief base *is a tuple* $(\mathcal{B}, \mathcal{R})$ *such that*

- $\mathcal{B}$ *is a finite* belief set*:* $\{(\chi_j, \gamma_j) \mid \chi_j \in \mathcal{L}_C$ *and* $\gamma_j \in \,]0, 1]\}_{j \in [\![1,k]\!]}$
- $\mathcal{R}$ *is a set of* belief rules $\{(\varphi_i \rightsquigarrow \psi_i, \rho_i, \alpha_i)\}_{i \in [\![1,n]\!]}$*, where*
  - $\varphi_i, \psi_i \in \mathcal{L}_x$*, they share the same free variable* x*,* $\varphi_i \rightsquigarrow \psi_i$ *is called a* default rule*,*

- $\rho_i \in]0,1]$ *encodes the* reliability *of the rule, i.e. the certainty of the conclusion when the rule is applied (in a non-exceptional context),*
- $\alpha_i \in [0,1]$ *encodes the* anchorage *of the rule, i.e. how willing the agent is to update the reliability of the rule.*[2]

Example 1 (All birds fly). Consider a child who has only seen birds that fly and therefore is convinced that all birds fly, but who is open to learning that ostriches and penguins are birds that do not fly. Then the stereotype (before the arrival of the evidence about ostriches and penguins) could be encoded as follows $(\texttt{Bird}(x) \rightsquigarrow \texttt{Fly}(x), 1, 0)$.

Example 2 (French people are rude). Consider an American moving to France who has heard that French people are rude and holds a strong belief in that statement while being open to being convinced otherwise if given enough evidence. Then the stereotype could be encoded as follows $(\texttt{French}(x) \rightsquigarrow \texttt{Rude}(x), 0.7, 0.4)$.

Example 3. An example of dogmatism is illustrated on "Flat Earth Society" group which currently has 200,000 members. This community believes that the Earth is a flat disk, surrounded by an enormous wall of ice (Antarctica) which would prevent us from falling when we reach the end. The flatists are convinced that we are trying to impose fake belief on them. Then the stereotype could be encoded as follows $(\neg\texttt{FlatEarther}(x) \rightsquigarrow \texttt{Wrong}(x), 1, 1)$. This belief rule is characterized by maximum reliability and anchorage what is specific to dogmatism.

2.3 Semantics of Belief Bases

Translation of Belief Bases into Classical Propositional Logic. Let $(\mathcal{B}, \mathcal{R})$ be a belief base. Assume that $\mathcal{L}$ contains a constant symbol a_d for each $d \in \mathcal{D}$. Therefore, given a structure $\mathbb{S} = (\mathcal{D}, \mathcal{I}_\mathcal{C}, \mathcal{I}_\mathcal{P})$ over $\mathcal{L}$ such that $\mathcal{I}_\mathcal{C}(a_d) = d$, every closed formula of the knowledge base can be equivalently encoded into a formula of classical propositional logic as follows. Let the set $\texttt{Prop} := \{z_{P(a_1,\ldots,a_n)} \mid P \in \mathcal{P}$ of arity n and $a_1, \ldots, a_n \in \mathcal{C}\}$ be a finite set of propositional variables on which is defined a propositional language denoted $\mathcal{L}_{\mathrm{CL}}$. Let us define the translation $\tau : \mathcal{L}_\mathrm{C} \to \mathcal{L}_{\mathrm{CL}}$ as follows: $\tau(\forall x.\phi) = \bigwedge_{d \in \mathcal{D}} \tau(\phi[a_d/x])$, $\tau(\neg\phi) = \neg\tau(\phi)$, $\tau(\phi \wedge \psi) = \tau(\phi) \wedge \tau(\psi)$ and $\tau(P(a_1, \ldots, a_n)) = z_{P(a_1,\ldots,a_n)}$, where $\phi[a_d/x]$ is the formula ϕ in which every occurrence of x has been replaced by a_d. Since we only consider closed formulas, $\tau(P(a_1, \ldots, a_n))$ is only applied in situations where $a_i \in \mathcal{C}$. The valuation $v_\mathbb{S} : \texttt{Prop} \to \{0, 1\}$ is defined as follows: $v_\mathbb{S}(z_{P(a_1,\ldots,a_n)}) = 1$ iff $\mathbb{S} \models P(a_1, \ldots, a_n)$. $v_\mathbb{S}$ is extended to every propositional formula in the standard way. We have $\mathbb{S} \models \phi$ iff $v_\mathbb{S}(\tau(\phi)) = 1$ for every $\phi \in \mathcal{L}_\mathrm{C}$. Belief rules can be translated in a similar way in classical propositional logic rules, by considering that the rule $\varphi_i \rightsquigarrow \psi_i$ encodes the statement $\forall x, \varphi_i \rightsquigarrow \psi_i$.

[2] If $\alpha = 0$, then the agent reasons very scientifically about the rule and will agree to update the reliability ρ of the rule when given new trusted evidence. If $\alpha = 1$, then the agent's opinion about the reliability of the rule cannot be changed.

Semantics of Belief Bases. Let Ω denote the set of valuations associated with the propositional language $\mathcal{L}_{\text{CL}}$. Let $\omega, \omega_i, \ldots$ denote the elements of Ω, and $[\omega]$ be the set of formulas of $\mathcal{L}_{\text{CL}}$ satisfied by ω.

We interpret a set of default rules Δ in terms of the *Lexicographic-ordering* $\succ_\Delta$ (read "is more plausible than") on valuations as defined in [3]. Given a set of default rules of the form $\alpha \rightsquigarrow \beta$, it is possible to compute a stratification of this set according to their specificity: here a rule has more (or less) specificity when it describes a more (or less) exceptional case, the rules that have no exception being in the most specific stratum. Note that a set of default rules may not admit a stratification. We assume that this is not the case here, i.e., that Δ can always be stratified which is called *consistent* by [8]. Pearl in [8] provides an algorithm named System Z[3] to compute automatically the strata, i.e., the subsets of formulas with the same rank. As shown in [2], the same ordering on valuations can be obtained by interpreting each default rule $\varphi_i \rightsquigarrow \psi_i$ by a constraint $\Pi(\varphi_i \wedge \psi_i) > \Pi(\varphi_i \wedge \neg\psi_i)$ on a possibility measure Π.

Definition 2 (Lexicographic ordering $\succ_\Delta$ *on* Ω). *Let $\Delta = \Delta_1 \cup \cdots \Delta_n$ be a stratified default base with n strata ordered from the most specific stratum Δ_1 to the least one Δ_n, let α, β be two formulas of $\mathcal{L}_{\text{CL}}$, let $\omega, \omega' \in \Omega$,*

- *Notations: str (for "strict") is a function that translates a set of default rules into a set of formulas of $\mathcal{L}_{CL}$ as follows $str(E) = \bigcup_{\varphi_i \rightsquigarrow \psi_i \in E}\{\neg\varphi_i \vee \psi_i\}$.*
- $\omega \succ_\Delta \omega'$ *iff there exists* $k \in [1, n]$ *s.t.* $\{\, |str(\Delta_k) \cap [\omega]| > |str(\Delta_k) \cap [\omega']|$ *and* $i < k, |str(\Delta_i) \cap [\omega]| = |str(\Delta_i) \cap [\omega']|$

The last item of Definition 2 explains the conditions for $\omega \succ_\Delta \omega'$: we compare by lexicographic order the tuples obtained by computing the number of formulas satisfied by each valuation (ω and ω') in each stratum.

Definition 3 (Models associated to a consistent belief base). *When $sk(\mathcal{B})$ is* consistent, *the set of models $\mathbb{M}$ associated with the belief base $(\mathcal{B}, \mathcal{R})$ (in which Δ is the set of rules of $\mathcal{R}$ without considering their reliability and anchorage) is $\mathbb{M} = \{v \mid v$ is a valuation such that $v(\phi) = 1$ for all $\phi \in sk(\mathcal{B})\}$. Moreover, the rules enable us to compare any valuation thanks to $\succ_\Delta$ defined in Definition 2.*

Example 4. We consider the first-order language $\mathcal{L}$ such that the predicates (indexed with their arity) are $\mathcal{P} = \{\texttt{Bird}_1, \texttt{Penguin}_1, \texttt{Fly}_1, \texttt{French}_1, \texttt{Rude}_1\}$ and with no constants and the following belief base $(\mathcal{B}, \mathcal{R})$:

$\mathcal{B} = \{(\forall x.(\texttt{Penguin}(x) \rightarrow \texttt{Bird}(x)), 0.8); (\forall x.(\texttt{Bird}(x) \rightarrow \neg\texttt{French}(x)), 1)\}$,
$\mathcal{R} = \{(\texttt{Bird}(x) \rightsquigarrow \texttt{Fly}(x), 0.9, 0), (\texttt{Penguin}(x) \rightsquigarrow \neg\texttt{Fly}(x), 0.99, 0), (\texttt{French}(x) \rightsquigarrow \texttt{Rude}(x), 0.7, 0.5)\}$.

First, we fix a finite domain of interpretation $\mathcal{D}$, then we translate $\mathcal{B}$ into $\mathcal{L}_{\text{CL}}$ over the variables $\texttt{Prop} = \{\texttt{Bird}(d), \texttt{Penguin}(d), \texttt{Fly}(d), \texttt{French}(d), \texttt{Rude}(d) \mid$

[3] In this paper, the indices of the strata are considered in reversed order compared to system Z.

$d \in \mathcal{D}\}$. From this, we build the set $\mathbb{M}$ of valuations satisfying the formulas of the belief base $\mathcal{B}$. Now we translate $\mathcal{R}$ (we drop the anchoring for the moment) into $\mathcal{L}_{\mathrm{CL}}$. First, notice that $\mathtt{Bird}(x) \rightsquigarrow \mathtt{Fly}(x)$ is less specific than $\mathtt{Penguin}(x) \rightsquigarrow \neg\mathtt{Fly}(x)$ since $\forall x, (\mathtt{Penguin}(x) \rightarrow \mathtt{Bird}(x))$. We get $\mathcal{R}_{\mathrm{CL}} = \Delta_1 \cup \Delta_2$ with

$\Delta_1 = \{\mathtt{Penguin}(d) \rightsquigarrow \neg\mathtt{Fly}(d) \mid d \in D\}$,

$\Delta_2 = \{\mathtt{Bird}(d) \rightsquigarrow \mathtt{Fly}(d) \mid d \in D\} \cup \{\mathtt{French}(d) \rightsquigarrow \mathtt{Rude}(d) \mid d \in D\}$.

Consider $\mathcal{D} = \{a, b, c\}$ and $v_1, v_2 \in \mathbb{M}$ such that $v_1(\mathtt{Penguin}(a)) = v_1(\mathtt{Bird}(b)) = v_1(\mathtt{French}(c)) = v_1(\mathtt{Fly}(b)) = v_1(\mathtt{Rude}(c)) = 1$, $v_1(\mathtt{Fly}(a)) = 0$, and $v_2(\mathtt{Penguin}(a)) = v_2(\mathtt{Bird}(b)) = v_2(\mathtt{French}(c)) = v_2(\mathtt{Fly}(b)) = 1, v_2(\mathtt{Fly}(a)) = v_2(\mathtt{Rude}(c)) = 0$.

Then, we get $v_1 \succ_\Delta v_2$, since

$|str(\Delta_1) \cap [v_1]| = |\{\neg\mathtt{Penguin}(x) \vee \neg\mathtt{Fly}(x)\}_{x\in\{a,b,c\}}| = 3$

$|str(\Delta_2) \cap [v_1]| = |\{\neg\mathtt{Bird}(x) \vee \mathtt{Fly}(x)\}_{x\in\{b,c\}} \cup \{\neg\mathtt{French}(x) \vee \mathtt{Rude}(x)\}_{x\in\{a,b,c\}}\}| = 5$

$|str(\Delta_1) \cap [v_2]| = |\{\neg\mathtt{Penguin}(x) \vee \neg\mathtt{Fly}(x)\}_{x\in a,b,c}\}| = 3$

$|str(\Delta_2) \cap [v_2]| = |\{\neg\mathtt{Bird}(x) \vee \mathtt{Fly}(x)\}_{x\in\{b,c\}} \cup \{\neg\mathtt{French}(x) \vee \mathtt{Rude}(x)\}_{x\in\{a,b\}}\}| = 4$.

In [7], an algorithm is proposed to translate a set of uncertain default rules $\mathrm{U}\Delta$ of the form $(\varphi \rightsquigarrow \psi, \rho)$ into a set of classic formulas associated with weights representing a lower bound of the necessity of the rule. Since default rules enable us to reason by assuming a non-exceptional situation, first each rule is rewritten into a formula $(\varphi \wedge \bigwedge_{i\in[1,k]} \neg e_i \rightarrow \psi, \rho)$ where $e_1, \ldots, e_k$ are exceptions to this rule. For computing exceptions, the algorithm starts from the second most specific stratum $\mathrm{U}\Delta_2$ (since the rules of $\mathrm{U}\Delta_1$ have no exception). At stratum $\mathrm{U}\Delta_s$ each rule can only admit exceptions in strata lower than s. Adapted to our context where we consider also a belief base $\mathcal{B}$, exceptions to a rule $(\varphi \rightsquigarrow \psi, \rho)$ are rules $(e_i \rightsquigarrow \psi_i, \rho_i) \in \mathrm{U}\Delta$ with compatible premises but incompatible conclusions (i.e., $\mathcal{B} \cup T_{s-1} \cup \{(\varphi \wedge e_i, \min(\rho, \rho_i))\} \nvdash_{>\mathrm{Inc}} \bot$ and $\mathcal{B} \cup T_{s-1} \cup \{(\psi \wedge \psi_i, \min(\rho, \rho_i)\} \vdash_{>\mathrm{Inc}} \bot$[4]), where T_{s-1} is the set of strict rules with explicit exceptions (and with certainty levels) coming from the translation of the rules of all the previous strata $\mathrm{U}\Delta_1 \cup \ldots \cup \mathrm{U}\Delta_{s-1}$. In addition to the translation of a rule, $(\varphi \rightsquigarrow \psi, \rho)$, k formulas of the form $\varphi \rightarrow \neg e_i$ are added in order to impose that when nothing says the contrary, the situation is not exceptional. [7] showed that each new formula $\varphi \rightarrow \neg e_i$ can be attributed a degree equal to $\min(\rho, \rho_i)$. Let $tr(\mathcal{B}, \mathrm{U}\Delta)$ be the translation of a set of uncertain rules $\mathrm{U}\Delta$ given a possibilistic base $\mathcal{B}$. In our setting, rules have also an anchorage, we transfer the anchorage to the new formulas in the same way that the degree is transferred, more precisions are given in the next section.

Example 5 (Example 4 continued). For the moment we forget the anchoring degrees $\mathcal{B} = \left\{\begin{matrix}(\mathtt{Penguin}(x) \rightarrow \mathtt{Bird}(x), 0.8)\\(\mathtt{Bird}(x) \rightarrow \neg\mathtt{French}(x), 1)\end{matrix}\right\}$, $\mathcal{R} = \left\{\begin{matrix}(\mathtt{Penguin}(x) \rightsquigarrow \neg\mathtt{Fly}(x), 0.99) & \mathrm{U}\Delta_1\\ \hline (\mathtt{Bird}(x) \rightsquigarrow \mathtt{Fly}(x), 0.9) & \\ (\mathtt{French}(x) \rightsquigarrow \mathtt{Rude}(x), 0.7) & \mathrm{U}\Delta_2\end{matrix}\right\}$ In order to rewrite $\mathcal{R}$ it is enough to consider the two rules of $\mathrm{U}\Delta_2$ and find their exceptions, here only $Bird(x) \rightsquigarrow Fly(x)$ admits exceptions (in stratum $\mathrm{U}\Delta_1$)

[4] Or equivalently $\mathrm{Inc}(\{(\varphi \wedge e_i, \min(\rho, \rho_i))\} \cup \mathcal{B} \cup T_{s-1}) = \mathrm{Inc}(B \cup T_{s-1})$ and $\mathrm{Inc}(\{(\psi \wedge \psi_i, \min(\rho, \rho_i))\} \cup \mathcal{B} \cup T_{s-1}) > \mathrm{Inc}(B \cup T_{s-1})$

which gives:

$$tr(\mathcal{B},\mathcal{R}) = \begin{cases} (\texttt{Penguin}(x) \rightarrow \neg\texttt{Fly}(x), 0.99, 0) \ (\texttt{Bird}(x) \wedge \neg\texttt{Penguin}(x) \rightarrow \texttt{Fly}(x), 0.9, 0) \\ (\texttt{French}(x) \rightarrow \texttt{Rude}(x), 0.7, 0.5) \ \ (\texttt{Bird}(x) \rightarrow \neg\texttt{Penguin}(x), 0.9, 0) \end{cases}$$

Note that here, forgetting anchoring, $\mathcal{B} \cup tr(\mathcal{B},\mathcal{R})$ is consistent: $\text{Inc}(B \cup tr(\mathcal{B},\mathcal{R})) = 0$.

3 Arrival of a New Piece of Information

In this section, we propose a strategy for considering the arrival of a new piece of information $\phi \in \mathcal{L}_{\text{C}}$ with a certainty degree γ. Since the anchorage reinforces the certainty of a rule, when ϕ is incompatible with a rule, we compare its certainty level γ with the aggregation of the level of certainty and the anchorage of this rule and react accordingly either to reject ϕ or to modify the rule. This aggregation, denoted $\oplus$, can be a simple addition or a more sophisticated operation. We leave its study for further research. Let us introduce the transformation *tra* of the belief base $(\mathcal{B},\mathcal{R})$ into a possibilistic base $\{(\phi_i,\gamma_i)\}_i \in [1,m]$ by aggregating certainty and anchorage levels:

$$tra(\mathcal{B},\mathcal{R}) = \left\{ (\phi_i,\gamma_i) \in \mathcal{L} \times [0,1] \,\middle|\, \begin{array}{l} (\phi_i,\gamma_i) \in \mathcal{B} \text{ or} \\ (\phi_i,\rho_i,\alpha_i) \in tr(\mathcal{B},\mathcal{R}) \text{ and } \gamma_i = \rho_i \oplus \alpha_i \end{array} \right\}$$

Given a belief base $(\mathcal{B},\mathcal{R})$, compute $\mathcal{B}' = tra(\mathcal{B},\mathcal{R})$

1. If ϕ is *inconsistent* with the 1-cut of $\mathcal{B}'$, $\mathcal{B}'_1$, i.e., $sk(\mathcal{B}'_1) \cup \{\phi\} \vdash \bot$, then ϕ is not integrated
2. If (ϕ,γ) is *compatible* with $(\mathcal{B},\Delta)$, in other words, if the level of inconsistency do not increase when adding (ϕ,γ), i.e., $\text{Inc}(\mathcal{B}' \cup \{(\phi,\gamma)\}) = \text{Inc}(\mathcal{B}')$ then (ϕ,γ) is added to $\mathcal{B}'$ giving $(\mathcal{B}' \cup (\phi,\gamma),\mathcal{R})$
3. Otherwise (when the new information is more certain than some contradicting formula of $(\mathcal{B},\Delta)$), let us consider the rule (i.e., the formula of $\mathcal{B}'$ that comes from a rule $(\varphi_i \rightsquigarrow \psi_i,\rho_i,\alpha_i) \in \mathcal{R}$) with a maximum certainty in $\mathcal{B}'$, violated by ϕ i.e., $\mathcal{B}' \vdash_{>Inc} \phi \rightarrow (\varphi_i \wedge \neg\psi_i)$[5] and $\rho_i \oplus \alpha_i$ is maximum among such rules
   (a) if $\gamma < \rho_i \oplus \alpha_i$ the piece of information ϕ is rejected (but the anchoring of the rule can be decreased $\varphi_i \rightsquigarrow \psi_i$, because even rejected information may influence the reasoner)
   (b) otherwise a rule $(\varphi_i \wedge \phi \rightsquigarrow \neg\psi_i,\gamma,0)$ is added encoding this exception. The certainty and anchorage of $\varphi_i \rightsquigarrow \psi_i$ are lowered yielding $(\varphi_i \rightsquigarrow \psi_i,\rho'_i,\alpha'_i)$ such that $(\rho'_i \oplus \alpha'_i < \rho_i \oplus \alpha_i)$. The same exception adding and degree lowering is done for all the remaining rules violated by ϕ.

This strategy proposes a way to take anchorage into account when revising a belief base. In our opinion, anchorage must reinforce the barrier to the addition of an exception. Hence, to determine whether to add an exception or not, we must compare the certainty level (γ) of the new information with the certainty of the intern rule combined with its anchorage $(\rho \oplus \alpha)$. In the case where

[5] When $sk(\mathcal{B}')$ is consistent, it amounts to check whether $\mathbb{M} \models \phi \rightarrow (\varphi_i \wedge \neg\psi_i)$.

the information is stronger, the anchorage must decrease and the exception be added. The operator $\oplus$ should translate an effect of reinforcing certainty (thus preventing a rational reasoning which would possibly allow for a revision). As soon as a rule dismantles the information then it is discarded, hence it is not necessary to look at less important rules violated by this information.

Example 6 (Example 4 continued). Assume that we take $\oplus = \max$, then the rewriting of the belief base gives a new possibilistic base $\mathcal{B}' = tra(\mathcal{B}, \mathcal{R})$ which contains:

$(\texttt{Bird}(x) \rightarrow \neg\texttt{French}(x), 1)$ $(\texttt{Bird}(x) \rightarrow \neg\texttt{Penguin}(x), 0.9)$
$(\texttt{Penguin}(x) \rightarrow \neg\texttt{Fly}(x), 0.99)$ $(\texttt{Penguin}(x) \rightarrow \texttt{Bird}(x), 0.8)$
$(\texttt{Bird}(x) \wedge \neg\texttt{Penguin}(x) \rightarrow \texttt{Fly}(x), 0.9)$ $(\texttt{French}(x) \rightarrow \texttt{Rude}(x), 0.7)$

We first consider the piece of information $\texttt{French}(Jeanne) \wedge \texttt{Bird}(Jeanne)$ meaning that we have an individual Jeanne that is both a human (who is French) and a bird. Then this information is not integrated (case 1) because it is contradictory to the only formula of certainty 1. Now, assume that $\phi_0 = (\texttt{Rude}(x) \rightarrow \texttt{RudeUS}(x) \wedge \texttt{RudeF}(x)$ with certainty 1 arrives, meaning that any rude person is rude for everyone, i.e., she is rude for French people and for US citizens. This formula being consistent with all the formulas of $\mathcal{B}'$, thus $Inc(\mathcal{B}' \cup (\phi_0, 1)) = Inc(\mathcal{B}') = 0$. Assume now, that the user learns that Jeanne is French and not considered rude by French people: $\phi_1 = French(Jeanne) \wedge \neg RudeF(Jeanne)$ with a certainty degree of 0.8. Then we apply the case 3b, ϕ_1 violates the rule $(French(x) \rightsquigarrow Rude(x), 0.7, 0.4)$ and $0.8 > 0.7 = 0.7 \oplus 0.4$ this leads us to add the rule $(French(x) \wedge French(Jeanne) \wedge \neg RudeF(Jeanne) \rightsquigarrow \neg Rude(x), 0.8, 0)$, i.e., $(French(Jeanne) \wedge \neg RudeF(Jeanne) \rightsquigarrow \neg Rude(Jeanne), 0.8, 0)$, namely, Jeanne who is not considered rude in France is not rude at all (here, we propose to consider the anchorage as null due to the novelty of this information). Additionally we have to reduce the certainty and anchorage of the initial rule. We could get for instance $(French(x) \rightsquigarrow Rude(x), 0.65, 0.3)$ in such a way that $0.65 \oplus 0.4 < 0.7 \oplus 0.4$ (witch holds when $\oplus = \max$).

Note that this is a preliminary example, to explain what we have planned, but for the moment we have only considered the decrease in anchorage and not its increase, since this amounts to studying the initial creation of prejudice, which is a complex, social phenomenon that we are leaving aside for the moment.

4 Concluding Remark

This article explores the way to encompass the handling of stereotypes in a logical framework. We showed that disposing of a framework where we can express both defeasibility, certainty and anchorage strength would be suitable for this purpose. The possibilistic setting seems suitable for dealing with these three notions. A lot of work remains to be done in order to consolidate the foundations of this new line of research. Two crucial points are the study of the different ways of defining the aggregation of certainty and anchorage, and the management of the reduction of these degrees.

References

1. Allport, G.W., Clark, K., Pettigrew, T.: The Nature of Prejudice. Addison-Wesley, Reading (1954)
2. Benferhat, S., Dubois, D., Prade, H.: Nonmonotonic reasoning, conditional objects and possibility theory. Artif. Intell. **92**(1–2), 259–276 (1997)
3. Benferhat, S., Cayrol, C., Dubois, D., Lang, J., Prade, H.: Inconsistency management and prioritized syntax-based entailment. In: Proceedings of the International Joint Conference on Artificial Intelligence (IJCAI 1993), vol. 93, pp. 640–645 (1993)
4. Dubois, D., Faux, F., Prade, H.: Prejudice in uncertain information merging: pushing the fusion paradigm of evidence theory further. Int. J. Approximate Reasoning **121**, 1–22 (2020)
5. Dubois, D., Lang, J., Prade, H.: Possibilistic logic. In: Handbook of Logic in Artificial Intelligence and Logic Programming, vol. 3, pp. 439–513. Oxford University Press (1994)
6. Dubois, D., Prade, H.: Possibility Theory - An Approach to Computerized Processing of Uncertainty. Plenum Press, New York (1988)
7. Dupin De Saint Cyr, F., Prade, H.: Handling uncertainty and defeasibility in a possibilistic logic setting. Int. J. Approximate Reasoning, Special Section on Logical Approaches to Imprecise Probabilities **49**(1), 67–82 (2008)
8. Pearl, J.: System Z: a natural ordering of defaults with tractable applications to nonmonotonic reasoning. In: Proceedings of the 3rd Conference on Theoretical Aspects of Reasoning about Knowledge, pp. 121–135 (1990)
9. Rokeach, M.: The nature and meaning of dogmatism (1954)
10. Dupin de Saint-Cyr, F., Faux, F.: Integrating evolutionary prejudices in belief function theory. In: Bouraoui, Z., Vesic, S. (eds.) ECSQARU 2023. LNCS, vol. 14294, pp. 400–414. Springer, Cham (2024). https://doi.org/10.1007/978-3-031-45608-4_30

SDF-FuzzIA: A Fuzzy-Ontology Based Plug-in for the Intelligent Analysis of Geo-Thematic Data

Giuseppe Filippone[1](✉), Gianmarco La Rosa[1], and Marco Elio Tabacchi[1,2]

[1] Dipartimento di Matematica e Informatica (DMI), Università degli Studi di Palermo, Palermo, Italy
{giuseppe.filippone01,gianmarco.larosa,marcoelio.tabacchi}@unipa.it
[2] Istituto Nazionale di Ricerche Demopolis, Palermo, Italy

Abstract. This short paper presents a description of SDF-FuzzIA, a Fuzzy-Ontology LLM-based system for the intelligent analysis of geo-thematic data that serves as a plug-in to the Sustainability Decision Framework (SDF) Decision Support System (DSS). A description of the components implemented in the system is given, followed by an explanation of the interaction between the components and the main system. As this still is a work in progress, future directions and possible hurdles are explored.

Keywords: Fuzzy Logic · Fuzzy Ontology · Soft Computing

1 Introduction

SDF (Sustainability Decision Framework) is a work-in-progress Decision Support System that deals with geo-thematic data varying from territorial information, to geopolitical quantitative analysis, to open access information on economies and productive parameters of all the countries past and present, developed by Dipartimento di Matematica e Informatica (DMI) of the Università degli Studi di Palermo (UniPa) and TD Group.

An essential part of the system is SDF-FuzzIA, a Fuzzy-Ontology Large Language Model based (LLM-based) system for the intelligent analysis of such data, whose task is to continually oversee data generated by the system about users' prompts and queries, to find unexpected correlations and suggest better courses of action in the exploration of results and further queries. Furthermore, by using LLMs, the system should be able to supplement and support decisions using natural language.

More specifically, among the aims of SDF-FuzzIA:

- A **Fuzzy-based system**, as input data gathered by diverse sources are often vague, incomplete, and sometimes contradictory.

S. Destercke et al. (Eds.): SUM 2024, LNAI 15350, pp. 163–169, 2025.
https://doi.org/10.1007/978-3-031-76235-2_13

- The ability to interact with the system using LLMs or other affine generative algorithms, which is expected for a modern DSS by the current trends in artificial intelligence (AI).
- **Explainability** of the results and Introspection of the algorithms used, as the user should always be able to find the source of the suggestions and evaluate its trustworthiness and applicability [5].

In the rest of this short paper, a description of the components implemented in the system is given, followed by an explanation of the interaction between the components and the main system. As this is still ongoing work, future directions and possible hurdles are explored as well. Moreover, to the best of our knowledge, there are no similar works, hence we do not present any comparison to related works.

2 SDF Project

The SDF Project is designed to be a comprehensive tool for strategic decision-making towards sustainable development, addressing significant technical challenges through a modular and iterative development approach using modern techniques.

The proposed research and development project aims to create a prototype of a digital platform called the Sustainability Decision Framework. The platform consists of a web portal for stakeholder communication and an application solution offering digital information services. Moreover, it utilizes a Data Lakehouse architecture for storing and analysing structured, semi-structured, and unstructured data. SDF supports thousands of indicators and millions of values from international institutional and non-institutional entities, collected annually over a 10-year period, relevant to geography, environment, demographics, economy, labour, health, social issues, culture, politics, governance, and development cooperation for all countries, such as those relative to sustainable development as outlined by the United Nations Agenda 2030.

The project aims to be accessible via the web, support strategic decision-making, produce customizable country profiles, promote research activities, and facilitate dialogue among stakeholders through a shared information heritage.

Implementing the platform provides several technical and scientific challenges. Among them identifying and analysing international data sources and indicators, overcoming differences in taxonomy and metadata, creating a global database with uniform criteria despite heterogeneous taxonomies and metadata, and correcting value errors, and descriptive, predictive, and prescriptive models by correlating numerous indicators and sectors.

Characteristics and Performance. The SDF Platform encompasses a web portal for stakeholder communication and an application solution with Master Data and Transactions environments. It features advanced search capabilities via parameter selection and natural language queries, demand-based informational outputs for decision-making, and supports user-related dataset integration. The platform responds into multiple formats and languages, with a high querying capacity exceeding 10 queries per second.

3 SDF-FuzzIA

The system is mainly a hybrid Explainable Artificial Intelligence (XAI), which is composed of a high-performance and high-accuracy AI, and a Fuzzy-based system, linked as in Fig. 1.

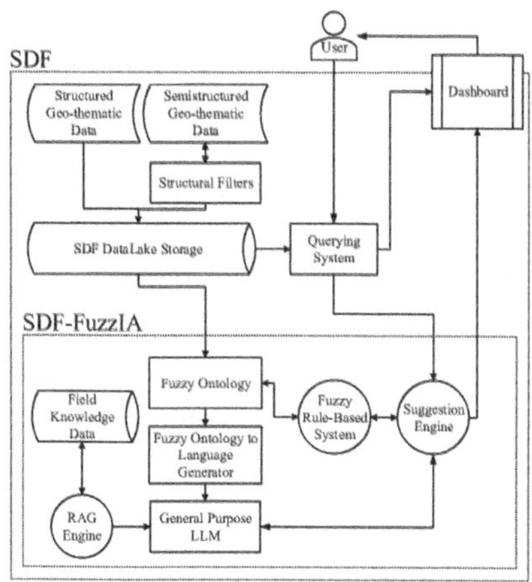

Fig. 1. The SDF-FuzzIA component embedded in the main SDF Decision Support System

The system is an open-access LLM based on Llama3 [13][1] with 70 billion parameters. This component is fine-tuned with semi-structured and unstructured open data from freely accessible sources. In particular, it will be trained on about 100 informative sources, 300 geographic areas, about 100 sectors (i.e., public administration, non-profit organization, and so on), about 800 topics (i.e., agriculture), about 25,000 indicators (i.e., the rate of female/male employer for each country), and more than 15 million indicators values.

It receives queries in natural language and provides answers in natural language together with an explanation to increase the model exportability and introspection. In particular, the explanation provided by the system is enriched with the knowledge given by the Fuzzy Ontology (described in detail in Paragraph *Fuzzy Ontology*) and the Fuzzy Rule-Based System (FRBS) components. Furthermore, the explanation is further enriched with the sources of the information used by the LLM to infer the explanation itself. Specifically, the system is instructed with fine-grained prompt engineering whose task is to provide the

[1] Introducing Meta Llama 3: The most capable openly available LLM to date.

user with traceable and searchable information. Prompt engineering is a technique used for generative AI to direct the system to produce outcomes that align with the desired specifications and exhibit a high level of quality.

General Purpose LLM. An LLM is a large-scale computational model, known for its capability to generate general-purpose language and perform other natural language processing tasks such as classification and question answering.

In particular, for the SDF-FuzzIA system, the LLM taken into account is Llama3, which is a decoder-only transformer with a tokenizer with a vocabulary of 2^{17} tokens, and a context window of 2^{12} tokens. We address the reader to [14] for the original definition of the transformer model, and to [8] for a general survey on transformers.

Taking into account the nowadays main publicly available LLMs (Gemma2[2], Llama3 [13][3], and Phi-3[4]), the Llama3 model was chosen after careful analysis and a set of qualitative comparisons with the domain expert in the geo-thematic field. In particular, the domain experts with the project provided a set of pairs of questions and answers in natural language ($n = 100$) related to geo-thematic data. Llama3 was the LLM that seemed to provide answers that were qualitatively more meaningful and better aligned with the answers provided for the purposes of the project. Specifically, the gold answers (base truth) provided by the domain experts were compared to the answers given by the different LLMs and LLama3 was the better performer.

In our system, the LLM works in a two-step phase. First, the LLM receives the user queries in natural language and generates a possible answer by using its internal knowledge. Then, the LLM uses the answer given by the FRBS in natural language to enrich and improve its initial answer. In both cases, the LLM is instructed by fine-grainer prompts aimed at providing explorable answers to the user. Finally, the enriched answer is provided to the Suggestion Engine (see Paragraph *Suggestion Engine*), which redirects the answer to the user Dashboard.

Field Knowledge Data and RAG Engine. Retrieval-Augmented Generation (RAG) [6][5] is an optimization process that enhances the capabilities of LLMs by referencing an authoritative knowledge base outside their training data before generating a response. This approach ensures that the output remains relevant, accurate, and useful in various contexts without the need for retraining, thereby enhancing the already advanced capabilities of LLMs in specific domains or an organization's internal knowledge base.

In our system, the database given by the Field Knowledge Data is used by the RAG Engine to optimize the LLM concerning the geo-thematic field. In particular, the LLM is refined with geo-thematic data (primarily semi-structured and unstructured data) from freely accessible sources selected by the domain

[2] Gemma 2 is now available to researchers and developers.
[3] Introducing Meta Llama 3: The most capable openly available LLM to date.
[4] Introducing Phi-3: Redefining what's possible with SLMs.
[5] What is Retrieval-Augmented Generation (RAG)?.

experts. This refinement enhances the model's capacity to provide context-aware information concerning the geo-thematic field and improves the intelligibility and trustworthiness of the responses provided to the user.

Fuzzy Ontology. Ontology can be defined as a structured representation of knowledge, comprising concepts, properties, relations, and axioms, which facilitates the utilisation of information by both humans and machines. It facilitates data integration, semantic research, system interoperability, and decision support. Classical ontologies frequently prove inadequate in addressing vagueness, a shortcoming that is addressed by fuzzy ontologies. Fuzzy ontologies build upon the foundations of classical ontologies by incorporating the principles of fuzzy logic. In particular, rules and constraints of fuzzy logic define (using axioms) the elements of the ontology. The application of Fuzzy ontologies allows for the representation of items with degrees of membership, rather than the binary classification typical of classical ontologies. For a more comprehensive treatment of these techniques, we direct the reader to [12]. Additionally, for an overview of the state of the art on this topic, we direct the reader to references [9] and [4].

The implementation of the SDF-FuzzIA ontology was carried out using Protégé (see [10]) with the help of two plugins: Cellfie (a Protégé Desktop plugin for importing spreadsheet data into OWL ontologies)[6] and Fuzzy OWL2 (a Protégé plugin for creating Fuzzy OWL 2 ontologies, see [3])[7]. In this context, the use of a fuzzy reasoner (e.g., DL2, see [2]) is considered necessary in such a scenario.

In light of the objectives put forth by SDF, the classes present in the ontology encompass both elements intrinsic to semantics about these themes (e.g., City, Country, Currency, etc.) and classes that facilitate the modelling of n-ary relationships between classes. As an example, the class "Distance" in the ontology represents items that possess the object properties "distance_from" and "distance_to," which link countries based on their distance. This distance is recorded as a data property "hasDistanceValue." The utility of a fuzzy ontology is demonstrated by the introduction of three subclasses, "LongDistance," "MidDistance," and "ShortDistance," which categorise distances as long, medium, or short, indicating the relative proximity of two countries.

Fuzzy Ontology to Language Generator. The Fuzzy Ontology to Language Generator (FO2L) component is used to transform the Fuzzy Ontology relations in natural language phrases that are used to supplement the LLM with context-aware information (see RAG in Paragraph *Field Knowledge Data and RAG Engine*) concerning the geo-thematic field. In particular, the rules used to make these transformations are given by the domain experts and are incorporated in this component. Moreover, the idea behind this component is given by [11], where the authors developed heuristics to convert user queries to ontology triples used to construct database queries.

[6] Cellfie plugin.
[7] FuzzyOWL2 plugin.

Fuzzy Rule-Based System. Fuzzy Rule-Based Systems (FRBSs) are rule-based systems that employ fuzzy sets and fuzzy logic for representing diverse forms of knowledge and for modelling the interactions and relationships existing between their variables. In the context of SDF-FuzzIA, the FRBS is used to provide semantically reliable answers to the user by employing the knowledge given by the Fuzzy Ontology. In particular, the rules employed by the FRBS exploit the indicators used in the Fuzzy Ontology to build a knowledge system.

The importance of this component is given by the necessity to develop an interpretable and reliable system (see, e.g., [7] for further details on the importance of reliable AI). Currently, AI models are still considered unreliable due to their inability to perform common-sense reasoning consistently. Even advanced models like GPT-4 exhibit these issues, as highlighted by the authors in their report [1]. Therefore, for the project's purposes, appropriate analyses are required to verify the reliability of the system's inference capabilities. Notably, these models can exhibit "hallucinations", providing grammatically correct and coherent responses that are factually incorrect or illogical. This phenomenon occurs because the LLMs predict the next most probable word based on the text statistics learned during training. While hallucinations allow LLMs to generate text across various application domains, enhancing the overall user experience depending on the context, they pose a threat to the reliability of such models in critical contexts. Hence, in scenarios requiring high reliability, it is necessary to develop hybrid systems where a LLM component provides highly accurate responses, and another component, such as a highly reliable model (e.g., Fuzzy Systems), explains why the LLM produced a particular response.

Suggestion Engine. The Suggestion Engine component receives the user queries in natural language and interacts with the FRBS and the LLM to build an explorable answer. In particular, the Suggestion Engine asks the FRBS for a semantically reliable answer, and the LLM for a high-accuracy answer. Then, it asks the LLM for a new answer based on the answer given by the FRBS and the previous answer given by the LLM itself. In this way, the final answer provided by the LLM is enriched with semantically reliable information concerning the geo-thematic context. Specifically, since the FRBS offers a structured set of rules and logic grounded in Fuzzy Ontology, it is possible to guide the LLM to conform to these principles when producing responses. By integrating the FRBS, the system can validate the LLM's outputs against established fuzzy rules, thereby ensuring that the responses are not only grammatically correct but also semantically and logically coherent.

4 Conclusions

SDF is an innovative decision support system in the geo-thematic field that utilizes SDF-FuzzIA, a hybrid AI system combining methodologies like LLM, Fuzzy Ontology, RAG, and Fuzzy Rule-Based systems. This hybrid approach aims to deliver accurate, explainable, and near real-time results by enhancing natural language interaction, domain knowledge integration, and semantic relevance. Although still under development, initial tests have shown promising results in replicating expert answers to specific questions. Challenges include managing computational complexity and maintaining response consistency as the data set expands.

References

1. OpenAI et al.: GPT-4 Technical Report (2024). arXiv:2303.08774
2. Bobillo, F., Delgado, M., Gómez-Romero, J.: DeLorean: a reasoner for fuzzy OWL 2. Expert Syst. Appl. **39**(1), 258–272 (2012). https://doi.org/10.1016/j.eswa.2011.07.016. ISSN 0957-4174
3. Bobillo, F., Straccia, U.: Fuzzy ontology representation using OWL 2. Int. J. Approximate Reasoning **52**(7), 1073–1094 (2011). https://doi.org/10.1016/j.ijar.2011.05.003. Selected Papers - Uncertain Reasoning Track - FLAIRS 2009. ISSN 0888-613X
4. Cross, V., Chen, S.: Fuzzy ontologies: state of the art revisited. In: Barreto, G.A., Coelho, R. (eds.) NAFIPS 2018. CCIS, vol. 831, pp. 230–242. Springer, Cham (2018). https://doi.org/10.1007/978-3-319-95312-0_20
5. Fernandez, A., et al.: Evolutionary fuzzy systems for explainable artificial intelligence: why, when, what for, and where to? IEEE Comput. Intell. Mag. **14**(1), 69–81 (2019). https://doi.org/10.1109/MCI.2018.2881645
6. Gao, Y., et al.: Retrieval-Augmented Generation for Large Language Models: A Survey (2024). arXiv:2312.10997
7. Landgrebe, J.: Certifiable AI. Appl. Sci. **12**(3) (2022). https://doi.org/10.3390/app12031050. ISSN 2076-3417
8. Lin, T., et al.: A survey of transformers. AI Open **3**, 111–132 (2022). https://doi.org/10.1016/j.aiopen.2022.10.001. ISSN 2666-6510
9. Manikandabalaji, Sivakumar, R.: Knowledge representation using fuzzy ontologies: a survey. IJCSNS **23**(12), 199 (2023)
10. Musen, M.A., Protégé Team: The protégé project: a look back and a look forward. AI Matters **1**(4), 4–12 (2015)
11. Ochieng, P.: PAROT: Translating natural language to SPARQL. Expert Syst. Appl. **176**, 114712 (2021). https://doi.org/10.1016/j.eswa.2021.114712
12. Straccia, U.: Foundations of Fuzzy Logic and Semantic Web Languages, 1st edn. Chapman and Hall/CRC (2013). https://doi.org/10.1201/b15460
13. Touvron, H., et al.: LLaMA: Open and Efficient Foundation Language Models (2023). arXiv:2302.13971
14. Vaswani, A., et al.: Attention is All You Need (2023). arXiv:1706.03762

On the Completeness and Complexity of Lifted Temporal Inference

Marcel Gehrke(✉)

Institute for Humanities-Centred Artificial Intelligence, University of Hamburg, Hamburg, Germany
marcel.gehrke@uni-hamburg.de

Abstract. For static lifted inference algorithms, completeness, i.e., domain liftability, is extensively studied. However, so far no domain liftability results for temporal lifted inference algorithms exist. In this paper, we contribute the first completeness and complexity analysis for a temporal lifted algorithm, the so-called lifted dynamic junction tree algorithm (LDJT), which is the only exact lifted temporal inference algorithm out there. To handle temporal aspects efficiently, LDJT uses conditional independences to proceed in time, leading to restrictions w.r.t. elimination orders. We show that these restrictions influence the domain liftability results and show that one particular case while proceeding in time, has to be excluded from FO^2. Additionally, for the complexity of LDJT, we prove that the lifted width is in even more cases smaller than the corresponding treewidth in comparison to static inference.

Keywords: Temporal Relational Inference · Completeness · Complexity

1 Introduction

For static lifted inference algorithms, completeness, which investigates domain liftability, i.e., an inference algorithm runs in time polynomial w.r.t. domain sizes [5], is extensively studied. Static lifted inference algorithms such as weighted first-order model counting (WFOMC), lifted variable elimination (LVE), or the lifted junction tree algorithm (LJT) are domain liftable for the FO^2 fragment (all models with at most two logical variables per parametric factor (parfactor)) [2,5,24]. To the best of our knowledge, so far no theoretical analyses w.r.t completeness and complexity results for temporal lifted inference algorithms exist. In this paper, we close this gap, by analysing the lifted dynamic junction tree algorithm (LDJT) [8] and identifying model classes for which temporal lifted query answering is guaranteed and give bounds for the query answering runtime.

Poole proposes LVE as an exact inference algorithm on relational models [20], which has been extended by de Salvo Braz by generalising lifted summing out [21] and by Milch et al. by introducing counting to lift certain computations where lifted summing out does not apply [16]. Taghipour et al. extend LVE into

S. Destercke et al. (Eds.): SUM 2024, LNAI 15350, pp. 170–185, 2025.
https://doi.org/10.1007/978-3-031-76235-2_14

the state of the art [23]. Taghipour et al. introduce completeness results for LVE with generalised counting [24]. Braun shows that the results also hold for LJT while answering multiple queries efficiently [2]. The completeness results also hold for WFOMC [5]. For WFOMC, these results have been extended in recent years mostly to include counting into the logic-based WFOMC [7,13,14,25,26]. Taghipour et al. present complexity results for LVE [23]. Braun extends these results to the case of answering multiple queries and LJT. There are approximate temporal relational algorithms [1,12,19], unfortunately without theoretical guarantees. Therefore, for static lifted algorithms, completeness and complexity analyses exist, but not for temporal lifted algorithms.

We contribute completeness and complexity results for LDJT, which uses (temporal) conditional independences to proceed in time. Specifically, we analyse: 1) the influence of using temporal conditional independences on the elimination order w.r.t. liftable models, 2) the completeness of LDJT, 3) the complexity of LDJT, and 4) how domain sizes influence the lifted width of LDJT in comparison to the treewidth of the propositional interface algorithm [17].

We show that using temporal conditional independences for inference leads to restrictions w.r.t. elimination orders of inference algorithms. Compared to known completeness results for static approaches, we show that these restrictions lead to an adjustment in the class of liftable models. For the completeness of LDJT, we prove that the liftability class has to be adjusted and prove a sufficient adjustment. In our complexity analysis, we show that compared to static inference, lifted temporal inference has even more advantages over the ground case. Increasing domain sizes may increase the treewidth, an exponential term in the complexity of the interface algorithm [17], while the lifted width of LDJT remains constant. Overall, with our completeness and complexity analysis, we give theoretical guarantees and show how crucial lifting is for temporal inference.

In the following, we recap parameterised probabilistic dynamic models (PDMs) as a representation for temporal probabilistic relational models and LDJT for query answering. Then, we show a type of model, where using temporal conditional independences leads to groundings. We use these insights to derive our completeness results. Further, we investigate the complexity of LDJT and set it into relation to propositional temporal algorithms and LJT. Lastly, we conclude.

2 Preliminaries

We shortly present all parts of LDJT and present the important bits in more detail in the appendix. LDJT uses PDMs as a temporal relational representation [11], which in turn are based on parameterised probabilistic models (PMs) [2].

2.1 Parameterised Probabilistic Models

We introduce parameterised random variables (PRVs) and parfactors to represent sets of indistinguishable random variable and factors.

Definition 1. *Let* $\mathbf{R}$ *be a set of random variable names,* $\mathbf{L}$ *a set of logical variable names,* Φ *a set of factor names, and* $\mathbf{D}$ *a set of constants (universe). All sets are finite. Each logical variable* L *has a domain* $\mathcal{D}(L) \subseteq \mathbf{D}$*. A* constraint *is a tuple* $(\mathcal{X}, C_{\mathbf{X}})$ *of a sequence of logical variables* $\mathcal{X} = (X^1, \ldots, X^n)$ *and a set* $C_{\mathcal{X}} \subseteq \times_{i=1}^{n} \mathcal{D}(X_i)$*. The symbol* $\top$ *for* C *marks that no restrictions apply, i.e.,* $C_{\mathcal{X}} = \times_{i=1}^{n} \mathcal{D}(X_i)$*. A* PRV $R(L^1, \ldots, L^n), n \geq 0$ *is a syntactical construct of a random variable name* $R \in \mathbf{R}$ *possibly combined with logical variable names* $L^1, \ldots, L^n \in \mathbf{L}$ *to represent a set of random variables. If* $n = 0$*, the PRV is parameterless and forms a propositional random variable. A PRV* A *or logical variable* L *under constraint* C *is given by* $A_{|C}$ *or* $L_{|C}$*, and yields a set of random variables or a set of constants respectively. We may omit* $|\top$ *in* $A_{|\top}$ *or* $L_{|\top}$*. The term* $\mathcal{R}(A)$ *denotes the possible values (range) of a PRV* A*. An* event $A = a$ *denotes the occurrence of PRV* A *with range value* $a \in \mathcal{R}(A)$*.*

Assume random variable names L, M, N, and O as well as $\mathbf{L} = \{X, Y\}$ with $\mathcal{D}(X) = \{x_1, x_2, x_3\}$ and $\mathcal{D}(Y) = \{y_1, y_2\}$. Using the names, we build boolean PRVs N, $L(X)$, $O(Y)$, and $M(X, Y)$. To set PRVs into relation, we use parfactors. A parfactor describes a function, mapping argument values to real values, of which at least one is non-zero.

Definition 2. *We denote a* parfactor g *by* $\phi(\mathcal{A})_{|C}$ *with* $\mathcal{A} = (A^1, \ldots, A^n)$ *a sequence of PRVs,* $\phi : \times_{i=1}^{n} \mathcal{R}(A^i) \mapsto \mathbb{R}^+$ *a function with name* $\phi \in \Phi$*, and* C *a constraint on the logical variables of* $\mathcal{A}$*, identical for all possible groundings of* $\mathcal{A}$ *w.r.t.* C *We may omit* $|\top$ *in* $\phi(\mathcal{A})_{|\top}$*. The term* $lv(Y)$ *refers to the logical variables in some element* Y*, a PRV, a parfactor or sets thereof. The term* $gr(Y_{|C})$ *denotes the set of all instances of* Y *w.r.t. constraint* C*. We call this process grounding. A set of parfactors forms a* model $G := \{g^i\}_{i=1}^{n}$*. The semantic of* G *is given by grounding and building a full joint distribution* P_G*. With* Z *as the normalisation constant,* G *represents* $P_G = \frac{1}{Z} \prod_{f \in gr(G)} f$*.*

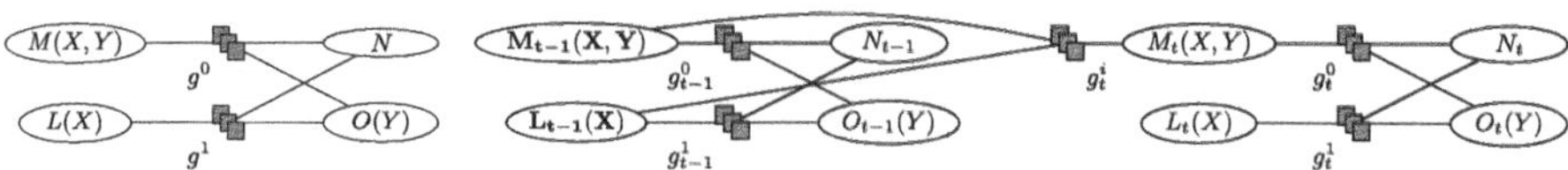

Fig. 1. Parfactor graph for our example G^{ex}

Fig. 2. $G^{ex}_{\rightarrow}$ the two-slice temporal parfactor graph for our example model G^{ex} (temporal copy pattern)

Let $G_{ex} = \{g^i\}_{i=0}^{1}$ with $g^0 = \forall y, x \in \mathcal{D}(Y) \times \mathcal{D}(X) : \phi^0(M(x, y), N, O(y))_{|\top}$ and $g^1 = \forall y, x \in \mathcal{D}(Y) \times \mathcal{D}(X) : \phi^1(L(x), N, O(y))_{|\top}$. The concrete input-output pairs are omitted. Figure 1 shows the graphical representation of our example PM G^{ex}, where ellipses are PRVs (representing a set of random variables) and boxes are parfactors (representing a set of factors). Each PRV represent multiple (ground) random variables and each parfactor represents multiple (ground) factors. Next, we add time to the model.

2.2 Adding Time to the Representation

Similar to a dynamic Bayesian network (DBN), there are two assumptions for PDMs, namely that the underlying process is stationary and, that the first-order Markov assumption holds. Thus, we define a PDM analogously to an DBN.

Definition 3. *A PDM G is a pair of PMs $(G_0, G_\rightarrow)$ where G_0 is a PM for the first time step and $G_\rightarrow$ is a two-slice temporal PM representing a PM for time slice $t-1$ and the same PM for time slice t. The time slices are connected by so-called* inter-*slice parfactors to model the temporal behaviour. G_0 can be extended using the temporal copy pattern $G_\rightarrow$ for any number of time steps. The semantics of G is to unroll G for T time steps resulting in a PM as defined above.*

Assume the initial model G_0 is the PM from Fig. 1 with time step 0 (and possibly priors) added. Figure 2 illustrates the temporal copy pattern, which consists of an PM for $t-1$ and one for t as well as an *inter*-slice parfactor g^i_t.

Having a PDM, we can ask queries on the model.

Definition 4. *Given a PDM G, a query term Q (ground PRV), and events $\mathbf{E}_{0:t} = \{E^i_t = e^i_t\}_{i,t}$, the expression $P(Q_t|\mathbf{E}_{0:t})$ denotes a* query *w.r.t. P_G.*

The problem of answering a query $P(A^i_\pi|\mathbf{E}_{0:t})$ w.r.t. the model is called *filtering* for $\pi = t$, *prediction* for $\pi > t$, and *hindsight* for $\pi < t$. In this paper, we analyse under which circumstances such queries are guaranteed to have a lifted solution as well as the complexity of such queries.

2.3 Query Answering Algorithm: LDJT

LDJT is an exact algorithm to efficiently answer multiple queries in a PDM [8]. LDJT constructs a so-called first-order junction tree (FO jtree) [3] from a PDM to answer multiple queries using LVE operations. An FO jtree has parameterised clusters (parclusters) as nodes. Similar to the propositional case, where the size of the clusters corresponds to the treewidth [6], we use parclusters in our complexity analysis for the lifted width. Parclusters divide a model into submodels using conditional independencies. LDJT performs a so-called message passing to distribute local information to prepare parclusters for query answering. Then, queries can be answered on these smaller parclusters.

Given the first-order Markov assumption of a PDM, LDJT can identify PRVs, which make one time step conditionally independent from the next. We call the set of these PRVs interface. The interface $\mathbf{I}_{t-1}$ consists of all PRVs from time step $t-1$, which occur in an *inter*-slice parfactors. In Fig. 2, the highlighted PRVs $M_{t-1}(X, Y)$ and $L_{t-1}(X)$ make up the interface $\mathbf{I}_{t-1}$. During the construction of LDJT's FO jtree structures, it ensures that one parcluster contains the PRVs of $\mathbf{I}_{t-1}$, which is called *in-cluster*, and constructs a parcluster containing $\mathbf{I}_t$, which is called *out-cluster*. To proceed in time, LDJT computes the (unnormalised) joint distribution over $\mathbf{I}_t$, which we call α_t message, and enters α_t in J_{t+1}.

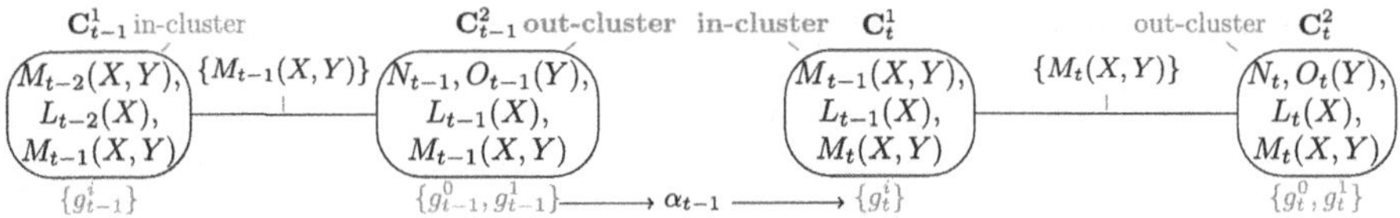

Fig. 3. J_{t-1} on the left and J_t on the right with the parclusters and messages depicted

Figure 3 depicts an FO jtree J_{t-1} for time step $t-1$ on the left and an FO jtree J_t for time step t on the right as well as the α_{t-1} message to make these two FO jtrees conditionally independent. LDJT computes the α_{t-1} message over $M_{t-1}(X,Y)$ and $L_{t-1}(X)$ using the *out-cluster* of J_{t-1} and sends α_{t-1} to the *in-cluster* of J_t. α_{t-1} makes $\mathbf{C}^2_{t-1}$ and $\mathbf{C}^1_t$ conditionally independent. Therefore, accounting for α_{t-1} in J_t makes J_t conditionally independent of all previous time steps. Similar arguments hold for messages within an FO jtree. During all computations LDJT tries to avoid groundings. To obtain a so-called lifted solution, LDJT aims to eliminate PRVs using lifted summing out, which eliminates PRVs using representatives and then efficiently accounts for ground eliminations, yielding tractability through exchangeability [18].

To be able to follow the completeness analysis, let us have a look at operators for lifted computations and briefly introduce when they are applicable as their applicability is crucial for the completeness results. Lifted summing out is the main operator to eliminate PRVs and besides other preconditions is applicable on a PRV A from a parfactor g if $lv(A) \supseteq lv(g)$ holds and if A only occurs in g. The latter precondition can easily be established using lifted multiplication. Further, counting can enable preconditions for lifted summing out [22].

To provide an idea of lifted summing out as well as counting, assume we only have g^1_{t-1} and want to eliminate $O_{t-1}(Y)$. To eliminate $O_{t-1}(Y)$ from g^1_{t-1}, LDJT cannot apply lifted summing out, as $L_{t-1}(X)$ is over the logical variable X ($\{Y\} \not\supseteq \{Y, X\}$). Thus, $O_{t-1}(Y)$ is not over all logical variables of g^1_{t-1}. In such a situation, LDJT can still obtain a lifted solution by counting the logical variable X. Instead of grounding, count-converting counts indistinguishable cases of the underlying random variables into a set of histograms to only have a polynomial instead of an exponential blowup. By counting, LDJT binds the counted logical variable and transforms PRVs, which are parameterised with that logical variable, into a set of histograms using the encoded symmetries, i.e., for a boolean PRV with a domain of two one would have three histograms. The three histograms would be: [2, 0], all true, [1, 1], one true and one false, and [0, 2], all false and each histogram maps to one potential. Hence, counting exploits indistinguishability to only have a polynomial blowup instead of an exponential. Counting has two preconditions. The first precondition is that the logical variable does not appear inside a counting formula in the parfactor. The second precondition is that the logical variable does not appear inside a constraint associated with a counting formula. So the logical variable is not allowed to occur in an inequality constraint with an already counted logical variable. For the second precondition, Taghipour introduces a merge count (merging a counted PRV with a PRV, whose logical variables are in an inequality constraint) operator to handle the second case. While PRVs are combined using lifted multiplication, counted (P)RVs are combined using a so-called merging operator. As the count-conversion binds X, the Y of $O_{t-1}(Y)$ is the only remaining logical variable in g^1_{t-1}. Thus, LDJT applies lifted summing out to eliminate $O_{t-1}(Y)$. For a more detailed introduction, as well as pre- and post-conditions, for lifted summing out and count-conversion, please refer to [22].

LJT and LDJT have to compute messages over sets of PRVs to make two parclusters conditionally independent from each other. Preserving PRVs for a message influences the elimination order, similar as in MAP queries, which in turn might lead to groundings. Having just one submodel consisting of the temporal copy pattern and all PRVs from $t-1$ in the interface is for us the case with the fewest restrictions, given our first-order Markov assumption. Actually, LJT and LDJT have operators to check for algorithm induced groundings, the so-called *fusion* [4] and *extension* [9,10] steps. These steps enlarge submodels and the interface in case conditional independencies lead to groundings and thereby can result in the case with the fewest restrictions. Thus, for the completeness analysis, we can just assume the worst case, which is computing the joint distribution of the the maximal interface on the complete temporal copy pattern, even if in reality the computations could use some conditional independences.

3 Temporal Elimination Order and Lifted Solutions

Next, we look at a case where LDJT has to ground for a model from FO^2. We use our example from Fig. 3 to derive a pattern, which might lead to groundings.

Example 1 (Groundings LDJT cannot prevent). Figure 3 depicts FO jtrees of $G^{ex}_{\rightarrow}$ for two time steps. Let us start by trying to proceed in time as depicted in Fig. 3. To compute α_{t-1}, the joint distribution over $M_{t-1}(X,Y)$ and $L_{t-1}(X)$, in $\mathbf{C}^2_{t-1}$, LDJT multiplies g^0_{t-1} and g^1_{t-1} as well as the message from $\mathbf{C}^1_{t-1}$. In the message from $\mathbf{C}^1_{t-1}$ the Y in $M_{t-1}(X,Y)$ is counted (the Y is counted to eliminate $L_{t-2}(X)$ using lifted summing out). Thus, to multiply the message with g^0_{t-1} and g^1_{t-1}, the Y also has to be counted in g^0_{t-1} and g^1_{t-1}. Unfortunately, LDJT cannot eliminate either N_{t-1} or $\#_Y O_{t-1}(Y)$ without having to ground. N_{t-1} or $\#_Y O_{t-1}(Y)$ do not have any free logical variables but M_{t-1} and L_{t-1} still have the free logical variable X, which cannot be counted as it appears in a counting formula in M_{t-1}. Thus, using these submodels and this interface, LDJT has to ground to compute α_{t-1} using temporal conditional independences.

Consider the fewest restrictions, the worst case from the previous section, i.e., use the maximal interface, $M_{t-1}(X,Y)$, $L_{t-1}(X)$, $O_{t-1}(Y)$ and N_{t-1}, as well as no submodels, which can be seen in our temporal copy pattern in Fig. 2. Given that model, LDJT has to compute a query $P(M_t(X,Y), L_t(X), O_t(Y), N_t)$, which is our α_t. The idea here is to delay the eliminations of N_{t-1} and $O_{t-1}(Y)$. In case LDJT can eliminate all PRVs from $t-1$ without having to ground, it is able to compute α_t in a lifted fashion. LDJT can eliminate $M_{t-1}(X,Y)$ by multiplying g^0_{t-1} and g^i and then applying lifted summing out. To eliminate either of N_{t-1}, $O_{t-1}(Y)$, or $L_{t-1}(X)$ LDJT first has to multiply the previous result with g^1_{t-1} Afterwards, LDJT has a parfactor over N_{t-1}, $O_{t-1}(Y)$, $L_{t-1}(X)$, and $M_t(X,Y)$. From that parfactor, LDJT needs to eliminate N_{t-1}, $O_{t-1}(Y)$, and $L_{t-1}(X)$. Lifted summing out can only be applied on $M_t(X,Y)$ as it is the only PRV with all the logical variables of the parfactor. Recall counting is only defined for one logical variable at a time. So LDJT cannot count X and Y of $M_t(X,Y)$. Nonetheless, LDJT can count either X or Y. In case X is counted, LDJT eliminates $O_{t-1}(Y)$ using lifted summing out. In case Y is counted, LDJT eliminates $L_{t-1}(X)$ using lifted summing out. However, in both cases, the remaining two (P)RVs cannot be eliminated in a lifted fashion, as in both cases $M_t(X,Y)$ still has one free logical variable. Therefore, LDJT has to ground $M_t(X,Y)$ before it can proceed with eliminating the remaining (P)RVs from $t-1$. Thus, to proceed in time, using the temporal conditional independences, LDJT has to ground.

The problem in Example 1 is that the PRV $M(X,Y)$ occurs for $t-1$ and t in the inter-slice parfactors. To compute the α_{t-1} message, LDJT has to preserve $M_{t-1}(X,Y)$. However, to eliminate other PRVs without groundings, LDJT has to eliminate $M_{t-1}(X,Y)$ first. Delaying the elimination of the other PRVs to J_t does not help either. Here, LDJT can eliminate $M_{t-1}(X,Y)$, but $M_t(X,Y)$ comes into play. Again, LDJT would have to eliminate $M_t(X,Y)$ before it can eliminate the remaining PRVs from time step $t-1$. However, $M_t(X,Y)$ cannot be eliminated as LDJT needs it to compute α_t.

The general problem here is that in the inter-slice parfactors there is a PRV with two logical variables for time step t and for time step $t+1$. Such a case might result in groundings. In the case that such a pattern leads to groundings, a lifted solution is possible by ignoring the temporal aspects, e.g., one can provide the unrolled PDM to LJT. That would result in a parcluster, which contains the PRV $M(X,Y)$ for all T time steps and therefore has a large lifted width. Nonetheless, by ignoring the temporal aspects of the model, a lifted solution is possible. LJT starts by eliminating $M(X,Y)$ for all time steps and then proceeds with other PRVs. Depending on domain sizes and time steps, either a lifted solution or accounting for temporal aspects is advantageous.

4 Completeness

In the following, we show completeness results for LDJT. Completeness investigates domain liftability of model classes, i.e., whether an inference algorithm runs in time polynomial w.r.t. domain sizes for certain model classes [5]. Models from the FO^2 fragment have at most two logical variables in each parfactor. These models are in this paper either PMs or PDMs. Braun shows that LJT is complete for models from FO^2 [2] and Taghipour et al. show that LVE with generalised counting is complete for models from FO^2 [24]. The completeness results also hold for other static lifted inference algorithms [5].

Corollary 1. *LVE and LJT are complete for any PDM G from* FO^2.

Proof. Unrolling G into a static model yields again a model from FO^2 and LVE as well as LJT are complete for models from FO^2. □

Hence, by not accounting for temporal conditional independences, LVE and LJT answer queries in polynomial time w.r.t. the domain size for any PDMs from FO^2. Let us now include temporal conditional independences for the completeness analysis and have a look at LDJT. For our completeness analysis of LDJT, we begin with a negative result, which directly follows from Example 1. Afterwards, we specify the model classes for which LDJT is complete and therefore can guarantee query answering results in a reasonable time, leading to the knowledge of tractable inference for (temporal) model classes.

Theorem 1. *LDJT is not complete for* FO^2.

Proof. Example 1 shows a model from FO^2 and LDJT cannot compute a lifted solution for that model. Hence, LDJT is not complete for all FO^2 models. □

So where is the difference between LVE and LDJT? In the static case, there is no restriction w.r.t. the elimination order and thus LVE can start by eliminating all PRVs with two logical variables, then proceed with PRVs with one logical variable and finally eliminate random variables. However, LDJT aims at handling temporal aspects efficiently, which is not given anymore by performing LVE on the unrolled model as temporal independences are not accounted for. For a large T, the unrolled model will be enormous. In our toy example, we have four PRVs. Assuming $1,000,000$ time steps, the unrolled model would have to store $4,000,000$ PRVs at once in memory instead of 8 at most. Additionally, if one would apply LJT on the unrolled model, the heuristic to construct its FO jtree, would most likely cluster all occurrences of $M_t(X, Y)$ together. Hence, it would have a single cluster with at least $1,000,000$ PRVs and the lifted width is dependent on the largest cluster. A large lifted width then also directly has a great influence on the corresponding runtime to perform inference. Therefore, LDJT trades in completeness to handle temporal aspects efficiently.

In maximum a posteriori (MAP) queries a similar problem occurs. In MAP queries, some PRVs have to be summed out and the remaining ones have to be maxed out, also leading to restrictions on the elimination order as some PRVs have to be preserved. MAP queries are also not complete for FO^2 [2].

Nonetheless, for the completeness of LDJT, only one distinct pattern in the *inter-slice* parfactors has to be excluded. Now, we prove for which model classes LDJT is complete. Let us call the class of models from FO^2 with inter-slice parfactors in which PRVs *from at most one time slice* are parameterised with at most two logical variables TFO^2. In the proof, we focus solely on the computations between time steps as the computations within time steps are complete.

Theorem 2. *LDJT is complete for models from TFO^2.*

Proof. We consider the three remaining cases, with logical variables, namely:

i) only PRVs with at most one logical variable in inter-slice parfactors,
ii) only PRVs with two logical variables for $t-1$ in inter-slice parfactors, and
iii) only PRVs with two logical variables for t in inter-slice parfactors.

Case i) is similar to the proof that LVE is complete for FO^2. To proceed in time, LDJT does not have to preserve any PRVs with two logical variables. Thus, LDJT can start by eliminating all PRVs with two logical variables from time step $t-1$. Afterwards, there are only PRVs with at most one logical variable from time step $t-1$. Now, LDJT can eliminate the remaining PRVs from time step $t-1$ that are not in the interface using the operators count-conversion (to ensure lifted summing out), merge-count (to merge a counted PRV and a PRV with inequality constraints), and merge (to merge two counted PRVs). The temporal message also does not cause to groundings in the next time step. The operators multiply (to multiply two PRVs) and merge ensure that multiplying α_{t-1} with the inter slice parfactors does not yield groundings. Next, all PRVs from $t-1$ can be eliminated without groundings, as the product contains only PRVs with at most one logical variable. Thus, LDJT and is complete for case i).

Case ii) means that at least one PRV from $t-1$ with two logical variables is in the interface. For the proof, we consider the worst case computational wise, i.e., all PRVs from $t-1$ in the interface. Therefore, LDJT does not have to eliminate any PRVs from $t-1$ to compute α_{t-1}. In time step t, α_{t-1} is initially multiplied with the inter-slice parfactors. Here, LDJT starts by eliminating all PRVs with two logical variables from time slice $t-1$ using lifted summing out (as we only consider FO^2 the preconditions of lifting summing out have to hold here). Given our case, we know that there are no PRVs with two logical variables for t in the inter-slice parfactors. Thus, LDJT can eliminate the remaining PRVs from time-slice $t-1$ with generalised counting. Now, there are only PRVs from t remaining. These PRVs from t also make up α_t. Thus, LDJT could eliminate all PRVs from $t-1$ without grounding and thereby compute α_t without having to ground. LDJT can use those to answer queries and here LDJT performs computations within a time step, which is proven to be complete for FO^2. Additionally, it can pass the remaining parfactors as the temporal message α_t to $t+1$, which as shown will not cause any grounding and therefore, LDJT solves this case in at most polynomial time w.r.t. domain sizes. Hence, LDJT is complete for case ii).

Case iii) is similar to case i). To compute temporal messages α_{t-1} there are no algorithm-induced groundings as LDJT does not have to preserve a PRV with two logical variables. Thus, there are no PRVs with two logical variables in α_{t-1}. In t, LDJT again starts by multiplying α_{t-1} with the inter-slice parfactors. Here, LDJT cannot directly eliminate all PRVs from $t-1$ as there is at least one PRV with two logical variables from t in the inter-slice parfactors, let us call this PRV A. However, A has not to be preserved to compute the next temporal message α_t. Thus, to compute α_t, LDJT can first eliminate A using lifted summing out. Afterwards, LDJT can eliminate all PRVs not in the interface, which includes all PRVs from $t-1$, using generalised counting as each of them has at most one logical variable. Hence, LDJT also computes the temporal messages in the third case without having to ground and is also complete for case iii).

So far, we only considered the forward pass to proceed in time. For a backward pass, to answer hindsight queries, the argumentation for case ii) and case iii) are exchanged. Therefore, overall LDJT is complete for all three cases, which means that LDJT is complete for models from TFO^2. □

Knowing that LDJT is complete for TFO^2, we now analyse another interesting model class, which has at most 1 logical variable in each PRV ($\mathcal{M}^{1prv}$).

Corollary 2. *LDJT is complete for* $\mathcal{M}^{1prv}$.

Proof. The proof directly follows from Theorem 2 and [22, Thm. 7.2]. Here, generalised counting ensures that LDJT is complete for $\mathcal{M}^{1prv}$. The rough idea of the proof is that all PRVs can be counted. Counting only leads to a polynomial and not an exponential blowup. After counting no free variables remain and from the polynomial representation they can be directly eliminated. With at most 1 logical variable for each PRV the problematic temporal pattern cannot occur. Thus, LDJT is complete for $\mathcal{M}^{1prv}$. □

In general, completeness results for relational inference algorithms assume liftable evidence. In case evidence breaks symmetries, query answering might not run in polynomial time but in exponential time w.r.t. domain sizes. Especially, for temporal models small differences in observations slowly ground a model. However, for conditioning with liftable evidence, the problem of evidence breaking symmetries over time has been dealt with by approximating symmetries [11]. Further, even if an algorithm is not complete for a certain class, the algorithm might still compute a lifted solution for some models of that class. Similar to static algorithms [23], LDJT can calculate a lifted solution for a 3-logical variable model, even though LDJT is not complete for 3-logical variable models. The same holds for models from FO^2 with inter-slice parfactors in which PRVs from both time slices are parameterised with two logical variables.

5 Complexity

For the complexity analysis, we have a look at the complexity of each step of LDJT, before we compile the overall complexity of LDJT. Each of these steps can also be found in the complexity analysis for LJT [2]. Therefore, we compare the complexity of LDJT to LJT and only analyse the differences.

5.1 LDJT Complexity

LDJT consists of the following steps: (i) FO jtree construction, (ii) evidence entering, (iii) message passing, and (iv) query answering. For LDJT, the complexity of the FO jtree construction is negligible. Compared to LJT [2], the complexity of the FO jtree construction of LDJT only differs in constant factors. The construction of the FO jtree structures J_0 and J_t is the very same as for LJT. The difference is that two FO jtrees are constructed. For *extension*, LDJT performs checks on two messages and twice the *fusion* step of LJT. The constant factors do not change the negligible complexity of the FO jtree construction [2].

For the complexity analysis of LDJT, we assume that the FO jtree structures of LDJT are minimal, i.e., they cease to be an FO jtree if anything was removed [3], and do not induce groundings. Further, we slightly change the definition of *lifted width* [2,24], as we now consider a PDM G and two FO jtrees, J_0 and J_t.

Definition 5. *Let* $w_{J_0} = (w_g^0, w_\#^0)$ *be the* lifted width *of* J_0 *and let* $w_{J_t} = (w_g^t, w_\#^t)$ *be the* lifted width *of* J_t. *The* lifted width w_J *of a pair* (J_0, J_t) *is a pair* $(w_g, w_\#)$, *where* $w_g = \max(w_g^0, w_g^t)$ *and* $w_\# = \max(w_\#^0, w_\#^t)$.

Further, T is the maximum number of time steps, k the largest lag[1], n is the largest domain size among $lv(G)$, $n_\#$ is the largest domain size of the counted logical variables, r is the largest range size in G, $r_\#$ is the largest range size among the PRVs in the counted RVs, and $n_J = max(n_{j_0}, n_{j_t})$ being the number of parclusters. The largest possible factor is given by $r^{w_g} \cdot n_\#^{w_\# \cdot r_\#}$. Hence, we always consider the largest width from either J_0 or J_t.

Evidence entering consists of absorbing evidence at each applicable node.

Lemma 1. *The complexity of absorbing an evidence parfactor is in*

$$O(T \cdot n_J \cdot \log_2(n) \cdot r^{w_g} \cdot n_\#^{w_\# \cdot r_\#}). \tag{1}$$

LDJT enters evidence in T FO jtrees, as there is evidence for up to T time steps.

Passing messages consists of calculating messages with LJT for every time step. Here, we consider the worst case, i.e., for each time step querying the first and last time step, the average case, i.e., *hindsight* and *prediction* queries with a constant lag, and the best case, i.e., only *filtering* queries.

Lemma 2. *The worst case complexity of passing messages is in*

$$O(T^2 \cdot n_J \cdot \log_2(n) \cdot r^{w_g} \cdot n_\#^{w_\# \cdot r_\#}). \tag{2}$$

The average case complexity of passing messages is in

$$O(k \cdot T \cdot n_J \cdot \log_2(n) \cdot r^{w_g} \cdot n_\#^{w_\# \cdot r_\#}). \tag{3}$$

The best case complexity of passing messages is in

$$O(T \cdot n_J \cdot \log_2(n) \cdot r^{w_g} \cdot n_\#^{w_\# \cdot r_\#}). \tag{4}$$

---

[1] Normally, the term lag is used for time differences in hindsight queries. We also use it for prediction queries.

The complexity of one complete message pass in an FO jtree, consists of calculating $2 \cdot (n_J - 1)$ messages and each message has a complexity of $O(\log_2 n \cdot r^{w_g} \cdot n_{\#}^{w_{\#} \cdot r_{\#}})$ [2]. One difference in LDJT compared to LJT is that LDJT needs to calculate $2 \cdot (n_J - 1) + 2$ messages for the current FO jtree, because LDJT calculates an α and a β message in addition to the normal message pass. For the FO jtrees used to answer *prediction* or *hindsight* queries, LDJT calculates at most $2 \cdot (n_J - 1) + 1$ messages, as LDJT calculates either an α or β message, respectively. In case during a backward or forward pass a particular FO jtree, J_t, is not used to answer any query, but is solely used to compute an α or respectively a β message, then computing $(n_J - 1)$ messages suffice for J_t. Hence, an efficient query answering plan reduces the complexity of message passing with constant factors. Next, we investigate the worst, average, and best case complexity of message passing as the complexity can drastically vary depending on the setting.

The worst case for LDJT is that for each time step, there is a query for the first and the last time step. Therefore, for each of the T time steps, LDJT needs to perform a message pass in all T FO jtrees, leading to $T \cdot T$ message passes. Hence, LDJT would perform a message pass for the current time step t, a backward pass from t to the first time step and a forward pass from t to the last time step. Thus, LDJT performs T message passes for each t, which overall leads to $T \cdot T$ message passes. The complexity of Eq. 2 is also the complexity of LJT given an unrolled FO jtree constructed by LDJT and evidence for each time step. However, with a huge T this scenario is rather far-fetched.

The average case for LDJT is that for each time step, LDJT answers a constant number of *hindsight* and *prediction* queries. Assume a set of queries is answered for each time step. Such a set could for example include queries with the largest difference $t - 10$ and $t + 15$. From this set the maximum lag k is 15. In general, *prediction* and *hindsight* queries are often close to the current time step and T can be huge ($k << T$). Overall, as an upper bound LDJT passes messages $2 \cdot k \cdot T$ times. The best case for LDJT are only *filtering* queries. That is to say, LDJT needs to perform exactly one message pass as well as compute the α message on each of the T FO jtrees.

The last step is *query answering*, which consists of finding a parcluster and answering a query, given a set of all queries, on an assembled submodel.

Lemma 3. *The complexity of answering a set of queries* $\{Q_k\}_{k=1}^m$ *is*

$$O(m \cdot \log_2(n) \cdot r^{w_g} \cdot n_{\#}^{w_{\#} \cdot r_{\#}}). \tag{5}$$

The complexity for query answering in LDJT is equal to the complexity in LJT.

We now combine the stepwise complexities to arrive at the complexity of LDJT by adding up the complexities in Eq. (1) to (5).

Theorem 3. *The worst case complexity of LDJT is*

$$O(((T^2 + T) \cdot n_J + m) \cdot \log_2(n) \cdot r^{w_g} \cdot n_{\#}^{w_{\#} \cdot r_{\#}}). \tag{6}$$

The average case complexity of LDJT is

$$O((k \cdot T \cdot n_J + m) \cdot \log_2(n) \cdot r^{w_g} \cdot n_{\#}^{w_{\#} \cdot r_{\#}}). \tag{7}$$

The best case complexity of LDJT is

$$O((T \cdot n_J + m) \cdot \log_2(n) \cdot r^{w_g} \cdot n_{\#}^{w_{\#} \cdot r_{\#}}). \tag{8}$$

In the complexity T^2, r^{w_g}, and $n_{\#}^{w_{\#} \cdot r_{\#}}$ can become rather huge. The r^{w_g} corresponds to the largest ground intermediate result, which one just has to compute. But LDJT just has to compute it once for every indistinguishable group instead of computing it for each object. For huge T's always querying the first and the last time step is really far-fetched in practice (predicting far into the future is as good as guessing). Further, predictions could be implemented more efficiently in LDJT by just computing the current state and projecting into the future by only once computing the temporal behaviour, see [15]. However, we decided against including this in this analysis, as LDJT is not implemented to use this fast prediction and it would make the complexity analysis even more involved. Thus, the complexity of the average case should be a good upper bound in most practical settings.

The term $n_{\#}^{w_{\#} \cdot r_{\#}}$ is only polynomial w.r.t. domain sizes (instead of exponential in the ground case). Normally, the domains become large, while the range values are constant and relative small, so only the base case increases, but not the exponent. Thus, even though the terms T^2 and $n_{\#}^{w_{\#} \cdot r_{\#}}$ can become quite large, they are still relative small in comparison to the complexity in the ground case. Let us now compare the complexity of LDJT to LJT [2] as well as to the propositional interface algorithm [17].

5.2 LJT Comparison

In this subsection, we compare the complexity of LJT and LDJT for a given PDM. In case LJT uses the very same heuristic to construct its FO jtree as LDJT does, i.e., unroll LDJT's structure for T time steps, LJT has the worst case complexity of LDJT. Each message pass of LJT, after entering new evidence, is over all T FO jtrees. Therefore, the worst case complexity of LDJT is the best LJT can achieve. Further, for LDJT for *hindsight* and *prediction* queries, computing (n_J-1) messages instead of $2\cdot(n_J-1)$ messages sometimes suffices. A highly adapted and efficient implementation of LJT could achieve that reduction at best only for *prediction* queries, but cannot achieve that for *hindsight* queries. Thus, LJT can only be as efficient as LDJT in the worst case.

The other possibility is that LJT uses any static heuristic to build its FO jtree. In this case, LJT would most likely have fewer, but larger parcluster. The number of parclusters is only a linear factor for LJT. However, the size of the parclusters directly influences the lifted width. Increasing the lifted width results in increasing an exponential term. Thus, the complexity of LJT, using a random (static) heuristic, will most likely be higher than the worst case complexity of LDJT.

5.3 Interface Algorithm Comparison

In this subsection, we have a closer look at the lifted width for temporal inference. We begin by looking at the speed-ups in the static case and then continue by showing that there is an additional speed-up for the temporal case. For LJT compared to the junction tree algorithm, the speed up is twofold. The first speed-up is that LJT has fewer nodes in an FO jtree J than the corresponding junction tree has, i.e., $n_{gr(J)} >> n_J$. The other speed up is that under the presence of counting, the lifted width is smaller than the treewidth.

Taghipour et al. show that in models that do not require count-conversions, the lifted width is equal to the treewidth [23]. Therefore, without count-conversions the w_g of the lifted width is equal to the treewidth of a junction tree. For temporal inference, the lifted width of LDJT often is much smaller than the treewidth of the interface algorithm [17], even without counting.

Theorem 4. *The lifted width remains constant while the treewidth increases given increasing domain sizes with at least one PRVs in the interface.*

Proof. The interface algorithm ensures that all interface random variables occur together in one cluster of a jtree. Therefore, the corresponding jtree has at least $|gr(\mathbf{I}_t)|$ random variables in a cluster. Thus, the treewidth depends on the domain sizes of interface PRVs. The w_g (which corresponds to the treewidth) of the lifted width w_J is independent of the domain sizes. Hence, the lifted width remains constant while the treewidth increases with domain sizes. □

For LVE and LJT without counting the lifted width and the treewidth are identical [23]. For LDJT even without counting, the lifted width can be drastically smaller than the treewidth and propositional inference is exponential w.r.t. the treewidth [6]. For LDJT, increasing domain sizes (without counting) only leads to increasing a logarithmic factor. Thus, LDJT can answer queries for large(r) domain sizes, which is infeasible in the propositional case due to an exponential blowup. Overall, lifting is necessary for temporal inference.

6 Conclusion

In this paper, we analyse completeness and complexity for LDJT. To the best of our knowledge, this is the first such analysis for a temporal lifted inference algorithm. LDJT uses temporal conditional independences, which induces restrictions on the elimination order. Based on these restrictions, we show that an adjustment w.r.t. the liftable model classes is necessary and prove a sufficient adjustment to the classes of liftable models w.r.t. completeness for static lifted inference algorithms. We only have to exclude one special case in the inter-slice parfactors. Thereby, we show that LDJT can guarantee a lifted solution for many models and thereby compute solutions in a reasonable time. For nearly all (realistic) scenarios, LDJT has a complexity linear to the maximum number of time steps, which is the desired behaviour for an exact temporal inference algorithm. Additionally, we also show how crucial lifting is for exact temporal inference. Here, even without counting, the lifted width is much smaller compared to the corresponding treewidth. For static algorithms, without counting, the lifted width is equal to the treewidth. With recent advances in preserving lifted models over time [11], LDJT preserves the advantages of lifting while answering conditioning queries.

Knowing the benefits of a lifted temporal inference algorithm, the next steps contain investigating how other sequential problems such as decision making can benefit from lifting. Further, we want to find an approach to generalise counting even further to make LDJT complete for FO^2.

References

1. Ahmadi, B., Kersting, K., Mladenov, M., Natarajan, S.: Exploiting symmetries for scaling loopy belief propagation and relational training. Mach. Learn. **92**(1), 91–132 (2013)
2. Braun, T.: rescued from a sea of queries: exact inference in probabilistic relational models. Ph.D. thesis (2020)
3. Braun, T., Möller, R.: Lifted junction tree algorithm. In: Friedrich, G., Helmert, M., Wotawa, F. (eds.) KI 2016. LNCS (LNAI), vol. 9904, pp. 30–42. Springer, Cham (2016). https://doi.org/10.1007/978-3-319-46073-4_3
4. Braun, T., Möller, R.: Preventing groundings and handling evidence in the lifted junction tree algorithm. In: Kern-Isberner, G., Fürnkranz, J., Thimm, M. (eds.) KI 2017. LNCS (LNAI), vol. 10505, pp. 85–98. Springer, Cham (2017). https://doi.org/10.1007/978-3-319-67190-1_7
5. Van den Broeck, G.: On the completeness of first-order knowledge compilation for lifted probabilistic inference. In: Proceedings of the NIPS11 (2011)
6. Darwiche, A.: Modeling and Reasoning with Bayesian Networks. Cambridge University Press, Cambridge (2009)
7. Dilkas, P., Belle, V.: Synthesising recursive functions for first-order model counting: challenges, progress, and conjectures. In: Proceedings of the KR23 (2023)
8. Gehrke, M., Braun, T., Möller, R.: Lifted dynamic junction tree algorithm. In: Chapman, P., Endres, D., Pernelle, N. (eds.) ICCS 2018. LNCS (LNAI), vol. 10872, pp. 55–69. Springer, Cham (2018). https://doi.org/10.1007/978-3-319-91379-7_5
9. Gehrke, M., Braun, T., Möller, R.: Preventing unnecessary groundings in the lifted dynamic junction tree algorithm. In: Mitrovic, T., Xue, B., Li, X. (eds.) AI 2018. LNCS (LNAI), vol. 11320, pp. 556–562. Springer, Cham (2018). https://doi.org/10.1007/978-3-030-03991-2_51
10. Gehrke, M., Braun, T., Möller, R.: Towards preventing unnecessary groundings in the lifted dynamic junction tree algorithm. In: Trollmann, F., Turhan, A.-Y. (eds.) KI 2018. LNCS (LNAI), vol. 11117, pp. 38–45. Springer, Cham (2018). https://doi.org/10.1007/978-3-030-00111-7_4
11. Gehrke, M., Möller, R., Braun, T.: Taming reasoning in temporal probabilistic relational models. In: Proceedings of the ECAI 2020, pp. 2592–2599 (2020)
12. Geier, T., Biundo, S.: Approximate online inference for dynamic Markov logic networks. In: Proceedings of the ICTAI11, pp. 764–768. IEEE (2011)
13. Kazemi, S.M., Kimmig, A., Van den Broeck, G., Poole, D.: New liftable classes for first-order probabilistic inference. In: Proceedings of NIPS16 (2016)

14. Kuzelka, O.: Counting and sampling models in first-order logic. In: Proceedings of the IJCAI-23, pp. 7020–7025 (2023)
15. Marwitz, F.A., Möller, R., Gehrke, M.: PETS: predicting efficiently using temporal symmetries in temporal PGMs. In: Bouraoui, Z., Vesic, S. (eds.) ECSQARU 2023. LNCS, vol. 14294, pp. 322–333. Springer, Cham (2023). https://doi.org/10.1007/978-3-031-45608-4_24
16. Milch, B., Zettlemoyer, L.S., Kersting, K., Haimes, M., Kaelbling, L.P.: Lifted probabilistic inference with counting formulas. In: Proceedings of the AAAI08 (2008)
17. Murphy, K.P.: Dynamic Bayesian networks: representation, inference and learning. Ph.D. thesis, University of California, Berkeley (2002)
18. Niepert, M., Van den Broeck, G.: Tractability through Exchangeability: a new perspective on efficient probabilistic inference. In: Proceedings of the AAAI14 (2014)
19. Papai, T., Kautz, H., Stefankovic, D.: Slice normalized dynamic Markov logic networks. In: Proceedings of the NIPS12, pp. 1907–1915 (2012)
20. Poole, D.: First-order probabilistic inference. In: Proceedings of the IJCAI03 (2003)
21. de Salvo Braz, R.: Lifted first-order probabilistic inference. Ph.D. thesis, Ph. D. Dissertation, University of Illinois at Urbana Champaign (2007)
22. Taghipour, N.: Lifted probabilistic inference by variable elimination. Ph.D. thesis, Ph. D. Dissertation, KU Leuven (2013)
23. Taghipour, N., Davis, J., Blockeel, H.: First-order decomposition trees. In: Proceedings of the NIPS13, pp. 1052–1060. Curran Associates Inc. (2013)
24. Taghipour, N., Fierens, D., Van den Broeck, G., Davis, J., Blockeel, H.: Completeness results for lifted variable elimination. In: Proceedings of AISTATS13 (2013)
25. Van Bremen, T., Kuželka, O.: Faster lifting for two-variable logic using cell graphs. In: Uncertainty in Artificial Intelligence, pp. 1393–1402. PMLR (2021)
26. Van Bremen, T., Kuželka, O.: Lifted inference with tree axioms. Artif. Intell. **324**, 103997 (2023)

Bel Coalitional Games

Michel Grabisch[1] and Silvia Lorenzini[2(✉)]

[1] Paris School of Economics, Université Paris I Panthéon-Sorbonne Centre d'Economie de la Sorbonne, Paris, France
michel.grabisch@univ-paris1.fr

[2] Department of Economics, University of Perugia, Perugia, Italy
silvia.lorenzini@dottorandi.unipg.it

Abstract. We introduce Bel coalitional games, that generalize classical coalitional games, where uncertainty is modelled through the Dempster-Shafer theory and every agent can have different knowledge. We propose the notion of contract in our framework, that specifies how agents divide the values of the coalitions and we use the Choquet integral to model the agents' preferences between contracts. Next, we study the core under two different moments of the game by defining the ex-ante core and the ex-t-interim core, where, in the latter, we need the Dempster conditional rule to update the mass functions of agents. In particular, in the last step of the ex-t-interim case and when the set of states reduces to singleton, i.e. when there is no uncertainty, we recover the classical definition of the core. Finally, we show some results about the ex-ante and the ex-t-interim core of Bel coalitional games, following the well-known results about classical coalitional games.

Keywords: Belief functions · Choquet expectation · Bel Coalitional Games

1 Introduction

Bayesian coalitional games, introduced in [8], generalize classical coalitional games, by assuming uncertainty in the framework. Our aim is to define a more general framework using the Dempster-Shafer theory [4,13], i.e., by considering belief functions instead of a probability measure. The choice of the Dempster-Shafer theory is motivated by the fact that it offers a well-established and flexible framework for uncertainty representation, suitable for cardinal information (in contrast to possibility theory, which is more suitable to ordinal information). In classical coalitional games, we assume that the payoff to each coalition is given by a fixed deterministic value and that these values are common knowledge among all agents, but both assumptions often fail to hold for real-world problems. In order to study problems of cooperation under uncertainty, we may use Bayesian coalitional games. However, the Bayesian hypothesis is strong and requires a good knowledge of the environment. For example, if there is a lack of information, the Bayesian approach may result in a model that does not align

S. Destercke et al. (Eds.): SUM 2024, LNAI 15350, pp. 186–200, 2025.
https://doi.org/10.1007/978-3-031-76235-2_15

with the agents' behavior, as illustrated by Ellsberg's paradox [6]. In the present paper, we propose a new kind of game with ambiguous information, which we call Bel coalitional game, where agents can have their own knowledge that is represented by belief functions. For non-cooperative games we have that Bel games [12] generalize Bayesian games, which are recovered when the mass function is a probability distribution. In this case Bel coalitional games do not constitute a true generalization of Bayesian coalitional games, in fact if we assume that all agents have same belief function and such belief function is a probability measure, then we obtain a particular case of Bayesian coalitional game, in which agents have same information sets composed of singletons. The paper is structured as follows. Section 2 recalls the basic notions of Dempster-Shafer theory, Choquet integration and coalitional games. The definition of Bel coalitional games is provided in Sect. 3, followed by an introduction to the notions of contract and core at two different stages of the game (ex-ante and ex-t-interim) in Sect. 4. Finally, Sect. 5 studies the non-emptiness of the core in Bel coalitional games by analyzing different scenarios of agents' knowledge, while Sect. 6 draws our conclusions.

Proofs are omitted due to the limited number of pages.

2 Preliminaries

2.1 Dempster-Shafer Theory

Let $\Omega = \{\omega_1, \ldots, \omega_d\}$ be a finite non-empty set of states of the world and denote by 2^Ω and $\mathbb{R}^\Omega$ the power set of Ω and the set of all random variables on Ω, respectively. A *belief function* (see [4,13]) is a set function $\nu : 2^\Omega \to [0,1]$ satisfying:

(i) $\nu(\emptyset) = 0$ and $\nu(\Omega) = 1$;

(ii) $\nu\left(\bigcup_{i=1}^{k} E_i\right) \geq \sum_{\emptyset \neq I \subseteq \{1,\ldots,k\}} (-1)^{|I|+1} \nu\left(\bigcap_{i\in I} E_i\right)$, for all $k \geq 2$ and $E_1, \ldots, E_k \in 2^\Omega$.

Notice that a probability measure π on 2^Ω is a particular belief function where *(ii)* holds as an equality.

A belief function ν is associated with a dual set function $\overline{\nu}$ on 2^Ω called *plausibility function* and defined, for all $A \in 2^\Omega$, as $\overline{\nu}(A) = 1 - \nu(A^c)$. Both belief function and plausibility function are completely characterized by their *Möbius inverse* (or *mass function*), that is, a set function $m_\nu : 2^\Omega \to [0,1]$ such that

(i) $m_\nu(\emptyset) = 0$;

(ii) $\sum_{A\subseteq\Omega} m_\nu(A) = 1$.

A subset of Ω with strictly positive mass function is called a *focal element* and we denote by $\mathcal{F}_\nu = \{E \in 2^\Omega : m_\nu(E) > 0\}$ the set of focal elements of ν. We have that, for all $A \in 2^\Omega$, it holds:

$$\nu(A) = \sum_{B \subseteq A} m_\nu(B) \quad \text{and} \quad \overline{\nu}(A) = \sum_{B \cap A \neq \emptyset} m_\nu(B).$$

Given a belief function ν and $X \in \mathbb{R}^\Omega$, the *Choquet expectation* of X with respect to ν is defined through the Choquet integral [7]

$$\mathbb{C}_\nu(X) = \oint X \, d\nu = \sum_{i=1}^{d} [X(\omega_{\sigma(i)}) - X(\omega_{\sigma(i+1)})] \nu(E_i^\sigma),$$

where σ is a permutation of $\{1, \dots, d\}$ such that $X(\omega_{\sigma(1)}) \geq \dots \geq X(\omega_{\sigma(d)})$, $E_i^\sigma = \{\omega_{\sigma(1)}, \dots, \omega_{\sigma(i)}\}$, for $i = 1, \dots, d$, and $X(\omega_{\sigma(d+1)}) = 0$. When ν reduces to a probability π, we have that $\mathbb{C}_\pi(X) = \mathbb{E}_\pi(X)$. We also have that every belief function ν is in one-to-one correspondence with the closed and convex set called *core* and defined by

$$\text{core}(\nu) = \{\pi : \pi \text{ is a probability measure on } \Omega, \, \pi(A) \geq \nu(A), \, \forall \, A \in 2^\Omega\}.$$

Finally, we recall that the Choquet expectation in the framework of Dempster-Shafer theory can be expressed both in terms of core(ν) and m_ν (see, e.g., [7]): for all $X \in \mathbb{R}^\Omega$

$$\mathbb{C}_\nu(X) = \min_{\pi \in \text{core}(\nu)} \int X \, d\pi = \sum_{B \in \mathcal{F}_\nu} \left(\min_{\omega \in B} X(\omega) \right) m_\nu(B)$$

$$\mathbb{C}_{\overline{\nu}}(X) = \max_{\pi \in \text{core}(\nu)} \int X \, d\pi = \sum_{B \in \mathcal{F}_\nu} \left(\max_{\omega \in B} X(\omega) \right) m_\nu(B).$$

We conclude this subsection with some useful properties of the Choquet integral with respect to a belief function ν:

- *positive homogeneity* (it holds also for $\overline{\nu}$): $\oint \alpha X \, d\nu = \alpha \oint X \, d\nu$, with $\alpha \geq 0$;
- *asymmetry*: $\oint - X \, d\nu = - \oint X \, d\overline{\nu}$;
- *monotonicity* (or *nondecreasingness*) with respect to the integrand (it holds also for $\overline{\nu}$):
$$X \leq Y \implies \oint X \, d\nu \leq \oint Y \, d\nu;$$
- *subadditivity*: $\oint X \, d\overline{\nu} + \oint Y \, d\overline{\nu} \geq \oint X + Y \, d\overline{\nu}$;
- *superadditivity*: $\oint X \, d\nu + \oint Y \, d\nu \leq \oint X + Y \, d\nu$;
- *comonotonic additivity* (it holds also for $\overline{\nu}$): if X and Y are comonotonic, i.e., there is no $\omega, \omega' \in \Omega$ such that $X(\omega) < X(\omega')$ and $Y(\omega) > Y(\omega')$, then
$$\oint (X + Y) \, d\nu = \oint X \, d\nu + \oint Y \, d\nu.$$

In particular we say that a random vector $X = (X_1, \ldots, X_n)$ is comonotonic if its components are pairwise comonotonic. In this case we have that

$$\oint \sum_{i=1}^{n} X_i \, d\nu = \sum_{i=1}^{n} \oint X_i \, d\nu.$$

2.2 Coalitional Games

Coalitional games' aim is the study of payoff division within groups of agents. A game assigns to each group of agents, referred to as a coalition, a set of possible payoffs.

Definition 1. *A coalitional game is a pair (N, v) where*

- $N = \{1, 2, \ldots, n\}$ *is a set of agents;*
- $v : 2^N \to \mathbb{R}$ *is a function that assigns to each group of agents $S \subseteq N$ a real-valued payoff.*

If the context makes N clear, we can refer to the coalitional game simply as v. An outcome or payoff vector $\boldsymbol{x}$ is a vector in $\mathbb{R}^N$ defining the payoff given to any agent in N.

Definition 2. *The core of a coalitional game (N, v) is defined by*

$$C(N, v) = \{\boldsymbol{x} \in \mathbb{R}^N : \sum_{i \in S} \boldsymbol{x}_i \geq v(S), \, \forall\, S \subseteq N, \, \sum_{i \in N} \boldsymbol{x}_i = v(N)\}.$$

Let us underline that the core of a coalitional game can be empty, meaning that there may be no payoff vector that satisfies the stated condition.

3 Bel Coalitional Games

To start with, let us recall the definition of Bayesian coalitional games [8]. First, we need to discuss what it means for a coalition to prefer one distribution of payoff to another. Let us underline that with distribution of payoff $\boldsymbol{x} = (\boldsymbol{x}_j)_{j \in S}$ we mean $\boldsymbol{x} : \Omega \to \mathbb{R}^S$.

Definition 3. *Let $\boldsymbol{x} = (\boldsymbol{x}_j)_{j \in S}$ and $\boldsymbol{y} = (\boldsymbol{y}_j)_{j \in S}$, with $\boldsymbol{x}, \boldsymbol{y} : \Omega \to \mathbb{R}^S$, be two distributions of payoffs for agents in coalition S. Let the preferences of the agents be $(\succcurlyeq_j)_{j \in S}$. We say that S weakly prefers $\boldsymbol{x}$ to $\boldsymbol{y}$, $\boldsymbol{x} \succcurlyeq_S \boldsymbol{y}$, if for all agents j, $\boldsymbol{x}_j \succcurlyeq_j \boldsymbol{y}_j$. We say that S strictly prefers $\boldsymbol{x}$ to $\boldsymbol{y}$, $\boldsymbol{x} \succ_S \boldsymbol{y}$, if the preferences are strict for all agents.*

Now we can present Bayesian coalitional games.

Definition 4. *A Bayesian coalitional game is given by $(N, \Omega, \pi, (\mathcal{I}_j)_{j \in N}, (\succcurlyeq_j)_{j \in N})$ where*

- $N = \{1, 2, \ldots, n\}$ *is a set of agents;*
- $\Omega = \{\omega_1, \omega_2, \ldots, \omega_d\}$ *is a set of possible worlds, where each world specifies a coalitional game defined over* N*;*
- π *is a common prior over the set of worlds* Ω*;*
- $\mathcal{I}_j$ *is agent* j*'s partition of the worlds in* Ω*;*
- $\succcurlyeq_j$ *describes agent* j*'s preference over distributions of payoffs, for each agent* j.

We can understand the Bayesian framework as follows. Imagine there is a set of potential scenarios or worlds, denoted as Ω, in which a coalitional game may unfold. These worlds are sampled from a probability distribution, π, which is shared knowledge among all participants. However, from each agent's perspective, they have specific subsets of worlds that are indistinguishable based on their available information. These subsets are contained in the information set $\mathcal{I}_j$ and collectively they form a partition of Ω. Importantly, these information sets are known by all agents and constitute common knowledge. To better model the belief of players, that in the real-word problems can be different and ambiguous, we consider a mass function m_{ν_j} for each player j, given by a belief function ν_j, defined on 2^Ω, and we do not impose that the information set $\mathcal{I}_j$ of player j has to be a partition of Ω, for every player j, but we just assume that Ω has to be covered.

Definition 5. *A Bel coalitional game is given by* $(N, \Omega, (m_{\nu_j})_{j\in N}, (\mathcal{I}_j)_{j\in N}, v_\omega)$ *where*

- $N = \{1, 2, \ldots, n\}$ *is a set of agents;*
- $\Omega = \{\omega_1, \omega_2, \ldots, \omega_d\}$ *is a set of possible worlds, where each world specifies a coalitional game* v_ω *defined over* N*;*
- $m_{\nu_j} : 2^\Omega \to [0, 1]$ *is agents* j*'s beliefs over the worlds* Ω*;*
- $\mathcal{I}_j$ *is agent* j*'s information set of the worlds* Ω *and we impose that it has to cover* Ω *and* $\mathcal{I}_j = \mathcal{F}_{\nu_j}$*, where* $\mathcal{F}_{\nu_j}$ *is the set of focal elements of* ν_j*;*
- $v_\omega : 2^N \to \mathbb{R}$ *is a coalitional game that assigns to each coalition* $S \subseteq N$ *its value in world* ω.

In our case, the agent's preferences are induced by the Choquet expectation, which is fully defined by m_{ν_j}, therefore we do not explicitly include it in the previous definition.

4 Contracts

In Bayesian and Bel coalitional games we have that agents may not know the exact value of a coalition, so it is not possible to specify a precise payoff to an agent in advance before uncertainty is resolved. Instead, as described in [8], we suppose that agents enter into agreements (assumed to be exogenously given), called *contracts*, about how to divide the values of the coalitions once the true state of the world is revealed to all agents.

Definition 6. *A contract among agents of coalition $S \subseteq N$ (S-contract) is a mapping from the set of worlds to payoff vectors, $\mathbf{c}^S : \Omega \to \mathbb{R}^S$, such that $\mathbf{c}_j^S(\omega)$ denotes the payoff to agent $j \in S$ in world ω. A contract is feasible if for all worlds ω*

$$\sum_{j \in S} \mathbf{c}_j^S(\omega) \leq v_\omega(S).$$

For individual agents contracts play no role and they simply receive their own payoff in each world. As in [8], we focus only on feasible contracts and we study the stability of N-contracts (grand contracts). The agent's perception of the desirability of a contract varies depending on when this assessment occurs. In our case we use the Choquet integral to model agents' preferences, hence we have that $\mathbf{c}^S \succcurlyeq_S \mathbf{c'}^S$, if for all agents $j \in S$, $\mathbb{C}_{\nu_j}(\mathbf{c}_j^S) \geq \mathbb{C}_{\nu_j}(\mathbf{c'}_j^S)$. Of course, we have that $\mathbf{c}^S \succ_S \mathbf{c'}^S$, if the preferences are strict for all agents. We find interesting to examine two scenarios: when the contract is evaluated before a world is drawn (we call the situation *ex-ante*) and when the contract is evaluated after a world is drawn and each agent is made aware of what possible world it could be (we call it *ex-t-interim*). More precisely, at each step t, a potential world is eliminated until only one remains and, consequently, we have an update of the mass functions of players at every step $t = 1, \ldots, T$, where $T = |\Omega| - 1$ is the maximal number of steps. Hence, at the end, players know the true state of world.

4.1 Ex-ante

In the ex-ante contract the distribution of payoffs to agent j is equal to $\mathbf{c}_j^S$. Let us give the notion of blocking for this particular contract, i.e., when a coalition is dissatisfied with a grand contract.

Definition 7. *Given a Bel coalitional game $(N, \Omega, (m_{\nu_j})_{j \in N}, (\mathcal{I}_j)_{j \in N}, v_\omega)$ and a grand contract $\mathbf{c}^N$, a coalition S ex-ante blocks $\mathbf{c}^N$ if there exists an S-contract $\mathbf{c}^S$ such that*

$$\mathbf{c}^S \succ_S \mathbf{c}_S^N,$$

where with $\mathbf{c}_S^N$ we mean $(\mathbf{c}_j^N)_{j \in S}$.

The ex-ante core of Bel coalitional games is defined using the notion of blocking.

Definition 8. *The ex-ante core of a Bel coalitional game is the set of all feasible grand contracts $\mathbf{c}^N$ for which no coalition $S \subseteq N$ ex-ante blocks $\mathbf{c}^N$.*

4.2 Ex-t-Interim

First, let the set of possible worlds, which contains the true one, at stage t be I^t, then we have that $w^* \in I^1 \supset I^2 \supset \cdots \supset I^T$. In the ex-$t$-interim stages, after agents get the information I^t, we have a sequential update of the information sets of players. To model this, we need the notion of conditioning. Literature proposed various definitions of conditional capacities that remain valid when we apply them to a belief function. In this paper we present the two most commonly used ones. The first one, known as the Dempster conditioning rule, was formulated by Dempster [4] in terms of the plausibility function $\overline{\nu}$, and, assuming $\overline{\nu}(B) > 0$, is defined as follows:

$$\overline{\nu}^D(A|B) = \frac{\overline{\nu}(A \cap B)}{\overline{\nu}(B)}.$$

Its conjugate belief function is

$$\nu^D(A|B) = 1 - \overline{\nu}^D(A^c|B) = 1 - \frac{\overline{\nu}(A^c \cap B)}{\overline{\nu}(B)} = \frac{\nu(A \cup B^c) - \nu(B^c)}{1 - \nu(B^c)}.$$

The Dempster conditional belief (plausibility) function ν^D ($\overline{\nu}^D$) continues to be a belief (plausibility) function. The respective mass function, for $\overline{\nu}(B) > 0$, is given by

$$m_\nu^D(A|B) = K_B \sum_{B \cap C = A} m_\nu(C),$$

where $K_B = \frac{1}{\overline{\nu}(B)}$ is a normalization factor, constant for a given subset $B \subseteq \Omega$. The conditional ν^D is the lower envelope of a set of conditional probabilities denoted as core($\nu^D(\cdot|B)$). In fact, starting from the conditional belief $\nu^D(\cdot|B)$ and taking its core, that is defined as

$$\text{core}(\nu^D(\cdot|B)) = \\ \{\pi : \pi \text{ is a probability measure on } \Omega,\ \pi(A) \geq \nu^D(A|B),\ \forall A \in 2^\Omega\},$$

we have that $\nu^D(\cdot|B)$ can be written as (see [5])

$$\nu^D(A|B) = \min_{\pi \in \text{core}(\nu^D(A|B))} \pi(A|B).$$

A decision-theoretic justification of an axiomatic version of this conditioning rule can be found in [9], while the issue of consistency of a conditional numerical assessment has been faced in [2,10].

The second rule we look at is the general conditional belief function of A given B, firstly recognized in Dempster [4], that is defined as

$$\nu^G(A|B) = \frac{\nu(A \cap B)}{\nu(A \cap B) + \overline{\nu}(A^c \cap B)},$$

where $\nu(A \cap B) + \overline{\nu}(A^c \cap B) > 0$. A general conditional belief function continues to be a belief function and it is the lower envelope of the set of all conditional probabilities defined by its core (see, also, [3,11])

$$\nu^G(A|B) = \min_{\pi \in \text{core}(\nu)} \pi(A|B).$$

Hence, the general conditional belief is the lower envelope of a set of probability measures and then all information have a relevant role. On the contrary, with

Table 1. Values at $t = 1$ for player 1, conditioning by $\{\omega_1, \omega_2\}$.

	$\{\omega_1\}$	$\{\omega_2\}$	$\{\omega_3\}$	$\{\omega_1, \omega_2\}$	$\{\omega_1, \omega_3\}$	$\{\omega_2, \omega_3\}$	Ω
ν_1	0	0	$\frac{1}{6}$	$\frac{1}{3}$	$\frac{1}{6}$	$\frac{1}{6}$	1
m_{ν_1}	0	0	$\frac{1}{6}$	$\frac{1}{3}$	0	0	$\frac{1}{2}$
ν_1^G	0	0	0	1	0	0	1
$m_{\nu_1}^G$	0	0	0	1	0	0	0
ν_1^D	0	0	0	1	0	0	1
$m_{\nu_1}^D$	0	0	0	1	0	0	0

Table 2. Values at $t = 1$ for player 2, conditioning by $\{\omega_1, \omega_2\}$.

	$\{\omega_1\}$	$\{\omega_2\}$	$\{\omega_3\}$	$\{\omega_1, \omega_2\}$	$\{\omega_1, \omega_3\}$	$\{\omega_2, \omega_3\}$	Ω
ν_2	$\frac{1}{5}$	0	$\frac{1}{5}$	$\frac{1}{5}$	$\frac{2}{5}$	$\frac{4}{5}$	1
m_{ν_2}	$\frac{1}{5}$	0	$\frac{1}{5}$	0	0	$\frac{3}{5}$	0
ν_2^G	$\frac{1}{4}$	0	0	1	$\frac{1}{4}$	0	1
$m_{\nu_2}^G$	$\frac{1}{4}$	0	0	$\frac{3}{4}$	0	0	0
ν_2^D	$\frac{1}{4}$	$\frac{3}{4}$	0	1	$\frac{1}{4}$	$\frac{3}{4}$	1
$m_{\nu_2}^D$	$\frac{1}{4}$	$\frac{3}{4}$	0	0	0	0	0

the Dempster conditioning rule, just the local information restricted to the conditioning event has a role. We decide to work with the Dempster conditional because it better models our situation, while, if we consider the general (Bayes) conditioning rule, we can show that it does not fit well our case. This is illustrated in the following example.

Example 1. Let us consider $N = \{1, 2\}$ and $\Omega = \{\omega_1, \omega_2, \omega_3\}$. However let $\mathcal{I}_1 = \{\{\omega_1, \omega_2\}, \{\omega_3\}, \Omega\} = \mathcal{F}_{\nu_1}$ and $\mathcal{I}_2 = \{\{\omega_1\}, \{\omega_3\} \{\omega_2, \omega_3\}\} = \mathcal{F}_{\nu_2}$ be the information sets of player 1 and 2, respectively. The mass function m_{ν_1} of player 1 is such that $m_{\nu_1}(\{\omega_1, \omega_2\}) = \frac{1}{3}$, $m_{\nu_1}(\{\omega_3\}) = \frac{1}{6}$, $m_{\nu_1}(\Omega) = \frac{1}{2}$ while the mass function m_{ν_2} of player 2 is such that $m_{\nu_2}(\{\omega_1\}) = \frac{1}{5}$, $m_{\nu_2}(\{\omega_3\}) = \frac{1}{5}$, $m_{\nu_2}(\{\omega_2, \omega_3\}) = \frac{3}{5}$. Let us suppose that at $t = 1$ we have that $\omega^* \in \{\omega_1, \omega_2\}$, so conditioning by $\{\omega_1, \omega_2\} = I^1$ for player 1 we get the values present in Table 1 while for player 2 we get the values present in Table 2.

For player 2 we can observe that with the Dempster conditioning rule, the mass is concentrated between ω_1 and ω_2 and ν_2^D becomes a probability measure. With this latter rule in fact we have that when we condition with respect to a focal element $B \in \mathcal{F}_\nu$ the mass of any focal element $A \in \mathcal{F}_\nu$ is transferred to the subset $A \cap B$ if it is non-empty and discarded otherwise. Furthermore, if we suppose that, at $t = 2$, we have that $I^2 = \{\omega_2\}$, i.e., $\omega^* = \omega_2$, with the Dempster rule we have, for both players, a belief function with mass function equal to 1 on $\{\omega_2\}$. However, with the general conditioning rule, for both players, we cannot calculate $\nu_1^G(A|\{\omega_2\})$ and $\nu_2^G(A|\{\omega_2\})$, for some $A \subseteq \mathcal{F}_{\nu_i^G}$, $I = 1, 2$. ♦

From now on we indicate the Dempster conditioned belief function at stage t as $\nu_j^t = \nu_j^{t-1}(\cdot|I^t)$, where $\nu_j^0 = \nu_j$. Unlike [8] we suppose that at interim stage players possess only their private information and they do not know anything about other players' information sets. Hence, also in this case, we give the definition of ex-t-interim blocking and ex-t-interim core.

Definition 9. *Given a Bel coalitional game $(N, \Omega, (m_{\nu_j})_{j\in N}, (\mathcal{I}_j)_{j\in N}, v_\omega)$ and a grand contract $\mathbf{c}^N$, a coalition S ex-t-interim blocks $\mathbf{c}^N$ if, at stage t, there exists an S-contract $\mathbf{c}^S$ such that*

$$\mathbf{c}^S_{|I^t} \succ_S \mathbf{c}^N_{S|I^t},$$

where with $\mathbf{c}^N_{S|I^t}$ we mean $(\mathbf{c}^N_{j|I^t})_{j\in S}$.

Definition 10. *The ex-t-interim core of a Bel coalitional game is the set of all feasible grand contracts $\mathbf{c}^N$ for which no coalition $S \subseteq N$ ex-t-interim blocks $\mathbf{c}^N$.*

Remark 1. Notice that at the last stage T the ex-T-interim core of a Bel coalitional game coincides with the core of a classical coalitional game.

5 Core of Bel Coalitional Games

In this section we show some results about the core. It is important to emphasize that in the particular case where all agents have the same belief function $\nu = \pi$, with π a probability measure, we cannot simply consider a classical coalitional game (N, v) such that $v = \mathbb{E}_\pi(v_\omega(S))$ and get that all classical game results hold too for ex-ante core and the ex-t-interim core cases. This is because the definitions of the ex-ante core and the ex-t-interim core coincide with the classical definition of the core only when there is no uncertainty, i.e., when $|\Omega| = 1$ (for the ex-t-interim case also when $t = T$ (see Remark 1)). First, we characterize the ex-ante core and the ex-t-interim core through the Choquet integral.

Theorem 1. *Given a Bel coalitional game $(N, \Omega, (m_{\nu_j})_{j\in N}, (\mathcal{I}_j)_{j\in N}, v_\omega)$, if a grand contract $\mathbf{c}^N$ is in the ex-ante core of the game then, for all $S \subseteq N$, there exists $j \in S$ such that*

$$\mathbb{C}_{\nu_j}(v_\omega(S)) \leq \mathbb{C}_{\overline{\nu}_j}\left(\sum_{i\in S} \mathbf{c}_i^N\right).$$

Remark 2. One can notice the presence of the conjugate belief function $\overline{\nu}_j$ in the right hand side, hence the corresponding plausibility function. This is due to the asymmetric property of the Choquet integral. Recall that $\nu_j \leq \overline{\nu}_j$, and by monotonicity of the Choquet integral, this means that the right hand side is greater than it would be with ν_j instead of $\overline{\nu}_j$, i.e., the upper bound is less tight.

The other implication is not true, let us see a counterexample.

Table 3. Values of the coalitions with respect to a world ω.

ω	ω_1	ω_2	ω_3
$v_\omega(\{1\})$	1	4	1
$v_\omega(\{2\})$	3	2	1
$v_\omega(\{3\})$	2	2	2
$v_\omega(\{1,2\})$	4	7	3
$v_\omega(\{1,3\})$	3	8	3
$v_\omega(\{2,3\})$	4	5	3
$v_\omega(\{1,2,3\})$	6	11	5

Example 2. Let us consider $N = \{1,2,3\}$ and $\Omega = \{\omega_1, \omega_2, \omega_3\}$. Let $\mathcal{I}_1 = \{\{\omega_1\}, \{\omega_2, \omega_3\}\} = \mathcal{F}_{\nu_1}$, $\mathcal{I}_2 = \{\{\omega_1\}, \{\omega_2\}, \{\omega_3\}\} = \mathcal{F}_{\nu_2}$ and, finally, $\mathcal{I}_3 = \{\{\omega_1, \omega_2\}, \{\omega_3\}\} = \mathcal{F}_{\nu_3}$ be the information sets of player 1, 2 and 3, respectively. The mass function m_{ν_1} of player 1 is given by $m_{\nu_1}(\{\omega_1\}) = \frac{1}{3}$, $m_{\nu_1}(\{\omega_2, \omega_3\}) = \frac{2}{3}$, the mass function m_{ν_2} of player 2 is $m_{\nu_2}(\{\omega_1\}) = \frac{1}{3}$, $m_{\nu_2}(\{\omega_2\}) = \frac{1}{3}$, $m_{\nu_2}(\{\omega_3\}) = \frac{1}{3}$, while the mass function m_{ν_3} of player 3 is $m_{\nu_3}(\{\omega_1, \omega_2\}) = \frac{2}{3}$, $m_{\nu_3}(\{\omega_3\}) = \frac{1}{3}$. The values of coalitions are given by Table 3.

By considering the following grand contract (payoffs appear in order of worlds)

$$\mathbf{c}^N = [(1,3,2), (4,5,2), (1,1,3)]$$

it is possible to show that for all $S \subseteq N$, there exists $j \in S$ such that

$$\mathbb{C}_{\nu_j}(v_\omega(S)) \leq \mathbb{C}_{\overline{\nu}_j}\left(\sum_{i \in S} \mathbf{c}_i^N\right).$$

Now, let us consider the coalition $S = \{1,2\}$ and the contract

$$\mathbf{c}^S = [(2,2), (1,6), (1,2)].$$

To determine if $\mathbf{c}^N$ is in the ex-ante core we compute the utilities:

$$\mathbb{C}_{\nu_1}(\mathbf{c}_1^N) = 1,\ \mathbb{C}_{\nu_1}(\mathbf{c}_1^S) = 1.3,\ \mathbb{C}_{\nu_2}(\mathbf{c}_2^N) = 3 \text{ and } \mathbb{C}_{\nu_2}(\mathbf{c}_2^S) = 3.3.$$

Hence, $\mathbb{C}_{\nu_1}(\mathbf{c}_1^N) < \mathbb{C}_{\nu_1}(\mathbf{c}_1^S)$ and $\mathbb{C}_{\nu_2}(\mathbf{c}_2^N) < \mathbb{C}_{\nu_2}(\mathbf{c}_2^S)$, i.e., there exists an S-contract $\mathbf{c}^S$, for the coalition S, such that $\mathbf{c}^S \succ_S \mathbf{c}^N$, so $\mathbf{c}^N$ is not in the ex-ante core of the game. ♦

By reinforcing the hypothesis of Theorem 1 we get the following result for the other implication, in the particular case where agents have the same belief function ν.

Theorem 2. *Let us consider that agents have the same belief function ν. Given a Bel coalitional game $(N, \Omega, m_\nu, \mathcal{I}, v_\omega)$, if for all $S \subseteq N$*

$$\mathbb{C}_\nu(v_\omega(S)) \leq \sum_{j \in S} \mathbb{C}_\nu(\mathbf{c}_j^N),$$

Table 4. Values of the coalitions with respect to a world ω.

ω	ω_1	ω_2	ω_3	ω_4	ω_5
$v_\omega(\{1\})$	3	5	2	1	3
$v_\omega(\{2\})$	2	6	1	3	3
$v_\omega(\{1,2\})$	6	12	4	6	7

then the grand contract $\mathbf{c}^N$ is in the ex-ante core of the game.

Let us see an example.

Example 3. Let us consider $N = \{1,2\}$ and $\Omega = \{\omega_1,\omega_2,\omega_3,\omega_4,\omega_5\}$. Let $\mathcal{I} = \{\{\omega_5\},\{\omega_1,\omega_2\},\{\omega_3,\omega_4\},\{\omega_1,\omega_3,\omega_4\}\} = \mathcal{F}_\nu$ be the information set of players. The mass function m_ν is given by $m_\nu(\{\omega_5\}) = \frac{1}{5}$, $m_\nu(\{\omega_1,\omega_2\}) = \frac{2}{5}$, $m_\nu(\{\omega_3,\omega_4\}) = \frac{1}{5}$, $m_\nu(\{\omega_1,\omega_3,\omega_4\}) = \frac{1}{5}$. In Table 4 we give the values of the coalitions in the five possible worlds.

To analyze this game, consider the following feasible grand contract

$$\mathbf{c}^N = [(3,3),(9,3),(1,3),(3,3),(4,3)].$$

As already mentioned, $\mathbf{c}_1^1$ and $\mathbf{c}_2^2$ are simply equal to rows 1 and 2 in the Table 4, respectively. To determine if $\mathbf{c}^N$ is in the ex-ante core of the game we compute the utilities:

$$\mathbb{C}_\nu(v_\omega(\{1\})) = 2.2,\ \mathbb{C}_\nu(v_\omega(\{2\})) = 1.8,\ \mathbb{C}_\nu(v_\omega(\{1,2\})) = 5.4,$$
$$\mathbb{C}_\nu(\mathbf{c}_1^N) = 2.4 \text{ and } \mathbb{C}_\nu(\mathbf{c}_2^N) = 3.$$

Hence, by Theorem 2, $\mathbf{c}^N$ is in the ex-ante core of the game. ♦

Finally, if ν is a probability measure π, as a consequence of previous results we have this useful corollary.

Corollary 1. *Given a Bel coalitional game $(N,\Omega,m_\nu,\mathcal{I},v_\omega)$, with $\nu = \pi$, then a grand contract $\mathbf{c}^N$ is in the ex-ante core of the game if and only if, for all $S \subseteq N$*

$$\mathbb{E}_\pi(v_\omega(S)) \leq \mathbb{E}_\pi\left(\sum_{i\in S} \mathbf{c}_i^N\right).$$

Remark 3. If ν is a probability measure we can see the Bel coalitional game as a particular Bayesian coalitional game and so we can get the latter result from Theorem 1 of [8].

Let us see the analogous results for the ex-t-interim case. To simplify the notation we denote $\mathbb{C}_{\nu_j^t}(v_{\omega\in I^t}(S))$ and $\mathbb{C}_{\nu_j^t}(\mathbf{c}^S_{|I^t})$ by $\mathbb{C}_{\nu_j^t}(v_\omega(S)|I^t)$ and $\mathbb{C}_{\nu_j^t}(\mathbf{c}^S|I^t)$, respectively.

Theorem 3. *Given a Bel coalitional game* $(N, \Omega, (m_{\nu_j})_{j\in N}, (\mathcal{I}_j)_{j\in N}, v_\omega)$, *if the grand contract* $\mathbf{c}^N$ *is in the ex-t-interim core of the game, then for all* $S \subseteq N$, *there exists* $j \in S$ *such that*

$$\mathbb{C}_{\nu_j^t}(v_\omega(S)|I^t) \leq \mathbb{C}_{\overline{\nu}_j^t}\left(\sum_{i\in S} \mathbf{c}_i^N|I^t\right).$$

Also in this case the other implication is not true, let us see a counterexample.

Example 4. We are under the hypothesis of Example 2 with the same grand contract $\mathbf{c}^N$. Let us assume that $I^1 = \{\omega_1, \omega_2\}$. It is possible to show that for all $S \subseteq N$, there exists $j \in S$ such that

$$\mathbb{C}_{\nu_j^1}(v_\omega(S)|I^1) \leq \mathbb{C}_{\overline{\nu}_j^1}\left(\sum_{i\in S} \mathbf{c}_i^N|I^1\right).$$

Now, let us consider the coalition $S = \{1, 3\}$ and the contract

$$\mathbf{c}^S = [(0, 3), (5, 3), (1, 2)].$$

To determine if $\mathbf{c}^N$ is in the ex-1-interim core we compute the utilities:

$$\mathbb{C}_{\nu_1^1}(\mathbf{c}_1^N|I^1) = 3,\ \mathbb{C}_{\nu_1^1}(\mathbf{c}_1^S|I^1) = 3.3,\ \mathbb{C}_{\nu_3^1}(\mathbf{c}_3^N|I^1) = 2 \text{ and } \mathbb{C}_{\nu_3^1}(\mathbf{c}_3^S|I^1) = 3.$$

Hence, $\mathbb{C}_{\nu_1^1}(\mathbf{c}_1^N|I^1) < \mathbb{C}_{\nu_1^1}(\mathbf{c}_1^S|I^1)$ and $\mathbb{C}_{\nu_3^1}(\mathbf{c}_3^N|I^1) < \mathbb{C}_{\nu_3^1}(\mathbf{c}_3^S|I^1)$, i.e., there exists an S-contract $\mathbf{c}^S$, for the coalition S, such that $\mathbf{c}^S_{|I^1} \succ_S \mathbf{c}^N_{S|I^1}$, so $\mathbf{c}^N$ is not in the ex-1-interim core of the game. ♦

Theorem 4. *Let us consider that agents have the same belief function* ν. *Given a Bel coalitional game* $(N, \Omega, m_\nu, \mathcal{I}, v_\omega)$, *if for all* $S \subseteq N$

$$\mathbb{C}_{\nu^t}(v_\omega(S)|I^t) \leq \sum_{j\in S} \mathbb{C}_{\nu^t}(\mathbf{c}_j^N|I^t),$$

then the grand contract $\mathbf{c}^N$ *is in the ex-t-interim core of the game.*

Example 5. We are under the hypothesis of Example 3. Let us suppose that at $t = 1$ we have that $\omega^* \in I^1 = \{\omega_1, \omega_3, \omega_4, \omega_5\}$. Also this time we compute the utilities:

$$\mathbb{C}_{\nu^1}(v_\omega(\{1\})|I^1) = 2.2,\ \mathbb{C}_{\nu^1}(v_\omega(\{2\})|I^1) = 1.8,\ \mathbb{C}_{\nu^1}(v_\omega(\{1,2\})|I^1) = 5.4,$$

$$\mathbb{C}_{\nu^1}(\mathbf{c}_1^N|I^1) = 2.4 \text{ and } \mathbb{C}_{\nu^1}(\mathbf{c}_2^N|I^1) = 3.$$

So, by Theorem 4, $\mathbf{c}^N$ is in the ex-1-interim core of the game. Table 5 shows utilities for $t = 2, 3, 4$.

Table 5. Utilities at each step t.

t	I^t	$\mathbb{C}_{\nu^t}(v_\omega(\{1\})\|I^t)$	$\mathbb{C}_{\nu^t}(v_\omega(\{2\})\|I^t)$	$\mathbb{C}_{\nu^t}(v_\omega(\{1,2\})\|I^t)$	$\mathbb{C}_{\nu^t}(\mathbf{c}_1^N\|I^t)$	$\mathbb{C}_{\nu^t}(\mathbf{c}_2^N\|I^t)$
2	$\{\omega_3,\omega_4,\omega_5\}$	1.67	1.67	4.99	1.99	3
3	$\{\omega_4,\omega_5\}$	1.67	3	6.3	3.3	3
4	$\{\omega_4\}$	1	3	6	3	3

Hence, again by Theorem 4, $\mathbf{c}^N$ is in the ex-t-interim core of the game, for $t = 2, 3, 4$. ♦

Also for the ex-t-interim case we have the following result. Since we use a different definition of ex-t-interim core, this result is different from the one present in [8].

Corollary 2. *Given a Bel coalitional game $(N, \Omega, m_\nu, \mathcal{I}, v_\omega)$, with $\nu = \pi$, then a grand contract $\mathbf{c}^N$ is in the ex-t-interim core of the game if and only if, for all $S \subseteq N$*

$$\mathbb{E}_{\pi^t}(v_\omega(S)|I^t) \leq \mathbb{E}_{\pi^t}\left(\sum_{i\in S} \mathbf{c}_i^N|I^t\right). \tag{1}$$

Now, we want a similar result of the well-known Bondareva [1] and Shapley [14] theorem for our cases. For any $S \subseteq N$, its characteristic vector $\mathbf{1}^S \in \mathbb{R}^N$ is defined by $\mathbf{1}_i^S = 1$ if $i \in S$ and $\mathbf{1}_i^S = 0$ otherwise.

Definition 11. *A collection $\mathcal{B}$ of non-empty subsets of N is balanced if there exist $\lambda_S > 0$, $S \in \mathcal{B}$, such that*

$$\sum_{S\in\mathcal{B}} \lambda_S \mathbf{1}^S = \mathbf{1}^N.$$

The λ_S are called balancing weights.

To simplify the problem we examine only the case in which agents have the same belief function ν.

Definition 12. *A Bel coalitional game $(N, \Omega, m_\nu, \mathcal{I}, v_\omega)$ is ex-ante balanced if for every balanced collection $\mathcal{B}$ it holds*

$$\sum_{S\in\mathcal{B}} \lambda_S \mathbb{C}_\nu(v_\omega(S)) \leq \mathbb{C}_{\overline{\nu}}(v_\omega(N)).$$

Definition 13. *A Bel coalitional game $(N, \Omega, m_\nu, \mathcal{I}, v_\omega)$ is ex-t-interim balanced if for every balanced collection $\mathcal{B}$ it holds*

$$\sum_{S\in\mathcal{B}} \lambda_S \mathbb{C}_{\nu^t}(v_\omega(S)|I^t) \leq \mathbb{C}_{\overline{\nu}^t}(v_\omega(N)|I^t).$$

Remark 4. Notice that at the last stage T the definition of ex-T-interim balanced game coincide with the one of balanced game for a classical coalitional game.

Theorem 5. *Let us consider a Bel coalitional game* $(N, \Omega, m_\nu, \mathcal{I}, v_\omega)$. *We have that:*

(i) if there exists a comonotonic $\mathbf{c}^N$ *in the ex-ante core of the game, then the Bel coalitional game is ex-ante balanced;*
(ii) if $\nu = \pi$, *then the ex-ante core of the game is non-empty if and only if the Bel coalitional game is ex-ante balanced;*
(iii) if there exists a comonotonic $\mathbf{c}^N$ *in the ex-t-interim core of the game, then the Bel coalitional game is ex-t-interim balanced;*
(iv) if $\nu = \pi$, *then the ex-t-interim core of the game is non-empty if and only if the Bel coalitional game is ex-t-interim balanced.*

It is well known that not all balanced collections are necessary to check that a game is balanced. In fact, for classical coalitional games minimal balanced collections suffice to characterize the class of coalitional games with a non-empty core. A balanced collection is *minimal* if it does not contain a proper subcollection that is balanced (see Chap. 3 of [7] for more details). For Bel coalitional games we have the following result.

Theorem 6. *Let us consider a Bel coalitional game* $(N, \Omega, m_\nu, \mathcal{I}, v_\omega)$. *We have that:*

(i) if there exists a comonotonic $\mathbf{c}^N$ *in the ex-ante core of the game, then for any minimal balanced collection* $\mathcal{B}$ *with system of balancing weights* $(\lambda_A)_{A\in\mathcal{B}}$, *we have*

$$\sum_{A\in\mathcal{B}} \lambda_A \mathbb{C}_\nu(v_\omega(A)) \leq \mathbb{C}_{\overline{\nu}}(v_\omega(N));$$

(ii) if $\nu = \pi$, *then the ex-ante core of the game is non-empty if and only if for any minimal balanced collection* $\mathcal{B}$ *with system of balancing weights* $(\lambda_A)_{A\in\mathcal{B}}$, *we have*

$$\sum_{A\in\mathcal{B}} \lambda_A \mathbb{E}_\pi(v_\omega(A)) \leq \mathbb{E}_\pi(v_\omega(N));$$

(iii) if there exists a comonotonic $\mathbf{c}^N$ *in the ex-t-interim core of the game, then for any minimal balanced collection* $\mathcal{B}$ *with system of balancing weights* $(\lambda_A)_{A\in\mathcal{B}}$, *we have*

$$\sum_{A\in\mathcal{B}} \lambda_A \mathbb{C}_{\nu^t}(v_\omega(S)|I^t) \leq \mathbb{C}_{\overline{\nu}^t}(v_\omega(N)|I^t);$$

(iv) if $\nu = \pi$, *the ex-t-interim core of the game is non-empty if and only if for any minimal balanced collection* $\mathcal{B}$ *with system of balancing weights* $(\lambda_A)_{A\in\mathcal{B}}$, *we have*

$$\sum_{A\in\mathcal{B}} \lambda_A \mathbb{E}_{\pi^t}(v_\omega(S)|I^t) \leq \mathbb{E}_{\pi^t}(v_\omega(N)|I^t).$$

6 Conclusion

We defined Bel coalitional games, a generalization of classical coalitional games and of Bayesian coalitional games. We investigated payoff division through contracts, by considering the Choquet integral to model agents' preferences, and we gave two different definitions of the core. Finally, we stated some results about the non-emptiness of the core, by using the notion of balanced and minimal balanced collection. The aim of future research is to use balanced collections also in the case in which agents have different belief functions and find efficient algorithms to verify the non-emptiness of the core.

Acknowledgements. The second author has been supported by the project Fondo Ricerca Ateneo WP4.1 esercizio 2022 - RATIONALISTS, funded by University of Perugia.

References

1. Bondareva, O.: The theory of the core in an n-person game. In: Vestnik LGU13, pp. 141–142 (1962)
2. Coletti, G., Petturiti, D., Vantaggi, B.: A Dutch book coherence condition for conditional completely alternating Choquet expectations. Boll. dell'Unione Mat. Ital. **13**, 585–593 (2020)
3. Coletti, G., Petturiti, D., Vantaggi, B.: Conditional belief functions as lower envelopes of conditional probabilities in a finite setting. Inf. Sci. **339**, 64–84 (2016)
4. Dempster, A.P.: Upper and lower probabilities induced by a multivalued mapping. Ann. Math. Stat. **38**, 325–339 (1967)
5. Denneberg, D.: Conditioning (updating) non-additive measures. Ann. Oper. Res. **52**, 21–42 (1994)
6. Ellsberg, D.: Risk, ambiguity, and the savage axioms. Q. J. Econ. **75**, 643–669 (1961)
7. Grabisch, M.: Set Functions, Games and Capacities in Decision Making. Theory and Decision Library C, Springer, Heidelberg (2016). https://doi.org/10.1007/978-3-319-30690-2
8. Ieong, S. and Shoham, Y.: Bayesian coalitional games. In: Proceedings of the Twenty-Third AAAI Conference on Artificial Intelligence, pp. 95–100 (2008)
9. Petturiti, D., Vantaggi, B.: Conditional decisions under objective and subjectuve ambiguity in Dempster-Shafer theory. Fuzzy Sets Syst. **447**, 155–181 (2022)
10. Petturiti, D., Vantaggi, B.: The extent of partially resolving uncertainty in assessing coherent conditional plausibilities. Fuzzy Sets Syst. **458**, 26–49 (2023)
11. Petturiti, D., Vantaggi, B.: Upper and lower conditional probabilities induced by a multivalued mapping. J. Math. Anal. Appl. **458**(2), 1214–1235 (2018)
12. Pomeret-Coquot, P., Fargier, H., Martin-Dorel, E.: Games of incomplete information: a framework based on belief functions. Int. J. Approximate Reasoning **151**, 182–204 (2022)
13. Shafer, G.: A Mathematical Theory of Evidence. Princeton University Press, Princeton (1976)
14. Shapley, L.S.: On balanced sets and cores. Naval Res. Logist. (NRL) **14**, 453–460 (1967)

Lifting Partially Observable Stochastic Games

Nazlı Nur Karabulut and Tanya Braun(✉)

Computer Science Department, University of Münster, Münster, Germany
{nnur.karabulut,tanya.braun}@uni-muenster.de

Abstract. Partially observable stochastic games (POSGs) are a Markovian formalism used to model a set of agents acting in a stochastic environment, in which each agent has its own reward function. As is common with multi-agent decision making problems, the model and runtime complexity is exponential in the number of agents, which can be prohibitively large. Lifting is a technique that treats groups of indistinguishable instances through representatives if possible, yielding tractable inference in the number of objects in a model. This paper applies lifting to the agent set in POSGs, yielding so-called isomorphic POSGs that have a model complexity no longer dependent on the number of agents, and presents a lifted solution approach that exploits this lifted agent set for space and runtime gains.

1 Introduction

Multi-agent decision making lies at the heart of artificial intelligence (AI) and is one of the most challenging problems in AI research. A common formalisation lies in partially observable stochastic games (POSGs). In POSGs, a set of agents acts in an environment modelled as a stochastic Markovian process, each with the goal to optimise its own reward function. POSGs are a generalisation of other formalisms based on Markovian modelling, from single-agent decision making in Markov decision processes (MDPs) and partially observable MDPs (POMDPs) to multi-agent decision making in decentralised POMDPs (DecPOMDPs), in which the agent set shares a single reward function. Both POSGs and DecPOMDPs have in common that their space and runtime complexity depends exponentially on the number of agents, making them all but unsolvable except for small problem sizes.

Unfortunately, there are applications that have an agent set size in the thousands and still require working with such a model-based approach to be able to argue about its reasoning. One such example are nanoscale medical systems, e.g., used for diagnosis or treatment [13]. Here, the idea is that a network of nanoagents is used to detect markers for a specific disease or deliver medicine to locations in the cardio-vascular system. Given its application in the medical domain, one needs to find accurate solutions [24] that can be justified and allows for arguing about agent set sizes as to not poison a patient. While the agent sets are typically very large, an upside is that they rarely come without any structure.

S. Destercke et al. (Eds.): SUM 2024, LNAI 15350, pp. 201–216, 2025.
https://doi.org/10.1007/978-3-031-76235-2_16

Typically and true for nanoscale systems, there are a small number of different types of agents to fulfil specific purposes within the system.

One promising direction that actually uses this structure has been explored when lifting the agent set of DecPOMDPs [3]. Lifting is a technique from probabilistic inference in probabilistic graphical models (PGMs), in which indistinguishable random variables are encoded using logical variables and treated through a representative [17], which leads to tractability in terms of domain sizes of those logical variables [15]. Indistinguishability of random variables is characterised by subgraph isomorphisms in the PGM between the variables and identical factors describing the influences between a variable and its neighbours in those subgraphs. In terms of the probabilistic behaviour, lifting assumes conditional independences between these indistinguishable instances, which yields the isomorphic structure in the graph. Lifting the agent set in DecPOMDPs leads to conditions for the transition, sensor, and reward function, under which representatives for groups of agents can be used during computations, yielding tractability in terms of agent numbers [3], which partitions the agent set.

However, lifting DecPOMDPs stops one step short of fully applying the lifting idea to multi-agent decision making, by still having a joint reward function. In this paper, we go this next step in providing each agent partition with its own reward function, which makes the formalisation a POSG. While lifting POSGs as an extension of lifted DecPOMDPs can be thought of as a theoretical endeavour, doing so has practical value as lifting POSGs can be helpful in modelling, e.g., nanoscale medical systems, where different types of agents may interact with each other or external factors and may have competing or complementary objectives. Additionally, we can further reduce the complexity of these models while appropriately modelling diverse goals of the agents, ensuring accurate and efficient decision-making in dynamic medical environments. Specifically, the contributions are twofold: (i) isomorphic POSGs based on a partitioning of the agent set for an explicit encoding of structure in the agent set and (ii) a lifted version of the multi-agent dynamic programming operator [7] to solve isomorphic POSGs. To the best of our knowledge, we are the first to apply the lifting idea to POSGs and show its potential for solving problem instances of new scale.

The remainder of the paper is structured as follows: We start with a recap of POSGs as well as dynamic programming for POSGs. Then, we define indistinguishability between agents and present isomorphic POSGs. Afterwards, we provide a lifted dynamic programming operator and end with a discussion.

2 Preliminaries

This section defines POSGs and recaps dynamic programming for POSGs. We look through a PGM lens for definitions, which are based on [7] and [18], using random variables, S, which can take discrete values, referred to as range, $ran(S) = \{s_1, \dots, s_n\}$. So-called *decision random variables* have actions as ranges. Setting S to a value $s \in ran(S)$ (an *event*) is denoted as $S = s$ or s for short if S is clear from its context. We denote sequences over a discrete time interval $[t_s, t_e]$ with subscript $t_s{:}t_e$, e.g., $(S_{t_s}, \dots, S_{t_e}) = S_{t_s:t_e}$. We use π to denote projection and θ to denote permutation.

2.1 Partially Observable Stochastic Game

A POSG encodes a set of agents working towards their own goals, represented through individual reward functions.

Definition 1. *A POSG M is a tuple $(\boldsymbol{I}, S, \boldsymbol{A}, \boldsymbol{O}, T, \boldsymbol{R})$, with*

- $\boldsymbol{I}$ *a set of N agents,*
- S *a random variable with a set of states as range,*
- $\boldsymbol{A} = \{A_i\}_{i \in \boldsymbol{I}}$ *a set of decision random variables A_i, one for each agent $i \in \boldsymbol{I}$, each with a set of local actions as range, with $ran(\boldsymbol{A}) = \times_{i \in \boldsymbol{I}} ran(A_i)$ the set of* joint actions,
- $\boldsymbol{O} = \{O_i\}_{i \in \boldsymbol{I}}$ *a set of random variables O_i, each with a set of local observations as range, with $ran(\boldsymbol{O}) = \times_{i \in \boldsymbol{I}} ran(O_i)$ the set of* joint observations,
- $T(S', S, \boldsymbol{A}, \boldsymbol{O}) = P(S', \boldsymbol{O} \mid S, \boldsymbol{A})$, *called transition function to give it a unique name, which is a probability distribution denoting the probability of moving from state s with joint action $\boldsymbol{a}$ to state s' and making the joint observation $\boldsymbol{o}$, with $T(S_0, ., ., .) = P(S_0)$ referring to a state prior, and*
- $\boldsymbol{R} = \{R_i(S, \boldsymbol{A})\}_{i \in \boldsymbol{I}}$ *a set of reward functions, one for each agent $i \in \boldsymbol{I}$.*

Optional are a finite horizon τ, a discount factor $\gamma \in [0, 1]$ (default 1), and an error margin $\epsilon > 0$. Each agent $i \in \boldsymbol{I}$ has a local policy $\pi_i : ran((O_{i,0:t})) \mapsto ran(A_i)$ mapping observation histories $o_{i,0:t}, t \leq \tau - 1$, to actions a, with $\boldsymbol{\pi} = (\pi_i)_{i \in \boldsymbol{I}}$ a joint policy. *A set of policies for agent i is denoted by Π_i. A set or sequence of elements over all agents except agent i is denoted by subscript $-i$.*

Each agent has its own set of actions and observations and its own reward function[1] whereas the state is joint. An agent's belief $b_i(s, \boldsymbol{\pi}_{-i})$ is a probability distribution over the state space $ran(S)$ and the other agents' policies $\boldsymbol{\Pi}_{-i}$. The value V_i of a belief b_i is defined by

$$V_i(b_i) = \max_{\pi_i \in \Pi_i} \sum_{s \in ran(S)} \sum_{\boldsymbol{\pi}_{-i} \in \boldsymbol{\Pi}_{-i}} b_i(s, \boldsymbol{\pi}_{-i}) V_i'(s, \pi_i, \boldsymbol{\pi}_{-i})$$

where $V_i'(s, \pi_i, \boldsymbol{\pi}_{-i})$ refers to the value of acting according to joint policy $\pi_i, \boldsymbol{\pi}_{-i}$ in state s and incorporates the reward functions in each step. Modelling-wise, DecPOMDPs only differ from POSGs by having a single reward function for all agents. However, having self-interested agents has consequences for potential solution methods: Whereas a DecPOMDP allows for brute-force joint policy evaluation, POSGs no longer enable such computation as it is no longer possible to define what an optimal policy is, which is why some of the most common DecPOMDP solution methods do not translate to the POSG setting [16]. In POSGs, one usually aims for a Nash equilibrium and possibly a Pareto optimal one, meaning that in an equilibrium, no agent has an incentive to change its policy without any agent changing its policy. The one solution method that both formalisms share is dynamic programming, which we recap next.

[1] $A_i = A_j$, $O_i = O_j$, or $R_i = R_j$, $i, j \in \boldsymbol{I}$, are possible but not mandatory.

Algorithm 1. Multi-agent Dynamic Programming Operator

function MA-DP(set of policies Π_i^{t-1} for each agent $i \in \boldsymbol{I}$ with value vectors $\boldsymbol{V}_i^{t-1}$)
  $\Pi_i^t \leftarrow$ Perform exhaustive backup using Π_i^{t-1} for each agent $i \in \boldsymbol{I}$
  $\boldsymbol{V}_i^t \leftarrow$ Calculate new value vectors for each agent $i \in \boldsymbol{I}$
  while $\exists \pi_{i,j}^t \in \Pi_i^t$: Eq. (1) holds **do**
    $\Pi_i^t \leftarrow \Pi_i^t \setminus \{\pi_{i,j}^t\}$, $\boldsymbol{V}_i^t \leftarrow \boldsymbol{V}_i^t \setminus \{v_{i,j}^t\}$
  return $\{(\Pi_i^t, \boldsymbol{V}_i^t)\}_{i \in \boldsymbol{I}}$

2.2 Dynamic Programming for POSGs

Solving a POSG with horizon τ can be done through a combination of dynamic programming and pruning [7]. Such a routine involves an iterative application of a dynamic programming operator where in each iteration $t \in \{0, \ldots, \tau - 1\}$, the operator, first, does an exhaustive backup that generates all possible policies of depth t for each agent given the existing policies of depth $t-1$, and second, prunes policies that are very weakly dominated until no more policies can be pruned. In the following, we define the dynamic programming operator as well as what constitutes weak domination between policies.

Algorithm 1 shows the dynamic programming operator. It takes a set of policies $\Pi_i^{t-1} = \{\pi_{i,j}^{t-1}\}_{j=1}^m$ of depth $t-1$ for each agent i. For each policy, there exists a value vector $v_{i,j}^{t-1}$ that denotes the value of that policy in each possible combination of state $s \in ran(S)$ and policies of the other agents $\boldsymbol{\pi}_{-i} \in \boldsymbol{\Pi}_{-i}^{t-1}$. We denote the set of corresponding value vectors by $\boldsymbol{V}_i^{t-1}$. The first step involves generating all possible policies of depth t given the policies of depth $t-1$, expanding all existing policies with all possible combinations of observations and actions. Pruning involves finding policies that are very weakly dominated by other policies, meaning that over the complete space of $S \times \boldsymbol{\Pi}_{-i}^t$ there is always another strategy with higher value. Formally, a policy $\pi_{i,j}^t$ is pruned if the following holds, which can be solved using a linear programme:

$$\forall s \in ran(S), \boldsymbol{\pi}_{-i}^t \in \boldsymbol{\Pi}_{-i}^t \exists \pi_{i,j'}^t \in \Pi_i^t : V_i(s, \pi_{i,j}^t, \boldsymbol{\pi}_{-i}^t) \leq V_i(s, \pi_{i,j'}^t, \boldsymbol{\pi}_{-i}^t), \quad (1)$$

3 Lifting Partially Observable Stochastic Games

POSGs are characterised by a model and runtime complexity for exact solution methods that is exponential in the number of agents. However, large multi-agent systems such as nanoscale medical systems often exhibit some form of structure among its agent set that we can exploit for efficiency gains. If assuming that there are groups of agents that are essentially indistinguishable, i.e., one expects them to behave identically if faced with the same situation, it is possible to reduce model and runtime complexity by being able to work with representatives for groups. A complexity reduction from exponential to logarithmic is possible for so-called isomorphic DecPOMDPs [3] if assuming that the number of groups K is much smaller than the overall number of agents N, i.e., $K \ll N$. In the following, we apply the idea to POSGs.

3.1 Indistinguishable Agents for Agent Tractability

Before considering agents that are indistinguishable within a group that is distinguishable from the next group, we consider indistinguishability between two agents. Intuitively, two agents are indistinguishable if they act the same way given the same situation and the possible effect is identical, which also means that they have the same set of policies available. This translates into the following aspects, which we explain below: (i) The available actions and observations as well as the reward function are identical. (ii) The behaviour in functions T and R is symmetric (what that means we see below). (iii) The two agents are conditionally independent in T and R given the state and the other agents' actions. The first item is essentially a precondition for the other two items as two agents cannot be indistinguishable in any regard if they have a different set of actions and observations and a different reward function, making them automatically distinguishable. Also, if the available actions and observations differ, the histories cannot be the same and the agents cannot perform the same action. The second item describes that in T and R, those two agents could switch their actions, while the resulting distribution over state and joint observation as well as the resulting reward would be the same, making them indistinguishable. The last item is a consequence of saying that the two agents have the same policies available to them. The same set of policies implies that both agents would act identically given the same situation, which requires conditional independence between those two agents. Next, we formalise these conditions for two agents before generalising the conditions for groups of agents.

The first item, which is rather straight-forward, can be formalised by stating that for two indistinguishable agents $i, j \in \boldsymbol{I}$,

$$\begin{aligned} ran(A_i) = ran(A_j) \wedge ran(O_i) = ran(O_j) \\ \wedge \forall s \in ran(S), \boldsymbol{a} \in ran(\boldsymbol{A}) : R_i(s, \boldsymbol{a}) = R_j(s, \boldsymbol{a}). \end{aligned} \tag{2}$$

The second item, symmetric behaviour, manifests itself in the functions T and R by being able to exchange the actions and observations of two indistinguishable agents and still receiving the same probability and reward. Formally, this behaviour between two indistinguishable agents $i, j \in \boldsymbol{I}$ can be characterised as follows, with $R_h = R_i = R_j$ due to Eq. (2):

$$\begin{aligned} \forall \boldsymbol{a} \in ran(\boldsymbol{A}), \boldsymbol{a} = (a_i, a_j, \boldsymbol{a}_{-i,-j}) : \forall \boldsymbol{o} \in ran(\boldsymbol{O}), \boldsymbol{o} = (o_i, o_j, \boldsymbol{o}_{-i,-j}) : \\ T(s', s, a_i, a_j, \boldsymbol{a}_{-i,-j}, \boldsymbol{o}) = T(s', s, a_j, a_i, \boldsymbol{a}_{-i,-j}, \boldsymbol{o}) \\ T(s', s, \boldsymbol{a}, o_i, o_j, \boldsymbol{o}_{-i,-j}) = T(s', s, \boldsymbol{a}, o_j, o_i, \boldsymbol{o}_{-i,-j}) \\ R_h(s, a_i, a_j, \boldsymbol{a}_{-i,-j}) = R_h(s, a_j, a_i, \boldsymbol{a}_{-i,-j}), h \in \{i, j\} \end{aligned} \tag{3}$$

The third item, conditional independence means distributions factorise according to $P(A, B \mid C) = P(A \mid C) \cdot P(B \mid C)$. Conditional independence between two indistinguishable agents $i, j \in \boldsymbol{I}$ affects the observations o_i, o_j and actions a_i, a_j, which are conditionally independent given the state and the other agents' actions. This independence yields a factorisation of T, where o_i, o_j as well as o_i (o_j) and a_j (a_i) are conditionally independent, and a factorisation of R regarding a_i, a_j, which looks as follows:

$$\begin{aligned}
&\forall \boldsymbol{a} \in ran(\boldsymbol{A}), \boldsymbol{a} = (a_i, a_j, \boldsymbol{a}_{-i,-j}) : \forall \boldsymbol{o} \in ran(\boldsymbol{O}), \boldsymbol{o} = (o_i, o_j, \boldsymbol{o}_{-i,-j}) : \\
&T(s', s, a_i, a_j, \boldsymbol{a}_{-i,-j}, o_i, o_j, \boldsymbol{o}_{-i,-j}) = P(s', o_i, o_j, \boldsymbol{o}_{-i,-j} \mid s, a_i, a_j, \boldsymbol{a}_{-i,-j}) \\
&= P_i(s', o_i, \boldsymbol{o}_{-i,-j} \mid s, a_i, \boldsymbol{a}_{-i,-j}) \cdot P_j(s', o_j, \boldsymbol{o}_{-i,-j} \mid s, a_j, \boldsymbol{a}_{-i,-j}) \\
&R_h(s, a_i, a_j, \boldsymbol{a}_{-i,-j}) = R_h^i(s, a_i, \boldsymbol{a}_{-i,-j}) + R_h^j(s, a_j, \boldsymbol{a}_{-i,-j}), h \in \{i, j\}
\end{aligned} \tag{4}$$

Note the dimension reductions in the factorisations compared to T, R, which we use for efficiency when applied to groups of agents. Equations (2) to (4) together actually yield $P_i = P_j$ and $R_h^i = R_h^j$, which we show next.

Lemma 1. *If Eqs. (2) to (4) hold, then the factors P_i, P_j and summands R_h^i, R_h^j in Eq. (4) are identical.*

Proof. As mentioned above, Eq. (2) is a prerequisite for indistinguishability and, specifically, for Eq. (3). If Eq. (3) holds, permutations of actions and observations for indistinguishable agents map to the same probability p in T and reward r in R. Let us first show that $P_i = P_j$ for the first line of Eq. (3). Inserting Eq. (4) leads to:

$$\begin{aligned}
&T(s', s, a_i, a_j, \boldsymbol{a}_{-i,-j}, \boldsymbol{o}) = T(s', s, a_j, a_i, \boldsymbol{a}_{-i,-j}, \boldsymbol{o}) = p \\
\Leftrightarrow &P_i(s', \boldsymbol{o} \mid s, a_i, \boldsymbol{a}_{-i,-j}) \cdot P_j(s', \boldsymbol{o} \mid s, a_j, \boldsymbol{a}_{-i,-j}) = p^{a_i} \cdot p^{a_j} \\
&= P_j(s', \boldsymbol{o} \mid s, a_j, \boldsymbol{a}_{-i,-j}) \cdot P_i(s', \boldsymbol{o} \mid s, a_i, \boldsymbol{a}_{-i,-j}) = p^{a_j} \cdot p^{a_i} = p
\end{aligned}$$

There are two cases to consider, one where $a_i = a_j = a$ (or a') and one where $a_i = a \neq a_j = a'$, which leads to the following set of equations:

$$\begin{aligned}
&P_i(. \mid s, a, \boldsymbol{a}_{-i,-j}) \cdot P_j(. \mid s, a, \boldsymbol{a}_{-i,-j}) = p_i^a \cdot p_j^a = p \Leftrightarrow p_j^a = \frac{p}{p_i^a} \Leftrightarrow p_i^a = \frac{p}{p_j^a} \\
&P_i(. \mid s, a, \boldsymbol{a}_{-i,-j}) \cdot P_j(. \mid s, a', \boldsymbol{a}_{-i,-j}) = p_i^a \cdot p_j^{a'} = p' \\
&P_i(. \mid s, a', \boldsymbol{a}_{-i,-j}) \cdot P_j(. \mid s, a, \boldsymbol{a}_{-i,-j}) = p_i^{a'} \cdot p_j^a = p' \\
&P_i(. \mid s, a', \boldsymbol{a}_{-i,-j}) \cdot P_j(. \mid s, a', \boldsymbol{a}_{-i,-j}) = p_i^{a'} \cdot p_j^{a'} = p'' \Leftrightarrow p_j^{a'} = \frac{p''}{p_i^{a'}} \Leftrightarrow p_i^{a'} = \frac{p''}{p_j^{a'}}
\end{aligned}$$

The two lines in the middle say that $p_i^{a'} \cdot p_j^a = p_i^a \cdot p_j^{a'} = p'$, which yields the following if replacing p_i^a and $p_i^{a'}$ as well as p_j^a and $p_j^{a'}$ with the corresponding fractions from above:

$$\begin{aligned}
&p_i^{a'} \cdot \frac{p}{p_i^a} = p_i^a \cdot \frac{p''}{p_i^{a'}} = p' \Leftrightarrow (p_i^{a'})^2 \cdot p = (p_i^a)^2 \cdot p'' = p_i^{a'} \cdot p_i^a \cdot p' \\
&\frac{p''}{p_j^{a'}} \cdot p_j^a = \frac{p}{p_j^a} \cdot p_j^{a'} = p' \Leftrightarrow (p_j^a)^2 \cdot p'' = (p_j^{a'})^2 \cdot p = p_j^{a'} \cdot p_j^a \cdot p'
\end{aligned}$$

Since the first halves before the equivalence sign both equal p', we have that the parts with p in them are equal, which we can further reformulate:

$$\frac{p_i^{a'}}{p_i^a} \cdot p = \frac{p_j^{a'}}{p_j^a} \cdot p \Leftrightarrow \frac{p_i^{a'}}{p_i^a} = \frac{p_j^{a'}}{p_j^a} \Leftrightarrow p_j^a = \frac{p_i^a}{p_i^{a'}} \cdot p_i^{a'}$$

The second halves after the equivalence sign are also equal, which yields:

$$p_i^{a'} \cdot p_i^a \cdot p' = p_j^{a'} \cdot p_j^a \cdot p' \Leftrightarrow p_j^a = \frac{p_i^{a'} \cdot p_i^a}{p_j^{a'}}$$

Taking both expressions for p_j^a, we get

$$\frac{p_i^a}{p_i^{a'}} \cdot p_j^{a'} = \frac{p_i^{a'} \cdot p_i^a}{p_j^{a'}} \Leftrightarrow p_i^a \cdot (p_j^{a'})^2 = p_i^a \cdot (p_i^{a'})^2 \Leftrightarrow p_j^{a'} = p_i^{a'}$$

which leads to $P_i = P_j$. The same reformulations can be done for the other two lines of Eq. (3), showing that $P_i = P_j$ as well as $R_i = R_j$ given Eq. (4). □

3.2 Partitions of Indistinguishable Agents

The previous section groups two agents together. The idea is of course to have a small number of groups with many indistinguishable agents in each group. As such, Eq. (2), (3), (4) have to hold for all agents within a group, which leads to a partitioning of the agent set.

The set of agents $\boldsymbol{I}$ partitions into K sets $\mathfrak{I}_k$, i.e., $\boldsymbol{I} = \bigcup_{k=1}^{K} \mathfrak{I}_k$, $\mathfrak{I}_k \neq \emptyset$, and $\forall k, l \in \{1, \ldots, K\}, k \neq l : \mathfrak{I}_k \cap \mathfrak{I}_l = \emptyset$, and it holds for each partition $\mathfrak{I}_k$ of indistinguishable agents that:

$$\begin{aligned} \forall i, j \in \mathfrak{I}_k : ran(A_i) = ran(A_j) \wedge ran(O_i) = ran(O_j) \\ \wedge \forall s \in ran(S), \boldsymbol{a} \in ran(\boldsymbol{A}) : R_i(s, \boldsymbol{a}) = R_j(s, \boldsymbol{a}), \end{aligned} \tag{5}$$

which is a straightforward generalisation of Eq. (2). The equation denotes that each agent in a partition needs to have the same actions and observations available as well as the same reward function. As such, it is sufficient to keep K decision and observation random variables A_k, O_k and reward functions R_k instead of N variables A_i and O_i and reward functions R_i.

Generalising Eq. (3) for partitions means that all permutations θ of those actions $\boldsymbol{a}_{-k}$ that belong to a partition $\mathfrak{I}_k$ in a joint action $\boldsymbol{a}$, i.e., $\boldsymbol{a}_k = \pi_{\mathfrak{I}_k}(\boldsymbol{a})$, map to the same numbers. The same holds for permutations of observations $\boldsymbol{o}_k = \pi_{\mathfrak{I}_k}(\boldsymbol{o})$. Formally, with $\boldsymbol{a}_{-k}$ ($\boldsymbol{o}_{-k}$) referring to the remaining actions (observations), i.e., $\boldsymbol{a}_{-k} = \boldsymbol{a} \setminus \boldsymbol{a}_k$ ($\boldsymbol{o}_{-k} = \boldsymbol{o} \setminus \boldsymbol{o}_k$), the condition changes to:

$$\begin{aligned} \forall \boldsymbol{a} \in ran(\boldsymbol{A}), \boldsymbol{a} = (\boldsymbol{a}_k, \boldsymbol{a}_{-k}) : \forall \theta(\boldsymbol{a}_k) : \forall \boldsymbol{o} \in ran(\boldsymbol{O}), \boldsymbol{o} = (\boldsymbol{o}_k, \boldsymbol{o}_{-k}) : \forall \theta(\boldsymbol{o}_k) : \\ T(s', s, \boldsymbol{a}_k, \boldsymbol{a}_{-k}, \boldsymbol{o}) = T(s', s, \theta(\boldsymbol{a}_k), \boldsymbol{a}_{-k}, \boldsymbol{o}) \\ T(s', s, \boldsymbol{a}, \boldsymbol{o}_k, \boldsymbol{o}_{-k}) = T(s', s, \boldsymbol{a}, \theta(\boldsymbol{o}_k), \boldsymbol{o}_{-k}) \\ R_k(s, \boldsymbol{a}_k, \boldsymbol{a}_{-k}) = R_k(s, \theta(\boldsymbol{a}_k), \boldsymbol{a}_{-k}) \end{aligned} \tag{6}$$

Generalising Eq. (4) to the partition case means that all agents within a partition are conditionally independent from each other, leading to factorising the functions T and R as follows, with with $a_i = \pi_{A_i}(\boldsymbol{a}_k)$ and $o_i = \pi_{O_i}(\boldsymbol{o}_k)$:

$$\begin{aligned} &\forall \boldsymbol{a}_{-k} \in ran(\boldsymbol{A}_{-k}) : \forall \boldsymbol{a}_k \in ran(\boldsymbol{A}_k) : \forall \boldsymbol{o}_{-k} \in ran(\boldsymbol{O}_{-k}) : \forall \boldsymbol{o}_k \in ran(\boldsymbol{O}_k) : \\ &\qquad T(s', s, \boldsymbol{a}_k, \boldsymbol{a}_{-k}, \boldsymbol{o}_k, \boldsymbol{o}_{-k}) = \prod_{i \in \mathfrak{I}_k} P_i(s', o_i, \boldsymbol{o}_{-k} \mid s, a_i, \boldsymbol{a}_{-k}) \\ &\qquad R_k(s, \boldsymbol{a}_k, \boldsymbol{a}_{-k}) = \sum_{i \in \mathfrak{I}_k} R_k^i(s, a_i, \boldsymbol{a}_{-k}). \end{aligned} \tag{7}$$

Given the generalisations in Eqs. (5) to (7), Lemma 1 applies to the group setting with the same argument as above.

Corollary 1. *If Eqs. (5) to (7) hold, then the factors P_i and summands R_k^i in Eq. (7) are identical.*

Corollary 1 allows for a more compact representation of T and R as both functions factorise into a set of functions, one for each partition. Additionally, the factorised functions have lower dimensionality, because they no longer require the joint action (and observation) as input. With $\pi_{\boldsymbol{k}}(\boldsymbol{a})$ denoting the projection of $\boldsymbol{a}$ onto the agents in $\boldsymbol{k}$ (the same for $\boldsymbol{o}$), the functions boil down to:

$$\begin{aligned} T(s', s, \boldsymbol{a}, \boldsymbol{o}) &= \prod_{\boldsymbol{k} \in \times_{k=1}^K \mathfrak{I}_k} \prod_{k \in \boldsymbol{k}} P_k(s', \pi_{\boldsymbol{k}}(\boldsymbol{o}) \mid s, \pi_{\boldsymbol{k}}(\boldsymbol{a})) \\ R(s, \boldsymbol{a}) &= \sum_{\boldsymbol{k} \in \times_{k=1}^K \mathfrak{I}_k} R_k(s, \pi_{\boldsymbol{k}}(\boldsymbol{a})) \end{aligned} \tag{8}$$

Corollary 2. *The complexity of Eq. (8) is no longer exponential in the number of agents N in a POSG M, in which Eqs. (5) to (7) hold.*

Proof. While there are $n^K \cdot K, n = \max_{k \in \{1,\dots,K\}} |\mathfrak{I}_k|$ many factors in T (n^K summands in R), the complexity of each T_k (R_k) is down to $s \cdot o^K \cdot a^K$ ($s \cdot a^K$), which is no longer exponential in the overall number of agents N.

With indistinguishability defined on a partition level, we next define isomorphic POSGs that compactly encode this indistinguishability.

3.3 Isomorphic POSGs

POSGs that have a partitioned agent set and additionally exhibit symmetric behaviour in the functions T and R as well as conditional independence among the members of the same partition allow for a more compact representation that no longer depends on the number of agents, but only the number of partitions as seen in Eq. (8). We call such a POSG isomorphic and define it as follows:

Definition 2. *An* isomorphic *POSG $\bar{M}$ is a tuple $(\bar{\boldsymbol{I}}, S, \bar{\boldsymbol{A}}, \bar{\boldsymbol{O}}, \bar{\boldsymbol{T}}, \bar{\boldsymbol{R}})$, with*

- $\bar{\boldsymbol{I}}$ *a partitioning* $\{\mathfrak{I}_k\}_{k=1}^K$ *of agents,* $n_k = |\mathfrak{I}_k|$ *and* $|\bar{\boldsymbol{I}}| = \sum_k n_k = N$*,*
- S *a random variable with a set of states as range as in Definition 1,*
- $\bar{\boldsymbol{A}} = \{\bar{A}_k\}_{k=1}^K$ *a set of decision random variables* $\bar{A}_k$*, each with possible* actions *as range,* $ran(A_i) = ran(\bar{A}_k) \forall i \in \mathfrak{I}_k$*, and* $ran(\boldsymbol{A}_k) = \times_{i \in \mathfrak{I}_k} ran(A_k)$*,*
- $\bar{\boldsymbol{O}} = \{\bar{O}_k\}_{k=1}^K$ *a set of random variables* $\bar{O}_k$*, each with a set of* observations *as range,* $ran(O_i) = ran(\bar{O}_k) \forall i \in \mathfrak{I}_k$*, and* $ran(\boldsymbol{O}_k) = \times_{i \in \mathfrak{I}_k} ran(O_k)$*,*
- $\bar{\boldsymbol{T}} = \{\bar{T}_k(S, S', \bar{\boldsymbol{A}}, \bar{\boldsymbol{O}})\}_{k=1}^K$ *a set of probability distributions* $\bar{T}_k(S, S', \bar{\boldsymbol{A}}, \bar{\boldsymbol{O}}) = P(S', \bar{\boldsymbol{O}} \mid S, \bar{\boldsymbol{A}})$*, and*
- $\bar{\boldsymbol{R}} = \{\bar{R}_k(S, \bar{\boldsymbol{A}})\}_{k=1}^K$ *a set of reward functions.*

Before presenting model complexity results, we show that an isomorphic POSG is equivalent to a standard (ground) POSG that fulfils Eq. (5), (6), (7).

Theorem 1. *An isomorphic POSG* $\bar{M}$ *has an equivalent POSG* M*, in which Eqs. (5) to (7) hold.*

Proof. We transform the isomorphic POSG $\bar{M}$ into a first-order representation by adding logical variables for each partition to represent the agents of each partition, leading to parameterised random variables $\bar{A}_k = A_k(X_k), \bar{O}_k = O_k(X_k)$ as decision and observation random variables, with the domains of the logical variables X_k being the partitions of agents, i.e., $dom(X_k) = \mathfrak{I}_k$. Ranges are not affected. The probability distributions $T_k(S, S', \bar{\boldsymbol{A}}, \bar{\boldsymbol{O}})$ and reward functions $R_k(S, S', \bar{\boldsymbol{A}})$ then can be written as

$$T_k(S, S', \bar{\boldsymbol{A}}, \bar{\boldsymbol{O}}) = T_k(S, S', A_1(X_1), \ldots, A_K(X_K), O_1(X_1), \ldots, O_K(X_K))$$
$$R_k(S, S', \bar{\boldsymbol{A}}) = R_k(S, S', A_1(X_1), \ldots, A_K(X_K))$$

which is only a syntactic transformation with no consequence for the semantics of the functions. We now turn $\bar{M}$ into M by essentially grounding $\bar{M}$, which means we expand the compact representations of $\bar{M}$ into the joint representations of M: Grounding a parameterised random variable replaces each logical variable with a constant, here an agent, leading to a set of random variables $A_k(x_{k,i}), O_k(x_{k,i})$, $i \in \mathfrak{I}_k$, which is equivalent to the random variables A_i, O_i of M with Eq. (5). Grounding a function that contains parameterised random variables yields a set of functions with identical mappings and the logical variables replaced with all possible combinations of constants (agents) over the partitions, i.e., $\times_{k=1}^K dom(X_k)$, which follow multiplicative semantics in T as T is a probability distribution and additive semantics in R as R is an additive reward function. As such, Eq. (7) holds, with $P_i = P_j = P_k$ and $R_i = R_j = R_k$ within each partition $\mathfrak{I}_k$. When multiplying identical functions per partition, symmetric behaviour as formalised in Eq. (6) emerges, since the same probabilities from P_k (rewards from R_k) are multiplied (added) when different permutations of a partition action (observation) are handled. □

Model Complexity. Given the more compact representation of the functions in a POSG, we get a drastic model space reduction:

Corollary 3. *The model complexity of an isomorphic POSG $\bar{M}$ is no longer dependent on the number of agents N.*

Proof. The sizes of the transition function T and reward functions $\boldsymbol{R}$, which follow from their input arguments, dominate the model complexity. Formally, the sizes $\mathbb{T}$ and $\mathbb{R}$ of the functions $T(S', S, \boldsymbol{A}, \boldsymbol{O})$ and $\{R_i(S, \boldsymbol{A})\}_{i \in \boldsymbol{I}}$, respectively, of a POSG M of Definition 1 equivalent to an isomorphic POSG $\bar{M}$ of Definition 2 lie in $\mathbb{T} \in O(s^2 a^N o^N)$ and $\mathbb{R} \in O(Nsa^N)$, with $a = \max_i |ran(A_i)|$ and $o = \max_i |ran(O_i)|$, $i \in \boldsymbol{I}$, which has N as an exponent. The sizes $\bar{\mathbb{T}}$ and $\bar{\mathbb{R}}$ of the functions $\boldsymbol{T}$ and $\boldsymbol{R}$ of an isomorphic POSG $\bar{M}$ of Definition 2 lie in $\bar{\mathbb{T}} \in O(Ks^2 a^K o^K)$ and $\bar{\mathbb{R}} \in O(Ksa^K)$, which does not reference N anymore. □

This indistinguishability only translates into smaller functions than in a ground POSG of Definition 1 if $K \ll N$. Given a particular ground POSG, one could check for Eqs. (5) to (6) to hold and then compactly encode the POSG as an isomorphic one, which yields K and the respective factorisations of T and $\boldsymbol{R}$. However, in such a scenario, one still has to store the ground POSG, which requires exponential space. Then, one has to check Eqs. (5) and (7), which can be especially cumbersome for Eq. (7), and compute the factors referenced in Eq. (6). Thus, the modelling scenario would be to directly formulate an isomorphic POSG for the cases where one assumes Eqs. (5) to (6) to hold to bypass the effort. In such a case, one could even consider tasks such as how many agents are necessary per partition to fulfil certain conditions, where the overall number of agents might no longer be known.

Of course, a compact representation does not help much if a solution algorithm has to revert to the grounded representation. Thus, next, we present a lifted version of the multi-agent dynamic programming operator [7] that works with isomorphic POSGs.

4 Lifting Dynamic Programming for Isomorphic POSGs

Multi-agent dynamic programming solves a finite-horizon POSG by iteratively eliminating weakly-dominated strategies [7] using a dynamic programming operator. For a lifted version that works with isomorphic POSGs, we have to adapt the operator to work with partitions. Given the results of the previous section, we prove in this section that all agents of the same partition use the same set of policies. Thus, it is sufficient to consider a representative agent for each partition and let the operator perform an exhaustive backup, compute corresponding value vectors, and then prune policies once per partition.

4.1 Weakly Dominated Policies Among Indistinguishable Agents

Before setting up the lifted dynamic programming operator, we argue why we can work with representatives for each partition of indistinguishable agents.

Lemma 2. *Indistinguishable agents share the same set of available policies as well as the same belief and state values.*

Algorithm 2. Lifted Multi-Agent Dynamic Programming Operator

function LIFT-MA-DP(Set of sets of policies with corresponding value vectors $\{(\Pi_k^{t-1}, \boldsymbol{V}_k^{t-1})\}_{k=1}^K$)
  $\Pi_k^t \leftarrow$ Perform exhaustive backup using Π_k^{t-1} for each partition $\mathfrak{I}_k \in \bar{\boldsymbol{I}}$
  $\boldsymbol{V}_k^t \leftarrow$ Calculate new value vectors for each partition $\mathfrak{I}_k \in \bar{\boldsymbol{I}}$
  while $\exists \pi_{k,l}^t \in \Pi_k^t$: Eq. (9) holds **do**
    $\Pi_k^t \leftarrow \Pi_k^t \setminus \{\pi_{k,l}^t\}$, $\boldsymbol{V}_k^t \leftarrow \boldsymbol{V}_k^t \setminus \{v_{k,l}^t\}$
  return $\{(\Pi_k^t, \boldsymbol{V}_k^t)\}_{k=1}^K$

Proof. With agents i, j of a partition $\mathfrak{I}_k$ sharing the same actions and observations, the same reward function R_k, and the same transition function T_k due to conditional independences among them, the same possible policies are available to agents of the same partition by construction. Additionally, the values V_i, V_j of a belief $b_i = b_j$ are identical, i.e., $V_i = V_j = V_k$ given a policy $\pi_i = \pi_j$, a state s, and the other agents' policies $\boldsymbol{\pi}_{-i} = \boldsymbol{\pi}_j$, which are identical due to $\pi_i = \pi_j$. The same holds for the values V_i', V_j' of $s, \pi_i = \pi_j, \boldsymbol{\pi}_i = \boldsymbol{\pi}_j$, which can be computed with the factorised versions of T and R, meaning $V_i' = V_j' = V_k'$. □

Theorem 2. *Policies that are weakly dominated for agent i are also weakly dominated for an indistinguishable agent j.*

Proof. As a direct consequence of Lemma 2, the same policies can be pruned if Eq. (1) holds. □

The effect of Theorem 2 is that we only need to back up policies once for each partition using a representative agent and then perform a pruning on those policies. Formally, the condition for pruning a partition policy $\pi_{k,l}^t$ becomes:

$$\forall s \in ran(S), \boldsymbol{\pi}_{-k}^t \in \boldsymbol{\Pi}_{-k}^t \exists \pi_{k,l}^t \in \Pi_k^t : V_k(s, \pi_{k,l}^t, \boldsymbol{\pi}_{-k}^t) \leq V_k(s, \pi_{k,m}^t, \boldsymbol{\pi}_{-k}^t), \quad (9)$$

The linear programme can be adapted accordingly (and is left out due to space restrictions). Next, we present the dynamic programming operator.

4.2 Lifted Dynamic Programming Operator

Given the results of the previous sections, we set up the lifted dynamic programming operator that operates on representatives for the partitions in an isomorphic POSG.

Algorithm 2 shows the lifted version of the operator, which works exactly as before, just on a partition level. So, it takes a set of policies $\Pi_k^{t-1} = \{\pi_{k,l}^{t-1}\}_{l=1}^m$ of depth $t-1$ for partition $\mathfrak{I}_k$. For each policy, there exists a value vector $v_{k,l}^{t-1}$ that denotes the value of that policy in each possible combination of state $s \in ran(S)$ and policies of the other partitions $\boldsymbol{\pi}_{-k} \in \boldsymbol{\Pi}_{-k}^{t-1}$. The operator then proceeds as before with first performing an exhaustive backup and second pruning weakly dominated policies per partition.

We briefly argue for correctness of Algorithm 2 based on the correctness of Algorithm 1:

Algorithm 3. Solving isomorphic POSGs

function LIFTED-SOLVE(POSG $\bar{M}$, horizon τ)
  $\Pi_k^0 \leftarrow \emptyset$ for each partition $\mathfrak{I}_k \in \bar{\boldsymbol{I}}$
  $\boldsymbol{V}_k^0 \leftarrow \emptyset$ for each partition $\mathfrak{I}_k \in \bar{\boldsymbol{I}}$
  while $t \in \{1, \dots, \tau\}$ **do**
    $\{(\Pi_k^t, \boldsymbol{V}_k^t)\}_{k=1}^K \leftarrow$ LIFT-MA-DP($\{(\Pi_k^{t-1}, \boldsymbol{V}_k^{t-1})\}_{k=1}^K$)
  return $\{(\Pi_k^\tau, \boldsymbol{V}_k^\tau)\}_{k=1}^K$

Theorem 3. *Using Algorithm 2 on an isomorphic POSG is equivalent to using Algorithm 1 on a POSG, in which Eqs. (5) to (7) hold.*

Proof. Given the equivalence between an isomorphic POSG and a POSG, in which Eqs. (5) to (7) hold, and Theorem 2, we conclude that Algorithm 2 only prunes policies that Algorithm 1 would prune.

Algorithm 3 shows how to solve an isomorphic POSG using Algorithm 2, which repeatedly calls Algorithm 2 for increasingly deeper policy trees until the horizon is reached. The resulting set of policies can then be used to select the best policies given an initial state distribution.

5 Discussion

Before we consider related work, we briefly consider the nanoscale medical system as an isomorphic POSG and discuss the expressivity of isomorphic POSGs, which touches upon the assumptions made and limitations imposed.

5.1 Nanoscale Medocal Systems as Isomorphic POSGs

When simulating a nanoscale medical system based on [2], its components are divided into two categories of agents, nanosensors and nanobots, based on how they respond to particular markers and so-called messages in a blood stream. Together, these agents build the set $\boldsymbol{I}$ of agents, which are divided into K partitions, one partition for each marker / message that should be detected in the blood stream. According to early research in such a nanoscale medical system [2], each partition may contain about 64,000 agents, which means that the agent set must be at least $K \cdot 64,000$, with K being single-digit. Each agent has the ability to observe its marker/message (or not) and to output a marker of its own (or not), yielding two possible observations and actions for each agent. If describing the environment through the presence and absence of such markers and messages, there are 2^K possible states. Considering concrete numbers, e.g., with four markers and one message type, the agent set is of size $N = 5 \cdot 64,000 = 320,000$ partitioned into $K = 5$ partitions and the state space is of size $s = 2^5 = 32$, which leads to model sizes of the following in the ground case:

$$\mathbb{T} \in \mathcal{O}(32^2 \cdot 2^{320,000} \cdot 2^{320,000}), \tag{10}$$

$$\mathbb{R} \in \mathcal{O}(320,000 \cdot 32 \cdot 2^{320,000}) \tag{11}$$

as well as model sizes of the following in the isomorphic case:

$$\bar{\mathbb{T}} \in \mathcal{O}(5 \cdot 32^2 \cdot 2^5 \cdot 2^5), \tag{12}$$

$$\bar{\mathbb{R}} \in \mathcal{O}(5 \cdot 32 \cdot 2^5). \tag{13}$$

Expressivity. Isomorphic POSG use the fact that we have some form of structure in the agent set for efficiency gains. Having a small number of different types of agents with a set of predetermined actions and observations available to them is a common occurrence and encoded in Eq. (5). However, the prerequisites regarding symmetric behaviour in Eq. (6) and conditional independence in Eq. (7) do pose some strong assumptions on the transition and reward function T and R in a trade-off with a drastic model complexity reduction and a much more efficient solution approach that is no longer exponentially dependent on N but only K, with $K \ll N$. However, isomorphic POSGs can encode more complex settings than a set of K independent POMDPs as the interaction between the different partitions is still a part of T and R.

Regarding the logical variables used in the proof for Theorem 1, this work introduces constructs of first-order logic into the agent set of a POSG, which allows for using this information explicitly for efficiency gains, which is not possible if using a standard POSG, which has this information only implicitly available.

Related Work. We look at related work for POSGs as well as lifting in the context of MDP-based formalisms.

Next to dynamic programming for POSGs [7], there exist approximate pruning methods as an extension [11]. To work on scalability, research has focused on approximations for specific variants of POSGs, such as zero-sum games (e.g., [8,22,25]) and one-sided POSGs (e.g., [4,9,10]), and POSGs with common payoffs [5]. On a more practical level, partially observable game-theoretic Golog extends Golog with game-theoretic multi-agent planning in POSGs [6].

In offline decision making using MDP-based frameworks, lifting has been used in calculations to exploit relational structures in the state space of (PO)MDPs: In first-order MDPs (FOMPDs) [1], the situation calculus [14] is used to describe the state space. Factorised FOMDPs additionally assume a factorised representation of the state space [19]. Although not lifting-adjacent, object-oriented POMDPs similarly factorise the state space based on objects in the state space [23]. In FO-POMDPs, lifting is applied to policies, pruning policies that are indistinguishable [20]. In open-universe FO-POMDPs [21], the open-universe assumption that often comes with first-order representations is added using Bayesian logic as a basis. Most recently, lifting has been applied to the agent set in DecPOMDPs [3]. On a propositional level, probabilistic inference techniques have been used for multi-agent decision making in DecPOMDPs, factorising the state space using dynamic Bayesian networks, a temporal PGM [12].

To the best of our knowledge, we are the first to apply techniques from lifted probabilistic inference to the general formalism of POSGs, providing groundwork for new solution methods and avenues for approximations.

6 Conclusion

This paper presents a new compact encoding of POSGs, called isomorphic POSGs, which uses a partitioning of the agent set based on indistinguishable agents and conditional independences among indistinguishable agents, which allows for a model complexity that is no longer dependent (exponential or otherwise) on the number of agents, but only dependent on the number of partitions. A lifted multi-agent dynamic programming operator works on isomorphic POSGs, using the compact encoding for efficiency gains, computing and evaluating policies for representatives of each partition, avoiding doing repetitive work for each agent in a partition. As such, isomorphic POSGs have great potential for applications that have a huge number of agents such as nanoscale medical systems.

The next step in future work lies in a case study on nanoscale medical systems, combining the research and work of nanotechnology, medicine, and artificial intelligence through simulation runs. Future work also includes working on loosening the assumptions made for isomorphic POSGs, thereby increasing the expressivity of such POSGs, inspired by the lifting tool of counting (in contrast to isomorphism), as well as additionally using lifting for the state space.

References

1. Boutilier, C., Reiter, R., Price, B.: Symbolic dynamic programming for first-order MDPs. In: IJCAI-01 Proceedings of the 17th International Joint Conference on Artificial Intelligence, pp. 690–697. IJCAI Organization (2001)
2. Braun, T., Fischer, S., Lau, F., Möller, R.: Lifting DecPOMDPs for nanoscale systems—a work in progress. In: 10th International Workshop on Statistical Relational AI at the 1st International Joint Conference on Learning and Reasoning (2021). https://arxiv.org/abs/2001.02021
3. Braun, T., Gehrke, M., Lau, F., Möller, R.: Lifting in multi-agent systems under uncertainty. In: UAI-22 Proceedings of the 38th Conference on Uncertainty in Artificial Intelligence, pp. 1–8. AUAI Press (2022)
4. Carr, S., Jansen, N., Bharadwaj, S., Spaan, M.T.J., Topcu, U.: Safe policies for factored partially observable stochastic games. In: RSS-21 Proceedings of Robotics: Science and Systems XVII, pp. 1–11. RSS Foundation (2021)
5. Emery-Montemerlo, R., Gordon, G., Schneider, J., Thrun, S.: Approximate solutions for partially observable stochastic games with common payoffs. In: AAAMAS-04 Proceedings of the 3rd International Joint Conference on Autonomous Agents and Multiagent Systems, pp. 136–143. IEEE (2004)
6. Finzi, A., Lukasiewicz, T.: Partially observable game-theoretic agent programming in Golog. Int. J. Approximate Reasoning **119**, 220–241 (2020)

7. Hansen, E.A., Bernstein, D.S., Zilberstein, S.: Dynamic programming for partially observable stochastic games. In: AAAI-04 Proceedings of the 19th National Conference on Artificial Intelligence, vol. 4, pp. 709–715 (2004)
8. Horák, K., Bošanský, B.: Solving partially observable stochastic games with public observations. In: Proceedings of the AAAI Conference on Artificial Intelligence, pp. 547–552. AAAI Press (2019)
9. Horák, K., Bošanský, B., Kiekintveld, C., Kamhoua, C.: Compact representation of value function in partially observable stochastic games. In: IJCAI-19 Proceedings of the 28th International Joint Conference on Artificial Intelligence, pp. 350–356. IJCAI Organisation (2019)
10. Horák, K., Bošanský, B., Pěchouček, M.: Heuristic search value iteration for one-sided partially observable stochastic games. In: AAAI-17 Proceedings of the 31st AAAI Conference on Artificial Intelligence, pp. 558–564 (2017)
11. Kumar, A., Zilberstein, S.: Dynamic programming approximations for partially observable stochastic games. In: FLAIRS-09 Proceedings of the 22nd International Florida Artificial Intelligence Research Society Conference. AAAI Press (2009)
12. Kumar, A., Zilberstein, S., Toussaint, M.: Probabilistic inference techniques for scalable multiagent decision making. J. Artif. Intell. Res. **53**, 223–270 (2015)
13. Lau, F., Wendt, R., Fischer, S.: DNA-based molecular communication as a paradigm for multi-parameter detection of diseases. In: ACM NanoCom-17 Proceedings of the 4th ACM International Conference on Nanoscale Computing and Communication 2017. ACM (2021)
14. McCarthy, J.: Situations, actions, and causal laws. Technical report, Standford University (1963)
15. Niepert, M., Van den Broeck, G.: Tractability through exchangeability: a new perspective on efficient probabilistic inference. In: AAAI-14 Proceedings of the 28th AAAI Conference on Artificial Intelligence, pp. 2467–2475. AAAI Press (2014)
16. Oliehoek, F.A., Amato, C.: A Concise Introduction to Decentralised POMDPs. Springer, Cham (2016). https://doi.org/10.1007/978-3-319-28929-8
17. Poole, D.: First-order probabilistic inference. In: IJCAI-03 Proceedings of the 18th International Joint Conference on Artificial Intelligence, pp. 985–991. IJCAI Organization (2003)
18. Russell, S., Norvig, P.: Artificial Intelligence: A Modern Approach. Pearson (2021)
19. Sanner, S., Boutilier, C.: Approximate solution techniques for factored first-order MDPs. In: ICAPS-07 Proceedings of the 17th International Conference on Automated Planning and Scheduling, pp. 288–295. AAAI Press (2007)
20. Sanner, S., Kersting, K.: Symbolic dynamic programming for first-order POMDPs. In: AAAI-10 Proceedings of the 24th AAAI Conference on Artificial Intelligence, pp. 1140–1146. AAAI Press (2010)
21. Srivastava, S., Russell, S., Ruan, P., Cheng, X.: First-order open-universe POMDPs. In: UAI-14 Proceedings of the 30th Conference on Uncertainty in Artificial Intelligence, pp. 742–751. AUAI Press (2014)
22. Tomášek, P., Horák, K., Aradhye, A., Bošanský, B., Chatterjee, K.: Solving partially observable stochastic shortest-path games. In: IJCAI-21 Proc. of the 30th International Joint Conference on Artificial Intelligence, pp. 4182–4189. IJCAI Organisation (2021)
23. Wandzel, A., Oh, Y., Fishman, M., Kumar, N., Wong, L.L., Tellex, S.: Multi-object search using object-oriented POMDPs. In: ICRA-19 Proceedings of the 2019 International Conference on Robotics and Automation, pp. 7194–7200. IEEE (2019)

24. Wemmenhove, B., Mooij, J.M., Wiegerinck, W., Leisink, M., Kappen, H.J., Neijt, J.P.: Inference in the promedas medical expert system. In: Bellazzi, R., Abu-Hanna, A., Hunter, J. (eds.) AIME 2007. LNCS (LNAI), vol. 4594, pp. 456–460. Springer, Heidelberg (2007). https://doi.org/10.1007/978-3-540-73599-1_61
25. Wiggers, A.J., Oliehoek, F.A., Roijers, D.M.: Structure in the value function of two-player zero-sum games of incomplete information. In: ECAI-16 Proceedings of the 22nd European Conference on Artificial Intelligence, pp. 1628–1629. IOS Press (2016)

Inconsistency Measurement in $\mathrm{LTL_f}$ Based on Minimal Inconsistent Sets and Minimal Correction Sets

Isabelle Kuhlmann[1(✉)] and Carl Corea[2]

[1] Artificial Intelligence Group, University of Hagen, Hagen, Germany
isabelle.kuhlmann@fernuni-hagen.de
[2] Institute for IS Research, University of Koblenz, Koblenz, Germany
ccorea@uni-koblenz.de

Abstract. We investigate the problem of measuring inconsistency in linear temporal logic on finite traces ($\mathrm{LTL_f}$). In particular, we present Answer Set Programming-based approaches to compute a selection of traditional inconsistency measures w.r.t. $\mathrm{LTL_f}$ knowledge bases. In contrast to existing works (mostly on propositional logic), these approaches are novel in the sense that they allow to assess logical inconsistency in presence of temporal operators, as offered by $\mathrm{LTL_f}$. In an experimental evaluation on real-world data from the area of business process management, we show that our approaches are practically feasible.

Keywords: Inconsistency measurement · Linear temporal logic · Answer set programming

1 Introduction

Linear Temporal Logic (LTL) [36] is an increasingly popular logic for specifying *temporal* system behavior. The core idea is that time is represented as a linear sequence of states, where at every state some statements may be true. In regard to this sequence, a set of temporal operators can be used to confine the allowed behavior by means of declarative constraints—for example, the operator $\mathbf{X}$ (*next*) means that a certain formula must hold at the next state, and the operator $\mathbf{G}$ (*globally*) means a formula must hold in every state. In this work, we focus on LTL over *finite traces* ($\mathrm{LTL_f}$), meaning that the considered sequence of states (and any corresponding interpretation) is finite. A core problem for $\mathrm{LTL_f}$ is that of determining whether a set of formulas, also called a *knowledge base* (KB), is unsatisfiable (*inconsistent*). For example, consider the KBs $\mathcal{K}_1 = \{\mathbf{G}a, \mathbf{G}\neg a\}$ and $\mathcal{K}_2 = \{\mathbf{G}a, \mathbf{G}\neg a, b \wedge \mathbf{X}a\}$. Obviously, both KBs are inconsistent, however, the granularity of the inconsistency is different—for $\mathcal{K}_1$, there is only one conflict, namely $\{\mathbf{G}a, \mathbf{G}\neg a\}$, but for $\mathcal{K}_2$ we have two conflicts, namely $\{\mathbf{G}a, \mathbf{G}\neg a\}$ and $\{\mathbf{G}\neg a, b \wedge \mathbf{X}a\}$.

S. Destercke et al. (Eds.): SUM 2024, LNAI 15350, pp. 217–232, 2025.
https://doi.org/10.1007/978-3-031-76235-2_17

Especially in the scope of inconsistency resolution, being able to assess the *severity* of inconsistency is a crucial asset, e. g., for comparing different specifications. As an example, it may be reasonable to argue that $\mathcal{K}_2$ is "more" inconsistent than $\mathcal{K}_1$, due to the different nature of the conflicts. The vast majority of works in the context of LTL_f in combination with inconsistency are on (binary) satisfiability checking and do not focus on quantifying the severity—or, degree—of inconsistency. Therefore, in the work at hand, we focus on the problem of *measuring* inconsistency in LTL_f. To this end, we present novel algorithmic approaches based on *Answer Set Programming* (ASP) for computing a selection of traditional inconsistency measures. Most notably, these approaches are able to analyze inconsistency under a correct consideration of temporal operators, e. g., the KB $\{a, \mathbf{X}\neg a\}$ is consistent in our setting.

In a previous work [3], we focused on state-based inconsistency measures, which can assess the number of states (in the linear sequence) that are affected by the inconsistency. This allows to understand inconsistency in a time-sensitive way, but is not formula-centric. In this work, we present approaches for formula-centric inconsistency measures based on minimal inconsistent subsets and minimal correction sets. This is useful for formula-driven analysis, and pin-pointing the causes of inconsistencies. Specifically, we present computational approaches for the MI-inconsistency measure ($\mathcal{I}_{\mathsf{MI}}$) [20], the *MUS-variable-based* inconsistency measure ($\mathcal{I}_{mv}$) [42], the *problematic* inconsistency measure ($\mathcal{I}_p$) [20] and the *repair* inconsistency measure ($\mathcal{I}_r$) [15] (see Sect. 2.2). The contributions of this work are as follows.

- We introduce the use of inconsistency measurement based on minimal inconsistent subsets, as well as minimal correction sets, in LTL_f.
- We present algorithmic approaches (based on ASP) for computing the inconsistency measures $\mathcal{I}_{\mathsf{MI}}, \mathcal{I}_{mv}, \mathcal{I}_p$ and $\mathcal{I}_r$ for LTL_f KBs.
- For $\mathcal{I}_{mv}$ and $\mathcal{I}_p$ we present two alternative approaches; one based on minimal inconsistent subsets, and one based on minimal correction sets.
- We empirically evaluate our approaches with real-life data sets of the *business process intelligence challenge.* Our experiments show that our approaches are feasible in many settings, allowing to efficiently compute the considered measures for KBs of industrial complexity (Sect. 4).

2 Preliminaries

We continue with preliminaries on LTL_f, the considered inconsistency measures, and answer set programming, in addition to an overview of related work.

2.1 LTL_f, Inconsistency

For LTL_f, we consider a linear sequence of states $t_0, ..., t_m$, where t_0 is the designated starting point. LTL_f formulas are then built over a fixed set of propositional symbols At, and are closed under the Boolean connectives, the unary operator $\mathbf{X}$ (*next*), and the binary operator $\mathbf{U}$ (*until*), using the grammar rule

$$\phi ::= a|(\neg\phi)|(\phi_1 \wedge \phi_2)|(\phi_1 \vee \phi_2)|(\mathbf{X}\phi)|(\phi_1 \mathbf{U} \phi_2)$$

with $a \in \mathsf{At}$.

An LTL$_f$ *knowledge base* (KB) $\mathcal{K}$ is a set of LTL$_f$ formulas. Let $\mathbb{K}$ be the set of all LTL$_f$ KBs. We denote the signature of a formula ϕ (or a KB $\mathcal{K}$) as $\mathsf{At}(\phi)$ (resp. $\mathsf{At}(\mathcal{K})$).

Semantics are given in form of LTL$_f$ interpretations. Concretely, an LTL$_f$ interpretation ω w.r.t. At is a function mapping each state and proposition to 0 or 1, meaning that $\omega(t, a) = 1$ if proposition a is assigned 1 (*true*) in state t. The satisfiability of a formula ϕ by an interpretation ω, denoted by $\omega \models \phi$, is defined via $\omega \models \phi \Leftrightarrow \omega, t_0 \models \phi$, where $\omega, t_i \models \phi$ for any interpretation ω as above and for every $t_i \in \{t_0, ..., t_m\}$ is inductively defined as follows:

$\omega, t_i \models a$ iff $\omega(t_i, a) = 1$ for $a \in \mathsf{At}$

$\omega, t_i \models \neg\phi$ iff $\omega, t_i \not\models \phi$

$\omega, t_i \models \phi_1 \wedge \phi_2$ iff $\omega, t_i \models \phi_1$ and $\omega, t_i \models \phi_2$

$\omega, t_i \models \phi_1 \vee \phi_2$ iff $\omega, t_i \models \phi_1$ or $\omega, t_i \models \phi_2$

$\omega, t_i \models \mathbf{X}\phi$ iff $i < m$ and $\omega, t_{i+1} \models \phi$

$\omega, t_i \models \phi_1 \mathbf{U} \phi_2$ iff either $\omega, t_i \models \phi_1 \wedge \phi_2$ or

$\quad (\omega, t_j \models \phi_2) \wedge (\omega, t_k \models \phi_1) \forall k \in \{i, \ldots, j-1\}$ with $j \in \{i+1, \ldots, m\}$

An interpretation ω satisfies a KB $\mathcal{K} \in \mathbb{K}$ iff $\omega \models \phi$ for all $\phi \in \mathcal{K}$. Further, we make use of the additional operators $\mathbf{F}$ (*eventually*) and $\mathbf{G}$ (*globally*), with $\mathbf{F}\phi$ being defined as $\top\mathbf{U}\phi$, and $\mathbf{G}$ as $\neg\mathbf{F}\neg\phi$.

Note that in some related works, the notion of "simple traces" is adopted, meaning that at every point in time only one proposition should be true [5]. In this work, we do not restrict ourselves to this notion. A simple corollary is that we allow for the satisfiability of formulas such as $a \wedge b$.

Let $\Omega(\mathsf{At})$ denote the set of all LTL$_f$ interpretations for At. Then, for a set of formulas $\mathcal{K}$, define $\mathsf{Mod}(\mathcal{K}) = \{\omega \in \Omega(\mathsf{At}) \mid \omega \models \mathcal{K}\}$ as the set of interpretations that satisfy $\mathcal{K}$. A KB $\mathcal{K}$ is then called consistent (or satisfiable) iff $\mathsf{Mod}(\mathcal{K}) \neq \emptyset$. Otherwise, we say $\mathcal{K}$ is inconsistent (or unsatisfiable), denoted $\mathcal{K} \models \perp$.

Definition 1. *A set $M \subseteq \mathcal{K}$ (with $\mathcal{K} \in \mathbb{K}$) is called a* minimal inconsistent subset *(MIS) of $\mathcal{K}$ if $M \models \perp$ and there is no $M' \subset M$ with $M' \models \perp$. Let $\mathsf{MI}(\mathcal{K})$ be the set of all MISes of $\mathcal{K}$.*

In addition to the notion of an MIS, we require the definition of a *minimal correction set*, which intuitively represents a (minimal) fragment of a KB $\mathcal{K}$ whose removal restores consistency. Moreover, we make use of the notion of *maximal satisfiable subsets*, which, as the name suggests, are satisfiable sets of which each superset is unsatisfiable.

Definition 2. *A set $M \subseteq \mathcal{K}$ (with $\mathcal{K} \in \mathbb{K}$) is a* minimal correction set *(MCS) if $\mathcal{K} \setminus M$ is consistent and for all $M' \subsetneq M$, it holds that $\mathcal{K} \setminus M'$ is inconsistent. Let $\mathsf{MC}(\mathcal{K})$ be the set of all MCSes of $\mathcal{K}$.*

Definition 3. *A set $M \subseteq \mathcal{K}$ (with $\mathcal{K} \in \mathbb{K}$) is a* maximal satisfiable subset *(MSS) if M is consistent and each $M' \supsetneq M$ is inconsistent.*

We further harness two properties stating that (i) the set-complement of an MSS M (i. e., $\mathcal{K} \setminus M$) is an MCS, and (ii) due to *hitting set duality* [37], we have $\bigcup \mathsf{MI}(\mathcal{K}) = \bigcup \mathsf{MC}(\mathcal{K})$.

2.2 Inconsistency Measures

An inconsistency measure is a function $\mathcal{I} : \mathbb{K} \to \mathbb{R}^{\infty}_{\geq 0}$ assigning to a KB a non-negative numerical value. Conventionally, a higher value indicates a larger degree of inconsistency, and the value 0 is reserved for consistent KBs (i. e., $\mathcal{I}(\mathcal{K}) = 0$ iff $\mathcal{K}$ is consistent). A variety of concrete measures has been proposed, cf. [40] for an overview. In this work, we focus on formula-centric inconsistency measures [19], which consider the *formulas* of a KB as the unit of inconsistency. In the following, let $\mathcal{K} \in \mathbb{K}$. The considered inconsistency measures $\mathcal{I}_{\mathsf{MI}}, \mathcal{I}_p$ [20], $\mathcal{I}_{mv}$ [42] and $\mathcal{I}_r$ [15] are defined as follows. The MI-measure $\mathcal{I}_{\mathsf{MI}}$ counts the number of MISes of $\mathcal{K}$, i. e., $\mathcal{I}_{\mathsf{MI}}(\mathcal{K}) = |\mathsf{MI}(\mathcal{K})|$. The *problematic* inconsistency measure counts the number of distinct formulas over all MISes (or, alternatively, over all MCSes, due to hitting set duality). Thus, $\mathcal{I}_p(\mathcal{K}) = |\bigcup \mathsf{MI}(\mathcal{K})| = |\bigcup \mathsf{MC}(\mathcal{K})|$. Similarly, the *MUS-variable-based* inconsistency measure $\mathcal{I}_{mv}$ counts the fraction of the atoms involved in $\mathsf{MI}(\mathcal{K})$ (or $\mathsf{MC}(\mathcal{K})$), divided by the total number of atoms: $\mathcal{I}_{mv}(\mathcal{K}) = \frac{|\mathsf{At}(\bigcup \mathsf{MI}(\mathcal{K}))|}{|\mathsf{At}(\mathcal{K})|} = \frac{|\mathsf{At}(\bigcup \mathsf{MC}(\mathcal{K}))|}{|\mathsf{At}(\mathcal{K})|}$. The *repair* measure $\mathcal{I}_r$ is based on MCSes, and counts the smallest number of formulas that must be deleted to make $\mathcal{K}$ consistent. In other words, $\mathcal{I}_r(\mathcal{K}) = \min\{|M| \mid M \in \mathsf{MC}(\mathcal{K})\}$.

Example 1. Consider $\mathcal{K}_3 = \{\mathbf{G}\neg a, b, \mathbf{G}a\}$. Then $\mathsf{MI}(\mathcal{K}_3) = \{\{\mathbf{G}\neg a, \mathbf{G}a\}\}$ and $\mathsf{MC}(\mathcal{K}_3) = \{\{\mathbf{G}\neg a\}, \{\mathbf{G}a\}\}$, and thus $\mathcal{I}_{\mathsf{MI}}(\mathcal{K}_3) = 1$, $\mathcal{I}_p(\mathcal{K}_3) = 2$, $\mathcal{I}_{mv}(\mathcal{K}_3) = 1/2$, and $\mathcal{I}_r(\mathcal{K}_3) = 1$.

2.3 Answer Set Programming

In general, *answer set programming* [11,31] is a declarative approach aimed at difficult search problems. The use of ASP is motivated by recent work on the topic of computing inconsistency degrees in LTL_f, as well as solving related problems [4,22,23]. The disjunctive extension of ASP [14] is furthermore capable of solving problems in Σ_2^p, which makes it a suitable approach for the task at hand [41]. In short, a *disjunctive logic program* is a set of rules of the form $h_1 \mid \ldots \mid h_m \leftarrow b_{m+1}, \ldots, b_n, \textbf{not}\, b_{n+1}, \ldots, \textbf{not}\, b_k.$, with $0 \leq m \leq n \leq k$, each h_i and b_i $(0 \leq i \leq k)$ being atoms, and "**not**" denoting default negation. The *head* of a rule r is $\mathsf{head}(r) = \{h_1, \ldots, h_m\}$ and the body is $\mathsf{body}(r) = \{b_{m+1}, \ldots, b_n, \textbf{not}\, b_{n+1}, \ldots, \textbf{not}\, b_k\}$. If $\mathsf{body}(r) = \emptyset$ and the head consists of a single atom, r is a *fact.* If $\mathsf{head}(r) = \emptyset$, then r is a *constraint.* We further divide the body elements of r into $\mathsf{body}^+(r) = \{b_{m+1}, \ldots, b_n\}$ and

$\mathsf{body}^-(r) = \{b_{n+1}, \ldots, b_k\}$. An atom/rule/program is *ground* if it does not contain any variables.[1] Let X be a set of ground atoms. We define X to be a *model* of a ground logic program P if for all $r \in P$, $\mathsf{head}(r) \cap X = \emptyset$ whenever $\mathsf{body}^+(r) \subseteq X$ and $\mathsf{body}^-(r) \cap X = \emptyset$. The *reduct* [13] of P w.r.t. X is defined as $P^X = \{\mathsf{head}(r) \leftarrow \mathsf{body}^+(r) \mid \mathsf{body}^-(r) \cap X \neq \emptyset \text{ with } r \in P\}$. Further, X is an *answer set* of P if it is a subset-minimal model of P^X. Note that we refer to the arity n of a predicate p as p/n. For details regarding the ASP syntax, as well as further language constructs, we refer the reader to [2].

2.4 Related Work

This work is related to algorithmic approaches for computing inconsistency measures. In this domain, the main focus has been on approaches for propositional logic, yielding a multitude of algorithmic approaches [21,24–27,33,34,39,42]. In contrast, this work proposes approaches for handling KBs specified in LTLf, which is not generally possible with existing approaches from propositional logic.

Moreover, several other works have recently addressed inconsistency measurement in formalisms other than propositional logic, such as spatio-temporal databases [16], classical and disjunctive databases [35], graph databases [18], and general information spaces [17].

Regarding inconsistency in LTLf, the main focus of related works has been on satisfiability checking, with various approaches based on automata-based reasoning [6,29], or, recently, also SAT-based approaches [10,28,38] and even subsymbolic approaches [32]. In this work, we go beyond the binary satisfiability notion and provide means for assessing the severity of inconsistency, which can provide many valuable insights for tasks such as debugging inconsistent KBs and belief revision. In [38], the authors show how to compute *unsatisfiable cores* in LTLf, which, in our terms, would be (possibly non-minimal) inconsistent subsets. Some measures presented in this work are based on a similar notion (minimal inconsistent subsets) and are therefore vaguely related. Yet, as stated above, the work in [38] does not enumerate MISes, which is however incorporated in an encoding in the work at hand (cf. $\mathcal{I}_{\mathsf{MI}}$). Regardless of this, note that, as introduced, many measures (such as $\mathcal{I}_r$) are not based on MISes in general, and therefore require different approaches as we will present at a later point.

3 Algorithmic Approaches Based on ASP

Regarding the computation of the considered inconsistency measures, we propose an ASP-based approach. To be precise, we describe one method to compute $\mathcal{I}_{\mathsf{MI}}$, and, respectively, $\mathcal{I}_r$, as well as two methods for $\mathcal{I}_{mv}$ and $\mathcal{I}_p$. Our motivation for this approach is twofold: First, computing the measures requires to conduct computationally expensive search problems (some complexity results are

[1] Note that, following the Clingo syntax, all variable names we use start with an uppercase letter, and all constant names start with a lowercase letter.

provided in [3]), so logic programming in general appears to be a suitable candidate. Second, ASP allows to encode the measure definitions in a well-readable form, which may prove to be beneficial when working with these formalizations. We acknowledge that there exist other possible solution approaches, however, the investigation of such is subject of future work.

With regard to $\mathcal{I}_{\mathsf{MI}}$, we need to know the number of MISes, so we introduce an ASP encoding which allows us to extract, and therefore also to count, MISes in LTL_f. In order to compute $\mathcal{I}_r$, on the other hand, we need to find the smallest MCS; to this end, we demonstrate how to encode the notion of MCSes in ASP. The fact that $\mathcal{I}_{mv}$ and $\mathcal{I}_p$ can be defined using either MISes or MCSes (see Sect. 2.2) allows us to present (and later on compare) two different approaches.

3.1 MIS-Based Approaches

In the following, we define an encoding $P^{\mathsf{MIS}}(\mathcal{K})$, which models the notion of an MIS w.r.t. a given KB $\mathcal{K}$ in ASP. First, we encode the formulas in $\mathcal{K}$ by using the same approach as in [4,22,23]. To be precise, each atom a in the signature of a formula $\phi \in \mathcal{K}$ is represented as $\mathit{atomInFormula}(a, \phi)$ (KB1). Further, atoms $a \in \mathsf{At}(\mathcal{K})$ and formulas $\phi \in \mathcal{K}$ are encoded as *atom*/1 (KB2) and *kbElement*/1 (KB3), respectively.[2] Moreover, we inductively represent each formula and its subformulas according to their connectives. Thus, a conjunction $\phi = \phi_1 \wedge \phi_2$ is modeled as $\mathit{conjunction}(\phi, \phi_1, \phi_2)$ (KB4). In the same fashion, we model disjunctions as *disjunction*/3 (KB5), negations as *negation*/2 (KB6), and formulas consisting of individual atoms as *formulaIsAtom*/2 (KB7). The temporal operators are encoded analogously, i. e., a formula $\phi = \mathbf{X}\phi_1$ is encoded as $\mathit{next}(\phi, \phi_1)$ (KB8), $\phi = \phi_1\mathbf{U}\phi_2$ as $\mathit{until}(\phi, \phi_1, \phi_2)$ (KB9), $\phi = \mathbf{G}\phi_1$ as $\mathit{globally}(\phi, \phi_1)$ (KB10), and $\phi = \mathbf{F}\phi_1$ as $\mathit{eventually}(\phi, \phi_1)$ (KB11). The temporal states are modeled by adding the fact $\mathit{finalState}(m)$ (KB12), with regard to a given final state t_m, in addition to the rule "$\mathit{state}(0\,..\,\mathrm{X}) \leftarrow \mathit{finalState}(\mathrm{X})$." (KB13). Based on Proposition 2 in [3], it follows that setting m to the maximal depth d over all formulas[3] suffices to obtain an interpretation that correctly characterizes (in)consistency of the KB. Thus, in the following, the length of interpretations is fixed to the maximal depth as described.

The two truth values 1 and 0 are represented as $\mathit{tv}(t)$ (KB14), and $\mathit{tv}(f)$ (KB15), respectively. We summarize the set of rules specifying the contents of the given KB $\mathcal{K}$ and the temporal states, as well as the truth values (i. e., the rules defined by (KB1)–(KB15)) as $P^{\text{KB}}(\mathcal{K})$, and initialize $P^{\mathsf{MIS}}(\mathcal{K}) := P^{\text{KB}}(\mathcal{K})$.

After representing the given KB itself in ASP, we still need to encode how the formulas should be evaluated. To this end, we introduce *truthValue*/4, and utilize it to construct rules for each operator. This overall approach is the same

[2] In the actual implementation, a and ϕ are represented by uniquely defined constants, i. e., strings starting with lowercase letters (e. g., ϕ_1 could be represented as *phi_1*).

[3] For any formula ϕ, $d(\phi)$ is inductively defined via $d(a) = 0$ for $a \in \mathsf{At}$, $d(\neg\phi) = d(\phi)$, $d(\phi_1 \wedge \phi_2) = d(\phi_1 \vee \phi_2) = \max\{d(\phi_1), d(\phi_2)\}$, $d(\mathbf{X}\phi) = 1 + d(\phi)$, and $d(\phi_1\mathbf{U}\phi_2) = 1 + \max\{d(\phi_1), d(\phi_2)\}$.

as the one described in [4,22,23], however, because in the work at hand we make use of the saturation technique [9], the use of the default negation (**not**) needs to be avoided, which consequently leads to a slightly different, more explicit representation. For instance, the conjunction $\phi = \phi_1 \wedge \phi_2$ could be interpreted as "ϕ evaluates to 1 (w.r.t. a given state) if both ϕ_1 and ϕ_2 are 1, and it evaluates to 0 otherwise". However, since "otherwise" would intuitively be modeled in ASP by means of default negation, we need to use the more explicit description "ϕ evaluates to 0 if ϕ_1 is 0 or if ϕ_2 is 0. The actual ASP representation of the conjunction operation is shown in Listing 1, and added to $P^{\mathsf{MIS}}(\mathcal{K})$.

```
truthValue(F, T, I, t) ← conjunction(F, G, H), truthValue(G, T, I, t), truthValue(H, T, I, t),
    state(T), interpretation(I).
truthValue(F, T, I, f) ← conjunction(F, G, _), truthValue(G, T, I, f), state(T), interpretation(I).
truthValue(F, T, I, f) ← conjunction(F, _, G), truthValue(G, T, I, f), state(T), interpretation(I).
```

Listing 1. Encoding of the $\wedge$ operation.

Note that the ASP representation of the evaluation of a formula refers not only to a temporal state (represented by *state*/1), but also to a specific truth assignment (*interpretation*/1). The purpose of the latter is to check multiple sets of formulas for satisfiability individually. As for the other classical logic operators ($\vee$ and $\neg$) as well as the evaluation of formulas consisting of individual atoms, are modeled in the same way and the corresponding rules are added to $P^{\mathsf{MIS}}(\mathcal{K})$.

The idea behind the encoding of the LTL-specific operators is essentially the same as for the classical operators, however, we need to pay closer attention to how the temporal states are interconnected. As explained in Sect. 2.1, a formula $\mathbf{X}\phi$ evaluates to 0 in state t_i if $t_i \geq m$, with m being the index of the final state (see line 1 in Listing 2). Otherwise, $\mathbf{X}\phi$ evaluates to the same truth value in state t_i as ϕ in $t_j := t_i + 1$ (line 2). With regard to $\mathbf{U}$, we have to distinguish between two cases in which a formula $\phi_1\mathbf{U}\phi_2$ can evaluate to 1 in a given state t_i (see Sect. 2.1). In the first case, $\phi_1 \wedge \phi_2$ evaluates to 1 in t_i (line 3 in Listing 2), in the second case, there exists a state t_j, with $j > i$ and $j \leq m$, such that ϕ_2 evaluates to 1 in t_j, and ϕ_1 evaluates to 1 in each $t_i, \ldots, t_{j-1}$. The latter point is achieved by means of a cardinality constraint (line 4). Further, there are two cases in which $\phi_1\mathbf{U}\phi_2$ evaluates to 0 w.r.t. t_i. The first case is that ϕ_2 never evaluates to 1 in any state t_j with $j \geq i$ (line 5). The other case is that ϕ_2 is 1 in some state t_j (and $j > i$), however, ϕ_1 becomes 0 in a state t_k, with $i \leq k < j$ (line 6). A formula $\mathbf{G}\phi$ evaluates to 0 in state t_i if ϕ is 0 in t_i itself or at least one of the following states (see line 7 in Listing 2). Thus, $\mathbf{G}\phi$ can only evaluate to 1 in t_i if it is 1 in each $t_i, \ldots, t_m$ (line 8). The $\mathbf{F}$ operator is encoded in the same fashion as the $\mathbf{G}$ operator, but in reverse—a formula $\mathbf{F}\phi$ evaluates to 0 in state t_i if it is false in each $t_i, \ldots, t_m$, and it evaluates to 1 if ϕ is 1 in t_i itself or in one of the subsequent states. All rules concerning the evaluation of the temporal operators are added to $P^{\mathsf{MIS}}(\mathcal{K})$.

```
truthValue(F, Ti, I, f) ← next(F, _), state(Ti), Ti ≥ M, finalState(M), interpretation(I).
truthValue(F, Ti, I, V) ← next(F, G), state(Ti), tv(V), Tj = Ti + 1, Ti < M, finalState(M),
        truthValue(G, Tj, I, V), interpretation(I).
truthValue(F, Ti, I, t) ← until(F, G, H), state(Ti), truthValue(G, Ti, I, t), truthValue(H, Ti, I, t),
        interpretation(I).
truthValue(F, Ti, I, t) ← until(F, G, H), state(Ti), state(Tj), Tj > Ti, Tj ≤ M, finalState(M),
        X{truthValue(G, T, I, t) : state(T), T ≥ Ti, T < Tj}X, X = Tj − Ti, truthValue(H, Tj, I, t),
        interpretation(I).
truthValue(F, Ti, I, f) ← until(F, G, H), state(Ti), X{truthValue(H, T, I, f) : state(T), T ≥ Ti}X,
        finalState(M), X = M − Ti + 1, interpretation(I).
truthValue(F, Ti, I, f) ← until(F, G, H), state(Ti), state(Tj), truthValue(H, Tj, I, t), Tj > Ti,
        Tj ≤ M, finalState(M), 1{truthValue(G, T, I, f) : state(T), T ≥ Ti, T < Tj},
        interpretation(I).
truthValue(F, Ti, I, f) ← globally(F, G), state(Ti), 1{truthValue(G, T, I, f) : state(T), T ≥ Ti},
        interpretation(I).
truthValue(F, Ti, I, t) ← globally(F, G), state(Ti), X{truthValue(G, T, I, t) : state(T), T ≥ Ti}X,
        finalState(M), X = M − Ti + 1, interpretation(I).
```

Listing 2. Encoding of the **X**, **U**, and **G** operations.

In order to actually retrieve an MIS, we first guess a candidate set S_{cs} consisting of at least one formula $\phi \in \mathcal{K}$ (see line 1 in Listing 3), and define some auxiliary predicates (lines 2–4). The first one (line 2) determines whether an atom is contained in S_{cs}. The second one (line 3) gets the number of elements in S_{cs}, i. e., $|S_{cs}|$. The last one (line 4) computes $|S_{cs}| - 1$, which is the size of the subsets of S_{cs} that need to be checked for satisfiability. To encode that each subset needs to be satisfiable, we define $|S_{cs}|$ subsets (line 5), and state that they need to consist of exactly $|S_{cs}| - 1$ formulas from S_{cs} (line 6). In addition, we need to ensure that no two subsets are the same (lines 7–8). Since each subset could be satisfiable under a different truth value assignment, we define one interpretation for each of them (line 9), and assign each atom exactly one truth value w.r.t. each interpretation (line 10). Note that u is a separate interpretation that we later require to check S_{cs} for unsatisfiability, which is why it is explicitly excluded from the satisfiability checks of the subsets. We can now model that a subset is satisfiable if all of its formulas can be evaluated to 1 under the same interpretation (line 11). Moreover, we add an integrity constraint (line 12) which ensures that all subsets must be satisfiable (otherwise, S_{cs} could not be an MIS) in order to derive an answer set.

```
 1 1{inCs(X) : kbElement(X)}.
 2 atomInCs(A) ← atomInFormula(A, F), inCs(F).
 3 numElsInCs(X) ← X = #count{F : inCs(F)}.
 4 numElsInSubset(X) ← numElsInCs(Y), X = Y − 1.
 5 subset(1..X) ← numElsInCs(X), X > 1.
 6 Y{inSubset(F, S) : inCs(F)}Y ← subset(S), numElsInSubset(Y).
 7 subsetEq(S1, S2) ← subset(S1), subset(S2), S1 ≠ S2, X = #count{F : inSubset(F, S1),
        inSubset(F, S2)}, numElsInSubset(Y), X == Y.
 8 ← subsetEq(S1, S2).
 9 interpretation(1..X) ← numElsInCs(X), X > 1.
10 1{truthValue(A, T, I, V) : tv(V)}1 ← atom(A), state(T), interpretation(I), I ≠ u.
11 subsetIsSat(S) ← subset(S), numElsInSubset(X), X{truthValue(F, 0, I, t) : inSubset(F, S),
        kbElement(F)}X, interpretation(I), I ≠ u.
12 ← not subsetIsSat(S), subset(S).
```

Listing 3. Satisfiability check of each subset of S_{cs}.

At this point, all that is left to do is to make sure the candidate set S_{cs} is unsatisfiable. To this end, we apply the saturation technique [9]. More precisely, by means of a disjunctive rule, we allow each atom in S_{cs} to be 1 *or* 0 (see line 2 of Listing 4) w.r.t. the aforementioned interpretation u (introduced in line 1). Thus, under this interpretation, each atom in $\mathsf{At}(S_{cs})$ can be 1 and 0 at the same time. The idea behind this approach is that u serves as a "universal" interpretation, meaning that we implicitly check all possible interpretations. If both 1 and 0 are derived for each atom (lines 5–6), but there is still a formula in S_{cs} that evaluates to 0 (line 3), it means that S_{cs} cannot be satisfied under any interpretation and is consequently unsatisfiable. The integrity constraint in line 4 enforces that S_{cs} is "not not unsatisfiable", i.e., it must be unsatisfiable. All rules defined in Listings 3–4 are added to $P^{\mathsf{MIS}}(\mathcal{K})$.

```
1 interpretation(u).
2 truthValue(A, T, u, t) | truthValue(A, T, u, f) ← atomInCs(A), state(T).
3 isUnsat ← truthValue(F, 0, u, f), inCs(F), kbElement(F).
4 ← not isUnsat.
5 truthValue(A, T, u, t) ← isUnsat, atomInCs(A), state(T).
6 truthValue(A, T, u, f) ← isUnsat, atomInCs(A), state(T).
```

Listing 4. Check whether S_{cs} is unsatisfiable.

We now discuss how the above approach can be used for inconsistency measurement. To compute $\mathcal{I}_{\mathsf{MI}}(\mathcal{K})$ (the number of MISes of a KB $\mathcal{K}$), we use the *projection* reasoning mode [12] to enumerate all possible MISes, and count them. More precisely, we add "**#project** *inCs*/1." to the encoding, and use the flag `--project` when calling Clingo to get all answer sets that are unique w.r.t. the predicate *inCs*/1. We denote $P^{\mathsf{MIS}}(\mathcal{K}) \cup \{$**#project** *inCs*/1.$\}$ as $P^{\mathsf{MIS}}_{\mathrm{proj}}(\mathcal{K})$.

Theorem 1. *Let $\mathcal{K}$ be a KB. The number of answer sets of $P^{\mathsf{MIS}}_{\mathrm{proj}}(\mathcal{K})$ under the projection reasoning mode corresponds exactly to $\mathcal{I}_{\mathsf{MI}}(\mathcal{K})$.*[4]

As for $\mathcal{I}_p$ and $\mathcal{I}_{mv}$, we do not need to consider all MISes, since we are only interested in which formulas, or, respectively, atoms, are included in at least one MIS. Therefore, we initialize a set $Q := \mathcal{K}$ (we also refer to this as the "query set"), and a set $R = \emptyset$, and search for an MIS that includes at least one $\phi_q \in Q$. If such an MIS is found, we add each formula contained in it (in the case of $\mathcal{I}_p$) or each atom contained in its signature (in the case of $\mathcal{I}_{mv}$) to R, and remove each formula in it from Q. This process is repeated until either Q is empty or no MIS overlapping with Q exists anymore. We denote the problem of finding an MIS M with $M \cap Q \neq \emptyset$ as MIS-OVERLAP$(\mathcal{K}, Q)$ (see [25] for a related approach in the context of propositional logic). This problem can be solved by extending the above ASP encoding $P^{\mathsf{MIS}}(\mathcal{K})$ by an additional encoding $P_{\text{MIS-OL}}(\mathcal{K}, Q)$ as defined in Listing 5. In addition, for each $\phi_q \in Q$, we add *queryFormula*(ϕ_q) as a fact to $P_{\text{MIS-OL}}(\mathcal{K}, Q)$. If no answer set is found, there are no overlapping MISes left, and we can stop the procedure. If a solution is found, we extract the formulas in the MIS, update Q and R accordingly, and start the next iteration (given that $|Q| > 0$). Once the procedure is done, we return $|R|$ in the case of $\mathcal{I}_p$, and $|R|/|\mathsf{At}(\mathcal{K})|$ in the case of $\mathcal{I}_{mv}$.

[4] Proofs for technical results are provided in the supplementary material; see https://fernuni-hagen.sciebo.de/s/yZIwvNsraOZKA8o.

```
validCS ← 1{inCs(F) : queryFormula(F)}.
← not validCS.
```

Listing 5. Check if S_{cs} contains at least one $\phi_q \in Q$.

Theorem 2. *Let $\mathcal{K}$ be a KB, $Q \subseteq \mathcal{K}$, and let $P^{\mathsf{MIS}}_{\text{MIS-OL}}(\mathcal{K}, Q) := P^{\mathsf{MIS}}(\mathcal{K}) \cup P_{\text{MIS-OL}}(\mathcal{K}, Q)$. If there exists an answer set of A of $P^{\mathsf{MIS}}_{\text{MIS-OL}}(\mathcal{K}, Q)$, the set $\{\phi_A \mid inCs(\phi_A) \in A\}$ corresponds to an* MIS *M of $\mathcal{K}$, such that $M \cap Q \neq \emptyset$. If no answer set exists, there exists no* MIS *M of $\mathcal{K}$ with $M \cap Q \neq \emptyset$.*

3.2 MCS-Based Approaches

We continue by describing an answer set program $P^{\mathsf{MCS}}(\mathcal{K})$, which models the concept of an MCS. The overall idea behind the approach of modeling MCSes is broadly similar to the previously described approach for MISes. To be precise, we guess a candidate set for an MSS, and check whether it is consistent and has no satisfiable superset. This allows us to retrieve the complement of the MSS, i. e., an MCS. However, at first we need to represent the given KB, which is again achieved by $P^{\text{KB}}(\mathcal{K})$, and we initialize $P^{\mathsf{MCS}}(\mathcal{K}) := P^{\text{KB}}(\mathcal{K})$. The evaluation of the operators $(\wedge, \vee, \neg, \mathbf{X}, \mathbf{U}, \mathbf{G}, \mathbf{F})$, and of formulas consisting of single atoms generally functions in the same manner as for MISes. However, in the context of MCSes, we use a slightly different predicate to check the candidate set S_{cs} for satisfiability than we do to check its supersets for unsatisfiability. To be precise, to ensure that each individual superset is unsatisfiable we use the predicate *truthValueSet/*4, which works exactly as *truthValue/*4 for MISes, but refers to a specific superset instead of a specific interpretation. E. g., *truthValueSet*$(\phi, \text{Sup}(S_{cs}), t_i, \theta)$ refers to a formula ϕ, some $\text{Sup}(\text{S}_{\text{cs}}) \supset S_{cs}$, a state t_i, and a truth value $\theta \in \{1, 0\}$. To check S_{cs} for satisfiability, we use the predicate *truthValueCS/*3, which once again is defined analogously to *truthValue/*4, but omits the reference to an interpretation or superset. The sets of rules defining the evaluation of the operators via *truthValueCS/*3, and *truthValueSet/*4 are added to $P^{\mathsf{MCS}}(\mathcal{K})$.

Again, we guess a candidate set S_{cs} (line 1 in Listing 6), and add auxiliary rules to represent the atoms in S_{cs} (line 2), as well as the number of formulas in S_{cs} (line 3). Since our goal is to find MCSes, we additionally need to represent the complement of S_{cs} (line 4). We further add an integrity constraint which ensures that the MCS is not empty (line 5). To check S_{cs} for satisfiability, we first assign each atom in each state exactly one truth value (line 6). We then define that the candidate set is consistent if each of its formulas evaluate to 1 in state t_0 (line 7), and enforce that only solutions in which S_{cs} is in fact consistent can be derived by adding the integrity constraint shown in line 8.

```
1 1{inCs(X) : kbElement(X)}.
2 atomInCs(A) ← atomInFormula(A, F), inCs(F).
3 numElsInCs(X) ← X = #count{F : inCs(F)}.
4 inComplement(F) ← kbElement(F), not inCs(F).
5 ← not 1{inComplement(X) : kbMember(X)}.
6 1{truthValueCS(A, S, V) : tv(V)}1 ← atom(A), state(T).
7 csIsSat ← numElementsInCs(X), X{truthValueCS(F, 0, t) : inCs(F), kbElement(F)}X.
8 ← not csIsSat.
```

Listing 6. Satisfiability check of candidate set S_{cs}.

We define the number of supersets of S_{cs} of size $|S_{cs}|+1$ (line 1 of Listing 7), as well as an instance of *superset*/1 for each superset (line 2). To construct the supersets, we define the formulas in $\mathcal{K} \setminus S_{cs}$ as "additional elements" (line 3). Note that we need exactly $|\mathcal{K}| - |S_{cs}|$ supersets, with each of them consisting of S_{cs} (line 4) and one additional formula from $\mathcal{K} \setminus S_{cs}$ (line 5). Moreover, lines 6 and 7 enforce that no two supersets are equal, and lines 8–9 contain auxiliary predicates that represent which atoms are in each superset's signature. At last, we check each superset for inconsistency. Again, we apply saturation to achieve this. We create a disjunctive rule that allows each atom to be both 1 and 0 w.r.t. a state and a superset (line 10). Overall, the inconsistency check functions in the same way as for the MIS variant, only with an additional reference to a specific superset in each rule. Thus, we model that a superset is inconsistent if at least one formula evaluates to 0 (line 11), even though all possible interpretations are implicitly considered (lines 13–14), and enforce that each superset must be inconsistent (line 12). Further, we add all rules introduced in Listings 6–7 to $P^{\mathsf{MCS}}(\mathcal{K})$.

```
1 numSupersets(X) ← numElementsInCs(Y), numKbElements(Z), X = Z − Y.
2 superset(1..X) ← numSupersets(X), X > 0.
3 1{additionalElement(F, S) : inComplement(F)}1 ← superset(S).
4 inSuperset(F, S) ← inCs(F), superset(S).
5 inSuperset(F, S) ← additionalElement(F, S), superset(S).
6 supersetEq(S1, S2) ← superset(S1), superset(S2), S1 ≠ S2, additionalElement(F1, S1),
        additionalElement(F2, S2), F1 == F2.
7 ← supersetEq(S1, S2).
8 atomInSuperset(A, S) ← atomInCs(A), superset(S).
9 atomInSuperset(A, S) ← additionalElement(F, S), atomInFormula(A, F).
10 truthValueSet(A, S, T, t) | truthValueSet(A, S, T, f) ← atomInSuperset(A, S), superset(S),
        state(T).
11 isUnsat(S) ← truthValueSet(F, S, 0, f), inSuperset(F, S), kbMember(F).
12 ← not isUnsat(S), superset(S).
13 truthValueSet(A, S, T, t) ← isUnsat(S), atomInSuperset(A, S), superset(S), state(T).
14 truthValueSet(A, S, T, f) ← isUnsat(S), atomInSuperset(A, S), superset(S), state(T).
```

Listing 7. Check whether the supersets of S_{cs} are unsatisfiable.

With regard to MCS-based inconsistency measurement, we first address $\mathcal{I}_{mv}$ and $\mathcal{I}_p$. For the MIS variant of these measures, we described an iterative procedure which solves the MIS-OVERLAP problem w.r.t. a given KB $\mathcal{K}$ and a query set $Q \subseteq \mathcal{K}$ at each step. In the same fashion, we define the MCS-OVERLAP problem, denoted as the problem of finding an MCS M such that $M \cap Q \neq \emptyset$ w.r.t. a KB $\mathcal{K}$ [25]. To solve this problem, we use encoding $P^{\mathsf{MCS}}(\mathcal{K})$ as defined above[5],

[5] Note that line 5 from Listing 6 can be omitted, since the rules in Listing 8 already ensure that the MCS is not empty.

and extend it by the program $P_{\text{MCS-OL}}(\mathcal{K}, Q)$. We add a fact *queryFormula*(ϕ_q) for each $\phi_q \in Q$ to $P_{\text{MCS-OL}}(\mathcal{K}, Q)$, as well as the two rules presented in Listing 8, which ensure that at least one formula from Q is contained in the complement of S_{cs} (i. e., the corresponding MCS).

```
validCS ← 1{inComplement(F) : queryFormula(F)}.
← not validCS.
```

Listing 8. Check if $\mathcal{K} \setminus S_{\text{cs}}$ contains at least one $\phi_q \in Q$.

Theorem 3. *Let $\mathcal{K}$ be a KB, $Q \subseteq \mathcal{K}$, and $P^{\mathsf{MCS}}_{\text{MCS-OL}}(\mathcal{K}, Q) := P^{\mathsf{MCS}}(\mathcal{K}) \cup P_{\text{MCS-OL}}(\mathcal{K}, Q)$. If there exists an answer set of A of $P^{\mathsf{MCS}}_{\text{MCS-OL}}(\mathcal{K}, Q)$, the set $\{\phi_A \mid inComplement(\phi_A) \in A\}$ corresponds to an MCS M of $\mathcal{K}$, such that $M \cap Q \neq \emptyset$. If no answer set exists, there exists no MCS M of $\mathcal{K}$ with $M \cap Q \neq \emptyset$.*

Regarding $\mathcal{I}_r$, recall that we are looking for the smallest MCS. To achieve this, we simply add the optimization statement $P_{\min} :=$ **#minimize**$\{1, \text{F} : inComplement(\text{F})\}$, which minimizes the number of elements in the complement of S_{cs}. We denote $P^{\mathsf{MCS}}_r(\mathcal{K}) := P^{\mathsf{MCS}}(\mathcal{K}) \cup P_{\min}$, and refer to a set that complies with the minimization as an *optimal* answer set.

Theorem 4. *Let $\mathcal{K}$ be a KB and let A_o be an optimal answer set of $P^{\mathsf{MCS}}_r(\mathcal{K})$. Then $|\{\phi_{A_o} \mid inComplement(\phi_{A_o}) \in A_o\}| = \mathcal{I}_r(\mathcal{K})$.*

4 Experimental Evaluation

Experimental Setup. We performed runtime experiments with data sets of the *Business Process Intelligence Challenge* (BPIC)[6], which is an international scientific challenge series that provides real-life (process) data from various domains such as healthcare, government or industry. The data is offered in the form of *event logs*, which are (multi-)sets of activity sequences stemming from industrial processes. We applied the state-of-the-art tool MINERful [8] to obtain sets of LTL_f formulas from the BPIC event logs. More precisely, the output of MINERful is in the *Declare* language (see [7]), which was then transformed to LTL_f formulas. Note that Declare offers a set of predefined templates for defining constraints, and each template has a clearly defined LTL_f semantics.

As mining parameters for our data generation, we selected standard values as suggested in [6], namely a support factor of 75%, as well as confidence and interest factors of 95%. For our evaluation, we used the BPIC event logs of all available challenges (2012–2020) (see the supplementary material for an overview of the characteristics of the considered event logs, as well as the code and data). The event logs stem from a variety of different domains, with the 2019 event log containing over 250,000 traces (sequences). If a KB $\mathcal{K}$ was found to be consistent, we added a formula $\mathbf{G}a$, with a being a randomly chosen atom from $\mathsf{At}(\mathcal{K})$. Moreover, we set a timeout of 60 s, and set the index of the final state m to the maximum depth of the given KB (see Sect. 3.1). All experiments were run on an Intel Xeon E5-2643 v3 3.40-GHz CPU with 192 GB RAM.

[6] https://icpmconference.org/2020/bpi-challenge/.

Table 1. Runtime results for MIS-based vs. MCS-based approaches w.r.t. $\mathcal{I}_{mv}$ and $\mathcal{I}_p$, as well as $\mathcal{I}_r$ and $\mathcal{I}_{\mathsf{MI}}$. Timeout: 60 s

Data set	Runtime (s)					
	MIS-based		MCS-based			
	$\mathcal{I}_{mv}$	$\mathcal{I}_p$	$\mathcal{I}_{mv}$	$\mathcal{I}_p$	$\mathcal{I}_r$	$\mathcal{I}_{\mathsf{MI}}$
BPIC2012	0.087	0.221	0.055	0.093	0.060	0.170
BPIC2015_1	14.881	–	0.774	8.221	0.167	–
BPIC2015_2	8.315	–	0.277	5.967	0.143	–
BPIC2015_3	–	–	7.032	–	0.278	–
BPIC2015_4	0.930	0.936	0.172	0.174	0.235	0.991
BPIC2016	1.867	1.842	0.248	0.240	0.305	1.419
BPIC2017	–	–	–	–	–	–
BPIC2018	–	–	–	–	–	–
BPIC2019	–	–	1.162	1.123	0.225	–
BPIC2020_1	7.874	6.437	0.808	0.779	0.265	7.065
BPIC2020_2	2.167	–	0.334	8.507	0.334	–
BPIC2020_3	–	–	0.518	4.777	0.409	–
BPIC2020_4	7.248	4.175	0.522	0.507	0.254	7.475
BPIC2020_5	7.458	7.412	0.727	0.706	0.238	8.213

Results. We begin by discussing the runtime results regarding the MIS-based and MCS-based approaches for $\mathcal{I}_{mv}$ and $\mathcal{I}_p$, which are shown in Table 1 (note that "–" marks a timeout). We can observe that (i) the instances BPIC2017 and BPIC2018 could not be solved by any approach, (ii) the MCS-based approaches overall outperform the MIS-based ones, and (iii) $\mathcal{I}_{mv}$ is in most cases computed faster than $\mathcal{I}_p$. The reason for (i) is most likely the fact that the two instances in question contain significantly more formulas than the other ones. Observation (ii) is in line with other findings from the literature which state that the computation of MCSes is often easier than the computation of MISes (see, e. g., [1,30]). As for (iii), since, in practice, it usually holds that $|\mathcal{K}| > |\mathsf{At}(\mathcal{K})|$, it is likely that the computation of $\mathcal{I}_p(\mathcal{K})$ requires more iterations than $\mathcal{I}_{mv}(\mathcal{K})$.

The rightmost part of Table 1 shows the runtimes regarding $\mathcal{I}_r$ and $\mathcal{I}_{\mathsf{MI}}$. While $\mathcal{I}_{\mathsf{MI}}$ times out w.r.t. the same instances as the MIS variant of $\mathcal{I}_p$, $\mathcal{I}_r$ could be computed in < 1 s in all cases (except for BPI2017 and BPI2018). Thus, overall, we did observe some limitations of our approaches—in particular the MIS-based ones—however, we could also see that it is feasible to use them in many practical application cases.

5 Conclusion

We presented algorithmic approaches based on ASP for computing a selection of inconsistency measures w.r.t. LTL$_f$ knowledge bases. Our results allow to compute the considered measures $\mathcal{I}_{\mathsf{MI}}, \mathcal{I}_p, \mathcal{I}_{mv}$ and $\mathcal{I}_r$ with a correct consideration of the temporal operators of LTL_f, which is not possible with existing solutions for propositional logic. As stated, inconsistency is a core problem of many domains dealing with LTL_f formulas, e.g., program verification or declarative process specifications. Thus, being able to compute the measures here can provide useful insights for debugging or re-modelling LTL_f specifications.

Our approach extends previous works on inconsistency measurement in LTL_f [3,4,22], but, in contrast to those works, focuses not on the number of states (in time) that are affected by inconsistency, but allows to look "into" MISes, respectively, MCSes. This provides new means for formula-driven analysis, such as pin-pointing highly problematic formulas, which could be useful in the scope of debugging. In future works, we aim to explore other algorithmic approaches for our use-case, e.g., it might be of interest to develop formula-centric measures for LTL_f that also take into account the number of temporal states that are affected. To investigate generalizability, we also aim to investigate other inconsistency measures based on entirely different principles, such as semantic inconsistency measures.

Acknowledgements. The research reported here was supported by the Deutsche Forschungsgemeinschaft (grants 506604007 (IK) and DE 1983/9-2 (CC)).

References

1. Bendík, J., Černá, I.: Replication-guided enumeration of minimal unsatisfiable subsets. In: Simonis, H. (ed.) CP 2020. LNCS, vol. 12333, pp. 37–54. Springer, Cham (2020). https://doi.org/10.1007/978-3-030-58475-7_3
2. Calimeri, F., et al.: ASP-core-2 input language format. Theory Pract. Logic Program. **20**(2), 294–309 (2020)
3. Corea, C., Grant, J., Thimm, M.: Measuring inconsistency in declarative process specifications. In: Di Ciccio, C., Dijkman, R., del Río Ortega, A., Rinderle-Ma, S. (eds.) BPM 2022. LNCS, vol. 13420, pp. 289–306. Springer, Cham (2022). https://doi.org/10.1007/978-3-031-16103-2_20
4. Corea, C., Kuhlmann, I., Thimm, M., Grant, J.: Paraconsistent reasoning for inconsistency measurement in declarative process specifications. Inf. Syst. **122**, 102347 (2024)
5. De Giacomo, G., Vardi, M.Y.: Linear temporal logic and linear dynamic logic on finite traces. In: Ijcai, vol. 13, pp. 854–860 (2013)
6. Di Ciccio, C., Maggi, F.M., Montali, M., Mendling, J.: Resolving inconsistencies and redundancies in declarative process models. Inf. Syst. **64**, 425–446 (2017)
7. Di Ciccio, C., Montali, M.: Declarative process specifications: reasoning, discovery, monitoring. In: van der Aalst, W.M.P., Carmona, J. (eds.) Process Mining Handbook. LNBIP, vol. 448, pp. 108–152. Springer, Cham (2022). https://doi.org/10.1007/978-3-031-08848-3_4
8. Di Ciccio, C., Schouten, M.H., De Leoni, M., Jan, M., et al.: Declarative process discovery with MINERful in ProM. In: CEUR Workshop Proceedings, vol. 1418, pp. 60–64. CEUR-WS.org (2015)
9. Eiter, T., Gottlob, G.: On the computational cost of disjunctive logic programming: propositional case. Ann. Math. Artif. Intell. **15**, 289–323 (1995)

10. Geatti, L., Gigante, N., Montanari, A., Venturato, G.: SAT meets tableaux for linear temporal logic satisfiability. J. Autom. Reason. **68**(2), 1–32 (2024)
11. Gebser, M., Kaminski, R., Kaufmann, B., Schaub, T.: Answer set solving in practice. Synthesis Lect. Artif. Intell. Mach. Learn. **6**(3), 1–238 (2012)
12. Gebser, M., Kaufmann, B., Schaub, T.: Solution enumeration for projected Boolean search problems. In: van Hoeve, W.-J., Hooker, J.N. (eds.) CPAIOR 2009. LNCS, vol. 5547, pp. 71–86. Springer, Heidelberg (2009). https://doi.org/10.1007/978-3-642-01929-6_7
13. Gelfond, M., Lifschitz, V.: The stable model semantics for logic programming. In: Proceedings of the ICLP/SLP, pp. 1070–1080. MIT Press (1988)
14. Gelfond, M., Lifschitz, V.: Classical negation in logic programs and disjunctive databases. N. Gener. Comput. **9**, 365–385 (1991)
15. Grant, J.: Measuring inconsistency in some branching time logics. J. Appl. Non-Classical Logics **31**(2), 85–107 (2021)
16. Grant, J., Martinez, M.V., Molinaro, C., Parisi, F.: Dimensional inconsistency measures and postulates in spatio-temporal databases. J. Arti. Intell. Res. **71**, 733–780 (2021)
17. Grant, J., Parisi, F.: General information spaces: measuring inconsistency, rationality postulates, and complexity. Ann. Math. Artif. Intell. **90**(2), 235–269 (2022)
18. Grant, J., Parisi, F.: On measuring inconsistency in graph databases with regular path constraints. Artif. Intell. **335**, 104197 (2024)
19. Hunter, A., Konieczny, S.: Approaches to measuring inconsistent information. In: Bertossi, L., Hunter, A., Schaub, T. (eds.) Inconsistency Tolerance. LNCS, vol. 3300, pp. 191–236. Springer, Heidelberg (2005). https://doi.org/10.1007/978-3-540-30597-2_7
20. Hunter, A., Konieczny, S.: Measuring inconsistency through minimal inconsistent sets. KR **8**(358-366), 42 (2008)
21. Jabbour, S., Sais, L.: Exploiting MUS structure to measure inconsistency of knowledge bases. In: European Conference on AI, pp. 991-998. IOS Press (2016)
22. Kuhlmann, I., Corea, C., Grant, J.: An ASP-based framework for solving problems related to declarative process specifications. In: Proceedings of the 21st International Workshop on Non-Monotonic Reasoning (NMR), pp. 129–132 (2023)
23. Kuhlmann, I., Corea, C., Grant, J.: Non-automata based conformance checking of declarative process specifications based on ASP. In: De Weerdt, J., Pufahl, L. (eds.) BPM 2023. LNBIP, vol. 492, pp. 396–408. Springer, Cham (2023). https://doi.org/10.1007/978-3-031-50974-2_30
24. Kuhlmann, I., Gessler, A., Laszlo, V., Thimm, M.: A comparison of ASP-based and SAT-based algorithms for the contension inconsistency measure. In: Dupin de Saint-Cyr, F., Öztürk-Escoffier, M., Potyka, N. (eds.) SUM 2022. LNCS, vol. 13562, pp. 139–153. Springer, Cham (2022). https://doi.org/10.1007/978-3-031-18843-5_10
25. Kuhlmann, I., Niskanen, A., Järvisalo, M.: Computing MUS-based inconsistency measures. In: Gaggl, S., Martinez, M.V., Ortiz, M. (eds.) JELIA 2023. LNCS, vol. 14281, pp. 745–755. Springer, Cham (2023). https://doi.org/10.1007/978-3-031-43619-2_50
26. Kuhlmann, I., Thimm, M.: An algorithm for the contension inconsistency measure using reductions to answer set programming. In: Davis, J., Tabia, K. (eds.) SUM 2020. LNCS (LNAI), vol. 12322, pp. 289–296. Springer, Cham (2020). https://doi.org/10.1007/978-3-030-58449-8_23

27. Kuhlmann, I., Thimm, M.: Algorithms for inconsistency measurement using answer set programming. In: Proceedings of the 19th International Workshop on Non-Monotonic Reasoning (NMR), pp. 159–168 (2021)
28. Li, J., Pu, G., Zhang, Y., Vardi, M.Y., Rozier, K.Y.: SAT-based explicit LTLf satisfiability checking. Artif. Intell. **289**, 103369 (2020)
29. Li, J., Zhang, L., Pu, G., Vardi, M.Y., He, J.: LTLf satisfiability checking. arXiv preprint arXiv:1403.1666 (2014)
30. Liffiton, M.H., Previti, A., Malik, A., Marques-Silva, J.: Fast, flexible MUS enumeration. Constraints An Int. J. **21**(2), 223–250 (2016)
31. Lifschitz, V.: Answer set Programming. Springer, Berlin (2019). https://doi.org/10.1007/978-3-030-24658-7
32. Luo, W., et al.: Teaching LTLf satisfiability checking to neural networks. In: IJCAI, pp. 3292–3298 (2022)
33. McAreavey, K., Liu, W., Miller, P.: Computational approaches to finding and measuring inconsistency in arbitrary knowledge bases. Int. J. Approximate Reasoning **55**(8), 1659–1693 (2014)
34. Niskanen, A., Kuhlmann, I., Thimm, M., Järvisalo, M.: MaxSAT-based inconsistency measurement. In: Proceedings of the 26th European Conference on Artificial Intelligence, pp. 1779–1786. IOS Press (2023)
35. Parisi, F., Grant, J.: On measuring inconsistency in definite and indefinite databases with denial constraints. Artif. Intell. **318**, 103884 (2023)
36. Pnueli, A.: The temporal logic of programs. In: 18th Annual Symposium on Foundations of Computer Science (SFCS 1977), pp. 46–57. IEEE (1977)
37. Reiter, R.: A theory of diagnosis from first principles. Artif. Intell. **32**(1), 57–95 (1987)
38. Roveri, M., Di Ciccio, C., Di Francescomarino, C., Ghidini, C.: Computing unsatisfiable cores for LTLf specifications. arXiv preprint arXiv:2203.04834 (2022)
39. Thimm, M.: Stream-based inconsistency measurement. Int. J. Approximate Reasoning **68**, 68–87 (2016)
40. Thimm, M.: Inconsistency measurement. In: Ben Amor, N., Quost, B., Theobald, M. (eds.) SUM 2019. LNCS (LNAI), vol. 11940, pp. 9–23. Springer, Cham (2019). https://doi.org/10.1007/978-3-030-35514-2_2
41. Thimm, M., Wallner, J.P.: On the complexity of inconsistency measurement. Artif. Intell. **275**, 411–456 (2019)
42. Xiao, G., Ma, Y.: Inconsistency measurement based on variables in minimal unsatisfiable subsets. In: European Conference on Artificial Intelligence, pp. 864–869 (2012)

Contribution of Subsets of Variables in Global Sensitivity Analysis with Dependent Variables

Christophe Labreuche[1,2(✉)]

[1] Thales Research & Technology, Palaiseau, France
christophe.labreuche@thalesgroup.com
[2] SINCLAIR AI Lab, Palaiseau, France

Abstract. Global Sensitivity Analysis aims at explaining how much each random variable contributes to the variance of the output of a black-box model. The standard approach – namely Sobol indices – computes the contribution of each subset of variables but requires that the variables are independent. The Shapley effect (based on the Shapley value) has been defined for dependent variables, but gives the contribution of each variable individually instead of the contribution of subsets of variables. The aim of this work is to propose a novel approach for dependent variables that defines the level of contribution of each subset of variables so that they sum up to the total variance of the output of the model. We show that we come up with known concepts – namely the Banzhaf values and interaction indices, up to a multiplicative factor.

Keywords: Global Sensitivity Analysis · Banzhaf values · Dependent Variables · Hoeffding Decomposition

1 Introduction

Explaining a black-box function is an important aspect of AI [13]. Existing approaches can be distinguished between local and global methods. The first ones are used to explain a particular instance. Model-agnostic approaches can be applied to any model, while model-specific methods are dedicated to a particular class of models. In Machine Learning (ML), *Feature Attribution* (FA) methods indicate the level to which each feature contributes to the model outcome on the instance. Well-known model-agnostic methods cover the Shapley value [20] or LIME [27]; saliency maps are examples of FA techniques used for more particular model architectures as the ones used in image processing. The asset of the Shapley value comes from its justification through axiomatic characterizations originally defined in the context of Cooperative Game Theory (CGT) [30]. In CGT, the Shapley value fairly shares among players (aka variables) the worth gained by all players together. It uses as input a *game*, which returns a worth to each subset of players. Note that CGT also proposes other values

S. Destercke et al. (Eds.): SUM 2024, LNAI 15350, pp. 233–248, 2025.
https://doi.org/10.1007/978-3-031-76235-2_18

such as the Banzhaf index [7]. This latter is originally a power index in a vote computed from the proportion of times a given voter would reverse the result of the scrutiny when changing her vote. A local linearity on the model is assumed with these indices as the explanation only returns one attribution index for each features. In order to account for interaction among features, some indices have also been defined on subsets of variables (rather than only on single feature) in the context of ML [2,26,34,36,37], including extensions of the Shapley value to coalitions of variables [14,32,34]. Note that the Shapley interaction indices were already known in CGT [12]. The Banzhaf index has also been extended to coalitions – known as Banzhaf interaction indices [12].

By contrast, global explanation methods aim at providing an interpretation of the model in general, by analysing the behaviour of the model over the complete range of input features, depicted by a dataset. *Global Sensitivity Analysis (GSA)* in statistics is a standard approach to produce global explanations. It examines the relationship between inputs and outputs, and more precisely how the uncertainty arising in the output of the model can be attributed to the various uncertainties in the inputs [28,38]. In *variance-based GSA*, it is assumed that the uncertainty in the output can be entirely described by the variance [8].

The focus of this paper is on variance-based GSA. The variance on the model output is then allocated among variables seen as random variables. According to the Sobol method, the variance of the output – called *total variance* – is decomposed into fractions attributed to subsets of input variables. In other words, unlike feature attribution in local explanation, the allocation is not done among the single variables, but among all subsets of features. The idea is that features are interacting in the model and the contribution allotted to a subset of variables indicates how these variables together interact and contribute to the output. This standard approach is depicted in Fig. 1–Left. More precisely, the Hoeffding decomposition of the model is first performed. It consists in writing the initial function f as a sum of functions f_S whose support is a different subset of the variables, with $S \subseteq N$ and N being the set of variable indices. These functions are orthogonal, which implies that the variance of function f is the sum of the variances over all sub-functions f_S. This yields Sobol indices [31]. These indices sum-up to one (property called *efficiency*), meaning that the total variance is shared among all subsets of variables. In the case of independent inputs, the interpretation of Sobol indices follows the efficiency property, resulting in a percentage of the total variance attributed to each subset of variables.

In practice, variables are often dependent due to some physical constraints [18]. In such situation, the Sobol indices fail to satisfy efficiency and other properties. As a result, the Sobol indices loose their interest as a well-founded explanation tool [29]. To circumvent this drawback, Owen proposed to turn the problem of sharing the variance to a CGT setting, and in particular the use of the Shapley value, which satisfies by design the desired property of efficiency [23]. More precisely, a cooperative game v is computed from the contribution of each subset of variables to the total variance, and then the Shapley value is applied on this game – see Fig. 1–Middle. However, there is a major difference between the Sobol

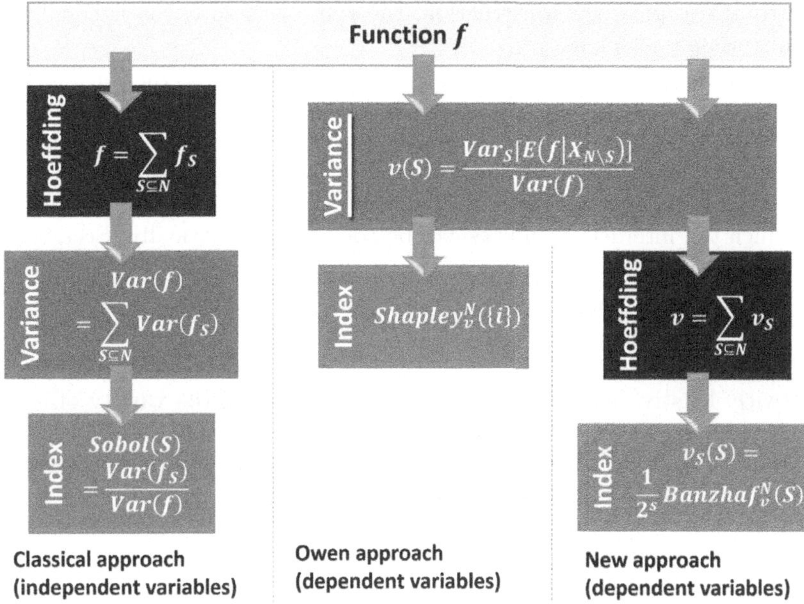

Fig. 1. Three approaches for variance-based GSA

and the Shapley indices, as the Shapley value allocates the global worth among the individual inputs and not their coalitions. The drawback of the Shapley value is thus to provide information only on the relative importance of single variables but not at all on the way they interact.

1.1 Contributions of the Paper

The main contribution of this paper is to propose a variance-based GSA approach applicable for dependent variables, for which the total variance is shared among all subsets of variables. To our knowledge, this is new. The approach is depicted in Fig. 1–Right. We start with a cooperative game defined from the proportion of the variance induced by a subset of variables compared to the total variance, as in the Owen approach [23]. In order to obtain indices defined on each subset of variables, which satisfy by design efficiency, we perform the Hoeffding decomposition on this game (rather than on the original function). Applying this decomposition on the Grand coalition N, each term corresponds to the influence index for a coalition of variables. The main result of this paper shows that the terms in this decomposition correspond to the Banzhaf interaction indices up to a multiplicative coefficient. Banzhaf interaction indices are already known in CGT [12], but have never been used in ML to our knowledge.

As summary, the main contributions of the paper are:

1. The main result shows the Hoeffding decomposition of a game (Sect. 3). Each term in this decomposition is the contribution of a subset S of variables to

the total variance. By construction, the sum of these terms is equal to the total variance (efficiency).
2. Each term in the decomposition corresponds to the interaction index of a coalition S of variables. The second contribution of the paper shows that the so-obtained interaction coefficients are not new as they are equal to the Banzhaf interaction indices up to a multiplicative constant (Sect. 4).
3. The Hoeffding decomposition that we obtain is illustrated on two examples in which the interaction indices can be computed analytically (Sect. 5).

1.2 Importance of the Contributions

The difficulty of using concepts from CGT in the context of explaining ML models is twice. Firstly, one has to choose the index among the variety of existing indices, including the Shapley and Banzhaf indices and their extensions to interaction among variables. Secondly, one has to choose a definition of the game on which the index is computed. Regarding this latter point, there are for instance several methods using the Shapley value, which only differ on the definition of the game. The game can indeed be obtained by averaging over all elements in the dataset as in Kernel-SHAP [20], can use a conditional distribution over the value of the instance on the coalition, as in SHAP [20] and TreeSHAP [19], can use a joint-marginal distribution, as in QII, or can use a uniform distribution and discard the training data, as in IME [33]. All this variability on the choice of the index and of the game induces some confusion to the practitioner. When comparing all these methods (choice of the game and the index) on datasets, it is hard to assess which combination works best. The standard metrics used to compare the methods, progressively inserts (deletes respectively) the most important features, which raises the question of applying a model with some missing values.

This paper does not come up with a novel interaction index. Instead of proposing another index, which intrinsically does not solve the choice problem of the practitioner, we consider variance-based GSA, which brings a general framework in which (1) the choice of the game is not arbitrary and has an intrinsic definition; (2) the choice of the expression of the value follows from the Hoeffding decomposition, which is at the heart of variance-based GSA. Hence the paper provides a nice interpretation of the already known Banzhaf interaction indices in the context of variance-based GSA.

2 Related Work and Background

2.1 Local and Global XAI Techniques

There exists a wide variety of XAI techniques [13]. Feature Attribution aims at assessing the contribution level of each input variable to the output for a particular instance [13]. *LIME* constructs a linear model around the instance to explain, where the weights serve as a measure of importance of features [27].

The Shapley value has been widely used for feature attribution [19,20,33]. It requires to construct a game in the sense of Game Theory, indicating for each subset of variable, an assessment of the output value when only this subset of input variables is used. Different definition of games have been proposed [1]: SHAP [20] and TreeSHAP [19] use a conditional distribution, Kernel-SHAP [20] averages over all dataset points. Alternative attribution methods have also been proposed such as the use of the Banzhaf value instead of the Shapley value [35].

In contrast, global explanation does not focus on a particular instance and aims to produce an overall understanding of the model. Partial Dependence Plots (PDP) and Accumulated Local Effects (ALE) show how the output of the model is influenced on average by each feature. In [3], the contribution of each variable to the output is also shown, but it better accounts for dependent variables thanks to an entropic projection. SAGE (Shapley Additive Global importancE) uses the Shapley value to provide the overall contribution level of features [6].

2.2 Sobol Indices

Let $N := \{1, \ldots, n\}$ denote a finite set of indices of variables $\mathbf{X} = (X_1, \ldots, X_n)$. Values of these variables are denoted by $\mathbf{x} = (x_1, \ldots, x_n)$. For convenience, subsets of N will be denoted by upper case letters, e.g. $S, T \subseteq N$, and their cardinality by the corresponding lowercase, e.g. $s := |S|, t := |T|$.

The aim of GSA is to study how perturbations on variables $\mathbf{X}$ generate perturbations on an output $Y = f(\mathbf{X})$ through a function f [28]. Variables $X_1, \ldots, X_n$ are assumed to be random variables. Variance-based sensitivity analysis consists in allocating the variance $\text{Var}(Y)$ among all subsets of variables. The Hoeffding decomposition is first applied on function f [28]

$$f(\mathbf{x}) = \sum_{S \subseteq N} f_S(\mathbf{x}_S), \tag{1}$$

where the support of f_S is variables S, and

$$f_S(\mathbf{x}_S) = \sum_{K \subseteq S} (-1)^{s-k} \, \mathbb{E}_{N \setminus K} \left[f(\mathbf{X}) | \mathbf{X}_K = \mathbf{x}_K \right]. \tag{2}$$

Then the Sobol index relatively to subset S of variables is

$$\text{Sob}(S) = \frac{\text{Var}(f_S)}{\text{Var}(Y)}.$$

One can readily see that when the variables are independent that [23]

$$\text{Non-negativity:} \qquad 0 \leq \text{Sob}(S) \leq 1 \quad \forall S \subseteq N, \tag{3}$$

$$\text{Efficiency:} \qquad \sum_{S \subseteq N} \text{Sob}(S) = 1. \tag{4}$$

Thanks to these properties, $\text{Sob}(S)$ measures the degree of contribution of subset S to the total variance $\text{Var}(Y)$.

Several proposals have been made in the case of dependent variables, for structures exhibiting partial independence [16], or e.g. by transforming the dependent variables into independent inputs [21]. However, none of these extensions [15] provide clearly interpretable indices with admitted properties.

The Sobol indices have also been used for local explanation of image processing [8]. Using the prediction of a perturbation of the initial image through several masks, the Sobol indices are deduced, providing the effect of individual pixels or a region, and of interactions among pixel regions.

2.3 Shapley Value in Cooperative Game Theory

A *game* (with transferable utility) is a set function $v : 2^N \to \mathbb{R}$ such that $v(\varnothing) = 0$. In *Cooperative Game Theory* (CGT), N is the set of players, and $v(S)$ (for $S \subseteq N$) is the amount of wealth produced by S when they cooperate. A player $i \in N$ is said to be *dummy* for v if $\forall S \subseteq N \setminus \{i\}$, $v(S \cup \{i\}) = v(S) + v(\{i\})$. A player $i \in N$ is said to be *null* for v if $\forall S \subseteq N \setminus \{i\}$, $v(S \cup \{i\}) = v(S)$.

The Shapley value fairly shares among players what is gained by all players [23]

$$\phi_i^{\mathrm{Sh}}(N, v) := \sum_{S \subseteq N \setminus \{i\}} \frac{s!(n-s-1)!}{n!} \left[v(S \cup \{i\}) - v(S) \right], \tag{5}$$

where $\phi_i^{\mathrm{Sh}}(N, v)$ is given to player i. The Shapley value is uniquely defined by four properties [30]: *Additivity* (for all games v, w, $\phi_i^{\mathrm{Sh}}(N, v + w) = \phi_i^{\mathrm{Sh}}(N, v) + \phi_i^{\mathrm{Sh}}(N, w)$), *Null player* (if player i is null for v, then $\phi_i^{\mathrm{Sh}}(N, v) = 0$)), *Symmetry* (for all v, for all permutations π on N, $\phi_{\pi(i)}^{\mathrm{Sh}}(\pi N, \pi v) = \phi_i^{\mathrm{Sh}}(N, v)$), and *Efficiency* ($\sum_{i \in N} \phi_i^{\mathrm{Sh}}(N, v) = v(N)$).

As explained before, the Shapley values have been recently used in GSA as an alternative to the Sobol indices to better account for dependent inputs [10,15,23]. When variables are dependent, an interesting approach consists in defining a set function $v : 2^N \to \mathbb{R}$ by, for all $S \subseteq N$ [23]

$$v(S) = \frac{\mathrm{Var}_S\left(\mathbb{E}_{N \setminus S}[Y | X_S]\right)}{\mathrm{Var}(Y)} \tag{6}$$

where $Y = f(X)$. Note that $v(\emptyset) = 0$. Then the Shapley value is computed on v to provide the contribution of variable i to the overall variance $v(N)$. In particular, each attribution score is non-negative (property non-negativity (3)) and they sum-up to 1 (property efficiency (4)), which is consitent with the idea of sharing the variance (a non-negative number) among variables.

2.4 Banzhaf Value

The Banzhaf value [7] of a player $i \in N$ in a game $v \in \mathcal{G}(2^N)$ is defined by

$$\phi_i^{\mathrm{Ban}}(N, v) = \sum_{S \subseteq N \setminus i} \frac{1}{2^{n-1}} \left(v(S \cup i) - v(S) \right).$$

A coalition $S \subseteq N$ is called a *partnership* when they act as a single player denoted by $[S]$ in the game. The associated game is $v_{[S]}$ defined on $N_{[S]} := (N \setminus S) \cup \{[S]\}$ by, for any $K \subseteq N \setminus S$, $v_{[S]}(K) = v(K)$ and $v_{[S]}(K \cup \{[S]\}) = v(K \cup S)$.

The Banzhaf value is uniquely axiomatized by a set of four axioms [7,17]: *Linearity* ($v \mapsto \phi_i^{\text{Ban}}(N, v)$ is linear), *Dummy player* (If $i \in N$ is a dummy player for v, then $\phi_i^{\text{Ban}}(N, v) = v(\{i\})$), *Symmetry*, and *2-efficiency* (For any game v, $\phi_i^{\text{Ban}}(N, v) + \phi_j^{\text{Ban}}(N, v) = \phi_{[ij]}^{\text{Ban}}(N_{[ij]}, v_{[ij]}), \forall i, j \in N$).

Note that the Shapley value also satisfies linearity and dummy player. We see the Shapley value and the Banzhaf values essentially satisfy to the same axioms, except the last one, where the Shapley value satisfies the Efficiency axiom whereas the Banzhaf value satisfies the 2-efficiency property. The 2-efficiency property says that the importance allocated to a partnership $[ij]$ of two players i and j is equal to the sum of the importance for these two players.

2.5 Interaction Indices

In the context of ML, there is an increasing interest in measuring the level of interaction among features on top of individual importance of features [37]. It quantifies the contribution of a subset of features to the model output. Several interaction indices have been proposed in ML. A first approach consists in replacing in the Shapley value, a single variable by a subset of variables [2]. In [26], the Shapley interaction indices are used in sensitivity analysis. When the user fixes the maximum cardinality ℓ of interaction that are computed, the *efficiency* property means that the sum of all interactions up to cardinality ℓ is equal to the game for the Grand coalition. It is satisfied by the Shapley Taylor interaction index [34], the Faithful Interaction index [36] and the n-Shapley values [4]. The Shapley Taylor interaction index of a unanimity game of coalition S is zero for all strict subsets of S [34]. The Faithful Interaction index is a solution to a weighted quadratic regression problem, where the interaction corresponds to a ℓ-additive approximation of the game [36]. The n-Shapley value is defined recursively from the Shapley interaction indices, using the Bernoulli numbers [4]. A unified approach to assess these interaction indices is proposed in [9] for Cardinal Interaction Indices. They are rewritten directly as a sum over terms of the game rather than over the derivative $\delta_S^v(\cdot)$, and are estimated with Monte Carlo integration. Note that, even though the Banzhaf interaction indices have been defined in CGT [12], they have not been considered in this ML literature.

3 Hoeffding Decomposition on a Game

A game $v : 2^N \rightarrow \mathbb{R}$ can be equivalently be represented by pseudo-Boolean function $b : \{0,1\}^N \rightarrow \mathbb{R}$ defined by

$$v(S) = b(1_S, 0_{N \setminus S}),$$

where $(1_S, 0_{N \setminus S})$ denotes an alternative taking value 1 on variables in S and value 0 otherwise. We consider a set of random variables $\mathbf{Z} = (Z_1, \ldots, Z_n)$ on

$\{0,1\}^N$ which are i.i.d. and uniformly distributed over $\{0,1\}$. This amounts to assigning the same probability for a variable $j \in N$ to belong to the coalition, or not to belong to the coalition.

Remark 1. In Sect. 2.3, we have seen that the Shapley value satisfied the symmetry axiom – meaning (5) handles players in a symmetric manner. Hence our assumption of a similar probability for a variable $j \in N$ to belong to the coalition, or not to belong to the coalition is already included in (5). Hence we can make this assumption.

Theorem 1. *Assuming that the presence or absence of the variables to a coalition are i.i.d. and given by a uniform distribution, the Hoeffding decomposition of game v is given by for all $P \subseteq N$*

$$v(P) = \sum_{S \subseteq N} v_S(P \cap S), \tag{7}$$

where $v_S : 2^S \to \mathbb{R}$ is given by

$$v_S(P) = \frac{(-1)^{s-p}}{2^n} \times \sum_{T_1 \subseteq S} \sum_{T_2 \subseteq N \setminus S} (-1)^{s-t_1}\, v(T_1 \cup T_2). \tag{8}$$

We note that $v_S(P)$ only marginally depends on P, though $(-1)^p$, and mainly depends on S.

Proof. We start with the Hoeffding decomposition of a Boolean function over independent uniform distributions over $\{0,1\}^N$, which reads (see (1) and (2)):

$$b(\mathbf{z}) = \sum_{S \subseteq N} b_S(\mathbf{z}_S),$$

where

$$b_S(\mathbf{z}_S) = \sum_{K \subseteq S} (-1)^{s-k} \mathbb{E}[b(\mathbf{Z}) | \mathbf{Z}_K = \mathbf{z}_K] = \sum_{K \subseteq S} \frac{(-1)^{s-k}}{2^{n-k}} \sum_{\mathbf{z}_{N \setminus K} \in \{0,1\}^{N \setminus K}} b(\mathbf{z}).$$

Transforming the Boolean function into a game, we obtain

$$v(A) := b(1_A, 0_{N \setminus A}) = \sum_{S \subseteq N} b_S(1_{A \cap S}, 0_{S \setminus A}) =: \sum_{S \subseteq N} v_S(S \cap A),$$

where $v_S(P) = b_S(\mathbf{z}_S)$ for $\mathbf{z}_S = (1_P, 0_{S \setminus P})$, with $P \subseteq S$. Thus $\mathbf{z}_K = (1_{P \cap K}, 0_{K \setminus P})$ for $K \subseteq S$. Then $\mathbf{z}_{N \setminus K} = (1_R, 0_{N \setminus (K \cup R)})$ for $R \subseteq N \setminus K$, then $\mathbf{z} = (\mathbf{z}_K, \mathbf{z}_{N \setminus K}) = (1_{(P \cap K) \cup R}, 0_{N \setminus [(P \cap K) \cup R]})$ and

$$v_S(P) = b(1_P, 0_{S \setminus P}) = \sum_{K \subseteq S} \frac{(-1)^{s-k}}{2^{n-k}} \sum_{R \subseteq N \setminus K} v(R \cup (P \cap K)).$$

We set $T = R \cup (P \cap K)$. We can write $T = T_1 \cup T_2$ with $T_1 \subseteq S$ and $T_2 \subseteq N \setminus S$. We have $K \subseteq (T_1 \cap P) \cup [S \setminus (T_1 \cup P)] =: M$ and $R = [T_1 \setminus K] \cup T_2$.

Hence

$$v_S(P) = \sum_{T_1 \subseteq S} \sum_{T_2 \subseteq N \setminus S} v(T_1 \cup T_2) \sum_{K \subseteq M} \frac{(-1)^{s-k}}{2^{n-k}}$$
$$= \sum_{T_1 \subseteq S} \sum_{T_2 \subseteq N \setminus S} v(T_1 \cup T_2) \frac{(-1)^{s-m}}{2^{n-m}} \sum_{K \subseteq M} \left(-\frac{1}{2}\right)^{m-k}.$$

As $\sum_{K \subseteq M} \left(-\frac{1}{2}\right)^{m-k} = \sum_{k=0}^{m} \binom{m}{k} \left(-\frac{1}{2}\right)^{m-k} = \left(1 - \frac{1}{2}\right)^m = \frac{1}{2^m}$, we obtain

$$v_S(P) = \sum_{T_1 \subseteq S} \sum_{T_2 \subseteq N \setminus S} v(T_1 \cup T_2) \frac{(-1)^{s-m}}{2^n}.$$

But (see Fig. 2)

$$m = |T_1 \cap P| + |S \setminus (T_1 \cup P)|$$
$$= [t_1 - |(S \setminus P) \cap T_1|] + [s - p - |(S \setminus P) \cap T_1|] = (s-p) + t_1 - 2\,|(S \setminus P) \cap T_1|$$

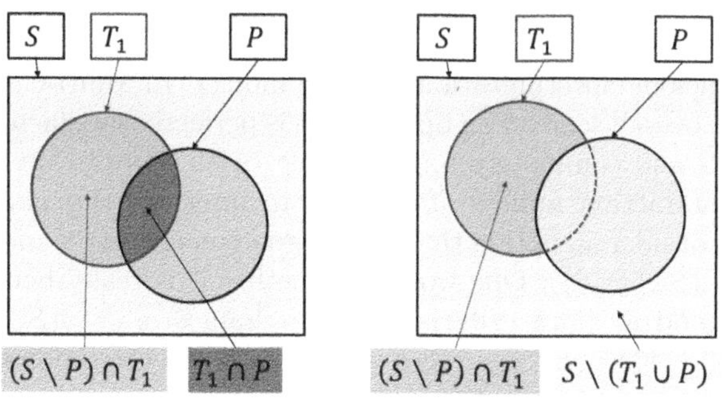

Fig. 2. Notation

Hence $(-1)^{s-m} = (-1)^{s-t_1}(-1)^{-(s-p)}(-1)^{2\,|(S \setminus P) \cap T_1|} = (-1)^{s-t_1}(-1)^{s-p}$ and

$$v_S(P) = \sum_{T_1 \subseteq S} \sum_{T_2 \subseteq N \setminus S} v(T_1 \cup T_2) \frac{(-1)^{s-t_1}\,(-1)^{s-p}}{2^n}$$
$$= \frac{(-1)^{s-p}}{2^n} \times \sum_{T_1 \subseteq S} \sum_{T_2 \subseteq N \setminus S} (-1)^{s-t_1}\, v(T_1 \cup T_2).$$

■

4 GSA for Dependent Variables Using Banzhaf Interaction Indices

In Theorem 1, we recognize the expression of the Banzhaf value for $v_S(P)$ when S is a a singleton. In order to interpret expression 8, we first recall the Banzhaf interaction index.

4.1 Banzhaf Interaction Indices for TU Games

Some known facts on the Banzhaf interaction indices are recalled in this section. The Banzhaf interaction indices are defined for $S \subseteq N$ with $S \neq \varnothing$ by [12]

$$I^{\text{Ban}}_{N,v}(S) = \frac{1}{2^{n-s}} \sum_{T \subseteq N\setminus S} \sum_{K \subseteq S} (-1)^{s-k} v(K \cup T). \tag{9}$$

In particular, for a pair $S = \{i,j\}$, we obtain

$$I^{\text{Ban}}_{N,v}(\{i,j\}) = \frac{1}{2^{n-2}} \sum_{T \subseteq N\setminus\{i,j\}} \delta_{i,j} v(T), \tag{10}$$

where $\delta_{i,j} v(S) := v(S \cup \{i,j\}) - v(S \cup \{i\}) - v(S \cup \{j\}) + v(S)$. We note that

$$I^{\text{Ban}}_{N,v}(\{i\}) = \phi^{\text{Ban}}_i(N,v).$$

Let us recall some interpretation of these indices. In Multi-Criteria Decision Aid, $v(S)$ is the overall score of an option that is perfectly satisfactory (with score 1) on criteria S and completely unacceptable (with score 0) on the remaining criteria. The interaction indices are crucial to interpret this game v [11]. For $S \subseteq N\setminus\{i,j\}$, consider an option that is very good on criteria S and unacceptable on criteria $N \setminus (S \cup \{i,j\})$. One wonders whether it is really beneficial for this option to be good at both criteria i and j. Values $\delta_i = v(S \cup \{i\}) - v(S)$, $\delta_j = v(S \cup \{j\}) - v(S)$ and $\delta_{ij} = v(S \cup \{i,j\}) - v(S)$ correspond to the added value for this option to be good at criterion i but not at criterion j, to be good at criterion j but not at criterion i and to be good at both criteria i and j respectively. When $\delta_{ij} > \delta_i + \delta_j$, improving both i and j gives strictly more than improving i and j separately. In this case, criteria i and j deserve to be well-satisfied together. We say in this case that there is *complementarity* among criteria i and j. When $\delta_{ij} < \delta_i + \delta_j$, it is not interesting to improve both criteria i and j together. We say in this case that there is *substitutability* among criteria i and j.

The concept of an interaction among pairs of players is not very well-known in CGT [22]. The first appearance of some interaction in CGT is due to Owen [25].

Expression 9 is uniquely characterized by the following axioms [12]: *Linearity* ($I^{\text{Ban}}_{N,v}(S)$ is linear on $\mathcal{G}(2^N)$ for every $S \subseteq N$), *Dummy player* (If $i \in N$ is a dummy player for $v \in \mathcal{G}(2^N)$, then (i) $I^{\text{Ban}}_{N,v}(\{i\}) = v(\{i\})$, and (ii) for every

$S \subseteq N \setminus \{i\}$, $S \neq \varnothing$, $I^{\text{Ban}}_{N,v}(S \cup \{i\}) = 0$), *Symmetry*, (for all $v \in \mathcal{G}(2^N)$, for all permutation π on N, $I^{\text{Ban}}_{N,v}(S) = I^{\text{Ban}}_{\pi N,\pi v}(\pi S)$), *2-efficiency*, and *Recursivity* (For any game v and any $S \subseteq N$, $s > 1$, $I^{\text{Ban}}_{N,v}(S) = I^{\text{Ban}}_{N\setminus\{j\},v^{-j}_{\cup j}}(S\setminus\{j\}) - I^{\text{Ban}}_{N\setminus\{j\},v^{-j}}(S\setminus\{j\})$ $,\forall j \in S$, where v^{-j} is the game v restricted to elements in $N \setminus j$, defined by $v^{-j}(S) = v(S), \forall S \subseteq N \setminus j$, and $v^{-j}_{\cup j}$ is the game on $N \setminus j$ in the presence of j defined by $v^{-j}_{\cup j}(S) = v(S \cup j) - v(S), \forall S \subseteq N \setminus j$).

When there will be no confusion, $I^{\text{Ban}}_{N,v}(S)$ will be simply denoted by $I^{\text{Ban}}(S)$.

4.2 Main Result

We directly obtain from Theorem 1 the following result.

Corollary 1. *Assuming that the presence or absence of the variables to a coalition are i.i.d. and given by a uniform distribution, the Hoeffding decomposition of game v is given by for all $P \subseteq N$*

$$v(P) = \sum_{S \subseteq N} v_S(P \cap S), \tag{11}$$

where $v_S : 2^S \to \mathbb{R}$ is given by

$$v_S(P) = \frac{(-1)^{s-p}}{2^s} I^{\text{Ban}}(S) \qquad \forall P \subseteq S. \tag{12}$$

Corollary 1 provides the Hoeffding decomposition of a game v. Taking 11 with $P = N$, we obtain

$$\sum_{S \subseteq N} \text{Index}(S) = v(N), \tag{13}$$

where Index : $2^N \to \mathbb{R}$ is given by

$$\text{Index}(S) := v_S(S) = \frac{1}{2^s} I^{\text{Ban}}(S). \tag{14}$$

In this expression, each term corresponds to a Banzhaf interaction index up to a multiplicative factor. Note that term $v_S(S)$ contains what regards interactions among S and not subsets of S, according to the definition of the Hoeffding decomposition. Property (13) means that the total variance $v(N)$ is shared among all subsets of variables. In this view, $v_S(S)$ is the contribution of subset S of variables in the total variance $v(N)$ of function f. The only major difference with the Sobol indices is that $v_S(S)$ may be negative, whereas all Sobol indices are necessarily non-negative.

5 Sobol-Like Indices for Dependent Variables

This section further develops the Banzhaf indices for GSA. We first define first-order and total indices, as for Sobol indices. The idea is that one cannot compute the influence of each subset of variable when the dimension is significant. One then computes the contribution of a variable taking into account more or less interaction terms (Sect. 5.1). Then Sect. 5.2 presents the illustration of our proposal on two examples.

5.1 1st Order and Total Indices

For the Sobol indices, it is convenient to compute two indices related to variable i. The first one is the 1st order index that simply considers the term of the decomposition for singleton $\{i\}$:

$$\text{Index}_{1\text{st}-\text{order}}(i) = v_{\{i\}}(\{i\}) = \frac{1}{2} I^{\text{Ban}}(\{i\}).$$

The *total* index for variable i is defined by:

$$\text{Index}_{\text{Total}}(i) = \sum_{S \ni i} v_S(S).$$

It sums-up the contribution of all subsets including variable i.

Lemma 1. *We have*

$$\text{Index}_{\text{Total}}(i) = \frac{1}{2}(v(N) - v(N \setminus \{i\})).$$

Proof. We have

$$\begin{aligned}
\text{Index}_{\text{Total}}(i) &= \sum_{S \ni i} v_S(S) = \sum_{S \ni i} \frac{1}{2^s} I^{\text{Ban}}(S) \\
&= \frac{1}{2^n} \sum_{S \ni i} \sum_{T \subseteq N \setminus S} \sum_{K \subseteq S} (-1)^{s-k} v(T \cup K) \\
&= \frac{1}{2^n} \sum_{M \subseteq N} v(M) \sum_{S \ni i; T \subseteq N \setminus S, K \subseteq S : M = T \cup K} (-1)^{s-k} \\
&= \frac{1}{2^n} \sum_{M \subseteq N} v(M) \sum_{S \ni i} (-1)^{s-|S \cap M|} \\
&= \frac{1}{2^n} \sum_{M \subseteq N \setminus \{i\}} v(M) \sum_{S \ni i} (-1)^{s-|S \cap M|} + \frac{1}{2^n} \sum_{M \subseteq N \setminus \{i\}} v(M \cup \{i\}) \sum_{S \ni i} (-1)^{s-|S \cap M|-1} \\
&= \frac{1}{2^n} \sum_{M \subseteq N \setminus \{i\}} (v(M \cup \{i\}) - v(M))\, \psi(M),
\end{aligned}$$

where for $M \subseteq N \setminus \{i\}$

$$\psi(M) := \sum_{S \ni i} (-1)^{s-|S\cap M|-1} = \sum_{R \subseteq N\setminus\{i\}:S=R\cup\{i\}} (-1)^{r+1-|R\cap M|-1}$$

$$\sum_{R_1 \subseteq M, R_2 \subseteq N\setminus(M\cup\{i\})} (-1)^{r_1+r_2-r_1} = \sum_{R_1\subseteq M} 1 \times \sum_{R_2 \subseteq N\setminus(M\cup\{i\})} (-1)^{r_2}.$$

We have $\sum_{R_1\subseteq M} 1 = 2^m$, and $\sum_{R_2\subseteq N\setminus(M\cup\{i\})}(-1)^{r_2}$ is equal to 0 if $M \neq N\setminus\{i\}$ and is equal to 1 if $M = N \setminus \{i\}$. In $\text{Index}_{\text{Total}}(i)$, there remains thus only term $M = N \setminus \{i\}$:

$$\text{Index}_{\text{Total}}(i) = \frac{1}{2^n} \times (v(N) - v(N \setminus \{i\})) \times 2^{n-1}.$$

■

5.2 Illustrative Examples

Example 1 (see sect 4.1 in [24]). Consider first a function of two variables which depends only on the first one: $f(x_1, x_2) = x_1$. The two variables are governed by the following distribution showing correlation:

$$\mathbf{X} \sim \mathcal{N}\left(0, \begin{pmatrix} 1 & \rho \\ \rho & 1 \end{pmatrix}\right).$$

We have $\text{Var}(Y) = 1$ with $Y = f(X_1, X_2)$. The game 6 is given by:

$$v(\emptyset) = 0 \ , \ v(\{1\}) = 1 \ , \ v(\{2\}) = \rho^2 \ , \ v(\{1,2\}) = 1.$$

Hence

$$\text{Index}(\emptyset) = \frac{2+\rho^2}{4} \qquad \text{Index}(\{1\}) = \frac{1}{2} - \frac{\rho^2}{4}$$

$$\text{Index}(\{2\}) = \frac{\rho^2}{4} \qquad \text{Index}(\{1,2\}) = -\frac{\rho^2}{4}$$

The sum of these indices is indeed $v(N) = 1$. This decomposition clearly separate the contribution of each term. $\text{Index}(\emptyset)$ holds the constant term, that is the average value of the game over all coalitions. More precisely, $v(S)$ measures the part the total variance hold by subset S, and $\text{Index}(\emptyset)$ is the average contribution of the variance over all subsets of N. $\text{Index}(\{1\})$ and $\text{Index}(\{2\})$ contain the sole contribution of variables 1 and 2 alone. Lastly, $\text{Index}(\{1,2\})$ measures the contribution of the interaction of the two variables. There is a negative interaction between the two variables.

Note that in comparison, the Shapley value of v is

$$\phi_1^{\text{Sh}}(N, v) = 1 - \frac{\rho^2}{2} \qquad \phi_2^{\text{Sh}}(N, v) = \frac{\rho^2}{2}$$

The contributions of single variables are identical with the two methods up to a multiplicative coefficient 2. ■

Example 2. Consider the following function $f(x_1, x_2) = x_1 + \alpha\, x_1\, x_2$, which exhibits interaction between these variables . The two variables are now assumed independent:

$$\mathbf{X} \sim \mathcal{N}\left(0, \begin{pmatrix} 1 & 0 \\ 0 & 1 \end{pmatrix}\right).$$

We note that $\text{Var}(Y) = 1 + \alpha^2$. The game (6) is given by:

$$v(\emptyset) = 0 \ , \ v(\{1\}) = \frac{1}{1+\alpha^2} \ , \ v(\{2\}) = 0 \ , \ v(\{1,2\}) = 1.$$

Hence

$$\text{Index}(\emptyset) = \frac{1+\alpha^2/2}{2(1+\alpha^2)} \qquad \text{Index}(\{1\}) = \frac{1+\alpha^2/2}{2(1+\alpha^2)}$$

$$\text{Index}(\{2\}) = \frac{\alpha^2/2}{2(1+\alpha^2)} \qquad \text{Index}(\{1,2\}) = \frac{\alpha^2/2}{2(1+\alpha^2)}$$

The sum of these indices is indeed $v(N) = 1$. There is a positive interaction between the two variables.

Note that in comparison, the Shapley value of v is

$$\phi_1^{\text{Sh}}(N, v) = \frac{2+\alpha^2}{2(1+\alpha^2)} \qquad \phi_2^{\text{Sh}}(N, v) = \frac{\alpha^2}{2(1+\alpha^2)}$$

Note that $\phi_1^{\text{Sh}}(N, v) = 2\,\text{Index}(\{1\})$ and $\phi_2^{\text{Sh}}(N, v) = 2\,\text{Index}(\{2\})$. ■

6 Conclusion

We have proposed an alternative approach to the Shapley effect for GSA. We first compute the game as in Owen's paper. The game quantifies the part to which each coalition contributes to the overall variance. We then compute the Hoeffding decomposition of this game. The main result of this paper is that this decomposition yield known factors, which are the Banzhaf interaction indices. This gives a finer analysis of the contribution of variables than with the Shapley effects.

For future works, we will develop estimation methods to compute these indices from a dataset. In the literature, the construction of the game defined by 6 from a dataset (i.e. "*given data*") is done with an estimation algorithm based on the K-means [5]. However, this estimation is biased. We will work on other estimations in future works.

Acknowledgements. This work has received support from the FaRADAI project (ref. 101103386) funded by the European Commission under the European Defence Fund (EDF-2021-DIGIT-R).

References

1. Aas, K., Jullum, M., Loland, A.: Explaining individual predictions when features are dependent: more accurate approximations to Shapley values. Artif. Intell. **298**, 103502 (2021)
2. Au, Q., Herbinger, J., Stachl, C., Bischl, B., Casalicchio, G.: Grouped feature importance and combined features effect plot. Data Min. Knowl. Disc. **36**(4), 1401–1450 (2022)
3. Bachoc, F., Gamboa, F., Halford, M., Loubes, J., Risser, L.: Explaining machine learning models using entropic variable projection. Inf. Inference **12**, 1686–1715 (2023)
4. Bordt, S., von Luxburg, U.: From Shapley values to generalized additive models and back. In: ICAIS, pp. 709–745. PMLR (2023)
5. Broto, B., Bachoc, F., Depecker, M.: Variance reduction for estimation of Shapley effects and adaptation to unknown input distribution. SIAM/ASA J. Uncertain. Quantif. **8**, 693–716 (2020)
6. Covert, I., Lundberg, S., Lee, S.: Understanding global feature contributions with additive importance measures. In: NEURIPS, pp. 17212–17223 (2020)
7. Dubey, P., Shapley, L.S.: Mathematical properties of the Banzhaf power index. Math. Oper. Res. **4**, 99–131 (1979)
8. Fel, T., Cadene, R., Chalvidal, M., Cord, M., Vigouroux, D., Serre, T.: Look at the variance! Efficient black-box explanations with Sobol-based sensitivity analysis. In: NeurIPS (2022)
9. Fumagalli, F., Muschalik, M., Kolpaczki, P., Hüllermeier, E., Hammer, B.: SHAP-IQ: unified approximation of any-order Shapley interactions. In: NeurIPS 2023 (2023)
10. Giudici, P., Raffinetti, E.: Shapley-Lorenz explainable artificial intelligence. Expert Syst. Appl. (2021)
11. Grabisch, M., Labreuche, C.: A decade of application of the Choquet and Sugeno integrals in multi-criteria decision aid. Ann. Oper. Res. **175**, 247–286 (2010)
12. Grabisch, M., Roubens, M.: An axiomatic approach to the concept of interaction among players in cooperative games. Int. J. Game Theory **28**, 547–565 (1999)
13. Guidotti, R., Monreale, A., Ruggieri, S., Turini, F., Giannotti, F., Pedreschi, D.: A survey of methods for explaining black box models. ACM Comput. Surv. **51**(5), Article 93 (2018)
14. Harris, C., Pymar, R., Rowat, C.: Joint shapley values: a measure of joint feature importance. arXiv preprint arXiv:2107.11357 (2021)
15. Iooss, B., Prieur, C.: Shapley effects for sensitivity analysis with correlated inputs: comparisons with Sobol' indices, numerical estimation and applications. Int. J. Uncertain. Quantif. **9**, 493–514 (2019)
16. Jacques, J., Lavergne, C., Devictor, N.: Sensitivity analysis in presence of model uncertainty and correlated inputs. Reliab. Eng. Syst. Saf. **91**, 1126–1134 (2006)
17. Lehrer, E.: An axiomatisation of the Banzhaf value. Int. J. Game Theory **17**, 89–99 (1988)
18. Lopez-Benito, A., Bolado-Lavin, R.: A case study on global sensitivity analysis with dependent inputs: the natural gas transmission model. Reliab. Eng. Syst. Saf. **165**, 11–21 (2017)
19. Lundberg, S., Enrion, G., Lee, S.: Consistent individualized feature attribution for tree ensembles. arXiv preprint arXiv:1802.03888 (2018),

20. Lundberg, S., Lee, S.: A unified approach to interpreting model predictions. In: NIPS 2017, pp. 4768–4777 (2017)
21. Mara, T., Tarantola, S.: Variance-based sensitivity indices for models with dependent inputs. Reliab. Eng. Syst. Saf. **107**, 115–121 (2012)
22. Murofushi, T., Soneda, S.: Techniques for reading fuzzy measures (III): interaction index. In: 9th Fuzzy System Symposium, Sapporo, Japan, pp. 693–696 (1993)
23. Owen, A.: Sobol' indices and Shapley value. SIAM/ASA J. Uncertain. Quantif. **2**, 245–251 (2014)
24. Owen, A., Prieur, C.: On Shapley value for measuring importance of dependent inputs. SIAM/ASA J. Uncertain. Quantif. **5**(1), 986–1002 (2017)
25. Owen, G.: Multilinear extensions of games. Management Sci. **18**, 64–79 (1972)
26. Rabitti, G., Borgonovo, E.: A Shapley-Owen index for interaction quantification. SIAM/ASA J. Uncertain. Quantif. **7**, 1060–1075 (2019)
27. Ribeiro, M., Singh, S., Guestrin, C.: "Why should i trust you?": explaining the predictions of any classifier. In: ACM KDD 2016, San Francisco, California, USA, pp. 1135–1144 (2016)
28. Saltelli, A., et al.: Global Sensitivity Analysis: The Primer. Wiley, New York (2008)
29. Saltelli, A., Tarantola, S.: On the relative importance of input factors in mathematical models: safety assessment for nuclear waste disposal. J. Am. Stat. Assoc. **97**, 702–709 (2002)
30. Shapley, L.S.: A value for n-person games. In: Kuhn, H.W., Tucker, A.W. (eds.) Contributions to the Theory of Games, vol. II, pp. 307–317. Princeton University Press (1953)
31. Sobol, I.: Sensitivity estimates for nonlinear mathematical models. Math. Model. Comput. Exp. **1**, 407–414 (1993)
32. Stoian, M.: Fast joint shapley values. In: Proceedings of the International Conference on Management of Data (SIGMOD 2023), pp. 285–287 (2023)
33. Strumbelj, E., Kononenko, I.: An efficient explanation of individual classifications using game theory. J. Mach. Learn. Res. **11**, 1–18 (2010)
34. Sundararajan, M., Dhamdhere, K., Agarwal, A.: The Shapley Taylor interaction index. In: ICML 2020, pp. 9259–9268. PMLR (2020)
35. Sundararajan, M., Najmi, A.: The many Shapley values for model explanation. In: ICML 2020, pp. 9269–9278 (2020)
36. Tsai, C., Yeh, C., Ravikumar, P.: Faith-Shap: the faithful Shapley interaction index. J. Mach. Learn. Res. **24**(94), 1–42 (2023)
37. Tsang, M., Rambhatla, S., Liu, Y.: How does this interaction affect me? Interpretable attribution for feature interactions. In: NeurIPS 2020, pp. 6147–6159 (2020)
38. Wei, P., Lu, Z., Song, J.: Variable importance analysis: a comprehensive review. Reliab. Eng. Syst. Saf. **142**, 399–432 (2015)

An Ontology-Based Approach for Handling Inconsistency in Explainable and Prioritized Access Control Models

Ahmed Laouar[1(✉)], Toky Raboanary[2], and Salem Benferhat[1]

[1] CRIL, Univ. Artois & CNRS, Lens, France
{laouar,benferhat}@cril.fr
[2] University of Cape Town, Cape Town, South Africa
traboanary@cs.uct.ac.za

Abstract. The development of secure and efficient solutions for access control is an important issue in a variety of applications. One of the main challenges is to avoid situations that make access control decision-making impossible. However, avoiding such situations hampers the evolution of the model, as it means either adding a large set of constraints or dealing with each conflict situation. It is, therefore, important to explore methods that deal with conflicts as they arise while providing explanations of the decision taken. In this work, we develop an ontology to manage dynamic and abstract access control rules based on the OrBAC (Organization-Based Access Control) model and integrate an ordering relation over instances of the ontology. Our method takes advantage of the application of inconsistency-tolerant semantics to resolve conflicts and generate explanations for transparency in decisions made. Our results show that the approach efficiently preserves the consistency of the decision taken and provides potentially useful and human-friendly explanations.

Keywords: Access Control · Ontology · Inconsistency · Explainability

1 Introduction

In computer security, various mechanisms are needed to make sure that information systems are properly accessed by authorized users. Some of these mechanisms focus on user authentication and identification, while other mechanisms focus on the modelling and the specification of access rights according to the required security policy [19]. Access control models are the main mechanisms to achieve this objective. They serve as a tool of implementing and enforcing security policies, by providing an access decision (grant or deny) based on some predefined access control security rules. A lot of research efforts focused on proposing different types of models [15,19], like the role-based model (RBAC) [21] and the organisation-based model (OrBAC) [16].

In this paper, we focus on dealing with conflicts that arise in generic and dynamic access control models. We chose the OrBAC model, mainly because it

S. Destercke et al. (Eds.): SUM 2024, LNAI 15350, pp. 249–264, 2025.
https://doi.org/10.1007/978-3-031-76235-2_19

is a dynamic model that has the capacity of inferring access rights and also because uncertainty in the model is natural and can be represented in multiple ways, like using possibilistic logic weights [2]. We propose to deal with conflicts on their occurrence rather than defining rules and constraints to avoid them as it is usually used in the literature [8].

This can be achieved by applying inconsistency-tolerant semantics [4], which are the formal way to reason in the presence of conflicts in ontologies [5].

Therefore, OrBAC must be adapted (or encoded) in some formal ontology fragment in order to apply such methods. More precisely, the main contributions of this paper are as follows:

- we provide an ontology that encodes the OrBAC model, alongside the use of hierarchies and a specific representation of uncertainty;
- we show how a partial order is propagated in the ontology using SWRL rules;
- we apply repair methods inspired from partially ordered possibilistic knowledge base repairs [17] to handle conflicts;
- we provide explanations for the model decisions, with and without conflicts;
- we show that reasoning services are done with simple SPARQL queries.

Section 2 provides a refresher on OrBAC. Section 3 introduces the OrBAC ontology and the representation of uncertainty. Section 4 introduces an efficient conflict handling method. Section 5 provides mechanisms to generate explanations of the decisions taken, before concluding the paper.

2 Refresher on OrBAC

The basic concepts in access control are subjects, which represent users, objects, which are passive entities like resources and files, and actions which are the basic operations performed by a subject on an object, like read and write. OrBAC [16] considers these concepts as concrete entities and abstracts each of them into an abstract entities. Subjects that fulfill same functions are grouped in roles, objects that satisfy common properties are grouped in views and actions that share the same principles are grouped in activities. In addition, OrBAC considers organisations as concrete concepts, and always defines rules in an organisation.

Concrete entities are linked to abstract entities using the connection relations Employ, Use and Consider. Employ(org,s,r) means that the subject "s" plays the role "r" within the organization "org". Use(org,o,v) means that the object "o" is used in the view "v" within the organization "org". Consider(org,α,a) means that the organization "org" considers that the action "α" fall within the activity "a". Moreover, OrBAC is said to be dynamic because it uses contexts. A privilege may be granted in a context and denied in another. It can represent environmental variables, like timeOfDay, or a particular state like emergency. Contexts are defined between a subject, an action and an object in an organisation using the relation Define such that Define(org,s,α,o,c) means that in the organization "org", the context "c" holds between the subject "s", the object "o" and the action "α".

Policy rules, which are also called abstract privileges, are defined using the relations Permission, Prohibition, Obligation and Recommendation. Each rule admits five arguments; Permission(org, r, a, v, c) means that the organization "org" grants the role "r" the permission to perform the activity "a" on the view "v" if the context "c" holds. The remaining rules are defined similarly.

Access rights, or concrete privileges, are derived for the concrete entities, namely subjects, actions and objects. They are represented with the relations Is-permitted, Is-prohibited, Is-obliged, and Is-recommended. The concrete privilege Is-permitted(s,α,o) means that the subject "s" is allowed to perform the action "α" on the object "o". The derivation is given as follows:

Definition 1 (Derivation of concrete rules). *Concrete permissions are derived from an OrBAC policy using the following rule:*
$\forall$org, $\forall$s, $\forall\alpha$, $\forall$o, $\forall$r, $\forall$a, $\forall$v, $\forall$c,
Permission(org,r,a,v,c) $\wedge$ Employ(org,s,r) $\wedge$ Use(org,o,v) $\wedge$ Consider(org, α, a) $\wedge$ Define(org,s,α,o,c) $\models$ Is-permitted(s,α,o)
The privileges Is-prohibited, Is-obliged, and Is-recommendedare derived similarly.

3 The OrBAC Ontology

3.1 Preliminaries

An ontology can be encoded in OWL [9] using multiple syntaxes. In this paper, RDF triples are used to represent the axioms of the proposed ontology, in addition to some properties from OWL. Note that OWL is not limited to the elements described in this paper, as it allows representing other notions like complex concepts, subproperties, etc. The ontology is constructed from finite sets of classes C, object properties P and individuals I. The class triples are of the form $\langle x$,rdf:type,owl:Class$\rangle$, where $x \in C$ and relationships between classes are of the form $\langle x$,rdfs:subClassOf,$y\rangle$, where $x, y \in C$. Object property triples have the form $\langle x$,rdf:type,owl:ObjectProperty$\rangle$, where $x \in P$ and $\langle x, p, y\rangle$ s.t. $p \in \{$rdfs:domain,rdfs:range$\}$, $x \in P$ and $y \in C$. Transitive object property triples are of the form $\langle x$,rdf:type,owl:TransitiveProperty$\rangle$. It should be noted that properties like TransitiveProperty can be encoded in RDF triples, but are not part of RDF/RDFS. Axioms about individuals are called assertions, and have the form $\langle x, p, y\rangle$, where $x \in I$ is an individual, p is a property and y can be an individual or a literal. In addition, Semantic Web Rule Language (SWRL) rules [13] and SPARQL [12] queries are used to implement reasoning services: SWRL rules can be represented in the form $\varphi \rightarrow \phi$, where φ and ϕ are conjunction of atoms involving classes and object properties. SPARQL ASK queries check for the existence of triples that match a specified query pattern and return True if the pattern matches, and False otherwise. Then, SELECT queries retrieve specific data based on the provided patterns, and the result is a set of variable bindings matching the patterns. UNION queries combine results of multiple patterns, and the result set includes all matches from each pattern, eliminating duplicates by default.

3.2 Ontology Classes and Object Properties

Representing OrBAC Concrete and Abstract Entities. Objects, subjects and actions are represented using the classes orbac:Object, orbac:Subject and orbac:Action, respectively. The orbac:Organisation class represents organisations in the model, and is also a sub class of the orbac:Subject class. They are all sub classes of the class orbac:ConcreteConcept. Activities, views and roles are represented using the classes orbac:Activity, orbac:View and orbac:Role, respectively. They are all sub classes of the class orbac:AbstractConcept. Moreover, the orbac:Context class represents contexts.

Representing Policy Rules. The abstract policy rules, called also access types in the following, are represented using the orbac:Permission, orbac:Prohibition, orbac:Obligation and orbac:Recommendation classes, which are all sub classes of the class orbac:AccessType. Each access type is connected with its constructing elements using the object properties: orbac:accessTypeOrg to link an access type with an organisation. orbac:accessTypeRole to link an access type with a role. orbac:accessTypeActivity to link an access type with an activity. orbac:accessTypeView to link an access type with a view. orbac:accessTypeContext to link an access type with a context. Each access type instance must be connected with exactly one entity for each of the above object properties.

Representing Connection Rules. The rules linking concrete entities with their abstractions, and the rule that defines whether a context holds or not for a subject, an action and an object in an organisation are represented using the classes orbac:Consider, orbac:Use, orbac:Employ and orbac:Define, which are all sub classes of orbac:ConnectionRule. These rules are connected with their entities using the following object properties. The class orbac:Consider uses the orbac:considersActivity, orbac:considersAction and orbac:considersOrg object properties to connect its instances to an activity, an action and an organisation respectively. The class orbac:Use uses the object properties orbac:usesView, orbac:usesObject and orbac:usesEmployer to connect its instances to a view, an object and an organisation, respectively. The class orbac:Employ uses the orbac:employesRole, orbac:employesEmployee and orbac:employesEmployer object properties to connect its instances to a role, a subject and an organisation, respectively. The class orbac:Define uses the orbac:definesSubject, orbac:definesAction, orbac:definesObject and orbac:definesContext properties to connect its instances to a subject, an action, an object and a context, respectively.

Running Example. In collaborative research projects, sharing resources involves access to sensitive data by users from multiple organisations, which requires proper access control mechanisms. We consider an example of a consortium of universities. The consortium defines a permission in the secondment context, a secondee can perform modification to the reports view, which contains secondment reports. It also defines a prohibition for a staff member to not modify the same reports, in the default context. Moreover, the consortium considers that the action edit falls under the activity modify and the file report1 is used within the view reports. Assume a user called Bob, who is employed within the consortium as a staff member. In addition, Bob is on a secondment in univ1, thus they are employed as a secondee there. The context default holds between Bob, edit and report1, in the consortium, since Bob is a staff member. Moreover, the context secondment also holds for Bob, edit and report1 since Bob is on a secondment. The representation of this example using the proposed orbac ontology is given in Fig. 1.

```
:perm1 rdf:type orbac:Permission.
:perm1 :accessTypeOrg :consortium.
:perm1 :accessTypeContext :secondment.
:perm1 :accessTypeActivity :modify.
:perm1 :accessTypeView :reports.
:perm1 :accessTypeRole :secondee.

:emp1 rdf:type orbac:Employ.
:emp1 :employesEmployer :univ1.
:emp1 :employesRole :secondee.
:emp1 :employesEmployee :Bob.

:use1 rdf:type orbac:Use.
:use1 :usesEmployer :univ1.
:use1 :usesView :reports.
:use1 :usesObject :report1.

:def2 rdf:type orbac:Define.
:def2 :definesOrg :univ1.
:def2 :definesContext :secondment.
:def2 :definesSubject :Bob.
:def2 :definesAction :edit.
:def2 :definesObject :report1.
```

```
:prohi1 rdf:type orbac:Prohibition.
:prohi1 :accessTypeOrg :consortium.
:prohi1 :accessTypeContext :default.
:prohi1 :accessTypeActivity :modify.
:prohi1 :accessTypeView :reports.
:prohi1 :accessTypeRole :staffMember.

:emp2 rdf:type orbac:Employ.
:emp2 :employesEmployer :consortium.
:emp2 :employesRole :staffMember.
:emp2 :employesEmployee :Bob.

:cons1 rdf:type orbac:Consider.
:cons1 :considersOrg :consortium.
:cons1 :considersActivity :Modify.
:cons1 :considersAction :edit.

:def1 rdf:type orbac:Define.
:def1 :definesOrg :consortium.
:def1 :definesContext :default.
:def1 :definesSubject :Bob.
:def1 :definesAction :edit.
:def1 :definesObject :report1.
```

Fig. 1. Running example triples.

Hierarchies in the OrBAC Ontology. In access control models, hierarchies define how privileges are inherited in the model. We propose the use of hierarchies of organisations and roles, as defined in [7]. For roles, two types of relationships are implemented: a sub-role inherits both permissions and prohibitions of a parent role, and a senior-role inherits permissions of a parent role and the parent role inherits its prohibitions. This separation results from the difference in significance of a relationship between roles. A senior role is considered as more privileged, thus inheritance goes upward.

- orbac:SubRole and orbac:SeniorRole classes as sub classes of the orbac:Role class.
- The orbac:hasParent property links a sub-role or a senior-role to a parent role.
- Sub-roles and senior-roles are defined in an organisation using the property orbac:subRoleOrg, which links a sub-role or a senior-role to an organisation.
- Organisational hierarchy is represented using the orbac:subOrganisationOf object property, which links two organisations. Every rule or relation that is defined in a parent organisation holds for the sub-organisation, for example: if ⟨perm1, orbac:accessTypeOrg, consortium⟩ $\wedge$ ⟨univ1, orbac:subOrganisationOf, consortium⟩ $\rightarrow$ ⟨perm1, orbac:accessTypeOrg, univ1⟩. The inverse does not hold.

We use SPARQL queries to check if a rule is inferred from the hierarchy, without adding the inferred axioms (triples) to the ontology.

Example 1. The following are examples of role hierarchies: the three triples on the left indicate that a secondee is a sub-role of employee in univ1. Similarly, on the right, the consortium considers that an employee is a sub-role of staffMember.

`:secondee rdf:type orbac:subRole.`
`:secondee orbac:hasParent :employee.`
`:secondee orbac:subRoleOrg :univ1.`
	`:employee rdf:type orbac:subRole.`
	`:employee orbac:hasParent :staffMember.`
	`:employee orbac:subRoleOrg :consortium.`

3.3 Representation of Uncertainty

The representation of priorities in OrBAC has been widely studied [3,8]. However, the proposed solutions focus on providing a preference of application between policy rules. The automatic assignment of preferences has been discussed in [8], but the authors did not elaborate on how it can be achieved. In this paper, a new representation of uncertainty is suggested, making the following separation:

- Fully certain rules: the policy rules (or access types) are established by policy experts and must be enforced, hence, they are given the highest priority.
- Uncertain rules: are the rules connecting concrete and abstract concepts.

Uncertainty is encoded using a preference relation defined between the individuals of the connection relations, namely, Employ, Consider, Use, and Define. Policy rules are assumed to be fully certain. Defining a preference between individuals of the uncertain rules is achieved using the transitive object property $\langle x$, orbac:isPreferredTo, $y\rangle$, which means x is more preferred to y. Equivalence holds when both $\langle x$, orbac:isPreferredTo, $y\rangle$ and $\langle y$, orbac:isPreferredTo, $x\rangle$ hold and incomparability is the absence of both relations.

The following describes how preferences are automatically assigned to the connection rules. The main method proceeds by ordering abstract concepts (e.g. roles). The order is then propagated using SWRL rules. Rule R1 in Table 1 is an example on how ordering is propagated from orbac:Role individuals to orbac:Employ individuals, other rules are defined similarly. Another method relies on hierarchies to define preferences. Mainly, a sub-role and a senior-role are considered more preferred than their parent roles. In addition, rules defined in sub-organisations have a higher preference. Rules R2 and R3 in Table 1 are examples of rules used to propagate ordering from hierarchies. All the rules are provided in the appendix. SWRL rules are executed efficiently using Drools [20].

Example 2. In our example, if we have ⟨secondment, orbac:isPreferredTo,default⟩ and ⟨secondee, orbac:isPreferredTo, staffMember⟩, then the SWRL rules result in ⟨emp1, orbac:isPreferredTo, emp2⟩ and ⟨def2, orbac:isPreferredTo, def1⟩. In addition, ⟨univ1, orbac:subOrganisationOf, consortium⟩, ⟨use1, orbac:usesEmployer, univ1⟩ and ⟨emp2, orbac:employesEmployer, consortium⟩, result in ⟨use1, orbac:isPreferredTo,emp2⟩.

Definition 2. *An* orbac *KB is a tuple* $\mathcal{K} = \langle \mathcal{T}, \mathcal{R}, \mathcal{T}_I \rangle$ *s.t.* $\mathcal{T}$ *is a finite set of axioms of the* orbac *ontology.* $\mathcal{R}$ *is a finite set of* SWRL *order propagation rules.*

Table 1. SWRL rules for order propagation. From R2. consider that r1 and r2 are roles, if r2 is a parent role of r1, then r1 is preferred to r2.

R1: orbac:isPreferredTo(?role1,?role2) ∧ orbac:employesRole(?employ1,?role1) ∧ orbac:employesRole(?employ2,?role2) → orbac:isPreferredTo(?employ1,?employ2)
R2: orbac:hasParent(?role1, ?role2) → orbac:isPreferredTo(?role1, ?role2)
R3: orbac:usesEmployer(?use1, ?org1) ∧ orbac:employesEmployer(?employ2, ?org2) ∧ orbac:subOrganisationOf(?org1, ?org2) → orbac:isPreferredTo(?use1, ?employ2)

$\mathcal{T}_I$ is a finite set of assertions (dataset) and I is the set of individuals. In the following, $\mathcal{T}_I = \mathcal{T}_U \cup \mathcal{T}_F$ is used to separate the assertions involving uncertain (U) and fully certain (F) individuals. $\mathcal{T}_U = \{\langle x, p, y\rangle \in \mathcal{T}_I | x = i \text{ or } y = i \text{ s.t. } \langle i, \mathsf{rdf:type}, t\rangle \in \mathcal{T}_I \text{ and } t \in U\}$, where $U =$ {orbac:Employ, orbac:Consider, orbac:Use, orbac:Define}.

The derivation of access rights, given by Definition 1, is encoded using SPARQL queries, called Is-permitted(s, α, o) and Is-prohibited(s, α, o), for a concrete permission and prohibition, respectively, where s, α, o are a subject, action, and object, respectively. Due to space limitations, the orbac ontology, alongside all the queries used in this paper are provided in the appendix[1].

Example 3. In the running example, both Is-permitted(Bob,edit,report1) and Is-prohibited(Bob,edit,report1) return true.

4 Inconsistency Handling

The main objective of encoding OrBAC as an ontology is to allow inconsistency handling as defined for formal ontologies [4]. The separation between fully certain and uncertain rules plays a key role in this application. In this section, we introduce the different reasoning services applied to the orbac ontology. Then, we discuss the application of partially ordered possibilistic repairs [17] to it.

4.1 Consistency Checking and Computing Conflicts

In access control, a knowledge base is considered as inconsistent if both a permission and a prohibition are inferred for the same subject, object and action. In the access control literature, inconsistency is also referred to by the presence of redundant rules [8]. The SPARQL query InconsistencyChecking$(\mathcal{K})$ checks if at least for one of the classes orbac:Employ, orbac:Consider, orbac:Use and orbac:Define, two different instances are defined for the same subject, action and object, respectively. And whether instances of both orbac:Permission and orbac:Prohibition are defined for their consecutive orbac:Role, orbac:Activity, orbac:View and orbac:Context. The query also checks if every instance is linked with the same organisation, or

[1] All paper appendices are available at https://github.com/ahmedlaouar/orbac.owl.

with one of its parent organisations and whether the permission is linked with a parent role or the prohibition is linked with a senior role. The definition following establishes inconsistency in the sense of the orbac ontology:

Definition 3. *Let $\mathcal{K}$ be an* orbac *KB, $\mathcal{K}$ is inconsistent if both of the queries* Is-permitted(s, α, o) and Is-prohibited(s, α, o) return true for the same s, α, and o (both a concrete permission and prohibition are derived).

Moreover, a conflict is the minimal set of individuals deriving both a permission and a prohibition for the same subject, action and object.

Definition 4. *Let $\mathcal{K} = \langle \mathcal{T}, \mathcal{R}, \mathcal{T}_U \cup \mathcal{T}_F \rangle$ be an* orbac *KB. Let I_U denote the uncertain individuals involved in $\mathcal{T}_U$. $\mathcal{C} \subseteq I_U$ is a conflict of $\mathcal{K}$ if:*

- InconsistencyChecking$(\langle \mathcal{T}, \mathcal{R}, \mathcal{T}_C \cup \mathcal{T}_F \rangle)$ = True, and;
- $\forall i \in \mathcal{C}$, InconsistencyChecking$(\langle \mathcal{T}, \mathcal{R}, \mathcal{T}_{C \setminus \{i\}} \cup \mathcal{T}_F \rangle)$ = False.

Where $\mathcal{T}_{C \setminus \{i\}} = \mathcal{T}_C \setminus \{\langle x, p, y \rangle \in \mathcal{T}_C | x = i \text{ or } y = i\}$.

The individuals in $\mathcal{I}_U$ ($I_U = \{i \in I | \langle i, \text{rdf:type}, t \rangle \in \mathcal{T}_U$ and $t \in$ {orbac:Employ, orbac:Consider, orbac:Use, orbac:Define}}) are considered uncertain by the representation in Sect. 3.3. These individuals are those included in conflicts. The remaining individuals are fully certain or irrelevant to conflicts. Computing all the conflicts of a KB is achieved using a SPARQL query called ComputeConflicts$(\mathcal{K})$, which is, to some extent, similar to the InconsistencyChecking$(\mathcal{K})$ query.

Example 4. The query InconsistencyChecking$(\mathcal{K})$ returns True from our example, because it infers both a permission and prohibition for (Bob, edit, report1). The conflict resulting from the query ComputeConflicts$(\mathcal{K})$ contains the following individuals: $\mathcal{C}$ = {emp1, use1, cons1, emp2, def1, def2}.

Proposition 1. *Checking the consistency and computing the conflicts of an* orbac *KB $\mathcal{K}$ using the queries* InconsistencyChecking$(\mathcal{K})$ and ComputeConflicts$(\mathcal{K})$ are both tractable. This follows from the efficiency of the used queries.

4.2 Repair-Based Method

An orbac knowledge base, as given by Definition 2 is partially preordered, thanks to the use of the property orbac:isPreferredTo. To resolve inconsistency in the orbac ontology, we propose to adapt the notion of accepted assertion used for DL-Lite$_{\mathcal{R}}$ partially preordered possibilistic knowledge bases in [1,17]. To achieve that, we start by introducing two notions: support and dominance. Schematically, a request for a permission is accepted if it admits at least one support which dominates all the supports for a prohibition [18].

Support: a support of a privilege for a given subject, action and object is the minimal set of individuals participating in its derivation. A privilege may have multiple supports. To compute the supports of a privilege, a SPARQL query is used. ComputeSupports$_p(\mathcal{K},s,\alpha,o)$, s.t. $p \in \{\mathsf{Permission}, \mathsf{Prohibition}\}$, and s, α and o are a subject, action and object, respectively, returns all the supports of a permission involving s, α and o. A similar query to compute supports of a prohibition is achieved by changing the requested access type with Prohibition, considering the fact that querying for a prohibition is slightly different, as it must check for senior roles prohibitions.

Definition 5. *Let $\mathcal{K} = \langle \mathcal{T}, \mathcal{R}, \mathcal{T}_U \cup \mathcal{T}_F \rangle$ be an* orbac *KB and s, α and o be a subject, an action and an object respectively. Let I_U denote the uncertain individuals in $\mathcal{T}_U$. $\mathcal{S} \subseteq I_U$ is a support of a privilege involving s, α and o if:*

- ComputeSupports$_p(\langle \mathcal{T}, \mathcal{R}, \mathcal{T}_\mathcal{S} \cup \mathcal{T}_F \rangle, s, \alpha, o) \neq \emptyset$, and;
- $\forall i \in \mathcal{S}$, ComputeSupports$_p(\langle \mathcal{T}, \mathcal{R}, \mathcal{T}_{\mathcal{S}\setminus\{i\}} \cup \mathcal{T}_F \rangle, s, \alpha, o) = \emptyset$.

Where $p \in \{\mathsf{Permission}, \mathsf{Prohibition}\}$ and $\mathcal{T}_{\mathcal{S}\setminus\{i\}} = \mathcal{T}_\mathcal{S} \setminus \{\langle x, p, y \rangle \in \mathcal{T}_\mathcal{S} | x = i \text{ or } y = i\}$. and $\mathcal{T}_\mathcal{S}$ is the set of triples involving elements of $\mathcal{S}$.

Example 5. A support of the permission of (Bob, edit, report1) contains the individuals $\mathcal{S}_{perm}$ = {emp1, def2, use1, cons1}, and a support of the prohibition of (Bob, edit, report1) contains the individuals $\mathcal{S}_{proh}$ = {emp2, def1, use1, cons1}.

Dominance: the object property orbac:isPreferredTo encodes a partial preorder over a knowledge base. The dominance relation as defined in [17] extends the partial preorder defined over the elements of a set into a partial preorder defined over its subsets. We adapt this definition to the relation defined in this paper.

Definition 6. *Let $\mathcal{K} = \langle \mathcal{T}, \mathcal{R}, \mathcal{T}_U \cup \mathcal{T}_F \rangle$ be an* orbac *KB. Let I_U denote the uncertain individuals in $\mathcal{T}_U$. Let $\mathcal{B}_1$ and $\mathcal{B}_2$ be two subsets of I_U.*
$\mathcal{B}_1$ dominates $\mathcal{B}_2$ if $\forall x \in \mathcal{B}_1$, $\exists y \in \mathcal{B}_2$ s.t. ⟨x, orbac:isPreferredTo, y⟩ holds and ⟨y, orbac:isPreferredTo, x⟩ does not hold.

Accepted Permission. In the case where, for a given query, both concrete permission and prohibition are derived, we adapt the notion of acceptance defined in [17] to decide whether a permission is granted or not. The idea is to verify for each support of a prohibition, if there exists a support of a permission dominating it. This way, multiple supports of a permission can participate in granting an access, while ensuring that there is no support of a prohibition that is more preferred to all the supports of a permission. Our approach is local and query-driven, and does not involve repairing the knowledge base. It is clearly in the spirit of argumentation methods. The method relies on the notions of support and dominance from previous section, and avoids inferring all privileges. Instead it checks if an access is granted using the following definition.

Definition 7. *Let $\mathcal{K} = \langle \mathcal{T}, \mathcal{R}, \mathcal{T}_U \cup \mathcal{T}_F \rangle$ be an inconsistent* orbac *KB. An access is accepted for a given subject s, action α and object o, if for each $\mathcal{S}_{proh}$ a support of a prohibition for (s, α, o) (as per Definition 5), $\exists \mathcal{S}_{perm}$ a support of a permission for (s, α, o) s.t. $\mathcal{S}_{perm}$ dominates $\mathcal{S}_{proh}$ (as per Definition 6).*

Example 6. Let us now complete our example. The consortium agreed that the connection rule Consider is fully certain. For the privileges of Bob to edit the object report1, in Example 5, we derived a single support for a permission $\mathcal{S}_{perm}$ and a single support $\mathcal{S}_{proh}$. The new supports are: $\mathcal{S}_{perm}$ = {emp1, use1, def2}, and $\mathcal{S}_{proh}$ = {emp2, use1, def1}. From Example 2, ⟨emp1, orbac:isPreferredTo, emp2⟩, ⟨def2, orbac:isPreferredTo, def1⟩ and ⟨use1, orbac:isPreferredTo, emp2⟩. Therefore, according to Definition 6, $\mathcal{S}$ dominates $\mathcal{C}$, hence a permission is accepted for Bob to edit the object report1.

Since the computation of supports is done in polynomial time, and under the assumption that the size of the support sets of permissions and prohibitions (for an access request) is polynomial, the decision whether or not to grant a permission (Definition 7) is also done in polynomial time w.r.t. the size of the dataset (data complexity).

5 Text-Based Generation of Explanations

Explainability in AI systems is becoming crucial nowadays because transparency is needed so that we can trust AI prediction. In Example 6, not only knowing that Bob can edit the object report1 is important, but also understanding the reasons for that decision made by the AI system. Different types of explanations can be provided, such as trace-based, counterfactual, contrastive and scientific explanations [6]. Our interest is trace-based and contrastive explanations. The former consists of finding the rationale that leads to the decision made by following the steps that were taken by the AI system [6], and the latter focuses on emphasising the features that make the differences between facts [6]. Our method leverages InferenceQuery($\mathcal{K}$) (presented in Sect. 3.3), ComputeConflicts($\mathcal{K}$) (Definition 4), ComputeSupports($\mathcal{K}$,s, α, o) (Definition 5) and the notion of dominance (Definition 6) to generate the explanations. It makes use of SimpleNLG [11] and nltk[2] libraries for natural language generation. This last section briefly presents our method for generating the explanations.

Algorithm 1 presents our method for generating the logic-based explanations (L) and the explanations in natural language (V) by receiving in inputs: OrBAC knowledge base $\mathcal{K}$, subject s, action α and object o. It starts with querying $\mathcal{K}$ for (s, α, o) to determine the different privileges with their justifications (stored in t), and it computes the conflicts (lines 1–2). If there is no conflict, the access is either permitted or prohibited where the contents of L and V explain it (lines 3–6). Otherwise, the process assumes that the access is denied at the beginning. In this case, the method continues with separating the 2 opposite traces: t_{perm}

[2] https://www.nltk.org/.

Algorithm 1: Generating explanations for the access control

Require: $\mathcal{K}$, s, α, o {$\mathcal{K}$ = orbac $\mathcal{KB}$, s = subject, α = action and o = object}
$t \leftarrow$ InferenceQuery($\mathcal{K}, s, \alpha, o$) {$t$ traces the inferred privileges.}
$C \leftarrow$ ComputeConflicts($\mathcal{K}, s, \alpha, o$)
if $C = \emptyset$ **then**
{**NO access conflict**}
$Access \leftarrow$ GetAccess(t) {$Access$ is either permitted or prohibited.}
$L \leftarrow$ Formulate($\mathcal{K}, t$) {L: Logic-based explanations}
$V \leftarrow V_{\text{simple}}(\mathcal{K}, t)$ {V: Verbalised explanations, V_{simple}: Trace verbalisation (t)}
else
{**Access conflict**}
$Access \leftarrow$ false {Access is prohibited.}
{***Trace-based explanations through the traces:*** t_{perm} ***and*** t_{perm}}
$t_{perm} \leftarrow$ SelectPermission($\mathcal{K}, t$) {$t_{perm} \in t$ where t_{perm} concerns permission.}
$t_{proh} \leftarrow$ SelectProhibition($\mathcal{K}, t$) {$t_{proh} \in t$ where t_{proh} concerns prohibition.}
$L \leftarrow$ Formulate($\mathcal{K}, t_{perm}$) $\cup$ Formulate($\mathcal{K}, t_{proh}$) {L from t_{perm} and t_{proh}}
$V \leftarrow$ append($V_{\text{simple}}(\mathcal{K}, t_{perm}), V_{\text{Simple}}(K, t_{proh})$) {Verbalisation}
{***Trace-based and contrastive explanations from supports***}
$\mathcal{S}_{perm} \leftarrow$ ComputeSupports($\mathcal{K}, t_{perm}$) {Set of supports of permission}
$\mathcal{S}_{proh} \leftarrow$ ComputeSupports($\mathcal{K}, t_{proh}$) {Set of supports of prohibition}
$Cont \leftarrow \mathcal{S}_{perm} \Delta \mathcal{S}_{proh}$ {$Cont$: contrasts between permission and prohibition}
$L \leftarrow L \cup \mathcal{S}_{perm} \cup \mathcal{S}_{proh} \cup Cont$ {$Cont$ presents the contrastive explanation.}
$V \leftarrow$ append($V_{\text{support}}(Cont)$) {V_{support}: Verbalisation for supports}
{***Presenting the preference relations***}
$L \leftarrow L \cup$ pref($\mathcal{K}, s, \alpha, o$) {Presenting all preferences related to (s, α, o)}
$V \leftarrow$ append(V_{pref}(pref($\mathcal{K}, s, \alpha, o$))) {$V_{\text{pref}}$: Verbalisation for preferences}
{***Trace-based explanations based on dominance***}
$i \leftarrow 0$, $L_2 \leftarrow \emptyset$, $V_2 \leftarrow \emptyset$
for all $s_{proh} \in \mathcal{S}_{proh}$ **do**
for all $s_{perm} \in \mathcal{S}_{perm}$ **do**
if dominates(s_{perm}, s_{proh}) {Following Definition 5} **then**
$L_2 \leftarrow L_2 \cup$ Trace(dominates(s_{perm}, s_{proh})) {Presenting the preferences}
$V_2 \leftarrow$ append(Trace(V_{pref}(dominates(s_{perm}, s_{proh})))) {Verbalisation}
$i \leftarrow i + 1$
break
end if
end for
end for
if $i = |\mathcal{S}_{proh}|$ **then**
$Access \leftarrow$ true {By following Definition 6, the access is granted.}
$L \leftarrow L \cup L_2$, $V \leftarrow$ append(V_2) {Providing the explanations}
end if
end if
Ensure: $Access$, L, V {$Access$ = Access control, L = logic-based explanations, V = explanations in natural language}

for permission and t_{proh} for prohibition, which form L and are verbalised in V (lines 9–12). Thereafter, it computes the supports of permission and prohibition, emphasises the contrasts regarding the conflict by means of $Cont = \mathcal{S}_{perm} \Delta \mathcal{S}_{proh}$ and shows this using logic-based explanations and natural language explanations (lines 13–17). The preferences related to (s, α, o) are computed to be part of the

Table 2. Example of a conflict of access. The explanations show why access is permitted and prohibited and why the final outcome is granted. The presentation of preferences is omitted for the sake of space.

Access	**Conflict**: Bob is permitted and prohibited to edit report1. **Outcome**: Bob can edit report1.
Logic-based explanation	perm1(consortium, secondee, modify, reports, secondment) $\wedge$ emp1(univ1, bob, secondee) $\wedge$ use1(univ1, report1, reports) $\wedge$ cons1(univ1, edit, modify) $\wedge$ def2(univ1, bob, edit, report1, secondment) $\wedge$ subOrganisationOf(univ1, consortium) $\models$ Is-permitted(Bob, edit, report1) prohib1(consortium, staffMember, modify, reports, default) $\wedge$ emp2(consortium, bob, staffMember) $\wedge$ use1(univ1, report1, reports) $\wedge$ cons1(univ1, edit, modify) $\wedge$ def1(consortium, bob, edit, report1, default) $\wedge$ subOrganisationOf(univ1, consortium) $\models$ Is-prohibited(Bob, edit, report1) **Supports** $\mathcal{S}_{perm}$ = {emp1, def2, use1} $\mathcal{S}_{proh}$ = {emp2, def1, use1} $Cont = \mathcal{S}_{perm} \Delta \mathcal{S}_{proh}$ = {emp1, emp2, def1, def2} **Outcome** Is-permitted(Bob, edit, report1) because: ⟨emp1, orbac:isPreferredTo, emp2⟩, ⟨def2, orbac:isPreferredTo, def1⟩, ⟨use1, orbac:isPreferredTo, emp2⟩
Explanation in English	There is a conflict. Bob, a secondee at consortium, can edit report1. Bob is permitted to modify reports in a secondment context, where report1 is considered as reports, edit it is classified as a modify activity, and univ1 is part of consortium. Bob, a staffMember at consortium, cannot edit report1. Bob is prohibited to modify reports in a default context, where report1 is considered as reports, edit it is classified as a modify activity, and univ1 is part of consortium. There are contrasts: (a) Bob is a secondee at consortium, and Bob is a staffMember at univ1. (b) In univ1, the context secondment holds between Bob, report1 and edit, and in consortium, the default context holds between Bob, report1 and edit. Bob can edit report1 because: (1) 'Bob is a secondee at consortium' is preferred to 'Bob is a staff member at univ1'. (2) 'In univ1, the secondment context holds between Bob, report1 and edit' is preferred to 'In consortium, the default context holds between Bob, report1 and edit'. (3) 'report1 is used in reports at univ1' is preferred to 'Bob is a staffMember at univ1'.

explanations (lines 18–19) to help understand the automatic resolution of the conflict. Lines 21–34 determine the access following Definition 6 and Definition 7. The access for (s, α, o) is granted if for each $s_{proh} \in \mathcal{S}_{proh}$, a support of prohibition (line 21), $\exists s_{perm} \in \mathcal{S}_{perm}$, a support of permission for (s, α, o), s.t. s_{perm} dominates s_{proh} (line 23). In line 24, Trace selects the preferences responsible for the dominance, which is used to constitute L_2 and V_2. Then, the access is permitted if $i = |\mathcal{S}_{proh}|$ holds, making L_2 and V_2 be part of the explanations.

Logic-Based Explanations. Formulate$(\mathcal{K}, t)$ translates the results t from querying $\mathcal{K}$ to logic-based representation. See the second line in Table 2 as an illustration. $\mathcal{S}_{perm} \cup \mathcal{S}_{proh} \cup Cont$, where $Cont = \mathcal{S}_{perm} \Delta \mathcal{S}_{proh}$ shows the contrast between the supports. pref$(\mathcal{K}, s, \alpha, o)$ presents the list of preference relations that are involved in determining the privilege for (s, α, o).

Explanations in Natural Language. Concerning verbalisation, we used a template-based approach to generate explanations in natural language because the structure of an ontology is fixed. A template is a linguistic structure with gaps designed to be completed to form a sentence [10]. A linguistic template has been developed for each case, and the selected individuals from $\mathcal{K}$ fill in the gaps. See the third line, in Table 2 as an illustration. $V_{\text{simple}}(\mathcal{K}, t)$ uses the following template: "[subject], [role] at [Organisation], [ability] [action] the [object], and [Organisation] is part of [Organisation2]. This is because [subject] is [decision] to [activity] [view] in [context] context, where [object] is considered as [view], and [action] it is classified as a [activity] activity.", where "[x]" represent a gap to be replaced. "[ability]" is either 'can' or 'cannot' if the access is permission or prohibition, respectively. "[decision]" is either 'permitted' or 'prohibited' if the access is permission or prohibition, respectively. The other gaps are replaced by the individuals in t. V_{support} and V_{pref} follow the same principle. $V_{\text{support}}(\mathcal{V}) = V_{\text{ind}}(v_1) + \cdots + V_{\text{ind}}(v_n)$, where $v_1, \ldots, v_n \in \mathcal{V}$ are individuals, and V_{ind} is a template-based verbaliser like V_{simple}. We defined a template for each entity, and the algorithm selects the individuals to render the explanations. As an illustration, the template "[subject] is [role] at [Organisation]" is a template for the entity Employ. Then, using it, V_{ind}(empl1) generates 'Bob is a secondee at univ1'. In the case of contrast (line 17), the supports with the same classes are verbalised in parallel. Lastly, $V_{\text{pref}}(\mathcal{P}) = V_{\text{pref'}}(p_1(a_1, b_1)) + \cdots + V_{\text{pref'}}(p_m(a_m, b_m))$, where $p_1(a_1, b_1), \ldots, p_m(a_m, b_m) \in \mathcal{P}$ are the preferences responsible for the dominance, $a_1, \ldots, a_m, b_1, \ldots, b_m$ are individuals, and $V_{\text{pref'}}(p_i(a_i, b_i)) = V_{\text{ind}}(a_i) +$ 'is preferred to' $+ V_{\text{ind}}(b_i)$, with $1 \leq i \leq m$.

Discussions. We generated all privileges from our knowledge base, and the authors judge that the generated explanations are useful to help understand the outcome. We used a grammar checker: language_tool_python[3] for getting an insight into the grammatically, and it is positive. We also used automatic metrics to know the readability of the explanations in natural language. As a

[3] https://pypi.org/project/language-tool-python/.

result, the reader needs to have at least some college education, which is in line with the education of administrators. However, the best method to evaluate the quality of generated texts is through human evaluation [14], and the usefulness of the explanations has been checked by the authors. We can just say that the method has the potential to be useful and human-friendly. Details of this simple evaluation are available in the appendix[4].

6 Concluding Discussions

In this paper, we propose an access control system relying on an ontology of the OrBAC model, on a SPARQL query engine to perform reasoning services and on a SWRL rules engine to automatically assign priorities in the ontology. We showed the efficiency of the different reasoning services and provided a new method to resolve conflicts, relying on inconsistency-tolerant semantics, which were initially defined for lightweight ontologies. Furthermore, trace-based and contrastive explanations are generated to support the decision made. Through a simple study, we found that the generated explanations are potentially useful and human-friendly. There is room for improvement. The importance of permission and prohibition are equal in our explanation generation, whereas prohibition might be essential. In addition, discriminating supports and conflicts may reduce the number of explanations while focusing on the relevant ones.

Acknowledgments. This research was supported by the European Union's Horizon research and innovation programme under the MSCA-SE (Marie Skłodowska-Curie Actions Staff Exchange); Call: HORIZON-MSCA-2021-SE-01; Project title: STARWARS (STormwAteR and WastewAteR networkS heterogeneous data AI-driven management) [grant agreement 101086252]. This research has also received support from the ANR project EXPIDA (EXplainable and parsimonious Preference models to get the most out of Inconsistent DAtabases), [grant number ANR-22-CE23-0017]. THR acknowledges support from the Hasso Plattner Institute for Digital Engineering through the HPI Research School at UCT. A. Laouar's PhD is supported by the ANR project Vivah (Vers une intelligence artificielle à visage humain) [grant number ANR-20-THIA-0004].

References

1. Belabbes, S., Benferhat, S.: Computing a possibility theory repair for partially preordered inconsistent ontologies. IEEE Trans. Fuzzy Syst. **30**(8), 3237–3246 (2021)
2. Benferhat, S., Bouriche, K., Ouzarf, M.: On the possibilistic handling of priorities in access control models. In: Sun, F., Li, T., Li, H. (eds.) Foundations and Applications of Intelligent Systems. AISC, vol. 213, pp. 275–285. Springer, Heidelberg (2014). https://doi.org/10.1007/978-3-642-37829-4_23
3. Benferhat, S., El Baida, R.: A prioritized-based approach to handling conflicts in access control. In: 16th IEEE International Conference on Tools with Artificial Intelligence, pp. 286–293. IEEE (2004)

[4] All paper appendices are available at https://github.com/ahmedlaouar/orbac.owl.

4. Bienvenu, M., Bourgaux, C.: Inconsistency-tolerant querying of description logic knowledge bases. In: Pan, J.Z., et al. (eds.) Reasoning Web 2016. LNCS, vol. 9885, pp. 156–202. Springer, Cham (2017). https://doi.org/10.1007/978-3-319-49493-7_5
5. Calvanese, D., De Giacomo, G., Lembo, D., Lenzerini, M., Rosati, R.: Tractable reasoning and efficient query answering in description logics: the DL-Lite family. J. Autom. Reason. **39**(3), 385–429 (2007)
6. Chari, S., Seneviratne, O., Gruen, D.M., Foreman, M.A., Das, A.K., McGuinness, D.L.: Explanation ontology: a model of explanations for user-centered AI. In: Pan, J.Z., et al. (eds.) ISWC 2020. LNCS, vol. 12507, pp. 228–243. Springer, Cham (2020). https://doi.org/10.1007/978-3-030-62466-8_15
7. Cuppens, F., Cuppens-Boulahia, N., Miège, A.: Inheritance hierarchies in the Or-BAC model and application in a network environment. In: Proceedings of the Foundations of Computer Security (FCS 2004), pp. 41–60 (2004)
8. Cuppens, F., Cuppens-Boulahia, N., Ghorbel, M.B.: High level conflict management strategies in advanced access control models. Electron. Notes Theor. Comput. Sci. **186**, 3–26 (2007). Proceedings of the First Workshop in Information and Computer Security (ICS 2006)
9. Dean, M., Schreiber, G. (eds.): OWL Web Ontology Language Reference. World Wide Web Consortium, Recommendation REC-owl-ref-20040210 (2004)
10. Gatt, A., Krahmer, E.: Survey of the state of the art in natural language generation: core tasks, applications and evaluation. J. Artif. Intell. Res. **61**, 65–170 (2018)
11. Gatt, A., Reiter, E.: SimpleNLG: a realisation engine for practical applications. In: Proceedings of the 12th European Workshop on Natural Language Generation (ENLG 2009), pp. 90–93 (2009)
12. Harris, S., Seaborne, A. (eds.): SPARQL 1.1 Query Language. World Wide Web Consortium, Recommendation REC-sparql11-query-20130321 (2013)
13. Horrocks, I., Patel-Schneider, P.F., Boley, H., Tabet, S., Grosof, B., Dean, M.: SWRL: A Semantic Web Rule Language Combining OWL and RuleML. World Wide Web Consortium, Member Submission SUBM-SWRL-20040521 (2004)
14. Howcroft, D.M., et al.: Twenty years of confusion in human evaluation: NLG needs evaluation sheets and standardised definitions. In: 13th International Conference on Natural Language Generation 2020, pp. 169–182. Association for Computational Linguistics (2020)
15. Jabal, A.A., et al.: Methods and tools for policy analysis. ACM Comput. Surv. (CSUR) **51**(6), 1–35 (2019)
16. Kalam, A.A.E., et al.: Organization based access control. In: Proceedings POLICY 2003. IEEE 4th International Workshop on Policies for Distributed Systems and Networks, pp. 120–131. IEEE (2003)
17. Laouar, A., Belabbes, S., Benferhat, S.: Tractable closure-based possibilistic repair for partially ordered DL-Lite ontologies. In: Gaggl, S., Martinez, M.V., Ortiz, M. (eds.) JELIA 2023. LNCS, vol. 14281, pp. 353–368. Springer, Cham (2023). https://doi.org/10.1007/978-3-031-43619-2_25
18. Laouar, A., Belabbes, S., Benferhat, S.: Conflict handling strategies for partially ordered access control security policies. In: 2024 IEEE Pacific Rim Conference on Communications, Computers and Signal Processing (PACRIM), pp. 1–6 (2024)
19. Mohamed, A.K.Y.S., Auer, D., Hofer, D., Küng, J.: A systematic literature review for authorization and access control: definitions, strategies and models. Int. J. Web Inf. Syst. **18**(2/3), 156–180 (2022)

20. Proctor, M.: Drools: a rule engine for complex event processing. In: Schürr, A., Varró, D., Varró, G. (eds.) AGTIVE 2011. LNCS, vol. 7233, pp. 2–2. Springer, Heidelberg (2012). https://doi.org/10.1007/978-3-642-34176-2_2
21. Sandhu, R., Ferraiolo, D., Kuhn, R.: The NIST model for role-based access control: towards a unified standard. In: Proceedings of the Fifth ACM Workshop on Role-Based Access Control. RBAC 2000, pp. 47–63. Association for Computing Machinery, New York (2000)

Estimating Causal Effects in Partially Directed Parametric Causal Factor Graphs

Malte Luttermann[1(✉)], Tanya Braun[2], Ralf Möller[3], and Marcel Gehrke[3]

[1] German Research Center for Artificial Intelligence (DFKI), Lübeck, Germany
malte.luttermann@dfki.de

[2] Data Science Group, University of Münster, Münster, Germany
tanya.braun@uni-muenster.de

[3] Institute for Humanities-Centered Artificial Intelligence, University of Hamburg, Hamburg, Germany
{ralf.moeller,marcel.gehrke}@uni-hamburg.de

Abstract. Lifting uses a representative of indistinguishable individuals to exploit symmetries in probabilistic relational models, denoted as parametric factor graphs, to speed up inference while maintaining exact answers. In this paper, we show how lifting can be applied to causal inference in partially directed graphs, i.e., graphs that contain both directed and undirected edges to represent causal relationships between directed and undirected edges to represent causal relationships between random variables. We present partially directed parametric causal factor graphs (PPCFGs) as a generalisation of previously introduced parametric causal factor graphs, which require a fully directed graph. We further show how causal inference can be performed on a lifted level in PPCFGs, thereby extending the applicability of lifted causal inference to a broader range of models requiring less prior knowledge about causal relationships.

Keywords: causal models · probabilistic relational models · lifted inference

1 Introduction

A fundamental problem for an intelligent agent performing reasoning under uncertainty is to compute the effect of an action on a certain random variable (randvar) on other randvars. When computing the effect of an action on a specific randvar, it is crucial to deploy the semantics of an intervention instead of performing a classical conditioning on that randvar [22, Chapter 4]. An intervention acting on a randvar R can be thought of as setting R to a fixed value and removing all incoming influences on the value of R. In practice, generally not all causal relationships in a given model are known and thus, only a partially directed graphical model is available. In such a partially directed graph,

S. Destercke et al. (Eds.): SUM 2024, LNAI 15350, pp. 265–280, 2025.
https://doi.org/10.1007/978-3-031-76235-2_20

directed edges represent cause-effect relationships and undirected edges represent causal relationships whose direction is unknown. In this paper, we solve the problem of efficiently estimating causal effects of actions in partially directed lifted probabilistic models, denoted as parametric factor graphs. Lifted representations combine relational logic (i.e., first-order logic with known universes) with probabilistic models. Thus, they are more expressive than propositional models such as factor graphs and allow for tractable probabilistic inference w.r.t. domain sizes of logical variables (logvars) by exploiting symmetries. We therefore provide a solution to enable first-order decision making with partial causal knowledge.

Previous Work. The estimation of causal effects using causal graphical models in form of directed acyclic graphs in combination with observational data has been extensively studied in the literature (see, e.g., [22,23,27]). Some works incorporate causal knowledge into (propositional) factor graphs (which are originally undirected graphical models) to enable the estimation of causal effects in factor graphs [6,29]. In practice, the underlying causal graph is often not fully known and hence, identifying and estimating causal effects when provided with observational data and a partially directed graph has been investigated [7,10,16,24]. However, all of these works perform causal effect estimation on a propositional level and thus lack the expressivity of relational logic, for example to capture the relationships between individual objects. To represent individual objects and the relationships between them, Poole [25] introduces parametric factor graphs as lifted representations, which combine relational logic and probabilistic models, thereby allowing to encode that certain properties hold for groups of indistinguishable objects. In probabilistic inference, lifting exploits symmetries to speed up inference while maintaining exact answers [20]. Over the past years, both algorithms for symmetry detection [1,8,11,13–15] allowing the construction of lifted representations such as parametric factor graphs as well as various lifted inference algorithms operating on parametric factor graphs have been developed and further refined [2–5,9,18,25,28]. More recently, Luttermann et al. [12] introduce parametric causal factor graphs (PCFGs) as an extension of parametric factor graphs allowing to incorporate causal knowledge into a lifted representation. Nevertheless, the authors assume that the causal relationships between the involved randvars are fully known, which is rarely the case in practical settings.

Our Contributions. We introduce partially directed parametric causal factor graphs (PPCFGs) as a generalisation of PCFGs to obtain a formalism that compactly encodes a full joint distribution over a set of randvars and at the same time incorporates causal knowledge in the model, if available. The major advantage of a PPCFG over an PCFG is that not all causal relationships between the involved randvars need to be known, thereby reducing the amount of prior knowledge required and thus making the model more suitable for many practical settings. We further define *d*-separation in PPCFGs to reason about conditional independence. In addition to that, we present an algorithm to efficiently estimate causal effects in PPCFGs on a lifted level, i.e., a representative of indistinguishable objects is used for computations to speed up inference. Our algorithm identifies whether a causal effect can be uniquely determined from the given PPCFG and if so, outputs the causal effect. If the undirected edges in the PPCFG lead to a causal effect being not uniquely identifiable, our algorithm efficiently enumerates all possible causal effects while operating on a lifted level.

Structure of This Paper. The remainder of this paper is structured as follows. We begin by introducing necessary background information and notations, starting with the introduction of PCFGs as a lifted representation of a fully directed graphical model. Afterwards, we present PPCFGs as a generalisation of PCFGs, allowing both for directed and undirected edges in the model and then define d-separation in PPCFGs. Thereafter, we provide an algorithm to efficiently estimate causal effects in PPCFGs before we conclude.

2 Preliminaries

We begin to introduce parameterised randvars (PRVs), which use logvars as parameters to represent sets of indistinguishable randvars.

Definition 1 (Parameterised Random Variable). *Let $\boldsymbol{R}$ be a set of andvar names, $\boldsymbol{L}$ a set of logvar names, and $\boldsymbol{D}$ a set of constants. All sets are finite. Each logvar L has a domain* $\mathrm{dom}(L) \subseteq \boldsymbol{D}$. *A* constraint *is a tuple $(\mathcal{X}, C_{\mathcal{X}})$ of a sequence of logvars $\mathcal{X} = (X_1, \dots, X_n)$ and a set $C_{\mathcal{X}} \subseteq \times_{i=1}^{n} \mathrm{dom}(X_i)$. The symbol $\top$ for C marks that no restrictions apply, i.e., $C_{\mathcal{X}} = \times_{i=1}^{n} \mathrm{dom}(X_i)$. A* PRV $R(L_1, \dots, L_n)$, $n \geq 0$, *is a syntactical construct of a randvar $R \in \boldsymbol{R}$ possibly combined with logvars $L_1, \dots, L_n \in \boldsymbol{L}$ to represent a set of randvar. If $n = 0$, the PRV is parameterless and forms a propositional randvar. A PRV A (or logvar L) under constraint C is given by $A_{|C}$ ($L_{|C}$), respectively. We may omit $|\top$ in $A_{|\top}$ or $L_{|\top}$. The term* $\mathrm{range}(A)$ *denotes the possible values of a PRV A. An* event *$A = a$ denotes the occurrence of PRV A with range value $a \in \mathrm{range}(A)$.*

Example 1. Consider $\boldsymbol{R} = \{Comp, Sal, Rev\}$ for competence, salary, and revenue, respectively, and $\boldsymbol{L} = \{E\}$ with $\mathrm{dom}(E) = \{alice, bob, charlie\}$ (employees), combined into PRVs $Comp(E)$, $Sal(E)$, and Rev with $\mathrm{range}(Comp(E)) = \mathrm{range}(Sal(E)) = \mathrm{range}(Rev) = \{\mathrm{low}, \mathrm{medium}, \mathrm{high}\}$.

A parametric factor (parfactor) describes a function, mapping argument values to positive real numbers, of which at least one is non-zero.

Definition 2 (Parfactor). *Let $\boldsymbol{\Phi}$ denote a set of factor names. We denote a* parfactor *g by $\phi(\mathcal{A})_{|C}$ with $\mathcal{A} = (A_1, \dots, A_n)$ being a sequence of PRVs, $\phi\colon \times_{i=1}^{n} \mathrm{range}(A_i) \mapsto \mathbb{R}^+$ being a function with name $\phi \in \boldsymbol{\Phi}$ mapping argument values to a positive real number called* potential, *and C being a constraint on the logvars of $\mathcal{A}$. We may omit $|\top$ in $\phi(\mathcal{A})_{|\top}$. The term $lv(Y)$ refers to the logvars in some element Y, a PRVs, a parfactor, or sets thereof. The term $gr(Y_{|C})$ denotes the set of all instances (groundings) of Y with respect to constraint C.*

Example 2. Take a look at the parfactor $g_1 = \phi_1(Comp(E), Rev)_{|\top}$. Assuming the same ranges of the PRVs and the same domains of the logvars as in Example 1, g_1 specifies $|\text{range}(Comp(E))| \cdot |\text{range}(Rev)| = 9$ input-output mappings $\phi_1(\text{low}, \text{low}) = \varphi_1$, $\phi_1(\text{low}, \text{medium}) = \varphi_2$, $\phi_1(\text{low}, \text{high}) = \varphi_3$, and so on with $\varphi_i \in \mathbb{R}^+$ for all $i = 1, \ldots, 9$. Further, $lv(g_1) = \{E\}$ and $gr(g_{1_\top}) = \{\phi_1(Comp(alice), Rev), \phi_1(Comp(bob), Rev), \phi_1(Comp(charlie), Rev)\}$.

A PCFG is then built from a set of parfactors $\{g_1, \ldots, g_m\}$ and the causal relationships between the PRVs, which are encoded by the direction of the edges in the graph structure of the PCFG [12]. In its original form, a PCFG is a fully directed graph, but we extend this definition to allow for both directed and undirected edges in the following section.

3 Partially Directed Parametric Causal Factor Graphs

So far, we have introduced PRVs and parfactors, which together form a parametric probabilistic model that combines probabilities and relational logic to allow for lifted probabilistic inference. We now move on to define PPCFGs as lifted models that are able to incorporate partial causal knowledge. By combining relational logic with probabilistic models, PPCFGs are more expressive than propositional models such as a partially directed Bayesian networks or factor graphs and allow us to exploit symmetries (in form of indistinguishable individuals) to speed up both probabilistic and causal inference.

Definition 3 (Partially Directed Parametric Causal Factor Graph). *A PPCFG is a graph $M = (\boldsymbol{A} \cup \boldsymbol{G}, \boldsymbol{E})$ that consists of variable nodes $\boldsymbol{A}$, factor nodes $\boldsymbol{G}$ ($\boldsymbol{A} \cap \boldsymbol{G} = \emptyset$), and a set of edges $\boldsymbol{E}$. Each variable node $A \in \boldsymbol{A}$ represents a PRV A and every factor node $g \in \boldsymbol{G}$ represents a parfactor $g = \phi(\mathcal{A})_{|C}$, where $\mathcal{A} = (A_1, \ldots, A_k)$ with $A_1 \in \boldsymbol{A}, \ldots, A_k \in \boldsymbol{A}$ is a sequence of PRVs, $\phi\colon \times_{i=1}^{k} \text{range}(A_i) \mapsto \mathbb{R}^+$ is a function, and C is a constraint on the logvars as of $\mathcal{A}$. We again may omit $|\top$ in $\phi(\mathcal{A})_{|\top}$. For a variable node $A \in \boldsymbol{A}$ and a factor node $g \in \boldsymbol{G}$, there is an undirected edge $g - A \in \boldsymbol{E}$ if A appears in the argument list of $g = \phi(\mathcal{A})$ and no information about the causal relationships between the PRVs in $\mathcal{A}$ is available. If it is known that all $A' \in \mathcal{A} \setminus \{A\}$ are causes of $A \in \mathcal{A}$ (or if $\mathcal{A} \setminus \{A\} = \emptyset$), the edge $g - A$ can be replaced by a directed edge $g \rightarrow A$.*

The semantics of M is given by grounding and building a full joint distribution. With Z as the normalisation constant and $\mathcal{A}_k$ denoting the PRVs appearing in the argument list of $\phi_k(\mathcal{A}_k)$, M represents the full joint distribution

$$P_M = \frac{1}{Z} \prod_{g \in G} \prod_{\phi_k \in gr(g)} \phi_k(\mathcal{A}_k).$$

Example 3. Following Examples 1 and 2, Fig. 1 depicts a PPCFG M modelling the interplay of a company's revenue and its employees' competences and salaries. Each parfactor represents a group of ground factors and thus, grounding M results in a partially directed factor graph (see Fig. 2). In this particular

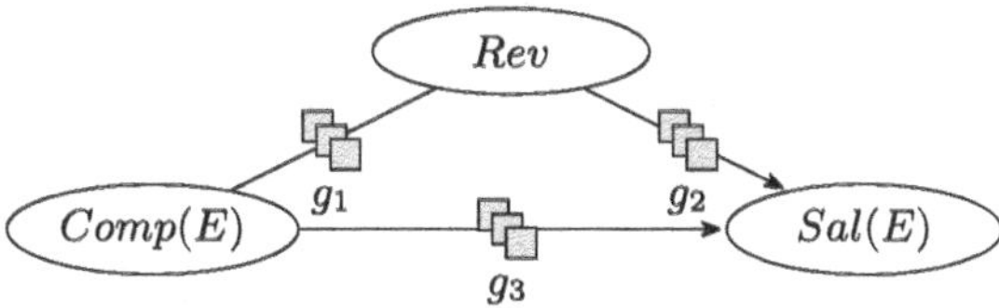

Fig. 1. A PPCFG modelling the interplay of a company's revenue and its employees' competences and salaries (without input-output mappings of parfactors).

example, the causal relationships encoded by M tell us that the revenue of the company influences the salary of each individual employee and the competence of a specific employee influences their salary. Moreover, there is a dependency between the competence of each individual employee and the revenue of the company, but the causal direction is not encoded in M. We humans expect that the competence of the employees influences the revenue of the company, but an autonomous agent might not have this information available, resulting in a partially directed graph.

In accordance with the literature, in this paper we consider PPCFGs that do not contain directed cycles. Throughout this paper, we denote the parents of a PRV A in a PPCFG $M = (\boldsymbol{A} \cup \boldsymbol{G}, \boldsymbol{E})$ by $\text{Pa}(A, M) = \{A' \in \boldsymbol{A} \mid \exists g \in \boldsymbol{G} \colon (g - A') \in \boldsymbol{E} \wedge (g \rightarrow A) \in \boldsymbol{E}\}$. Analogously, we define the children of A in M as $\text{Ch}(A, M) = \{A' \in \boldsymbol{A} \mid \exists g \in \boldsymbol{G} \colon (g \rightarrow A') \in \boldsymbol{E} \wedge (g - A) \in \boldsymbol{E}\}$ and the neighbours of A in M as $\text{Ne}(A, M) = \{A' \in \boldsymbol{A} \mid \exists g \in \boldsymbol{G} \colon (g - A') \in \boldsymbol{E} \wedge (g - A) \in \boldsymbol{E}\}$.

Note that the semantics of a PPCFG is defined with respect to the set of parfactors and thus is independent of the given causal relationships. A PPCFG hence simultaneously encodes a full joint probability distribution over a set of randvars (represented by PRVs) and a set of causal relationships between those randvars[1]. Another important remark is that grounding a PPCFG results in a partially directed factor graph as introduced by Frey [6]. The important advantage is that the PPCFG combines relational logic with probabilistic models, thereby being more expressive than a factor graph and allowing us to exploit symmetries (in form of indistinguishable individuals) to speed up inference.

Example 4. Take a look at Fig. 2, which presents the resulting model when grounding the PPCFG M given in Fig. 1. The model $gr(M)$ now contains the randvars *Comp(alice)*, *Comp(bob)*, *Comp(charlie)*, which are represented by a single PRV *Comp(E)* with a logvar E, $\text{dom}(E) = \{alice, bob, charlie\}$, in M (analogously for *Sal(alice)*, *Sal(bob)*, *Sal(charlie)* and *Sal(E)*). Moreover, each parfactor g_i in M represents a set of three (identical) ground factors ϕ_i in the grounded model $gr(M)$. In consequence, the lifted representation M is more compact than its grounded counterpart $gr(M)$, thereby reducing run times required to perform inference. The underlying assumption is that there are indistinguishable objects, here employees, which can be represented by a representative.

[1] Thus, the setting of having an PPCFG is equivalent to having a causal graph and observational data for the randvars occurring in the causal graph [22].

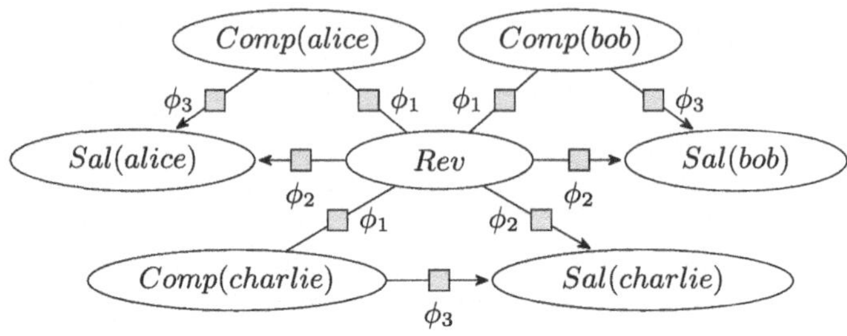

Fig. 2. A visualisation of the resulting model when grounding the PPCFG M given in Fig. 1, where $\text{dom}(E) = \{alice, bob, charlie\}$.

Since a PPCFG encodes a full joint probability distribution, a PPCFG can be used to compute marginal distributions of grounded PRVs given observations of specific events. A query asks for a probability distribution (or a specific probability) given a set of observed events or, in the case of an interventional query, given a set of interventions fixing the values of certain (grounded) PRVs.

Definition 4 (Query). *A* query $P(Q \mid E_1 = e_1, \ldots, E_k = e_k)$ *consists of a query term* Q *and a set of events* $\{E_j = e_j\}_{j=1}^k$ *with* Q *and* E_j *being grounded or parameterless PRVs. To query a specific probability instead of a probability distribution, the query term is an event* $Q = q$*. An* interventional query $P(Q \mid do(R_1 = r_1, \ldots, R_k = r_k))$ *asks for a probability distribution given that the grounded or parameterless PRVs* R_j *are set to a fixed value* r_j*.*

Example 5. The query $P(Rev \mid Comp(alice) = \text{high})$ asks for the probability distribution of Rev given that the event $Comp(alice) = \text{high}$ is observed. Keep in mind that, in general, observing and intervening are two different things, e.g., generally $P(Rev \mid Comp(alice) = \text{high}) \neq P(Rev \mid do(Comp(alice) = \text{high}))$.

Before we consider interventions in more detail, we first take a closer look at conditional independence in PPCFGs, which is needed to introduce the semantics of an intervention. Afterwards, we formally define the semantics of an intervention and show how causal effects, which rely on the notion of an intervention, can efficiently be estimated in PPCFGs.

4 Conditional Independence in PPCFGs

The notion of d-separation [21] provides a graphical criterion to test for conditional independence in directed acyclic graphs and is essential to compute the effect of an intervention in the sense that all non-causal paths, so-called backdoor paths, need to be blocked to remove spurious effects. Frey [6] extends the notion of d-separation to partially directed factor graphs and we build on this definition to define d-separation in PPCFGs (analogously to d-separation in PPCFGs [12]). Note that d-separation in PPCFGs is defined on a ground level.

Definition 5 (d-separation). *Let $M = (\boldsymbol{A} \cup \boldsymbol{G}, \boldsymbol{E})$ be a PPCFG. Given three disjoint sets of randvars $\boldsymbol{X}$, $\boldsymbol{Y}$, and $\boldsymbol{Z}$ (subsets of $gr(\boldsymbol{A})$), we say that $\boldsymbol{X}$ and $\boldsymbol{Y}$ are conditionally independent given $\boldsymbol{Z}$, written as $\boldsymbol{X} \perp\!\!\!\perp \boldsymbol{Y} \mid \boldsymbol{Z}$, if the nodes in $\boldsymbol{Z}$ block all paths between the nodes in $\boldsymbol{X}$ and the nodes in $\boldsymbol{Y}$ in $gr(M)$. A path is a connected sequence of edges (independent of their directions) and it is therefore also possible for a path to pass from a parent of a factor to another parent of that factor. A path is blocked by the nodes in $\boldsymbol{Z}$ if*

1. *the path contains the pattern $\phi_1 \rightarrow A \leftarrow \phi_2$ such that neither A nor any of its descendants are in $\boldsymbol{Z}$, or*
2. *the path passes from ϕ_1 through A to ϕ_2 such that it does not contain the pattern $\phi_1 \rightarrow A \leftarrow \phi_2$ and A is in $\boldsymbol{Z}$, or*
3. *the path passes from a parent of a factor ϕ to another parent of ϕ, and neither the child of ϕ nor any of its descendants are in $\boldsymbol{Z}$.*

Example 6. Consider the grounded PPCFG M depicted in Fig. 2. M encodes, for example, that the competence of *alice* is independent of *bob*'s salary given the revenue of the company, written as $\{Comp(alice)\} \perp\!\!\!\perp \{Sal(bob)\} \mid \{Rev\}$, because all paths from $Comp(alice)$ to $Sal(bob)$ pass through Rev.

Remark 1. Definition 5 is slightly different from the definition of d-separation provided by Frey [6] for partially directed factor graphs in the sense that Definition 5 is more flexible by allowing for variable nodes to have multiple parent factor nodes (e.g., in Fig. 1, $Sal(E)$ has two parent factor nodes g_2 and g_3 instead of a single parent factor node with two inputs from $Comp(E)$ and Rev).

Remark 2. In case a PPCFG M is constructed by hand (instead of being learned from observational data), it is possible to construct a mismatch of the conditional independence statements encoded in the graph structure of M and the conditional independence statements implied by the full joint probability distribution encoded by M. We therefore assume that the graph structure of a given PPCFG M encodes exactly the same conditional independence statements as the underlying full joint probability distribution encoded by M.

As we have seen, d-separation in PPCFGs is defined on a ground level. However, it is possible to check for d-separation on a lifted level without having to ground the entire PPCFG. For this purpose, the well-known Bayes-Ball algorithm [26] that allows us to check for conditional independence can be applied on a lifted level, that is, it can be run on the graph structure of the PPCFG instead of on its grounded graph structure [17]. The idea of the Bayes-Ball algorithm is that a bouncing ball is sent through the graph structure such that the ball can pass through a node, bounce back, or be blocked to determine blocked paths (more details can be found in [26]). It is also possible to check whether PRVs (instead of ground randvars) are conditionally independent in a highly efficient manner on a lifted level. In a PPCFG, each PRV A is represented by a single variable node and thus, checking for conditional independence statements that involve A can be done by looking at this single variable node instead of taking into account all groundings $gr(A)$ of A. In contrast, in a propositional setting (i.e., in a ground model), each ground randvar in $gr(A)$ must be looked at individually. For example, to check whether $\{Comp(E)\} \perp\!\!\!\perp \{Sal(E)\} \mid \{Rev\}$ holds, only three variable nodes are of relevance in the lifted representation (see Fig. 1) while $2 \cdot |\text{dom}(E)| + 1$ nodes are of relevance in the ground model (Fig. 2).

We next show how the estimation of causal effects, which rely on the notion of an intervention, can efficiently be realised in PPCFGs. The idea is again to avoid grounding the entire PPCFG (if possible) and to perform the causal effect estimation on a lifted level to speed up causal inference.

5 Efficient Estimation of Causal Effects in PPCFGs

To compute the effect of actions carried out on randvars on other randvars, we have to answer interventional queries [22]. An intervention on a grounded or parameterless PRV R, denoted as $do(R = r)$ where $r \in \text{range}(R)$, changes the structure of the underlying model by setting the value of R to r and removing all incoming influences on R. An intervention is defined on a fully directed graph.

Definition 6 (Intervention). *Let $M = (\boldsymbol{A} \cup \boldsymbol{G}, \boldsymbol{E})$ be a fully directed PPCFG and let $gr(\boldsymbol{A}) = \{R_1, \ldots, R_n\}$ denote the set of randvars obtained by grounding the PRVs in $\boldsymbol{A}$. Any probability distribution entailing the conditional independence statements encoded by M can be factorised as $P(R_1, \ldots, R_n) = \prod_{i=1}^{n} P(R_i \mid \text{Pa}(R_i, gr(M)))$. An intervention $do(R'_1 = r'_1, \ldots, R'_k = r'_k)$ changes the underlying probability distribution such that*

$$
\begin{aligned}
&P(R_1 = r_1, \ldots, R_n = r_n \mid do(R'_1 = r'_1, \ldots, R'_k = r'_k)) \\
&= \begin{cases} \displaystyle\prod_{R_i \in \{R_1, \ldots, R_n\} \setminus \{R'_1, \ldots, R'_k\}} P(r_i \mid \text{pa}(R_i, gr(M))) & \text{if } \forall j \in \{1, \ldots, k\} \colon r_j = r'_j \\ 0 & \text{otherwise,} \end{cases}
\end{aligned}
$$

where $\text{pa}(R_i, gr(M))$ denotes the values of $\text{Pa}(R_i, gr(M))$.

Observe that the definition of an intervention refers to a fully directed PPCFG. In general, we deal with PPCFGs that might contain undirected edges and thus represent a whole class of fully directed PPCFGs, namely all fully directed PPCFGs that entail the same conditional independence statements as the initial PPCFG[2]. To determine the semantics of an intervention, we thus need to know the parents of the PRVs on which we intervene. As we deal with PPCFGs that might contain undirected edges, however, we do not always know the real parents of a PRV.

Example 7. Given the PPCFG depicted in Fig. 1, the company might wonder whether it is worth it to send its employee *alice* to an expensive training

[2] For classical (propositional) directed acyclic graphs, the set of fully directed acyclic graphs representing identical conditional independence statements is known under the name of a *Markov equivalence class*.

course to increase the revenue of the company due to the increased competence of *alice*. Hence, the company is interested in computing the quantity $P(Rev \mid do(Comp(alice) = \text{high}))$. Since an intervention can be thought of as setting the value of *Comp(alice)* to high and removing all incoming edges of *Comp(alice)* in the graph, there are two possible scenarios when computing $P(Rev \mid do(Comp(alice) = \text{high}))$: (i) *Rev* is a parent of *Comp(alice)* in the true model and thus, the underlying probability distribution is changed, or (ii) *Comp(alice)* is a parent of *Rev* in the true model and therefore, the underlying probability distribution remains unchanged as *Comp(alice)* has no parents. Without further background information, we do not know which of these two scenarios is actually correct and thus, we need to consider both possibilities.

Fortunately, we do not always have to consider all possible fully directed PPCFGs represented by a given PPCFG when computing the effect of an intervention because there are settings in which the represented fully directed PPCFGs all yield the same effect of the intervention. In particular, in case all parents of the randvars on which we intervene are known, we can uniquely determine the effect of the intervention even if there are still undirected edges in the PPCFG[3].

Theorem 1. *Let* $M = (\boldsymbol{A} \cup \boldsymbol{G}, \boldsymbol{E})$ *denote a PPCFG and let* $P(Q \mid do(R'_1 = r'_1, \ldots, R'_k = r'_k))$ *be an interventional query with* $Q \in gr(\boldsymbol{A})$, $R'_1 \in gr(\boldsymbol{A}), \ldots, R'_k \in gr(\boldsymbol{A})$, *and* $\{Q\} \cap \{R'_1, \ldots, R'_k\} = \emptyset$. *If it holds that* $\text{Ne}(R'_1, gr(M)) = \emptyset, \ldots, \text{Ne}(R'_k, gr(M)) = \emptyset$, *then the result of* $P(Q \mid do(R'_1 = r'_1, \ldots, R'_k = r'_k))$ *is identical for all fully directed PPCFGs represented by* M.

Proof. Let $\{R_1, \ldots, R_\ell\} = gr(\boldsymbol{A}) \setminus \{Q, R'_1, \ldots, R'_k\}$ denote the set of all grounded randvars that do not occur in the given query. From Definition 6 we know that

$$\begin{aligned} &P(Q = q \mid do(R'_1 = r'_1, \ldots, R'_k = r'_k)) \\ &= \sum_{r_1 \in \text{range}(R_1)} \cdots \sum_{r_\ell \in \text{range}(R_\ell)} \prod_{R_i \in gr(\boldsymbol{A}) \setminus \{R'_1, \ldots, R'_k\}} P(r_i \mid \text{pa}(R_i, gr(M))). \end{aligned} \tag{1}$$

If it holds that $\text{Ne}(R'_1, gr(M)) = \emptyset, \ldots, \text{Ne}(R'_k, gr(M)) = \emptyset$, then the parents of $R'_1, \ldots, R'_k$ are fully known and identical in all fully directed PPCFGs represented by M. Hence, it remains to be shown that the factorisation of all ground randvars that are not in $\{R'_1, \ldots, R'_k\}$ is equivalent for all fully directed PPCFGs represented by M. We know that all fully directed PPCFGs represented by M entail exactly the same conditional independence statements as M and thus, the factorisations induced by these fully directed PPCFGs entail equivalent semantics (just as all Bayesian network structures over a fixed set of randvars entailing the same conditional independence statements induce equivalent factorisations of the underlying probability distribution). Consequently, Eq. (1) yields the same result for all fully directed PPCFGs represented by M.

[3] In fact, this has been shown for propositional models [16,19] and we now transfer this result to relational models.

Theorem 1 implies that we do not have to consider all possible edge directions of the undirected edges in a PPCFG when computing the effect of an intervention but just the possible directions of the undirected edges that are relevant for the intervention, that is, the directions of the undirected edges that are connected to the randvars on which we intervene. Note that all terms required to answer the interventional query can be computed by querying the PPCFG, as the PPCFG compactly encodes the full joint probability distribution over all ground randvars. The semantics of the PPCFG is well-defined even if there are undirected edges in the graph because the factors do not necessarily encode distributions conditioned on the parents but instead encode arbitrary local distributions that factorise the full joint probability distribution.

Intuitively, it becomes clear that the effect of an intervention cannot be uniquely determined if there are undirected edges connected to the randvars on which we intervene because there are different possible parent sets that might result in multiple disjoint effects of the intervention.

Theorem 2. *Let $M = (\boldsymbol{A} \cup \boldsymbol{G}, \boldsymbol{E})$ denote a PPCFG and let $P(Q \mid do(R'_1 = r'_1, \ldots, R'_k = r'_k))$ be an interventional query with $Q \in gr(\boldsymbol{A})$, $R'_1 \in gr(\boldsymbol{A})$, $\ldots, R'_k \in gr(\boldsymbol{A})$, and $\{Q\} \cap \{R'_1, \ldots, R'_k\} = \emptyset$. If there exists a randvar $R'_i \in \{R'_1, \ldots, R'_k\}$ such that $\mathrm{Ne}(R'_i, gr(M)) \neq \emptyset$, then the result of $P(Q \mid do(R'_1 = r'_1, \ldots, R'_k = r'_k))$ is not guaranteed to be uniquely determined.*

Proof. If there exists a randvar $R'_i \in \{R'_1, \ldots, R'_k\}$ such that $\mathrm{Ne}(R'_i, gr(M)) \neq \emptyset$ holds, there are undirected edges connected to R'_i, implying that the parents of R'_i are not guaranteed to be identical in all fully directed PPCFGs represented by M. Since we know that Eq. (1) gives us the result of the given query, the factors being removed from the product on the right hand side of the equation might differ depending on the actual parents of the ground randvars on which we intervene, thereby possibly yielding different results for the query.

Remark 3. There are scenarios in which it is possible to uniquely determine the result of an interventional query even if there are undirected edges connected to the randvars on which we intervene as not all undirected edges can be oriented in both directions (because they are not allowed to introduce a cycle or to change the conditional independence statements encoded in the graph structure)[4].

Combining the insights from Theorems 1 and 2 naturally leads to an algorithm to compute the effect of interventions in PPCFGs. The idea is that all possible parent sets of the intervention variables have to be considered. If there is just one possible set of parents, the effect of the intervention can be uniquely determined, otherwise there are multiple possible effects that are enumerated. This idea is incorporated in the IDA algorithm and its variants [7,10,16] for single interventions (i.e., interventions $do(R' = r')$) in propositional models. Algorithm 1 displays our proposed algorithm, which extends the idea of just

[4] An orientation of an undirected edge alters the conditional independence statements if the new orientation introduces the pattern $R_1 - \phi_1 \rightarrow R_2 \leftarrow \phi_2 - R_3$ such that R_1 and R_3 are not directly connected via a factor—see Item 1 in Definition 5.

Algorithm 1: Lifted Causal Inference in PPCFGs

Input : A PPCFG $M = (\boldsymbol{A} \cup \boldsymbol{G}, \boldsymbol{E})$ and an interventional query $P(Q \mid do(R'_1 = r'_1, \ldots, R'_k = r'_k))$ with $Q \in gr(\boldsymbol{A})$, $R'_1 \in gr(\boldsymbol{A}), \ldots, R'_k \in gr(\boldsymbol{A})$ and $\{Q\} \cap \{R'_1, \ldots, R'_k\} = \emptyset$.

Output: The set of all possible post-intervention distributions $P(Q \mid do(R'_1 = r'_1, \ldots, R'_k = r'_k))$ represented by M.

$\boldsymbol{P} \leftarrow \emptyset$;
$\{R_1, \ldots, R_\ell\} \leftarrow gr(\boldsymbol{A}) \setminus \{Q, R'_1, \ldots, R'_k\}$;
$M' \leftarrow$ PCFG after splitting parfactors in M based on each $R'_i \in \{R'_1, \ldots, R'_k\}$;
foreach $\boldsymbol{C}_1 \subseteq \mathrm{Ne}(R'_1, M'), \ldots, \boldsymbol{C}_k \subseteq \mathrm{Ne}(R'_k, M')$ *s.t. all* $\boldsymbol{C}_i$ *are cliques* **do**
  $M'' \leftarrow$ PCFG after orienting all edges $C - \phi - R'_i$ as $C - \phi \rightarrow R'_i$ in M' for all $C \in \boldsymbol{C}_i$ for every $i \in \{1, \ldots, k\}$;
  $M'' \leftarrow$ Arbitrary fully directed PPCFG represented by M'';
  if *no such* M'' *exists* **then**
    continue;
  $D \leftarrow \sum_{r_1 \in \mathrm{range}(R_1)} \cdots \sum_{r_\ell \in \mathrm{range}(R_\ell)} \prod_{R_i \in gr(\boldsymbol{A}) \setminus \{R'_1, \ldots, R'_k\}} P(r_i \mid \mathrm{pa}(R_i, M''))$;
  Add D to $\boldsymbol{P}$;
return $\boldsymbol{P}$

considering the possible parent sets of intervention variables to joint interventions (i.e., interventions $do(R'_1 = r'_1, \ldots, R'_k = r'_k)$ where $k \geq 1$) in relational models[5].

Given a PPCFG $M = (\boldsymbol{A} \cup \boldsymbol{G}, \boldsymbol{E})$ and an interventional query $P(Q \mid do(R'_1 = r'_1, \ldots, R'_k = r'_k))$, Algorithm 1 proceeds as follows to compute the set of all possible results for the given query. First, Algorithm 1 splits the parfactors in M based on each $R'_i \in \{R'_1, \ldots, R'_k\}$. Splitting the parfactors results in a modified PPCFG M' entailing equivalent semantics as M and works as follows [4]. Recall that $R'_i = A(L_1 = l_1, \ldots, L_j = l_j)$, $l_1 \in \mathrm{dom}(L_1), \ldots, l_j \in \mathrm{dom}(L_j)$, is a specific instance of a PRV $A(L_1, \ldots, L_j)$. The idea behind the splitting procedure is that we would like to separate $gr(A)$ into two sets $gr(A) \setminus \{R'_i\}$ and $\{R'_i\}$, as R'_i has to be treated differently than the remaining instances of A due to the intervention. Thus, every parfactor g for which there is an instance $\phi \in gr(g)$ such that R'_i appears in the argument list of ϕ is split. Formally, splitting a parfactor g replaces g by two parfactors $g'_{|C'}$ and $g''_{|C''}$ and adapts the constraints of $g'_{|C'}$ and $g''_{|C''}$ such that the inputs of $g'_{|C'}$ are restricted to all sequences that contain R'_i and the inputs of $g''_{|C''}$ are restricted to the remaining input sequences. After the splitting procedure, the semantics of the model remains unchanged as the groundings of M' are still the same as the groundings of the initial model M—they are just arranged differently across the sets of ground instances.

Algorithm 1 then iterates over all possible parent sets (i.e., over all subsets of undirected neighbours) of $R'_1, \ldots, R'_k$. When considering the subsets of undi-

[5] Algorithm 1 transfers the results for joint interventions in propositional models with undirected edges [19] to relational models.

rected neighbours, it is necessary that all subsets are jointly valid, that is, they are not allowed to alter the conditional independence statements encoded by the model and they must not introduce any cycles when oriented towards $R'_1, \ldots, R'_k$. To ensure the validity of these subsets, they are required to form a clique. A clique $\boldsymbol{C}$ is a subset of nodes such that all pairs of nodes in $\boldsymbol{C}$ are directly connected via a factor[6]. By ensuring that the subsets of undirected neighbours form cliques, the orientation of the undirected neighbours of $R'_1, \ldots, R'_k$ towards $R'_1, \ldots, R'_k$ does not introduce any pattern $R_1 - \phi_1 \rightarrow R_2 \leftarrow \phi_2 - R_3$ such that R_1 and R_3 are not directly connected via a factor, as due to the clique property R_1 and R_3 are always guaranteed to be directly connected via a factor. Thus, the conditional independence statements encoded by M'' are equivalent to those encoded by M. Having obtained the parent sets of $R'_1, \ldots, R'_k$, Algorithm 1 next extends the modified model M'' to any fully directed PPCFG represented by M'', if such a fully directed PPCFG exists. Afterwards, M'' is guaranteed to be a fully directed PPCFG represented by M and hence, the result of the provided query is given by Eq. 1, which requires us to know the parents of all non-intervention randvars to compute a product over conditional probability distributions. Since M encodes the full joint probability distribution over $gr(\boldsymbol{A})$, the conditional probability distributions are obtained by querying M, which allows us to compute the results for these queries using lifted probabilistic inference, thereby avoiding to ground the entire model if possible. The resulting post-intervention distribution is then added to the result set $\boldsymbol{P}$ and the algorithm continues the above steps for the next possible parent set until all possible parent sets have been taken into account. In case there is no causal explanation for the given PPCFG M (that is, M does not represent any fully directed PPCFG), Algorithm 1 returns an empty set. Such a situation might occur if M already contains a directed cycle or if there are undirected edges that cannot be oriented without introducing a new cycle or changing the conditional independence statements encoded by the model.

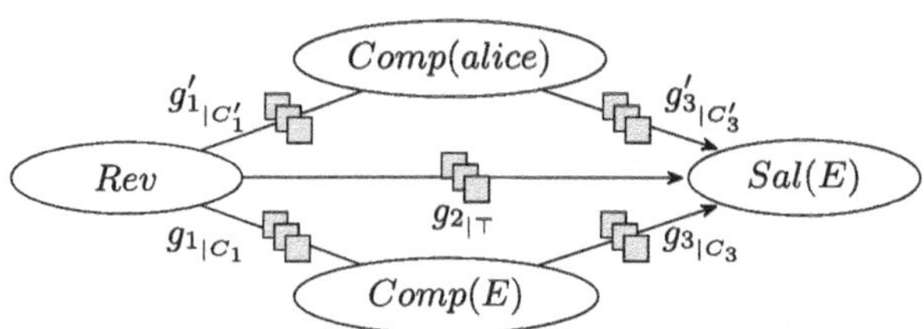

Fig. 3. The modified PPCFG obtained after splitting the parfactors in the PPCFG shown in Fig. 1 to separate *Comp(alice)* from *Comp(E)*.

---

[6] Formally, a clique $\boldsymbol{C}$ in a PPCFG $M = (\boldsymbol{A} \cup \boldsymbol{G}, \boldsymbol{E})$ is a subset of nodes such that for each pair of nodes $C_1 \in \boldsymbol{C}$, $C_2 \in \boldsymbol{C}$ with $C_1 \neq C_2$ it holds that there exists a factor ϕ such that there is an edge between C_1 and ϕ as well as an edge between C_2 and ϕ in $\boldsymbol{E}$ (either directed or undirected).

Example 8. Assume we want to compute $P(Rev \mid do(Comp(alice) = \text{high}))$ in the PPCFG M depicted in Fig. 1. As the intervention operates on $Comp(alice)$, first the parfactors in M are split on $Comp(alice)$ to obtain the modified PPCFG M' shown in Fig. 3. In M', $Comp(alice)$ is now a separate node in the graph, so its possible parents can be determined. Formally, g_1 has been replaced by two parfactors $g_{1_{|C_1}}$ and $g'_{1_{|C'_1}}$ with constraints $C_1 = (E, \{bob, charlie\})$ and $C'_1 = (E, \{alice\})$. Analogously, g_3 has been replaced by $g_{3_{|C_3}}$ and $g'_{3_{|C'_3}}$. Next, Algorithm 1 considers all possible parent sets of $Comp(alice)$, which are given by $\text{Pa}_1(Comp(alice), M'') = \emptyset$ (resulting from orienting $Comp(alice) - g'_{1_{|C'_1}} - Rev$ as $Comp(alice) - g'_{1_{|C'_1}} \rightarrow Rev$) and $\text{Pa}_2(Comp(alice), M'') = \{Rev\}$ (resulting from orienting $Comp(alice) - g'_{1_{|C'_1}} - Rev$ as $Comp(alice) \leftarrow g'_{1_{|C'_1}} - Rev$). Depending on whether $Comp(alice)$ actually has a parent or not, the post-intervention distribution is different and hence, Algorithm 1 returns a set containing two possible results for the query $P(Rev \mid do(Comp(alice) = \text{high}))$.

Note that the model shown in Fig. 3 is more compact than the fully grounded model in Fig. 2 as Algorithm 1 only grounds necessary parts of the model. Algorithm 1 can further benefit from interventions on multiple indistinguishable objects under the assumption that the graph structure is identical for all groundings. For example, when considering the effect of a training course given to multiple employees on the revenue of the company, even if there are hundreds of employees, after intervening on their competence there are still only two nodes representing their competences in the graph, namely one node for all employees that have received the training course and another node for the remaining employees. In this case, Algorithm 1 has to consider only two possible parent sets regardless of the number of employees while there are $2^{|\text{dom}(E)|}$ possible parent sets in an equivalent propositional model to consider (where $|\text{dom}(E)|$ is the number of employees).

Corollary 1. *Let $M = (\boldsymbol{A} \cup \boldsymbol{G}, \boldsymbol{E})$ be a PPCFG. When intervening on a non-grounded PRV $A(L_1, \dots, L_j) \in \boldsymbol{A}$, under the assumption that the graph structure is identical for all groundings, it holds that*

1. *Algorithm 1 considers $O(2^{|\text{Ne}(A,M)|})$ possible parent sets in the worst case, and*
2. *in a propositional model, $O(2^{\prod_{R \in gr(A)} |\text{Ne}(R, gr(M))|})$ possible parent sets have to be considered in the worst case.*

It is also possible to intervene on a non-grounded PRV. For example, the company might wonder whether it is worth it to send all its employees to a training course to increase the revenue of the company. Then, we are interested in computing the quantity $P(Rev \mid do(Comp(E) = \text{high}))$. Finally, note that Algorithm 1 can handle multiple query variables at once instead of a single query variable Q (the query variables are then not summed out in Line 9).

6 Conclusion

We introduce PPCFGs as probabilistic relational models that allow to incorporate partial causal knowledge, thereby enabling lifted causal inference without the need for a fully specified causal model. A lifted representation such as a PPCFG is more expressive than a propositional model and allows for tractable inference with respect to domain sizes of logvars as. We further present an algorithm to efficiently compute the effect of joint interventions in PPCFGs (i.e., on a lifted level). Our proposed algorithm is also able to efficiently deal with interventions on PRVs representing sets of indistinguishable randvars.

In future work, we aim to investigate the effect of interventions in a relational model with mutual interdependencies in form of bidirectional edges. We conjecture that randvars with mutual interdependencies can be collapsed into a single node in the graph such that our proposed algorithm can still be applied. Another interesting direction for future work is to allow for hidden confounders.

Acknowledgements. This work is funded by the BMBF project AnoMed 16KISA057.

References

1. Ahmadi, B., Kersting, K., Mladenov, M., Natarajan, S.: Exploiting symmetries for scaling loopy belief propagation and relational training. Mach. Learn. **92**, 91–132 (2013)
2. Braun, T., Möller, R.: Lifted junction tree algorithm. In: Friedrich, G., Helmert, M., Wotawa, F. (eds.) KI 2016. LNCS (LNAI), vol. 9904, pp. 30–42. Springer, Cham (2016). https://doi.org/10.1007/978-3-319-46073-4_3
3. Braun, T., Möller, R.: Parameterised queries and lifted query answering. In: Proceedings of the 27th International Joint Conference on Artificial Intelligence (IJCAI-2018), pp. 4980–4986. IJCAI Organization (2018)
4. De Salvo Braz, R., Amir, E., Roth, D.: Lifted first-order probabilistic inference. In: Proceedings of the 19th International Joint Conference on Artificial Intelligence (IJCAI-2005), pp. 1319–1325. Morgan Kaufmann Publishers Inc. (2005)
5. De Salvo Braz, R., Amir, E., Roth, D.: MPE and partial inversion in lifted probabilistic variable elimination. In: Proceedings of the 21th National Conference on Artificial Intelligence (AAAI-2006), pp. 1123–1130. AAAI Press (2006)
6. Frey, B.J.: Extending factor graphs so as to unify directed and undirected graphical models. In: Proceedings of the 19th Conference on Uncertainty in Artificial Intelligence (UAI-2003), pp. 257–264. Morgan Kaufmann Publishers Inc. (2003)
7. Guo, R., Perkovic, E.: Minimal enumeration of all possible total effects in a Markov equivalence class. In: Proceedings of the 24th International Conference on Artificial Intelligence and Statistics (AISTATS-2021), pp. 2395–2403. PMLR (2021)
8. Kersting, K., Ahmadi, B., Natarajan, S.: Counting belief propagation. In: Proceedings of the 25th Conference on Uncertainty in Artificial Intelligence (UAI-2009), pp. 277–284. AUAI Press (2009)
9. Kisyński, J., Poole, D.: Constraint processing in lifted probabilistic inference. In: Proceedings of the 25th Conference on Uncertainty in Artificial Intelligence (UAI-2009), pp. 293–302. AUAI Press (2009)

10. Liu, Y., Fang, Z., He, Y., Geng, Z.: Collapsible IDA: collapsing parental sets for locally estimating possible causal effects. In: Proceedings of the 36th Conference on Uncertainty in Artificial Intelligence (UAI-2020), pp. 290–299. PMLR (2020)
11. Luttermann, M., Braun, T., Möller, R., Gehrke, M.: Colour passing revisited: lifted model construction with commutative factors. In: Proceedings of the 38th AAAI Conference on Artificial Intelligence (AAAI-2024), pp. 20500–20507. AAAI Press (2024)
12. Luttermann, M., Hartwig, M., Braun, T., Möller, R., Gehrke, M.: Lifted causal inference in relational domains. In: Proceedings of the 3rd Conference on Causal Learning and Reasoning (CLeaR-2024), pp. 827–842. PMLR (2024)
13. Luttermann, M., Machemer, J., Gehrke, M.: Efficient detection of commutative factors in factor graphs. In: Proceedings of the Twelfth International Conference on Probabilistic Graphical Models (PGM-2024), pp. 38–56. PMLR (2024)
14. Luttermann, M., Machemer, J., Gehrke, M.: Efficient detection of exchangeable factors in factor graphs. In: Proceedings of the 37th International FLAIRS Conference (FLAIRS-2024) (2024). Florida Online Journal
15. Luttermann, M., Möller, R., Gehrke, M.: Lifting factor graphs with some unknown factors. In: Bouraoui, Z., Vesic, S. (eds.) ECSQARU 2023. LNCS, vol. 14294, pp. 337–347. Springer, Cham- (2023). https://doi.org/10.1007/978-3-031-45608-4_25
16. Maathuis, M.H., Kalisch, M., Bühlmann, P.: Estimating high-dimensional intervention effects from observational data. Ann. Stat. **37**, 3133–3164 (2009)
17. Meert, W., Taghipour, N., Blockeel, H.: First-order bayes-ball. In: Balcázar, J.L., Bonchi, F., Gionis, A., Sebag, M. (eds.) ECML PKDD 2010. LNCS (LNAI), vol. 6322, pp. 369–384. Springer, Heidelberg (2010). https://doi.org/10.1007/978-3-642-15883-4_24
18. Milch, B., Zettlemoyer, L.S., Kersting, K., Haimes, M., Kaelbling, L.P.: Lifted probabilistic inference with counting formulas. In: Proceedings of the 23th AAAI Conference on Artificial Intelligence (AAAI-2008), pp. 1062–1068. AAAI Press (2008)
19. Nandy, P., Maathuis, M.H., Richardson, T.S.: Estimating the effect of joint interventions from observational data in sparse high-dimensional settings. Ann. Stat. **45**, 647–674 (2017)
20. Niepert, M., Van den Broeck, G.: Tractability through exchangeability: a new perspective on efficient probabilistic inference. In: Proceedings of the 28th AAAI Conference on Artificial Intelligence (AAAI-2014), pp. 2467–2475. AAAI Press (2014)
21. Pearl, J.: Fusion, propagation, and structuring in belief networks. Artif. Intell. **29**, 241–288 (1986)
22. Pearl, J.: Causality: Models, Reasoning and Inference, 2nd edn. Cambridge University Press, Cambridge (2009)
23. Pearl, J., Glymour, M., Jewell, N.P.: Causal Inference in Statistics: A Primer, 1st edn. Wiley, Hoboken (2016)
24. Perkovic, E.: Identifying causal effects in maximally oriented partially directed acyclic graphs. In: Proceedings of the 36th Conference on Uncertainty in Artificial Intelligence (UAI-2020), pp. 530–539. PMLR (2020)
25. Poole, D.: First-order probabilistic inference. In: Proceedings of the 18th International Joint Conference on Artificial Intelligence (IJCAI-2003), pp. 985–991. Morgan Kaufmann Publishers Inc. (2003)
26. Shachter, R.D.: Bayes-ball: rational pastime (for determining irrelevance and requisite information in belief networks and influence diagrams). In: Proceedings of the

14th Conference on Uncertainty in Artificial Intelligence (UAI-1998), pp. 480–487. Morgan Kaufmann Publishers Inc. (1998)
27. Spirtes, P., Glymour, C., Scheines, R.: Causation, Prediction, and Search, 2nd edn. MIT Press, Cambridge (2000)
28. Taghipour, N., Fierens, D., Davis, J., Blockeel, H.: Lifted variable elimination: decoupling the operators from the constraint language. J. Artif. Intell. Res. **47**, 393–439 (2013)
29. Winn, J.: Causality with gates. In: Proceedings of the 15th International Conference on Artificial Intelligence and Statistics (AISTATS-2012), pp. 1314–1322. PMLR (2012)

Discrete Minimax Binary Relevance Classifier for Imbalanced Multi-label Classification

Salvador Madrigal, Vu-Linh Nguyen, Cyprien Gilet(✉), and Sébastien Destercke

Heudiasyc Laboratory, University of Technology of Compiègne, Compiègne, France
salvador.madrigal-castillo@etu.utc.fr,
{vu-linh.nguyen,cyprien.gilet,sebastien.destercke}@hds.utc.fr

Abstract. Multi-label classification (MLC) is a supervised learning problem where each instance can be associated with none, one, or multiple labels. MLC has received increasing attention due to its wide range of applications, such as text categorization and medical diagnosis. Despite a rich literature on MLC, handling imbalanced data, often encountered in real-world MLC datasets, has not been tackled satisfactorily. Based on a thorough literature review, it appears that the existing methods for imbalanced MLC are either hard to be coupled with sound theoretical guarantees or of limited scalability. This paper discusses the potential (dis)advantages of existing methods for imbalanced MLC, when being coupled with Binary relevance classifier (BRC), and introduces Discrete Minimax BRC (DMBRC), which would be a promising attempt to robustify the BRC by leveraging theoretically sound properties of the Discrete Minimax Classifier. We also provide empirical evidence to illustrate how DMBRC may be advantageous in balancing the label-wise error rates. Finally, we envision future works on further strengthening DMBRC in both label-wise error rates and conventional MLC evaluation metrics.

Keywords: Multi-label classification · Binary relevance · Imbalance data · Discrete Minimax Classifier

1 Introduction

This paper seeds in the context of supervised multi-label classification (MLC) for safety-critical detection, such as diagnosing pathologies in precision medicine, detecting anomalies, frauds, or failures of components in condition monitoring.

1.1 Problem Statements

Given observations of features which can be numeric, categorical, images, or any other kind of signals describing an instance, the purpose of MLC is to determine

S. Destercke et al. (Eds.): SUM 2024, LNAI 15350, pp. 281–296, 2025.
https://doi.org/10.1007/978-3-031-76235-2_21

the actual class labels of the sample in order to support the experts of the application domain in their diagnosis.

Differ from usual *single-label classification tasks*, which classify instances into one of several mutually exclusive classes, *MLC* allows one to assign multiple class labels to each instance [2,8,26,38]. An example of MLC problems is multi-cancer early detection [24], in which each patient may develop either non or multiple cancer types. Another MLC problem is fault detection, such as in aeroengines [42] and battery systems [43], where multiple faults can coexist.

While MLC is becoming increasingly promising in safety-critical detection, an important issue often occurs in such contexts: the presence of imbalanced labels. In MLC datasets that contain rare class labels, conventional classifiers, such as Multi-label k Nearest Neighbours (MLKNN) [46], tend to underestimate these class labels, predicting them less frequently, which results in a high number of false negatives class-wise (see Fig. 1(a) and (c)). Therefore, the goal is to reduce the number of false positives by balancing the error rates (see Fig. 1(b) and (d)). The issue of imbalanced labels generally harms classification performances associated with the most imbalanced labels, favouring the most represented situation. This may typically happen when some labels are important to detect but are scarcely observed and difficult to predict. For example, diagnosing rare pathologies in precision medicine, and predicting failures of a component in condition monitoring, are crucial but difficult to carry out.

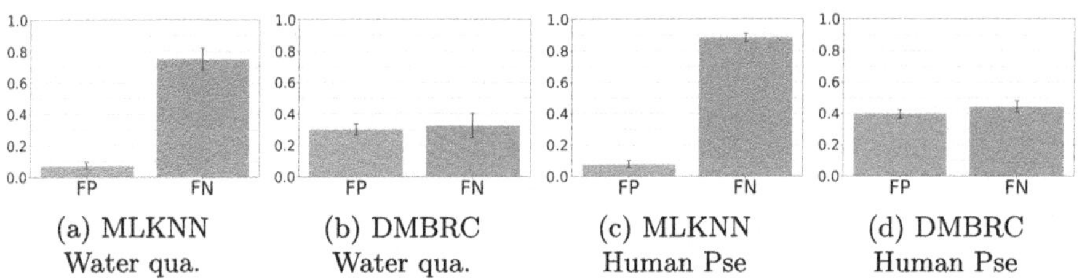

(a) MLKNN Water qua. (b) DMBRC Water qua. (c) MLKNN Human Pse (d) DMBRC Human Pse

Fig. 1. Mean False Positive (FP) and False Negative (FN) rates for the fourth label with a prior probability of 0.21 in the Water Quality dataset, and for the second label (0.26) in the Human PseAAC dataset, obtained after a 5-fold cross-validation procedure using MLKNN and our new method named DMBRC.

1.2 Related Works and State of the Art

The issue of imbalanced classes has become more and more studied in MLC from the past decades [4,7,20,38], while remaining a challenging problem. In [38], the authors proposed a complete and interesting survey of methods designed to handle imbalanced data in MLC. They notably highlight that the methods aiming to address the imbalance problem in MLC can be divided into four categories: resampling methods [4,6–8,20], classifier adaptation [9,35,45], ensemble methods

[23,29,37] and cost-sensitive approaches [3,11,40]. Let us note that commonly used single-label classifiers robust to imbalance data (such as Weighted Logistic Regression, Weighted Decision Trees [31]) can be straigthforwardly adapted to MLC when considering the binary relevance strategy, and with more efforts when considering more complex techniques such as classifier chains [34].

Classifier adaptation techniques handling imbalanced labels often either (1) re-define each binary classification problem as a multi-class classification problem, in which the instances in the majority class are partitioned into multiple (sub)-classes, to reduce the potential impact of the majority classes during the training time [45] or (2) partition the input space into multiple regions and solve the binary classification problems on the regions independently [9] or (3) only allow the neighbor training instances to have impact on the prediction of the query instance [35]. Clearly, such techniques require one to choose at least one sensitive hyper-parameter, such as the number of (sub)-classes, the number of smaller regions, and the threshold that determines the nearest neighbors. Moreover, it is unclear why/how splitting the majority class into multiple (sub)-classes and partitioning the input space into multiple regions without further processing may help to mitigate class imbalance.

Cost-sensitive approaches [3,11,40] explicitly increase the impact of the instances from the minority class by adjusting class weights in the training loss. This would be analogous to learning a classifier on a modified training data set. Therefore, it might be inconvenient if one wishes to use the classifier for other purposes, such as doing descriptive statistics, i.e., summarizing the characteristics of the data set, or accommodating emerging evaluation metrics at the prediction time.

Resampling methods often oversample to increase the impact of instances with minority classes or undersample to decrease the impact of instances within the majority classes. Yet, it may provide promising predictive results in practice [4,6–8,20]. Similar to cost-sensitive approaches [3,11,40], they also learn a classifier on a modified training data set, yet in a much less controlled way, as they typically rely on random processes.

Ensemble methods consist of ensemble generation and aggregation. During ensemble generation, one can optionally couple the data-generating/sampling process (to train the ensemble members) with other methods for class imbalance. The chosen data-generating/sampling process may also amplify the imbalance. For example, bagging seems to lower the chance of selecting instances from the minority class when creating training data sets, from which the ensemble members are trained, and thus may lower the impact of instances from the minority class in general. It is also known that different aggregators may be in favor of different evaluation metrics [27].

1.3 Towards the Minimax Classifier

The issue of imbalance classes is well known in single-label classification since the past century [1,16,32], and the *Minimax Classifier* have been analytically demonstrated to be an optimal approach to deal with imbalanced data in the context of single-label classification [1,32]. The minimax classifier is indeed the Bayesian classifier for which the risks per class are all minimized and balanced.

In statistical decision theory [1,32], minimax classifiers are usually fitted by maximizing the Bayes risk with respect to the prior probabilities over the simplex, which requires to know the conditional distributions of the features in each class in order to analytically calculate the Bayes risk. However, this task remains a challenge in Machine Learning since we do not know the class-conditional distributions of the features, especially when dealing with several classes, mixed features, and arbitrary loss function. Indeed, computing an accurate estimate of the feature joint distribution in each class to achieve a good estimate of the empirical Bayes risk over the simplex remains highly complicated. Furthermore, in most real-world applications, the estimation of the empirical Bayes risk over the simplex is generally intractable because of the curse of dimensionality. Nowadays, only a few minimax algorithms have been proposed to deal with these general cases in single-label classification [16,19,21].

In the past few years, the authors in [16,19] developed a new approach for computing a minimax classifier for single-label classification tasks, suitable to process both numeric and categorical attributes, that can process a large number of classes, that can be coupled with any pretrained deep neural network [18], and which has been applied to precision medicine [15] or condition monitoring [17]. The procedure partitions the feature space beforehand and learns the minimax classifier by using a closed-form expression of the empirical Bayes risk over the simplex. While the authors show that discrete empirical Bayes risk is a concave non-differentiable multivariate piecewise affine function concerning the priors, they provide an efficient algorithmic procedure to obtain the Discrete Minimax Classifier (DMC). This approach can outperform several other state-of-the-art methods to obtain guaranteed robustness against imbalanced class risks, even when dealing with a large number of classes (for example $K > 100$). However, the use of minimax classifiers has not been studied yet in the context of MLC. The objective of this paper is to propose an opening approach to introduce the minimax classifier in MLC based on the DMC proposed in [16,19].

The paper is organized as follows. Section 2 recalls the main concepts of imbalanced MLC. In Sect. 3, we propose a binary relevance strategy to consider the DMC in MLC. Our proposed classifier aims to minimize and balance all the risks per class and with respect to any kind of loss/cost function that penalizes the classification errors. Section 4 empirically assesses our proposed classifier, in comparison with 11 other MLC methods/algorithms, on six real data sets.

2 Multi-label Classification

This section recalls the main concepts of MLC. Let $\mathcal{X} = \mathbb{R}^d$ be a d-dimensional feature space, and let $\Lambda = \{\lambda_1, \dots, \lambda_K\}$ be a set containing K labels. A multi-label sample is a pair $(\boldsymbol{x}, \boldsymbol{y})$, where $\boldsymbol{x} \in \mathcal{X}$ is a d-dimensional feature vector and $\boldsymbol{y} \in \mathcal{Y} := \{0,1\}^K$, where, for any $1 \leq k \leq K$, $y^k = 1$ ($y^k = 0$) indicates that the label λ_k is relevant/present (irrelevant). A multi-label dataset (MLD),

$\mathcal{D} = \{(\boldsymbol{x}_n, \boldsymbol{y}_n)|1 \leq n \leq N\}$, is formed by N pairs of multi-label samples. A multi-label classifier, $\delta : \mathcal{X} \rightarrow \mathcal{Y}$, is a classifier for which given an unseen sample $\boldsymbol{x}$, the classifier returns a prediction $\hat{\boldsymbol{y}} \in \mathcal{Y}$.

There are two main approaches for dealing with MLC: problem transformation and algorithm adaptation [2]. The first one, problem transformation, aims to transform the MLC problem into one or more binary or multiclass classifications, some popular methods include: Binary relevance [44], classifier chains [33,34] and power label set [2]. On the other hand, algorithm adaptation, aims to adapt existing algorithms so that they can work with the MLC setting, like the MLKNN [46]. In this paper, we primarily explore the binary relevance (BR). The BR method consists in transforming the multi-label problem (MLP) into a collection of independent binary classification problems [44], one per label. Formally, given a multi-label data set $\mathcal{D} = \{(\boldsymbol{x}_n, \boldsymbol{y}_n)|1 \leq n \leq N\}$, the BR methods consist in creating K different data sets $\mathcal{D}_k = \{(\boldsymbol{x}_n, \boldsymbol{y}_n^k)|1 \leq n \leq N\}$, for $k = 1, \ldots, K$, and learning for each dataset $\mathcal{D}_k$ a binary classifier $\delta_k : \mathcal{X} \rightarrow \{1, 0\}$.

While BR methods, focusing in this paper, are among the most algorithmically simplest MLC methods, they are arguably sound methods for optimizing the commonly used decomposable losses [12]. Moreover, it seems to be a convenient choice regarding the trade-off between the theoretical soundness and computational expenses in various applications of MLC, where the training data can come with incomplete or missing data. A notable example of such scenarios would be predicting antimicrobial resistance phenotypes (susceptible, or resistant) of multiple drugs given genomic sequences of strains [13,25], where a significant proportion of training instances are partially annotated. Finally, BR can sometimes provide competitive empirical results, compared to power label sets and classifier chains, with respect to both decomposable and non-decomposable losses [27,41].

2.1 Imbalance in Multi-label Data

Many real-life applications suffer from label imbalance [38], highlighting the importance of determining whether an MLD is imbalanced. One straightforward method to determine label imbalance in an MLP is to observe the label distribution, as shown in the sequel by Fig. 3. This visualization provides an idea of how the labels are distributed, but is only useful when there are a few labels. For datasets with a large number of labels, it is beneficial to employ metrics to measure the imbalance. Different measures have been proposed [4] to quantify the imbalance present in a MLD.

Imbalance Ratio per Label (IRLbl) [4]. For any $\lambda_k \in \Lambda$, the IRLbl, defined as

$$\text{IRLbl}(\lambda_k|\mathcal{D}) = \frac{\max\limits_{k^* \in \{1,\ldots,K\}} \left(\sum_{n=1}^{N} [\![y_n^{k^*} = 1]\!] \right)}{\sum_{n=1}^{N} [\![y_n^k = 1]\!]}, \tag{1}$$

where $[\![\cdot]\!]$ is the indicator function, i.e., $[\![A]\!] = 1$ if the predicate A is true and $[\![A]\!] = 0$ otherwise. The $\text{IRLbl}(\lambda_k|\mathcal{D})$ is the ratio between the most frequent label and λ_k. It is 1 for the most frequent label and a higher value means a higher level of imbalance for λ_k. Note however that the measure is not upper-bounded and is relative.

Mean Imbalance Ratio (MeanIR) [4]. This score measures the average level of imbalance in an MLD and is defined as

$$\text{MeanIR}(\mathcal{D}) = \frac{1}{K}\sum_{k=1}^{K} \text{IRLbl}(\lambda_k|\mathcal{D}). \tag{2}$$

Of course, the same critic applies to MeanIR than to IRLbl.

Coefficient of Variation of IRLbl (CVIR) [4]. This score measures the variation of IRLbl and indicates if all labels have the same level of imbalance:

$$\text{CVIR}(\mathcal{D}) = \frac{1}{\text{MeanIR}(\mathcal{D})\sqrt{K-1}}\sqrt{\sum_{k=1}^{k}(\text{IRLbl}(\lambda_k|\mathcal{D}) - \text{MeanIR}(\mathcal{D}))^2}. \tag{3}$$

The higher the value the greater the level of imbalance between labels.

2.2 Performance Metrics for Multi-label Problems

For binary and multi-class classification, the performance of a classifier depends on whether the unseen samples are correctly classified. In the case of MLC, predictions can considered as correct, partially correct, or partially incorrect [22]. We shall detail 3 commonly used metrics used in this paper.

For any query instance $\boldsymbol{x}$, let $\boldsymbol{y}$ and $\hat{\boldsymbol{y}}$ be the true labels and the predicted labels, respectively. The subset 0/1 loss and the Hamming loss (respectively denoted by $\mathcal{L}_{0/1}$, $\mathcal{L}_{\text{Ham}}$ and reminded in Eq. (4)), which are the lower the better scores, assess the classifier δ according to its ability to exactly predict all the labels and accurately predict the labels on average, respectively.

$$\mathcal{L}_{0/1}(\delta|\mathcal{D}) = \frac{1}{N}\sum_{n=1}^{N}[\![\hat{\boldsymbol{y}}_n \neq \boldsymbol{y}_n]\!], \quad \mathcal{L}_{\text{Ham}}(\delta|\mathcal{D}) = \frac{1}{N}\sum_{n=1}^{N}\frac{1}{K}\sum_{k}^{K}[\![\hat{y}_n^k \neq y_n^k]\!]. \tag{4}$$

The F1 score given by Eq. (5) is a metric that combines precision and recall. It is calculated as the harmonic mean of precision and recall. Higher values indicate better performance of the model.

$$\text{F1}(\delta|\mathcal{D}) = 2\sum_{n=1}^{N}\frac{\sum_{k=1}^{K}\hat{y}_n^k \cdot y_n^k}{\sum_{k=1}^{K}\hat{y}_n^k + \sum_{k=1}^{K}y_n^k}. \tag{5}$$

3 Discrete Minimax Binary Relevance Classifier

The DMC [16] is a single-label classifier that aims to address the imbalance issue with statistical guarantees and when considering any kind of loss function L

(which allows one to penalize the classification errors). More precisely, the DMC aims to minimize and balance the risks per class during the training procedure. In particular, in the context of binary classification and when considering the 0/1 loss function, the DMC allows one to minimize and balance the false positive and false negative rates, especially when dealing with imbalance data.

The primary goal of this paper is to extend the statistical guarantees provided by the DMC [16] to the MLC using the Binary Relevance strategy. In other words, given a MLD, $\mathcal{D} = \{(\boldsymbol{x}_n, \boldsymbol{y}_n) | 1 \leq n \leq N\}$, we create K different datasets $\mathcal{D}_k = \{(\boldsymbol{x}_n, y_n^k) | 1 \leq n \leq N\}$, for $k = 1, \ldots, K$, and we aim to learn for each set $\mathcal{D}_k$ a binary DMC denoted by $\delta_k^M : \mathcal{X} \to \{1, 0\}$.

To this aim, given a training set $\mathcal{D}_k$, we first partition the feature space $\mathcal{X}$ into T different regions $\Omega = \{\phi_1, \ldots, \phi_T\}$, as proposed in [16], using for example the Kmeans partitioning or the decision tree partitioning. This defines a mapping $\Phi : \mathcal{X} \to \Omega$, which maps any instance $\boldsymbol{x} \in \mathcal{X}$ to a discrete profile $\Phi(\boldsymbol{x}) \in \Omega$. For all $t \in \{1, \ldots, T\}$ and for each binary class $\ell \in \{0, 1\}$, the estimated probability that an instance $\boldsymbol{x} \in \mathcal{X}$ has the discrete profile ϕ_t given its real class y^k is $\ell \in \{0, 1\}$ is given by

$$\hat{p}_{\ell t} := \frac{1}{|\mathcal{I}_\ell|} \sum_{i \in \mathcal{I}_\ell} [\![\Phi(\boldsymbol{x}_i) = \phi_t]\!]. \tag{6}$$

Here, $\mathcal{I}_\ell$ is the set containing all the training instances such that $y_i^k = \ell$.

For the following, let us define $L = \{L_{\ell j} : \ell, j \in \{0, 1\}\}$ the costs of classification errors such that $L_{\ell, j}$ is the cost of predicting the class j given that the real class is ℓ. For example, when considering the 0/1 loss function, we have $L_{0,0} = L_{1,1} = 0$ and $L_{0,1} = L_{1,0} = 1$. Furthermore, let us define $\mathbb{S} = \big\{\pi = [\pi_0, \pi_1] \in [0, 1]^2 : \pi_0 + \pi_1 = 1\big\}$ the 2-dimensional simplex.

Based on the partitioned feature space $\Omega = \Phi(\mathcal{X})$ and similarly to [16], we can demonstrate that the empirical Bayes risk $V_k : \mathbb{S} \to \mathbb{R}_+$ (as a function of the priors), associated with the binary-classification set $\mathcal{D}_k$, is analytically given by

$$V_k(\pi) = \sum_{\ell=0}^{1} \sum_{t=1}^{T} \sum_{z=0}^{1} L_{\ell z}\, \pi_\ell\, \hat{p}_{\ell t}\, [\![\sum_{j \in \{0,1\}} L_{jz} \pi_j \hat{p}_{jt} = \min_{q \in \{0,1\}} \sum_{j \in \{0,1\}} L_{jq} \pi_j \hat{p}_{jt}]\!]. \tag{7}$$

We can furthermore demonstrate that V_k is a concave multivariate piecewise affine function over $\mathbb{S}$ with a finite number of pieces.

Finally, similarly to [16], the DMC $\delta_k^M : \mathcal{X} \to \{1, 0\}$ associated with the binary classification dataset $\mathcal{D}_k$ is given by

$$\delta_k^M : \boldsymbol{x} \mapsto \arg\min_{j \in \{0,1\}} \sum_{t=1}^{T} \sum_{\ell=0}^{1} L_{\ell j}\, \bar{\pi}_\ell\, \hat{p}_{\ell t}\, [\![\Phi(\boldsymbol{x}_i) = \phi_t]\!]. \tag{8}$$

In Eq. (8), $\bar{\pi} = [\bar{\pi}_0, \bar{\pi}_1]$ corresponds to the least favorable priors that maximize the Bayes risk V_k over the simplex $\mathbb{S}$ and can be easily computed with a projected subgradient algorithm [16]. Note that the classifier obtained by Eq. (8) will balance class-wise risks, as illustrated by Fig. 1(b) and (d). Without entering into too much details (due to page limits), the DMC will seek the class prior distribution $\bar{\pi}$ that will balance the risks.

For the MLC, the idea of the **Discrete Minimax Binary Relevance Classifier** (DMBRC) is to extend the statistical guarantees from the multi-class problem to the multi-label problem. To achieve, first, the feature space of the MLP, $\mathcal{X}$, is discretized, and then following the idea of the BR method, a DMC, $\delta_{\bar{\pi}}^{B} : \mathcal{X} \rightarrow \{0, 1\}$ for the binary is trained per each label individually (Fig. 2).

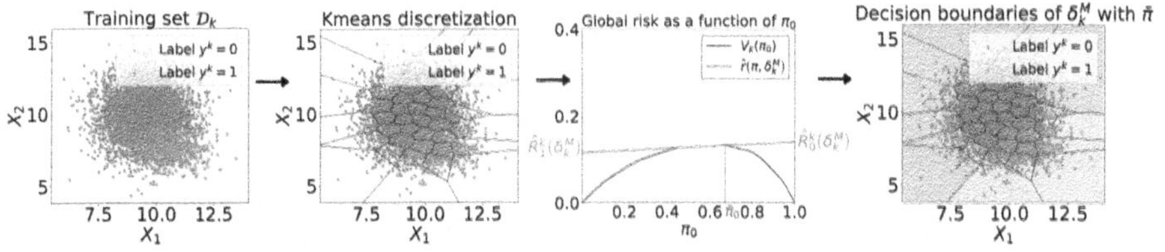

Fig. 2. Illustrative example of how to compute δ_k^M.

4 Experiments

In this section, we conduct an empirical study to assess the performance of our new Discrete Minimax Binary Relevance Classifier (DMBRC), compared to other methods suitable to imbalanced MLC.

4.1 Experimental Setting

The datasets used are the following: yeast, scene, CHD_49 and Tmc2007 from the Mulan repository [39]; Water-quality (water qua) from the repository from the University of Cordoba (see, https://www.uco.es/kdis/mllresources/) and HumanPseAAC (Human Pse) from the cometa repository [5]. Information about the different datasets can be viewed in Table 1 and their distributions per label are given in Fig. 3, with the exception of TMC2007 due to its large number of labels. Note that they display various levels of imbalance and disparity across labels: some data sets have labels that are highly imbalanced, such as yeast (label 9), CHD_49 (label 4) or Human Pse. (label 4), while others have labels whose percentage of appearance remain relatively high across labels (scene and water quality), with different degrees of disparity (scene percentage are all similar, while water quality has a higher disparity).

For each data set, a 5–fold-cross-validation is employed, and for each method, the mean and standard deviation (std) of the performance scores on the test sets are reported. The subset zero-one metric (4), F1 score (5), and Hamming metric (4) are employed as conventional MLC performance scores, which are not specially designed for assessing the ability to balance the class conditional

Table 1. Overview of datasets

Dataset	Instances	Features	Labels	MeanIR	CVIR
Yeast	2417	103	14	7.20	1.88
Scene	2407	294	6	1.25	0.12
CHD_49	555	49	6	5.77	1.75
Water qua	1060	16	14	15.29	1.08
Human Pse.	3106	440	14	1.77	0.30
Tmc2007	28596	49060	22	17.13	0.81

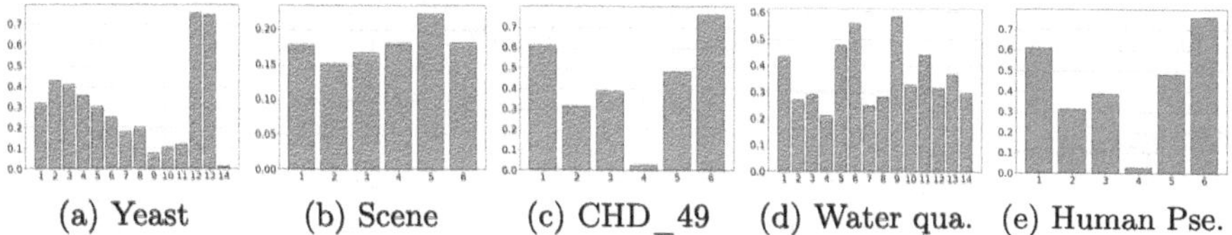

(a) Yeast (b) Scene (c) CHD_49 (d) Water qua. (e) Human Pse.

Fig. 3. Datasets distribution per label.

risk. To assess this aspect, for each given classifier δ, we quantify the ability to balance the false positive rate and false negative rate associated with each label $\lambda \in \Lambda$ using the following criterion

$$\psi(\delta) = \max_{\lambda \in \Lambda} |R_0^\lambda(\delta) - R_1^\lambda(\delta)|, \tag{9}$$

where $R_0^\lambda(\delta)$ denotes the false negative rate associated with the label λ for the classifier δ and $R_1^\lambda(\delta)$ the false positive rate. A smaller $\psi(\delta)$ indicates that the multi-label classifier δ provides a better balance of the false positive and false negative rates. A perfectly balanced classifier would have $\psi(\delta) = 0$, a value $\psi(\delta) = 1$ indicating that at least for one label, a classifier always predict either false or positive.

To seek an extensive empirical study, we implemented the DMBRC and 11 other competitors in Python. The competitors can be divided into three groups of different natures. The first group consists of three classifiers that do not take into account class imbalance: the MLKNN classifier and two binary relevance methods with logistic regression (LR) and decision tree (DT) as the base learners. The second group consists of two cost-sensitive binary relevance classifiers, weighted logistic regression (WLR) and weighted decision tree (WDT), which explicitly take into account class imbalance as part of the training loss. The third group consists of six resampling methods, which employ multi-label random oversampling [7] (ROS), multi-label random undersampling [7] (RUS), and multi-label SMOTE [8] (SM) to resample the datasets and then train binary relevance LR and DT. We refer to these methods as ROS+LR, ROS+DT, RUS+LR, RUS+DT, SM+LR, and SM+DT.

With the exception of MLKNN [46], whose implementation is provided in the Scikit-multilearn package [36], all the other methods are of binary relevance nature and only differ in their base learners. We use the implementation for binary relevance from the Scikit-multilearn package [36], which allows customization of the base learner using base learners from other packages such as the Scikit-learn library [31], from which we call functions for training LR and DT.

For the multi-label resampling methods, we use an unofficial implementation on GitHub (see https://github.com/Seal-Li/Multi-label-imbalance). For DMBRC, we employ the DMC using the source code provided by [16], with some modifications. The source code used in our experiments has been made public at https://github.com/SalvadorMadrigal/DMBRC-for-Imbalanced-MLC.

4.2 Results

The predictive performance, in terms of the subset zero-one metric (4), F1 score (5), and Hamming metric (4), and the ψ scores are given in Table 2 and Table 3. Overall, the predictive performance provided by DMBRC is on par with the ones provided by other cost-sensitive classifiers, i.e., WLR and WDT. This is a satisfying result since DMBRC, WLR, and WDT all take into account the class imbalance, but in different manners and are useful for different purposes (besides opting for the subset zero-one metric (4), F1 score (5), and Hamming metric (4)). Therefore, DMBRC complements the existing collection of classifiers/methods to balance accuracies meaningfully.

Considering the full set of methods/classifiers covered, DMBRC is never the worst, but also never the best. If one ranks the methods/classifiers covered by their performance, DMBRC consistently occupies middle positions in the ranks. This is perfectly expected: the additional constraint of being balanced will typically lower the average accuracy, compared to an unconstrained classifier optimizing average accuracy. It is therefore normal that the DMBRC cannot outperform, in average, methods such as MLKNN, LR or DT that are not also taking care of balancing the true positive and negative rates for each label. This should therefore not be considered as bad news as DMBRC and other cost-sensitive classifiers essentially scarify average scores to gain a better balance of the class-conditional risk, i.e., their primary focus.

However, when looking at the ψ scores (9), we clearly see that DMBRC and other cost-sensitive classifiers do their balancing jobs well, as they consistently outperform other competitors on this criterion. Interestingly, DMBRC provides the best scores on all the tested data sets, showing that its theoretical properties are paired with very good empirical results. This would be a strong motivation to further boost DMBRC in both its primary focus of balancing the false positive and false negative rates associated with each label.

Table 2. Evaluation of the subset zero-one metric ($\mathcal{L}_{0/1}$), F-1 score (F1), Hamming metric ($\mathcal{L}_{\text{Ham}}$), and ψ metric. The arrow next to the metric represents ↑ the higher the better and ↓ the lower the better. The results are presented as [mean ± std]. The best and the worst performances are given in **bold** and red, respectively.

Metric	Classifier	Yeast	Scene	CHD_49
$\mathcal{L}_{0/1}$ ↓	MLKNN	**0.81 ± 0.02**	**0.37 ± 0.02**	0.88 ± 0.05
	LR	**0.85 ± 0.01**	0.46 ± 0.02	**0.83 ± 0.05**
	DT	0.93 ± 0.01	0.60 ± 0.03	0.88 ± 0.04
	DMBRC	0.91 ± 0.02	0.68 ± 0.03	0.95 ± 0.02
	WLR	0.93 ± 0.02	0.57 ± 0.02	0.91 ± 0.04
	WDT	0.97 ± 0.01	0.76 ± 0.02	0.96 ± 0.02
	ROS+LR	0.86 ± 0.01	0.47 ± 0.02	0.86 ± 0.04
	ROS+DT	0.94 ± 0.02	0.60 ± 0.01	0.87 ± 0.04
	RUS+LR	0.85 ± 0.01	0.47 ± 0.02	0.84 ± 0.04
	RUS+DT	0.93 ± 0.01	0.64 ± 0.03	0.87 ± 0.03
	SM+LR	0.91 ± 0.01	0.54 ± 0.02	0.93 ± 0.02
	SM+DT	0.96 ± 0.01	0.63 ± 0.01	0.88 ± 0.02
F1 ↑	MLKNN	0.61 ± 0.02	**0.70 ± 0.02**	0.59 ± 0.04
	LR	0.61 ± 0.01	0.62 ± 0.01	0.64 ± 0.02
	DT	0.56 ± 0.02	0.47 ± 0.04	0.59 ± 0.04
	DMBRC	0.52 ± 0.01	0.64 ± 0.01	0.48 ± 0.04
	WLR	0.52 ± 0.02	**0.70 ± 0.02**	0.54 ± 0.01
	WDT	0.46 ± 0.01	0.59 ± 0.01	0.47 ± 0.04
	ROS+LR	0.60 ± 0.01	0.62 ± 0.01	0.61 ± 0.03
	ROS+DT	0.54 ± 0.03	0.48 ± 0.02	0.59 ± 0.04
	RUS+LR	0.60 ± 0.01	0.61 ± 0.02	0.64 ± 0.01
	RUS+DT	0.56 ± 0.01	0.44 ± 0.03	0.60 ± 0.03
	SM+LR	**0.64 ± 0.01**	0.67 ± 0.02	**0.67 ± 0.02**
	SM+DT	0.57 ± 0.01	0.56 ± 0.02	0.65 ± 0.03
$\mathcal{L}_{\text{Ham}}$ ↓	MLKNN	**0.20 ± 0.01**	**0.09 ± 0.01**	0.32 ± 0.02
	LR	**0.20 ± 0.01**	0.10 ± 0.00	**0.29 ± 0.02**
	DT	0.23 ± 0.01	0.13 ± 0.00	0.32 ± 0.02
	DMBRC	0.33 ± 0.01	0.18 ± 0.01	0.44 ± 0.02
	WLR	0.33 ± 0.01	0.13 ± 0.00	0.37 ± 0.01
	WDT	0.38 ± 0.01	0.20 ± 0.01	0.46 ± 0.02
	ROS+LR	**0.20 ± 0.00**	0.10 ± 0.00	0.31 ± 0.02
	ROS+DT	0.24 ± 0.01	0.13 ± 0.00	0.30 ± 0.02
	RUS+LR	**0.20 ± 0.01**	0.10 ± 0.00	**0.29 ± 0.01**
	RUS+DT	0.23 ± 0.00	0.14 ± 0.01	0.31 ± 0.02
	SM+LR	0.25 ± 0.01	0.12 ± 0.01	0.35 ± 0.01
	SM+DT	0.30 ± 0.01	0.16 ± 0.00	0.31 ± 0.02
ψ ↓	MLKNN	1.00 ± 0.00	0.49 ± 0.03	1.00 ± 0.00
	LR	1.00 ± 0.00	0.50 ± 0.05	1.00 ± 0.00
	DT	1.00 ± 0.00	0.62 ± 0.07	1.00 ± 0.00
	DMBRC	**0.66 ± 0.20**	**0.09 ± 0.02**	**0.45 ± 0.23**
	WLR	0.68 ± 0.19	0.11 ± 0.03	0.65 ± 0.17
	WDT	0.67 ± 0.11	0.13 ± 0.04	0.48 ± 0.15
	ROS+LR	1.00 ± 0.00	0.49 ± 0.04	0.92 ± 0.02
	ROS+DT	1.00 ± 0.00	0.62 ± 0.05	1.00 ± 0.00
	RUS+LR	1.00 ± 0.00	0.59 ± 0.06	0.97 ± 0.04
	RUS+DT	1.00 ± 0.00	0.68 ± 0.05	1.00 ± 0.00
	SM+LR	1.00 ± 0.00	0.25 ± 0.04	1.00 ± 0.00
	SM+DT	1.00 ± 0.00	0.35 ± 0.02	1.00 ± 0.00

Table 3. Evaluation of the subset zero-one metric ($\mathcal{L}_{0/1}$), F-1 score (F1), Hamming metric ($\mathcal{L}_{\text{Ham}}$), and ψ metric. The arrow next to the metric represents ↑ the higher the better and ↓ the lower the better. The results are presented as [mean ± std]. The best and the worst performances are given in **bold** and red, respectively.

Metric	Classifier	Water qua	Human Pse	Tmc2007
$\mathcal{L}_{0/1}$ ↓	MLKNN	**0.84 ± 0.01**	**0.98 ± 0.00**	0.74 ± 0.01
	LR	**0.84 ± 0.01**	**0.98 ± 0.01**	**0.69 ± 0.00**
	DT	0.89 ± 0.02	0.99 ± 0.01	0.83 ± 0.01
	DMBRC	0.97 ± 0.01	0.99 ± 0.01	0.96 ± 0.01
	WLR	0.92 ± 0.00	0.99 ± 0.01	0.83 ± 0.01
	WDT	0.99 ± 0.00	0.99 ± 0.00	0.98 ± 0.00
	ROS+LR	0.85 ± 0.01	0.99 ± 0.01	0.70 ± 0.01
	ROS+DT	0.91 ± 0.01	0.99 ± 0.01	0.83 ± 0.01
	RUS+LR	0.84 ± 0.02	0.99 ± 0.01	0.69 ± 0.00
	RUS+DT	0.92 ± 0.01	0.99 ± 0.01	0.83 ± 0.00
	SM+LR	0.90 ± 0.01	1.00 ± 0.00	0.76 ± 0.01
	SM+DT	0.90 ± 0.01	1.00 ± 0.00	0.84 ± 0.00
F1 ↑	MLKNN	0.20 ± 0.01	0.54 ± 0.01	0.62 ± 0.00
	LR	0.22 ± 0.01	0.47 ± 0.01	**0.68 ± 0.00**
	DT	0.19 ± 0.01	0.49 ± 0.01	0.52 ± 0.01
	DMBRC	0.30 ± 0.01	0.53 ± 0.02	0.45 ± 0.00
	WLR	**0.35 ± 0.01**	0.53 ± 0.01	0.63 ± 0.00
	WDT	0.24 ± 0.01	0.52 ± 0.01	0.41 ± 0.01
	ROS+LR	0.21 ± 0.01	0.48 ± 0.01	0.68 ± 0.00
	ROS+DT	0.11 ± 0.01	0.49 ± 0.01	0.53 ± 0.01
	RUS+LR	0.22 ± 0.02	0.49 ± 0.01	0.68 ± 0.00
	RUS+DT	0.11 ± 0.02	0.49 ± 0.01	0.53 ± 0.01
	SM+LR	0.13 ± 0.01	**0.57 ± 0.01**	0.69 ± 0.00
	SM+DT	0.13 ± 0.01	0.56 ± 0.01	0.60 ± 0.01
$\mathcal{L}_{\text{Ham}}$ ↓	MLKNN	0.09 ± 0.00	**0.29 ± 0.01**	0.07 ± 0.00
	LR	0.09 ± 0.00	**0.29 ± 0.01**	0.06 ± 0.00
	DT	0.09 ± 0.00	0.30 ± 0.00	0.08 ± 0.00
	DMBRC	0.23 ± 0.01	0.36 ± 0.01	0.20 ± 0.00
	WLR	0.17 ± 0.00	0.35 ± 0.01	0.10 ± 0.00
	WDT	0.27 ± 0.01	0.36 ± 0.01	0.22 ± 0.00
	ROS+LR	0.09 ± 0.00	**0.29 ± 0.01**	0.06 ± 0.00
	ROS+DT	0.09 ± 0.00	0.31 ± 0.01	0.08 ± 0.00
	RUS+LR	0.09 ± 0.00	**0.29 ± 0.01**	0.06 ± 0.00
	RUS+DT	0.09 ± 0.00	0.31 ± 0.00	0.08 ± 0.00
	SM+LR	**0.08 ± 0.00**	0.41 ± 0.01	0.07 ± 0.00
	SM+DT	0.09 ± 0.00	0.39 ± 0.01	0.08 ± 0.00
ψ ↓	MLKNN	1.00 ± 0.00	0.84 ± 0.03	0.90 ± 0.03
	LR	1.00 ± 0.00	0.97 ± 0.03	0.88 ± 0.04
	DT	1.00 ± 0.00	0.94 ± 0.08	1.00 ± 0.00
	DMBRC	**0.87 ± 0.04**	**0.19 ± 0.04**	**0.09 ± 0.03**
	WLR	0.97 ± 0.01	0.25 ± 0.05	0.22 ± 0.06
	WDT	0.90 ± 0.03	0.29 ± 0.03	0.16 ± 0.02
	ROS+LR	1.00 ± 0.00	0.96 ± 0.03	0.81 ± 0.06
	ROS+DT	1.00 ± 0.00	0.85 ± 0.02	1.00 ± 0.00
	RUS+LR	1.00 ± 0.00	0.96 ± 0.03	0.87 ± 0.04
	RUS+DT	1.00 ± 0.00	0.88 ± 0.07	1.00 ± 0.00
	SM+LR	1.00 ± 0.00	0.92 ± 0.03	0.92 ± 0.02
	SM+DT	1.00 ± 0.00	0.70 ± 0.09	1.00 ± 0.00

5 Conclusion

This paper complements the literature on handling imbalanced MLC data with a new approach based on the theoretical minimax strategy [1,14,16,32]. Our new classifier DMBRC attempts to robustify the binary relevance classification by leveraging theoretically sound properties provided by the DMC [16].

We moreover provide empirical evidence to illustrate how classifiers, which take into account class imbalance, may be advantageous when being assessed by the ability to balance the class conditional risk. The empirical evidence also suggests that DMBRC is the most promising classifier in this aspect.

Motivated by these promising pieces of evidence, we envision future works on further strengthening DMBRC in both label-wise error rates and conventional MLC evaluation metrics. We also plan to leverage the fact that DMC can be coupled with deep neural networks to handle images to broaden the application domain of DMBRC to high-stakes applications such as predicting multiple diseases given medical images [30]. In addition, since DMC theoretical guarantees are not limited to binary classification, we also think of extending the current approaches to the graded multi-label [10] setting or to multi-dimensional classification [28], for which approaches such as WLR or WDT.

References

1. Berger, J.O.: Statistical Decision Theory and Bayesian Analysis. Springer Series in Statistics, 2nd edn. Springer, New York (1985). https://doi.org/10.1007/978-1-4757-4286-2
2. Bogatinovski, J., Todorovski, L., Džeroski, S., Kocev, D.: Comprehensive comparative study of multi-label classification methods. Expert Syst. Appl. **203**, 117215 (2022)
3. Cao, P., Liu, X., Zhao, D., Zaiane, O.: Cost sensitive ranking support vector machine for multi-label data learning. In: Abraham, A., Haqiq, A., Alimi, A.M., Mezzour, G., Rokbani, N., Muda, A.K. (eds.) HIS 2016. AISC, vol. 552, pp. 244–255. Springer, Cham (2017). https://doi.org/10.1007/978-3-319-52941-7_25
4. Charte, F., Rivera, A., del Jesus, M.J., Herrera, F.: A first approach to deal with imbalance in multi-label datasets. In: Pan, J.-S., Polycarpou, M.M., Woźniak, M., de Carvalho, A.C.P.L.F., Quintián, H., Corchado, E. (eds.) HAIS 2013. LNCS (LNAI), vol. 8073, pp. 150–160. Springer, Heidelberg (2013). https://doi.org/10.1007/978-3-642-40846-5_16
5. Charte, F., Rivera, A.J., Charte, D., del Jesus, M.J., Herrera, F.: Tips, guidelines and tools for managing multi-label datasets: the mldr.datasets R package and the Cometa data repository. Neurocomputing **289**, 68–85 (2018)
6. Charte, F., Rivera, A.J., del Jesus, M.J., Herrera, F.: MLeNN: a first approach to heuristic multilabel undersampling. In: Corchado, E., Lozano, J.A., Quintián, H., Yin, H. (eds.) IDEAL 2014. LNCS, vol. 8669, pp. 1–9. Springer, Cham (2014). https://doi.org/10.1007/978-3-319-10840-7_1
7. Charte, F., Rivera, A.J., del Jesus, M.J., Herrera, F.: Addressing imbalance in multilabel classification: measures and random resampling algorithms. Neurocomputing **163**, 3–16 (2015)

8. Charte, F., Rivera, A.J., del Jesus, M.J., Herrera, F.: MLSMOTE: approaching imbalanced multilabel learning through synthetic instance generation. Knowl.-Based Syst. **89**, 385–397 (2015)
9. Chen, K., Lu, B.L., Kwok, J.T.: Efficient classification of multi-label and imbalanced data using min-max modular classifiers. In: The 2006 IEEE International Joint Conference on Neural Network Proceedings, pp. 1770–1775. IEEE (2006)
10. Cheng, W., Hüllermeier, E., Dembczynski, K.J.: Graded multilabel classification: the ordinal case. In: Proceedings of the 27th International Conference on Machine Learning (ICML-2010), pp. 223–230 (2010)
11. Daniels, Z., Metaxas, D.: Addressing imbalance in multi-label classification using structured hellinger forests. In: Proceedings of the AAAI Conference on Artificial Intelligence, vol. 31 (2017)
12. Dembczynski, K., Cheng, W., Hüllermeier, E.: Bayes optimal multilabel classification via probabilistic classifier chains. In: Proceedings of the 27th International Conference on Machine Learning (ICML), pp. 279–286 (2010)
13. Do, V.H., et al.: PanKA: leveraging population pangenome to predict antibiotic resistance. iScience **27**(9), 110623 (2024)
14. Gilet, C.: Discrete minimax classifier for personalized diagnosis in medicine. Ph.D. thesis, Université Côte d'Azur (2021). https://tel.archives-ouvertes.fr/tel-03553934
15. Gilet, C., Barbosa, S., Fillatre, L.: Minimax classifier with box constraint on the priors. In: Machine Learning for Health (ML4H) at NeurIPS 2019. Proceedings of Machine Learning Research (2019)
16. Gilet, C., Barbosa, S., Fillatre, L.: Discrete box-constrained minimax classifier for uncertain and imbalanced class proportions. IEEE Trans. Pattern Anal. Mach. Intell. **44**(6), 2923–2937 (2020)
17. Gilet, C., Fillatre, L.: Anomaly detection with discrete minimax classifier for imbalanced dataset or uncertain class proportions. In: Gelman, L., Martin, N., Malcolm, A.A., (Edmund) Liew, C.K. (eds.) Advances in Condition Monitoring and Structural Health Monitoring. LNME, pp. 179–187. Springer, Singapore (2021). https://doi.org/10.1007/978-981-15-9199-0_17
18. Gilet, C., Guyomard, M., Barbosa, S., Fillatre, L.: Multiclass minimax learning for deep neural networks. In: Proceedings of the 31st European Signal Processing Conference (EUSIPCO) (2023)
19. Gilet, C., Guyomard, M., Destercke, S., Fillatre, L.: Softmin discrete minimax classifier for imbalanced classes and prior probability shifts. Mach. Learn. **113**(2), 605–645 (2023). https://doi.org/10.1007/s10994-023-06397-8
20. Giraldo-Forero, A.F., Jaramillo-Garzón, J.A., Ruiz-Muñoz, J.F., Castellanos-Domínguez, C.G.: Managing imbalanced data sets in multi-label problems: a case study with the SMOTE algorithm. In: Ruiz-Shulcloper, J., Sanniti di Baja, G. (eds.) CIARP 2013. LNCS, vol. 8258, pp. 334–342. Springer, Heidelberg (2013). https://doi.org/10.1007/978-3-642-41822-8_42
21. Guerrero-Curieses, A., Alaíz-Rodríguez, R., Cid-Sueiro, J.: A fixed-point algorithm to minimax learning with neural networks. IEEE Trans. Syst. Man Cybern. Part C (Appl. Rev.) **34**, 383–392 (2004)
22. Han, M., Wu, H., Chen, Z., Li, M., Zhang, X.: A survey of multi-label classification based on supervised and semi-supervised learning. Int. J. Mach. Learn. Cybern. **14**(3), 697–724 (2023)
23. Liu, B., Tsoumakas, G.: Making classifier chains resilient to class imbalance. In: Asian Conference on Machine Learning, pp. 280–295. PMLR (2018)

24. Liu, M.C., et al.: Sensitive and specific multi-cancer detection and localization using methylation signatures in cell-free DNA. Ann. Oncol. **31**(6), 745–759 (2020)
25. Moradigaravand, D., Palm, M., Farewell, A., Mustonen, V., Warringer, J., Parts, L.: Prediction of antibiotic resistance in Escherichia coli from large-scale pan-genome data. PLoS Comput. Biol. **14**(12), e1006258 (2018)
26. Nguyen, V.L., Hüllermeier, E.: Multilabel classification with partial abstention: bayes-optimal prediction under label independence. J. Artif. Intell. Res. **72**, 613–665 (2021)
27. Nguyen, V.-L., Hüllermeier, E., Rapp, M., Loza Mencía, E., Fürnkranz, J.: On aggregation in ensembles of multilabel classifiers. In: Appice, A., Tsoumakas, G., Manolopoulos, Y., Matwin, S. (eds.) DS 2020. LNCS (LNAI), vol. 12323, pp. 533–547. Springer, Cham (2020). https://doi.org/10.1007/978-3-030-61527-7_35
28. Nguyen, V.L., Yang, Y., De Campos, C.: Probabilistic multi-dimensional classification. In: Uncertainty in Artificial Intelligence, pp. 1522–1533. PMLR (2023)
29. Pakrashi, A., Mac Namee, B.: Stacked-MLkNN: a stacking based improvement to multi-label k-nearest neighbours. In: First International Workshop on Learning with Imbalanced Domains: Theory and Applications, pp. 51–63. PMLR (2017)
30. Paul, H.Y., Kim, T.K., Siegel, E., Yahyavi-Firouz-Abadi, N.: Demographic reporting in publicly available chest radiograph data sets: opportunities for mitigating sex and racial disparities in deep learning models. J. Am. Coll. Radiol. **19**(1), 192–200 (2022)
31. Pedregosa, F., et al.: Scikit-learn: machine learning in Python. J. Mach. Learn. Res. **12**, 2825–2830 (2011)
32. Poor, H.V.: An Introduction to Signal Detection and Estimation, 2nd edn. Springer, New York (1994). https://doi.org/10.1007/978-1-4757-2341-0
33. Read, J., Pfahringer, B., Holmes, G., Frank, E.: Classifier chains for multi-label classification. Mach. Learn. **85**, 333–359 (2011)
34. Read, J., Pfahringer, B., Holmes, G., Frank, E.: Classifier chains: a review and perspectives. J. Artif. Intell. Res. **70**, 683–718 (2021)
35. Sun, K.W., Lee, C.H.: Addressing class-imbalance in multi-label learning via two-stage multi-label hypernetwork. Neurocomputing **266**, 375–389 (2017)
36. Szymański, P., Kajdanowicz, T.: A scikit-based Python environment for performing multi-label classification. arXiv preprint arXiv:1702.01460 (2017)
37. Tahir, M.A., Kittler, J., Bouridane, A.: Multilabel classification using heterogeneous ensemble of multi-label classifiers. Pattern Recogn. Lett. **33**(5), 513–523 (2012)
38. Tarekegn, A.N., Giacobini, M., Michalak, K.: A review of methods for imbalanced multi-label classification. Pattern Recogn. **118**, 107965 (2021)
39. Tsoumakas, G., Spyromitros-Xioufis, E., Vilcek, J., Vlahavas, I.: MULAN: a Java library for multi-label learning. J. Mach. Learn. Res. **12**, 2411–2414 (2011)
40. Wu, G., Tian, Y., Liu, D.: Cost-sensitive multi-label learning with positive and negative label pairwise correlations. Neural Netw. **108**, 411–423 (2018)
41. Wu, G., Zhu, J.: Multi-label classification: do hamming loss and subset accuracy really conflict with each other? In: Proceedings of the 34th International Conference on Neural Information Processing Systems (NeurIPS), pp. 3130–3140 (2020)
42. Yuan, Y., Liu, X., Ding, S., Pan, B.: Fault detection and location system for diagnosis of multiple faults in aeroengines. IEEE Access **5**, 17671–17677 (2017)
43. Zhang, K., Hu, X., Liu, Y., Lin, X., Liu, W.: Multi-fault detection and isolation for lithium-ion battery systems. IEEE Trans. Power Electron. **37**(1), 971–989 (2021)
44. Zhang, M.L., Li, Y.K., Liu, X.Y., Geng, X.: Binary relevance for multi-label learning: an overview. Front. Comput. Sci. **12**, 191–202 (2018)

45. Zhang, M.L., Li, Y.K., Yang, H., Liu, X.Y.: Towards class-imbalance aware multi-label learning. IEEE Trans. Cybern. **52**(6), 4459–4471 (2020)
46. Zhang, M.L., Zhou, Z.H.: ML-KNN: a lazy learning approach to multi-label learning. Pattern Recogn. **40**(7), 2038–2048 (2007)

Entropic Regularization Schemes for Learning Fuzzy Similarity Measures Based on the d-Choquet Integral

Christophe Marsala[1], Davide Petturiti[2(✉)], and Barbara Vantaggi[3]

[1] Sorbonne Université, CNRS, LIP6, Paris, France
christophe.marsala@lip6.fr
[2] Dip. Economia, Università degli Studi di Perugia, Perugia, Italy
davide.petturiti@unipg.it
[3] Dip. MEMOTEF, Sapienza Università di Roma, Rome, Italy
barbara.vantaggi@uniroma1.it

Abstract. We consider the problem of learning one of three possible fuzzy generalizations of the Jaccard similarity measure, based on the d-Choquet integral. Each of the resulting fuzzy similarity measures is parameterized by a capacity and by a real parameter. The capacity describes the weights assigned to groups of attributes and their interactions, while the real parameter is related to the restricted dissimilarity function used to evaluate differences among attributes. To face identifiability issues and in view of an XAI use of the learned capacity, the parameters' set is restricted to the set of (at most) 2-additive completely monotone capacities. Next, under a suitable definition of entropy for completely monotone capacities, we address different entropic regularization schemes to single out interactions between groups of attributes. This is done by taking as reference a local uniform Möbius inverse over sets of attributes with the same cardinality.

Keywords: Fuzzy similarity measure · d-Choquet integral · Similarity learning · Entropic regularization

1 Introduction

The ubiquitous use of similarity and dissimilarity measures in machine learning and decision support systems makes it necessary to develop models and methods that guarantee interpretability. This is particularly important in the context of eXplainable AI (XAI), when high stake decisions are involved [30]. At the same time, during the last years, many similarity measures have been proposed (see, e.g., [26]), mainly focusing on the particular nature of data and on the properties required by a similarity measure [4,13].

This abundance of choices opens the problem of what is the semantic behind a particular similarity measure. Indeed, a deeper understanding of the concept of similarity is required, especially if XAI is in view. For this reason, a series

S. Destercke et al. (Eds.): SUM 2024, LNAI 15350, pp. 297–311, 2025.
https://doi.org/10.1007/978-3-031-76235-2_22

of papers (see, e.g., [3,8–11]) coped with figuring out the comparative nature of similarity (and dissimilarity) measures on fuzzy description profiles. In parallel, another stream of research dealt with developing similarity (and dissimilarity) measures encoding a non-purely metric semantic (see, e.g., [2,12,31]). This paper contributes to this second line of research, referring to the similarity measures introduced in [27] which are based on the notion of *d-Choquet integral* [6] and are parameterized by a capacity ν and a restricted dissimilarity function δ.

The quoted similarity measures $\mathbf{S}_i^{\nu,\delta}(X,Y)$, for $i = 1,2,3$, generalize the classical Jaccard similarity measure [19] and embody three levels of semantics:

(i) The choice of the functional form $\mathbf{S}_i^{\nu,\delta}$ implies how common and different parts of fuzzy description profiles are aggregated: in the particular case $\mathbf{S}_1^{\nu,\delta}$, we get a symmetric fuzzy version of Tversky's contrast model [33].
(ii) The choice of the capacity ν expresses how interactions among attributes are weighted: the corresponding Möbius inverse μ_ν acts as a significance assessment that allows for positive or negative interactions.
(iii) The choice of the restricted dissimilarity function δ encodes how pointwise dissimilarities are evaluated: choosing one of the parametric forms $\delta_{p,p}, \delta_{1,p}, \delta_{p,1}$, a tuning on sample similarity comparisons can be performed.

The main difficulty connected to the above similarity measures is the elicitation of ν (or, equivalently, μ_ν), that can be faced with a *similarity learning* task, relying on a dataset of labeled data. Nevertheless, for a fixed functional form $\mathbf{S}_i^{\nu,\delta}$, the large number of parameters naturally raises the problem of identifiability, that has been recently addressed in learning Choquet functionals [5,22]. Here we show that focusing only on classification capabilities provided by ν as done in [27] can be too weak an approach to address identifiability: further constraints must be imposed, in order to achieve a meaningful ν, besides its classification capabilities.

In this paper, restricting to the family of (at most) 2-additive completely monotone capacities [17] and relying on the definition of entropy given in [29], we propose three different entropic regularization schemes to use in the learning of ν. Every such scheme is a multicriteria optimization problem, blending accuracy and information-theoretic properties, by referring to local uniform Möbius inverses over sets of attributes with the same cardinality. The goal is to restrict the set of optimal Möbius inverses in a way to address non-identifiability. Next, we perform a preliminary experimental analysis on two fuzzified versions of real datasets, showing the sensitivity of the learned ν to the entropic schemes' hyperparameters.

The paper is structured as follows. Section 2 recalls the necessary preliminaries together with the three similarity measures introduced in [27]. Section 3 formulates the similarity learning problem and proposes three entropic regularization schemes for enforcing information-theoretic properties to the capacity parameterizing each of the similarity measures. Section 4 carries out a preliminary experimental analysis on two fuzzyfied datasets. Finally, Sect. 5 draws our conclusions and future perspectives.

2 Fuzzy d-Choquet Similarity Measures

We present the fuzzy similarity measures introduced in [27] and recall their main properties, together with the necessary preliminary material.

In what follows we consider a finite set of objects of interest and assume that every object is described by a set of attributes indexed by the finite set $N = \{1, \ldots, n\}$. Every attribute $i \in N$ can be present in an object X with a different degree of membership $X(i) \in [0, 1]$, thus any object description is identified with a function $X : N \to [0, 1]$, that we call *fuzzy description profile.* Notice that every fuzzy description profile X can be considered as a fuzzy subset of N [34], that we identify with its *membership function.*

Denote by $\mathcal{F} = [0, 1]^N$ the set of all possible fuzzy description profiles and by $\mathcal{C} = \{0, 1\}^N$ the subset of crisp description profiles.

We consider a t-norm T together with its dual t-conorm S and the complement $(\cdot)^c = 1 - (\cdot)$ to perform fuzzy set-theoretic operations. As usual (see [24]), we denote the main t-norms and t-conorms, for every $x, y \in [0, 1]$, as

$$\begin{aligned} T_M(x,y) &= \min\{x,y\}, & S_M(x,y) &= \max\{x,y\},\\ T_P(x,y) &= x \cdot y, & S_P(x,y) &= x + y - x \cdot y,\\ T_L(x,y) &= \max\{x+y-1, 0\}, & S_L(x,y) &= \min\{x+y, 1\}. \end{aligned}$$

For every $X, Y \in \mathcal{F}$, we define $X \cap Y = T(X, Y)$, $X \backslash Y = T(X, Y^c)$, $Y \backslash X = T(Y, X^c)$, $X \Delta Y = S(X \backslash Y, Y \backslash X)$ and $X \cup Y = S(X, Y)$, where all operations are intended pointwise on the elements of N. All t-norms and t-conorms extend uniquely to k-ary operations, for $k \geq 2$, due to associativity [24], and so do the corresponding fuzzy set-theoretic operations.

2.1 Capacities and the d-Choquet Integral

Let N be endowed with the power set 2^N. As is well-known (see, e.g., [18]), a *(normalized) capacity* is a set function $\nu : 2^N \to [0, 1]$ satisfying:

(i) $\nu(\emptyset) = 0$ and $\nu(N) = 1$;
(ii) $\nu(A) \leq \nu(B)$ when $A \subseteq B$, for all $A, B \in 2^N$.

Moreover, every capacity ν is associated with a set function $\mu_\nu : 2^N \to \mathbb{R}$ called *Möbius inverse* such that, for all $A \in 2^N$, it holds that

$$\mu_\nu(A) = \sum_{B \subseteq A} (-1)^{|A \backslash B|} \nu(B) \quad \text{and} \quad \nu(A) = \sum_{B \subseteq A} \mu_\nu(B).$$

As shown in [7], a function $\mu_\nu : 2^N \to \mathbb{R}$ is the Möbius inverse of a capacity ν, if and only if it satisfies:

(i) $\mu_\nu(\emptyset) = 0$;
(ii) $\sum_{B \in 2^N} \mu_\nu(B) = 1$;
(iii) $\sum_{\{i\} \subseteq B \subseteq A} \mu_\nu(B) \geq 0$, for all $A \in 2^N$ and all $i \in A$.

The above properties imply that $\mu_\nu(\{i\}) \geq 0$, for all $i \in N$. Moreover, if $\mu_\nu(B) \geq 0$, for all $B \in 2^N$, then the corresponding ν is a *completely monotone capacity* [18].

A capacity ν is then called *k-additive* (with $1 \leq k \leq n$) if $\mu_\nu(A) = 0$, for all $A \in 2^N$ with $|A| > k$, and there exists $A \in 2^N$ with $|A| = k$ such that $\mu_\nu(A) \neq 0$ [17]. In particular, a 1-additive capacity reduces to a *probability measure*. In what follows, we denote by $\mathbf{A}_2(N)$ the set of (at most) 2-additive completely monotone capacities and refer to the following particular elements of this set:

- ν_{1u} is the *uniform probability measure* such that $\nu_{1u}(\{i\}) = \frac{1}{n}$, for all $i \in N$, whose Möbius inverse is $\mu_{\nu_{1u}}(\{i\}) = \frac{1}{n}$, for all $i \in N$, and 0 otherwise;
- ν_{2u} is the *2-additive uniform capacity* such that $\nu_{2u}(\{i,j\}) = \frac{1}{\binom{n}{2}}$, for all $\{i,j\} \in 2^N$, whose Möbius inverse is $\mu_{\nu_{2u}}(\{i,j\}) = \frac{1}{\binom{n}{2}}$, for all $\{i,j\} \in 2^N$, and 0 otherwise.
- ξ_B for $B \in 2^N \setminus \{\emptyset\}$ with $|B| \leq 2$ is the *vacuous capacity at B* such that $\xi_B(A) = 1$ if $B \subseteq A$ and 0 otherwise, whose Möbius inverse is $\mu_{\xi_B}(B) = 1$ and 0 otherwise.

Given a capacity ν, several notions of expectation are available in the literature (see, e.g., [18]). Here, we refer to a generalization of the classical Choquet integral introduced in [6]. Such a generalization relies on the choice of a function $\delta : [0,1]^2 \to [0,1]$ called a *restricted dissimilarity function* asked to satisfy, for all $x, y, z \in [0,1]$, the following conditions:

1. $\delta(x,y) = \delta(y,x)$;
2. $\delta(x,y) = 1$ if and only if $\{x,y\} = \{0,1\}$;
3. $\delta(x,y) = 0$ if and only if $x = y$;
4. if $x \leq y \leq z$, then $\delta(x,y) \leq \delta(x,z)$ and $\delta(y,z) \leq \delta(x,z)$.

In this paper we will restrict to the following parametric version with $p, q \in (0, +\infty)$:

$$\delta_{p,q}(x,y) = |x^p - y^p|^{\frac{1}{q}}. \tag{1}$$

Due to space limitations, we will analyze only the cases $\delta_{p,p}$, $\delta_{1,p}$, $\delta_{p,1}$, parameterized by $p \in (0, +\infty)$. These three particular δ's allow to investigate the internal-external, external-only, and internal-only deformation of the absolute difference, relying on a single parameter that acts in a reciprocal way.

Given a capacity ν and a restricted dissimilarity measure δ, the *d-Choquet integral* is the functional $\mathbb{C}_{\nu,\delta} : [0,1]^N \to [0,n]$ defined, for all $X \in [0,1]^N$, as

$$\mathbb{C}_{\nu,\delta}(X) = \sum_{i=1}^{n} \delta\left(X(\sigma(i)), X(\sigma(i-1))\right) \nu(\{\sigma(i), \ldots, \sigma(n)\}),$$

where σ is a permutation of N such that $X(\sigma(1)) \leq \cdots \leq X(\sigma(n))$ and $X(\sigma(0)) := 0$. In particular, if $X \in \{0,1\}^N$, then X reduces to the indicator $\mathbf{1}_A$ of a subset A of N, and so $\mathbb{C}_{\nu,\delta}(\mathbf{1}_A) = \nu(A)$.

2.2 Fuzzy Similarities

Different definitions of similarities have been given for fuzzy subsets [3,14,15] essentially based on the "common" and the "different" parts of the compared fuzzy subsets.

Referring to a fixed pair of dual t-norm T and t-conorm S for fuzzy set-theoretic operations, in [27] we introduced three classes of similarity measures $\mathbf{S}_i^{\nu,\delta} : \mathcal{F}^2 \to [0, +\infty)$, for $i = 1, 2, 3$, each parameterized by a capacity ν and by a restricted dissimilarity function δ, defined, for every $X, Y \in \mathcal{F}$, as:

$$\mathbf{S}_1^{\nu,\delta}(X,Y) = \frac{\mathbb{C}_{\nu,\delta}(X \cap Y)}{\mathbb{C}_{\nu,\delta}(X \backslash Y) + \mathbb{C}_{\nu,\delta}(Y \backslash X) + \mathbb{C}_{\nu,\delta}(X \cap Y)}, \tag{2}$$

$$\mathbf{S}_2^{\nu,\delta}(X,Y) = \frac{\mathbb{C}_{\nu,\delta}(X \cap Y)}{\mathbb{C}_{\nu,\delta}(X \Delta Y) + \mathbb{C}_{\nu,\delta}(X \cap Y)}, \tag{3}$$

$$\mathbf{S}_3^{\nu,\delta}(X,Y) = \frac{\mathbb{C}_{\nu,\delta}(X \cap Y)}{\mathbb{C}_{\nu,\delta}(X \cup Y)}. \tag{4}$$

In the context of similarity measures, ν can be seen as a non-additive weighting function related to a set of fuzzy attributes indexed by N. Under this interpretation, the Möbius inverse μ_ν is the actual weight attached to every set of attributes, allowing for modeling (positive or negative) interactions among fuzzy attributes. With this meaning in view, in [2,12] μ_ν has been called a *significance assessment*.

We assume throughout the paper that ν satisfies the following property

(P) $\nu(A) > 0$, for all $A \in 2^N \backslash \{\emptyset\}$,

which is equivalent to the fact that the Möbius inverse of ν is such that $\mu_\nu(\{i\}) > 0$, for all $i \in N$. This ensures that all attributes included in a description profile should be "significant".

Remark 1. In [27] we proved that, assuming a capacity ν which satisfies **(P)** and $T = T_M$, the ratios in (2)–(4) are always well-defined, except for the case $X = Y = \mathbf{1}_\emptyset$. In this limit case, we set $\mathbf{S}_i^{\nu,\delta}(\mathbf{1}_\emptyset, \mathbf{1}_\emptyset) := 1$, for $i = 1, 2, 3$. Under the quoted assumptions, for $\delta \in \{\delta_{p,p}, \delta_{1,p}, \delta_{p,1}\}$, some bounding conditions on the values of $\mathbf{S}_i^{\nu,\delta}(X,Y)$ have been given in [27]. In particular, there we showed that $\mathbf{S}_3^{\nu,\delta}(X,Y)$ can take values greater than 1 for $\delta = \delta_{p,p}$ or $\delta = \delta_{1,p}$.

Taking $\nu = \nu_{1u}$ and $\delta = \delta_{1,1}$, the restrictions of $\mathbf{S}_i^{\nu,\delta}(X,Y)$ to $\mathcal{C}$, for $i = 1, 2, 3$, reduce to the classical *Jaccard's index* [19]. More generally, for a probability measure ν and $\delta = \delta_{1,1}$ we get a weighted version of the Jaccard's index [2].

We notice that, the restrictions of $\mathbf{S}_i^{\nu,\delta}$ to $\mathcal{C}$, for $i = 1, 2, 3$, coincide with the similarity measures defined in [2], for any choice of δ and a completely monotone ν. On the other hand, if we take $\delta = \delta_{1,1}$, then $\mathbf{S}_i^{\nu,\delta}$, for $i = 1, 2, 3$, coincide with the similarity measures defined in [12]. This implies that, in general, $\mathbf{S}_i^{\nu,\delta}(X,X) < 1$, for $i = 1, 2$. In the particular case $\delta = \delta_{1,1}$ and ν is a probability measure, $\mathbf{S}_3^{\nu,\delta}$ is a special case of the similarity measure introduced in [31].

As a by-product, taking $\delta = \delta_{1,1}$, by [12] we derive that $\mathbf{S}_i^{\nu,\delta}$, for $i = 1, 2, 3$, do not generally satisfy T'-*transitivity*, where T' is a t-norm possibly different from the t-norm T used in the fuzzy set-theoretic operations, i.e., the property:

(T) $\mathbf{S}_i^{\nu,\delta}(X, Z) \geq T'(\mathbf{S}_i^{\nu,\delta}(X, Y), \mathbf{S}_i^{\nu,\delta}(Y, Z))$, for all $X, Y, Z \in \mathcal{F}$.

We notice that for $\delta = \delta_{p,p}$ or $\delta = \delta_{1,p}$ property **(T)** does not make sense for $\mathbf{S}_3^{\nu,\delta}$, since it may take values greater than 1. On the other hand, when ν satisfies **(P)** and is a probability measure, $T = T_M$ and $\delta = \delta_{p,1}$, then the similarity measure $\mathbf{S}_3^{\nu,\delta}$ satisfies **(T)** with $T' = T_L$, i.e., T_L-transitivity. In turn, this last fact was proven in [12,31] for the particular case $\delta = \delta_{1,1}$.

3 Similarity Learning Problem

The three similarity measures $\mathbf{S}_i^{\nu,\delta}$, for $i = 1, 2, 3$, essentially rely on the choice of ν and $\delta \in \{\delta_{p,p}, \delta_{p,1}, \delta_{1,p}\}$, for a suitable $p \in (0, +\infty)$. Surely, the most difficult part in getting an operative $\mathbf{S}_i^{\nu,\delta}$ is the elicitation of ν, due to its exponential size, so, in this section we face the learning of μ_ν and the tuning of p.

The result of the similarity learning contributes to XAI on three different layers: the chosen $\mathbf{S}_i^{\nu,\delta}$ tells us how common and different parts of profiles are aggregated; the Möbius inverse μ_ν of ν singles out interactions on the groups of attributes where it is different from zero; δ tells us how pointwise dissimilarities are evaluated and to which degree p.

In the rest of the paper we fix the t-norm $T = T_M$ and t-conorm $S = S_M$ to perform fuzzy set-theoretic operations. This choice guarantees that the similarity measures $\mathbf{S}_i^{\nu,\delta}$, for $i = 1, 2, 3$, are always well-defined (see Remark 1). We also notice that a different choice of T and S can impact on the final result of the similarity learning task, thus a systematic comparative study in the line of [1] can be envisaged.

The learning of ν is actually faced by learning the corresponding Möbius inverse μ_ν which, in turn, depends on the choice of the functional form $\mathbf{S}_i^{\nu,\delta}$ and of δ, the latter being tied to the choice of p. In the more general case of a capacity ν, negative interactions among groups of attributes are possible but the learning task is complicated by the set of constraints *(i)–(iii)* in Sect. 2.1, that restrict the feasible μ_ν's.

In this paper, in analogy with [2,27], we will limit ourselves to the case of completely monotone capacities satisfying **(P)**, for which the corresponding Möbius inverse is non-negative and strictly positive on the singletons. This choice turns out to be particularly relevant as it allows us to enforce information-theoretic properties for ν, as discussed below.

A first approach to face the similarity learning is to look for δ, ν and $\mathbf{S}_i^{\nu,\delta}$ that maximize accuracy in a classification problem. In this case, we aim at learning the capacity ν and tuning the parameter $p \in (0, +\infty)$ for each $\mathbf{S}_i^{\nu,\delta}$, by relying on a set of labeled fuzzy description profiles.

We consider a training dataset $\mathcal{D}_{\text{train}} = \{(X_1, y_1), \ldots, (X_m, y_m)\}$ of labeled fuzzy data, where $X_j \in \mathcal{F} = [0,1]^N$ is a fuzzy description profile, while y_j is the corresponding class.

For a fixed similarity $\mathbf{S}_i^{\nu,\delta}$, for $i = 1, 2, 3$, where $\delta \in \{\delta_{p,p}, \delta_{p,1}, \delta_{1,p}\}$, we define a *Nearest-Neighborhood (NN) classifier*: each fuzzy description profile X_j with $(X_j, y_j) \in \mathcal{D}_{\text{train}}$ is assigned to a class y_j^* solving the problem

$$(X_j^*, y_j^*) \in \underset{(X_h, y_h) \in \mathcal{D}_{\text{train}} \setminus \{(X_j, y_j)\}}{\arg\max} f_i^{\nu,\delta}(X_h, y_h),$$

where $f_i^{\nu,\delta}(X_h, y_h) := \mathbf{S}_i^{\nu,\delta}(X_j, X_h)$. Our goal is to find the significance assessment μ_ν that maximises the *Leave-One-Out (LOO) objective function*

$$N_{LOO}(\mu_\nu) = |\{y_j \,:\, y_j = y_j^*,\, (X_j, y_j) \in \mathcal{D}_{\text{train}}\}|,$$

which counts the number of correctly classified instances. The use of $N_{LOO}(\mu_\nu)$ is inspired to the Neighbourhood Component Analysis appearing in [16].

Remark 2. We point out that in place of a NN classifier we could use a k-NN classifier but this would require a longer time for computing the assigned class y_j^* corresponding to each X_j with $(X_j, y_j) \in \mathcal{D}_{\text{train}}$.

In [27] we addressed the similarity learning problem as the maximization of $N_{LOO}(\mu_\nu)$ in the space of non-negative Möbius inverses, which is a continuous optimization problem with a non-continuous objective function, for which classical optimization techniques cannot be used.

Besides the optimization difficulty, the function $N_{LOO}(\mu_\nu)$ shows an intrinsic identifiability problem since, for a fixed choice of $\mathbf{S}_i^{\nu,\delta}$ and δ, there can exists possibly infinitely many capacities ν's with Möbius inverses μ_ν's maximizing $N_{LOO}(\mu_\nu)$, as the following example shows. Incidentally, the quoted example shows that the issue of non-identifiability holds also if we restrict to the class $\mathbf{A}_2(N)$.

Example 1. Take $N = \{1, 2, 3, 4\}$ and let $\mathcal{D}_{\text{train}} = \{(X_1, y_1), \ldots, (X_4, y_4)\}$ be the dataset reported below

N	1	2	3	4	Class
X_1	0.8	0.6	0.1	0.1	0
X_2	0.7	0.7	0.1	0.1	0
X_3	0.1	0.2	0.8	0.9	1
X_4	0	0.2	0.9	1	1

For $\alpha \in (0, 1]$, we define the completely monotone capacity $\nu_\alpha = \alpha\nu_{1u} + (1-\alpha)\nu_{2u}$ with Möbius inverse μ_{ν_α}, where ν_{1u} and ν_{2u} are defined as in Sect. 2.1. Notice that ν_α satisfies **(P)** and is 2-additive, thus it belongs to $\mathbf{A}_2(N)$.

For each fixed $\mathbf{S}_i^{\nu,\delta}$, with $i = 1, 2, 3$, and each $\delta \in \{\delta_{p,p}, \delta_{p,1}, \delta_{1,p}\}$, with p ranging in $[0.5, 2.5]$ with a 0.5 step, we get the maximum value $N_{LOO}(\mu_{\nu_\alpha}) = 4$ for all α ranging in $[0.0001, 1]$ with a 0.0001 step.

In other terms, for this dataset, independently of the choice of the similarity measure and the restricted dissimilarity measure, all the Möbius inverses in $\{\mu_{\nu_\alpha} : \alpha \in [0.0001, 1]\}$ are global optimizers of N_{LOO}. ♦

The previous example highlights that the maximization of N_{LOO} is too weak a notion if we want to use ν and the related μ_ν for XAI, so, further constraints must be imposed, in order to achieve a meaningful ν, besides its classification capabilities.

In this light, a first restriction we impose is to limit to the set $\mathbf{A}_2(N)$ of (at most) 2-additive completely monotone capacities: in this way we have a computational simplification of the learning task since the number of unknowns reduces to $n + \binom{n}{2}$. Elements of $\mathbf{A}_2(N)$ also favor an easier interpretation of interactions among attributes as only pairwise interactions are considered. Besides the size reduction, the restriction to the set $\mathbf{A}_2(N)$ allows us to impose some meaningful information-theoretic properties to the learned ν through suitable entropic regularization schemes, which are the novelty of the present paper.

In general, given a completely monotone capacity ν with Möbius inverse μ_ν, we can introduce the following version of *entropy* proposed in [29], which is an immediate generalization of classical Shannon entropy [32]

$$H(\mu_\nu) = -\sum_{B \subseteq N} \mu_\nu(B) \log \mu_\nu(B), \tag{5}$$

under the usual convention $0 \log 0 = 0$. Notice that the term for $B = \emptyset$ is automatically dropped being it 0. Restricting to elements of $\mathbf{A}_2(N)$, Eq. (5) simplifies since the summation is limited to all $B \subseteq N$, $0 < |B| \leq 2$, and the maximum is achieved at the Möbius inverse $\mu_{\nu_{\alpha^*}}$ of the 2-additive completely monotone capacity $\nu_{\alpha^*} = \alpha^* \nu_{1u} + (1 - \alpha^*)\nu_{2u}$ with $\alpha^* = \frac{n}{n+\binom{n}{2}}$, which is such that $\mu_{\nu_{\alpha^*}}(B) = \frac{1}{n+\binom{n}{2}}$, for all $B \subseteq N$, $0 < |B| \leq 2$, and 0 otherwise.

The Möbius inverse $\mu_{\nu_{\alpha^*}}$ expresses a situation of maximal uncertainty on the significance of single attributes and interactions among pairs of attributes. Nevertheless, $\mu_{\nu_{\alpha^*}}$ treats in the same manner single attributes and pairs, i.e., it does not distinguish between single-attribute importance and pairwise interactions.

On the other hand, the meaning we give to μ_ν on singletons is different to that we give on pairs. For this, each μ_ν can be decomposed in two set functions μ_ν^1, μ_ν^2 such that $\mu_\nu = \mu_\nu^1 + \mu_\nu^2$ with

$$\mu_\nu^k(B) = \begin{cases} \mu_\nu(B), \text{ if } |B| = k, \\ 0, \qquad\quad \text{otherise,} \end{cases} \tag{6}$$

for all $B \in 2^N$ and $k = 1, 2$.

The decomposition $\mu_\nu = \mu_\nu^1 + \mu_\nu^2$, where the two components are defined as in (6), allows us to distinguish between the information content carried by single attributes and that carried by pairs of attributes. Referring to the normalization $\overline{\mu}_\nu^k = \frac{\mu_\nu^k}{\|\mu_\nu^k\|_1}$ where $\|\mu_\nu^k\|_1 = \sum_{B \subseteq N} |\mu_\nu^k(B)|$, for $k = 1, 2$, we can introduce the entropy related to the significance of single attributes and that related to pairwise interactions, defined, for $k = 1, 2$, as

$$H_k(\overline{\mu}_\nu^k) = - \sum_{\substack{B \subseteq N \\ |B|=k}} \overline{\mu}_\nu^k(B) \log \overline{\mu}_\nu^k(B). \tag{7}$$

In the particular case μ_ν is the Möbius inverse of a 1-additive capacity, that is $\mu_\nu^2 \equiv 0$, we set $\overline{\mu}_\nu^2 \equiv 0$ and $H_2(\overline{\mu}_\nu^2) = 0$.

It is easily shown that $\mu_{\nu_{1u}}$ is the maximizer of H_1 while $\{\mu_{\xi_{\{i\}}} : \{i\} \in 2^N\}$ is the set of minimizers of H_1. Similarly, $\mu_{\nu_{2u}}$ turns out to be the maximizer of H_2, while $\{\mu_{\xi_{\{i,j\}}} : \{i,j\} \in 2^N\}$ is the set of minimizers of H_2. We also notice that, for every $\alpha \in (0,1)$, the (at most) 2-additive completely monotone capacity $\nu_\alpha = \alpha\nu_{1u} + (1-\alpha)\nu_{2u}$ has Möbius inverse μ_{ν_α} whose decomposition according to (6) maximizes both H_1 and H_2, since $\overline{\mu}_{\nu_\alpha}^1 = \mu_{\nu_{1u}}$ and $\overline{\mu}_{\nu_\alpha}^2 = \mu_{\nu_{2u}}$.

The previous discussion leads us to the following three objective functions to maximize over $\mathbf{A}_2(N)$, where $\lambda, \lambda_1, \lambda_2, \lambda_3 > 0$:

$$g_1(\mu_\nu) = \frac{N_{LOO}(\mu_\nu)}{|\mathcal{D}_{\text{train}}|} + \lambda H(\mu_\nu), \tag{8}$$

$$g_2(\mu_\nu) = \frac{N_{LOO}(\mu_\nu)}{|\mathcal{D}_{\text{train}}|} + \lambda_1 H_1(\overline{\mu}_\nu^1) + \lambda_2 H_2(\overline{\mu}_\nu^2), \tag{9}$$

$$g_3(\mu_\nu) = \frac{N_{LOO}(\mu_\nu)}{|\mathcal{D}_{\text{train}}|} + \lambda_1 H_1(\overline{\mu}_\nu^1) - \lambda_3 H_2(\overline{\mu}_\nu^2). \tag{10}$$

The first term $\frac{N_{LOO}(\mu_\nu)}{|\mathcal{D}_{\text{train}}|}$ appearing in all functions g_j's is the *relative LOO objective function*, accounting for the classification accuracy provided by μ_ν in the LOO classification procedure over $\mathcal{D}_{\text{train}}$. Notice that $\frac{N_{LOO}(\mu_\nu)}{|\mathcal{D}_{\text{train}}|}$ ranges in $[0,1]$.

The three objective functions above express three different entropic regularization schemes of the relative LOO objective function:

Function g_1**:** The term $\lambda H(\mu_\nu)$ enforces a diffuse distribution of weights on the sets of attributes in $\{B \in 2^N \setminus \{\emptyset\} : |B| \leq 2\}$ as it is maximized by $\mu_{\nu_{\alpha^*}}$. This term ranges in $[0, \lambda \log(n + \binom{n}{2})]$ and does not distinguish the information carried by single attributes and pairs.

Function g_2**:** The terms $\lambda_1 H_1(\overline{\mu}_\nu^1)$ and $\lambda_2 H_2(\overline{\mu}_\nu^2)$ enforce a diffuse distribution of weights on the sets of attributes in $\{B \in 2^N \setminus \{\emptyset\} : |B| = 1\}$ and $\{B \in 2^N \setminus \{\emptyset\} : |B| = 2\}$, respectively, as they are maximized by $\mu_{\nu_{1u}}$ and $\mu_{\nu_{2u}}$, respectively. These two terms range in $[0, \lambda_1 \log n]$ and $[0, \lambda_2 \log \binom{n}{2}]$, respectively. The difference between g_2 and g_1 is that in g_2 we can differentiate the enforcing of the information carried by single attributes and by pairs.

Function g_3**:** The term $\lambda_1 H_1(\overline{\mu}_\nu^1)$ plays the same role played in g_2. On the other hand, the term $-\lambda_3 H_2(\overline{\mu}_\nu^2)$ enforces a sparse distribution of weights on the sets of attributes in $\{B \in 2^N \setminus \{\emptyset\} : |B| = 2\}$ as it is maximized by elements of $\{\mu_{\xi_{i,j}} : \{i,j\} \in 2^N\}$. This term ranges in $[-\lambda_3 \log \binom{n}{2}, 0]$. We notice that the third term is maximized also in the limit case where ν reduces to a probability measure, as it vacuously reduces to 0. The difference between g_3 and g_2 is that g_3 maximizes the information content carried by single attributes simultaneously promoting sparsity on pairs. This last fact could better highlight pairwise interactions.

We also notice that all g_j's are scalarized versions of three multicriteria optimization problems, blending accuracy and information-theoretic properties. We stress that the constants $\lambda, \lambda_1, \lambda_2, \lambda_3$ should be interpreted as hyper-parameters, that act on the structure of the learned μ_ν, thus they require a sensitivity analysis.

Given a dataset $\mathcal{D}$ we partition it into two disjoint sets $\mathcal{D} = \mathcal{D}_{\text{train}} \cup \mathcal{D}_{\text{test}}$ and, working with $\mathcal{D}_{\text{train}}$, for fixed $\mathbf{S}_i^{\nu,\delta}$, with $i = 1, 2, 3$, $\delta \in \{\delta_{p,p}, \delta_{p,1}, \delta_{1,p}\}$, and $p \in (0, +\infty)$, for $j = 1, 2, 3$, we solve the optimization problem

$$\begin{aligned} &\text{maximize } g_j(\mu_\nu) \\ &\text{subject to } \nu \in \mathbf{A}_2(N). \end{aligned} \tag{11}$$

Remark 3. The set $\mathbf{A}_2(N)$ is a closed convex set of capacities with extreme points $\{\xi_B : B \in 2^N \setminus \{\emptyset\}, |B| = 1, 2\}$. In turn, the strict concavity of entropic terms $\lambda H(\mu_\nu)$ and $\lambda_1 H_1(\overline{\mu}_\nu^1)$ favors the positivity of the Möbius inverse (at least) on the singletons and this enforces property **(P)** when solving (11).

4 Preliminary Experimental Analysis

Since the objective functions g_j's introduced in Sect. 3 are discontinuous due to the relative LOO objective function term, to solve (11) we adopt the *Particle Swarm Optimization (PSO)* technique, which is a stochastic incomplete method operating on a fixed number of candidate μ_ν's [23]. All the learning task is carried out in `Python` and for the PSO implementation we refer to the `PySwarms` library [28] version `1.3.0`.

Due to space limitations, we show some preliminary results by limiting the analysis to the `Iris` and `Penguins` datasets, which are popular datasets available in the `Kaggle` platform [21]. The datasets have been pre-processed, by normalizing attribute ranges in $[0, 1]$.

Both pre-processed `Iris` and `Penguins` datasets have 4 attributes and a class label taking 3 possible values, with 150 and 333 rows, respectively. Both datasets have been partitioned in a training and a test set with a 50%–50% ratio, adopting a stratified sampling. We choose the 50%–50% ratio since our main goal is to extract from $\mathcal{D}_{\text{train}}$ three explanations encoded in the Möbius inverses obtained by maximizing the three objective functions g_j's. Next, we measure the classification accuracy of the three explanations by working on a different but analogous dataset $\mathcal{D}_{\text{test}}$.

For $j = 1, 2, 3$, the optimization problem (11) relies on the choice of $\mathbf{S}_i^{\nu,\delta}$, with $i = 1, 2, 3$, $\delta \in \{\delta_{p,p}, \delta_{p,1}, \delta_{1,p}\}$, and $p \in (0, +\infty)$. For this reason, we proceed with an exploratory analysis by evaluating the performance in terms of the relative

LOO objective function (which measures accuracy) on $\mathcal{D}_{\text{train}}$, of the reference capacity $\nu_{\alpha^*} = \alpha^*\nu_{1u} + (1-\alpha^*)\nu_{2u}$, with $\alpha^* = \frac{n}{n+\binom{n}{2}}$. Notice that ν_{α^*} expresses the maximum parameter uncertainty, according to both g_1 and g_2.

Figure 1 shows the value of the relative LOO objective function as a function of p, the latter ranging in $[0.5, 4.5]$ with a 0.5 step, for all $\mathbf{S}_i^{\nu,\delta}$, with $i = 1, 2, 3$, and $\nu = \nu_{\alpha^*}$.

For the Iris dataset, Figs. 1a, 1c and 1e show that the maximum value of the relative LOO objective function for $\nu = \nu_{\alpha^*}$ on $\mathcal{D}_{\text{train}}$ is 96%, which is achieved with $\mathbf{S}_3^{\nu,\delta}$: $\delta = \delta_{p,1}$, $p \in \{3.5, 4, 4.5\}$.

For the Penguins dataset, Figs. 1b, 1d and 1f show that the maximum value of the relative LOO objective function for $\nu = \nu_{\alpha^*}$ on $\mathcal{D}_{\text{train}}$ is 96.99%, which is achieved with $\mathbf{S}_2^{\nu,\delta}$: $\delta = \delta_{p,p}$, $p = 1$; $\delta = \delta_{p,1}$, $p \in \{1, 2.5\}$; $\delta = \delta_{1,p}$, $p \in \{0.5, 1\}$; and also with $\mathbf{S}_3^{\nu,\delta}$: $\delta = \delta_{p,p}$, $p \in \{1.5, 2, 2.5, 3, 3.5\}$.

For the Iris dataset we fix $\mathbf{S}_3^{\nu,\delta}$, $\delta = \delta_{p,1}$, $p = 4$, while for the Penguins dataset we fix $\mathbf{S}_3^{\nu,\delta}$, $\delta = \delta_{p,p}$, $p = 3$, in agreement with the exploratory analysis.

Next, for both datasets we proceed in solving problem (11) on $\mathcal{D}_{\text{train}}$ for all the g_j's with the chosen similarity measure, restricted dissimilarity function and p. The optimization is performed with 200 iterations of PSO using 20 particles which are initialized with a random generation in a neighborhood of ν_{α^*}.

Concerning functions g_j's, we report the results of a sensitivity analysis by taking $\lambda \in \{0.05, 0.1, 0.15\}$, $\lambda_1 = \lambda_3 = 2\lambda$ and $\lambda_2 = \frac{\lambda}{2}$. Tables 1 and 2 report the Möbius inverse of the (at most) 2-additive completely monotone capacity ν_{g_j} learned by maximizing g_j, for $j = 1, 2, 3$, each highlighting the importance of single attributes or pairwise coalitions of attributes, in terms of the deviation from the Möbius inverses $\mu_{\nu_{\alpha^*}}$, $\mu_{\nu_{1u}}$ and $\mu_{\nu_{2u}}$ (up to normalization).

Table 1. Optimal Möbius inverse $\mu_{\nu_{g_j}}$ obtained maximizing g_j with PSO, for $j = 1, 2, 3$, on $\mathcal{D}_{\text{train}}$ and value of the relative LOO objective function (which measures accuracy) on $\mathcal{D}_{\text{test}}$ for Iris dataset with $\mathbf{S}_3^{\nu,\delta}$, $\delta = \delta_{p,1}$, $p = 4$.

	Sets	{1}	{2}	{3}	{4}	{1, 2}	{1, 3}	{1, 4}	{2, 3}	{2, 4}	{3, 4}	Acc. $\mathcal{D}_{\text{test}}$
$\lambda = 0.05$	$\mu_{\nu_{g_1}}$	0.103	0.099	0.103	0.100	0.099	0.100	0.099	0.099	0.100	0.099	96%
	$\mu_{\nu_{g_2}}$	0.094	0.097	0.091	0.089	0.109	0.119	0.110	0.105	0.084	0.103	96%
	$\mu_{\nu_{g_3}}$	0.207	0.142	0.092	0.164	0.172	0.023	0.036	0.003	0.001	0.161	94.67%
$\lambda = 0.1$	$\mu_{\nu_{g_1}}$	0.104	0.102	0.100	0.096	0.102	0.104	0.094	0.099	0.090	0.108	96%
	$\mu_{\nu_{g_2}}$	0.098	0.094	0.100	0.101	0.099	0.113	0.115	0.087	0.102	0.093	96%
	$\mu_{\nu_{g_3}}$	0.164	0.194	0.106	0.115	0.033	0.015	0.180	0.013	0.001	0.177	94.67%
$\lambda = 0.15$	$\mu_{\nu_{g_1}}$	0.102	0.098	0.102	0.104	0.103	0.094	0.096	0.100	0.101	0.101	96%
	$\mu_{\nu_{g_2}}$	0.088	0.086	0.087	0.089	0.111	0.102	0.111	0.114	0.106	0.106	96%
	$\mu_{\nu_{g_3}}$	0.125	0.231	0.206	0.130	0.000	0.189	0.002	0.040	0.046	0.032	98.67%

All learned capacities have been then tested on the test dataset $\mathcal{D}_{\text{test}}$, computing the relative LOO objective function (which measures accuracy).

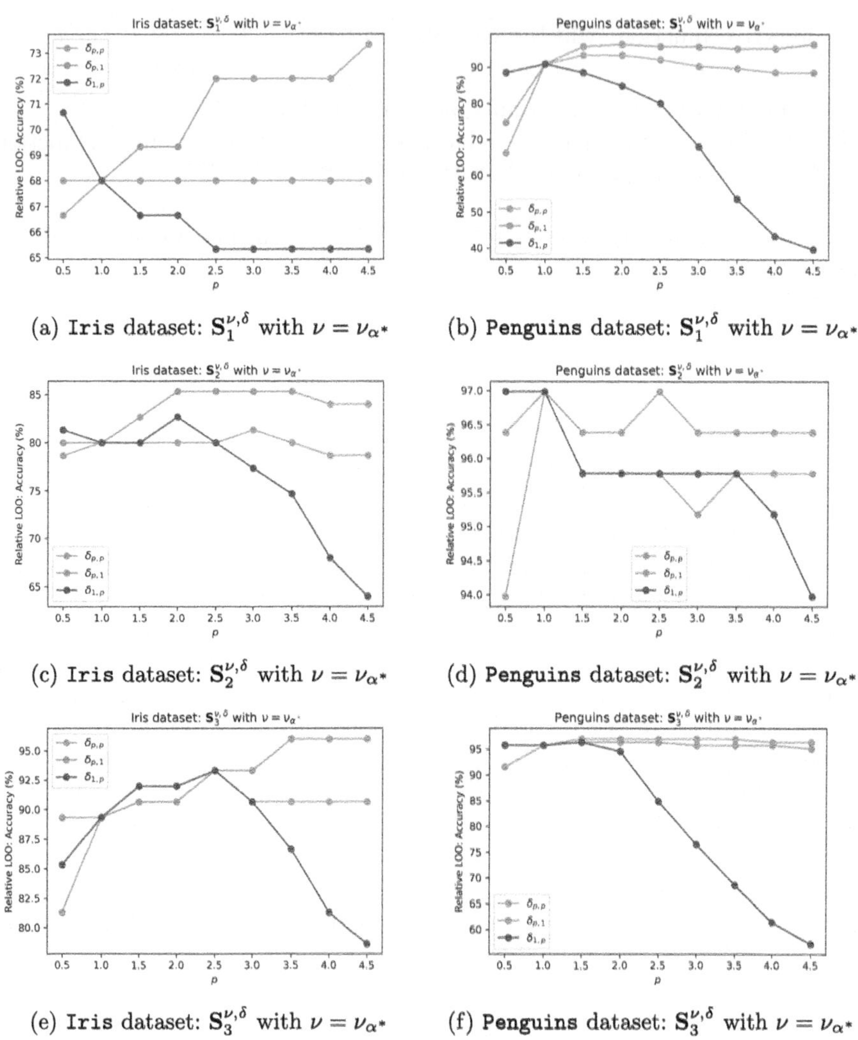

(a) `Iris` dataset: $\mathbf{S}_1^{\nu,\delta}$ with $\nu = \nu_{\alpha^*}$

(b) `Penguins` dataset: $\mathbf{S}_1^{\nu,\delta}$ with $\nu = \nu_{\alpha^*}$

(c) `Iris` dataset: $\mathbf{S}_2^{\nu,\delta}$ with $\nu = \nu_{\alpha^*}$

(d) `Penguins` dataset: $\mathbf{S}_2^{\nu,\delta}$ with $\nu = \nu_{\alpha^*}$

(e) `Iris` dataset: $\mathbf{S}_3^{\nu,\delta}$ with $\nu = \nu_{\alpha^*}$

(f) `Penguins` dataset: $\mathbf{S}_3^{\nu,\delta}$ with $\nu = \nu_{\alpha^*}$

Fig. 1. Relative LOO objective function measuring accuracy (%) on $\mathcal{D}_{\text{train}}$, seen as a function of p, for datasets `Iris` and `Penguins`, and $\mathbf{S}_i^{\nu,\delta}$, $i = 1, 2, 3$, with $\nu = \nu_{\alpha^*}$.

Table 2. Optimal Möbius inverse $\mu_{\nu_{g_j}}$ obtained maximizing g_j with PSO, for $j = 1,2,3$, on $\mathcal{D}_{\text{train}}$ and value of the relative LOO objective function (which measures accuracy) on $\mathcal{D}_{\text{test}}$ for `Penguins` dataset with $\mathbf{S}_3^{\nu,\delta}$, $\delta = \delta_{p,p}$, $p = 3$.

	Sets	{1}	{2}	{3}	{4}	{1,2}	{1,3}	{1,4}	{2,3}	{2,4}	{3,4}	Acc. $\mathcal{D}_{\text{test}}$
$\lambda = 0.05$	$\mu_{\nu_{g_1}}$	0.122	0.091	0.105	0.062	0.089	0.132	0.068	0.153	0.072	0.106	98.20%
	$\mu_{\nu_{g_2}}$	0.139	0.148	0.129	0.112	0.099	0.083	0.056	0.098	0.058	0.079	98.80%
	$\mu_{\nu_{g_3}}$	0.200	0.159	0.169	0.143	0.077	0.003	0.011	0.208	0.020	0.009	97.60%
$\lambda = 0.1$	$\mu_{\nu_{g_1}}$	0.115	0.076	0.106	0.058	0.106	0.126	0.087	0.149	0.058	0.118	98.80%
	$\mu_{\nu_{g_2}}$	0.116	0.094	0.129	0.098	0.099	0.131	0.049	0.138	0.053	0.093	97.60%
	$\mu_{\nu_{g_3}}$	0.137	0.109	0.145	0.116	0.244	0.037	0.039	0.147	0.002	0.025	97.60%
$\lambda = 0.15$	$\mu_{\nu_{g_1}}$	0.114	0.112	0.112	0.076	0.120	0.102	0.086	0.122	0.066	0.090	98.80%
	$\mu_{\nu_{g_2}}$	0.114	0.102	0.134	0.089	0.087	0.109	0.075	0.147	0.042	0.101	97.60%
	$\mu_{\nu_{g_3}}$	0.207	0.199	0.126	0.140	0.206	0.019	0.005	0.087	0.009	0.002	96.41%

5 Conclusion

The discussion carried out in Sect. 3 showed that identifiability is an important issue in the learning of similarity measures $\mathbf{S}_i^{\nu,\delta}$, for $i = 1,2,3$, that cannot be faced just by looking at classification capabilities. The three entropic regularization schemes we propose have exactly the purpose of restricting the set of parameters with a meaningful and interpretable set of constraints. In detail, this is done by blending classification capabilities with information-theoretic properties, i.e., by taking the locally uniform Möbius inverses $\mu_{\nu_{\alpha*}}$, $\mu_{\nu_{1u}}$ and $\mu_{\nu_{2u}}$ as references. We also performed a sensitivity analysis on the entropic regularization schemes' hyper-parameters to show their effect on the structure of the learned Möbius inverse.

Looking towards future research we aim at developing a more extensive experimental analysis based on a larger set of datasets and a more detailed ensemble of entropic schemes. This latter extension can be pursued in two directions: *(i)* switching to the set of (at most) k-additive completely monotone capacities, for $k \geq 2$; *(ii)* considering different notions of entropy for completely monotone capacities, so as to enforce different information-theoretic properties. Concerning point *(ii)*, two promising directions seem to explore other notions of entropy for completely monotone capacities [20] or to generalize the notion of extropy introduced in [25] for probability measures. We finally notice that, when δ is taken equal to $\delta_{p,q}$ as in (1), the tuning of both p, q can be structured as a searching problem over a grid decomposition of $(0, +\infty)^2$.

Acknowledgement. We acknowledge the support of the PRIN 2022 project "Models for dynamic reasoning under partial knowledge to make interpretable decisions" (Project number: `2022AP3B3B`, CUP Master: `J53D23004340006`, CUP: `B53D23009860006`) funded by the European Union – Next Generation EU (Missione 4 Componente 2).

References

1. Ahmad, K., Mesiarova-Zemankova, A.: Choosing t-norms and t-conorms for fuzzy controllers. In: Fourth International Conference on Fuzzy Systems and Knowledge Discovery (FSKD 2007), vol. 2, pp. 641–646 (2007)
2. Baioletti, M., Coletti, G., Petturiti, D.: Weighted attribute combinations based similarity measures. In: Greco, S., Bouchon-Meunier, B., Coletti, G., Fedrizzi, M., Matarazzo, B., Yager, R.R. (eds.) IPMU 2012. CCIS, vol. 299, pp. 211–220. Springer, Heidelberg (2012). https://doi.org/10.1007/978-3-642-31718-7_22
3. Bouchon-Meunier, B., Coletti, G., Lesot, M.-J., Rifqi, M.: Towards a conscious choice of a fuzzy similarity measure: a qualitative point of view. In: Hüllermeier, E., Kruse, R., Hoffmann, F. (eds.) IPMU 2010. LNCS (LNAI), vol. 6178, pp. 1–10. Springer, Heidelberg (2010). https://doi.org/10.1007/978-3-642-14049-5_1
4. Bouchon-Meunier, B., Rifqi, M., Bothorel, S.: Towards general measures of comparison of objects. Fuzzy Sets Syst. **84**(2), 143–153 (1996)
5. Bresson, R.: Neural learning and validation of hierarchical multi-criteria decision aiding models with interacting criteria. Ph.D. thesis, Université Paris-Saclay (2022)
6. Bustince, H., et al.: d-Choquet integrals: Choquet integrals based on dissimilarities. Fuzzy Sets Syst. **414**, 1–27 (2021)
7. Chateauneuf, A., Jaffray, J.Y.: Some characterizations of lower probabilities and other monotone capacities through the use of Möbius inversion. Math. Soc. Sci. **17**(3), 263–283 (1989)
8. Coletti, G., Bouchon-Meunier, B.: A study of similarity measures through the paradigm of measurement theory: the classic case. Soft. Comput. **23**(16), 6827–6845 (2019)
9. Coletti, G., Bouchon-Meunier, B.: A study of similarity measures through the paradigm of measurement theory: the fuzzy case. Soft. Comput. **24**(15), 11223–11250 (2020)
10. Coletti, G., Petturiti, D., Bouchon-Meunier, B.: A measurement theory characterization of a class of dissimilarity measures for fuzzy description profiles. In: Lesot, M.-J., et al. (eds.) IPMU 2020. CCIS, vol. 1238, pp. 258–268. Springer, Cham (2020). https://doi.org/10.1007/978-3-030-50143-3_20
11. Coletti, G., Petturiti, D., Bouchon-Meunier, B.: Weighted and Choquet L^p distance representation of comparative dissimilarity relations on fuzzy description profiles. Ann. Math. Artif. Intell. (2024). https://doi.org/10.1007/s10472-024-09924-y
12. Coletti, G., Petturiti, D., Vantaggi, B.: Fuzzy weighted attribute combinations based similarity measures. In: Antonucci, A., Cholvy, L., Papini, O. (eds.) ECSQARU 2017. LNCS (LNAI), vol. 10369, pp. 364–374. Springer, Cham (2017). https://doi.org/10.1007/978-3-319-61581-3_33
13. Couso, I., Garrido, L., Sánchez, L.: Similarity and dissimilarity measures between fuzzy sets: a formal relational study. Inf. Sci. **229**, 122–141 (2013)
14. De Baets, B., De Meyer, H.: Transitivity-preserving fuzzification schemes for cardinality-based similarity measures. Eur. J. Oper. Res. **160**(3), 726–740 (2005)
15. De Baets, B., Janssens, S., De Meyer, H.: On the transitivity of a parametric family of cardinality-based similarity measures. Int. J. Approximate Reasoning **50**(1), 104–116 (2009)
16. Goldberger, J., Hinton, G., Roweis, S., Salakhutdinov, R.: Neighbourhood components analysis. In: Saul, L., Weiss, Y., Bottou, L. (eds.) Advances in Neural Information Processing Systems, vol. 17. MIT Press (2004)

17. Grabisch, M.: K-order additive fuzzy measures. In: Proceedings of the 6th International Conference on Information Processing and Management of Uncertainty in Knowledge-Based Systems (IPMU), Granada, Spain, pp. 1345–1350 (1996)
18. Grabisch, M.: Set Functions, Games and Capacities in Decision Making. Springer, Cham (2016). https://doi.org/10.1007/978-3-319-30690-2
19. Jaccard, P.: Nouvelles recherches sur la distribution florale. Bull. de la Société Vaudoise des Sci. Nat. **44**, 223–270 (1908)
20. Jiroušek, R., Shenoy, P.: A new definition of entropy of belief functions in the Dempster-Shafer theory. Int. J. Approximate Reasoning **92**, 49–65 (2018)
21. Kaggle. https://www.kaggle.com
22. Kaldjob, P.K., Mayag, B., Bouyssou, D.: Study of the instability of the sign of the nonadditivity index in a Choquet integral model. In: Ciucci, D., et al. (eds.) IPMU 2022. CCIS, vol. 1602, pp. 197–209. Springer, Cham (2022). https://doi.org/10.1007/978-3-031-08974-9_16
23. Kennedy, J., Eberhart, R.: Swarm Intelligence. Morgan Kaufmann (2001)
24. Klement, E., Mesiar, R., Pap, E.: Triangular Norms, Trends in Logic, vol. 8. Kluwer Academic Publishers, Dordrecht/Boston/London (2000)
25. Lad, F., Sanfilippo, G., Agrò, G.: Extropy: complementary dual of entropy. Stat. Sci. **30**(1), 40–58 (2015)
26. Lesot, M.J., Rifqi, M., Benhadda, H.: Similarity measures for binary and numerical data: a survey. Int. J. Knowl. Eng. Soft Data Paradigms **1**(1), 63–84 (2009)
27. Marsala, C., Petturiti, D., Vantaggi, B.: Adding semantics to fuzzy similarity measures through the d-Choquet integral. In: Bouraoui, Z., Vesic, S. (eds.) ECSQARU 2023. LNCS, vol. 14294, pp. 386–399. Springer, Cham (2024). https://doi.org/10.1007/978-3-031-45608-4_29
28. Miranda, L.J.V.: PySwarms: a research toolkit for Particle Swarm Optimization in Python. J. Open Source Softw. **3**(21), 433 (2018)
29. Nguyen, H.: On entropy of random sets and possibility distributions. Anal. Fuzzy Inf. **1**, 145–156 (1987)
30. Rudin, C.: Stop explaining black box machine learning models for high stakes decisions and use interpretable models instead. Nat. Mach. Intell. **1**(5), 206–215 (2019)
31. Scozzafava, R., Vantaggi, B.: Fuzzy inclusion and similarity through coherent conditional probability. Fuzzy Sets Syst. **160**(3), 292–305 (2009)
32. Shannon, C.: A mathematical theory of communication. Bell Syst. Tech. J. **27**(3), 379–423 (1948)
33. Tversky, A.: Features of similarity. Psychol. Rev. **84**(4), 327–352 (1977)
34. Zadeh, L.: Fuzzy sets. Inf. Control **8**(3), 338–353 (1965)

Elicitation for Decision Problems Under Severe Uncertainties

Nawapon Nakharutai[1], Matthias C. M. Troffaes[3], and Sébastien Destercke[2(✉)]

[1] Department of Statistics, Faculty of Science, Chiang Mai University, Chiang Mai 50200, Thailand
nawapon.nakharutai@cmu.ac.th

[2] Department of Mathematical Sciences, Durham University, Durham, UK
sebastien.destercke@hds.utc.fr

[3] UMR CNRS 7253 Heudiasyc, Sorbonne Université, Université de Technologie de Compiègne, Compiègne, France
matthias.troffaes@durham.ac.uk

Abstract. In this paper, we investigate the problem of eliciting information from an expert, where the assumed uncertainty model is a coherent upper prevision (or equivalently a closed convex set of probabilities). The goal is to solve a decision problem under the maximality decision rule, with as few queries to the expert as possible. To address this, we study the range of coherent upper bounds an expert may give on a given query. In doing so, we provide new results and characterisations for this range. We then use these results to provide an algorithm of elicitation. We illustrate the algorithm on an example.

Keywords: decision · elicitation · coherence

1 Introduction

In many applications, to support effective decision making under uncertainty, expert elicitation is critical. Elicitation faces several important challenges [16]. These include how to ask queries to experts to avoid psychological biases [9], how to elicit beliefs in addition to payments to eliminate hedging opportunities [4], how to translate communicated information into mathematical models, and how to elicit both preferences and beliefs about decision problems [11,15].

While the existing literature on the topic is very rich, most authors focus on modelling expert knowledge in terms of precise probabilities (see for example Renooij [18] and Gartwaithe *et al.* [8]). This includes situations where the expert only provides partial information, as for instance in [14, Sec. 6.]. However, in case of severe uncertainty, it has been argued by many authors [2,22,27] that allowing the experts to provide bounded probability assessments rather than precise probabilities is more reasonable. Indeed, by allowing experts to provide bounded probability assessments, they can more honestly represent their uncertainty, especially in the face of severe uncertainty [27]. In this paper too, we will consider eliciting bounded probabilities.

S. Destercke et al. (Eds.): SUM 2024, LNAI 15350, pp. 312–324, 2025.
https://doi.org/10.1007/978-3-031-76235-2_23

Additionally, most existing elicitation procedures do not directly consider the decision problem for which the elicitation is eventually needed. This means that the elicitation procedure may be sub-optimal, i.e., we may ask the expert for information that is not needed to solve the decision problem at hand. While often elicitation of uncertainty is of interest by itself regardless of any decision problem, there are plenty of situations where uncertainties are elicited to solve a specific decision problem known at the time of elicitation. One aim of this paper is to explicitly consider the decision problem to more effectively target the elicitation, and thereby reduce the cognitive burden on the expert.

The problem of eliciting bounded probabilities has been studied extensively [21,26], including elicitation in related uncertainty theories such as evidence theory [29] and possibility theory [20]. Applications include environmental modelling [19] and medical diagnosis [6]. However, here too, most of these approaches focus on eliciting the models without considering a possible underlying decision problem. Two exceptions are the work of Jansen *et al.* [10] and T'Joens [23]. However, these works differ from ours in several aspects. Jansen *et al.* [10] aim to find a preference system as efficiently as possible, whilst we aim at finding an optimal solution to the decision problem as efficiently as possible. T'Joens [23] frames the problem within the framework of desirability whilst we use upper previsions, and additionally makes the strong assumption that the unknown uncertainty model elicited from the expert leads to a unique optimal recommendation which we do not assume.

In this paper, we study the problem of elicitation in the context of a given (but arbitrary) decision problem: what are the best questions we can ask an expert, to identify the best possible actions for a specific decision problem? To tackle this, we treat the elicitation problem as an optimization problem, similarly to what is sometimes done when eliciting multi-criteria preferences [1,3,5]. Specifically, we solve the following problem:

> Given current expert assessments, and a set of potential queries that we can ask the expert, which query should we present next to gain the most useful information for solving the decision problem?

The questions we ask, called queries, are in the form of uncertain rewards, also called gambles, and the expert answers us by giving us their upper bound on the expectation of such a gamble; such upper bound is sometimes also called an upper prevision [25,27,28].

To determine which query is most useful, we use an objective that directly relates to the decision problem: which query has the most potential for reducing the set of so-called maximal [24] actions? To determine this, we need to know in advance the range of upper bounds on the expectation that the expert might provide. For this purpose, we develop the theory concerning the range of coherent extensions, expanding on some results from Walley [27, §3.1.6, p. 126]. We note that verification of coherence of conditional probability bounds was also studied in [6].

The rest of the paper is organised as follows. Section 2 reviews the theory of upper previsions and the concept of maximality in decision making. Section 3, proposes a new algorithm for elicitation, along with a fully worked example. Section 4 concludes the paper.

2 Preliminaries

2.1 Upper Previsions

We denote by Ω a set of possible states. For simplicity, throughout this paper, we assume Ω is finite. We consider a decision problem where a decision maker can choose a finite number of acts from a set of acts $\mathcal{A}$. For each $a \in \mathcal{A}$, if $\omega \in \Omega$ happens to be the true state of nature, then $a(\omega)$ represents the reward to the decision maker, expressed as a real value on a utility scale. In this way, each act a is modelled as a real-valued function on a set of possible states. Such a function, representing an uncertain reward, is called a *gamble* [27]. Following [27], we will assume gambles are bounded, and $\mathcal{L}$ will denote the set of all gambles.

If a probability mass function p on Ω is available, the decision maker can simply pick an act that maximizes expected utility, i.e. $a^* = \arg\max_{a\in\mathcal{A}} \mathbb{E}_p(a)$, where $\mathbb{E}_p(a) := \sum_{\omega\in\Omega} p(\omega)a(\omega)$. However, a complete probability mass function may not be available. In that case, the decision maker may ask an expert to state their knowledge about the unknown true value of $\omega \in \Omega$. We assume here that we can elicit some information in terms of an upper prevision $\overline{P}$ which is a real-valued function defined on some subset $\operatorname{dom}\overline{P}$ of $\mathcal{L}$. Specifically, for each gamble $f \in \operatorname{dom}\overline{P}$, the value $\overline{P}(f)$ represents the expert's infimum selling price for f. So, by specifying $\overline{P}$, the expert declares that $\overline{P}(f) - f + \epsilon$ is acceptable uncertain reward for all $\epsilon > 0$ and $f \in \operatorname{dom}\overline{P}$. As we shall see, $\overline{P}(f)$ can also be interpreted as an upper bound on the expectation of f. Note that upper previsions generalize a very wide range of uncertainty models used in practical problems, including belief functions and possibility distributions [25].

In this study, we are interested in iteratively making queries to the expert in order to learn their uncertainty in the form of upper previsions, so that the decision maker can solve the underlying decision problem. To do so, we must first state some properties and consistency conditions for upper previsions.

We say that $\overline{P}'$ *dominates* $\overline{P}$, and write $\overline{P}' \leq \overline{P}$, when $\operatorname{dom}\overline{P}' \supseteq \operatorname{dom}\overline{P}$ and $\overline{P}'(f) \leq \overline{P}(f)$ for all $f \in \operatorname{dom}\overline{P}$. When $\overline{P}'$ dominates $\overline{P}$, all selling prices implied by $\overline{P}$ are also implied by $\overline{P}'$, and so $\overline{P}'$ is more informative compared to $\overline{P}$.

Definition 1 **([27, §3.3.1, p. 133]).** *The* credal set *of $\overline{P}$ is the set of probability mass functions on Ω whose expectations dominate $\overline{P}$, that is*

$$\mathcal{M}(\overline{P}) := \{p\colon \forall f \in \mathcal{L}, \mathbb{E}_p(f) \leq \overline{P}(f)\} \tag{1}$$

Note that $\mathcal{M}(\overline{P})$ is convex and closed. By $\mathcal{K} \Subset \operatorname{dom}\overline{P}$ we mean that $\mathcal{K}$ is a finite subset of $\operatorname{dom}\overline{P}$. We also define $\mathbb{R}_+ := \{x \in \mathbb{R}\colon x \geq 0\}$.

Definition 2 ([27, §**3.1.3(d)**], [25, **Cor. 8.38(ii)**]). *The* natural extension *of $\overline{P}$ is the upper prevision $\overline{\mathbb{E}}$ defined on every $f' \in \mathcal{L}$ by*

$$\overline{\mathbb{E}}(f') := \inf_{\substack{\mathcal{K} \Subset \operatorname{dom} \overline{P} \\ \lambda\colon \mathcal{K} \to \mathbb{R}_+}} \sup \left[f' + \sum_{f \in \mathcal{K}} \lambda_f (\overline{P}(f) - f) \right] \tag{2}$$

$$= \max_{p \in \mathcal{M}(\overline{P})} \mathbb{E}_p(f') \tag{3}$$

In Eq. (3), if $\mathcal{M}(\overline{P}) = \emptyset$, the maximum is taken to be $-\infty$. Equation (3) follows from Eq. (2) by duality [27, §3.3.3, p. 134], but it also has an obvious direct interpretation as the upper expectation induced by the credal set of $\overline{P}$.

Definition 3 ([27, §**2.4.4, p. 69**] **and** [25, **Cor. 8.38(i)**]). *We say that $\overline{P}$* avoids sure loss *when $\overline{\mathbb{E}}(0) = 0$, i.e. when*

$$\inf_{\substack{\mathcal{K} \Subset \operatorname{dom} \overline{P} \\ \lambda\colon \mathcal{K} \to \mathbb{R}_+}} \sup \left[\sum_{f \in \mathcal{K}} \lambda_f (\overline{P}(f) - f) \right] \geq 0. \tag{4}$$

or equivalently, when $\mathcal{M}(\overline{P}) \neq \emptyset$.

Definition 4 ([27, §**2.5.4, p. 75**], [25, **Cor. 8.38(iii)**]). *We say that $\overline{P}$ is* coherent *when $\overline{\mathbb{E}}(f') = \overline{P}(f')$ for all $f' \in \operatorname{dom} \overline{P}$, i.e.*

$$\inf_{\substack{f' \in \operatorname{dom} \overline{P}, \mathcal{K} \Subset \operatorname{dom} \overline{P} \\ \lambda' \in \mathbb{R}_+, \lambda\colon \mathcal{K} \to \mathbb{R}_+}} \sup \left[\sum_{f \in \mathcal{K}} \lambda_f (\overline{P}(f) - f) - \lambda' (\overline{P}(f') - f') \right] \geq 0. \tag{5}$$

or equivalently, when for every $f' \in \operatorname{dom} \overline{P}$ there is a $p \in \mathcal{M}(\overline{P})$ such that $\overline{P}(f') = \mathbb{E}_p(f')$.

We assume that the decision maker remains coherent throughout the process of elicitation, i.e. that their $\overline{P}$ is a coherent upper prevision. Otherwise, $\overline{P}$ must be fixed through natural extension.

As we shall see next, the natural extension is also an important tool for solving decision problems.

2.2 Decision Making

To select the best possible acts in $\mathcal{A}$, there are many decision criteria that can be used with upper previsions [24]. Here, we will consider *maximality*:

$$\operatorname{opt}(\overline{P}, \mathcal{A}) := \{a \in \mathcal{A} \colon \forall a' \in \mathcal{A}, \overline{\mathbb{E}}(a - a') \geq 0\} \tag{6}$$

The next property of maximality will be very useful later when we do elicitation.

Table 1. Total net profit of investing money under each stock fund and inflation rate for Example 1.

stock fund	L	M	H
1	2	5	7
2	3	7	3
3	5	2	7
4	7	4	2
5	5	6	5
6	3	8	4

Table 2. Values of $\overline{\mathbb{E}}(a-a')$ in Example 1.

$a \backslash a'$	1	2	3	4	5	6
1	0.0	1.5	0.0	0.33	−0.5	1.0
2	1.5	0.0	1.5	0.0	−.5	0.0
3	3.0	3.0	0.0	1.5	1.0	2.5
4	5.0	4.0	2.0	0.0	2.0	4.0
5	3.0	2.0	2.0	1.0	0.0	2.0
6	2.0	0.67	2.0	0.67	0.0	0.0

Theorem 1 ([24]). If $\overline{P}'$ dominates $\overline{P}$ then $\text{opt}(\overline{P}', \mathcal{A}) = \text{opt}(\overline{P}', \text{opt}(\overline{P}, \mathcal{A}))$.

Example 1. Consider a situation where we want to invest money in a stock fund. The return depends on the inflation rate across the next year. Suppose that next year's inflation rate could be either low (L), moderate (M) or high (H), that is, $\Omega = \{L, M, H\}$. There are six stock funds available, so $\mathcal{A} = \{1, \dots, 6\}$. The net profit under each fund and inflation rate is presented in Table 1. Suppose that, prior to investment, we elicit an expert's information about next year's inflation rate. The expert states:

- L is at least as probable as M.
- L is at least as probable as H.

This information can be phrased in terms of an upper prevision:

$$\overline{P}(\mathbb{I}_M - \mathbb{I}_L) = 0 \text{ and } \overline{P}(\mathbb{I}_H - \mathbb{I}_L) = 0, \tag{7}$$

Here, we use the notation $\mathbb{I}_{\omega'}$ to denote the gamble formed by the indicator function of the event $\{\omega'\}$, i.e. $\mathbb{I}_{\omega'}(\omega) = 1$ if $\omega' = \omega$, else 0. For every pair of acts a and a' in $\mathcal{A}$, we can calculate $\overline{\mathbb{E}}(a - a')$. These values are listed in Table 2. We can see that the set of maximal acts is $\text{opt}(\overline{P}, \mathcal{A}) = \{3, 4, 5, 6\}$ since these acts have corresponding rows in Table 2 with no negative entries.

As we can see in Example 1 the information from the expert allow us to suppress two non-maximal acts and leave us with four remaining maximal acts. Suppose we want to draw additional information from the expert by asking them to state an upper prevision of additional gambles which we will call *queries*. This new information may reduce the values of $\overline{\mathbb{E}}(a - a')$ and therefore the number of maximal acts (see Theorem 1). Specifically, we would like to optimally pick an additional query that maximize our chances to suppress as many additional acts as possible. To do so, we must first study the range of possible coherent extensions.

2.3 Range of Coherent Extensions

We will aim to extend the domain of an expert's coherent upper prevision by one gamble at a time. To do so, we introduce the following notation. Given any coherent upper prevision $\overline{P}$, $g \in \mathcal{L}$, and $\beta \in \mathbb{R}$ we define $\overline{P}_{g,\beta}$ on $\operatorname{dom}\overline{P} \cup \{g\}$ as follows:

$$\overline{P}_{g,\beta}(f) := \begin{cases} \overline{P}(f) & \text{if } f \in \operatorname{dom}\overline{P} \backslash \{g\}, \\ \beta & \text{if } f = g. \end{cases} \tag{8}$$

By $\overline{\mathbb{E}}_{g,\beta}$ we denote the natural extension of $\overline{P}_{g,\beta}$.

We are interested in knowing, for a given $g \notin \operatorname{dom}\overline{P}$, the range of β for which $\overline{P}_{g,\beta}$ is coherent. A formula for this was given in [27, §3.1.6, p. 126]. The next theorem provides new formulas for the lower bound of this range, which are more concise and easier to interpret compared to [27, §3.1.6, p. 126].

Theorem 2. *Let $g \notin \operatorname{dom}\overline{P}$ and $\beta \in \mathbb{R}$. Then $\overline{P}_{g,\beta}$ is coherent if and only if*

$$\widetilde{\mathbb{E}}(g) \leq \beta \leq \overline{\mathbb{E}}(g), \tag{9}$$

where

$$\widetilde{\mathbb{E}}(g) := \sup_{f' \in \operatorname{dom}\overline{P}} -\overline{\mathbb{E}}_{-f',-\overline{P}(f')}(-g) \tag{10}$$

$$= \sup_{f' \in \operatorname{dom}\overline{P}} \min_{\substack{\mu \in \mathcal{M}(\overline{P}) \\ \mathbb{E}_\mu(f') = \overline{P}(f')}} \mathbb{E}_\mu(g). \tag{11}$$

Note that $\overline{P}_{-f',-\overline{P}(f')}$ is not necessarily coherent, but it avoids sure loss due to the coherence of $\overline{P}$ [27, p. 123]. So, for any given $g \notin \operatorname{dom}\overline{P}$, any $\overline{P}_{g,\beta}$ is coherent only for $\beta \in [\widetilde{\mathbb{E}}(g), \overline{\mathbb{E}}(g)]$. Therefore, this is the range of coherent value that an expert can provide, assuming the expert is coherent (an assumption we make here). If the bounds coincide, then we already have full knowledge about the upper prevision of g, and querying it would be useless.

Note that the upper bound $\overline{\mathbb{E}}(g)$ does not add any information to what we already know since $\overline{\mathbb{E}} = \overline{\mathbb{E}}_{g,\overline{\mathbb{E}}(g)}$ [25, Cor. 4.32], whilst the lower bound $\widetilde{\mathbb{E}}(g)$ is the most informative value a subject could give if questioned about g, by Lemma 1.

Lemma 1. *$\overline{\mathbb{E}}_{g,\beta} \geq \overline{\mathbb{E}}_{g,\alpha}$ for all $\beta \geq \alpha$.*

Proof. $\overline{P}_{g,\alpha}$ dominates $\overline{P}_{g,\beta}$ whenever $\beta \geq \alpha$. So, by [25, Prop. 4.27], $\overline{\mathbb{E}}_{g,\alpha}$ must also dominate $\overline{\mathbb{E}}_{g,\beta}$.

3 Elicitation for Decision Problems

This section introduces and analyzes a new procedure to perform elicitation for decision problems under severe uncertainty. Unlike most elicitation procedures, this procedure tries to gain as much information as quickly as possible that is relevant to the decision problem at hand.

Algorithm 1. Elicitation

```
Require: 𝒜, P̄, 𝒬
Ensure: opt(P̄, 𝒜)
 1: while 𝒬 ≠ ∅ do
 2:     𝒜 ← opt(P̄, 𝒜)
 3:     for all q ∈ 𝒬 do
 4:         Compute Ẽ(q) and Ē(q)
 5:         Remove q from 𝒬 if Ẽ(q) = Ē(q)
 6:         Otherwise, calculate opt(P̄_{q,Ẽ(q)}, 𝒜)
 7:     end for
 8:     q* ← arg min |opt(P̄_{q,Ẽ(q)}, 𝒜)|
 9:     Present q* to the expert, who returns its upper prevision value, β
10:     P̄ ← P̄_{q*,β}
11:     Remove q* from 𝒬
12: end while
13: return opt(P̄, 𝒜)
```

3.1 Elicitation Procedure

The proposed procedure is given in Algorithm 1.

We used Theorem 1 in line 2 of the algorithm. Specifically, since all upper previsions considered later in the process will always dominate $\overline{P}$, implies that all non-maximal acts from $\mathcal{A}$ with respect to $\overline{P}$ can be removed because these can never become maximal at any later stage of the elicitation process. This can significantly reduce the number of pairs a and a' for which we have to calculate $\overline{\mathbb{E}}(a-a')$ (each $\overline{\mathbb{E}}(a-a')$ requires solving a linear program).

By coherence, the expert will never give a value for β that is less than $\widetilde{\mathbb{E}}(q)$. Consequently, by using $\overline{P}_{q,\widetilde{\mathbb{E}}(q)}$ to pick q^*, we adopt the most optimistic attitude towards the expert's possible answers, leading to the largest potential reduction in the number of maximal acts. Interestingly, such an optimistic view is at work in other selection settings such as racing algorithms [12].

To reduce cognitive burden on the expert, the condition $\mathcal{Q} \neq \emptyset$ could be further expanded, for instance, by simply limiting the total number of queries to be asked, or by considering that having almost indifferent options in $\mathcal{A} \leftarrow \text{opt}(\overline{P}, \mathcal{A})$ is sufficient to stop the procedure. As there are many sensible ways to do this, we have omitted this from the algorithm description.

Another aspect that is important is the set of queries $\mathcal{Q}$. One could for instance query the expert on the upper and lower probabilities of events $A \subseteq \Omega$, which correspond to indicator functions $\mathbb{I}_A$ in terms of gambles. For instance, asking about the lower probability of A is equivalent to asking about the upper prevision of $-\mathbb{I}_A$, and of $\mathbb{I}_A$ if one is interested in the upper probability. Other common queries include comparative ones between events A and B, e.g. $\mathbb{I}_A - \mathbb{I}_B$ [7,13], or the direct comparisons of two alternatives within $\mathcal{A}$. There is clearly a balance to find between the expressivity of $\mathcal{Q}$, and the cognitive accessibility of the queries for the decision maker. However, exploring such a question is out of the scope of the current initial study.

Regarding the complexity of Algorithm 1, computing Eq. (11) in line 4 requires solving one linear program per item in dom $\overline{P}$, and this for each remaining item in $\mathcal{Q}$, meaning that the number of linear programs to solve roughly increases quadratically as we collect queries. We also assume that we can compute $\mathrm{opt}(\overline{P}, \mathcal{A})$ and $\mathrm{opt}(\overline{P}_{q,\widetilde{\mathbb{E}}(q)}, \mathcal{A})$, meaning that we are assuming that we can enumerate acts explicitly, and that those remain in reasonable number (say, less than 100). Of course, a lot of those computations can be parallelised to ensure scalability, yet considering combinatorial optimisation problems would still require some adaptations.

3.2 Fully Worked Example

We now demonstrate Algorithm 1 through Examples 2 to 4, where we iteratively make queries to the expert.

Example 2. Recall the situation from Example 1 where we elicited an expert's knowledge about next year's inflection rate through an upper prevision $\overline{P}$. Based on this information, the maximal acts were 3, 4, 5 and 6. Consider the set of gambles $\mathcal{Q} = \{-\mathbb{I}_L, -\mathbb{I}_M, -\mathbb{I}_H\}$. We want to elicit the expert's upper prevision of one of these queries; this is equivalent to asking about lower probability of each atom. For each query, we can compute the coherent range using Eqs. (3) and (11):

$$\widetilde{\mathbb{E}}(-\mathbb{I}_L) = -0.5 \qquad \overline{\mathbb{E}}(-\mathbb{I}_L) = -0.33 \tag{12}$$

$$\widetilde{\mathbb{E}}(-\mathbb{I}_M) = -0.33 \qquad \overline{\mathbb{E}}(-\mathbb{I}_M) = 0.0 \tag{13}$$

$$\widetilde{\mathbb{E}}(-\mathbb{I}_H) = -0.33 \qquad \overline{\mathbb{E}}(-\mathbb{I}_H) = 0.0 \tag{14}$$

So, for each question $q \in \mathcal{Q} = \{-\mathbb{I}_L, -\mathbb{I}_M, -\mathbb{I}_H\}$, we assume that the expert will state a value $\beta \in [\widetilde{\mathbb{E}}(q), \overline{\mathbb{E}}(q)]$ as their upper prevision for q. For example, for the atom L, any coherent lower probability must be between 0.33 and 0.5. If the expert's answer is not in this range, then the expert is incoherent. If they avoid sure loss, we may adjust the expert's answer through natural extension to make them coherent. If they do not avoid sure loss, we must ask the expert to revise their answer until they do avoid sure loss.

Here are two extreme cases. If the expert gives $\beta = \overline{\mathbb{E}}(q)$, then $\overline{\mathbb{E}} = \overline{\mathbb{E}}_{q,\beta}$, then no information is gained to assist the decision maker. For the other extreme case, if the expert gives $\beta = \widetilde{\mathbb{E}}(q)$, then this is the most informative case that reduces as much as possible the credal set while still being coherent.

We can see from Table 3 that for $-\mathbb{I}_L$, all rows remain positive. Consequently, regardless of the expert answer, as long as the expert remains coherent, asking for their upper prevision of $-\mathbb{I}_L$ will not change the number of maximal elements. On the other hand, $-\mathbb{I}_M$ has more potential: Table 4 shows that stock fund 3 may no longer be maximal since there is a negative value in the corresponding row. Finally, $-\mathbb{I}_H$ has the most potential: Table 4 shows that stock funds 4 and 6 may no longer be maximal (Table 5).

Table 3. Values of $\overline{\mathbb{E}}_{-\mathbb{I}_L,-0.5}(a-a')$ in Example 2.

a \ a'	3	4	5	6
3	0.0	1.5	1.0	2.5
4	2.0	0.0	2.0	4.0
5	2.0	0.5	0.0	2.0
6	2.0	0.0	0.0	0.0

Table 4. Values of $\overline{\mathbb{E}}_{-\mathbb{I}_M,-0.33}(a-a')$ in Example 2.

a \ a'	3	4	5	6
3	0.0	0.33	−0.67	−0.33
4	2.0	0.0	0.67	1.33
5	2.0	1.0	0.0	0.67
6	2.0	0.67	0.0	0.0

Table 5. Values of $\overline{\mathbb{E}}_{-\mathbb{I}_H,-0.33}(a-a')$ in Example 2.

a \ a'	3	4	5	6
3	0.0	1.5	1.0	2.5
4	−0.33	0.0	0.33	2.0
5	0.67	1.0	0.0	1.67
6	0.33	0.67	−0.33	0.0

In conclusion, it is not worth asking about $-\mathbb{I}_L$, but asking about $-\mathbb{I}_M$ or $-\mathbb{I}_H$ can potentially reduce the number of maximal acts. Since adding $-\mathbb{I}_H$ to the domain has the potentially largest reduction in the number of maximal acts, we will ask the expert for their upper prevision of $-\mathbb{I}_H$.

Suppose the expert specifies a value $\beta = -0.2$ for the upper prevision of $-\mathbb{I}_H$ (so a lower probability of 0.2 for H). So, we now have $\operatorname{dom}\overline{P} = \{\mathbb{I}_M - \mathbb{I}_L, \mathbb{I}_H - \mathbb{I}_L, -\mathbb{I}_H\}$ with

$$\overline{P}(\mathbb{I}_M - \mathbb{I}_L) = 0, \qquad \overline{P}(\mathbb{I}_H - \mathbb{I}_L) = 0, \qquad \overline{P}(-\mathbb{I}_H) = -0.2 \tag{15}$$

Next, we can calculate $\overline{\mathbb{E}}(a-a')$ based on this updated $\overline{P}$; see Table 6. As we can see from Table 6, the maximal acts are stock funds 3, 4, and 5.

Table 6. Values of $\overline{\mathbb{E}}(a-a')$ after adding $-\mathbb{I}_H$ to $\operatorname{dom}\overline{P}$ in Example 2.

a \ a'	3	4	5	6
3	0.0	1.5	1.0	2.5
4	0.6	0.0	1.0	2.8
5	1.2	1.0	0.0	1.8
6	1.0	0.67	−0.2	0.0

Example 3. We can now continue the process by considering the remaining queries, that is, $\mathcal{Q} = \{-\mathbb{I}_L, -\mathbb{I}_M\}$. As before, we first compute the range of coherent extensions for these queries:

$$\widetilde{\mathbb{E}}(-\mathbb{I}_L) = -0.4 \qquad \overline{\mathbb{E}}(-\mathbb{I}_L) = -0.33 \tag{16}$$

$$\widetilde{\mathbb{E}}(-\mathbb{I}_M) = -0.33 \qquad \overline{\mathbb{E}}(-\mathbb{I}_M) = 0.0 \tag{17}$$

Again, for each $q \in \mathcal{Q}$, suppose that the expert will provide a value $\beta \in [\widetilde{\mathbb{E}}(q), \overline{\mathbb{E}}(q)]$. As before, we use the most optimistic case where $\beta = \widetilde{\mathbb{E}}(q)$ to choose a query. We compute the values for $\overline{\mathbb{E}}_{q,\widetilde{\mathbb{E}}(q)}(a - a')$, where $q = -\mathbb{I}_L$ and $q = -\mathbb{I}_M$ and present results in Tables 7 and 8. Since all rows in Table 7 are non-negative, it is not worth asking the expert about $-\mathbb{I}_L$. However, as there is a strictly negative number in Table 8, asking about $-\mathbb{I}_M$ might make stock fund 4 non-maximal.

Suppose that the expert says $\overline{P}(-\mathbb{I}_M) = -0.2$. So, we now have $\operatorname{dom} \overline{P} = \{\mathbb{I}_M - \mathbb{I}_L, \mathbb{I}_H - \mathbb{I}_L, -\mathbb{I}_H, -\mathbb{I}_M\}$ with

$$\overline{P}(\mathbb{I}_M - \mathbb{I}_L) = 0, \quad \overline{P}(\mathbb{I}_H - \mathbb{I}_L) = 0, \quad \overline{P}(-\mathbb{I}_H) = -0.2, \quad \overline{P}(-\mathbb{I}_M) = -0.2. \tag{18}$$

We then can calculate $\overline{\mathbb{E}}(a - a')$ for the remaining acts and present them in Table 9. Unfortunately, all entries in Table 9 are non-negative, so the maximal acts remain the same.

Table 7. Values of $\overline{\mathbb{E}}_{-L,-0.4}(a - a')$ in Example 3.

$a \backslash a'$	3	4	5
3	0.0	1.5	1.0
4	0.6	0.0	1.0
5	1.2	0.8	0.0

Table 8. Values of $\overline{\mathbb{E}}_{-M,-0.33}(a - a')$ in Example 3.

$a \backslash a'$	3	4	5
3	0.0	0.33	−0.67
4	0.6	0.0	−0.33
5	1.2	1.0	0.0

Table 9. Values of $\overline{\mathbb{E}}(a - a')$ after adding $-M$ to $\operatorname{dom} \overline{P}$ in Example 3.

$a \backslash a'$	3	4	5
3	0.0	0.8	0.0
4	0.6	0.0	0.2
5	1.2	1.0	0.0

Table 10. Values of $\overline{\mathbb{E}}_{-L,-0.4}(a - a')$ in Example 4.

$a \backslash a'$	3	4	5
3	0.0	0.8	0.0
4	0.6	0.0	0.2
5	1.2	0.8	0.0

Example 4. Finally, we consider the last query $-\mathbb{I}_L$ and compute its range of coherent extensions:

$$\widetilde{\mathbb{E}}(-\mathbb{I}_L) = -0.4 \quad \overline{\mathbb{E}}(-\mathbb{I}_L) = -0.33. \tag{19}$$

A coherent expert will give a value $\beta \in [\widetilde{\mathbb{E}}(-\mathbb{I}_L), \overline{\mathbb{E}}(-\mathbb{I}_L)]$. Again, we consider the most optimistic case where $\beta = \widetilde{\mathbb{E}}(-\mathbb{I}_L)$ and compute the values for $\overline{\mathbb{E}}_{-\mathbb{I}_L,-0.4}(a - a')$; see Table 10. Since all entries in Table 10 are non-negative, no answer from the expert about $-\mathbb{I}_L$ will change the number of maximal elements. Therefore, we can stop the elicitation process here.

4 Conclusion

In this paper, we studied the problem of elicitation in the context of decision making. We aimed to gain the most useful information to efficiently identify queries for a specific decision problem. To do so, we treated the expert's information as an upper prevision and proposed a new algorithm to most effectively pick a query to present to an expert. Assuming the expert remains coherent at all times, we used the lower bound of the coherent range to select the query that has most potential to reduce the number of maximal acts. A fully worked example demonstrated the algorithm.

In future work, we might explore additional stopping criteria to terminate elicitation. For example, we may consider to terminate the elicitation process when we can no longer remove any non-maximal acts regardless of the next query. This could save cognitive burden on the expert since we can present fewer queries. Moreover, instead of using the lower bound of the coherent range, we can look at other values, such for instance a mixture between the upper and lower bounds. Similarly, it would be desirable to test the impact of choosing different sets of possible queries, both in terms of convergence to the desired numerical outcome, and in terms of easiness for the decision maker. Such comparisons would necessitate extensive experiments.

Finally, it would be interesting to test the procedure on an actual decision problem with an actual expert, for instance to see if our assumption of coherence of the expert is reasonable in practice. In order to do this, we could take inspiration from previous experimental campaigns aiming at eliciting actual imprecise probabilities [17].

Acknowledgments. This project is funded by National Research Council of Thailand (NRCT): Contract number N42A660964. This research was supported by Chiang Mai University.

Declaration

Disclosure of Interests. The authors have no competing interests to declare that are relevant to the content of this article.

References

1. Adam, L., Destercke, S.: Possibilistic preference elicitation by minimax regret. In: Uncertainty in Artificial Intelligence, pp. 718–727. PMLR (2021)
2. Augustin, T., Coolen, F.P.A., de Cooman, G., Troffaes, M.C.M.: Introduction to Imprecise Probabilities, vol. 591. Wiley, Hoboken (2014)
3. Benabbou, N., Gonzales, C., Perny, P., Viappiani, P.: Minimax regret approaches for preference elicitation with rank-dependent aggregators. EURO J. Decis. Process. **3**, 29–64 (2015)

4. Blanco, M., Engelmann, D., Koch, A.K., Normann, H.T.: Belief elicitation in experiments: is there a hedging problem? Exp. Econ. **13**, 412–438 (2010). https://doi.org/10.1007/s10683-010-9249-1
5. Boutilier, C.: Computational decision support: regret-based models for optimization and preference elicitation (2013)
6. Coletti, G., Scozzafava, R.: The role of coherence in eliciting and handling imprecise probabilities and its application to medical diagnosis. Inf. Sci. **130**(1), 41–65 (2000). https://doi.org/10.1016/S0020-0255(00)00085-2
7. Erreygers, A., Miranda, E.: A graphical study of comparative probabilities. J. Math. Psychol. **104**, 102582 (2021)
8. Garthwaite, P.H., Kadane, J.B., O'Hagan, A.: Statistical methods for eliciting probability distributions. J. Am. Stat. Assoc. **100**(470), 680–701 (2005)
9. Gilovich, T., Griffin, D., Kahneman, D.: Heuristics and Biases: The Psychology of Intuitive Judgment. Cambridge University Press, Cambridge (2002)
10. Jansen, C., Blocher, H., Augustin, T., Schollmeyer, G.: Information efficient learning of complexly structured preferences: elicitation procedures and their application to decision making under uncertainty. Int. J. Approximate Reasoning **144**, 69–91 (2022). https://doi.org/10.1016/j.ijar.2022.01.016
11. Kadane, J.B., Winkler, R.L.: Separating probability elicitation from utilities. J. Am. Stat. Assoc. **83**(402), 357–363 (1988)
12. Maron, O., Moore, A.W.: The racing algorithm: model selection for lazy learners. Artif. Intell. Rev. **11**, 193–225 (1997)
13. Miranda, E., Destercke, S.: Extreme points of the credal sets generated by comparative probabilities. J. Math. Psychol. **64**, 44–57 (2015)
14. Oakley, J.E., O'Hagan, A.: Uncertainty in prior elicitations: a nonparametric approach. Biometrika **94**(2), 427–441 (2007)
15. O'Hagan, A., et al.: The elicitation of probabilities, pp. 61–96. Wiley (2006). https://doi.org/10.1002/0470033312.ch4
16. O'Hagan, A., et al.: Uncertain Judgements: Eliciting Experts' Probabilities. Wiley, Hoboken (2006)
17. Quaeghebeur, E., Wesseling, C., Beauxis-Aussalet, E., Piovesan, T., Sterkenburg, T.: The CWI world cup competition: eliciting sets of acceptable gambles. In: Proceedings of the Tenth International Symposium on Imprecise Probability: Theories and Applications, pp. 277–288. PMLR (2017)
18. Renooij, S.: Probability elicitation for belief networks: issues to consider. Knowl. Eng. Rev. **16**(3), 255–269 (2001). https://doi.org/10.1017/S0269888901000145
19. Rinderknecht, S.L., Borsuk, M.E., Reichert, P.: Bridging uncertain and ambiguous knowledge with imprecise probabilities. Environ. Model. Softw. **36**, 122–130 (2012)
20. Sandri, S.A., Dubois, D., Kalfsbeek, H.W.: Elicitation, assessment, and pooling of expert judgments using possibility theory. IEEE Trans. Fuzzy Syst. **3**(3), 313–335 (1995)
21. Silva, A.A., de Souza, F.C.: A protocol for the elicitation of imprecise probabilities. In: ISIPTA, vol. 1, pp. 315–321 (2005)
22. Smithson, M.: Elicitation, pp. 318–328. Wiley (2014). https://doi.org/10.1002/9781118763117.ch15
23. T'Joens, N.: Active elicitation of imprecise probability models. Master's thesis, Ghent University (2017). https://lib.ugent.be/catalog/rug01:002367397
24. Troffaes, M.C.M.: Decision making under uncertainty using imprecise probabilities. Int. J. Approximate Reasoning **45**(1), 17–29 (2007). https://doi.org/10.1016/j.ijar.2006.06.001

25. Troffaes, M.C.M., de Cooman, G.: Lower Previsions. Wiley Series in Probability and Statistics, Wiley, Hoboken (2014). https://doi.org/10.1002/9781118762622
26. Utkin, L.V.: Probabilities of judgments provided by unknown experts by using the imprecise Dirichlet model. Risk Decis. Policy **9**(4), 371–389 (2004)
27. Walley, P.: Statistical Reasoning with Imprecise Probabilities. Chapman and Hall, London (1991)
28. Williams, P.M.: Notes on conditional previsions. Int. J. Approximate Reasoning **44**(3), 366–383 (2007)
29. Yaghlane, A.B., Denoeux, T., Mellouli, K.: Constructing belief functions from qualitative expert opinions. In: 2006 2nd International Conference on Information & Communication Technologies, vol. 1, pp. 1363–1368. IEEE (2006)

Accelerate K-Mode Algorithms Using The Triangle Inequality

Vu-Linh Nguyen[1(✉)], Toan Nguyen-Mau[2], and Van-Nam Huynh[2]

[1] Heudiasyc Laboratory, University of Technology of Compiègne, Compiègne, France
vu-linh.nguyen@hds.utc.fr

[2] Japan Advanced Institute of Science and Technology, Nomi, Japan
{nmtoan,huynh}@jaist.ac.jp

Abstract. Clustering is an unsupervised machine learning task that aims to discover natural groups in the given dataset. K-mode algorithms, which are adaptions of K-mean algorithms for continuous data, are among the most popular algorithms for discovering clusters in categorical data. In this paper, we present some results on how to accelerate them using the triangle inequality, while still always computing exactly the same result as the original K-mode algorithms. We also provide some empirical evidence to illustrate the potential gains provided by leveraging the triangle inequality. Finally, we envision future work aimed at providing a comprehensive understanding of the use of triangle inequality in accelerating other clustering algorithms for categorical data.

Keywords: Clustering Categorical Data · K-Modes · Triangle Inequality · Decomposable Dissimilarities

1 Introduction

In the common setting of clustering, the primary goal is to partition N given objects into K clusters, where K is typically fixed before performing the clustering process, in which each object belongs to the corresponding (closest) cluster center [3,9,28]. K-mean algorithms are among the most popular methods for clustering continuous data, which aims to minimize the within-cluster variations. K-mode algorithms [2,11,13,20] are adaptions of K-mean algorithms and among the most popular clustering algorithms for discovering clusters in categorical data.

Yet, it has been shown that leveraging the triangle inequality may dramatically accelerate K-mean algorithms [6,7,9,10,15]. To the best of our knowledge, there is a lack of extensive investigation on similar aspects for K-mode algorithms. To complement the existing literature on clustering categorical data, this paper presents some first results on accelerating K-mode algorithms using the triangle inequality and further discussions on worthy follow-up work on this line of research. Yet, computational complexities in computing modes are not the primary focus of this work. This problem would not be decoupled from how to effectively update modes when considering the development of K-mode algorithms as

S. Destercke et al. (Eds.): SUM 2024, LNAI 15350, pp. 325–339, 2025.
https://doi.org/10.1007/978-3-031-76235-2_24

a whole. We therefore discuss such aspects for various distances that can be implemented with the K-mode algorithms to facilitate follow-up works.

We provide in the next section a detailed description of K-mode algorithms, discuss their computational complexity aspects, and then discuss the potential use of the triangle inequality in accelerating them in Sect. 3. The key idea is to eliminate redundant pairwise computations (informed by the triangle inequality) when performing the K-mode algorithms. Therefore, the accelerated K-mode algorithms still always return exactly the same result as the original K-mode algorithms, while often requiring one to compute a smaller number of pairwise dissimilarities. Our main contribution here would be insightful discussions on conditions under which some procedures to accelerate K-mean algorithms [6,7,9,10,15] can be adapted to accelerate K-mode algorithms.

Section 4 presents some empirical evidence to illustrate the potential gains by leveraging the triangle inequality from the computational aspects. As a first attempt, we shall exclusively focus on the basic K-mode algorithm [11,20], in which the dissimilarity is the Hamming dissimilarity, though our investigation can be enlarged to cover other decomposable distances straightforwardly. Section 5 envisions future work aimed at providing a comprehensive understanding of the use of triangle inequality in accelerating other clustering algorithms [2,11,13,18,19,27] for categorical data. Finally, Sect. 6 concludes this paper.

2 *K*-Mode Algorithms

This section provides a generic formulation of K-mode algorithms in the setting of categorical data, where each object is specified by P categorical attributes $\mathbf{A} = \{A^1, \ldots, A^P\}$. For each $p \in [P] := \{1, \ldots, P\}$, let $\mathcal{A}^p := \{a_1^p, \ldots, a_{Q_p}^p\}$ be the domain of A^p. Let $\mathcal{A} := \mathcal{A}^1 \times \ldots \times \mathcal{A}^P$ be the set of $\prod_{p=1}^P Q_p$ possible objects.

Let $\mathbf{D} = \{\boldsymbol{a}_1, \ldots, \boldsymbol{a}_N\} \subset \mathcal{A}$ be a dataset consisting of N objects, where $\boldsymbol{a}_n := (a_n^1, \ldots, a_n^P) \in \mathcal{A}$, $n \in [N]$. The primary goal of K-mode algorithms is to partition the N objects into a pre-defined number K of disjoint clusters/groups $\mathbf{G}_1, \ldots, \mathbf{G}_K$ which optimizes some evaluation criterion reflecting the clustering performance. Formally, the disjoint clusters should satisfy the following properties: $\mathbf{G}_k \neq \emptyset$, $\forall k \in [K]$, $\cup_{k=1}^K \mathbf{G}_k = \mathbf{A}$, and $\mathbf{G}_k \cap \mathbf{G}_{k'} = \emptyset$, $\forall k \neq k' \in [K]$.

Let $d : \mathcal{A} \times \mathcal{A} \longmapsto R_+$ be a dissimilarity. Ideally, a K-Mode algorithm should seek a partition $\mathbf{G} = \cup_{k=1}^K \mathbf{G}_k$ that minimizes the total within-cluster variation

$$V_W^d(\mathbf{G}) = \sum_{k=1}^{K} \sum_{\boldsymbol{a}_n \in \mathbf{G}_k} \sum_{\boldsymbol{a}_{n'} \in \mathbf{G}_k} d\left(\boldsymbol{a}_n, \boldsymbol{a}_{n'}\right) . \tag{1}$$

K-mode algorithms often start with the initialization step which generates K initial mode $\boldsymbol{c}_k$, one per cluster, and alternate between two following steps until some stopping criterion, such as the maximum number of iteration T, is met:

- **Assignment step:** Assign each object to the cluster with the nearest mode w.r.t. the dissimilarity d.
- **Update step:** Re-calculate the modes using the current assignment.

Algorithm 1. Lloyd's procedure reproduced from [15,17]

Input: A number K, a data set $\mathbf{D} = \{\boldsymbol{a}_1, \ldots, \boldsymbol{a}_N\}$, and initial modes $\boldsymbol{c}_k$, $k \in [K]$,
while Termination conditions are not met **do**
  for $n \leftarrow 1$ **to** N **do** # Assignment step
    for $k \leftarrow 1$ **to** K **do**
      Compute $d(\boldsymbol{a}_n, \boldsymbol{c}_k)$
    end for
    $I(n) \leftarrow \text{argmin}_{k \in [K]} d(\boldsymbol{a}_n, \boldsymbol{c}_k)$
  end for
  for $k \leftarrow 1$ **to** K **do** # Update step
    Determine $\mathbf{G}_k = \{\boldsymbol{a}_n \in \mathbf{D} | I(n) = k\}$
    Determine $\boldsymbol{c}_k$ using (3)
  end for
end while
Output: Final modes $\boldsymbol{c}_k$, $k \in [K]$ and final assignments $I(n)$, $n \in [N]$

The conventional Lloyd's procedure, for performing K-mode algorithms is summarized in Algorithm 1. The computational complexity of K-mode algorithms can greatly depend on the nature of the dissimilarity d. In the next sections, we discuss this aspect for the assignment and update steps, with a focus on decomposable dissimilarities, which are the main focus of this paper. From now on, we call any dissimilarity d that can be expressed in the form

$$d(\boldsymbol{a}_n, \boldsymbol{a}_{n'}) = \sum_{p=1}^{P} d^p(a_n^p, a_{n'}^p), \forall (a_n^p, a_{n'}^p) \in \mathcal{A} \times \mathcal{A}, \tag{2}$$

a decomposable dissimilarity, and call any dissimilarity that cannot be expressed in this form a non-decomposable dissimilarity. Examples of decomposable dissimilarities are the Hamming dissimilarity [20], the Distance Learning Dissimilarity for Categorical Data (DILCA) [12], and a few other dissimilarities [2,20].

2.1 Update Step

Given a cluster $\mathbf{G}_k$, a commonly used representation of the mode $\boldsymbol{c}_k$ is the object, which does not necessarily belong to $\mathbf{D}$, minimizing the average dissimilarity to the elements of $\mathbf{G}_k$. Formally, we have

$$\boldsymbol{c}_k \in \underset{\boldsymbol{c} \in \mathcal{A}}{\text{argmin}} \sum_{\boldsymbol{a}_n \in \mathbf{G}_k} d(\boldsymbol{a}_n, \boldsymbol{c}) . \tag{3}$$

The complexity of finding a mode $\boldsymbol{c}_k$ (3) may greatly depend on the nature of the dissimilarity d. Efficient algorithms to find $\boldsymbol{c}_k$ are known in the cases of the Hamming dissimilarity [11] and a few decomposable dissimilarities [2,20]. Interestingly, the problem of finding $\boldsymbol{c}_k$ (3) is computationally equivalent to finding Bayes-optimal predictions (BOPs) in probabilistic multi-dimensional classification (PMDC) [22]. This relation should be clear when considering the degree of membership of each $\boldsymbol{a}_n$ w.r.t. the cluster $\mathbf{G}_k$ as the conditional probability

$$\Pr(\boldsymbol{a}_n|\mathbf{G}_k) \approx \frac{1}{|\mathbf{G}_k|} \sum_{\boldsymbol{a}_{n'} \in \mathbf{G}_k} [\![\boldsymbol{a}_n = \boldsymbol{a}_{n'}]\!], \tag{4}$$

where $[\![\cdot]\!]$ is the indicator function, i.e., $[\![A]\!] = 1$ if the predicate A is true and $= 0$ otherwise. Under this perspective, an optimal mode $\boldsymbol{c}_k$ becomes

$$\boldsymbol{c}_k \in \underset{\boldsymbol{c} \in \mathcal{A}}{\operatorname{argmin}} \sum_{\boldsymbol{a}_n \in \mathbf{G}_k} d(\boldsymbol{a}_n, \boldsymbol{c}) = \underset{\boldsymbol{c} \in \mathcal{A}}{\operatorname{argmin}} \sum_{\boldsymbol{a}_n \in \mathcal{A}} d(\boldsymbol{a}_n, \boldsymbol{c}) \Pr(\boldsymbol{a}_n|\mathbf{G}_k). \tag{5}$$

This observation simply implies that all the (known) results on finding BOPs in PMDC [22] can be leveraged to develop efficient algorithms to find the mode $\boldsymbol{c}_k$ (3). In the cases of binary data, i.e., $Q_p = 2$, for any $p \in [P]$, finding $\boldsymbol{c}_k$ (3) is computationally equivalent to finding BOPs in probabilistic multi-label classification (PMLC) [5,21,30].

The Cases of Decomposable Dissimilarities. The following remark informs that when d is decomposable, finding a mode (3) can be solved reasonably fast.

Remark 1. Let d be any decomposable dissimilarity (2). The problem of finding a mode of cluster (3) can be done in time $O\left(\sum_{p=1}^{P} (Q_p)^2\right)$ given the marginal probabilities masses

$$\Pr(a_{q_p}^p|\mathbf{G}_k) \approx \frac{1}{|\mathbf{G}_k|} \sum_{\boldsymbol{a}_{n'} \in \mathbf{G}_k} [\![a_{q_p}^p = a_{n'}^p]\!], \forall p \in [P], \forall a_{q_p}^p \in \mathcal{A}^p. \tag{6}$$

Proof. A proof is simple and is given for completeness. For any decomposable dissimilarity d (2), the optimization problem (5) can be rewritten as

$$\begin{aligned}
\boldsymbol{c}_k &\in \underset{\boldsymbol{c} \in \mathcal{A}}{\operatorname{argmin}} \sum_{\boldsymbol{a}_n \in \mathcal{A}} \left(\sum_{p=1}^{P} d^p(a_n^p, c^p) \right) \Pr(\boldsymbol{a}_n|\mathbf{G}_k) \qquad (7)\\
&= \underset{\boldsymbol{c} \in \mathcal{A}}{\operatorname{argmin}} \sum_{p=1}^{P} \left(\sum_{\boldsymbol{a}_n \in \mathcal{A}} d^p(a_n^p, c^p) \Pr(\boldsymbol{a}_n|\mathbf{G}_k) \right)\\
&= \underset{\boldsymbol{c} \in \mathcal{A}}{\operatorname{argmin}} \sum_{p=1}^{P} \left(\sum_{a_{q_p}^p \in \mathcal{A}^p} \left(\sum_{\substack{\boldsymbol{a}_n \in \mathcal{A} \\ a_n^p = a_{q_p}^p}} d^p(a_{q_p}^p, c^p) \Pr(\boldsymbol{a}_n|\mathbf{G}_k) \right) \right)\\
&= \underset{\boldsymbol{c} \in \mathcal{A}}{\operatorname{argmin}} \sum_{p=1}^{P} \left(\sum_{a_{q_p}^p \in \mathcal{A}^p} \left(d^p(a_{q_p}^p, c^p) \sum_{\substack{\boldsymbol{a}_n \in \mathcal{A} \\ a_n^p = a_{q_p}^p}} \Pr(\boldsymbol{a}_n|\mathbf{G}_k) \right) \right)\\
&= \underset{\boldsymbol{c} \in \mathcal{A}}{\operatorname{argmin}} \sum_{p=1}^{P} \left(\sum_{a_{q_p}^p \in \mathcal{A}^p} d^p(a_{q_p}^p, c^p) \Pr(a_{q_p}^p|\mathbf{G}_k) \right).
\end{aligned}$$

Therefore, $\boldsymbol{c}_k = (c_k^1 \dots, c_k^P)$ is specified by

$$c_k^p \in \operatorname*{argmin}_{c^p \in \mathcal{A}^p} \sum_{a_{q_p}^p \in \mathcal{A}^p} d^p(a_{q_p}^p, c^p) \Pr(a_{q_p}^p | \mathbf{G}_k) \, , p \in [P] \, . \tag{8}$$

For each $p \in [P]$, finding c_k^p requires comparing the Q_p possible $c^p \in \mathcal{A}^p$ and can be done in time $O\left((Q_p)^2\right)$. □

As a sequel, in the case of Hamming dissimilarity, which is defined as

$$d_{\text{Ham}}(\boldsymbol{a}_n, \boldsymbol{a}_{n'}) = \sum_{p=1}^{P} [\![a_n^p \neq a_{n'}^p]\!] \, , \forall (a_n^p, a_{n'}^p) \in \mathcal{A} \times \mathcal{A} \, , \tag{9}$$

finding a mode can be done in $O\left(\sum_{p=1}^{P} Q_p\right)$ given the probabilities masses (6). This is because $\boldsymbol{c}_k = (c_k^1 \dots, c_k^P)$ is specified by

$$\begin{aligned} c_k^p &\in \operatorname*{argmin}_{c^p \in \mathcal{A}^p} \sum_{a_{q_p}^p \in \mathcal{A}^p} [\![a_{q_p}^p \neq c^p]\!] \Pr(a_{q_p}^p | \mathbf{G}_k) = \operatorname*{argmin}_{c^p \in \mathcal{A}^p} \sum_{a_{q_p}^p \in \mathcal{A}^p \setminus \{c^p\}} \Pr(a_{q_p}^p | \mathbf{G}_k) \\ &= \operatorname*{argmin}_{c^p \in \mathcal{A}^p} (1 - \Pr(c^p | \mathbf{G}_k)) = \operatorname*{argmax}_{c^p \in \mathcal{A}^p} \Pr(c^p | \mathbf{G}_k) \, , \end{aligned} \tag{10}$$

for any $p \in [P]$, and finding the most probable $c_k^p \in \mathcal{A}^p$ can be done in $O(Q_p)$.

2.2 Assignment Step

The assignment step is another source of computational complexity one may have to deal with. Naive solutions often require one to compute the dissimilarities between all the possible pairs $(\boldsymbol{a}_n, \boldsymbol{c}_k)$. This can be extremely expensive when the data set $\mathbf{D}$ is large and of high dimension, i.e., P is large, and the chosen dissimilarity d is complex. More precisely, the assignment step of Lloyd's procedure summarised in Algorithm 1 requires time $O(NK\delta(N, K))$, where $\delta(N, K)$ is an upper cost of computing $d(\boldsymbol{a}_n, \boldsymbol{c}_k)$.

In the next section, we shall explore the potential use of the triangle inequality in accelerating the assignment step. The key idea is to eliminate redundant calculations on pairwise dissimilarity, informed by the triangle inequality, when updating the modes when the K-Modes procedure goes along.

3 Accelerated K-Mode Algorithms

Yet, there are different procedures to accelerate K-mean algorithms by leveraging the triangle inequality [6,7,9,10,15], which are initiated by [9]. As shall be discussed in this section, not all of them can be (easily) adapted to accelerate K-mode algorithms.

3.1 Triangle Inequality Based Accelerated Procedures

The prerequisite is that the dissimilarity d satisfies the triangle inequality, i.e.,

$$d(\boldsymbol{a}_n, \boldsymbol{a}_{n"}) \le d(\boldsymbol{a}_n, \boldsymbol{a}_{n'}) + d(\boldsymbol{a}_{n'}, \boldsymbol{a}_{n"}), \tag{11}$$

for any possible triplet $(\boldsymbol{a}_n, \boldsymbol{a}_{n'}, \boldsymbol{a}_{n"}) \in \mathcal{A} \times \mathcal{A} \times \mathcal{A}$. The next remark shows that various decomposable dissimilarities d (2) satisfy the triangle inequality (11).

Remark 2. Let d be a decomposable dissimilarity of the form (2). Assume that for any $p \in [P]$, the basic dissimilarity d^p satisfies the triangle inequality (11). Then, d satisfies the triangle inequality (11).

Proof. A proof is trivial. Let $(\boldsymbol{a}_n, \boldsymbol{a}_{n'}, \boldsymbol{a}_{n"})$ be any possible triplet of objects in $\mathcal{A}$. For any $p \in [P]$, since d^p satisfies the triangle inequality (11), we have

$$d^p(a_n^p, a_{n"}^p) \le d^p(a_n^p, a_{n'}^p) + d^p(a_{n'}^p, a_{n"}^p) \tag{12}$$

Therefore, d satisfies the triangle inequality (11), i.e.,

$$\begin{aligned} d(\boldsymbol{a}_n, \boldsymbol{a}_{n"}) &= \sum_{p=1}^{P} d^p(a_n^p, a_{n"}^p) \le \sum_{p=1}^{P} (d^p(a_n^p, a_{n'}^p) + d^p(a_{n'}^p, a_{n"}^p)) \\ &= \sum_{p=1}^{P} d^p(a_n^p, a_{n'}^p) + \sum_{p=1}^{P} d^p(a_{n'}^p, a_{n"}^p) = d(\boldsymbol{a}_n, \boldsymbol{a}_{n'}) + d(\boldsymbol{a}_n, \boldsymbol{a}_{n"}). \quad \square \end{aligned}$$

As a sequel, the Hamming dissimilarity (9) satisfies the triangle inequality (11) since it is decomposable and its the basic dissimilarity $d^p_{\text{Ham}} = [\![\cdot \neq \cdot]\!]$ satisfies the triangle inequality. Discussions on some other (non-)decomposable dissimilarities that satisfy the triangle inequality (11) are deferred to Sect. 5.1.

Clearly, accelerated procedures which are purely based on the triangle inequality (11), such as Elkan's procedure [9], Drake's procedure [7,10] and Yinyang procedure [6], should be adapted to accelerate K-mode algorithms straightforwardly whenever the dissimilarity d satisfies the triangle inequality (11). However, there can be accelerated procedures, which can be hardly adapted to accelerate K-mode algorithms (even if d satisfies the triangle inequality (11)). An example of such procedures is the Annulus procedure [7,10], in which the notion of the origin, which is unclear how to define for categorical data, is required.

Yet, our discussions suggest that there may be (plenty of) room to accelerate K-Modes by leveraging the triangle inequality (11), even by simply choosing suitable existing accelerated procedures developed for K-mean algorithms and then adapting them for K-mode algorithms. Before investigating efforts on developing/implementing/adapting such accelerated procedures, which may be far from obvious (at least from the engineering aspects) [15], we think it would be beneficial to anticipate to what extent accelerated procedures may be useful.

To answer this question, we implement Elkan's procedure [9] whose details are recalled in the next section, and then empirically assess its advantages, compared to Lloyd's procedure, in Sect. 4. Elkan's procedure has been shown to be a strong procedure in detecting redundant pairwise dissimilarity calculations [15]. Therefore, it would also help to estimate the number of redundant pairwise dissimilarity calculations, which can be detected by related accelerated procedures, such as Drake's procedure [7,10] and Yinyang procedure [6].

3.2 Elkan's Procedure

To facilitate follow-up works (which may require one to re-implement things from scratch and parallelize algorithms to translate gains in the number of pairwise dissimilarities into gains in the running time), we provide a rather detailed explanation of Elkan's procedure reproduced from [9,15]. This might be beneficial to understand why Elkan's procedure may work and understand related procedures, such as Drake's procedure [7,10] and Yinyang procedure [6].

For any $\boldsymbol{a}_n$, let $I(n)$ be the index of the current closest modes. The idea is to maintain N upper bounds (13) and NK lower bounds (14):

$$u(n) \geq d(\boldsymbol{a}_n, \boldsymbol{c}_{I(n)})\,, n \in [N]\,, \tag{13}$$

$$l(n,k) \leq d(\boldsymbol{a}_n, \boldsymbol{c}_k)\,, n \in [N]\,, k \in [K]\,. \tag{14}$$

Assume the upper bounds and lower bounds are given. If $u(n) \leq l(n,k)$, then from (13) and (14), it is clear that

$$d(\boldsymbol{a}_n, \boldsymbol{c}_{I(n)}) \leq u(n) \leq l(n,k) \leq d(\boldsymbol{a}_n, \boldsymbol{c}_k)\,. \tag{15}$$

Therefore, we know that $\boldsymbol{a}_n$ can not belong to $\boldsymbol{c}_k$ without the need of computing $d(\boldsymbol{a}_n, \boldsymbol{c}_{I(n)})$ and $d(\boldsymbol{a}_n, \boldsymbol{c}_k)$.

For any $k \in [K]$, let $\boldsymbol{c}'_k$ and $\boldsymbol{c}_k$ be respectively the current and updated modes. During the update step, Elkan uses the triangle inequality (11) to update the bounds (13) and (14) using the $N(K+1)$ current bounds and K dissimilarities:

$$\delta(k) := d(\boldsymbol{c}'_k, \boldsymbol{c}_k), k \in [K]\,. \tag{16}$$

The updating rules are

$$l(n,k) = l'(n,k) - \delta(k) \text{ and } u(n) = u'(n) + \delta(I(n))\,. \tag{17}$$

The relation (13) is ensured because the triangle inequality (11) implies that

$$u(n) \geq d(\boldsymbol{a}_n, \boldsymbol{c}'_{I(n)}) + d(\boldsymbol{c}'_{I(n)}, \boldsymbol{c}_{I(n)}) \geq d(\boldsymbol{a}_n, \boldsymbol{c}_{I(n)})\,. \tag{18}$$

The relation (14) is ensured because the triangle inequality (11) implies that

$$l(n,k) \leq d(\boldsymbol{a}_n, \boldsymbol{c}'_k) - d(\boldsymbol{c}'_k, \boldsymbol{c}_k) \leq d(\boldsymbol{a}_n, \boldsymbol{c}_k)\,. \tag{19}$$

Elkan's procedure also use the triangle inequality (11) to avoid redundant pairwise dissimilarity calculations $d(\boldsymbol{a}_n, \boldsymbol{c}_k)$ whenever

$$d(\boldsymbol{a}_n, \boldsymbol{c}_k) \geq 2d(\boldsymbol{a}_n, \boldsymbol{c}_{I(n)})\,. \tag{20}$$

This is because the inequality (11) implies that

$$d(\boldsymbol{a}_n, \boldsymbol{c}_k) \geq d(\boldsymbol{c}_k, \boldsymbol{c}_{I(n)}) - d(\boldsymbol{a}_n, \boldsymbol{c}_k) \geq 2d(\boldsymbol{a}_n, \boldsymbol{c}_k) - d(\boldsymbol{a}_n, \boldsymbol{c}_k) = d(\boldsymbol{a}_n, \boldsymbol{c}_k)\,. \tag{21}$$

In other words, $\boldsymbol{a}_n$ can not belong to $\boldsymbol{c}_k$.

Altogether, Elkan's procedure can be implemented as in the Algorithm 2.

Algorithm 2. Elkan's procedure reproduced from [9,15]

```
Input: Initial modes $c_k$, $k \in [K]$ and objects $a_n$, $n \in [N]$
for $n \leftarrow 1$ to $N$ do                          # Initialize lower and upper bounds
    $I(n) \leftarrow 1$, $u(n) \leftarrow \infty$
    for $k \leftarrow 1$ to $K$ do $l(n,k) \leftarrow 0$
end for
while Termination conditions are not met do
    Compute $d(c_{k'}, c_k)$, $k, k' \in [K]$
    for $k \leftarrow 1$ to $K$ do
        $s(k) \leftarrow \frac{1}{2} \min_{k' \neq k} d(c_{k'}, c_k)$
    end for
    d = []
    for $n \leftarrow 1$ to $N$ do                                      # Assignment step
        if $u(n) > s(I(n))$ then
            $r \leftarrow$ true
            for $k \leftarrow 1$ to $K$ do
                $z = \max(l(n,k), d(c_{I(n)}, c_k)/2)$
                if $k = I(n)$ or $u(n) \leq z$ then continue
                if r then
                    $u(n) \leftarrow d(a_n, c_{I(n)})$, $r \leftarrow$ false
                    if $u(n) \leq z$ then continue
                end if
                $l(n,k) \leftarrow d(a_n, c_{I(n)})$
                if $l(n,k) < u(n)$ then
                    $I(n) \leftarrow k$, $u(n) \leftarrow l(n,k)$
                    continue
                end if
            end for
            d.append($n$)
        end if
    end for
    for $n \in$ d do                                     # Update lower and upper bounds
        $l(n,k) = d(a_n, c_k)$, $k \in [K]$
        $I(n) = \text{argmin}_k\, d(a_n, c_k)$
        $u(n) = d(a_n, c_{I(n))}$
    end for
    for $k \leftarrow 1$ to $K$ do                                         # Update step
        $c'_k \leftarrow c_k$, $c_k \leftarrow$ Update Step, $\delta(k) \leftarrow d(c'_k, c_k)$
    end for
    for $n \leftarrow 1$ to $N$ do                      # Update lower and upper bounds
        $u(n) \leftarrow u(n) + \delta(I(n))$
        for $k \leftarrow 1$ to $K$ do
            $l(n,k) \leftarrow l(n,k) - \delta(k)$
        end for
    end for
end while
Output: Final $c_k$, $k \in [K]$ and $I(n)$, $n \in [N]$
```

4 Experiments

Since we are mainly interested in the running time in diverse scenarios rather than opting for the clustering performance on specific datasets, we conduct experiments on several synthetic datasets, with controllable characteristics, including the number of objects N, number of attributes P, and compactness of the dataset, to assess the potential gains provided by the Elkan's procedure (See Algorithm 2), compared to the naive Lloyd's procedure (See Algorithm 1).

4.1 Experimental Setting

We simplify the data generation protocol by setting $Q = Q_1 = \ldots = Q_P$. We also introduce the parameter σ to control the compactness of the datasets. For each pair $(n, p) \in [N] \times [P]$, we randomly generate the value a_n^p = random. randint$(0, Q-1)$+int(random.gauss$(0, \sigma*Q)$)+Q, where random.randint$(0, Q-1)$ randomly draws a number in $\{0, 1, \ldots, Q-1\}$ and int(random.gauss$(0, \sigma * Q)$) returns the integer portion of the value randomly drawn by a Gaussian distribution with the mean 0 and the standard deviation $\sigma * Q$.

We exclusively focus on the basic K-mode algorithm [20] and its accelerated version implemented with the Hamming dissimilarity (9). The number of iterations is set to 10 in all the experiments. To make both algorithms a bit faster (while still comparable), we call the pairwise_distances function from Sklearn to compute the pairwise dissimilarities in line 3–8 of Algorithm 1 and line 31–35 of Algorithm 2, instead of running the for-loops.

We consider the clustering/running time as a function of the configuration (N, P, Q, σ, K), where some parameters shall be fixed to highlight the impact of the varied parameters. We set σ to 0.1 in all the experiments and defer an analysis of its impact to a further study[1]. The source code used in our experiments has been made public at https://github.com/nmtoan91/kmodes-triangle-inequality.

4.2 Results

Figure 1(a) suggests that when the number of objects N increases, the Elkan's procedure may help to accelerate the K-modes significantly. Notably, when N reaches $2^{20} = 1,048,576$, Elkan's procedure demonstrates a 56% faster execution time, compared to the original K-Modes. We guess that with a large N, the modes move slowly in the data space and some objects can stay far from a certain number of temporary modes when the iteration process goes along, thereby the Elkan's procedure may bring advantages by detecting more redundant pairwise dissimilarity calculations while paying an affordable cost of computing/maintaining the lower/upper bounds (given that the number K can be considered rather small in this configuration).

[1] Readers might wonder why the choices of parameters in our experiments seem a bit arbitrary. We think it is a common challenge when designing such a comparative study, which requires synthetic/real datasets reflecting desirable characteristics in diverse scenarios, and might call for further designs of sound experimental protocols.

The potential advantages provided the Elkan's procedure when the number of attributes P increases suggested by Fig. 1(b) seems reasonable because the cost of computing redundant pairwise dissimilarity calculations should (rapidly) increase when P increases. This might be a reasonable motivation to go with the Elkan's procedure (and accelerated procedures in general) because several real-world datasets may contain large numbers of attributes.

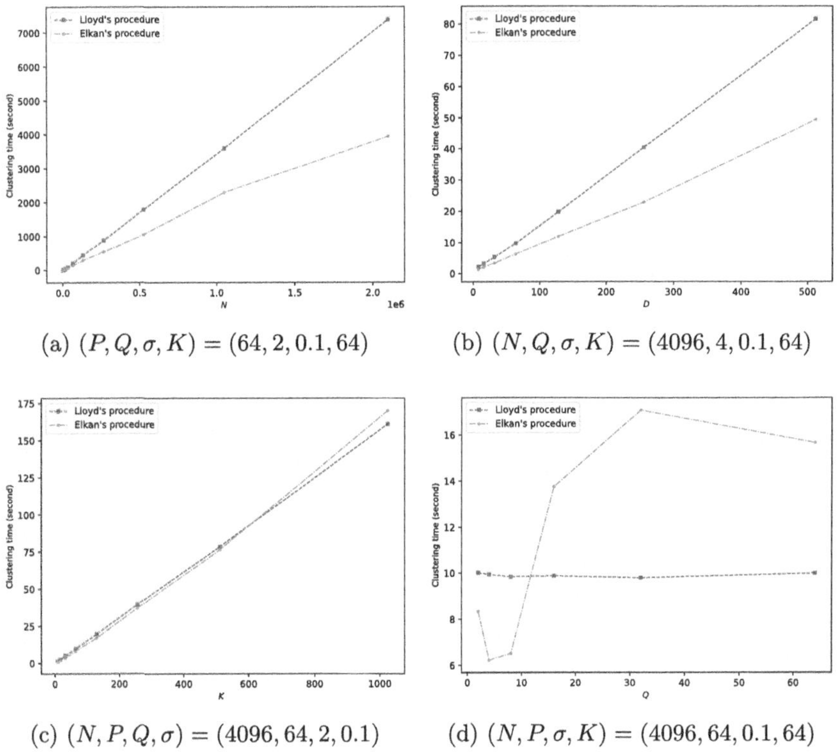

(a) $(P, Q, \sigma, K) = (64, 2, 0.1, 64)$ (b) $(N, Q, \sigma, K) = (4096, 4, 0.1, 64)$

(c) $(N, P, Q, \sigma) = (4096, 64, 2, 0.1)$ (d) $(N, P, \sigma, K) = (4096, 64, 0.1, 64)$

Fig. 1. Clustering time as a function of (N, P, Q, σ, K). In each scenario, the missing parameter is varied within the domain informed in the horizontal axis.

Figure 1(c) suggests that when the number K is small/moderate (i.e., less than 500), the Elkan's procedure may provide (tiny) gains, which are clearly dominated by its losses when the number K increases. We guess this is due to the rapidly increasing cost of computing/maintaining the lower/upper bounds the Elkan's procedure has to pay. This also suggests that implementing and assessing the Drake's procedure and the Yinyang procedure, which may provide promising trade-offs between detecting redundant pairwise dissimilarity calculations and the cost of computing/maintaining bounds, may be beneficial.

Finally, Fig. 1(d) suggests that Elkan's procedure may provide small or moderate gains when the attributes have small less than 10 possible values. We think this should not be considered as bad news as there would be several real-world datasets of that type.

5 Discussions

This section envisions potential follow-up works in developing and accelerating other clustering algorithms for categorical data.

5.1 K-Mode Algorithms with Other Dissimilarities

Note that K-mode algorithms can be implemented with different dissimilarities [2,11,13]. We shall discuss the opportunities and challenges of implementing and accelerating other variants of K-mode algorithms.

DILCA. For any $p \in [P]$, let $\text{Co}(p)$ be the set of indices of attributes that are highly correlated to A^p. For any $p \in [P]$, the basic dissimilarity is defined as

$$d^p_{\text{DILCA}}(a^p_n, a^p_{n'}) := \sqrt{\frac{\sum_{p' \in \text{Co}(p)} \sum_{q_{p'} \in [Q_{p'}]} \left(\Pr(a^p_n - a^p_{n'} | a^p_{q_{p'}})\right)^2}{\sum_{p' \in \text{Co}(p)} Q_{p'}}}, \tag{22}$$

where $\Pr(a^p_n - a^p_{n'} | a^p_{q_{p'}}) = \Pr(a^p_n | a^p_{q_{p'}}) - \Pr(a^p_{n'} | a^p_{q_{p'}})$. The DILCA is defined as

$$d_{\text{DILCA}}(\boldsymbol{a}_n, \boldsymbol{a}_{n'}) = \sum_{p=1}^{P} d^p_{\text{DILCA}}(a^p_n, a^p_{n'}), \forall (\boldsymbol{a}_n, \boldsymbol{a}_{n'}) \in \mathcal{A} \times \mathcal{A}. \tag{23}$$

Next remark ensures that the DILCA [12] satisfies the triangle inequality (11).

Remark 3. The DILCA (23) satisfies the triangle inequality (11).

Proof. Since the DILCA (23) is decomposable, we only need to show that basic dissimilarities (22) satisfy the triangle inequality (11), which can be verified easily. More precisely, for any $p \in [P]$ and any triplet $(a^p_n, a^p_{n'}, a^p_{n''})$, the inequality

$$d^p_{\text{DILCA}}(a^p_n, a^p_{n''}) \le d^p_{\text{DILCA}}(a^p_n, a^p_{n'}) + d^p_{\text{DILCA}}(a^p_{n'}, a^p_{n''}). \tag{24}$$

is ensured by the properties that $\sum_{p' \in \text{Co}(p)} Q_{p'}$ is a constant and the Euclidean distance satisfies the triangle inequality (11) (See, e.g., [29][page 357]). □

We think implementing K-mode algorithms with the DILCA (23) may result in another clustering algorithm with a relatively fast update step (ensured by the Remark 1), whose assignment step may be accelerated by leveraging the triangle inequality (as ensured by Remark 3).

F_1^d Dissimilarity. When dealing with binary data sets, i.e., $\mathcal{A} = \{0,1\}^P$, it might be also interesting to implement K-mode algorithms with F_1^d dissimilarity

$$F_1^d(\boldsymbol{a}_n, \boldsymbol{a}_{n'}) = 1 - \frac{2\sum_{p=1}^P a_n^p a_{n'}^p}{\sum_{p=1}^P a_n^p + \sum_{p=1}^P a_{n'}^p} \tag{25}$$

which is a popular dissimilarity in information retrieval (and in machine learning in general) [1,16], it seems that further efforts are required to build efficient and effective implementations. The connection between finding modes (3) and finding BOPs (5) in PMLC [25,30] informs that finding a mode (3) can be done polynomially in the number of attributes given suitable probabilities/frequencies. Such hints/results might be leveraged to implement the update step. Unfortunately, the F_1^d dissimilarity (25) dissatisfies the triangle inequality (11) [23,26]. This may lower the chances of accelerating the assignment step.

Jaccard Dissimilarity. For binary data sets, i.e., $\mathcal{A} = \{0,1\}^P$, it might be also interesting to implement K-mode algorithms with Jaccard Dissimilarity

$$J^d(\boldsymbol{a}_n, \boldsymbol{a}_{n'}) = 1 - \frac{\sum_{p=1}^P a_n^p a_{n'}^p}{\sum_{p=1}^P a_n^p + \sum_{p=1}^P a_{n'}^p - \sum_{p=1}^P a_n^p a_{n'}^p}, \tag{26}$$

which is another popular dissimilarity [8,24]. It is known that the Jaccard dissimilarity (26) satisfies the triangle inequality (11) (See [14] and references therein). Therefore, one might try to accelerate the assignment step by leveraging this inequality. Yet, finding modes (3) (or BOPs (5) in PMLC) is conjectured to be intractable [4,30]. Approximating modes (3) under the assumption that $\Pr(\boldsymbol{a}_n|\mathbf{G}_k) = \prod_{p=1}^P \Pr(a_n^p|\mathbf{G}_k)$ can be done polynomially in the number attributes given suitable probabilities/frequencies [21,25]. Such hints/results might be leveraged to implement the update step.

5.2 K-Representatives

K-representatives [18,27] is another promising clustering algorithm. This algorithm can be seen as instantiations of a generic clustering approach where the center $\boldsymbol{c}_k$ of a given cluster $\mathbf{G}_k$ is represented by a joint probability distribution $\Pr(\mathcal{A} \,|\, \mathbf{G}_k)$, instead of a member of $\mathcal{A}$, which roughly means each element $\boldsymbol{c} \in \mathcal{A}$ can be a possible center with the probability $\Pr(\boldsymbol{c} \,|\, \mathbf{G}_k)$. The distance from each object $\boldsymbol{a}_n \in \mathcal{A}$ to the center $\Pr(\mathcal{A} \,|\, \mathbf{G}_k)$ is then defined as the expectation of the chosen dissimilarity d, i.e., the average dissimilarity to the possible centers:

$$d(\boldsymbol{a}_n, \Pr(\mathcal{A} \,|\, \mathbf{G}_k)) = \sum_{c \in \mathcal{A}} d(\boldsymbol{a}_n, \mathbf{c})\Pr(\mathbf{c}|\mathbf{G}_k)\,. \tag{27}$$

Again, the computation of the dissimilarity (27) during the assignment step may benefit from known results on finding BOPs in PMDC [22] and PMLC [5,21,30] because it seems to be computationally equivalent to computing expected losses in finding BOPs. Yet, adapting triangle inequality-based accelerated procedures to accelerate K-representatives seems to be challenging because one may need to simultaneously deal with dissimilarity on object-center pairs and dissimilarity on center-center pairs, which may be of very different natures. One might try to adapt such accelerated procedures from the abstract level by checking the conditions under which a given object cannot either belong to a given cluster or move to another cluster rather than its current cluster.

6 Conclusion

This paper presents some results on accelerating K-mode algorithms by leveraging the triangle inequality, while still always computing exactly the same result as the standard procedure. We also provide some empirical evidence to illustrate the gains and losses provided by Elkan's procedure in several scenarios. This also suggests that implementing and assessing other accelerated procedures, such as Drake's procedure and the Yinyang procedure, which may provide promising trade-offs between the ability to detect redundant pairwise dissimilarity calculations and the cost of computing/maintaining bounds, may be beneficial. We also discuss the opportunities and challenges of implementing other variants of K-Modes and accelerating them. Finally, we provide similar discussions on the case of K-representatives, which is another promising clustering algorithm/approach.

Acknowledgments. This work was supported by the CPJ in Trustworthy AI (Ref. ANR-R311CHD). This work was also sponsored by the Office of Naval Research (ONR) and the Office of Naval Research Global (ONRG) under grant number N62909-23-1-2058. The views and conclusions contained herein are those of the authors only and should not be interpreted as representing those of the U.S. Government.

References

1. Beil, F., Ester, M., Xu, X.: Frequent term-based text clustering. In: Proceedings of the Eighth ACM SIGKDD International Conference on Knowledge Discovery and Data Mining (KDD), pp. 436–442 (2002)
2. Cao, F., Liang, J., Li, D., Bai, L., Dang, C.: A dissimilarity measure for the k-modes clustering algorithm. Knowl.-Based Syst. **26**, 120–127 (2012)
3. Cattell, R.B.: The description of personality: basic traits resolved into clusters. Psychol. Sci. Public Interest **38**(4), 476 (1943)
4. Chierichetti, F., Kumar, R., Pandey, S., Vassilvitskii, S.: Finding the jaccard median. In: Proceedings of the Twenty-First Annual ACM-SIAM Symposium on Discrete Algorithms (SODA), pp. 293–311. SIAM (2010)
5. Dembczyński, K., Waegeman, W., Cheng, W., Hüllermeier, E.: On label dependence and loss minimization in multi-label classification. Mach. Learn. **88**, 5–45 (2012)
6. Ding, Y., Zhao, Y., Shen, X., Musuvathi, M., Mytkowicz, T.: Yinyang k-means: a drop-in replacement of the classic k-means with consistent speedup. In: Proceedings of the 32nd International Conference on International Conference on Machine Learning (ICML), pp. 579–587 (2015)

7. Drake, J.: Faster k-means clustering. Ph.D. thesis (2013)
8. Dubes, R.C.: How many clusters are best?-An experiment. Pattern Recogn. **20**(6), 645–663 (1987)
9. Elkan, C.: Using the triangle inequality to accelerate k-means. In: Proceedings of the 20th International Conference on Machine Learning (ICML), pp. 147–153 (2003)
10. Hamerly, G., Drake, J.: Accelerating Lloyd's algorithm for k-means clustering. Partitional clustering algorithms, pp. 41–78 (2015)
11. Huang, Z.: Extensions to the k-means algorithm for clustering large data sets with categorical values. Data Min. Knowl. Disc. **2**(3), 283–304 (1998)
12. Ienco, D., Pensa, R.G., Meo, R.: From context to distance: learning dissimilarity for categorical data clustering. ACM Trans. Knowl. Discov. Data **6**(1), 1–25 (2012)
13. Khan, S.S., Ahmad, A.: Cluster center initialization algorithm for k-modes clustering. Expert Syst. Appl. **40**(18), 7444–7456 (2013)
14. Kosub, S.: A note on the triangle inequality for the Jaccard distance. Pattern Recogn. Lett. **120**, 36–38 (2019)
15. Kwedlo, W., Czochanski, P.J.: A hybrid MPI/OpenMP parallelization of k-means algorithms accelerated using the triangle inequality. IEEE Access **7**, 42280–42297 (2019)
16. Larsen, B., Aone, C.: Fast and effective text mining using linear-time document clustering. In: Proceedings of the Fifth ACM SIGKDD International Conference on Knowledge Discovery and Data Mining (KDD), pp. 16–22 (1999)
17. Lloyd, S.: Least squares quantization in PCM. IEEE Trans. Inf. Theory **28**(2), 129–137 (1982)
18. Mau, T.N., Huynh, V.N.: An LSH-based k-representatives clustering method for large categorical data. Neurocomputing **463**, 29–44 (2021)
19. Mau, T.N., Inoguchi, Y., Huynh, V.N.: A novel cluster prediction approach based on locality-sensitive hashing for fuzzy clustering of categorical data. IEEE Access **10**, 34196–34206 (2022)
20. Ng, M.K., Li, M.J., Huang, J.Z., He, Z.: On the impact of dissimilarity measure in k-modes clustering algorithm. IEEE Trans. Pattern Anal. Mach. Intell. **29**(3), 503–507 (2007)
21. Nguyen, V.L., Hüllermeier, E.: Multilabel classification with partial abstention: bayes-optimal prediction under label independence. J. Artif. Intell. Res. **72**, 613–665 (2021)
22. Nguyen, V.L., Yang, Y., de Campos, C.P.: Probabilistic multi-dimensional classification. In: Proceedings of the 39th Conference on Uncertainty in Artificial Intelligence (UAI), pp. 1522–1533 (2023)
23. Oksuz, K., Cam, B.C., Akbas, E., Kalkan, S.: Localization recall precision (LRP): a new performance metric for object detection. In: Proceedings of the European Conference on Computer Vision (ECCV), pp. 504–519 (2018)
24. Pandit, S., Gupta, S., et al.: A comparative study on distance measuring approaches for clustering. Int. J. Res. Comput. Sci. **2**(1), 29–31 (2011)
25. Pillai, I., Fumera, G., Roli, F.: Designing multi-label classifiers that maximize f measures: state of the art. Pattern Recogn. **61**, 394–404 (2017)
26. Powers, D.M.: What the f-measure doesn't measure: features, flaws, fallacies and fixes. arXiv preprint arXiv:1503.06410 (2015)
27. San, O.M., Huynh, V.N., Nakamori, Y.: An alternative extension of the k-means algorithm for clustering categorical data. Int. J. Appl. Math. Comput. Sci. **14**(2), 241–247 (2004)

28. Saxena, A., et al.: A review of clustering techniques and developments. Neurocomputing **267**, 664–681 (2017)
29. Strichartz, R.S.: The Way of Analysis. Jones & Bartlett Learning, Burlington (2000)
30. Waegeman, W., Dembczyński, K., Jachnik, A., Cheng, W., Hüllermeier, E.: On the bayes-optimality of f-measure maximizers. J. Mach. Learn. Res. **15**, 3333–3388 (2014)

A Reinforcement Learning Approach for Resolving Inconsistencies in Qualitative Constraint Networks

Anastasia Paparrizou, Michael Sioutis(✉), and Yoan Thomas

LIRMM UMR 5506, Université de Montpellier & CNRS, Montpellier, France
{Anastasia.Paparrizou,Michael.Sioutis,Yoan.Thomas}@lirmm.fr

Abstract. In this paper, we present a reinforcement learning approach for resolving inconsistencies in qualitative constraint networks (QCNs). QCNs are typically used in constraint programming to represent and reason about intuitive spatial or temporal relations like x {*is inside of* $\vee$ *overlaps*} y. Naturally, QCNs are not immune to uncertainty, noise, or imperfect data that may be present in information, and thus, more often than not, they are hampered by inconsistencies. We propose a multi-armed bandit approach that defines a well-suited ordering of constraints for finding a maximal satisfiable subset of them. Specifically, our learning approach interacts with a solver, and after each trial a reward is returned to measure the performance of the selected action (constraint addition). The reward function is based on the reduction of the solution space of a consistent reconstruction of the input QCN. Early experimental results obtained by our algorithm suggest that we can do better than the state of the art in terms of both effectiveness, viz., lower number of repairs obtained for an inconsistent QCN, and efficiency, viz., faster runtime.

Keywords: Spatial and Temporal Reasoning · Qualitative Constraints · Imperfect Data · Inconsistency Resolution · Maximizing Satisfiability · Reinforcement Learning · Multi-armed Bandit

1 Introduction

Representing and reasoning about spatial or temporal information in a natural, human-like manner, is the area of study of Qualitative Spatio-Temporal Reasoning (QSTR), a rich symbolic AI framework spanning various fields, such as constraint programming, logic, and mathematics [10,17]. As an example, within QSTR, we can consider a relation like x {*is inside of* $\vee$ *overlaps*} y, which does not involve any quantitative information and is rather intuitive. Such relations, and combinations thereof, can be modeled as a qualitative constraint network (QCN), a simplified example of which is provided in Fig. 1. Quite

S. Destercke et al. (Eds.): SUM 2024, LNAI 15350, pp. 340–353, 2025.
https://doi.org/10.1007/978-3-031-76235-2_25

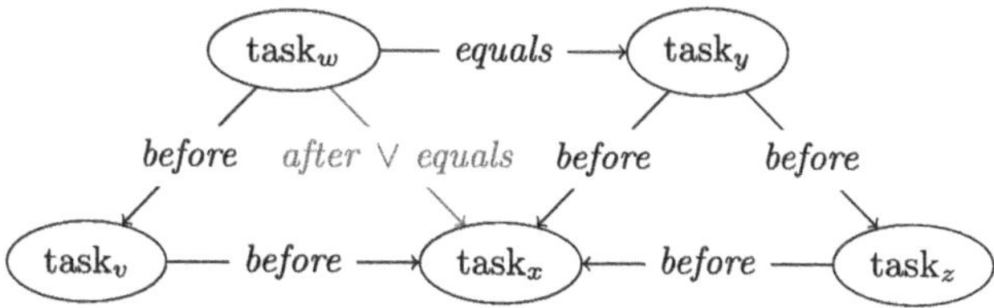

Fig. 1. An illustration of the MAX-QCN problem for a qualitative constraint network (QCN) [8] and the terminology used here; the simplified QCN, which can be viewed as a temporal plan, is inconsistent, and one solution of the MAX-QCN problem, viz., an *optimal* scenario, can be obtained by placing task$_w$ $\{$*before*$\}$ task$_x$ and hence repairing the respective constraint in the figure; note that, had we chosen to keep task$_w$ $\{$*after* ∨ *equals*$\}$ task$_x$, we would have had to make two repairs.

naturally, representing spatial or temporal information even in that symbolic, intuitive way, is not exempt from the presence of inconsistencies, as, more often than not, the information is infused with uncertainty, noise, and imperfect data among other things; we tackle this issue in this paper.

Context and Motivation

We focus on the MAX-QCN problem [8], which is the problem of maximizing satisfiability in a QCN; see Fig. 1. Concretely, given a QCN $\mathcal{N}$, solving the MAX-QCN problem of $\mathcal{N}$ is obtaining a deviating configuration that maximizes the number of satisfied constraints in $\mathcal{N}$. As mentioned earlier, representing spatial or temporal information in QSTR may inevitably lead to inconsistencies; this can be, for example, due to human error, noise in the data we abstract from, and/or inaccurate classifiers. As illustration, in scheduling inconsistencies may form due to the unavailability of resources for certain tasks [16]. In particular, timetabling, as an instance of scheduling, involves assigning temporal intervals to tasks that are subject to limited resources. For example, in the context of a railway network, an inconsistency can appear when two (or more) trains are allocated the same railway line in overlapping temporal intervals; to ensure safety of operations, the inconsistency must then be repaired by considering other railway lines, temporal intervals, and/or preferences alike, and minimizing changes so as to perturb the timetable to the least extent possible. Resolving inconsistencies in QSTR extends to the broader context of neuro-symbolic AI architectures dealing with textual entailment or image analysis, among other tasks [15]; indeed, resolving/minimizing inconsistency (equivalently, maximizing satisfiability) can be seen as an important step of logical reasoning in the neuro-symbolic cycle, see, e.g., Fig. 1 in [35], where minimizing inconsistency in a knowledge base forms the basis of logical abduction in abductive learning.

Related Work and Contribution

Solving (optimally) the MAX-QCN problem, or coming close to a solution of it, has been dealt with in various works in the literature, e.g., [8,9,28]. Of these works, [28] is the state of the art wrap-up of optimal encodings and (sub-)optimal heuristics for tackling this problem. Specifically, with respect to heuristics for approaching a solution of the MAX-QCN problem, [28] proposes a portfolio-style approach that combines several diverse procedures, each of which probes the constraints of a QCN in linear fashion, following its own ordering of constraints, and filters out the ones that fail the satisfiability check. Notably, each such procedure taken alone implements a simple instance of basic linear search for computing a maximal satisfiable subset of constraints, cf. [5,19], yet taken all together, hence considering different orderings of constraints, can produce a powerful approximation of the MAX-QCN problem. To a certain extent, finding a good initial ordering of constraints in this context is similar to starting from a configuration that violates as few constraints as possible, cf. [21], in the sense that we would ideally want to have all the constraints of a maxim**um** (cardinality) subset of constraints first in our ordering (as this would give the optimal solution). As illustration, let us revisit Fig. 1 and consider the case where task$_w$ $\{after \vee equals\}$ task$_x$ is the first constraint to be checked for inclusion in a (initially empty) satisfiable subset; clearly, it will be included, but it will result in a deviation from an optimal solution.

In this paper, with respect to the previous discussion, we leverage machine learning to improve inconsistency resolution in QSTR and make the following contributions:

(i) We propose a multi-armed bandit approach [14] to synthesize different orderings of constraints, with the aim to produce a new one that will outperform each of the former ones in the number of repairs performed for an inconsistent QCN (the lower, the better); to this end, we build upon the approach of [28] and we infuse it with a reinforcement learning aspect, where every action of processing the next constraint among different orderings of constraints is rewarded based on how much it reduces the solution space.

(ii) We experimentally compare an implementation of our multi-armed bandit approach to the implementation of [28] using a standard dataset of QCNs of Interval Algebra [1], and observe several pertinent properties of our method, the most important ones being that it is both more effective (fewer number of repairs) and more efficient (faster runtime) than the one of [28]; in addition, we employ an optimal Partial MaxSAT solver to show/reaffirm that solving (optimally) the MAX-QCN problem is, generally, impractical.

Organization. The rest of the paper is organized as follows. In Sect. 2 we provide definitions/notations regarding QSTR and the MAX-QCN problem that are necessary for following the paper. Then, in Sect. 3 we detail our reinforcement learning approach for tackling the MAX-QCN problem, and in Sect. 4 we experimentally evaluate it against the state of the art. Finally, we conclude with some discussion for future work in Sect. 5.

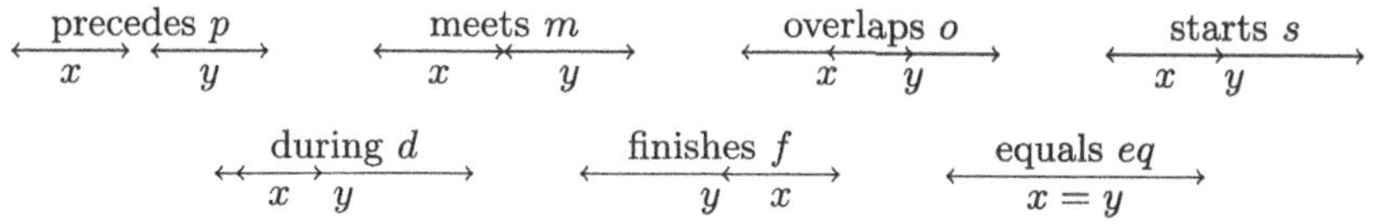

precedes p — x y | meets m — x y | overlaps o — x y | starts s — x y

during d — x y | finishes f — y x | equals eq — $x = y$

Fig. 2. A representation of the 13 base relations b of IA, each one relating two potential intervals x and y as in $x\ b\ y$; the converse of b, i.e., b^{-1}, can be denoted by bi and is omitted in the figure.

2 Preliminaries

A binary qualitative spatial or temporal constraint language is based on a finite set B of *jointly exhaustive and pairwise disjoint* relations, called *base relations* [17] and defined over an infinite domain D (e.g., $\mathbb{R}$). The base relations of a particular qualitative constraint language can be used to represent the definite knowledge between any two of its entities with respect to the level of granularity provided by the domain D (e.g., for D $= \mathbb{R}$, we could have B $= \{<, =, >\}$, with the usual semantics). The set B contains the identity relation Id, and is closed under the *converse* operation ($^{-1}$). Indefinite knowledge can be specified by a union of possible base relations, and is represented by the set containing them.

As illustration, consider the well-known qualitative temporal constraint language of Interval Algebra (IA) [1]. IA considers time intervals on the real line, and the set of base relations B $= \{eq\ (= \mathsf{Id}),\ p,\ pi,\ m,\ mi,\ o,\ oi,\ s,\ si,\ d,\ di,\ f,\ fi\}$ to encode knowledge about the temporal relations between such intervals, as described in Fig. 2.

Representing and reasoning about qualitative spatio-temporal information pertaining to a set of base relations B can be facilitated by a *qualitative constraint network* (QCN):

Definition 1. *A* qualitative constraint network *(*QCN*) is a tuple* (V, C) *where:*

- $V = \{v_1, \ldots, v_n\}$ *is a non-empty finite set of variables (representing entities in* D*);*
- *and* C *is a mapping* $C : V \times V \rightarrow 2^{\mathsf{B}}$ *such that,* $\forall v \in V$, $C(v, v) = \{\mathsf{Id}\}$, *and,* $\forall v, v' \in V$, $C(v, v') = (C(v', v))^{-1}$.

An example QCN of IA is shown in Fig. 3a; for conciseness, converse relations or Id loops are not shown in the figure.

Definition 2. *Let* $\mathcal{N} = (V, C)$ *be a* QCN, *then:*

- *a* solution *of* $\mathcal{N}$ *is a mapping* $\sigma : V \rightarrow$ D *such that,* $\forall (u, v) \in V \times V$, $\exists b \in C(u, v)$ *such that* $(\sigma(u), \sigma(v)) \in b$*; and* $\mathcal{N}$ *is* satisfiable *iff it admits a solution (see Fig. 3b);*
- *a sub-*QCN *(also known as* refinement*)* $\mathcal{N}'$ *of* $\mathcal{N}$, *denoted by* $\mathcal{N}' \subseteq \mathcal{N}$, *is a* QCN (V, C') *such that,* $\forall u, v \in V$, $C'(u, v) \subseteq C(u, v)$*;*

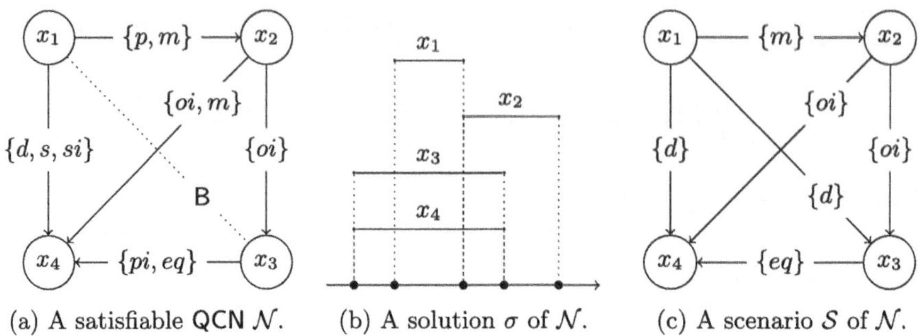

(a) A satisfiable QCN $\mathcal{N}$. (b) A solution σ of $\mathcal{N}$. (c) A scenario $\mathcal{S}$ of $\mathcal{N}$.

Fig. 3. Figurative examples of QCN terminology using IA.

- $\mathcal{N}$ *is* atomic *iff,* $\forall v, v' \in V$, $C(v, v')$ *is a* singleton relation, *i.e., a relation* $\{b\}$ *with* $b \in \mathsf{B}$*;*
- *a* scenario $\mathcal{S}$ *of* $\mathcal{N}$ *is an atomic satisfiable sub-*QCN *of* $\mathcal{N}$ *(see Fig. 3c);*
- *the* constraint graph *of* $\mathcal{N}$*, denoted by* $\mathsf{G}(\mathcal{N})$*, is the graph* (V, E) *where* $\{u, v\} \in E$ *iff* $C(u, v) \neq \mathsf{B}$ *and* $u \neq v$*;*
- $\mathcal{N}$ *is denoted by* $\mathcal{N}_\top$ *when each of its constraints is* universal, *i.e., iff,* $\forall v, v' \in V$ *with* $v \neq v'$, $C(v, v') = \mathsf{B}$.

The MAX-QCN Problem

The MAX-QCN problem has been introduced in the context of QSTR in [8]. Given a QCN $\mathcal{N}$ over a set of variables V, the MAX-QCN problem is the problem of finding a scenario over V that maximizes the number of satisfied constraints in $\mathcal{N}$, or, equivalently, the problem of finding a scenario over V that minimizes the number of unsatisfied constraints in $\mathcal{N}$. Such scenarios are called *optimal* scenarios of $\mathcal{N}$. Clearly, if a QCN $\mathcal{N}$ is satisfiable, any scenario of $\mathcal{N}$ is also an optimal scenario of $\mathcal{N}$. The reader is kindly asked to revisit Fig. 1 in the introduction for a simplified example of the MAX-QCN problem and a solution of it. Solving the MAX-QCN problem is clearly at least as difficult as solving the satisfiability checking problem of a QCN, which is NP-hard in general for most calculi [10].

3 A Multi-armed Bandit Approach

Our overall approach is presented in Algorithm 1, called POLYPUS; in what follows, we provide a description of this algorithm, focusing on its major components and analyzing its complexity.

Algorithm 1: POLYPUS($\mathcal{N}, \mathcal{A}, e$)

input : A QCN $\mathcal{N} = (V, C)$, a set $\mathcal{A}$ of bijections $\alpha : E \rightarrow \{0, 1, \ldots, |E| - 1\}$, where $E = E(\mathsf{G}(\mathcal{N}))$ (i.e., roughly, a set of orderings of the constraints in $\mathcal{N}$), and a number of epochs e with $e \in \mathbb{N}^+$

output : A subset $p \subseteq E(\mathsf{G}(\mathcal{N}))$ corresponding to a maximal satisfiable subset of constraints in $\mathcal{N}$

```
P ← ∅;
armPulls ← map();
meanReward ← map();
counter ← 0;
foreach α ∈ A do
    armPulls[α] ← 0;
    meanReward[α] ← 0;
for n from 1 to e do
    p ← ∅;
    N' = (V, C') ← N_T;
    unprocessedConstraints ← E(G(N)); // each constraint initiates a trial
    while unprocessedConstraints ≠ ∅ do
        α ← α ∈ argmax_{α'∈A} ( meanReward[α'] + sqrt( 2 · ln(counter + 1) / (armPulls[α'] + 1) ) );
            // UCB1
        c ← min({c' ∈ {0, 1, ..., |E(G(N))| − 1} | α^{-1}(c') ∈ unprocessedConstraints});
        {u, v} ← α^{-1}(c);
        C'(u, v) ← C(u, v); C'(v, u) ← C(v, u);
        (result, reward) ← SAT(N');    // see Equation 1 for reward description
        if result = true then
            p ← p ∪ {{u, v}};
        else
            C'(u, v) ← B; C'(v, u) ← B;
        meanReward[α] ← ((meanReward[α] · armPulls[α]) + reward) / (armPulls[α] + 1);
        armPulls[α] ← armPulls[α] + 1;
        unprocessedConstraints ← unprocessedConstraints \ {{u, v}};
        counter ← counter + 1;
    P ← P ∪ {p};
return p ∈ argmax_{p'∈P}(|p'|); // largest maximal satisfiable subset is returned
```

The task of obtaining an appropriate ordering of constraints, as detailed in Sect. 1, can be encoded in a *multi-armed* bandit problem: given a QCN $\mathcal{N}$ and a set of arms $\alpha_1, \cdots, \alpha_k$, the bandit problem consists in assessing the performance of the arms after a sequence of trials $1, \ldots, T$. An arm α_i corresponds to a given ordering of constraints defined under a certain criterion or heuristic (details on arms follow later on), and T is the number of constraints.

More formally, a *multi-armed* bandit problem is a sequential decision process where the learner interacts with its environment. During each trial t, the algorithm selects an arm α_i with $i \in \{1, \cdots, k\}$ and receives a reward $reward_t(\alpha_i)$ for this arm. The goal is to minimize the expected regret across trials, which is defined as the expectation of the difference between the total reward obtained by the best arm and the total reward obtained by the bandit algorithm. The algorithm observes the reward for the chosen arm after each trial, but not for the other arms that could have been selected. Therefore, the minimization of regret is achieved by balancing *exploration* (acquiring new information) and *exploitation* (using acquired information). The *exploration/exploitation* ratio is defined by the UCB1 policy [3] that among other policies displayed a good performance in several works related to constraint programming [6,18,22,31,34]. The ultimate goal is to find a strategy for mapping each trial t to a probability distribution over $\{\alpha_1, \cdots, \alpha_k\}$ in order to maximize cumulative rewards.

A trial t in our case is defined as the period where an arm is pulled until a solver—here, a native qualitative constraint-based solver was chosen—returns its output (lines 13–17 in Algorithm 1). During each trial t, a constraint is added to the network (initially empty) according to the order defined by the i arm selected at run t (line 16). If the added constraint provokes unsatisfiability, then the constraint is removed from the network (line 21) and we proceed to the next constraint. The ordering of constraints that results in the largest satisfiable subset of constraints is the best one (line 27). In the following, we analyze the set of arms we deploy and we assess the performance of the chosen arm via a reward function.

Arms

Given a QCN $\mathcal{N}$, an *arm* in Algorithm 1 is an ordering (permutation) of the (non-universal) constraints of $\mathcal{N}$, and is represented by a bijection $\alpha : E \rightarrow \{0, 1, \ldots, |E| - 1\}$, where $E = E(\mathsf{G}(\mathcal{N}))$. Thus, all the different arms form a set $\mathcal{A}$ of such bijections, which is provided as additional input to the algorithm. Each arm (i.e., ordering of constraints) is defined based on one of the following constraint ordering strategies, whose consecutive application constructs the ordering (a more detailed explanation of these strategies is provided in [29]):

- **max**: choose the constraint that contains the base relation with the most *local models* [29].[1]
- **avg**: choose the constraint with the highest average count of local models (i.e., each of its base relations contributes a count and we take the average of these counts).

[1] In sum, a *local model* of a constraint is a solution of the sub-network that solely involves the two variables of the constraint and a third neighboring variable; hence, the total number of local models of a constraint can be seen as an indicator of how much that constraint is supported (or tolerated) in the original network.

- **sum**: choose the constraint with the highest cumulative count of local models. (i.e., each of its base relations contributes a count and we take the sum of these counts).
- **weight**: choose the constraint with the largest weight, where the larger the weight, the more permissive the constraint is; see, e.g., Fig. 9 in [7].[2]
- **card + weight**: choose the constraint whose smallest decomposition into sub-relations of a (maximal) tractable subset $\mathcal{S} \in 2^{\mathsf{B}}$ [24] (e.g., the ORD-Horn set for IA [20]) is the largest one; use weight as a tie-breaker (this is typical in the literature, e.g., [24]).

The above strategies have been found to be the best-performing ones in the approach of [28], called GREEDUS, in that they help to produce a lower number of repairs for a given inconsistent QCN compared to other strategies.

Remark 1. GREEDUS [28] acts as a virtual best solver at the QCN level, in the sense that it tries each arm separately, for all the constraints of a given QCN collectively, and then records the best result among the arms. On the other hand, POLYPUS aims to combine different arms for a same given QCN, in an effort to operate as a virtual best solver at the constraint level of the QCN; obtaining the actual virtual best solver in this context is infeasible, as for m constraints we would need to explore $m!$ orderings to find the best one (i.e., the best arm).

Reward

A state-of-the-art solver for checking the satisfiability of a given QCN defined on a set of base relations B, be it SAT- or qualitative constraint-based, will typically return a locally consistent sub-QCN (refinement) of that QCN over a (maximal) tractable subset $\mathcal{S} \in 2^{\mathsf{B}}$ if the QCN is satisfiable [25,28]; a scenario (qualitative solution) of the QCN can then be extracted in a backtrack-free manner. Such subsets exist for most well-known qualitative constraint languages [10], and in what follows we assume the use of such a solver.

Given a QCN $\mathcal{N}$, a set of arms $\mathcal{A}$, and a solver as described earlier, let $\mathcal{N}'$ be the QCN that is returned by the solver when given $\mathcal{N}$ as input. Further, let $|\mathcal{N}|$ denote the *size* of a QCN $\mathcal{N} = (V, C)$, i.e., its total number of base relations; formally, $|\mathcal{N}| = \sum_{u,v \in V} |C(u,v)|$. A not straightforward and critical part in multi-armed bandits is to define an appropriate reward function that reflects, at each trial, the performance of the arm pulled, giving us a numeric feedback. In general, the theoretically most relevant reward is the one that optimises directly the objective function (i.e., to maximise the number of the constraints such that the resulting network is satisfiable). Indeed, the first reward we tried was the one that was promoting the arm that was generating the largest maximal satisfiable

[2] Permissiveness, or, equivalently, restrictiveness, is measured for each base relation by successively composing the base relation with every possible relation, and then summing up the cardinalities of the resulting compositions; the result is then suitably scaled and a weight is assigned to each base relation, and the weight of a constraint is then given by the sum of the weights of its base relations.

set of constraints among other arms. We were giving the value 1 as reward to the arm if the solver was producing a satisfiable instance (each time a constraint was added at line 17), and 0 otherwise. So the arm that was producing the largest wrt to the number of constraints satisfiable network had also the biggest mean reward. However, in practice this reward did not work well, as 0/1 values were not allowing for a good discrimination among the arms. Therefore, we tried functions with finer granularity, capturing the space reduction, the constraint weighting inspired from [21,31,32].

After conducting an empirical study on such possible measures, we define the reward for a pulled arm $\alpha \in \mathcal{A}$ as follows:

$$reward(\alpha) = \begin{cases} 1 - \left(\frac{|\mathcal{N}| - |\mathcal{N}'|}{|\mathcal{N}|}\right), & \text{if } \mathcal{N} \text{ is satisfiable} \\ 0, & \text{otherwise} \end{cases} \tag{1}$$

This particular reward performed well, since, as was also observed in the literature, the size of a QCN has a fairly strong positive relationship with the number of scenarios that the QCN contains, specifically, the Spearman's rank correlation coefficient between the two variables is ≈ 0.8 [32]. Intuitively, the more scenarios a QCN contains, the "more" satisfiable it is, and as the size of a QCN is very easy to compute (compared to counting an exponential number of scenarios), we found it to be a pertinent and useful value for our approach.

Computational Complexity

Given a QCN $\mathcal{N}$, the total time complexity of Algorithm 1 is $\Theta(|E(\mathsf{G}(\mathcal{N}))| \cdot \beta)$, where β is the time complexity of the solver that performs the satisfiability check in line 17 of the algorithm; in the case where a qualitative constraint-based solver is used, as is the case in this paper, $\beta = O(|\mathsf{B}|^{|E(\mathsf{G}(\mathcal{N}))|})$, however, in practice, a satisfiability check can be typically performed very fast. Specifically, with regard to the total time complexity, taking into account that the number of epochs e is a small constant, it can be seen that $\Theta(|E(\mathsf{G}(\mathcal{N}))|$ calls are performed to such a satisfiability checking solver (lines 11–17), and that all other operations take constant time. Thus, a total time complexity of $\Theta(|E(\mathsf{G}(\mathcal{N}))| \cdot \beta)$ is achieved.

4 Experimentation

In this section, with respect to resolving inconsistencies in QCNs, i.e., addressing the MAX-QCN problem, we experimentally compare an in-house implementation of POLYPUS introduced in Sect. 3 (Algorithm 1) to the greedy approach from [28], called GREEDUS. Interestingly, and in relation to Remark 1, GREEDUS can be encoded by POLYPUS if an arm is consecutively pulled for all constraints in a QCN, and then the best result among all arms (the result is defined as the number of repairs, where the best is lowest) is recorded for that QCN. We also use the RC2 MaxSAT solver [13] via the PySAT toolkit [12] to compute some ground truth, i.e., optimal number of repairs for our QCNs, to the extent possible.

In our evaluation campaign, we investigate the following hypotheses about the need for a more systematic learning approach than a naive greedy one.

H1 Does synthesizing different strategies for a same instance gives us a better result than choosing the best strategy (standalone) for that instance?

H2 Does this learning approach pay off in terms of runtime and constraints added to the network or does the overhead of learning outweigh its benefits?

In what follows, we show that we can give a positive answer to both of the aforementioned questions.

Note 1. All code, dataset, & logs are available at https://doi.org/10.5281/zenodo.11032284.

Dataset and Setup

We considered Interval Algebra (IA) network instances generated by the standard $\mathsf{A}(n, d, l)$ model [24], used extensively in the literature. In short, $\mathsf{A}(n, d, l)$ creates network instances of size n, average constraint graph degree d, and an average number l of base relations per constraint. We set $n = 30$ and $l = 6.5$, and we generated 100 network instances for *each* degree d between 11 and 15 with a 1-degree step; hence, 500 network instances in total. Typically, the larger the average degree d, the harder the MAX-QCN problem is to solve [8,9] (see also the number of timeouts of RC2 in Fig. 4d for the denser instances). For the experiments we used an Intel® Core^{TM} CPU i7-12700H @ 4.70 GHz, 16 GB of RAM, and the Ubuntu Linux 22.04 LTS OS, and one CPU core per network. All coding/running for our implementations was done in Python 3 and sped up with PyPy (https://www.pypy.org/).

Results and Remarks

All of the experimental results are concisely presented in Fig. 4. We make the following *four* observations, which positively answer hypotheses **H1** and **H2**:

(i) POLYPUS outperfoms GREEDUS (and RANDOM) on the denser of instances, viz., for $d = 15$ (and $d = 14$), with respect to the avg. number of repairs per instance, for a number of epochs $e \geq 5$ (Figs. 4a and 4d);
(ii) POLYPUS outperfoms GREEDUS on the densest of instances with respect to runtime, for a number of epochs $e \leq 7$ (Figure 4b);
(iii) POLYPUS learns and improves on a strategy to obtain fewer repairs in a QCN as the number of epochs e increases (Fig. 4a), outperforming RANDOM (and GREEDUS);
(iv) the strategy that is learned by POLYPUS is versatile, ranking and combining arms (orderings of constraints) differently as trials proceed (Fig. 4c).[3]

[3] For random selection each of the five arms would be selected $\approx$ 20% of the time across trials.

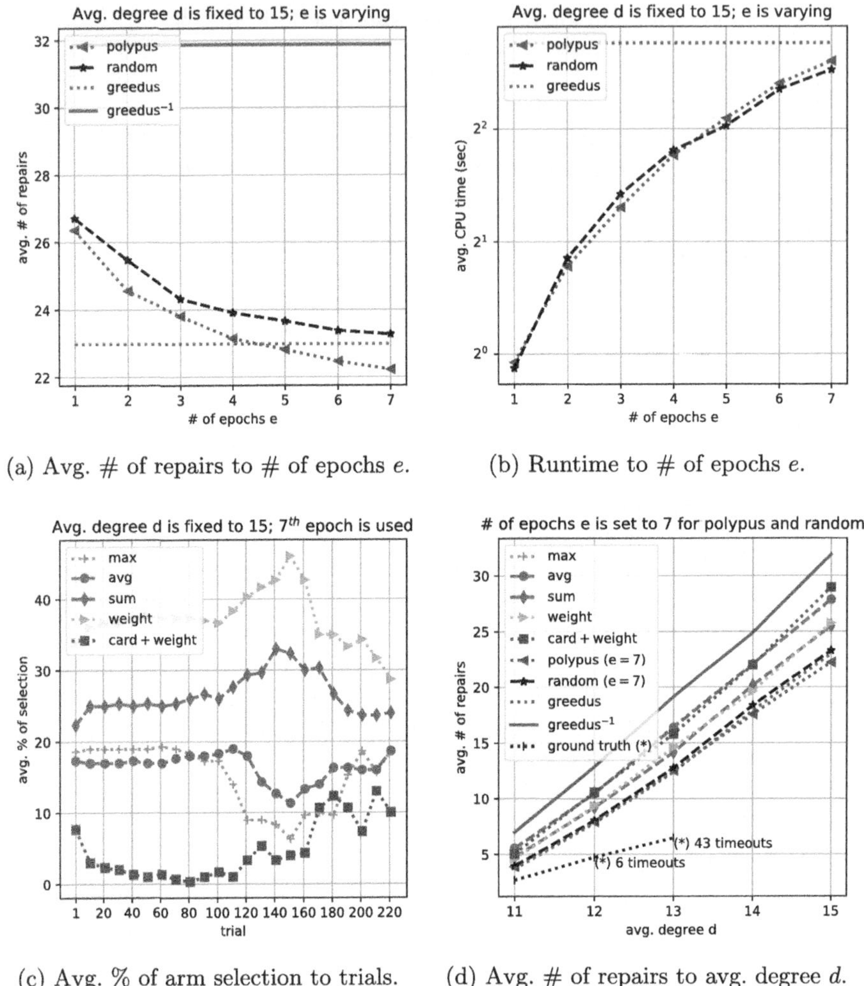

(a) Avg. # of repairs to # of epochs e.

(b) Runtime to # of epochs e.

(c) Avg. % of arm selection to trials.

(d) Avg. # of repairs to avg. degree d.

Fig. 4. Assessing the performance of an implementation of POLYPUS and GREEDUS [28], respectively, with Interval Algebra (IA) network instances of model $\mathsf{A}(n = 30, d, l = 6.5)$ [24]; epochs e do not apply to GREEDUS, i.e., e is always 1 (see Figs. 4a and 4b), GREEDUS^{-1} is GREEDUS with the worst (instead of the best) of the 5 strategies being chosen for every instance (see also Fig. 4d), RANDOM corresponds to a uniformly random arm selection (see line 13 in Algorithm 1), and the ground truth was (attempted to be) computed with the RC2 MaxSAT solver offering [13] of the PySAT toolkit [12] with a 3 600s timeout for every instance (only timely outputs were used).

Finally, we can note that there is a trade-off between efficiency (runtime) and effectiveness (number of repairs) for POLYPUS pertaining to the number of epochs e, as shown in Figs. 4a and 4b (and expected by algorithm design alone), and we can also *re*affirm the fact that obtaining optimal results for this type of problem is prohibitive in the general case (see the results of RC2 and also [8,9,28]).

Remark 2. With respect to the discussion so far, it might be tempting to question whether RANDOM could outperform POLYPUS for a high(er) number of epochs e (Fig. 4a). Even though this could be true for a very high number of e (in the sense that if one randomly tries different orderings of constraints for long enough, then they will eventually get the good result), we have not found it to be the case for $e \leq 1\,000$, at which point the runtime deteriorates significantly anyway, voiding any positive outcomes of the learning approach with regard to number of repairs obtained.

5 Conclusion and Future Directions

In this paper, we presented a reinforcement learning approach for resolving inconsistencies in qualitative constraint networks (QCNs). The use of such networks is typical in constraint programming when we want to represent and reason about intuitive constraints involving spatio-temporal information. Specifically, given a QCN $\mathcal{N}$, our multi-armed bandit approach aims to define a well-suited ordering of the constraints of $\mathcal{N}$ for finding a maximal (or even maximum cardinality) satisfiable subset of them. To this end, our learning approach interacts with a solver and after each trial a reward is returned to measure the performance of the selected action (constraint addition). The reward function is based on the reduction of the solution space of a consistent reconstruction of the input QCN. Our preliminary evaluation suggests that we can do better than the state of the art in terms of runtime and number of repairs. Also, our results reaffirm that the problem of resolving inconsistencies in an optimal manner is, generally, too costly, cf. [28] where smaller-sized instances were used.

This study opens up the following avenues for future research. Resolving inconsistencies in a QCN via filtering out problematic constraints and obtaining a maximal satisfiable subset of them, as we are doing here, lies at the heart of other recently introduced inconsistency-related tasks, such as decomposing a QCN into consistent components [26], or "freezing" some of its constraints to emulate a notion of paraconsistency [27] (where minimizing the number of paraconsistent constraints is critical). Thus, our work here can enhance the performance of the aforementioned approaches, and we aim to explore this in our next steps. Further, we introduced our reinforcement learning framework in the context of inconsistency resolution, but it could also prove useful when extracting solutions from already satisfiable QCNs, as it has been done in traditional constraint programming [6,18,22,31,34]. Moreover, the learning approach on its own provides many possibilities for future work in terms of defining reward functions and utilizing other exploration policies (such as Epsilon-greedy [30], EXP3 [3] and MOSS [2]), and making a cross-comparison analysis for uncovering more opportunities to tap into learning. Ultimately, it could be worth viewing a QCN as a CSP or SAT instance [33] and, hence, leveraging different arsenals of techniques that are particular to MaxCSP or MaxSAT respectively [4]. Finally, in this paper we considered a repair to be the act of replacing a hard constraint with another (hard) one, but it would be well-worth exploring cases where there exists a probability distribution over the constraints, as is the case in probabilistic spatiotemporal knowledge bases [11,23].

Acknowledgments. The work was partially funded by the Agence Nationale de la Recherche (ANR) for the "Hybrid AI" project that is tied to the chair of Dr. Sioutis, and the I-SITE program of excellence of Université de Montpellier that complements the ANR funding.

Disclosure of Interests. The authors have no competing interests to declare that are relevant to the content of this article.

References

1. Allen, J.F.: Maintaining knowledge about temporal intervals. Commun. ACM **26**, 832–843 (1983)
2. Audibert, J.Y., Bubeck, S.: Minimax policies for adversarial and stochastic bandits. In: COLT. Montreal, Canada (2009)
3. Auer, P., Cesa-Bianchi, N., Freund, Y., Schapire, R.: The nonstochastic multiarmed bandit problem. SIAM J. Comput. **32**, 48–77 (2002)
4. Bacchus, F., Järvisalo, M., Martins, R.: Maximum satisfiabiliy. In: Biere, A., Heule, M., van Maaren, H., Walsh, T. (eds.) Handbook of Satisfiability - Second Edition, vol. 336, pp. 929–991. IOS Press (2021)
5. Bailey, J., Stuckey, P.J.: Discovery of minimal unsatisfiable subsets of constraints using hitting set dualization. In: PADL (2005)
6. Balafrej, A., Bessiere, C., Paparrizou, A.: Multi-armed bandits for adaptive constraint propagation. In: IJCAI (2015)
7. van Beek, P., Manchak, D.W.: The design and experimental analysis of algorithms for temporal reasoning. J. Artif. Intell. Res. **4**, 1–18 (1996)
8. Condotta, J., Mensi, A., Nouaouri, I., Sioutis, M., Said, L.B.: A practical approach for maximizing satisfiability in qualitative spatial and temporal constraint networks. In: ICTAI (2015)
9. Condotta, J., Nouaouri, I., Sioutis, M.: A SAT approach for maximizing satisfiability in qualitative spatial and temporal constraint networks. In: KR (2016)
10. Dylla, F., et al.: A survey of qualitative spatial and temporal calculi: algebraic and computational properties. ACM Comput. Surv. **50**, 7:1–7:39 (2017)
11. Grant, J., Molinaro, C., Parisi, F.: Probabilistic spatio-temporal knowledge bases: capacity constraints, count queries, and consistency checking. Int. J. Approx. Reason. **100**, 1–28 (2018)
12. Ignatiev, A., Morgado, A., Marques-Silva, J.: PySAT: a python toolkit for prototyping with SAT oracles. In: SAT (2018)
13. Ignatiev, A., Morgado, A., Marques-Silva, J.: RC2: an Efficient MaxSAT Solver. J. Satisf. Boolean Model. Comput. **11**, 53–64 (2019)
14. Problem, T.M.-A.B., Katehakis, M.N., Jr., A.F.V.: Decomposition and computation. Math. Oper. Res. **12**, 262–268 (1987)
15. Lee, J.H., Sioutis, M., Ahrens, K., Alirezaie, M., Kerzel, M., Wermter, S.: Neurosymbolic spatio-temporal reasoning. In: Compendium of Neurosymbolic Artificial Intelligence, vol. 369, pp. 410–429. IOS Press (2023)

16. Leung, J.Y. (ed.): Handbook of Scheduling - Algorithms, Models, and Performance Analysis. Chapman and Hall/CRC, UK (2004)
17. Ligozat, G.: Qualitative Spatial and Temporal Reasoning. ISTE, Wiley, Hoboken (2013)
18. Loth, M., Sebag, M., Hamadi, Y., Schoenauer, M.: Bandit-based search for constraint programming. In: CP (2013)
19. Marques-Silva, J., Heras, F., Janota, M., Previti, A., Belov, A.: On computing minimal correction subsets. In: IJCAI (2013)
20. Nebel, B., Bürckert, H.: Reasoning about temporal relations: a maximal tractable subclass of Allen's interval algebra. J. ACM **42**, 43–66 (1995)
21. Nöhrer, A., Biere, A., Egyed, A.: Managing SAT inconsistencies with HUMUS. In: VaMoS (2012)
22. Paparrizou, A., Wattez, H.: Perturbing branching heuristics in constraint solving. In: CP (2020)
23. Parisi, F., Grant, J.: Knowledge representation in probabilistic spatio-temporal knowledge bases. J. Artif. Intell. Res. **55**, 743–798 (2016)
24. Renz, J., Nebel, B.: Efficient methods for qualitative spatial reasoning. J. Artif. Intell. Res. **15**, 289–318 (2001)
25. Renz, J., Nebel, B.: Qualitative spatial reasoning using constraint calculi. In: Aiello, M., Pratt-Hartmann, I., Van Benthem, J. (eds.) Handbook of Spatial Logics, pp. 161–215. Springer, Dordrecht (2007). https://doi.org/10.1007/978-1-4020-5587-4_4
26. Salhi, Y., Sioutis, M.: A decomposition framework for inconsistency handling in qualitative spatial and temporal reasoning. In: KR (2023)
27. Salhi, Y., Sioutis, M.: A Paraconsistency framework for inconsistency handling in qualitative spatial and temporal reasoning. In: ECAI (2023)
28. Sioutis, M.: Embarrassingly greedy inconsistency resolution of qualitative constraint networks. In: TIME (2023)
29. Sioutis, M., Wolter, D.: Dynamic branching in qualitative constraint-based reasoning via counting local models. Inf. Comput. **281**, 104787 (2021)
30. Sutton, R., Barto, G.A.: Reinforcement learning: an introduction. IEEE Trans. Neural Netw. Learn. Syst. **9**, 1054 (1998)
31. Wattez, H., Koriche, F., Lecoutre, C., Paparrizou, A., Tabary, S.: Learning variable ordering heuristics with multi-armed bandits and restarts. In: ECAI (2020)
32. Wehner, J., Sioutis, M., Wolter, D.: On robust vs fast solving of qualitative constraints. J. Heuristics **29**, 461–485 (2023)
33. Westphal, M., Wölfl, S.: Qualitative CSP, finite CSP, and SAT: comparing methods for qualitative constraint-based reasoning. In: IJCAI (2009)
34. Xia, W., Yap, R.H.C.: Learning robust search strategies using a bandit-based approach. In: AAAI (2018)
35. Zhou, Z.: Abductive learning: towards bridging machine learning and logical reasoning. Sci. China Inf. Sci. **62**, 76101:1–76101:3 (2019)

r-ERBFN: An Extension of the Evidential RBFN Accounting for the Dependence Between Positive and Negative Evidence

Frédéric Pichon[1(✉)], Serigne Diène[1], Thierry Denœux[2,3], Sébastien Ramel[1], and David Mercier[1]

[1] Univ. Artois, EA 3926 LGI2A, 62400 Béthune, France
{Frederic.Pichon,Serigne.Diene,Sebastien.Ramel, David.Mercier}@univ-artois.fr

[2] Université de technologie de Compiègne, CNRS, Heudiasyc, Compiègne, France
thierry.denoeux@utc.fr

[3] Institut universitaire de France, Paris, France

Abstract. Recently, it was shown that a radial basis function network (RBFN) with a softmax output layer amounts to pooling by Dempster's rule positive and negative evidence for each class, and approximating the resulting belief function by a probability distribution using the plausibility transform. This so-called latent belief function offers a richer uncertainty quantification than the probabilistic output of the RBFN. In this paper, we show that there exists actually a set of latent belief functions for a RBFN. This set is obtained by considering all possible dependence structures, which are described by correlations, between the positive and negative evidence for each class. Furthermore, we show that performance can be enhanced by optimizing the correlations brought to light.

Keywords: Belief function · Dempster's rule · Dependence · Evidential classification

1 Introduction

Evidential classifiers, the most well-known being arguably the evidential k-nearest neighbor classifier [4] and its prototype-based improvement [5], are classifiers whose predictive uncertainty about the unknown class $\boldsymbol{\theta} \in \Theta = \{\theta_1, \ldots, \theta_K\}$ of an instance with feature vector $\mathbf{x}$ is represented by a belief function $Bel_{\mathbf{x}}$ [3,21]. They allow the distinction between aleatory uncertainty and epistemic uncertainty [11], which is akin to the distinction between conflicting evidence and lack of evidence [17]. Such a distinction is important in situations where the final decision can be postponed (e.g. classification with a reject option) or where additional data can be gathered (e.g. active learning) [17]. Moreover, their fine uncertainty quantification can also be exploited to enhance the predictions of a deep neural network architecture, such as a CNN, as first shown in [25].

Of particular interest in this paper is the evidential classifier introduced recently in [9,16], as an alternative approach to the prototype-based evidential

S. Destercke et al. (Eds.): SUM 2024, LNAI 15350, pp. 354–368, 2025.
https://doi.org/10.1007/978-3-031-76235-2_26

classifier [5] having similar properties. This classifier was obtained by applying ideas developed in [7], to a radial basis function network (RBFN) with a softmax output layer (or with an output layer containing a single unit with logistic activation function in the case of binary classification). It was used in [16] to enhance the predictions of a UNet model [19] for a task of lymphoma segmentation from 3D PET-CT images.

In essence, this classifier, called hereafter the evidential RBFN (ERBFN), reveals a predictive, so-called latent, belief function $Bel_{\mathbf{x}}$ underlying the probabilistic prediction $P_{\mathbf{x}}$ of a given (trained) RBFN with a softmax output layer. This belief function underlies the probabilistic prediction in the sense that its transformation into a probability distribution using the plausibility transformation [2] is exactly $P_{\mathbf{x}}$. $Bel_{\mathbf{x}}$ is obtained by, first, defining positive and negative pieces of evidence for each class based on the parameters of the RBFN and on $\mathbf{x}$, and, then, pooling them by Dempster's rule.

In the ERBFN, positive and negative evidence for a given class are considered independent. However, they are obtained from the same set of values and therefore the independence assumption may be questioned. As shown in this paper, this assumption is actually inconsequential insofar as any possible dependence structure yields a predictive latent belief function, that is, a predictive belief function whose plausibility transformation is $P_{\mathbf{x}}$. However, this dependence structure, which as will be seen can be characterized following [14] by a correlation, does have an impact on the predictive belief function and therefore does matter.

To select the dependence structure, i.e., correlation, for each class, different approaches can be followed depending on the available information. When the only information available is the given RBFN with its (trained) parameters, then the best attitude is to be cautious, that is, one should select the correlations leading to the most uncertain (least informative) predictive belief function. This is known as following the least commitment (or maximum uncertainty) principle [18,24], which plays a role in Dempster-Shafer Theory (DST) similar to the principle of maximum entropy in probability theory. We note that such an approach leads to a simple and sound solution (if one uses the informational ordering considered in [7]), which is not reported here due to lack of space.

When, in addition to the trained RBFN, some learning data are available, which is a situation that is likely in practice and is the one considered in this paper, then it becomes possible to search for the correlations that will yield the best performance, according to some uncertainty quantification quality criterion. Classical prediction quality criteria, such as error rate, are not very well adapted as given some labelled data, they can only evaluate the quality of crisp (precise and certain) predictions. We propose to optimize the correlations with respect to the classification equivalent of the evidential uncertainty quantification quality criterion introduced recently by Denœux in regression [10,12]. Its rationale is that the uncertainty quantification is all the better if high degrees of belief tend to be assigned to the true classes and low degrees of belief are assigned to the complements of the true classes, i.e., high degrees of plausibility are assigned to the true classes. It generalizes the cross entropy loss in probability theory. As will be seen, such an optimization leads to predictive latent belief functions that tend to have better uncertainty quantification than the one of [9,16].

This paper is organized as follows. First, necessary background on the Dempster-Shafer theory of belief functions is provided in Sect. 2. Then, in Sect. 3, a means to represent the dependence structure between positive and negative evidence for a proposition by a correlation is presented and used to unveil a new result concerning so-called separable belief functions. This result is then exploited in Sect. 4 to introduce a new evidential classifier, called **r**-ERBFN, which is an extension of the ERBFN. Its additional parameters are a correlation r in the binary classification case and a vector $\mathbf{r}$ of K correlations in the multi-class classification case, allowing to account for the dependence between positive and negative evidence for each class. This classifier allows us to reveal alternative latent belief functions to that of the ERBFN for a given RBFN. A criterion for selecting a particular latent belief function among the available ones is described in Sect. 5. Experiments on real data are reported in Sect. 6. Finally, Sect. 7 concludes the paper. Proofs are omitted due to lack of space.

2 Background on Dempster-Shafer Theory

2.1 Evidence Representation

In Dempster-Shafer theory [3,21], a piece of evidence about the true (unknown) answer $\boldsymbol{\theta}$ to some question is represented by a *mass function*, which is a mapping $m : 2^{\Theta} \rightarrow [0,1]$ such that $m(\emptyset) = 0$ and $\sum_{A \subseteq \Theta} m(A) = 1$, with $\Theta = \{\theta_1, \ldots, \theta_K\}$ the set of possible answers to the question. The mass $m(A)$, for some $A \subseteq \Theta$, represents the probability that the evidence supports exactly the proposition $\boldsymbol{\theta} \in A$ (and nothing else more or less specific). Any subset $A \subseteq \Theta$ such that $m(A) > 0$ is called a *focal set* of m. If Θ is a focal set, then m is *non dogmatic*.

The *vacuous* mass function has Θ as only focal set; it corresponds to a totally uninformative piece of evidence. A mass function m whose focal sets are singletons only, is said to be *Bayesian*; it corresponds to the probability distribution $p : \Theta \rightarrow [0,1]$ such that $p(\theta) = m(\{\theta\})$ for all $\theta \in \Theta$.

A mass function that has the form $m(A) = 1-d$, $m(\Theta) = d$, for some $A \subset \Theta$ such that $A \neq \emptyset$ and some $d \in [0,1]$, is said to be *simple*. The quantity d is called the degree of *diffidence* in A [13]. The quantity $w := -\ln(d)$ is called the *weight of evidence* [21]. Such a mass function may be conveniently denoted by A^d or, equivalently, by A_w. It represents a piece of evidence that can be interpreted in two ways, with respective probabilities $1-d$ and d: according to the first interpretation, the evidence tells that $\boldsymbol{\theta} \in A$, and in the second interpretation, the evidence is useless, i.e., it tells $\boldsymbol{\theta} \in \Theta$.

More generally, a mass function "involves a probability model for the evidence bearing on [the] question" [22]. This model is the following (see, e.g., [8,23]). The piece of evidence can be interpreted in different ways with given probabilities, with Ω the (finite) set of interpretations and P the probability measure on Ω. If interpretation $\omega \in \Omega$ holds, the evidence tells that $\boldsymbol{\theta} \in \Gamma(\omega)$, with $\Gamma(\omega)$

a nonempty subset of Θ. The tuple $(\Omega, 2^\Omega, P, \Gamma)$ is called a source [6] and is formally a random set. It induces the mass function m such that $m(A) = P(\{\omega \in \Omega : \Gamma(\omega) = A\})$, for all $A \in 2^\Theta \backslash \{\emptyset\}$.

Given a mass function m and any $A \subseteq \Theta$, the probability that the evidence implies $\boldsymbol{\theta} \in A$ is $Bel(A) := \sum_{B \subseteq A} m(B)$ and that it does not contradict $\boldsymbol{\theta} \in A$ is $Pl(A) := \sum_{B \cap A \neq \emptyset} m(B)$. Functions $Bel : 2^\Theta \to [0,1]$ and $Pl : 2^\Theta \to [0,1]$ are called the belief and plausibility functions, respectively, and are in one-to-one correspondence with m. The contour function $\pi : \Theta \to [0,1]$ is the restriction of the plausibility function to singletons, i.e., $\pi(\theta) = Pl(\{\theta\})$, for all $\theta \in \Theta$.

2.2 Evidence Combination

Let $(\Omega_1, 2^{\Omega_1}, P_1, \Gamma_1)$ and $(\Omega_2, 2^{\Omega_2}, P_2, \Gamma_2)$, with $\Gamma_i : \Omega_i \to 2^\Theta \backslash \{\emptyset\}$, $i = 1, 2$, be two sources representing two pieces of evidence about $\boldsymbol{\theta}$ and inducing mass functions m_1 and m_2, respectively. Assume these sources to be independent, i.e., the joint probability $P_{12}(\omega_1, \omega_2)$ that the pair of interpretations $(\omega_1, \omega_2) \in \Omega_1 \times \Omega_2$ holds is equal to $P_1(\omega_1) \cdot P_2(\omega_2)$.

Let us make the subsequent assumption that the sources are reliable and let $\Gamma_\cap(\omega_1, \omega_2) := \Gamma_1(\omega_1) \cap \Gamma_2(\omega_2)$ for all $(\omega_1, \omega_2) \in \Omega_1 \times \Omega_2$. According to this assumption, if interpretations ω_1 and ω_2 both hold, then we know for sure that $\boldsymbol{\theta} \in \Gamma_\cap(\omega_1, \omega_2)$, and if $\Gamma_\cap(\omega_1, \omega_2) = \emptyset$, we know that ω_1 and ω_2 cannot hold simultaneously, and therefore the probability that a particular event in $\Omega_1 \times \Omega_2$ holds is obtained by conditioning P_{12} on the event $\Theta_\cap = \{(\omega_1, \omega_2) \in \Omega_1 \times \Omega_2 : \Gamma_\cap(\omega_1, \omega_2) \neq \emptyset\}$.

Let $P_\cap$ be the probability measure on $\Omega_1 \times \Omega_2$ resulting from the conditioning of P_{12} on the event $\Theta_\cap$. Under the assumptions that the pieces of evidence represented by mass functions m_1 and m_2 are independent and reliable, our knowledge about $\boldsymbol{\theta}$ can then be represented by the mass function denoted $m_1 \oplus m_2$, called the orthogonal sum of m_1 and m_2, and induced by the random set $(\Omega_1 \times \Omega_2, 2^{\Omega_1 \times \Omega_2}, P_\cap, \Gamma_\cap)$. It is easy to show that

$$(m_1 \oplus m_2)(A) = \frac{1}{1-\kappa} \sum_{B \cap C = A} m_1(B) m_2(C), \quad \forall A \subseteq \Theta, A \neq \emptyset,$$

and $(m_1 \oplus m_2)(\emptyset) = 0$, with $\kappa := \sum_{B \cap C = \emptyset} m_1(B) m_2(C)$ the *degree of conflict* between m_1 and m_2. The orthogonal sum is well defined if $\kappa < 1$.

The binary operation $\oplus$ is called Dempster's rule. It satisfies several properties. It is commutative, associative and has the vacuous mass function as only neutral element. Furthermore, given two simple mass functions A_{w_1} and A_{w_2}, their orthogonal sum is the simple mass function $A_{w_1 + w_2}$.

Another property of Dempster's rule is related to the plausibility transformation method [2], which allows us to approximate a mass function m by a Bayesian mass function p_m obtained by normalizing the contour function π of m:

$$p_m(\{\theta_k\}) := \frac{\pi(\theta_k)}{\sum_{\ell=1}^K \pi(\theta_\ell)}, \quad k = 1, \ldots, K.$$

Given two mass functions m_1 and m_2, we have $p_{m_1 \oplus m_2} = p_{m_1} \oplus p_{m_2}$ [27], i.e., combination and approximation commute. In the remainder of this paper, the approximation of a mass function according to the plausibility transformation method is simply referred to as, for short, its approximation.

Dempster's rule allows us to define the notion of a separable mass function: a mass function m is *separable* if it can be obtained as the combination by Dempster's rule of simple mass functions. Furthermore, if m is non dogmatic, then m can be canonically decomposed as [21]:

$$m = \bigoplus_{\emptyset \neq A \subset \Theta} A^{d(A)}, \tag{1}$$

with $d(\cdot)$ a mapping from $2^{\Theta} \setminus \{\emptyset, \Theta\}$ to $(0, 1]$ called diffidence function [13].

Finally, let us remark that the orthogonal sum $m_1 \oplus m_2$ of two mass functions m_1 and m_2 relies on the assumption that they are induced by independent sources, which amounts to specifying the joint probability measure P_{12} on $\Omega_1 \times \Omega_2$ to be the product measure $P_1 \otimes P_2$. However, in principle, any dependence structure, and thus any P_{12} having P_1 and P_2 as marginals, can be selected. This is illustrated by Shafer in [22, Example 3], which is a case of non independence between sources inducing simple mass functions. Another example of such a case is provided by Example 1[1].

Example 1. Assume m_1 and m_2 are simple mass functions, induced by sources $(\Omega_i, 2_i^{\Omega}, P_i, \Gamma_i)$, with $\Omega_i = \{0, 1\}$, $P_i(0) = 0.2$, $\Gamma_i(0) = A_i$ and $\Gamma_i(1) = \Theta$ for some $A_i \subset \Theta$, $i = 1, 2$. Let S_i be the random variable, with state space Ω_i, representing the interpretation for the i-th source. Then, specifying $P_{12}(0, 0) = 0.2$ and $P_{12}(1, 1) = 0.8$, models the dependency $S_2 = S_1$ (we have $P_{12}(S_2 = 0|S_1 = 0) = 1$ and $P_{12}(S_2 = 1|S_1 = 1) = 1$).

In Sect. 3, we will see that any possible dependence structure between two simple mass functions can be characterized by a correlation.

3 Dependence Between Positive and Negative Evidence

Positive and negative items of evidence with respect to a class, as defined in [7], and more generally with respect to a proposition $\theta \in A$, are nothing but simple mass functions with focal set A and focal set $\overline{A}$, respectively. Combining them by Dempster's rule corresponds to assuming that they are independent. In Sect. 3.2, we extend their combination to any possible dependence structure, which we characterize by a correlation. This is obtained as a particular case of the more general problem of combining two simple mass functions $A_1^{d_1}$ and $A_2^{d_2}$ having some dependence structure, which leads to a generalization of Dempster's rule for combining simple mass functions (Sect. 3.1). Then, in a second step (Sect. 3.3),

[1] Example 1 is based on the probabilistic dependence structure considered in [23, Example 1].

we use this rule to unveil a new result concerning the approximation of separable (non dogmatic) mass functions, which is instrumental for our extension of the ERBFN.

3.1 Correlation-Based Specification of the Dependence

Let us assume that we have two (non dogmatic) simple mass functions $m_1 = A_1^{d_1}$ and $m_2 = A_2^{d_2}$, for some $A_i \subset \Theta$ and $d_i \in (0,1]$, $i = 1,2$, induced by two sources $(\Omega_i, 2_i^{\Omega}, P_i, \Gamma_i)$, with $\Omega_i = \{0,1\}$, $P_i(1) = d_i$, $\Gamma_i(0) = A_i$ and $\Gamma_i(1) = \Theta$, $i = 1,2$. Let S_i be the random variable, with state space Ω_i, representing the interpretation for the i-th source. As explained in Sect. 2.2, specifying the dependence structure between these items of evidence amounts to specifying a joint probability measure P_{12} on $\Omega_1 \times \Omega_2$, with marginals P_1 and P_2.

It is easy to see that, given d_1 and d_2, P_{12} is completely characterized simply by providing $d_{12} := P_{12}(S_1 = 1, S_2 = 1)$. Indeed, we have

$$\begin{aligned}
P_{12}(S_1 = 1, S_2 = 1) &= d_{12}, \\
P_{12}(S_1 = 1, S_2 = 0) &= P_1(S_1 = 1) - P_{12}(S_1 = 1, S_2 = 1) \\
&= d_1 - d_{12}, \\
P_{12}(S_1 = 0, S_2 = 1) &= P_2(S_2 = 1) - P_{12}(S_1 = 1, S_2 = 1) \\
&= d_2 - d_{12}, \\
P_{12}(S_1 = 0, S_2 = 0) &= 1 - (d_{12} + d_1 - d_{12} + d_2 - d_{12}) \\
&= 1 - d_1 - d_2 + d_{12}.
\end{aligned} \tag{2}$$

Thanks to Fréchet [15], we know that $d_{12} \in [\max(0, d_1 + d_2 - 1), \min(d_1, d_2)]$, and thus any dependence structure between the two pieces of evidence can be specified by choosing a number in this latter interval. Moreover, specifying the probability $d_{12} = P_{12}(S_1 = 1, S_2 = 1)$, given $d_1 = P_1(S_1 = 1)$ and $d_2 = P_2(S_2 = 1)$, actually amounts to specifying the dependence between events $S_1 = 1$ and $S_2 = 1$. Following [14], this dependence can be completely characterized and without loss of information by a scalar $r \in [-1, 1]$, representing the correlation between the events. A model of correlation between two events of respective probabilities p_1 and p_2 with correlation $r \in [-1, 1]$ is provided in [14]: it is based on the Frank family of copulas and it is such that the probability p_{12} of their conjunction is equal for correlation r to $p_{12} = F(p_1, p_2, r)$ with

$$F(p_1, p_2, r) = \begin{cases} \min(p_1, p_2) & \text{if } r = 1, \\ p_1 \cdot p_2 & \text{if } r = 0, \\ \max(0, p_1 + p_2 - 1) & \text{if } r = -1, \\ \log_s[1 + (s^{p_1} - 1)(s^{p_2} - 1)/(s - 1)] & \text{otherwise,} \end{cases}$$

where $s = \tan(\pi(1 - r)/4)$. This family is continuous and strictly increasing in r, i.e. for $r < r'$, we have $F(p_1, p_2, r) \leq F(p_1, p_2, r')$ for all $(p_1, p_2) \in [0,1]^2$ and there exist $(p_1, p_2) \in [0,1]^2$ such that $F(p_1, p_2, r) < F(p_1, p_2, r')$. The cases

$r = -1$, $r = 0$ and $r = 1$ correspond to particular dependence structures: opposite dependence, independence, and perfect dependence, respectively [14]; we can notice that P_{12} in Example 1 is obtained for correlation $r = 1$.

In short, the dependence structure between two sources underlying two simple mass functions is characterized by a correlation $r \in [-1, 1]$. Now, if we assume further that these sources are reliable, and that their dependence is specified by r, our knowledge about $\boldsymbol{\theta}$ can be represented by the mass function denoted $A_1^{d_1} \oplus_r A_2^{d_2}$ and induced by the random set $(\Omega_1 \times \Omega_2, 2^{\Omega_1 \times \Omega_2}, P_\cap^r, \Gamma_\cap)$, where $P_\cap^r$ is the probability measure P_{12} defined by (2) with $d_{12} = F(d_1, d_2, r)$ and conditioned on the event $\Theta_\cap$. The binary operation $\oplus_r$ is a generalization of Dempster's rule for the combination of two simple mass functions ($\oplus$ is recovered for $r = 0$).

3.2 Dependent Positive and Negative Evidence

Consider the special case of Sect. 3.1, where $A_1 = A$ for some $A \subset \Theta$, $A \neq \emptyset$, and $A_2 = \overline{A_1}$, i.e., mass functions m_1 and m_2 represent positive and negative items of evidence, respectively, with respect to proposition $\boldsymbol{\theta} \in A$. Assuming these items of evidence to be reliable and their dependence to be specified by some correlation $r \in [-1, 1]$, our knowledge about $\boldsymbol{\theta}$ is then represented by mass function $A^{d_1} \oplus_r \overline{A}^{d_2}$.

Proposition 1. *We have*

$$(A^{d_1} \oplus_r \overline{A}^{d_2})(A) = \frac{d_2 - F(d_1, d_2, r)}{d_1 + d_2 - F(d_1, d_2, r)},$$

$$(A^{d_1} \oplus_r \overline{A}^{d_2})(\overline{A}) = \frac{d_1 - F(d_1, d_2, r)}{d_1 + d_2 - F(d_1, d_2, r)},$$

$$(A^{d_1} \oplus_r \overline{A}^{d_2})(\Theta) = \frac{F(d_1, d_2, r)}{d_1 + d_2 - F(d_1, d_2, r)},$$

and $(A^{d_1} \oplus_r \overline{A}^{d_2})(B) = 0$ *for all* $B \in 2^\Theta \setminus \{A, \overline{A}, \Theta\}$.

3.3 Introducing Dependence Between Positive and Negative Evidence in a Separable Mass Function

Consider the canonical decomposition (1) of a non dogmatic separable mass function m. Let A be some strict non empty subset of Θ. We have $d(A) \leq 1$ and $d(\overline{A}) \leq 1$. In other words, the mass function m involves a (possibly vacuous) positive evidence and a (possibly vacuous) negative evidence for the proposition $\boldsymbol{\theta} \in A$. More generally, it may be remarked that m involves (possibly vacuous) positive and negative evidence for $2^{|\Theta|-1} - 1$ propositions.

It is then clear that there exist $n \leq 2^{|\Theta|-1} - 1$ distinct, strict and non empty subsets $A_1, \ldots, A_n$ of Θ, with $A_i \neq \overline{A_j}$ for all $i \neq j$, such that m can be rewritten as

$$m = \bigoplus_{i=1}^{n} (A_i^{d_i^+} \oplus \overline{A_i}^{d_i^-}) \tag{3}$$

with $d_i^+ = d(A_i)$ and $d_i^- = d(\overline{A_i})$. Although this is inconsequential for our subsequent developments, we remark that expression (3) is obviously not unique. In particular, the list of subsets $A_1, \ldots, A_n$ can, or not, include a subset A_i such that $d(A_i) = 1$ and $d(\overline{A_i}) = 1$, without changing the fact that (3) holds, given that the vacuous mass function is a neutral element for Dempster's rule. Furthermore, if both $d(A) < 1$ and $d(\overline{A}) < 1$ for some subset A, then either A or $\overline{A}$ can arbitrarily be chosen to be one of the subsets A_i.

In any case, Eq. (3) brings to light that a non dogmatic separable mass function relies on the combination of independent positive and negative pieces of evidence for n propositions $\boldsymbol{\theta} \in A_i$, $i = 1, \ldots, n$.

Theorem 1. *Let m be the mass function given by* (3). *Let $\mathbf{r} := (r_1, \ldots, r_n) \in [-1, 1]^n$. Let $m_{\mathbf{r}} := \bigoplus_{i=1}^n (A_i^{d_i^+} \oplus_{r_i} \overline{A_i}^{d_i^-})$. We have $p_m = p_{m_{\mathbf{r}}}$ with p_m and $p_{m_{\mathbf{r}}}$ the approximations of m and $m_{\mathbf{r}}$, respectively.*

Theorem 1 shows that whatever the dependence structure, i.e., correlation r_i, chosen between the positive and negative evidence for proposition $\boldsymbol{\theta} \in A_i$, $i = 1, \ldots, n$, the approximation of the resulting mass function does not depend on this choice.

4 The r-ERBFN Classifier

In this section, we start by introducing an evidential classifier, which is an extension of the ERBFN classifier [9,16] accounting for the dependence between positive and negative evidence for each class (Sect. 4.1). Then, we show that, similarly as the ERBFN reveals a latent mass function for a given RBFN, this classifier also produces a latent mass function for this RBFN (Sect. 4.2).

4.1 Model

Let $\mathbf{x} \in \mathcal{X}$ be the feature vector of some instance with unknown class $\boldsymbol{\theta} \in \Theta = \{\theta_1, \ldots, \theta_K\}$. Let $\mathbf{p}_j \in \mathcal{X}$, $j = 1, \ldots, J$, be J prototypes. Let $s_j = \exp(-\gamma_j d_j)$ be the degree of similarity between $\mathbf{x}$ and $\mathbf{p}_j$, where $d_j = ||\mathbf{x} - \mathbf{p}_j||$ is the Euclidean distance between $\mathbf{x}$ and $\mathbf{p}_j$ and $\gamma_j > 0$ is a parameter.

Case $K = 2$. Let $v_j \in \mathbb{R}$ be a parameter associated to prototype $\mathbf{p}_j$. Let $w_j = s_j v_j$. Let $w_j^+ = \max(0, w_j)$ and $w_j^- = \max(0, -w_j)$ be the positive and negative parts, respectively, of w_j. Let m_j^+ be the simple mass function with focal set $\{\theta_1\}$ and weight of evidence w_j^+, i.e., $m_j^+ = \{\theta_1\}_{w_j^+}$. Let m_j^- be the simple mass function with focal set $\{\theta_2\}$ and weight of evidence w_j^-, i.e., $m_j^- = \{\theta_2\}_{w_j^-}$. In other words, prototype $\mathbf{p}_j$ induces a positive evidence m_j^+ and a negative evidence m_j^- for class θ_1.

Let $m^+ := \bigoplus_{j=1}^J m_j^+$ be the overall, i.e., given all prototypes, positive evidence for class θ_1. Similarly, let $m^- := \bigoplus_{j=1}^J m_j^-$ be the overall negative evidence for class θ_1. We have $m^+ = \{\theta_1\}^{d^+}$, with $d^+ = \exp(-w^+)$ where $w^+ = \sum_{j=1}^J w_j^+$, and $m^- = \{\theta_2\}^{d^-}$, with $d^- = \exp(-w^-)$ where $w^- = \sum_{j=1}^J w_j^-$.

It can be remarked that mass functions m_j^+, $j = 1, \ldots, J$, are completely determined by distinct values w_j, $j = 1, \ldots, J$, i.e., changing the value w_j for some j does not affect mass functions m_k^+, $k \neq j$, therefore it seems reasonable to assume that they are independent between themselves, hence the definition of m^+. The same can be said about mass functions m_j^-, $j = 1, \ldots, J$. On the contrary, we can remark that changing the value w_j for some j will affect in general *both* m^+ and m^-. Hence, when pooling the overall positive and negative evidence for θ_1 in order to obtain our overall evidence – represented by some mass function $m_{\mathbf{x}}$ – with respect to the class of the instance, it seems safer to assume that there is some dependence between them. As we have seen, such a dependence can be characterized by a correlation $r \in [-1, 1]$, leading to the following definition:

Definition 1 (r-ERBFN). *The output of the r-ERBFN classifier is the mass function $m_{\mathbf{x},r}$ defined as*

$$m_{\mathbf{x},r} := m^+ \oplus_r m^-. \tag{4}$$

Remark 1. The 0-ERBFN is nothing but the ERBFN classifier introduced in [16, Section 3.2]. It corresponds to assuming that the overall positive and negative evidence for θ_1 are independent.

Case $K > 2$. Let $v_{jk} \in \mathbb{R}$ be a parameter associated to prototype $\mathbf{p}_j$ and to class θ_k. Let $w_{jk} = s_j v_{jk}$. Let $w_{jk}^+ = \max(0, w_{jk})$ and $w_{jk}^- = \max(0, -w_{jk})$. Let $m_{jk}^+ = \{\theta_k\}_{w_{jk}^+}$ and $m_{jk}^- = \overline{\{\theta_k\}}_{w_{jk}^-}$. In other words, prototype $\mathbf{p}_j$ induces a positive evidence m_{jk}^+ and a negative evidence m_{jk}^- for class θ_k.

Let $m_k^+ := \bigoplus_{j=1}^J m_{jk}^+$, respectively $m_k^- := \bigoplus_{j=1}^J m_{jk}^-$, be the overall positive, respectively negative, evidence for class θ_k. We have $m_k^+ = \{\theta_k\}^{d_k^+}$, with $d_k^+ = \exp(-w_k^+)$ where $w_k^+ = \sum_{j=1}^J w_{jk}^+$, and $m_k^- = \overline{\{\theta_k\}}^{d_k^-}$, with $d_k^- = \exp(-w_k^-)$ where $w_k^- = \sum_{j=1}^J w_{jk}^-$.

Using a similar reasoning as that in the case $K = 2$, we can safely assume that: mass functions m_{jk}^+ (resp. m_{jk}^-), $j = 1, \ldots, J$, are independent; mass functions m_k^+ and m_k^- are not independent. The dependence between these latter mass functions can be characterized by a correlation r_k. Our overall evidence for class θ_k is then represented by mass function $m_k := m_k^+ \oplus_{r_k} m_k^-$.

If we make the assumption that the prototypes $\mathbf{p}_j$ (together with their associated parameters γ_j) have been identified, i.e., are fixed, then we can remark that mass functions m_k, $k = 1, \ldots, K$, are determined by distinct sets of values: m_k is determined by the set $\{v_{jk} : 1 \leq j \leq J\}$ whereas $m_{k'}$, $k' \neq k$, is determined by the set $\{v_{jk'} : 1 \leq j \leq J\}$. Hence, under this assumption, mass functions m_k, $k = 1, \ldots, K$, can be considered independent, leading to the following definition:

Definition 2 (r-ERBFN). *The output of the* $\mathbf{r}$*-ERBFN classifier, with* $\mathbf{r} = (r_1 \dots, r_K)$*, is the mass function* $m_{\mathbf{x},\mathbf{r}}$ *defined as*

$$m_{\mathbf{x},\mathbf{r}} := \bigoplus_{k=1}^{K} \left(m_k^+ \oplus_{r_k} m_k^- \right). \tag{5}$$

Remark 2. The $\mathbf{0}$-ERBFN is nothing but the classifier described in [9]. It corresponds to assuming that the overall positive and negative evidence for each class are independent.

4.2 Latent Mass Function

Case $K = 2$. We recall that a RBFN with a logistic output unit is a probabilistic classifier for a binary classification problem ($\Theta = \{\theta_1, \theta_2\}$). It is a neural network composed of a hidden layer containing J hidden units, each hidden unit j, $j = 1, \dots, J$, being characterized by a prototype $\mathbf{p}_j$ and a scale parameter $\gamma_j > 0$. The activation of hidden unit j is $s_j = \exp(-\gamma_j d_j)$, where $d_j = ||\mathbf{x} - \mathbf{p}_j||$ with $\mathbf{x}$ the feature vector of an instance. Furthermore, let v_j be the weight of the connection between hidden unit j and the logistic output unit. Then, the probabilistic prediction $P_{\mathbf{x}}$ of this classifier is

$$P_{\mathbf{x}}(\theta_1) = \frac{1}{1 + \exp(-\sum_{j=1}^{J} s_j v_j)}. \tag{6}$$

Now, consider a r-ERBFN, for some $r \in [-1, 1]$, whose parameters $\mathbf{p}_j$, γ_j and v_j, have been identified to that of a given RBFN with a logistic output unit. Let $m_{\mathbf{x},r}$ denote the output mass function defined by (4) of this r-ERBFN.

Theorem 2. *For all* $r \in [-1, 1]$*, the approximation* $p_{m_{\mathbf{x},r}}$ *of* $m_{\mathbf{x},r}$ *satisfies*

$$p_{m_{\mathbf{x},r}}(\{\theta_1\}) = P_{\mathbf{x}}(\theta_1). \tag{7}$$

Theorem 2 shows that the output $m_{\mathbf{x},r}$ of a r-ERBFN, whose parameters have been identified to that of a given RBFN, is a latent mass function for the probabilistic prediction $P_{\mathbf{x}}$ of this RBFN, for all $r \in [-1, 1] \setminus \{0\}$, in the same way as is the output of the 0-ERBFN.

Case $K > 2$. A RBFN with a softmax output layer is a probabilistic classifier for a multi-class classification problem, whose parameters are prototypes $\mathbf{p}_j$ and scale parameters $\gamma_j > 0$, $j = 1, \dots, J$, as well as weights v_{jk} connecting hidden unit j and output unit k, $j = 1, \dots, J$, $k = 1, \dots, K$. Its probabilistic prediction $P_{\mathbf{x}}$ is

$$P_{\mathbf{x}}(\theta_k) = \frac{\exp(\sum_{j=1}^{J} s_j v_{jk})}{\sum_{\ell=1}^{K} \exp(\sum_{j=1}^{J} s_j v_{j\ell})}. \tag{8}$$

Consider a $\mathbf{r}$-ERBFN, for some $\mathbf{r} := (r_1, \dots, r_K) \in [-1, 1]^K$, whose parameters $\mathbf{p}_j$, γ_j and v_{jk}, have been identified to that of a given RBFN with a softmax output layer. Let $m_{\mathbf{x},\mathbf{r}}$ denote the output mass function defined by (5) of this $\mathbf{r}$-ERBFN.

Theorem 3. *For all* $\mathbf{r} \in [-1,1]^K$, *the approximation* $p_{m_{\mathbf{x},\mathbf{r}}}$ *of* $m_{\mathbf{x},\mathbf{r}}$ *satisfies*

$$p_{m_{\mathbf{x},\mathbf{r}}}(\{\theta_k\}) = P_{\mathbf{x}}(\theta_k), \quad \forall \theta_k \in \Theta. \tag{9}$$

Proof. It has been established in [9] that Eq. (9) holds for $\mathbf{r} = \mathbf{0}$. The theorem follows from Theorem 1. □

We have thus proved that in the multi-category case also, there exists a set of latent mass functions for the probabilistic prediction of a RBFN, of which the one identified in [9] is a particular member, obtained for $\mathbf{r} = \mathbf{0}$.

5 Identification of the Correlations

Assume a RBFN having a softmax output layer[2] and whose parameter values are given. Let us consider the $\mathbf{r}$-ERBFN and identify its prototypes, parameters γ_j and v_{jk} to those of this RBFN. To compute the output mass function of the $\mathbf{r}$-ERBFN, it remains to identify the correlations $\mathbf{r}$. Note that this amounts to selecting a particular latent mass function among the set of latent mass functions brought to light in Sect. 4.2.

In order to select a given $\mathbf{r}$, one can consider its prediction error (or loss). When a prediction is probabilistic, i.e., in the form of a probability distribution $P_{\mathbf{x}}$, its loss is typically evaluated by the negative log-likelihood (or cross-entropy)

$$\mathcal{L}(y, P_{\mathbf{x}}) = -\ln P_{\mathbf{x}}(y), \tag{10}$$

with y the true class of the instance with feature vector $\mathbf{x}$. Minimizing (10) is equivalent to maximizing the probability of the true class.

In the case of the $\mathbf{r}$-ERBFN, the prediction is evidential, i.e., in the form of a mass function $m_{\mathbf{x}}$. Following [10,12], since in this case we no longer have a single probability for the true class but two numbers - a degree of belief $Bel_{\mathbf{x}}(\{y\})$ and a degree of plausibility $Pl_{\mathbf{x}}(\{y\})$ - we can consider the following generalized negative log-likelihood (GNLL)

$$\mathcal{L}(y, m_{\mathbf{x}}) = -\frac{1}{2}\ln Bel_{\mathbf{x}}(\{y\}) - \frac{1}{2}\ln Pl_{\mathbf{x}}(\{y\}). \tag{11}$$

Minimizing (11) amounts to seeking high degrees of belief and of plausibility for the true class. Moreover, we may notice that if $m_{\mathbf{x}}$ is Bayesian, i.e., corresponds to a probability distribution, then we have $Bel_{\mathbf{x}}(\{y\}) = Pl_{\mathbf{x}}(\{y\}) = m_{\mathbf{x}}(\{y\})$, and loss (11) reduces, as may be required, to (10).

Given a loss of the form (11) and some learning data $\{\mathbf{x}_i, y_i\}_{i=1}^n$ where $\mathbf{x}_i$ is the feature vector of instance i and y_i is its true class, we may then fit over this learning set, the correlation vector $\mathbf{r}$, i.e., we can search for the vector of correlations $\hat{\mathbf{r}}$ that optimizes the total GNLL over this learning data:

$$\hat{\mathbf{r}} = \arg \min_{\mathbf{r} \in [-1,1]^K} \sum_{i=1}^{n} \mathcal{L}(y_i, m_{\mathbf{x}_i,\mathbf{r}}).$$

[2] We focus for short on the case $K > 2$ in this section, but our developments also hold for $K = 2$.

The resulting optimized correlation vector $\hat{\mathbf{r}}$ may then be used to compute the predictive latent mass function for any test feature vector $\mathbf{x}$.

6 Experiments

The previous section has put forward the $\hat{\mathbf{r}}$-ERBFN, as a classifier producing sensible predictive latent mass functions for a RBFN, according to the principle of minimizing the loss. The purpose of this section is to illustrate using some numerical experiments, the interest of this classifier with respect to the original proposal from [9,16], which corresponds to the $\mathbf{0}$-ERBFN. First, we describe how we trained the RBFN in our experiments (Sect. 6.1). Then, we provide the remainder of our experimental protocol and the results obtained (Sect. 6.2).

6.1 Training of the RBFN

In all of our experiments, the parameters $\mathbf{p}_j$, $\gamma_j > 0$ and v_{jk} of the considered $\mathbf{r}$-ERBFN classifiers were identified to that of a RBFN with a softmax output layer (or a logistic output unit in the case of a binary classification problem) learnt over the training dataset following two phase learning as described in [20]. Precisely, in the first phase, for each class, three prototypes were obtained as the centers of the clusters resulting from applying the (constrained[3] [1, Algorithm 2.2]) K-means clustering procedure to the examples of the class. Furthermore, the scale parameter γ_j associated to prototype $\mathbf{p}_j$ was set to $\gamma_j = 1/(2\sigma_j^2)$ where (kernel width) σ_j was the mean of the distances between the prototype $\mathbf{p}_j$ and the training examples in its associated cluster. In the second phase, the connection weights v_{jk} between hidden units and output units were learnt by minimisation of the $L2$ regularized cross entropy loss, using gradient descent (with learning rate and regularization coefficient both set to 10^{-3} and with 10^3 epochs).

6.2 Experimental Settings and Results

We used four real datasets[4] considered in related work [11]: Pima (7 features, 2 classes, 532 instances), Ionosphere (33 features, 2 classes, 351 instances), Glass identification (9 features, 6 classes, 214 instances), Vowel identification (10 features, 6 classes, 540 instances). For each dataset, we proceeded similarly as in [12, Section 5]. Specifically, the data were split randomly (using stratified random sampling) into training, validation and test sets containing, respectively, 60%, 20% and 20% of the instances. The training set was used to learn the RBFN as presented in Sect. 6.1, the validation set was used to optimize $\mathbf{r}$ as described in Sect. 5, and the test set was used to evaluate the performance, according to the average GNLL, of $\mathbf{r} = \hat{\mathbf{r}}$ as well as of $\mathbf{r} = \mathbf{0}$. This process was repeated 50 times.

Figure 1 shows boxplots of test GNLL values for the four datasets.

[3] Enforcing at least two training examples of the given class per cluster.

[4] Pima is available from the `R` package `MASS` [26]. Ionosphere, Glass and Vowel are available from the UCI ML repository https://archive.ics.uci.edu. For Vowel, we considered only the first six classes, as in [11].

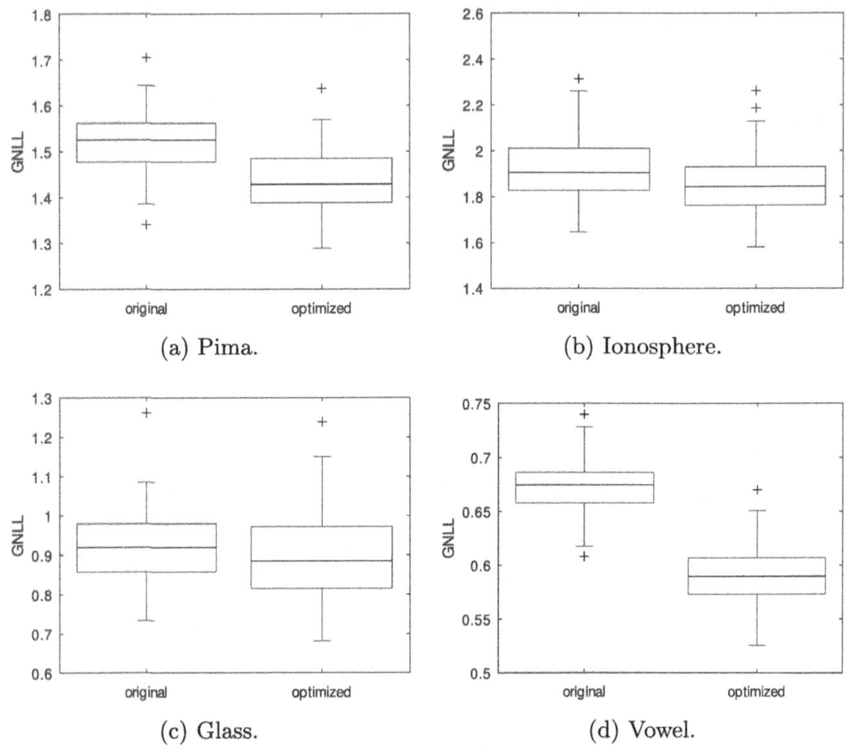

(a) Pima. (b) Ionosphere. (c) Glass. (d) Vowel.

Fig. 1. Generalized negative log-likelihood for the Pima (1a), Ionosphere (1b), Glass (1c) and Vowel (1d) datasets for $\mathbf{r} = \mathbf{0}$ (original) and $\mathbf{r} = \hat{\mathbf{r}}$ (optimized).

We can see that for all four datasets, the $\hat{\mathbf{r}}$-ERBFN outperforms the **0**-ERBFN; the differences are highly significant (p-values of paired t-tests for the comparison of GNLL values were at most[5] 2.6×10^{-11} over all datasets.)

7 Conclusion

This paper has brought to light a set of latent belief functions for a RBNF, extending the latent belief function identified in [9,16] to all possible dependence structures between positive and negative evidence for each class. These latent belief functions allow some performance improvement in terms of uncertainty quantification. A singular – least informative – belief function exists in this set; it will be described in a future publication.

Acknowledgments. Serigne Diène's PhD work is funded by the Hauts-de-France region and Artois University.

[5] P-value obtained for the Glass dataset.

References

1. Bradley, P.S., Bennett, K.P., Demiriz, A.: Constrained k-means clustering. Technical report. MSR-TR-2000-65, Microsoft Research, Redmond (2000). www.microsoft.com/en-us/research/wp-content/uploads/2016/02/tr-2000-65.pdf
2. Cobb, B.R., Shenoy, P.P.: On the plausibility transformation method for translating belief function models to probability models. Int. J. Approx. Reason. **41**(3), 314–330 (2006)
3. Dempster, A.P.: Upper and lower probabilities induced by a multivalued mapping. Ann. Math. Stat. **38**, 325–339 (1967)
4. Denœux, T.: A k-nearest neighbor classification rule based on Dempster-Shafer theory. IEEE Trans. Syst. Man Cybern. **25**(5), 804–213 (1995)
5. Denœux, T.: A neural network classifier based on Dempster-Shafer theory. IEEE Trans. Syst. Man Cybern. - Part A **30**(2), 131–150 (2000)
6. Denœux, T.: Quantifying predictive uncertainty using belief functions: different approaches and practical construction. In: Kreinovich, V., Sriboonchitta, S., Chakpitak, N. (eds.) TES 2018. SCI, vol. 753, pp. 157–176. Springer, Cham (2018). https://doi.org/10.1007/978-3-319-70942-0_8
7. Denœux, T.: Logistic regression, neural networks and Dempster-Shafer theory: a new perspective. Knowl.-Based Syst. **176**, 54–67 (2019)
8. Denœux, T.: Belief functions induced by random fuzzy sets: a general framework for representing uncertain and fuzzy evidence. Fuzzy Sets Syst. **424**, 63–91 (2021)
9. Denœux, T.: Théorie des fonctions de croyance et apprentissage automatique, (2022). journée Apprentissage automatique multimodal et fusion d'informations (2ème édition), GdR ISIS, virtual, 19th January 2022
10. Denœux, T.: Quantifying prediction uncertainty in regression using random fuzzy sets: the ENNreg model. IEEE Trans. Fuzzy Syst. **31**(10), 3690–3699 (2023)
11. Denœux, T.: Uncertainty quantification in logistic regression using random fuzzy sets and belief functions. Int. J. Approx. Reason. **168**, 109159 (2024)
12. Denœux, T.: Combination of dependent and partially reliable Gaussian random fuzzy numbers. Inf. Sci. **681**, 121208 (2024)
13. Dubois, D., Faux, F., Prade, H.: Prejudice in uncertain information merging: pushing the fusion paradigm of evidence theory further. Int. J. Approx. Reason. **121**, 1–22 (2020)
14. Ferson, S., et al.: Dependence in probabilistic modeling, Dempster-Shafer theory, and probability bounds analysis. Technical report. SAND2004-3072, Sandia Nat. Lab., Albuquerque, New Mexico (2004)
15. Fréchet, M.: Généralisations du théorème des probabilités totales. Fundam. Math. **25**, 379–387 (1935)
16. Huang, L., Ruan, S., Decazes, P., Denœux, T.: Lymphoma segmentation from 3D PET-CT images using a deep evidential network. Int. J. Approx. Reason. **149**, 39–60 (2022)
17. Hüllermeier, E., Waegeman, W.: Aleatoric and epistemic uncertainty in machine learning: an introduction to concepts and methods. Mach. Learn. **110**, 457–506 (2021)
18. Klir, G.J.: Uncertainty and Information: Foundations of Generalized Information Theory. Wiley-IEEE Press, Hoboken (2005)
19. Ronneberger, O., Fischer, P., Brox, T.: U-Net: convolutional networks for biomedical image segmentation. In: Navab, N., Hornegger, J., Wells, W.M., Frangi, A.F. (eds.) MICCAI 2015. LNCS, vol. 9351, pp. 234–241. Springer, Cham (2015). https://doi.org/10.1007/978-3-319-24574-4_28

20. Schwenker, F., Kestler, H.A., Palm, G.: Three learning phases for radial-basis-function networks. Neural Netw. **14**(4), 439–458 (2001)
21. Shafer, G.: A Mathematical Theory of Evidence. Princeton University Press, Princeton, N.J. (1976)
22. Shafer, G.: Probability judgment in artificial intelligence. In: Kanal, L.N., Lemmer, J.F. (eds.) Uncertainty in Artificial Intelligence, Machine Intelligence and Pattern Recognition, vol. 4, pp. 127–135. North-Holland (1986)
23. Shenoy, P.: On distinct belief functions in the Dempster-Shafer theory. In: Miranda, E., Montes, I., Quaeghebeur, E., Vantaggi, B. (eds.) Proceedings of the Thirteenth International Symposium on Imprecise Probability: Theories and Applications, vol. 215, pp. 426–437. PMLR (2023)
24. Smets, P.: Belief functions: the disjunctive rule of combination and the generalized Bayesian theorem. Int. J. Approx. Reason. **9**(1), 1–35 (1993)
25. Tong, Z., Xu, P., Denœux, T.: An evidential classifier based on Dempster-Shafer theory and deep learning. Neurocomputing **450**, 275–293 (2021)
26. Venables, W.N., Ripley, B.D.: Modern Applied Statistics with S, fourth edn. Springer, New York (2002). https://doi.org/10.1007/978-0-387-21706-2
27. Voorbraak, F.: A computationally efficient approximation of Dempster-Shafer theory. Int. J. Man-Mach. Stud. **30**(5), 525–536 (1989)

Frank's Triangular Norms in Piaget's Logical Proportions

Henri Prade(✉) and Gilles Richard

Institut de Recherche en Informatique de Toulouse (IRIT) - CNRS,
118, route de Narbonne, 31062 Toulouse Cedex 9, France
{henri.prade,gilles.richard}@irit.fr

Abstract. Starting from the Boolean notion of logical proportion in Piaget's sense, which turns out to be equivalent to analogical proportion, this note proposes a definition of analogical proportion between numerical values based on triangular norms (and dual co-norms). Frank's family of triangular norms is particularly interesting from this perspective. The article concludes with a comparative discussion with another very recent proposal for defining analogical proportions between numerical values based on the family of generalized means.

1 Introduction

In the appendix to a 1952 French book [7], in which he continued his systematic study of the transformations of logical statements, psychologist Jean Piaget introduced a quaternary operation he called *logical proportion*. He considered the notion important enough to devote a few pages to it in his introductory book [8] (pp. 35–37) the following year. It turns out that this so-called *logical proportion* is one of the possible (equivalent) expressions of a Boolean analogical proportion [10]. However, Piaget apparently never made the connection between his logical proportions and analogy.

This is the starting point of this short paper for proposing a new extension of analogical proportions to numerical values, based on triangular norms [4]. The paper is organized as follows. Section 2 recalls the link between Piaget's logical proportions and Boolean analogical proportions. Section 3 presents a new way of extending analogical proportions to numerical data using a pair of dual triangular norm and co-norm in place of conjunction and disjunction connectives in Piaget's logical proportions. This preserves the expected properties of analogical proportions. Moreover the Frank's family of triangular norms and co-norms [3] seems especially appropriate due to a particular relation linking the triangular norm and the dual co-norm which makes sense in an analogical proportion perspective. Sect. 4 examines the differences and potential links with an interesting, recent, proposal based on a parameterized family of means, also used for extending analogical proportions to numerical values.

S. Destercke et al. (Eds.): SUM 2024, LNAI 15350, pp. 369–377, 2025.
https://doi.org/10.1007/978-3-031-76235-2_27

2 Piaget's Logical Proportion and Analogy

Let a, b, c, d denote four logical propositions (in this section, $\wedge$, $\vee$, $\neg$, $\equiv$ denote the usual Boolean coonectives for conjunction, disjunction, negation and equivalence respectively). Then a, b, c, d make a *logical proportion* in the sense of Piaget [7,8] if the two following conditions hold:

$$a \wedge d = b \wedge c \text{ and } a \vee d = b \vee c.$$

When a, b, c, d are propositional truth values, the two conditions underlying Piaget's logical proportions define the following quaternary connective $Pia(a, b, c, d) = [(a \wedge d) \equiv (b \wedge c)] \wedge [(a \vee d) \equiv (b \vee c)]$ obviously equivalent to:

$$[(a \wedge d) \equiv (b \wedge c)] \wedge [(\neg a \wedge \neg d) \equiv (\neg b \wedge \neg c)] \quad (1)$$

This Pia connective, when true, emphasizes *similarity* and can be read as "what a and d have in common (positively or negatively), b and c have it also". It also says that "when a and d differs (one is true, the other is false) then b and c also differ". $Pia(a, b, c, d)$ is true only for the 6 patterns exhibited in the table below (and thus false for the $2^4 - 6 = 10$ remaining patterns).

Table 1. Valuations that make $Pia(a, b, c, d)$ true

a	b	c	d
0	0	0	0
1	1	1	1
0	0	1	1
1	1	0	0
0	1	0	1
1	0	1	0

According to [6], Table 1 is also known to correspond to the truth table of a Boolean *ana*logical proportion. An analogical proportion is a statement involving four items, usually denoted $a : b :: c : d$, which reads "a is to b as c is to d". One of the logical expressions of a Boolean analogical proportion, emphasizing *dissimilarity* inside pairs (a, b) and (c, d), is:

$$[(a \wedge \neg b) \equiv (c \wedge \neg d)] \wedge [(\neg a \wedge b) \equiv (\neg c \wedge d)] \quad (2)$$

In other words, $Pia(a, b, c, d)$ is equivalent to (2) and is one of the logical expressions of an analogical proportions [6,10]. Formula (2) precisely expresses that "a differs from b as c differs from d and b differs from a as d differs from c" (and "when a and b do not differ, c and d do not differ"). It can be checked that $a : b :: c : d$ thus defined satisfies the postulates usually assumed for analogical expressions:

- *reflexivity*: $a : b :: a : b$;
- stability under *symmetry*: $a : b :: c : d \Rightarrow c : d :: a : b$;
- stability under *central permutation*: $a : b :: c : d \Rightarrow a : c :: b : d$.

As a consequence, an analogical proportion also satisfies

- $a : a :: b : b$ (sameness) ;
- $a : b :: c : d \Rightarrow d : b :: c : a$ (external permutation);
- $a : b :: c : d \Rightarrow b : a :: d : c$ (internal reversal);
- $a : b :: c : d \Rightarrow d : c :: b : a$ (complete reversal).

Note that reflexivity postulate $a : b :: a : b$, true for all a and b, forces a Boolean analogical proportion to be true for the valuations $(0, 0, 0, 0)$, $(0, 1, 0, 1)$, $(1, 0, 1, 0)$, and $(1, 1, 1, 1)$. The minimal Boolean model that satisfies the three above postulates is true for the 6 patterns for Boolean variables a, b, c, d exhibited in Table 1 [11]. Besides it can be checked that Boolean analogical proportions are *transitive*, namely

$$(a : b :: c : d) \wedge (c : d :: e : f) \Rightarrow a : b :: e : f$$

Moreover there are *code independent*, i.e.,

$$(a : b :: c : d) \Rightarrow \neg a : \neg b :: \neg c : \neg d$$

which means that the proportion is stable exchanging 0 and 1 in Table 1.

Analogical proportions have been extended to numerical values a, b, c, d (assumed to be normalized between 0 and 1) by starting from (1) or (2) and by replacing connectives $\wedge$, $\vee$, $\neg$, $\equiv$ by multiple-valued connectives, the choice of these latter connectives being dictated by the preservation of a maximum of desirable properties [2,9]. This has led in (1) to take i) min for the central $\wedge$, ii) $\min(s \rightarrow_L t, t \rightarrow_L s) = 1 - |s - t|$ for the two equivalence connectives, where $s \rightarrow_L t = \min(1, 1 - s + t)$ is Łukasiewicz implication, iii) the bounded difference $\max(0, s - t) = 1 - (s \rightarrow_L t)$, (using $1 - (\cdot)$ as negation) for the four expressions of the form $s \wedge \neg t$. For (2), one has kept the same choices for the central $\wedge$ and for the equivalences, and used min and max for the inside $\wedge$ and $\vee$.

This graded view leads to privilege the connectives of MV algebras. In what follows, we explore another avenue.

3 A Triangular Norm-Based Definition of an Analogical Proportion

Triangular norms are associative operations of the unit interval [12], which have been extensively studied in fuzzy set theory [4]. They appear to be the proper operations for defining fuzzy set intersections, just as their dual co-norms are for defining fuzzy set unions. We start with a brief refresher on triangular norms, before introducing them in a new definition of analogical proportions for numerical values.

3.1 A Brief Refresher on Triangular Norms

Triangular norms (t-norms for short) and triangular conorms (t-conorms for short) were invented by Schweizer and Sklar [4,12], in the framework of probabilistic metric spaces, for the purpose of expressing the triangular inequality. They also turn out to be the most general binary operations on [0, 1] that meet natural and intuitive requirements for conjunction and disjunction operations. Namely, a t-norm T is a binary operation on $[0,1]$, i.e., $T : [0,1] \times [0,1] \to [0,1]$, that satisfies the following conditions:

- commutative: $T(a,b) = T(b,a)$;
- associative: $T(a,T(b,c)) = T(T(a,b),c)$;
- non-decreasing in both arguments: $T(a,b) \leq T(a',b')$ if $a \leq a'$ and $b \leq b'$;
- boundary conditions: $T(a,1) = T(1,a) = a$.

It can be proved that $T(a,0) = T(0,a) = 0$. It is known that the minimum operation is the greatest t-norm, i.e., for any t-norm T, $T(a,b) \leq \min(a,b)$ holds for all $a,b \in [0,1]$. Typical basic examples of t-norms are

- the minimum: $T(a,b) = \min(a,b)$,
- the product: $T(a,b) = a \cdot b$,
- the linear (or Łukasiewicz) t-norm: $T(a,b) = \max(0, a+b-1)$.

Note the inequalities: $\max(0, a+b-1) \leq a \cdot b \leq \min(a,b)$.

These three cases are important since any *continuous* t-norm is definable as an *ordinal sum*[1] of copies of Łukasiewicz, minimum and product t-norms (see, e.g.,. [4]).

In the following, we only consider continuous t-norms. A t-norm is said to be Archimedian if $\forall a, 0 < a < 1, T(a,a) < a$. Any continuous Archimedian t-norm has an additive generator f, and can be written as $T(a,b) = f^{(-1)}(f(a)+f(b))$ where f is a strictly decreasing function from [0, 1] to $[0,+\infty)$ and $f^{(-1)}$ is its pseudo-inverse defined by $f^{(-1)}(x) = 1$ if $x \in [0, f(1)]$, $f^{(-1)} = f^{-1}(x)$ if $x \in [f(1), f(0)]$, $f^{(-1)} = 0$ if $x \in [f(0), +\infty)$. If f is such that $f(1) = 0$ and $lim_{x \to 0^+} f(x) = +\infty$, the t-norm is said to be *strict*. The product is a typical example of strict t-norm (for which $f(x) = -ln(x)$). When T is strict, $f^{(-1)}(x) = f^{-1}(x)$. If $f(1) = 0$ and $f(0)$ is finite, the t-norm is said to be *nilpotent*; the Łukasiewicz t-norm is a typical example ($f(x) = 1 - x$).

The De Morgan-like dual notion of a t-norm (w.r.t. negation $n(a) = 1 - a$, or a more general strong negation) is that of a t-conorm $S(a,b) = n(T(n(a), n(b)))$, where a strong negation n is such that $n(0) = 1$; $n(1) = 0$ (it coincides with Boolean negation on $\{0, 1\}$); $n(a) \geq n(b), \text{if} a \leq b$, and $n(n(a)) = a$ (n is decreasing and involutive). In the following we use $n(a) = 1 - a$, for all $a \in [0,1]$.

[1] An ordinal sum is defined in the following way. Let $\{\mathcal{I}_n\}_n$ be a countable family of pairwise disjoint sub-intervals in [0, 1]. To each $\mathcal{I}_n = [x_n, y_n]$, we associate an Archimedean t-norm T_n. Then the ordinal sum is the triangular norm $T(a,b) = x_n + (y_n - x_n)T_n(\frac{a-x_n}{y_n-x_n}, \frac{b-x_n}{y_n-x_n})$ if $a,b \in [x_n, y_n]$; $T(a,b) = \min(a,b)$ otherwise [4].

Thus, a binary operation S on $[0,\ 1]$ is called a t-conorm if it satisfies the same properties as the ones of a t-norm except for the boundary conditions, namely, here 0 is an identity and 1 is absorbent:

$$\text{boundary conditions: } S(0,a) = S(a,0) = a.$$

Hence $S(a,1) = S(1,a) = 1$. Dually, the maximum operation is the smallest t-conorm ($S(a,b) \geq \max(a,b)$). Typical basic examples of t-conorms are the dual of minimum, product and Łukasiewicz' t-norms, namely the maximum $S(a,b) = max(a,b)$, the so-called probabilistic sum $S(a,b) = a + b - a\dot{b}$ and the bounded sum $S(a,b) = min(1, a+b)$. Note now the inequalities $\max(a,b) \leq a+b-a \cdot b \leq \min(1, a+b)$.

3.2 Definition and Properties

Definition 1. *Let a, b, c, d be four numerical values in [0, 1]. a, b, c, d are said to form an* analogical proportion *based on the t-norm T (and its dual t-conorm S_T), which is denoted $A_T(a,b,c,d)$, if and only if $T(a,d) = T(b,c)$ and $S_T(a,d) = S_T(b,c)$.*

Note that this definition is all-or-nothing: $A_T(a,b,c,d)$ is true or false. This contrasts with the multiple-valued logic extensions of $a : b :: c : d$, where the multiple-valued quaternary connective obtained becomes a matter of degree when a, b, c, d are numerical values [2]. Definition 1 might be weakened by requiring only equality for T (or for S_T), but in the following we keep it as it is for symmetry reason.

It is easy to check that

- when $a, b, c, d \in \{0, 1\}$, $A_T(a,b,c,d)$ is true only for the 6 valuations of the table in Sect. 2 that make $a : b :: c : d$ true.
- $A_T(a,b,c,d)$ satisfies the three postulates of analogical proportions: i) reflexivity, ii) stability under symmetry, and iii) stability under central permutation, thanks to the commutativity of T (and thus of S_T). It is clear that $A_T(a,b,c,d)$ also satisfies the consequences of the three postulates.

Let us consider the cases of the three main t–norms.

- $T(a,b) = \min(a,b)$ and $S_T(a,b) = \max(a,b)$. Then $A_T(a,b,c,d)$ means $\min(a,d) = \min(b,c)$ and $\max(a,d) = \max(b,c)$. So $A_T(a,b,c,d)$ holds only if $a = b$ (resp. $a = c$) and then $c = d$ (resp. $b = d$).
- $T(a,b) = ab$ and $S_T(a,b) = a+b-ab$. Then $A_T(a,b,c,d)$ means $ad = bc$ and $a + d - ad = b + c - bc$, which is equivalent to $ad = bc$ and $a + d = b + c$. Thus, a, b, c, d form both an arithmetic proportion (i.e., $a - b = c - d$), and a geometric proportion (i.e., $\frac{a}{b} = \frac{c}{d}$ provided that $b \neq 0$ and $d \neq 0$). In the case $a = 0$, this imposes b or c to be 0 as well. In case $b = 0$, then $d = c$, if $c = 0$, then $d = b$. This leads to the only candidate proportions $A_T(0,0,c,c)$ and $A_T(0,b,0,b)$. More generally, letting $a + d = S$ and $ad = P$, leads to $d = S - a$, then $a(S - a) = P$ and a should be solution of the quadratic equation $x^2 - Sx + P = 0$, thus $x = (S \pm \sqrt{S^2 - 4P})/2$, leading to $a = b$ (and $c = d$) or $a = c$ (and $b = d$) (since $S^2 - 4P = (b-c)^2$).

- $T(a,b) = \max(0, a+b-1)$ and $S_T(a,b) = \min(1, a+b)$. Then $A_T(a,b,c,d)$ means $\max(0, a+d-1) = \max(0, b+c-1)$ and $\min(1, a+d) = \min(1, b+c)$, i.e., $a+d = b+c$, or if we prefer $a-b = c-d$.

Thus with Definition 1, we obtain two different views of an analogical proportion between numerical values:

- a *conservative* view with T equal to minimum or product in $A_T(a,b,c,d)$, where the only accepted analogical proportions are of the form $a : a :: c : c$ or $a : b :: a : b$.
- a *liberal* view with T in $A_T(a,b,c,d)$ equal to Łukasiewicz' t-norm where $a : b :: c : d$ holds if and only if $a - b = c - d$.

These two options correspond to the cases where the two multiple-valued logic extensions of $a : b :: c : d$ [2] are equal to 1 respectively. The conservative view is clearly much more drastic than the liberal one.

One can also observe that $a = b \Rightarrow c = d$ for the three main t-norms, one may conjecture that it remains true for any continuous t-norm.

Besides, it is clear that $A_T(a,b,c,d)$ is transitive (i.e.,$A_T(a,b,c,d), A_T(c,d,g,h) \Rightarrow A_T(a,b,g,h)$) when T is one of the three main triangular norms. But the property is more general. For instance, when T is strict, using its additive generator, we have $f(a)+f(d) = f(b)+f(c)$ and $f(c)+f(h) = f(d)+f(g)$ that entails $f(a)+f(h) = f(b)+f(g)$, and similarly for the dual co-norm equality.

Moreover, the code independency property, valid in the Boolean model of analogical proportions and for the arithmetic proportion, is still valid for A_T: for instance, $T(a,d) = T(b,c)$ implies $S(1-a, 1-d) = 1-T(a,d) = 1-T(b,c) = S(1-b, 1-c)$ (similar reasoning to prove $T(1-a, 1-d) = T(1-b, 1-c)$).

Lastly, the analogical proportion $a : b :: 1-b : 1-a$, true in the Boolean case, is true with numerical values for Łukasiewicz t-norm, but not for the product or the min.

3.3 Appropriateness of Frank's Triangular Norms

There exist several parameterized families of triangular norms [4]. One is the Frank's family [3], which includes the main triangular norms. These are defined by

$$T_F^p(a,b) = \begin{cases} \min(a,b) & \text{if } p = 0 \\ a \cdot b & \text{if } p = 1 \\ \max(0, a+b-1) & \text{if } p = +\infty \\ \log_p(1 + \frac{(p^a-1)(p^b-1)}{p-1}) & \text{otherwise.} \end{cases}$$

and their additive generator is

$$f_F^p(x) = \begin{cases} -\log x & \text{if } p = 1 \\ 1 - x & \text{if } p = +\infty \\ \log(\frac{p-1}{p^x-1}) & \text{otherwise.} \end{cases}$$

More importantly for our topic, Frank's triangular norms (and their ordinal sums) are the only triangular norms that satisfy the following remarkable property

$$\forall p, T_F^p(a,b) + S_{T_F}^p(a,b) = a + b \tag{3}$$

This echoes the following canonical analogical proportion [10] that holds in any lattice structure [1]:

$$a \wedge b : a :: b : a \vee b$$

since it is clear that (3) expresses that $T_F^p(a,b) : a :: b : S_{T_F}^p(a,b)$ is an arithmetic proportion.

A candidate option to estimate a p such that $A_{T_F^p}(a,b,c,d)$ holds is to minimize the sum of the difference $|T_F^p(a,d) - T_F^p(b,c)| + |S_F^p(a,d) - S_F^p(b,c)|$. As a preliminary, we have experimented with $p = 2$, $p = 10$ and $p = 100$, with fixed values for a, b, c with $a \leq b \leq c \leq d$: e.g., $a = 0.01, b = 0.2, c = 0.3$. In Fig. 1, we provide the curves of $diff_p(x) = |T_F^p(a,x) - T_F^p(b,c)| + |S_F^p(a,x) - S_F^p(b,c)|$, for $p = 2, 10, 100$ and x varying linearly from 0.3 to 1. We observe that the value $x = d$ such that $diff_p(x) = 0$ slowly increases with p to 0.49 (obtained for $p = +\infty$, i.e., for $a + d = b + c$).

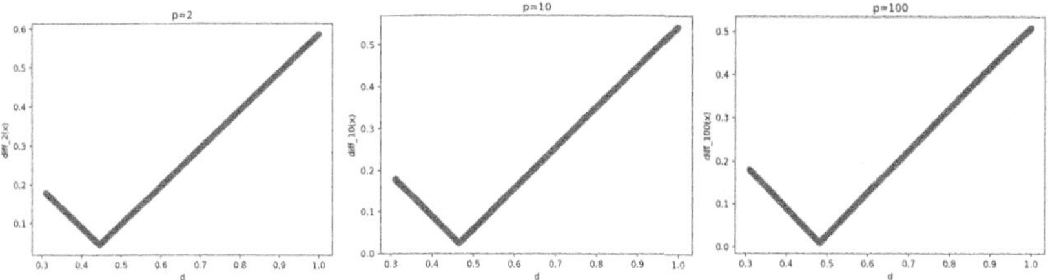

Fig. 1. $p = 2$; $p = 10$; $p = 100$

4 Related Work and Discussion

One can observe that the arithmetic proportion $a - b = c - d$ is equivalent to $\frac{1}{2}(a+d) = \frac{1}{2}(b+c)$, i.e., to the equality of the arithmetic means of the extremes and means in $a : b :: c : d$. This idea was recently extended in an interesting article [5], where the authors proposed a definition of analogical proportion for numerical values a, b, c, d (assuming without loss of generality that they are in the unit interval $[0,1]$). It is based on *generalized means*, as follows:

Definition 2. *Let $a, b, c, d \in [0,1]$. a, b, c, d are said to satisfy an analogical proportion if and only if $M_r(a,d) = M_r(b,c)$, where $M_r(x,y) = [\frac{1}{2}(x^r + y^r)]^{1/r}$.*

For $r = -1, 0, 1, 2$, we obtain the harmonic, geometric, arithmetic, and quadratic mean respectively. We have $\forall r, \min(x,y) \leq M_r(x,y) \leq \max(x,y)$, and the bounds are reached for $r = -\infty$ and $r = +\infty$ respectively.

This definition satisfies the 3 postulates of an analogical proportion. It also satisfies $a = b \Rightarrow c = d$, as well as $a \wedge b : a :: b : a \vee b$ for $\wedge = \min$ and $\vee = \max$. Moreover, $a : b :: 1 - b : 1 - a$ holds for $r = 1$.

Nevertheless it does not satisfy some additional property such as code independency. Take for instance $p = 2$: $M_2(a, d) = M_2(b, c)$ implies $a^2 + d^2 = b^2 + c^2$, but developing $(1 - a)^2 + (1 - d)^2$ and $(1 - b)^2 + (1 - c)^2$ shows that the 2 expressions will be equal if and only if $a + d = b + c$, which is not necessarily the case. Staying in $[0, 1]$ with $a = 0.1, b = 0.5, c = 0.5, d = 0.7$, we have that $M_r(a, d) = M_r(b, c)$ but we do not have $M_r(1 - a, 1 - d) = M_r(1 - b, 1 - c)$.

Besides, it was established [5] that given $a < b < c < d$, $\exists! r$ such that $M_r(a, d) = M_r(b, c)$.

There seems to be an obvious parallel between Definition 1 and Definition 2, up to the point that the conditions are respectively stated in terms of t-norms and dual t-conorms on the one hand and in terms of mean on the other hand. The two definitions coincides if T is the Łukasiewicz t-norm and $M_p = M_1$. Definition 1 is closer to logic and to multiple-valued logic extensions of analogical proportion [2] than Definition 2 that uses means which are not multiple-valued logic operations strictly speaking. The use of the parameterized family of Frank's t-norms enables us to modulate the conditions defining $A_{T_F^p}(a, b, c, d)$; we may wonder if when $0 \leq a < b < c < d \leq 1$, $\exists p$ such that $A_{T_F^p}(a, b, c, d)$ holds. This is an open question.

Besides, a long time ago, a multiple-valued logic extension of (1) based on Goguen's implication ($a \rightarrow b = 1$ if $a = 0$ and $a \rightarrow b = \min(1, b/a)$ otherwise) was proposed [6] which is equal to 1 if and only if $ad = bc$, thus coinciding with Definition 2 for M_0. This suggests that the agreement of Definition 2 with logic may be greater than it appears at first glance.

5 Concluding Remarks

The ambition of this note is to show the possibility of a definition, based on t-norms, of an analogical proportion between numerical values that makes sense and merits to be studied further. Some families of t-(co)norms are closer to generalized averages than Frank's, such as Schweizer and Sklar's family [4,12], also including the main t-norms, but they do not satisfy (3). Further research should lead to a unifying view on the topic.

Acknowledgements. This research was supported by the ANR project "Analogies: from Theory to Tools and Applications" (AT2TA), ANR-22-CE23-0023.

References

1. Barbot, N., Miclet, L., Prade, H.: Analogy between concepts. Artif. Intell. **275**, 487–539 (2019)
2. Dubois, D., Prade, H., Richard, G.: Multiple-valued extensions of analogical proportions. Fuzzy Sets Syst. **292**, 193–202 (2016)

3. Frank, M.J.: On the simultaneous associativity of f(x, y) and $x+y-f$ (x, y). Aequ. Math. **19**, 194–226 (1979)
4. Klement, E., Mesiar, R., Pap, E.: Triangular Norms. Kluwer Academic Pub, Boston (2000)
5. Lepage, Y., Couceiro, M.: Analogie et moyenne généralisée. In: Actes des 18èmes Journées d'Intelligence Artificielle Fondamentale, La Rochelle, pp. 114–124, 1–3 July 2024
6. Miclet, L., Prade, H.: Handling analogical proportions in classical logic and fuzzy logics settings. In: Sossai, C., Chemello, G. (eds.) ECSQARU 2009. LNCS (LNAI), vol. 5590, pp. 638–650. Springer, Heidelberg (2009). https://doi.org/10.1007/978-3-642-02906-6_55
7. Piaget, J.: Essai sur les Transformations des Opérations Logiques: les 256 Opérations Ternaires de la Logique Bivalente des Propositions. Presses Univ. de France, Paris (1952)
8. Piaget, J.: Logic and Psychology. Manchester Univ, Press (1953)
9. Prade, H., Richard, G.: Multiple-valued logic interpretations of analogical, reverse analogical, and paralogical proportions. In: Proceedings of the 40th IEEE International Symposium on Multiple-Valued Logic (ISMVL'10), Barcelona, pp. 258–263 (2010)
10. Prade, H., Richard, G.: From analogical proportion to logical proportions. Log. Univers. **7**(4), 441–505 (2013)
11. Prade, H., Richard, G.: Analogical proportions: from equality to inequality. Int. J. Approx. Reason. **101**, 234–254 (2018)
12. Schweizer, B., Sklar, A.: Associative functions and abstract semi-groups. Publ. Math. Debrecen **10**, 69–180 (1963)

Social Ranking Under Incomplete Knowledge: Elicitation of the Lex-Cel Necessary Winners

Ariane Ravier[1(✉)], Sébastien Konieczny[2], Stefano Moretti[1], and Paolo Viappiani[1]

[1] LAMSADE, CNRS, Université Paris-Dauphine, Université PSL, Paris 75016, France
{ariane.ravier,stefano.moretti,paolo.viappiani}@lamsade.dauphine.fr
[2] CRIL - CNRS - Université d'Artois, Lens, France
konieczny@cril.fr

Abstract. In this paper, we introduce two elicitation approaches for the determination of *lex-cel necessary winners* (*i.e.*, the elements that are ranked highest according to *lex-cel*) in a social ranking problem, when the knowledge about preferences over the coalitions is incomplete, and the initially accessible input is a subset of an existing total preorder. The first approach is preorder-driven, and elicitates enough of the underlying total preorder to determine the lex-cel necessary winners. The second approach is element-driven and guides comparisons based on strategically-located coalitions. Finally, we present experimental results and discuss the performance of each approach depending on various parameters and scenarios.

1 Introduction

The concept of social ranking has been recently introduced in the literature to address the problem of ranking individual elements based on their contribution in establishing the position of groups or coalitions within a society [1]. To generate a social ranking, a ranking over sets of objects (such as groups, team's squads, winning and losing coalitions within a voting body, etc.) is taken as input. This generates a ranking over individual elements (such as group members, team's players, voters, etc.) that reflects their overall influence within the ranking over sets.

In [2], a social ranking solution called *lexicographic excellence* (lex-cel) is identified as the only one satisfying a set of appealing properties. According to this rule, individual elements are ranked based on their occurrences in the highest positions of the ranking over coalitions (note that this ranking may contain equivalences) as follows. First, one counts the number of occurrences of each element (i.e., the number of sets containing the element) in every equivalence class of the ranking over sets. When comparing two elements, one must focus to the

S. Destercke et al. (Eds.): SUM 2024, LNAI 15350, pp. 378–393, 2025.
https://doi.org/10.1007/978-3-031-76235-2_28

number of occurrences in the best equivalence class of sets, the greater number of occurrences being the better element. In case the number of occurrences in the best equivalence class is the same, attention shifts to the number of occurrences in the second-best equivalence class of sets, with the higher number determining the better element. If parity persists, the third-best equivalence class is considered, and so on, until all the equivalence classes are examined (of course, in case of a tie in each equivalence class of sets, the two elements are declared indifferent). For instance, in a situation with only two individual elements 1 and 2 and a ranking over sets such that the set $\{1, 2\}$ and the singleton $\{2\}$ are in the best equivalence class and both are strictly better than the singleton $\{1\}$, the lex-cel places 2 above 1, as 2 occurs twice in the best equivalence class, while 1 occurs just once.

Generalizations of lex-cel have been proposed in [3,4], considering the size of coalitions in addition to the occurence of elements in the sets, and in [5], under variable domains of coalitions. Other social ranking solutions, based on different criteria, have been proposed, for instance, in [6–9]. As noticed in [10], however, generating a complete ranking over individual elements may not be necessary. In such cases, the objective may be to select only the most important ones. As a motivating example, the authors of [10] consider the problem of identifying the most influential scientists within an academic association based on the number and quality of their publications and taking into account the contributions of scientists to multi-authored publications. They first propose to establish a ranking over groups of scientists according to a predefined bibliometric criterion, then they adopt a property-driven approach to argue that most influential scientists across different groups are the elements with the highest position according to lex-cel computed on the ranking of groups.

In this paper we apply the lex-cel method to identify the most significant elements (also called "winners") in the presence of partial information. Indeed, we observe that the process of acquiring the necessary information to compute the lex-cel is often highly time-consuming, as one can argue that collecting and processing data to compare any possible pair of groups of a finite set becomes quickly unaffordable in practice. So, the main contribution of this paper is to provide elicitation procedures aiming at identifying winners (according to lex-cel) using a subset of the ranking relation over coalitions. In other words, our goal is to find a procedure to determine the winner(s) of the lex-cel by asking a minimum number of questions to an expert (*e.g.*, the director of the academic association) about pairwise comparisons among groups.

Our approach bears similarity with works in computational social choice that deal with determining possible and necessary winners [11–13] when only a partial preference profile, composed of partial orders, is given in input. Adaptive elicitation approaches have been proposed that refine the partial profile in an iterative manner until a necessary winner is identified or regret drops below a threshold [14–16].

We start by presenting some preliminary notions and concepts, then introduce the problem at hand. We describe, first, a preorder-driven elicitation method to determine the lex-cel necessary winners by reconstituting the underlying total preorder, and, second, an element-driven method, based on a strategic choice of coalitions to compare. Finally, we present experimental results and discuss the performance of each method depending on various scenarios.

2 Preliminaries

Given a finite set $\mathcal{E} = \{1, \ldots, n\}$, a **binary relation** $\succsim_{\mathcal{E}}$ over $\mathcal{E}$ is a subset of the Cartesian product $\mathcal{E} \times \mathcal{E}$. In the following, given two elements $i, j \in \mathcal{E}$, the fact that $(i, j) \in \succsim_{\mathcal{E}}$ is also denoted by $i \succsim_{\mathcal{E}} j$. A binary relation $\succsim_{\mathcal{E}}$ over $\mathcal{E}$ that is reflexive and transitive is called **preorder** over $\mathcal{E}$, and the pair $(\mathcal{E}, \succsim_{\mathcal{E}})$ is a **preordered set**. The set of all total preorders on $\mathcal{E}$ is denoted by $\boldsymbol{\mathcal{R}}(\mathcal{E})$. In the remaining of the paper a preorder may also be referred to as a **preference relation**. A total preorder that is also antisymmetric is called a **linear order**. An **extension** of a preorder $\succsim_{\mathcal{E}}$ is a total preorder $\succsim_{\mathcal{E}}^{*} \in \boldsymbol{\mathcal{R}}(\mathcal{E})$ such that $\succsim_{\mathcal{E}} \subseteq \succsim_{\mathcal{E}}^{*}$. If an extension is a linear order it is also called **linear extension**.

Given a total preorder $\succsim_{\mathcal{E}}$ over $\mathcal{E}$, we denote by $\sim_{\mathcal{E}}$ the symmetric part of $\succsim_{\mathcal{E}}$ (*i.e.* $i \sim_{\mathcal{E}} j$ if $i \succsim_{\mathcal{E}} j$ and $j \succsim_{\mathcal{E}} i$) and by $\succ_{\mathcal{E}}$ its asymmetric part (*i.e.* $i \succsim_{\mathcal{E}} j$ and not $j \succsim_{\mathcal{E}} i$). A **maximal element** $i \in \mathcal{E}$ of a preordered set $(\mathcal{E}, \succsim_{\mathcal{E}})$ is such that there exists no $j \in \mathcal{E}$ verifying $j \succ_{\mathcal{E}} i$; we denote by $\mathcal{M}^{\succsim_{\mathcal{E}}}$ the set of all maximal elements of $(\mathcal{E}, \succsim_{\mathcal{E}})$. Given a preordered set $(\mathcal{E}, \succsim_{\mathcal{E}})$, the preordered set restricted to a subset $C \subseteq \mathcal{E}$ is defined as the pair $(C, \succsim_{\mathcal{E}}^{C})$, where $\succsim_{\mathcal{E}}^{C} \subseteq C \times C$ is such that $i \succsim_{\mathcal{E}}^{C} j \Leftrightarrow i \succsim_{\mathcal{E}} j$ for all $i, j \in C$; if $\succsim_{\mathcal{E}}^{C}$ is total, the preordered set $(C, \succsim_{\mathcal{E}}^{C})$ is called a **chain**. A **maximal chain** is a chain $(C, \succsim_{\mathcal{E}}^{C})$ such that there exists no $D \subseteq \mathcal{E}$ with $C \subseteq D$ and such that $(D, \succsim_{\mathcal{E}}^{D})$ is a chain.

Let X be a finite set. We denote by $\mathcal{P}(X)$ the set of non-empty subsets of X, *i.e.* $\mathcal{P}(X) = \{A \subseteq X : A \neq \emptyset\}$. The elements of $\mathcal{P}(X)$ are called **coalitions**. A total preorder $\succsim_{\mathcal{P}(X)} \in \boldsymbol{\mathcal{R}}(\mathcal{P}(X))$ is called a **power relation** (when the set $\mathcal{P}(X)$ is clear from the context, a power relation on $\mathcal{P}(X)$ is simply denoted by $\succsim$). Consider a power relation $\succsim \in \boldsymbol{\mathcal{R}}(\mathcal{P}(X))$ and two coalitions $S, T \in \mathcal{P}(X)$ such that $S \succsim T$ (interpreted as "S is at least as powerful as T w.r.t. the power relation $\succsim$"). The **quotient order** of $\succsim$ is denoted by $\Sigma_1 \succ \Sigma_2 \succ \cdots \succ \Sigma_l$ where each **equivalence class** Σ_k, $k \in \{1, \ldots, l\}$ is generated by the symmetric part of $\succsim$. This means that all coalitions in Σ_1 are "equally powerful" and are "strictly more powerful" than the sets in Σ_2 and so on. We denote by $x_k = |\{S \in \Sigma_k : x \in S\}|$ the number of occurrences of element x in Σ_k, for any $k = 1, \ldots, l$. Let us denote by $\theta^{\succsim}(x)$, for any $x \in X$, the l-dimensional vector $\theta^{\succsim}(x) = (x_1, \ldots, x_l)$ associated with $\succsim$. This vector is called the **occurrence vector** of x, as it indicates the number of occurrences of x in each equivalence class of $\succsim$.

A **social ranking** $R : \boldsymbol{\mathcal{R}}(\mathcal{P}(X)) \to \boldsymbol{\mathcal{R}}(X)$ is a mapping that associates a power relation $\succsim \in \boldsymbol{\mathcal{R}}(\mathcal{P}(X))$ over coalitions with a total preorder $R^{\succsim} \in \boldsymbol{\mathcal{R}}(X)$ over single elements. Let **i**,**j** be two vectors of the same size, the lexicographic total preorder $\geq_L$ is such that $\mathbf{i} \geq_L \mathbf{j}$ if either $\mathbf{i} = \mathbf{j}$ or there exists t such that $i_t > j_t$ and $i_r = j_r$ for all $r \in \{1, \ldots, t-1\}$. The *lexicographic excellence (lex-cel)*

[2] is a social ranking $R_{le}^{\succsim} : \mathcal{R}(\mathcal{P}(X)) \rightarrow \mathcal{R}(X)$ such that for all $\succsim \in \mathcal{R}(\mathcal{P}(X))$ and all $x, y \in X$: $x \; R_{le}^{\succsim} \; y \iff \theta^{\succsim}(x) \;\geq_L\; \theta^{\succsim}(y)$.

The symmetric and the asymmetric part of $R_{le}^{\succsim}$ are denoted by $I_{le}^{\succsim}$ and $P_{le}^{\succsim}$, respectively. An element $x \in X$ such that $x \; R_{le}^{\succsim} \; y$ for all $y \in X$ is a *winner* in $R_{le}^{\succsim}$. There may be several winners; an axiomatic approach to the problem of selecting the winners provided by the lex-cel has been studied in [10].

Remark 1. It is easy to show that if $\succsim$ is a linear order over $\mathcal{P}(X)$, then $\theta^{\succsim}(\mathrm{x})$ is a 0, 1-vector and $\theta^{\succsim}(x) \neq \theta^{\succsim}(y)$ for all $x, y \in X$ with $x \neq y$; so, $R_{le}^{\succsim}$ is a linear order over X and there is a unique winner in $R_{le}^{\succsim}$.

Example 1. Let $X = \{1, 2, 3, 4\}$, and $\succsim_{\mathcal{P}(X)}$ (or, simply, $\succsim$) be a preorder represented by the preference graph of Fig. 1 (left-side) and its maximal chains (right-side).

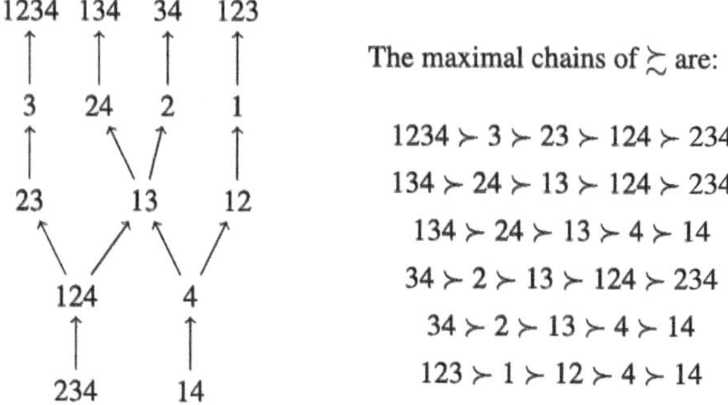

Fig. 1. Preference graph and maximal chains of $\succsim$

The following power relation $\succsim^*$ is an extension of $\succsim$: $1234 \succ^* 134 \succ^* 123 \succ^* 24 \succ^* 3 \succ^* 34 \succ^* 2 \succ^* 1 \succ^* 13 \succ^* 23 \succ^* 124 \succ^* 12 \succ^* 234 \sim^* 4 \succ^* 14$, from which we determine the following occurrence vectors: $\theta^{\succsim^*}(1) = (1,1,1,0,0,0,0,1,1,0,1,1,0,1)$, $\theta^{\succsim^*}(2) = (1,0,1,1,0,0,1,0,0,1,1,1,1,0)$, $\theta^{\succsim^*}(3) = (1,1,1,0,1,1,0,0,1,1,0,0,1,0)$, $\theta^{\succsim^*}(4) = (1,1,0,1,0,1,0,0,0,0,1,0,2,1)$.

According to lex-cel, we have $3 \; P_{le}^{\succsim^*} \; 1 \; P_{le}^{\succsim^*} \; 4 \; P_{le}^{\succsim^*} \; 2$, and the winner in $R_{le}^{\succsim^*}$ is 3.

3 Problem Description

We consider social rankings situations where only partial information about the coalitions is available; instead of a power relation in $\mathcal{R}(\mathcal{P}(X))$, we are given a preorder $\succsim_{\mathcal{P}(X)}$. In a way similar to what is done in social choice [11], we define possible and necessary winners, by considering winners in extensions of $\succsim_{\mathcal{P}(X)}$.

Definition 1. *An element $x \in X$ is a* ***lex-cel possible winner*** *on $\succsim_{\mathcal{P}(X)}$ if there exists an extension $\succsim^*_{\mathcal{P}(X)}$ of $\succsim_{\mathcal{P}(X)}$ such that x is a winner in $R_{le}^{\succsim^*_{\mathcal{P}(X)}}$.*

Definition 2. *An element $x \in X$ is a* ***lex-cel necessary winner*** *on $\succsim_{\mathcal{P}(X)}$ if x is a winner in $R_{le}^{\succsim^*_{\mathcal{P}(X)}}$ for every extension $\succsim^*_{\mathcal{P}(X)}$ of $\succsim_{\mathcal{P}(X)}$.*

Example 2. Let $\succsim_{\mathcal{P}(X)}$ be the preorder introduced in Example 1. Possible extensions of $\succsim_{\mathcal{P}(X)}$ (other than $\succeq^*$) are $\succeq^a$ and $\succeq^b$ such that $134 \succ^a 34 \succ^a 2 \succ^a 24 \succ^a 1234 \succ^a 3 \succ^a 13 \succ^a 123 \succ^a 23 \succ^a 124 \succ^a 1 \succ^a 12 \succ^a 234 \succ^a 4 \succ^a 14$ and $123 \succ^b 134 \succ^b 1 \succ^b 34 \succ^b 12 \succ^b 1234 \succ^b 2 \succ^b 3 \succ^b 24 \succ^b 23 \succ^b 13 \succ^b 4 \succ^b 14 \succ^b 124 \succ^b 234$.

One can check that 4 is winning in $R_{le}^{\succeq^a}$, 1 is winning in $R_{le}^{\succeq^b}$, but there is no extension of $\succeq$ such that 2 is winning. Actually, only 1, 3 and 4 are lex-cel possible winners of $\succeq$.

Remark 2. Since an extension of $\succsim_{\mathcal{P}(X)}$ must exist, the set of lex-cel possible winners on a preorder $\succsim_{\mathcal{P}(X)}$ is non-empty.

The problem of generating linear extensions of a partially ordered set (poset) has been widely studied in the literature [17,18]. However, as it will be better clarified later, we are not interested in generating the set of lex-cel possible winners. Our goal is to find a "parsimonious" procedure to elicit the winners in $R_{le}^{\succsim^*_{\mathcal{P}(X)}}$, where $\succsim^*_{\mathcal{P}(X)}$ is the (unique) true extension of $\succsim_{\mathcal{P}(X)}$, and the concept of lex-cel possible winner is only used to guide our elicitation procedures.

Proposition 1. *Consider*[1] *the quotient order $\Sigma_1 \succ_{\mathcal{P}(X)} \cdots \succ_{\mathcal{P}(X)} \Sigma_l$. Each winner in $R_{le}^{\succsim_{\mathcal{P}(X)}}$ belongs to Σ_1.*

Now, let $\succsim_{\mathcal{P}(X)}$ be a preorder and let $\succsim^*_{\mathcal{P}(X)} \in \mathcal{R}(\mathcal{P}(X))$ be an extension of $\succsim_{\mathcal{P}(X)}$. Notice that the maximal coalitions in the (totally) preordered set $(\mathcal{P}(X), \succsim^*_{\mathcal{P}(X)})$ are also maximal in the preordered set $(\mathcal{P}(X), \succsim_{\mathcal{P}(X)})$, *i.e.* $\mathcal{M}^{\succsim^*_{\mathcal{P}(X)}} \subseteq \mathcal{M}^{\succsim_{\mathcal{P}(X)}}$.

Let $\succsim^*_{\mathcal{P}(X)} \in \mathcal{R}(\mathcal{P}(X))$ be a power relation, and suppose that it is only partially known, *i.e.* only a subset $\succsim_{\mathcal{P}(X)} \subseteq \succsim^*_{\mathcal{P}(X)}$ is initially revealed. Our problem is to gather, through an elicitation process, sufficient information over $\succsim^*_{\mathcal{P}(X)} \setminus \succsim_{\mathcal{P}(X)}$ to determine the set of winners in $R_{le}^{\succsim^*_{\mathcal{P}(X)}}$. As we are mainly concerned with the cognitive effort provided by the user, we base our evaluation of an elicitation method's performance on the number of queries it submits. Starting from the (input) preorder $\succsim_{\mathcal{P}(X)} \subseteq \succsim^*_{\mathcal{P}(X)}$, we investigate different elicitation procedures aimed at submitting queries to a user (expert) about unknown relations between pairs of coalitions. We assume that each query asked to the user over an arbitrary pair of coalitions (let's say, coalitions C_1 and C_2 in $\mathcal{P}(X)$) receives in answer a set of pairs expressing the true value of the relation between

[1] Proofs for all propositions are available in the Appendix, available at this url.

the two coalitions, *i.e.*, $C_1 \succsim C_2$ or $C_2 \succsim C_1$ (or both, if $C_1 \sim C_2$). A tuple (C_i, C_j) indicates that $C_i \succsim^*_{\mathcal{P}(X)} C_j$.

Our objective is to minimize the number of queries submitted to the user to determine a non-empty set of lex-cel necessary winners on the set of revealed relations, that is finding a set $Q \subseteq \left(\succsim^*_{\mathcal{P}(X)} \setminus \succsim_{\mathcal{P}(X)} \right)$ such that:

$$Q \in \arg \min_{R \subseteq \left(\succsim^*_{\mathcal{P}(X)} \setminus \succsim_{\mathcal{P}(X)}\right)} \{|R| : \mathcal{NW}^{\succsim_{\mathcal{P}(X)} \cup R} \neq \emptyset\} \tag{1}$$

where $\mathcal{NW}^{\succsim_{\mathcal{P}(X)} \cup R}$ is the set of lex-cel necessary winners on $\succsim_{\mathcal{P}(X)} \cup R$. Problem (1) is called the Necessary Winner in Social Ranking (NWSR) problem. It is an extension of the social ranking problem, in a context where the user's preferences are only partially known. This can be because the user has not yet proceeded to the comparison of all possible coalitions, as this can prove cognitively taxing. However, it is assumed that there exists an underlying total preorder over $\mathcal{P}(X)$, which is obviously compatible with the preorder given as input. As the total preorder represents the user's preferences, we assume it may be discovered through queries. From the above definitions and propositions, we can now present the different steps of our two elicitation approaches for NWSR.

4 Elicitation by Reconstitution: RECO

A first preorder-based approach aims at reconstituting sufficiently many of the most preferred coalitions in the underlying total preorder so as to determine the lex-cel necessary winners with certainty. The process is iterative: at each step, we present the user with two coalitions over which they must express their preference.

Recall that the restriction of a preorder $\succsim_{\mathcal{P}(X)}$ over $\mathcal{P}(X)$ to a collection $\mathcal{S} \subseteq \mathcal{P}(X)$ is the preorder $\succsim^{\mathcal{S}}_{\mathcal{P}(X)}$ over $\mathcal{S}$ such that $C_1 \succsim^{\mathcal{S}}_{\mathcal{P}(X)} C_2 \Leftrightarrow C_1 \succsim_{\mathcal{P}(X)} C_2$ for all $C_1, C_2 \in \mathcal{S}$. With a slight abuse of notation, given a collection of coalitions $\mathcal{C} \subseteq \mathcal{P}(X)$, we denote by $x_{\mathcal{C}} = |\{C \in \mathcal{C} : x \in C\}|$ the number of occurrences of element x in the collection $\mathcal{C}$.

Algorithm 1 takes as input a preorder $\succsim_{\mathcal{P}(X)}$ and returns an updated preorder $\succsim \supseteq \succsim_{\mathcal{P}(X)}$, as well as a set of winners W. The procedure starts by considering all elements in X as possible winners. As long as there are still several possible winners and not all the coalitions have been considered and ranked (this is to address the scenario in which there may be several necessary winners, due to equivalences in the total preorder), two coalitions are selected using `SELECT`($\succsim$), a subroutine which, given a preorder $\succsim$ over a set $\mathcal{S}$, selects two coalitions $C_1, C_2 \in \mathcal{S}$ such that $(C_1, C_2) \notin \succsim$ and $(C_2, C_1) \notin \succsim$. We study different approaches to coalition selection, *i.e.* different definitions of `SELECT`($\succsim$). Using the function `QUERY`(C_1,C_2), the user is then asked to indicate the set of preference relations over C_1 and C_2 as a set of tuples. The function `Transitive_Closure` returns the smallest (with respect to the number of related pairs) transitive superset

Algorithm 1: Elicitation by reconstitution

Input: A preorder $\succsim_{\mathcal{P}(X)}$;
Output: A preorder $\succsim \supseteq \succsim_{\mathcal{P}(X)}$ and a set W of lex-cel necessary winners on $\succsim$;

```
C ← P(X) ; W ← X ; ≿ ← ≿^C_P(X) ;
while |W| > 1 and C ≠ ∅ do
    if S ~ T for all S,T ∈ M^≿ then
        W ← {x ∈ W : x_{M^≿} ≥ y_{M^≿}  for each  y ∈ W};
        C ← C \ M^≿;
        ≿ ← ≿^C_P(X);
    else
        C1, C2 ← SELECT(≿^C_P(X)) ;
        Newcomp ← QUERY(C1, C2);
        ≿ ← Transitive_Closure(≿ ∪ Newcomp);
return ≿;
return W.
```

of the updated $\succsim$. If all maximal elements from the maximal chains (*i.e.* all coalitions in $\mathcal{M}^{\succsim}$) have been evaluated as equivalent, they must belong to the same equivalence class Σ, which can no longer be extended: we therefore update our set of possible winners to contain only those who are most represented in Σ (in accordance with the lexcel), and remove from consideration the elements in $\mathcal{M}^{\succsim}$, as they are considered to have been added to the reconstructed order and their impact on the set of possible winners has been taken into account.

Proposition 2. *Algorithm 1 returns a set W of lex-cel necessary winners for the power relation $\succsim$.*

The performance of this approach depends heavily on the manner in which the two coalitions C_1 and C_2 are selected at each iteration. We therefore study several different manners of selecting coalitions, which we will refer to as *variants.*[2]

The *Complete randomness variant* (C-RAND) is a first naïve approach consists in simply selecting two uncompared coalitions randomly within $\mathcal{P}(X)$. However, it quickly appears that this approach is far from efficient, as we aim at reconstituting the preorder over the top coalitions, and the complete randomness variant can compare coalitions regardless of their likelihood to be among the most preferred. As the top coalition in $\succsim^*_{\mathcal{P}(X)}$ is necessarily a maximal element of $\succsim_{\mathcal{P}(X)}$, restricting the pool of candidate coalitions to $\mathcal{M}^{\succsim_{\mathcal{P}(X)}}$ significantly improves the relevance of the queries.

The *Restricted randomness variant* (R-RAND) consists in selecting at random two uncompared coalitions within $\mathcal{M}^{\succsim^{\mathcal{C}}_{\mathcal{P}(X)}}$.

[2] The pseudo-code for each variant is available in the Appendix.

We wish to reduce the impact of randomness on the selection process, and therefore introduce new strategies, the first of which shares the lexicographic approach of lex-cel. The *Restricted maximal length variant* (R-MAX_LEN) makes the hypothesis that longer maximal chains are more likely to be informative. It uses a vector $\Gamma = (\gamma_1, \ldots, \gamma_s)$, where s is the number of maximal chains in $\succsim_{\mathcal{P}(X)}$ and γ_i is the i-th longest maximal chain in $\succsim_{\mathcal{P}(X)}$. We define $M^{\Gamma} = (m_1, \ldots, m_s)$ an s-sized vector such that m_i is one of the maximal elements of γ_i.

Remark 3. Note that randomness may still play a part in this variant. Indeed, let $\succsim_{\mathcal{P}(X)}$ be the preorder presented in Example 1, the associated set of maximal elements is then such that $\mathcal{M}^{\succsim_{\mathcal{P}(X)}} = \{1234, 134, 34, 123\}$. As all maximal chains are of identical length, the tie-breaker is the order in which they have been given as input. Therefore, in this particular instance, the first comparison will be between 1234 and 134.

Finally, we try to focus on a maximization of a query's informativity by exploiting properties specific to lex-cel; this gives rise to the *Restricted minimal nonempty intersection* (R-MIN_INT) variant. We can define a category of queries we call *informative* with regards to a set of lex-cel possible winners. Given a preorder $\succsim_{\mathcal{P}(X)}$ over which we can determine a set of lex-cel possible winners W, a query over a pair of coalitions $C_1, C_2 \in \mathcal{P}(X)$ is said to be informative if, provided C_1 and C_2 are the top coalitions among $\mathcal{C}$, the recovered preference over them allows for the reduction of the set W. A query is maximally informative when the preference it uncovers can reduce W to a single element: the lex-cel necessary winner has then been uncovered.

For an element x to be removed from W in favour of another element $y \in W$, simply by comparing C_1 and C_2 as top coalitions among $\mathcal{C}$, implies that $\exists C \in \{C_1, C_2\}$ such that $x \notin C$ yet $y \in C$. This means the intersection of C_1 and C_2 must be smaller than the largest of the two coalitions. The result of the query must then lead to the placement of the coalition containing y before any containing x. We therefore seek to minimize the intersection of pairs of candidate coalitions, in order to minimize the number of elements that may remain in W after submission of the query. Note that if this intersection set is empty, then the query is considered non-informative, for when none of the candidates are present in the k-th equivalence class, the k-th component of both their occurrence vectors will be 0, and will not allow the lexicographic relation to discriminate between the candidates.

Example 3. Let $\succsim_{\mathcal{P}(X)}$ be the preorder presented in Example 1. The set $\mathcal{M}^{\succsim_{\mathcal{P}(X)}}$ is therefore such that $\mathcal{M}^{\succsim_{\mathcal{P}(X)}} = \{1234, 134, 34, 123\}$. We have that $|1234 \cap 134| = |1234 \cap 123| = 3$, $|1234 \cap 34| = |134 \cap 34| = |134 \cap 123| = 2$ and $|123 \cap 34| = 1$. The minimal non-empty intersection is of size 1, and the relative comparison will therefore be between 123 and 34.

The existence of informative queries highlighted in the restricted lexicographic variant leads us to observe that eliciting by reconstituting the total preorder may not always be an efficient approach. Indeed, there exist coalitions which will bring no information to lex-cel, and any query about their placement in the total preorder is therefore superfluous. A second approach to elicitation suggests to focus directly on the elements themselves, and only submit to the user targeted queries over key informative coalitions which will help determining the preference over elements without needing to discover the entire preorder.

5 Elicitation Through Prospection: PROSP

A second element-based approach aims at directly discerning the preference over the individual elements themselves. To do so, it considers the likelihood of certain elements within the population to be a lex-cel necessary winner, then determines key coalitions, the preference between which will be decisive during the application of lex-cel to assess the dominance of an element over another.[3]

Given a total preorder $\succsim^*_{\mathcal{P}(X)}$, and $x, y \in X$, we denote by $C_{x,-y}$ the most highly ranked coalition in $\succsim^*_{\mathcal{P}(X)}$ verifying that $x \in C_{x,-y}$ and $y \notin C_{x,-y}$.

Proposition 3. *Given a total preorder* $\succsim^*_{\mathcal{P}(X)}$, *let* $x, y \in X$ *be present in a best coalition* $C^* \in \succsim^*_{\mathcal{P}(X)}$. *It holds that* $x \; R_{le}^{\succsim^*_{\mathcal{P}(X)}} y$ *if* $C_{x,-y} \succ^*_{\mathcal{P}(X)} C_{y,-x}$.

Remark 4. Given a total preorder $\succsim^*_{\mathcal{P}(X)}$, let $x, y \in X$ be present in a best coalition $C^* \in \succsim^*_{\mathcal{P}(X)}$. If $C_{x,-y} \sim^*_{\mathcal{P}(X)} C_{y,-x}$, then neither $C_{x,-y}$ nor $C_{y,-x}$ will be decisive in the comparison of x and y. As such, $C_{x,-y}$ and $C_{y,-x}$ may be removed from consideration, and their values updated.

For any $x \in X$, let $\mathcal{P}_x = \{C \in \mathcal{P}(X) : x \in C\}$. For any pair $x, y \in X$, $x \neq y$, let $\mathcal{P}_{x,y} = \{\mathcal{P}_x \cup \mathcal{P}_y\} \setminus \{\mathcal{P}_x \cap \mathcal{P}_y\}$ be the set of coalitions in $\mathcal{P}(X)$ containing either x or y, but not both elements. $\mathcal{M}^{\succsim}_{x,y}$ denotes the set of maximal elements in $\succsim^{\mathcal{P}_{x,y}}_{\mathcal{P}(X)}$, that can then be partitioned into two disjoint subsets $\mathcal{M}^{\succsim}_{x,-y} = \{C \in \mathcal{M}^{\succsim}_{x,y} : x \in C\}$ and $\mathcal{M}^{\succsim}_{y,-x} = \{C \in \mathcal{M}^{\succsim}_{x,y} : y \in C\}$.

At each step, the procedure determines the set W of lex-cel possible winners in the current preorder, then selects the two deemed most likely to be more preferred amongst them, and determines coalitions which would play a key role in establishing a preference between the two elements. The user is only presented queries relative to these key coalitions.

From Proposition 1, we know that, in order to be a winner, an element $x \in X$ must be present in a top coalition of the total preference relation. While the exact set of the top coalitions is unknown, we proceed using $\mathcal{M}^{\succsim}_{\mathcal{P}(X)}$, as we know all top coalitions must belong to that set.

Determining the Most Likely Lex-cel Possible Winners: The Global Lex-cel. In order to minimize the number of queries to the user, we try and

[3] The pseudo-code for PROSP is presented in the Appendix.

determine the possible winners most likely to perform better. As a lex-cel necessary winner is one of the most present elements in the top coalitions of the total preorder, and since the higher a coalition is placed within a maximal chain, the most likely it is to be among the top coalitions, we study the presence of each lex-cel possible winner in each position. To do so, we present a generalisation of lex-cel, applicable to any set of maximal chains.

Let $\mathcal{A}^{\succsim_{\mathcal{P}(X)}}$ be the set of all maximal chains in $\succsim_{\mathcal{P}(X)}$. We denote by l the size of a longest maximal chain in $\mathcal{A}^{\succsim_{\mathcal{P}(X)}}$, and by A_k the k-th ranked equivalence class in a maximal chain $A \in \mathcal{A}^{\succsim_{\mathcal{P}(X)}}$, $\forall k \in \{1, \dots, |A|\}$. For any maximal chain $A \in \mathcal{A}^{\succsim_{\mathcal{P}(X)}}$ and for any element $i \in X$, we denote by $i_k^A = |\{C \in A_k \mid i \in C\}|$ the number of occurrences of i in the k-th equivalence class of the maximal chain A. We then denote by $\vartheta^{\succsim_{\mathcal{P}(X)}}(i)$ the l-dimensional vector

$$\vartheta^{\succsim_{\mathcal{P}(X)}}(i) = \Big(\sum_{A \in \mathcal{A}^{\succsim_{\mathcal{P}(X)}}} i_1^A, \dots, \sum_{A \in \mathcal{A}^{\succsim_{\mathcal{P}(X)}}} i_l^A\Big).$$

Definition 3. *The global lexicographic-excellence (global lex-cel) is the binary relation $R_{gle}^{\succsim_{\mathcal{P}(X)}}$ such that, for any preorder $\succsim_{\mathcal{P}(X)}$, and for any $i, j \in X$:*

$$i \; R_{gle}^{\succsim_{\mathcal{P}(X)}} \; j \iff \vartheta^{\succsim_{\mathcal{P}(X)}}(i) \geq_L \vartheta^{\succsim_{\mathcal{P}(X)}}(j)$$

Example 4. Let $\succsim_{\mathcal{P}(X)}$ be the preorder presented in Example 1. All maximal chains in $\succsim_{\mathcal{P}(X)}$ are of length at most $l = 5$, therefore we determine the l-sized vectors associated with every element in the population: $\vartheta^{\succsim_{\mathcal{P}(X)}}(1) = (4, 1, 5, 3, 3), \vartheta^{\succsim_{\mathcal{P}(X)}}(2) = (2, 4, 2, 3, 3), \vartheta^{\succsim_{\mathcal{P}(X)}}(3) = (6, 1, 5, 0, 3)$, and $\vartheta^{\succsim_{\mathcal{P}(X)}}(4) = (5, 2, 0, 6, 6)$, from which we can determine the following total preorder over X:

$$3 \; R_{gle}^{\succsim_{\mathcal{P}(X)}} \; 4 \; R_{gle}^{\succsim_{\mathcal{P}(X)}} \; 1 \; R_{gle}^{\succsim_{\mathcal{P}(X)}} \; 2.$$

Using the global lex-cel, we determine a preliminary total preorder over X, from which we select the two best elements as the pair driving the current iteration of the procedure.

Querying the User. We submit queries to determine the top coalition $C \in \mathcal{M}_{x,y}^{\succsim_{\mathcal{P}(X)}}$. Depending on whether C is in $\mathcal{M}_{x,-y}^{\succsim_{\mathcal{P}(X)}}$ or $\mathcal{M}_{y,-x}^{\succsim_{\mathcal{P}(X)}}$, we are then able to determine the preference relation between x and y.

Proposition 4. *Given a total preorder $\succsim_{\mathcal{P}(X)}^*$. Let $\succsim_{\mathcal{P}(X)}$ be a preorder such that $\succsim_{\mathcal{P}(X)} \subseteq \succsim_{\mathcal{P}(X)}^*$. Let $x, y \in X$ be the two best elements in X according to the global lex-cel. It holds that $x R_{le}^{\succsim_{\mathcal{P}(X)}^*} y$ iff there exists a coalition in $\mathcal{M}_{x,-y}^{\succsim_{\mathcal{P}(X)}}$ preferred to every coalition in $\mathcal{M}_{y,-x}^{\succsim_{\mathcal{P}(X)}}$ or if $\mathcal{M}_{y,-x}^{\succsim_{\mathcal{P}(X)}} = \emptyset$ and $\mathcal{M}_{x,-y}^{\succsim_{\mathcal{P}(X)}} \neq \emptyset$.*

We consider that a coalition in $\mathcal{M}_{x,y}^{\succsim_{\mathcal{P}(X)}}$ is more likely to be the most preferred in the set when its worst position in any maximal chain is still a low position.

For this reason, in stead of the sets $\mathcal{M}_{x,-y}^{\succsim_{\mathcal{P}(X)}}$ and $\mathcal{M}_{y,-x}^{\succsim_{\mathcal{P}(X)}}$, the procedure uses the vector of coalitions ordered in ascending worst position in a maximal chain of $\succsim_{\mathcal{P}(X)}$, denoted by $\hat{M}_{x,-y}^{\succsim_{\mathcal{P}(X)}}$ and $\hat{M}_{y,-x}^{\succsim_{\mathcal{P}(X)}}$ respectively.

We submit to the user a query over the first coalition from each vector. Without loss of generality, when it is found that a coalition $C_2 \in \hat{M}_{x,-y}^{\succsim_{\mathcal{P}(X)}}$ is dominated by $C_1 \in \hat{M}_{y,-x}^{\succsim_{\mathcal{P}(X)}}$, the value of C_2 is updated to the next coalition in $\hat{M}_{x,-y}^{\succsim_{\mathcal{P}(X)}}$. In the event that C_1 and C_2 are equivalent, they are both replaced in $\mathcal{M}_{x,y}^{\succsim_{\mathcal{P}(X)}}$ by the first following coalition C' in their respective maximal chains such that $|\{x,y\} \cap C'| = 1$. If, without loss of generality, there is no such coalition placed after C_1 in any maximal chain, then C_1 is simply removed from $\mathcal{M}_{x,y}^{\succsim_{\mathcal{P}(X)}}$. The sets $\hat{M}_{x,-y}^{\succsim_{\mathcal{P}(X)}}$ and $\hat{M}_{y,-x}^{\succsim_{\mathcal{P}(X)}}$ are then redetermined accordingly. This procedure is repeated until a coalition from one of the sets has been established to be preferred to all others in the complementary set, in which case, from Proposition 4, we can establish the preference relation between x and y. If $\mathcal{M}_{x,y}^{\succsim_{\mathcal{P}(X)}}$ becomes empty without having uncovered any preference between any coalition from $\hat{M}_{x,-y}^{\succsim_{\mathcal{P}(X)}}$ and any coalition from $\hat{M}_{y,-x}^{\succsim_{\mathcal{P}(X)}}$, we gather that x and y must be equivalent. In this case, we flag x as being equivalent to y, and artificially remove y from the pool of possible winners. If x turns out to be the necessary winner, its flag will serve as a reminder that it is not the only necessary winner: PROSP will return the set of elements having been found to be equivalent to x.

Example 5. Let $X = \{1,2,3\}$. Let the total preorder $\succsim_{\mathcal{P}(X)}^{*}$ be such that $1 \sim 2 \succ 12 \succ 13 \succ 23 \succ 123 \succ 3$, and let the available preorder $\succsim_{\mathcal{P}(X)} \subseteq \succsim_{\mathcal{P}(X)}^{*}$ contain two maximal chains $1 \succ 12 \succ 13 \succ 23 \succ 123 \succ 3$ and $2 \succ 12 \succ 13 \succ 23 \succ 123 \succ 3$, from which we obtain the set of lex-cel possible winners $W = \{1,2\}$. It is unnecessary to apply the global lex-cel to W, as we can only compare the elements 1 and 2. We easily determine that $\mathcal{M}_{1,2}^{\succsim_{\mathcal{P}(X)}} = \{\{1\},\{2\}\}$. The first query is between $\{1\}$ and $\{2\}$, to which the answer will be that $\{1\} \sim \{2\}$. Both elements are removed from consideration for the remainder of the comparison between elements 1 and 2. The set $\mathcal{M}_{1,2}^{\succsim_{\mathcal{P}(X)}}$ is then computed again using the preorder $\succsim_{\mathcal{P}(X)}$ from which all preferences concerning either $\{1\}$ or $\{2\}$ have been removed. The coalition $\{1\}$ is replaced by $\{1,3\}$; the coalition $\{2\}$ is replaced by $\{1,3\}$ as well. Consequently, the updated set is now $\mathcal{M}_{1,2}^{\succsim_{\mathcal{P}(X)}} = \{\{1,3\}\}$ which, in turn, means that $\mathcal{M}_{1,-2}^{\succsim_{\mathcal{P}(X)}} = \{\{1,3\}\}$ and $\mathcal{M}_{2,-1}^{\succsim_{\mathcal{P}(X)}} = \emptyset$. By construction of the algorithm, this means that 1 is preferred to 2, *i.e.* $1 \; R_{le} \; 2$, from which we update the set L containing elements found, in previous steps, to be dominated by another possible winner. In this case, we have $L = \{2\}$. Since $W \setminus L = \{1\}$, there is only one possible winner left, which, by definition, means that it is the necessary winner. The procedure then returns 1 as the necessary winner, which is indeed the winner for the preorder $\succsim_{\mathcal{P}(X)}$.

6 Experimental Results

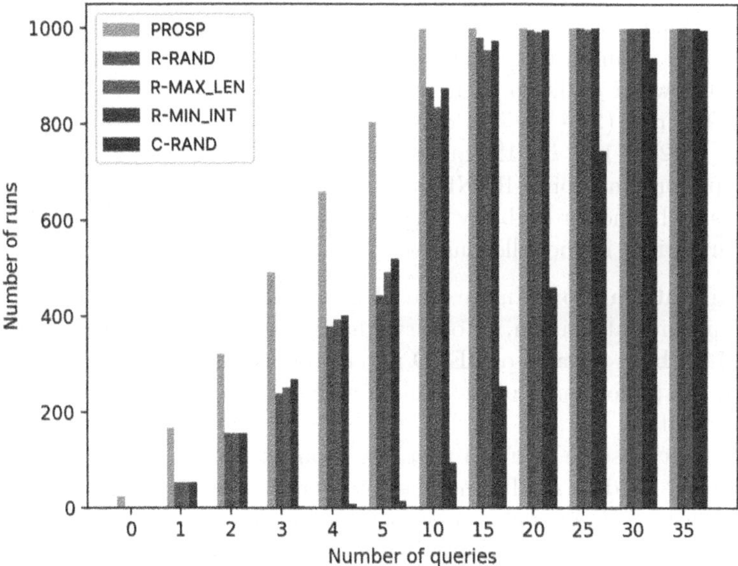

Fig. 2. Number of runs (out of 1000) for which each method submits at most x queries over a population of size 4

We compare the performance of both elicitation methods based on the number of queries submitted over preorders defined under different constraints. All presented results are from experiments over 100 to 1000 different preorders (referred to as "runs"). Unless specified otherwise, all preorders derived from the total preorder, which serve as input information for our problem, are generated uniformly at random. By this, we mean a total order over all coalitions is generated uniformly at random. This order is then turned into a preorder by coursing through all coalitions from best to worst, and making each equivalent to it next preferred with a probability 0.2. Then, coursing through coalitions of the preorder from most to least preferred, each is added to one of m maximal chains, with m the number of maximal chains also determined uniformly at random within the set $\{1, \ldots, \frac{2^n}{2}\}$, with n the size of the population. Once a coalition has been added to an i-th maximal chain, $\forall i \in \{1, \ldots, m-1\}$, it may be added to an additional chain with probability 0.2.[4]

In any scenario, we find that the C-RAND variant of RECO is always inefficient, as it most often submits far too many queries to the user, most of which can be deemed uninformative. Furthermore, we find that its restricted version, R-RAND, systematically outperforms it. Indeed, Fig. 2 shows that, even when

[4] Additional results under particular parameters are available in the Appendix.

the population contains only 4 elements, C-RAND is significantly outperformed by every other approach over all 1000 runs, as it always requires more queries than any other method: at the threshold of 20 questions, all other variants have solved NWSR on nearly all runs, yet C-RAND has only solved it for half of the runs. Note that, for a population of size $n = 4$, there are $2^n = 16$ coalitions to rank in total, and a worst-case scenario in terms of submitted queries would compare the worst coalition to all others ($2^n - 1 = 15$ queries), then the second worst to the rest ($2^n - 2 = 14$ queries), and so on, leading to a maximum of $\sum_{i \in \{0, 2^n - 1\}} (2^n - 1) - i = 120$ queries.

The performance of C-RAND only worsens as the size of the population grows. As such, and for legibility purposes, we will not include the results of this particular variant in the following sections.

Uniformly at Random. In the case where the underlying total preorder is generated uniformly at random, we observe that the performance of the R-MIN_INT and R-MAX_LEN variants of RECO never produce the least amount of queries. However, the performance of the remaining two approaches varies with the size of the population.

Indeed, we observe in Fig. 3 that, for a population of size smaller or equal to 6, PROSP yields a smaller number of queries than the R-RAND variant of RECO. However, for a population of size greater than or equal to 7, the R-RAND variant outperforms PROSP more and more significantly.

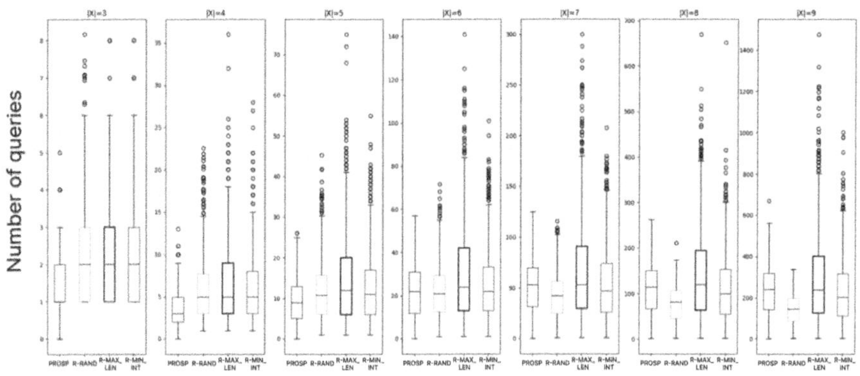

Fig. 3. Number of queries submitted by each method based on the size of the population.

Additive Preferences. Preferences expressed by a ranking $\succsim^*$ are called **additive** if there exists a utility function $f : \mathcal{P}(X) \rightarrow \mathbb{N}$ such that $\forall C \subseteq X, f(C) = \sum_{x \in C} f(x)$ verifying that, $\forall C, D \subseteq X, C \succsim^* D \Leftrightarrow f(C) \geq f(D)$. Under the hypothesis of an underlying total preorder expressing additive preferences, Fig. 4, testing over 100 runs for each size of X, shows that PROSP yields systematically better results than any other variant of RECO.

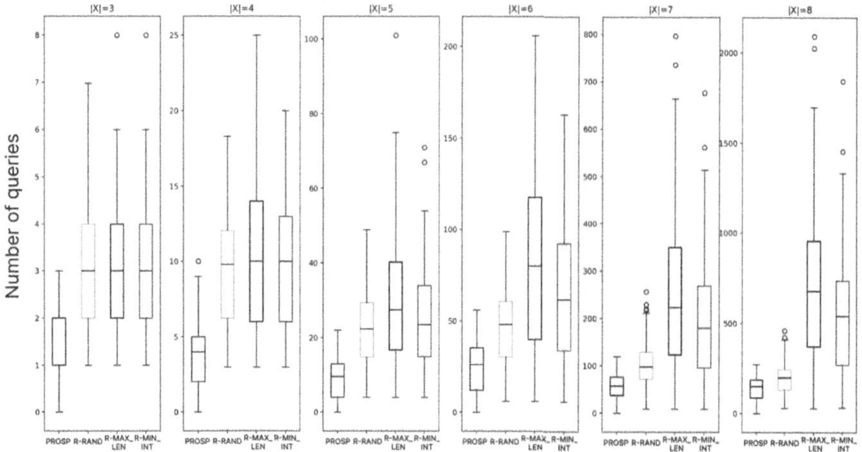

Fig. 4. Number of queries submitted by each method over additive total preorders

Best-case Scenario: A Unique Lex-cel Possible Winner. Under any restriction applied to the studied preorder, we find that PROSP maintains an enticing quality. Indeed, there exists what may be referred to as a "best-case scenario", in which the preorder given as input is such that one could directly determine the lex-cel necessary winner without soliciting the user. This scenario arises when there is only one true lex-cel possible winner (an example is provided in the Appendix[5]).

By construction of RECO, it is essential to determine at least the first coalition of the total preorder to assess any preference over elements of X. This means that, with the exception of the very specific scenario in which all maximal chains in the initial preorder start with the same singleton, we know that at least one query will always be submitted to the user using any variant of RECO. On the other hand, because prospecting leads us to directly confront key coalitions, PROSP does not need any additional input from the user in order to determine the lex-cel necessary winner when there is only one lex-cel possible winner from the start. In simpler terms, PROSP never submits a query when there is no need for additional information from the user's expressed partial preferences, whereas the RECO will (nearly) always submit at least one query.

Discussion of the Results. Under no restriction over the type of underlying total preorder, the R-RAND variant appears more the more efficient out of all proposed methods. However, PROSP remains enticing, as it submits fewer queries when the underlying preorder is additive, and maintains the notable advantage of not submitting any query when none is necessary.

[5] The Appendix will be published as a technical report and is available at this url.

Conclusions and Perspectives

We introduced two elicitation approaches to determine the lex-cel necessary winners when provided with a preorder over the set of a population's coalitions. One approach aims at reconstituting enough of the underlying total preorder to be able to determine with certainty which elements are the true winners; the second determines key coalitions in the comparison of elements deemed most likely to be winners by the global lex-cel. Our experiments suggest that the performance of the different elicitation methods varies depending on the type of power relation being examined, as the random variant of RECO seems to fare better when there is no hypothesis on the total preorder's type, but PROSP yields better results on additive underlying total preorders. Furthermore, PROSP has the attractive quality to not submit queries when none is necessary.

Additionally, as both approaches aim to determine the lex-cel necessary winners with certainty, they require a great number of queries as the size of the population increases, which may still prove strenuous for the user. We believe it would also be pertinent to study approximated resolution methods for this problem, which could guarantee that the number of submitted queries would remain manageable to the user.

Finally, both approaches operate under the hypothesis that the underlying total preorder is available, and any query will provide a fixed preference relation. This is a strong assumption, which could be made more flexible, and we believe it could be interesting to study scenarios in which not all queries may lead to a definite preference relation.

Acknowledgements. We acknowledge a financial support from the project THEMIS ANR-20-CE23-0018 of the French National Research Agency (ANR).

References

1. Moretti, S., Öztürk, M.: Some axiomatic and algorithmic perspectives on the social ranking problem. In: Rothe, J. (ed.) ADT 2017. LNCS (LNAI), vol. 10576, pp. 166–181. Springer, Cham (2017). https://doi.org/10.1007/978-3-319-67504-6_12
2. Bernardi, G., Lucchetti, R., Moretti, S.: Ranking objects from a preference relation over their subsets. Soc. Choice Welf. **52**, 589–606 (2019)
3. Algaba, E., Moretti, S., Rémila, E., Solal, P.: Lexicographic solutions for coalitional rankings. Soc. Choice Welf. 1–33 (2021)
4. Béal, S., Rémila, E., Solal, P.: Lexicographic solutions for coalitional rankings based on individual and collective performances. J. Math. Econ. 102738 (2022)
5. Suzuki, T., Horita, M.: Consistent social ranking solutions. Soc. Choice Welf. **62**, 549–569 (2024)
6. Allouche, T., Escoffier, B., Moretti, S., Ozturk, M.: Social ranking manipulability for the CP-majority, Banzhaf and lexicographic excellence solutions. In: Proceedings of IJCAI'20, pp. 17–23, 2020
7. Haret, A., Khani, H., Moretti, S., Ozturk, M.: Ceteris paribus majority for social ranking. In: Proceedings of IJCAI'18, pp. 303–309, 2018

8. Khani, H., Moretti, S., Ozturk, M.: An ordinal banzhaf index for social ranking. In: Proceedings of IJCAI'19, pp. 378–384, 2019
9. Suzuki, T., Horita, M.: Which set of agents plays a key role? An impossibility in transforming binary relations. Math. Soc. Sci. **129**, 12–19 (2024)
10. Konieczny, S., Moretti, S., Ravier, A., Viappiani, P.: Selecting the most relevant elements from a ranking over sets. In: Proceedings of the SUM'22, pp. 172–185, 2022
11. Konczak, K., Lang, J.: Voting procedures with incomplete preferences. In: Proceedings of the IJCAI'05 Workshop on Advances in Preference Handling, January 2005
12. Xia, L., Conitzer, V.: Determining possible and necessary winners given partial orders. J. Artif. Intell. Res. **41**, 25–67 (2011)
13. Lang, J.: Collective decision making under incomplete knowledge: possible and necessary solutions. In: Proceedings of IJCAI'20, pp. 4885–4891 (2020)
14. Lu, T., Boutilier, C.: Preference elicitation and robust winner determination for single- and multi-winner social choice. Artif. Intell. **279** (2020)
15. Benabbou, N., Di Sabatino Di Diodoro, S., Perny, P., Viappiani, P.: Incremental preference elicitation in multi-attribute domains for choice and ranking with the borda count. In: Schockaert, S., Senellart, P. (eds.) SUM 2016. LNCS (LNAI), vol. 9858, pp. 81–95. Springer, Cham (2016). https://doi.org/10.1007/978-3-319-45856-4_6
16. Napolitano, B., Cailloux, O., Viappiani, P.: Simultaneous elicitation of scoring rule and agent preferences for robust winner determination. In: Fotakis, D., Ríos Insua, D. (eds.) ADT 2021. LNCS (LNAI), vol. 13023, pp. 51–67. Springer, Cham (2021). https://doi.org/10.1007/978-3-030-87756-9_4
17. Brightwell, G., Winkler, P.: Counting linear extensions. Order **8**, 225–242 (1991)
18. Pruesse, G., Ruskey, F.: Generating linear extensions fast. SIAM J. Comput. **23**(2), 373–386 (1994)

User's Preference Modelling with Gödel Integral

Agnès Rico[1], Marie-Jeanne Lesot[2], and Christophe Marsala[2](✉)

[1] Université Lyon 1, Lyon, France
Agnes.Rico@univ-lyon1.fr
[2] Sorbonne Université CNRS, LIP6, Paris, France
{Marie-Jeanne.lesot,Christophe.Marsala}@lip6.fr

Abstract. The Gödel integral is a variant of the Sugeno integral. It is an expressive aggregation operator to compute a global evaluation combining local values taken for a set of considered criteria and a capacity, a set function that represents the importance of these criteria and their interactions. A crucial question is then the elicitation of this capacity so as to match the user preferences. Such a task is common in many fields, such as XAI, where counterfactual explanations can be generated in the form of counterfactual examples by aggregating a pair of objective and subjective criteria. In this paper, we focus on approaches to elicit such a capacity when the user preferences are expressed as an ordering relation between classes of equivalent objects. We first propose a theoretical characterisation of the set of admissible capacities: it allows to determine lower and upper bounds of possible capacities when it is not empty and it provides explanations when it is empty. Then, we introduce the algorithm GICEP (for *Gödel Integral for Capacity Elicitation from Preference relations*), to compute these bounds and to provide explanations. Experiments conducted on synthetic data show the relevance and efficiency of GICEP.

Keywords: Gödel integral · Capacity elicitation · Preference relation · eXplainable AI

1 Introduction

Given some examples of user preferences on objects evaluated according to some criteria, a common task in decision making theory aims at finding a representation of these preferences. It is often addressed through parameterised aggregation operators, such as fuzzy integrals, e.g. Choquet integrals, Sugeno integrals or their variants, used to model the decision maker behaviour [9]. The preference representation then takes the form of a parameter elicitation issue [12]. In particular, it has been discussed in the case of Choquet integrals [1,4,10] or Sugeno integrals [7,16].

This paper considers the case of the Gödel integrals [6] which are a variant of the classical Sugeno integral [18]: relying on a different link between its parameter and its output characterisation, the Gödel integral does not allow the direct use of previous results in the literature.

S. Destercke et al. (Eds.): SUM 2024, LNAI 15350, pp. 394–407, 2025.
https://doi.org/10.1007/978-3-031-76235-2_29

Beside the type of considered aggregation operator and its parameters, existing works also differ depending on the chosen expression for the preference relation. A preference relation can be given by a user who provides examples of her preferences as a dataset containing pairs (x, α) where x is an object evaluated according to some criteria, represented as a vector of local values, and α is the expected global evaluation, corresponding to the desired result, see e.g. [3] or [17].

A preference relation can also be given in a less demanding manner from the user, asking her only to state relative order relations. More precisely, the considered input is an order on the objects x, represented as a totally ordered sequence of object equivalence classes: it can for instance take the form $\{a, b\} \prec \{c\} \prec \{d, e\}$, meaning that the user has no preference between d and e, that are considered equivalent, but prefers them to c that, in turn, is preferred to the equivalent objects a and b. We consider this case in this paper.

Furthermore, the paper considers the case when the objects are evaluated by two evaluation criteria, i.e. are associated to 2D vectors (for the illustrative example, $\forall x \in \{a, b, c, d, e\}$, $x = (x_1, x_2)$). This issue for instance occurs in the eXplainable Artificial Intelligence (XAI) domain, when generating explanations in the form of counterfactual examples (see e.g. [11]): candidate counterfactual examples are evaluated according to two criteria, namely an objective one and a subjective one.

The first criterion, the objective one, only depends on the considered machine learning task, it includes for instance the proximity between the considered candidate and the data point whose prediction must be explained, see e.g. [19], or the candidate realism as compared to the data distribution [2,13] to name a few. The subjective criterion depends on the user receiving the explanation, and for instance measures the compatibility of the candidate with the user knowledge [11]. The two criteria are then aggregated to define the global quality of the candidate; the use of Gödel integrals to that aim has been introduced in [11], exploiting their high expressive power to model the user preference. The motivating example for the contribution proposed in this paper is then the question of determining the parameters of this aggregation operator, based on relations where the user expresses whether two candidate explanations are equivalent or strictly preferred one to another.

Therefore, the aim of this paper is to identify the parameter of a Gödel integral, the so-called capacity, such that the induced Gödel integral represents a given preference relation. Section 2 reminds the definition of Gödel integral and existing results regarding the elicitation of its parameter in the case when the global evaluation scale is a totally ordered set and the user preferences are expressed as pairs (x, α). Additionally, it underlines the specificity of the Gödel capacity elicitation task as compared to the Sugeno integral case. Section 3 describes the theoretical results we establish about the Gödel integral when the user preferences are expressed as pairwise order relations for 2D evaluation vectors. Section 4 presents the GICEP algorithm we propose to implement the theoretical results into an efficient procedure. Section 5 illustrates the relevance of the proposed algorithm with a set of experiments on synthetic data. Section 6 concludes the paper and discusses some directions for future works.

2 Background and Notation

In this section, definitions of the Gödel integrals and their parameters (namely, the capacity) are recalled. Existing work on elicitation is presented. The difference with the Sugeno integrals is then underlined.

Let $\mathcal{C} = \{1, 2\}$ be a set of 2 criteria on which n objects from $\mathcal{E} = \{x^1, \cdots, x^n\}$ are evaluated. Each $x \in \mathcal{E}$ is a vector $x = (x_1, x_2) \in [0, 1]^2$. We denote 1_A the characteristic function of any subset $A \subseteq \mathcal{C}$, i.e. the vector x such that $x_i = 1$ if $i \in A$ and 0 otherwise.

All criteria use as evaluation scale the real interval $[0, 1]$ which constitutes a totally ordered set, i.e. a scale on which the Gödel integrals are defined [6].

2.1 Capacities

The Gödel integrals [6] constitute a family of aggregation operators that generalise the classical Sugeno integral [18] as detailed in [5]. It is parameterised by a capacity [8] (also called fuzzy measure) which determines the importance and interaction of the criteria under consideration.

Formally, a *capacity* is a function $\mu : 2^{\mathcal{C}} \to [0, 1]$ that associates each subset of criteria $A \subseteq \mathcal{C}$ with a weight $\mu(A)$. By definition, μ is increasing with respect to set inclusion and satisfies the boundary conditions $\mu(\emptyset) = 0$ and $\mu(\mathcal{C}) = 1$. In the following, we denote V the set of all capacities (for a given, omitted, $\mathcal{C}$). When $\mathcal{C} = \{1, 2\}$, in addition to the boundary conditions, the capacity is defined by two values, $\mu(\{1\})$ and $\mu(\{2\})$ denoted μ_1 and μ_2.

A partial order $\leq$ between capacities is defined as $\mu \leq \mu'$ iff $\mu(A) \leq \mu'(A)$ for all $A \subseteq \mathcal{C}$. The set V equipped with this partial order is a bounded lattice. It can also be identified by the tuple $(V, \wedge, \vee, \bot, \top)$ where the binary operators $\vee$ (join), $\wedge$ (meet) and the elements $\bot$ and $\top$ are defined as follows, for any $\mu, \mu' \in V$ and for any $A \subseteq \mathcal{C}$:

- $(\mu \wedge \mu')(A) = \min(\mu(A), \mu'(A))$;
- $(\mu \vee \mu')(A) = \max(\mu(A), \mu'(A))$;
- for all $A \subset \mathcal{C}$, $\bot(A) = 0$ and $\bot(\mathcal{C}) = 1$;
- for all $A \neq \emptyset$, $\top(A) = 1$ and $\top(\emptyset) = 0$.

Two capacities $\underline{\mu}$ and $\overline{\mu}$ define $[\underline{\mu}, \overline{\mu}]$ the subset $\{\gamma \in V \mid \underline{\mu} \leq \gamma \leq \overline{\mu}\}$; it is non-empty if and only if $\underline{\mu} \leq \overline{\mu}$.

2.2 Gödel Integrals

The *Gödel integral* (GI) is an aggregation function that reduces any object x to a single value. It relies on the *Gödel conjunction* [6] defined for any $\alpha, \beta \in [0, 1]$ by:

$$\alpha \otimes_G \beta = \begin{cases} 0 & \text{if } \beta \leq 1 - \alpha \\ \beta & \text{otherwise.} \end{cases}$$

This function $\otimes_G$ is increasing in its 2 arguments and satisfies: $1 \otimes_G \beta = \beta$, $\alpha \otimes_G 1 = 1$ if $\alpha \neq 0$, and $0 \otimes_G \beta = \alpha \otimes_G 0 = 0$.

With the notation introduced in Sect. 2.1, the Gödel integral of $x = (x_1, x_2) \in [0,1]^2$ with respect to the capacity $\mu \in V$ is defined as

$$G_\mu(x) = \max\left(\mu_1 \otimes_G x_1, \mu_2 \otimes_G x_2, \min(x_1, x_2)\right) \tag{1}$$

It can only take two values according to the four following possibilities:

$$G_\mu(x) = \begin{cases} \min(x_1, x_2) & \text{if } x_1 \leq 1 - \mu_1 \text{ and } x_2 \leq 1 - \mu_2; \\ x_1 & \text{if } x_1 > 1 - \mu_1 \text{ and } x_2 \leq 1 - \mu_2; \\ x_2 & \text{if } x_1 \leq 1 - \mu_1 \text{ and } x_2 > 1 - \mu_2; \\ \max(x_1, x_2) & \text{if } x_1 > 1 - \mu_1 \text{ and } x_2 > 1 - \mu_2. \end{cases}$$

Thus, for any $x = (x_1, x_2) \in [0,1]^2$, $G_\mu(x) \in \{x_1, x_2\}$.

2.3 Eliciting Gödel Integral

The task of eliciting a Gödel integral aims at identifying a capacity μ (more precisely a set of capacities) such that the Gödel integral it induces represents a given user preference relation. To the best of our knowledge, the only existing work on that topic [3] considers the case when the preferences to be satisfied are pairs (x, α) where x is a vector and α its expected global evaluation, as specified by the user. The scale L such that $\mu : 2^{\mathcal{C}} \to L$ is not necessarily $[0,1]$ but any totally ordered set with a bottom element and a top element, respectively denoted 0 and 1, and equipped with an inverse order map $u \mapsto 1 - u$.

When $|\mathcal{C}| = 2$, the results established in [3] state that for any $x \in \mathcal{E}$ and $\alpha \in L$, the set of capacities μ such that $G_\mu(x) = \alpha$ with $\alpha \in L$ equals $[\underline{\mu}^{x,\alpha}, \overline{\mu}^{x,\alpha}]$ where the boundary definitions depends on the value of α (see [3] for more details).

2.4 Difference Between Gödel Integrals and Sugeno Integrals

Sugeno integral (SI) is classically used in decision making as aggregation function [9,14]. The Sugeno integral of $x \in [0,1]^n$ with respect to capacity μ is defined as:

$$S_\mu(x) = \max_{A \subseteq \mathcal{C}} \min(\mu(A), \min_{i \in A} x_i).$$

Many authors study its elicitation (for instance [15,16]), using the fact that $S_\mu(1_A)$ is equal to $\mu(A)$. It is thus possible to elicit the capacity from the global evaluation of such vectors.

It is not the case with Gödel integral as it holds that

$$G_\mu(1_A) = \mu(A) \otimes 1 = \begin{cases} 1 \text{ if } \mu(A) > 0 \\ 0 \text{ otherwise.} \end{cases}$$

Note that Boolean objects 1_A can be classified into two classes, according to the value, 0 or 1, of their Gödel integral. The only information about those in class 1 is that their capacity is non-zero.

If we consider $\min(\alpha, 1_A)$ the vector whose i-th component is equal to α if $i \in A$ and 0 otherwise, according to [6], it holds that:

$$G_\mu(\min(\alpha, 1_A)) = \mu(A) \otimes \alpha = \begin{cases} \alpha \text{ if } \mu(A) > 1 - \alpha \\ 0 \text{ otherwise} \end{cases}$$

For the Gödel integral, a preference given on Boolean objects of type 1_A does not make it possible to obtain the value of $\mu(A)$ but an inequality regarding the possible values of A.

3 Gödel Integral Elicitation

This section presents the theoretical results of our proposed approach.

Let $\mathcal{P} = \{\mathscr{C}_1, \ldots, \mathscr{C}_p\}$ be a partition of $\mathcal{E}$, i.e. such that (i) $\bigcup_{i=1} \mathscr{C}_i = \mathcal{E}$, (ii) $\forall i, \forall j$, $\mathscr{C}_i \cap \mathscr{C}_j = \emptyset$, and (iii) $\forall i$, $\mathscr{C}_i \neq \emptyset$. This partition $\mathcal{P}$ induces an order $\prec$ and an equivalence relation $\sim$ on $\mathcal{E}$ such that:

1. $\forall i$, $\forall e, f \in \mathscr{C}_i$, $e \sim f$;
2. $\forall i, \forall j$, $\forall e \in \mathscr{C}_i, \forall f \in \mathscr{C}_j$, $i < j \implies e \prec f$.

Therefore, the $\mathscr{C}_i$ are equivalence classes and $\mathcal{P}$ is a preference relation among examples from $\mathcal{E}$. Our aim is to elicit a capacity μ such that

- $\forall i$, $\forall e, f \in \mathscr{C}_i$, $G_\mu(e) = G_\mu(f)$;
- $\forall i, \forall j$, $\forall e \in \mathscr{C}_i, \forall f \in \mathscr{C}_i$, $i < j \implies G_\mu(e) < G_\mu(f)$.

In the following, after a preliminary result, we characterise the equivalence class induced by a given capacity and, then, reciprocally, the capacity induced by an equivalence class.

3.1 Technical Result

This section first adapts the results reminded in Sect. 2.3 to a continuous evaluation scale L, i.e. when the considered examples are evaluated on $[0, 1]$. Our aim is to solve equations of the form, $G_\mu(x) = \alpha$, here for $\alpha \in [0, 1]$.

If $\alpha \notin \{x_1, x_2\}$, the equation has no solution. Otherwise, the solution depends on the relative position of x_1, x_2.

Proposition 1. *The equation $G_\mu(x) = x_i$, $i \in \{1, 2\}$ has the following solutions*

- *if $x_i = x_j$: any capacity μ is solution.*
- *if $x_i < x_j$, $\mu_i \in [0, 1]$ and $\mu_j \in [0, 1 - x_j]$.*
- *if $x_i > x_j$, $\mu_i \in (1 - x_i, 1]$ and $\mu_j \in [0, 1]$.*

Proof. It is easy to check that if $x_i = x_j$ any capacity is a solution. For all x, $G_\mu(x) = \max(\mu_i \otimes x_i, \mu_j \otimes x_j, \min(x_i, x_j))$.

- If x_i is the minimum, $G_\mu(x) = x_i$ if and only if $\mu_j \otimes x_j = 0$, i.e., $\mu_j \leq 1 - x_j$
- If x_i is the maximum $G_\mu(x) = x_i$ if and only if $\mu_i \otimes x_i = x_i$, i.e., $\mu_i > 1 - x_i$.

Example 1. Given $x = (0.5, 0.6)$, a capacity μ such that $G_\mu(x) = 0.5$ satisfies $\mu_1 \in [0, 1]$ and $\mu_2 \in [0, 0.4]$.
A capacity μ such that $G_\mu(x) = 0.6$ satisfies $\mu_1 \in [0, 1]$ and $\mu_2 \in (0.4, 1]$.

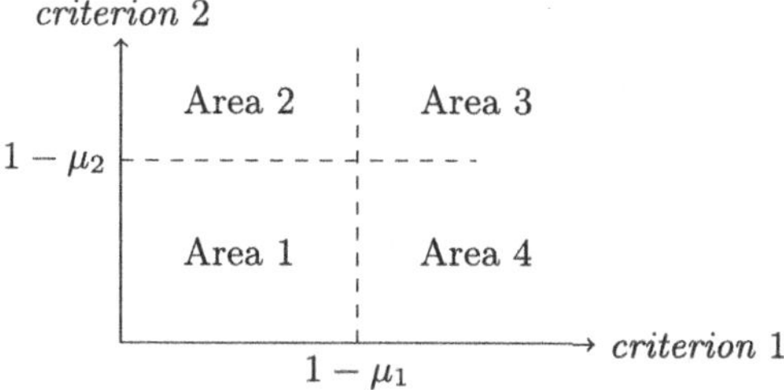

Fig. 1. Areas defined by a capacity μ.

3.2 Characterisation of the Capacity Induced Equivalence Class

Given a capacity μ and $\alpha \in [0,1]$, we identify the equivalence class of the objects whose GI is equal to α, this class is denoted $C_{\mu,\alpha} = \{x \in [0,1]^2 \mid G_\mu(x) = \alpha\}$. In this equivalence class, the objects can be ordered considering the partial order based on the coordinates: $(x_1, x_2) \leq (y_1, y_2)$ if and only if $x_1 \leq y_1$ and $x_2 \leq y_2$.

Knowing that the Gödel integral of an object x can take only 2 values (see Sect. 2.2), x belongs to $C_{\mu,\alpha}$ only if one of its component is equal to α, formally $C_{\mu,\alpha} \subseteq \{(\alpha, x_2),\ x_2 \in [0,1]\} \cup \{(x_1, \alpha),\ x_1 \in [0,1]\}$.

The values $1 - \mu_1$ and $1 - \mu_2$ define 4 area in the space of the criteria as shown in Fig. 1.

For each area, the boundaries of the objects in the class are defined as follows (lower bounds are denoted $\underline{x}, \underline{y}, \ldots$ and upper bounds are denoted $\overline{x}, \overline{y}, \ldots$).

Proposition 2. *The different bounds of $C_{\mu,\alpha}$ are:*

- *Area 1 ($\alpha \leq 1 - \mu_1$ and $\alpha \leq 1 - \mu_2$): $\underline{x} = (\alpha, \alpha)$, $\overline{x} = (\alpha, 1 - \mu_2)$ and $\overline{y} = (1 - \mu_1, \alpha)$.*
- *Area 2 ($\alpha \leq 1 - \mu_1$ and $\alpha > 1 - \mu_2$): $\underline{x} = (0, \alpha)$ and $\overline{x} = (1 - \mu_1, \alpha)$.*
- *Area 3 ($\alpha > 1 - \mu_1$ and $\alpha > 1 - \mu_2$): $\underline{x} = (0, \alpha)$, $\underline{y} = (\alpha, 0)$ and $\overline{x} = (\alpha, \alpha)$.*
- *Area 4 ($\alpha > 1 - \mu_1$ and $\alpha \leq 1 - \mu_2$): $\underline{x} = (\alpha, 0)$ and $\overline{x} = (\alpha, 1 - \mu_2)$.*

Proof. For each area, objects of the form (x_1, α) and of the form (α, x_2) are examined.

- Area 1 ($\alpha \leq 1 - \mu_1$ and $\alpha \leq 1 - \mu_2$): for $x = (x_1, \alpha)$: $G_\mu(x) = \max(\mu_1 \otimes x_1, \min(x_1, \alpha)) = \alpha$ iff $\alpha \leq x_1 \leq 1 - \mu_1$. for $x = (\alpha, x_2)$: $G_\mu(x) = \max(\mu_2 \otimes x_2, \min(\alpha, x_2)) = \alpha$ iff $\alpha \leq x_2 \leq 1 - \mu_2$.
- Area 2 ($\alpha \leq 1 - \mu_1$ and $\alpha > 1 - \mu_2$): for $x = (x_1, \alpha)$: $G_\mu(x) = \max(\mu_1 \otimes x_1, \alpha, \min(x_1, \alpha)) = \alpha$ iff $\mu_1 \otimes x_1 \leq \alpha$ (i.e., $x_1 \leq 1 - \mu_1$ since $\alpha \leq 1 - \mu_1$). for $x = (\alpha, x_2)$: $G_\mu(x) = \max(\mu_2 \otimes x_2, \min(\alpha, x_2)) = \alpha$ iff $x_2 = \alpha$.
- Area 3 ($\alpha > 1 - \mu_1$ and $\alpha > 1 - \mu_2$): for $x = (x_1, \alpha)$: $G_\mu(x) = \max(\mu_1 \otimes x_1, \alpha, \min(x_1, \alpha)) = \alpha$ iff $\mu_1 \otimes x_1 \leq \alpha$ i.e., $x_1 \leq \alpha$ (since $1 - \mu_1 < \alpha$). for $x = (\alpha, x_2)$: $G_\mu(x) = \max(\alpha, \mu_2 \otimes x_2, \min(\alpha, x_2)) = \alpha$ iff $\mu_2 \otimes x_2 \leq \alpha$ i.e., $x_2 \leq \alpha$ (since $1 - \mu_2 < \alpha$).

- Area 4 ($\alpha > 1 - \mu_1$ and $\alpha \le 1 - \mu_2$): for $x = (x_1, \alpha)$: $G_\mu(x) = \max(\mu_1 \otimes x_1, \min(x_1, \alpha)) = \alpha$ iff $x_1 = \alpha$. for $x = (\alpha, x_2)$: $G_\mu(x) = \max(\alpha, \mu_2 \otimes x_2, \min(x_1, \alpha)) = \alpha$ iff $\mu_2 \otimes x_2 \le \alpha$ i.e. $x_2 \le 1 - \mu_2$ since $\alpha \le 1 - \mu_2$.

Example 2. Let $x = (0.5, 0.6)$ and $y = (0.5, 0.4)$ be in the class $C_{\mu,0.5}$ for the capacity μ such that $\mu_1 = 0.6$ and $\mu_2 = 0.4$.

As $1 - \mu_1 = 0.4$ and $1 - \mu_2 = 0.6$, we have $\alpha = 0.5$ in Area 4. The bounds of the class $C_{\mu,0.5}$ are $\underline{x} = (0.5, 0)$ and $\overline{x} = (0.5, 0.6)$. The element $z = (0.4, 0.5)$ is not greater than $\underline{x} = (0.5, 0)$ so it is not in $\mathscr{C}_{0.5}$. It is easy to check that $G_\mu(z) = 0.4$.

3.3 Characterisation of an Equivalence Class Induced Capacity

In this section, we consider the reverse problem: given an equivalence class $\mathscr{C}$, we study the conditions that are required for the existence of a capacity μ and a real value $\alpha \in [0, 1]$ such that $G_\mu(x) = \alpha$ for all x in $\mathscr{C}$.

First, let us consider two objects x, y in the same equivalence class, i.e., such that $x \sim y$ and let us elicit μ such that $G_\mu(x) = G_\mu(y)$.

The boundary constraints required by a capacity μ in relation with the values of x and y when they have at least one value in common, are summarised in Table 1. When the 2 objects x and y have no common value there is no solution for the capacity. The other case with no capacity solution is characterised in Proposition 3.

Proposition 3. *For $x, y \in [0, 1]^2$ such that $\{x_1, x_2\} \cap \{y_1, y_2\} \neq \emptyset$, there does not exist a capacity μ such that $G_\mu(x) = G_\mu(y)$ in the two following cases:*

- $x_i < x_j = G_\mu(x) = G_\mu(y) = y_i < y_j$;
- $y_i < y_j = G_\mu(y) = G_\mu(x) = x_i < x_j$.

Proof. As it can be deduced from the study of the cases in Table 1, two cases can be highlighted for which it is not possible to find a solution for the capacity:

- Case 1: to solve $G_\mu(x) = x_i = y_j = G_\mu(y)$ where x_i is the minimum value of x and y_j is the maximum of y. In this case, to have a non empty set, it is required that $1 - y_j \le 1 - x_j$ i.e., $x_j \le y_j$. $x_i = y_j$, thus $x_j \le x_i$. By hypothesis $x_i \le x_j$, thus we have $x_i = x_j$.
- Case 2: to solve $G_\mu(x) = x_i = y_j = G_\mu(y)$ where x_i is the strict maximum value of x and y_j is the strict minimum of y. To have a non empty set we need to have $1 - x_i < 1 - y_i$ i.e. $y_i < x_i = y_j$ a contradiction with the fact that y_j is the minimum. So in this case the solution set is empty.

Proposition 3 can be generalised to equivalence classes containing more than two objects as follows.

Table 1. Constraints on μ such that $G_\mu(x) = \alpha$ and $G_\mu(y) = \alpha$

	Position of x related to y			
	$x_i \leq x_j$ $y_i < y_j$	$x_i \leq x_j$ $y_i \geq y_j$	$x_i > x_j$ $y_i \leq y_j$	$x_i > x_j$ $y_i > y_j$
	Case $G_\mu(x) = x_i$ and $G_\mu(y) = y_i$:			
μ_i	$[0, 1]$	$]1 - y_i, 1]$	$]1 - x_i, 1]$	$]1 - \min(x_i, y_i), 1]$
μ_j	$[0, 1 - \max(x_j, y_j)]$	$[0, 1 - x_j]$	$[0, 1 - y_j]$	$[0, 1]$
	Case $G_\mu(x) = x_i$ $G_\mu(y) = y_j$			
μ_i	$[0, 1]$	$[0, 1 - y_i]$	$]1 - x_i, 1]$	$]1 - x_i, 1 - y_i]$
μ_j	$]1 - y_j, 1 - x_j]$	$[0, 1 - x_j]$	$[1 - y_j, 1]$	$[0, 1]$

Proposition 4. *Let $\mathscr{C} = \{x^1, \dots, x^p\}$, there exists a capacity μ such that G_μ represents $\mathscr{C}$ if and only if $\nexists k, \nexists l \in \{1, \cdots, p\}$ such that $x_i^l = \alpha = \max(x_1^l, x_2^l)$ and $x_j^k = \alpha = \min(x_1^k, x_2^k)$ with $x_1^l \neq x_2^l$, $x_1^k \neq x_2^k$ and $i \neq j$.*

Proof. There exists a capacity μ solving the problem except in the case where the maximal value of one x^l is not greater than the minimal value of one x^k and they respectively correspond to different dimensions (or reciprocally).

Example 3. Let us consider two elements $x = (0.5, 0.6)$ and $y = (0.4, 0.5)$. The question is whether there exists a capacity μ such that $G_\mu(x) = G_\mu(y)$.

- Without Proposition 4. The common value for x and y is 0.5. So we have to solve $G_\mu(x) = G_\mu(y) = 0.5$. From x it is required that $\mu_1 \in [0, 1]$ and $\mu_2 \in [0, 0.4]$. From y, it is required that $\mu_1 \in [0, 1]$ and $\mu_2 \in [0.5, 1]$. These constraints are not compatible, thus there is no capacity μ such that G_μ represents $x \sim y$.
- With Proposition 4, the value 0.5 is the minimum for x and the maximum for y. However, the involved indexes are different, so there is no capacity solution.

4 Proposed GICEP Algorithm

In this section, we introduce the *Gödel Integral for Capacity Elicitation from Preference relations* (GICEP) algorithm whose pseudo-code is given in Algorithm 1. This algorithm takes as argument a partition $\mathcal{P} = \{\mathscr{C}_1, \dots, \mathscr{C}_p\}$, as

Algorithm 1.

function GICEP($\mathcal{P}$)
  $(\underline{\mu}, \overline{\mu}) \leftarrow ((0,0),(1,1))$
  for $i = 1, \ldots, |\mathcal{P}|$ **do**
    $\alpha_i \leftarrow$ CHECKCLASS($\mathscr{C}_i$)
    if i>1 **then**
      CHECKORDER(α_i, α_{i-1})
    $(\underline{\mu_i}, \overline{\mu_i}) \leftarrow$ CHECKBOUNDS($\mathscr{C}_i, \alpha_i$)
    $(\underline{\mu}, \overline{\mu}) \leftarrow$ AGGREGATE($(\underline{\mu}, \overline{\mu}), (\underline{\mu_i}, \overline{\mu_i})$)
  return $(\underline{\mu}, \overline{\mu})$ ▷ Boundaries for μ

defined in Sect. 3, that represents the preferences of a user over a set of examples. The algorithm then returns the boundaries $(\underline{\mu}, \overline{\mu})$ of the capacities that can be used to represent $\mathcal{P}$ with the Gödel integral. In cases where no solution exists, the algorithm gives an explanation to the user about the reason of the impossibility.

The next subsections describes in turn the functions called by the GICEP algorithm.

4.1 Function CHECKCLASS

Given an equivalence class $\mathscr{C}$, CHECKCLASS($\mathscr{C}$) checks whether there exists μ such that G_μ represents $\mathscr{C}$. This function returns a value α compatible with all examples from $\mathscr{C}$, thanks to the use of Proposition 4. More precisely this function checks that:

- there exists $\alpha \in [0,1]$ such that $\forall e \in \mathscr{C}$, $e = (\alpha, e_2)$ or $e = (e_1, \alpha)$;
- there does not exist $e, f \in \mathscr{C}$ such that $e_j = \alpha = \max(e_1, e_2)$ and $f_k = \alpha = \min(f_1, f_2)$ with $e_1 \neq e_2$, $f_1 \neq f_2$ and $j \neq k$.

If it is not possible to find such a compatible value α, the function fails and produces a message for the user explaining that it is not possible to find a value compatible with the Gödel integral aggregation function in the corresponding class of examples.

4.2 Function CHECKORDER

The function CHECKORDER(α_i, α_{i-1}) checks the compatibility of two equivalence classes with successive indices. More precisely, it checks whether class $\mathscr{C}_i$ can be considered as "greater" than class $\mathscr{C}_{i-1}$. To do this verification, the function checks if the given values α_i and α_{i-1}, previously associated by the CHECKCLASS function to the equivalence classes, are such that $\alpha_i > \alpha_{i-1}$.

If it is not the case, a message is produced to inform the user that there is an incompatibility for these two classes.

4.3 Function CheckBounds

From an equivalence class $\mathscr{C}$ that can be represented with a Gödel integral according to the CheckClass function and the value α it provides, CheckBounds($\mathscr{C}, \alpha$) returns the bounds of the set of the capacities that enable to obtain α for all objects from $\mathscr{C}$.

This function implements Proposition 1 for any object $x \in \mathscr{C}$ to determine the boundaries of the set of capacities μ that are solution of $G_\mu(x) = \alpha$. The obtained boundaries are iteratively aggregated with a call to the Aggregate function defined below.

4.4 Function Aggregate

The function Aggregate($(\underline{\mu}^1, \overline{\mu}^1), (\underline{\mu}^2, \overline{\mu}^2)$) combines two estimations of the lower and upper boundaries $(\underline{\mu}^1, \overline{\mu}^1)$ and $(\underline{\mu}^2, \overline{\mu}^2)$ to obtain an aggregated estimation $(\underline{\mu}, \overline{\mu})$: the value $\underline{\mu}$ is the maximum of the lower capacities and the value $\overline{\mu}$ is the minimum of the upper capacities.

If the obtained boundaries $\underline{\mu}$ and $\overline{\mu}$ do not satisfy $\underline{\mu} \leq \overline{\mu}$, the Aggregate function produces a message to inform the user, informing her that the two estimations are incompatible.

When the Aggregate function is called by the CheckBounds function, no incompatibility can appear as the given estimations $(\underline{\mu}^1, \overline{\mu}^1)$ and $(\underline{\mu}^2, \overline{\mu}^2)$ have been obtained from objects belonging to the same equivalence class whose compatibility has been previously checked by the CheckClass function.

On the other hand, the call of the Aggregate function in line 8 of GICEP algorithm aggregates estimates from different equivalence classes. However, the CheckOrder function only validated the compatibility of the α values associated with the classes and not that of the estimates of the limits of their capacities.

5 Illustrative Experiments

This section presents the experiments conducted with the GICEP algorithm on artificial data, describing the experimental protocol and the obtained results.

5.1 Experimental Protocol

In order to simulate user preferences that may or may not be represented by a Gödel integral, our experimental protocol consists of four steps detailed below.

In the first step, we generate randomly the data set $\mathcal{E}$. In the second step, a reference capacity $\mu^\star$ is randomly generated. In the third step, the partition $\mathcal{P}$ of $\mathcal{E}$ induced by the Gödel integral defined by $\mu^\star$ is built by defining the ordered equivalence classes as $\mathscr{C} = \{x \in \mathcal{E} \,|\, G_{\mu^\star}(x) = \alpha\}$ where α is chosen from a controlled set of values. The fourth step, which is optional, disturbs the partition by adding *a posteriori* inconsistencies.

Table 2. Experimental results in the case of compatible preferences

n	$\mu_1^\star$	$\mu_2^\star$	$[\underline{\mu}_1, \overline{\mu}_1]$	$[\underline{\mu}_2, \overline{\mu}_2]$
20	0.86	0.14	[0.70, 1.00]	[0.10, 0.30]
1000	0.47	0.53	[0.40, 0.50]	[0.50, 0.60]
1000	0.13	0.87	[0.10, 0.20]	[0.80, 0.90]
10000	0.45	0.55	[0.40, 0.50]	[0.50, 0.60]
10000	0.05	0.95	[0.00, 0.10]	[0.90, 1.00]

In order for the equivalence classes induced by these steps to be considered representative of user preferences and to be relevant, they must be limited in number and not reduced to singletons. To guarantee these properties, we use the fact that $G_{\mu^\star}(x) \in \{x_1, x_2\}$ to generate data in $[0,1]^2$, limiting the number of values for x_1 and x_2 by using their precision: with a precision of 1 digit for the decimal value for example, the possible values for $G_{\mu^\star}(x)$ are only limited to the 10 values 0.1, 0.2, ..., 1.0 and thus leading to, at most, to 10 equivalence classes.

The first step of the data generation depends on two parameters: (i) the total number n of examples in $\mathcal{E}$; and (ii) the chosen precision p_v. The results presented in this section are obtained with $p_v = 1$. The set of objects $\mathcal{E}$ is obtained by a random generation of the data in $[0,1]^2$, according to a uniform distribution and with a precision of p_v digits for the decimal value.

The second step builds the reference capacity $\mu^\star$ determined by its two parameters $\mu_1^\star$ and $\mu_2^\star$ randomly generated according to a uniform distribution on $[0,1]$.

The third step builds the set of classes $\mathscr{C}_i$ with $\alpha_i \in \{k/10^{p_v}, k = 1, \ldots, 10^{p_v}\}$, keeping only the non empty ones to define partition $\mathcal{P}$.

The fourth step is detailed in Sect. 5.3 which discusses the results obtained in the case where the user preferences are incompatible according to a Gödel integral.

The two next sections present the experimental results obtained with the GICEP algorithm on data generated applying this protocol.

5.2 Experiments with Compatible Preferences

Table 2 shows the experimental results obtained when preferences can be represented by a Gödel integral (i.e. compatible preference), for different parameter values. It highlights the fact that GICEP provides a correct result in all cases: the induced capacity sets always contain the reference $\mu^\star$ capacity. Moreover, it offers a good approximation quality, evaluated by the size of the interval $\overline{\mu}_i - \underline{\mu}_i$, $i \in \{1, 2\}$.

Note that the case $n = 10000$ can be considered as unrealistic because it does not seem possible for a user to provide such a large number of preferences. However, the aim in this case is to check the capacity of the GICEP algorithm to process a very large data set, that has been confirmed by the obtained results showed in Table 2.

5.3 Experiments with Incompatible Preferences

To check that the proposed GICEP algorithm is capable of detecting inconsistencies in a set of user's preferences, we also run a set of experiments on data corresponding to preferences that cannot be represented by a Gödel integral (i.e. incompatible preferences). To do this, two kinds of perturbation are introduced. The first one consists in moving an example from its class (i.e. the class it has been associated to in the third step) to another one. The second kind of perturbation consists in inverting the order of two classes of the partition obtained in the third step.

In each case, running the GICEP algorithm enables us to obtain the expected results, that is to say, the detection of the incompatibility of the preferences and the production of a message to inform the user. Here are the obtained results for the two kinds of perturbation:

- in the first case: the CheckClass function produces a message informing the user of the incompatibility and provides the explanation accordingly: " `not valid for class k : no compatible value` ".
- in the second case: the CheckOrder function produces a message informing the user of the incompatibility and provides the explanation related to the misplaced class i: " `no value compatible with the previous class` ".

6 Conclusion and Future Works

In this paper, a new approach is introduced to elicit the capacity (i.e. the main operator parameter) of the Gödel integral when the user's preferences are expressed as an ordering relation between classes of equivalent objects.

We first establish a theoretical characterisation of the set of admissible capacities, and we introduce an implementation as the *Gödel Integral for Capacity Elicitation from Preference relations* (GICEP) algorithm. This algorithm additionally offers functions that provide the user with explanations in the case when no capacity can be found to model her preferences. Illustrative experiments conducted on synthetic data show the relevance and efficiency of the proposed approach.

At a theoretical level, future works will aim at establishing guarantees regarding the quality of the output, e.g. the size of the boundaries. We will also aim at studying the general case when the objects are assessed with more than two criteria so as to establish a general elicitation procedure.

At an application level, future works will, in particular, consider the eXplainable Artificial Intelligence (XAI) domain, where the evaluated objects are candidate explanations, in particular counterfactual examples assessed by an objective criterion and a subjective one as introduced in [11]. In this case, the preferences expressed by the user provide indications regarding the desired combination of the two criteria: the elicitation procedure then offers tools to personalise the generated explanation, by the definition of a user-centred global quality criterion.

References

1. Angilella, S., Greco, S., Matarazzo, B.: Non-additive robust ordinal regression with Choquet integral, bipolar and level dependent Choquet integrals. In: IFSA/EUSFLAT Conference, pp. 1194–1199 (2009)
2. Artelt, A., Hammer, B.: Convex density constraints for computing plausible counterfactual explanations. In: Proceedings of the International Conference on Artificial Neural Networks and Machine Learning, ICANN2020, pp. 353–365 (2020)
3. Baaj, I., Rico, A.: Qualitative integrals with Gödel's implication and conjunction: elicitation and if-then rules extraction. In: IEEE International Conference on Fuzzy Systems, FUZZ-IEEE 2022, Padua, Italy, 18–23 July 2022, pp. 1–8. IEEE (2022)
4. Benabbou, N., Perny, P., Viappiani, P.: Incremental elicitation of Choquet capacities for multicriteria choice, ranking and sorting problems. Artif. Intell. **246**, 152–180 (2017)
5. Dubois, D., Prade, H., Rico, A.: Residuated variants of Sugeno integrals: towards new weighting schemes for qualitative aggregation methods. Inf. Sci. **329**, 765–781 (2016)
6. Dubois, D., Prade, H., Rico, A., Teheux, B.: Generalized qualitative Sugeno integrals. Inf. Sci. **415**, 429–445 (2017)
7. Gagolewski, M., James, S., Beliakov, G.: Supervised learning to aggregate data with the Sugeno integral. IEEE Trans. Fuzzy Syst. **27**(4), 810–815 (2019)
8. Grabisch, M.: Set functions games and capacities in decision making. In: Game theory, Social Choice, Decision Theory and Optimization. No. C 46 in Theory and Decision Library. Springer, Heidelberg (2016). https://doi.org/10.1007/978-3-319-30690-2
9. Grabisch, M., Labreuche, C.: A decade of application of the Choquet and Sugeno integrals in multi-criteria decision aid. Ann. Oper. Res. **175**, 247–286 (2010)
10. Hüllermeier, E., Tehrani, A.F.: Efficient learning of classifiers based on the 2-additive choquet integral. In: Moewes, C., Nürnberger, A. (eds.) Computational Intelligence in Intelligent Data Analysis, pp. 17–29. Springer, Heidelberg (2013). https://doi.org/10.1007/978-3-642-32378-2_2
11. Jeyasothy, A., Rico, A., Lesot, M.J., Marsala, C., Laugel, T.: Knowledge integration in XAI with Gödel integrals. In: 2023 IEEE International Conference on Fuzzy Systems (FUZZ) Incheon, South Korea, pp. 1–6 (2023)
12. Mousseau, V., Pirlot, M.: Preference elicitation and learning. EURO J. Decis. Process **3**, 1–3 (2015)
13. Poyiadzi, R., Sokol, K., Santos-Rodriguez, R., De Bie, T., Flach, P.: FACE: feasible and actionable counterfactual explanations. In: Proceedings of the AAAI/ACM Conference on AI, Ethics, and Society (2020)
14. Rico, A.: Discrete Sugeno integrals and their applications. In: Scalable Uncertainty Management - 12th International Conference, SUM, pp. 18–32 (2018)
15. Rico, A., Grabisch, M., Labreuche, C., Chateauneuf, A.: Preference modelling on totally ordered sets by the Sugeno integral. Disc. Appl. Math. **147**(1), 113–124 (2005)

16. Rico, A., Viappiani, P.: Incremental elicitation of capacities for the sugeno integral with a maximin approach. In: Davis, J., Tabia, K. (eds.) SUM 2020. LNCS (LNAI), vol. 12322, pp. 156–171. Springer, Cham (2020). https://doi.org/10.1007/978-3-030-58449-8_11
17. Rico, A., Lesot, M.J., Marsala, C.: A maximin approach to elicit gödel integral in an XAI context. In: IEEE World Congress on Computational Intelligence (IEEE WCCI 2024/Fuzz-IEEE 2024), Yokohama, Japan (2024)
18. Sugeno, M.: Theory of fuzzy integrals and its applications. Ph.D. thesis, Tokyo Institute of Technology (1974)
19. Van Looveren, A., Klaise, J.: Interpretable counterfactual explanations guided by prototypes. In: Proceedings of European Conference on Machine Learning (2021)

Extended Boltzmann Machine Generative Model

Lancelot Tullio(✉) and Maria Rifqi

LEMMA, University Paris–Panthéon–Assas, Paris, France
lancelot.tullio@etudiants.u-paris2.fr, maria.rifqi@u-paris2.fr

Abstract. The increase in computing power in recent years has brought generative models and the use of synthetic data back to the fore to solve a variety of previously unsolved problems, in particular when fields are subject to constraints linked to the sensitivity of the information processed. This article proposes a modified version of restricted Boltzmann machines (RBM), known as Bernoulli machines, to improve its ability to handle non-binary data without making the methodology more complex to understand and manipulate. To assess the performance of our algorithm, we compare it with various generative models that are well documented and have repeatedly proven their effectiveness in a variety of contexts. We also chose to use a large number of open source datasets with different types of features and different sizes in order the verify the generalization capacity and sclalability of our approach.

Keywords: Machine Learning · Generative approach · SDV · Restricted Boltzmann Machines · Generative Adversarial Network · Multidimensional relationship

1 Introduction

Personal data, commercial information, demographic statistics, our modern world generates an exceptional amount of data on a daily basis. However, even in the midst of this abundance, certain problems remain and require the use of synthetic data or data augmentation.

Lin and Tsai (2020) [16], for example, discuss the value of synthetic data in the management of data quality, which impacts the reliability of machine learning methods and requires making statistical assumptions that are rarely empirically justified. Cao et al. (2021) [17] propose data generation to deal with unbalanced databases, with a view to improving the robustness of predictions made by models. Some fields are also subject to additional constraints linked to the sensitivity of the information processed. This is particularly the case in the medical sector [4], or in the fight against fraud and money laundering, as Gascon et al. (2018) [20] point out. Generative AI techniques therefore appear to be a powerful way of solving these problems, as demonstrated by Open AI's GPT tool when it was launched in November 2022. Other generative methods are also

S. Destercke et al. (Eds.): SUM 2024, LNAI 15350, pp. 408–420, 2025.
https://doi.org/10.1007/978-3-031-76235-2_30

performing well. These include Bayesian neural networks developed by Chow et al. [3], variational autoencoders by Xu and et al. (2019) [22] and the Python SDV library proposed by Montanez and co-authors [18], based on antagonistic generative networks.

Among the Generative AI techniques mentioned above, we can still mention the Boltzmann Machine (RBM) developed by Hinton and Sejnowski (1983) [8] and revisited by Lezmi et al. (2020) [13] on problems of robustness of econometric models used in financial markets.

In this paper, we show that by adapting the structure of Bernoulli's Restricted Boltzmann Machine, we obtain a new synthetic data generator that is robust to a variety of variable types, while preserving the statistical properties of the original data. We also propose an experimental study on several databases with diverse characteristics, comparing the performances obtained with those of other generators implemented in the SDV Python library [19], and evaluating the quality of the data generated using a global metric based on the GAN model methodology, an implementation of which we propose, although it is also available within the SDV library.

2 Restricted Boltzmann Machines

2.1 Bernoulli RBMs

Fischer and Igel (2014) [6] propose to consider RBM as a special case of neural networks where **information has no particular direction** and therefore has the ability to move from hidden to visible layers and vice versa. This representation allows us to think of RBM as a Markov random field associated with an undirected bi-partite graph consisting of m visible units representing the observable features of the original database and n hidden units capturing the dependency structure (see Fig. 1).

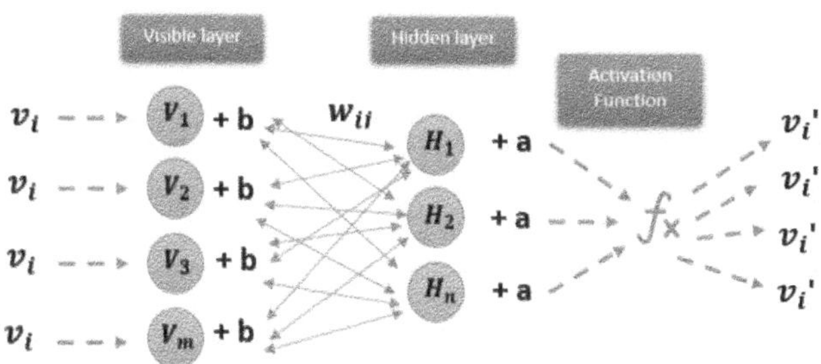

Fig. 1. RBM schema with m visible units and n hidden units

Let us denote by $v = (v_1, v_2, ..., v_m)$ and $h = (h_1, h_2, ..., h_n)$ the values of visibles $V = (V_1, V_2, ..., V_m)$ and hidden variables $H = (H_1, H_2, ..., H_n)$ respectively, where v_i and h_j are the binary states of the ith visible variable V_i and the

j^{th} hidden variable H_j. It is possible to define a joint probability distribution as a function of the neural network parameters. More precisely, this probability measure is based on a second intermediate function, the energy function, constructed from the neuron's characteristics. The latter is defined by Lezmi et al. (2020) [13] as follows:

$$E(v,h) = -\sum_{i=1}^{m}\sum_{j=1}^{n} W_{ij} * h_j * v_i - \sum_{i=1}^{m} b_i * v_i - \sum_{j=1}^{n} a_j * h_j \quad (1)$$

with i the number of visible neurons, j the number of hidden neurons in the RBM, w_{ij} representing a real weight associated with the edge between v_i and h_j, and b_i, a_j representing the real bias terms associated respectively with the ith visible and jth hidden variable. Note that this optimization problem can be solved iteratively using contrastive divergence, as demonstrated by Hinton et al. (2002) [7]. The Algorithm 1 presented below comes [5].

Algorithm 1. Bernoulli RBM algorithm

Require: $N \geq 1$, the number of observations, m and n the respective number of visible and hidden units
  Input: $(v_{(1)}, \ldots, v_{(N)})$
  Initialization: Set $\Delta a_i = \Delta b_j = \Delta w_{i,j} = 0$ for i=1,...,m, j=1,...,n
  for v from $v_{(1)}$ to $v_{(N)}$ **do**
    $v^0 \leftarrow v$
    for $t = 1$ to k **do**
      for $j = 1$ to n **do**
        compute $p(h_j^t)$=$\mathbb{P}(h_j^t|v^{t-1})$
        sample the jth hidden unit state: $h_j^t \sim \mathcal{B}(p(h_j^t))$
      end for
      for $i = 1$ to m **do**
        compute $p(v_i^t)$=$\mathbb{P}(v_i^t|h^t)$
        sample the ith visible unit state: $v_i^t \sim \mathcal{B}(p(v_i^t))$
      end for
    end for
    for $i = 1$ to m **do**
      for $j = 1$ to n **do**
        $\Delta a_i \leftarrow \Delta a_i + v_i^k - v_i^0$
        $\Delta b_j \leftarrow \Delta b_j + \mathbb{P}(h_j = 1|v^k) - \mathbb{P}(h_j = 1|v^0)$
        $\Delta w_{i,j} \leftarrow \Delta w_{i,j} + \mathbb{P}(h_j = 1|v^k) \times v_i^k - \mathbb{P}(h_j = 1|v^0) \times v_i^0$
      end for
    end for
  end for
  Output: Gradient estimation Δa_i,Δb_j and $\Delta w_{i,j}$

This generative model has been shown to perform well in the case of binary data, particularly for image generation, as demonstrated by Bidaurrazaga et al. (2023) [1]. However, Algorithm 1 remains limited for solving problems involving continuous quantitative or qualitative data. Indeed, as Borisov and co-authors (2023) [2] point out in their work, data transformation using a *One Hot Encoder* necessarily leads to dimensionality problems, also known as *curse of dimensionality* [21].

2.2 An Evolving Structure

Several modifications of RBM to extend its field of application have been proposed. For example, Kondratyev and Schwarz (2019) [10] propose a system of binary decomposition of continuous quantitative values, thus making it possible to deconstruct the original data to make them binary, processing them with RBM, before being reconstructed by doing the reverse operation. This idea was later complemented by Lailler and Cohen (2022) [12], where a discretization function based on the construction of a bounded value scale is introduced, with characteristics similar to normalization techniques. Finally, it should be noted that these different approaches focus mainly on problems that relate to the ability of RBM to generate data of different types. Lezmi et al. (2020) [13] show that the ability of RBM to preserve the dependency structure of the original data should not be neglected. These authors therefore propose an alternative solution, called *Split Sampling*, based on a copula function within the original RBM structure.

3 Extended Boltzmann Machines

For our part, we want to keep the original RBM structure while making some modifications to improve its ability to handle non-binary data. We propose an alternative to the RBM algorithm (see Algorithm 2), called *Extended RBM algorithm*. Starting from Bernoulli's RBM, we apply the following **three modifications**:

1. Initialization of weights and biases according to a uniform distribution,
2. Univariate means and standard deviation calculation of non binary variables,
3. Introduction of a Gaussian simulator for numerical variables.

The Gaussian generator assumes that each of the observed phenomena follows a normal distribution. Although this assumption may seem restrictive and unrealistic, we justify it by the Moivre-Laplace theorem which states that the succession of random variables independent of each other but following the same probability law can be approximated by a normal law. This theorem introduces little complexity to the RBM structure by considering the following generating function:

$$F(X) = \frac{1}{2}(1 + \mathrm{erf}(\frac{Z - \mu}{\sigma\sqrt{(2)}})) \tag{2}$$

with $\mathrm{erf}(X)$ the Gaussian error function derived from integral calculus and approximated by Maclaurin's formula.

Algorithm 2. Extended RBM algorithm

Require: $N \geq 1$, the number of observations, m and n the respective number of visible and hidden units
  Input: $(v_{(1)}, \ldots, v_{(N)})$
  Initialization: Set $\Delta a_i = \Delta b_j = \Delta w_{i,j} \sim U(0,1)$ for i=1,...,m and j=1,...,n
  for $p = 1$ to m **do** ▷ with (V_1,...,V_m) the visible variables
    append $\mu \leftarrow \overline{V_p}$
    append $\sigma \leftarrow \sigma_{V_p}$
  end for
    for v from $v_{(1)}$ to $v_{(N)}$ **do** ▷ iteration on observations
      $v^0 \leftarrow v$
      for $t = 1$ to k **do**
        for $j = 1$ to n **do** ▷ iteration on hidden elements
          compute $p(h_j^t)$=$\mathbb{P}(h_j^t|v^{t-1})$
          sample the jth hidden unit state: $h_j^t \sim \mathcal{B}(p(h_j^t))$
        end for
        for $i = 1$ to m **do** ▷ iteration on visible elements
          compute $p(v_i^t)$=$\mathbb{P}(v_i^t|h^t)$
          if $v_i \in \{0,1\}$ **then**
            sample the ith visible unit state: $v_i^t \sim \mathcal{B}(p(v_i^t))$
          else if $v_i \notin \{0,1\}$ **then**
            sample the ith visible unit state: $v_i^t \sim \mathcal{N}(p(v_i^t), \mu_i, \sigma_i)$
          end if
            end for
          end for
          for $i = 1$ to m **do**
            for $j = 1$ to n **do**
              $\Delta a_i \leftarrow \Delta a_i + v_i^k - v_i^0$
              $\Delta b_j \leftarrow \Delta b_j + \mathbb{P}(h_j = 1|v^k) - \mathbb{P}(h_j = 1|v^0)$
              $\Delta w_{i,j} \leftarrow \Delta w_{i,j} + \mathbb{P}(h_j = 1|v^k) \times v_i^k - \mathbb{P}(h_j = 1|v^0) \times v_i^0$
            end for
          end for
        end for
        Output: Gradient estimation Δa_i,Δb_j and $\Delta w_{i,j}$

By reversing this formula, we obtain the quantile function of the normal distribution as:

$$Z = Z(\mu, \sigma, \text{erf}^{-1}) \tag{3}$$

The quantile Z is a function of μ and σ, i.e. the mean and variance of the initial distribution of the continuous quantitative variable, as well as the inverse of the Gaussian error function. We therefore deduce that, to generate new data, we need to calculate the mean, variance and a probability of occurrence that allows us to **generate a number**. This probability of occurrence is derived from the calculation of Gibbs sampling, denoted $p(v_i^t)$ in the algorithmic representation.

4 Experimental Results

4.1 Quality Indicators

As we saw in the previous section, the modifications we are making to RBM don't make it any more complex to understand, but they do increase the potential range of applications. Nevertheless, it remains necessary to ensure that the data generated by Algorithm 2 is of good quality meaning that it is close to the original data.

To this end, we propose two ways to evaluate the quality of data generated: globally, by studying all data variables at the same time, and locally, by studying each variable independently of the others.

4.2 Global Study Results

We assess the global quality of the data generated by using the GAN systems methodology, along the lines of the functions available in the SDV library. The Python codes and datasets are available on the githbub https://github.com/Lancelot-T/Extended_RBM.git. The idea is that if the synthetic data is close to the original one, the controller cannot discriminate them by confusing original and sythetic data, and hence, its performance in accuracy is low. The datasets used all come from the UCI repository [14].

For each generative model, we build a database composed of synthetic and real data, with a binary target variable taking the value 1 when the data is synthetic and 0 otherwise. This database is then divided into two sub-populations using a 70/30 split, with a training sample on one side and a test sample on the other (see Fig. 2). We chose to calibrate a logistic regression as a controller, yet it learns a simple model, because we did not wish to introduce biases linked to the unavoidable optimisation of hyperparameters in alternative machine learning algorithms. Finally, we apply logistic regression to the test sample and calculate the accuracy rate using a default quantile of 50 percent. These values are reported in the next section.

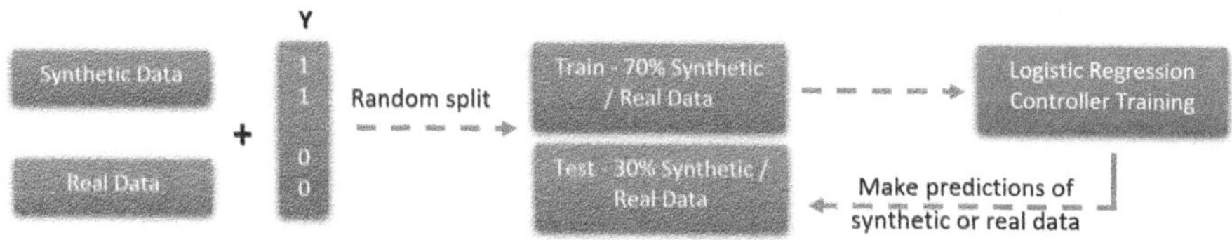

Fig. 2. Synthetic data quality GAN methodology

We divide the datasets in two categories: small datasets (with a number of variables between 5 and 33 and a number of observations between 150 and 1000), and large datasets (with a number of variables between 9 and 167 and a number of observations between 4227 and 178 081).

Table 1 gives the results we obtained on 10 datasets considered to be small. We can see that the Extended RBM we have proposed deceive the controller in 7 cases out of 10 comparing with other generative models, with accuracy rates averaging around 42%, compared with 60% for the Gaussian copula, which still has a good ability to deceive, with 2 wins out of 10.

Table 1. Comparisons of the quality of the obtained results by different methods for small datasets - The lower the accuracy rate, the closer the synthetic date are to real data - The lowest accuracy rate in bold indicates the best performance.

Means of accuracy rate calculated from 100 simulations for each dataset					
Dataset	RBM	Gaussian Copula	CTGAN	TVAE	Copula-GAN
Iris	**45 ± 1.9**	65 ± 5	85.8 ± 3.4	54.4 ± 4.4	72.5 ± 0.04
Wine	**36.5 ± 0.8**	52.7 ± 4.5	77 ± 3.34	64.4 ± 3.4	91.3 ± 1.8
Cancer	93 ± 4.2	**62 ± 2.2**	98 ± 0.70	71.2 ± 2.2	97 ± 0.85
FedEco	**42 ± 5.2**	61 ± 3.1	94 ± 1.3	71 ± 2.5	97.3 ± 1.1
Student	**45 ± 13.6**	100 ± 0	86 ± 1.9	98.5 ± 0.47	97.8 ± 0.2
Heart	**43.2 ± 0.8**	63.8 ± 2.6	77.4 ± 2.9	74.7 ± 2.1	71 ± 2.6
AutoMPG	**44.5 ± 1**	53.2 ± 3.2	92.7 ± 1.5	65.9 ± 2.3	73.6 ± 3.1
StatLog	97.6 ± 0	**47.2 ± 1.9**	75 ± 1.2	89 ± 0.8	70 ± 1.4
Ecoli	85 ± 2.4	100 ± 0	84 ± 2.1	**60 ± 2.4**	100 ± 0
Breast	**43.6 ± 1.2**	46.9 ± 3.9	52.4 ± 3.9	79.4 ± 1.9	55.6 ± 3.5

Table 2. Comparisons of the quality of the obtained results by different methods for large datasets - The lower the accuracy rate, the closer the synthetic date are to real data - The lowest accuracy rate in bold indicates the best performance.

Means of accuracy rate calculated from 10 simulations for each dataset					
Dataset	RBM	Gaussian Copula	CTGAN	TVAE	CopulaGAN
Bank	**48.6 ± 0.1**	77 ± 0.38	62 ± 0.4	85 ± 0.28	62 ± 0.3
El Nino	**49.8 ± 0.005**	62 ± 0.11	62 ± 0.09	76 ± 0.08	86 ± 0.92
Euronext	**50.9 ± 3.8**	54 ± 0.6	82 ± 0.67	76 ± 0.08	65 ± 0.13
H.California	**49.5 ± 0.08**	91 ± 0.3	91 ± 0.3	91 ± 0.3	91 ± 0.3
Marketing	96.2 ± 1	79 ± 0.2	**62 ± 0.24**	77 ± 0.1	63 ± 0.1
Shoppers	97.2 ± 1.1	91 ± 0.24	**63 ± 0.3**	86.9 ± 0.14	84.7 ± 0.3
Musk	100 ± 0	100 ± 0	95 ± 0.4	**71 ± 0.65**	100 ± 0
Spam	91.7 ± 0.12	99 ± 0.06	**78 ± 0.5**	90 ± 0.3	80.6 ± 0.5
AirQuality	**49.3 ± 0.11**	95 ± 0.3	58 ± 0.5	61.2 ± 0.67	95.3 ± 0.13
Adult	**51.5 ± 1.9**	87.6 ± 0.2	68 ± 0.2	70.1 ± 0.2	65.6 ± 0.2

Table 2, for its part, gives the results we obtained on 10 datasets considered to be large. We note a proportion of victories for the Extended RBM in terms of deception substantially equal to that calculated on small samples with 6 victories out of 10 study cases. It also appears that generative models of the GAN type seem to work better on large samples than on small ones.

4.3 Local Study Results

Drawing on the work of Dankar et al. (2022) [4], we propose to carry out variable-level analyses to study more precisely the statistical properties of real and synthetic data according to these two generative models on this database.

More specifically, 3 indicators are considered. The fidelity of synthetic attribute values to real values can be assessed using the Kullback-Leibler divergence [11]. The bivariate dependency structure of real and synthetic data is assessed by Kendall's tau [9], as this statistic is constructed from observation ranks. Finally, the shape of the distributions of real and synthetic data is analyzed by means of the Kolmogorov-Smirnov test [15], slightly modified by replacing the maximum of the distances by the mean of the distances, which is more in line with the desired objective.

The results presented in Table 3 highlight the ability of RBM to retain the statistical properties of real data for synthetic data. More specifically, the results obtained at the global level with the Kolmogorov-Smirnov and Kullback-Leibler statistics are weak for both generative models, but more particularly for the RBM. However, the Kendall's tau values obtained by the Gaussian Copula are more contrasted, showing an alteration in the dependency structure of the original database. On the other hand, this observation is not verified for the RBM,

Table 3. Comparisons of the quality of the obtained results by variables belonging to Dataset Breast - Best performances in bold.

Means of indicators calculated from 100 simulations for each feature of dataset Breast						
Variables	RBM τ	CG τ	RBM KS	CG KS	RBM KL	CG KL
Age	**0.98**	−0.009	**0.008**	0.125	**0.01**	0.01
Menopause	**0.91**	−0.002	0.036	**0.012**	0.18	**0.00**
TumorSize	**0.98**	−0.008	**0.021**	0.162	**0.00**	0.17
InvNodes	**0.98**	−0.004	**0.00**	0.03	**0.00**	0.01
NodeCaps	**0.99**	0.002	**0.00**	0.036	**0.00**	0.00
DegMalig	**0.84**	0.002	**0.138**	0.17	0.00	**0.00**
Breast	**1**	0.003	**0.00**	0.02	**0.00**	0.00
BreastQuad	**0.99**	0.002	**0.00**	0.055	**0.00**	0.01
Irradiat	**1**	−0.006	**0.00**	0.017	**0.00**	0.00
Class	**1**	0.002	**0.00**	0.21	**0.00**	0.00

whose Kendall's tau values are close to 1 for all variables, underlining its good behavior in relation to the dependency structure present in the original database.

We have carried out studies on the probability densities of real and synthetic data for each of the dataset variables. Figure 3 shows the results obtained for the *AutoMPG* dataset. The distributions represented in the first two columns are derived from the first six real variables of the original dataset. The third and fourth columns correspond to the distributions of these same variables according to the data generated by RBM.

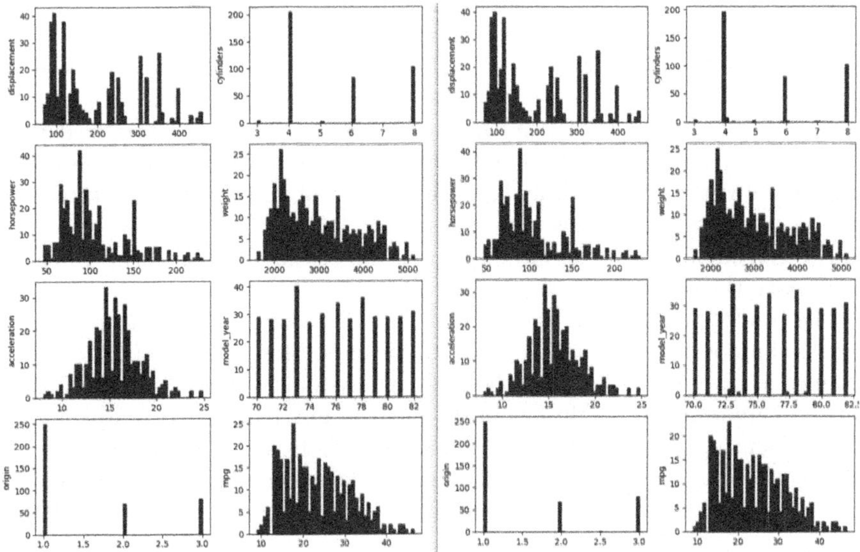

Fig. 3. Two dimensions analysis of AutoMPG dataset variables between real and synthetic data

The results show a high degree of similarity between the probability densities derived from real variables and those derived from synthetic variables. Furthermore, the specificities of the distributions are correctly taken into account, whether we're dealing with a continuous or a discrete quantitative variable. More, the proportions of extreme or outlier values present in the initial data are faithfully reproduced. It should also be noted that the satisfactory levels of Kendall's tau, combined with this representation of univariate distributions, reinforce the empirical evidence to guarantee adequate conservation of the initial dependency structure.

Figure 4 shows a three dimensional analysis of the distributions of several continuous quantitative variables in the *Wine* dataset. The first column shows a representation of the real data, while the second highlights the version of the data generated by RBM. Finally, the third column illustrates the result obtained on data from the generative Gaussian Copula model.

These results highlight the ability of generative models to preserve the dependency structure shown by a three-dimensional approach, and in particular for the RBM. More precisely, the different values of the axes used to construct the 3D plane, derived from the RBM, appear identical to those present on the axes of the 3D plane of the real data, which is not entirely the case for the 3D plane constructed from the data generated by the Gaussian Copula.

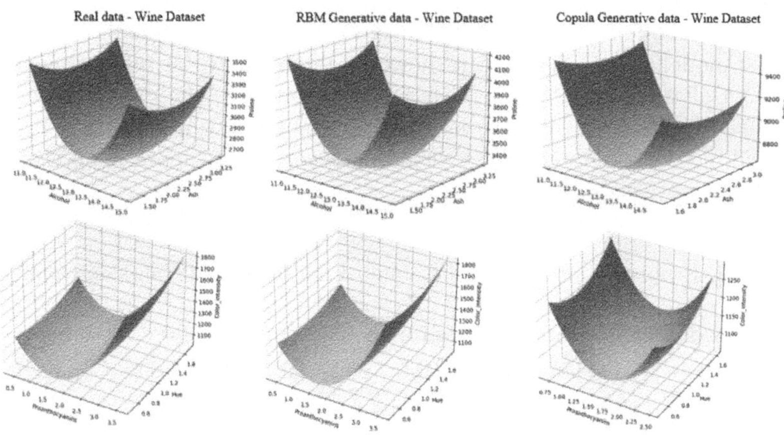

Fig. 4. Three dimensions analysis of certain variables belonging to the Wine dataset

On the other hand, the representations in the second row of Fig. 4 show an identical spatial object between the real data and those provided by RBM, but significantly different for the data generated by the Gaussian Copula. We can therefore reasonably conclude that the dependency structure of the initial data in the *Wine* dataset is better preserved by the RBM than by the Gaussian Copula for this dataset.

4.4 Auto-Generation Synthetic Data

We also investigated the ability of both the Extended RBM and the Gaussian Copula to faithfully generate data from synthetic data. Specifically, we used both models to generate synthetic data for each of the datasets presented in Table 1, and then generated new synthetic data, using the same models, from the first synthetic data. We then applied the strategy, illustrated in Fig. 3, of evaluating the overall quality of the synthetic data from the first synthetic data generation to ensure that the data obtained from the real distributions keep a stable dependency structure. The results are presented in Table 4.

We see a stability of the capacity to the RBM to generating new data from already synthetic distributions. Indeed, if we consider logistic regression as a controller, we can see that, on average, it is always as much deceived when it try to distinguish real data from synthetic data (Table 1) as when it wishes to distinguish first generation synthetic data from those second generation (Table 4). On the other hand, the results produced with the Gaussian Copula as a generative model are more contrasted, notably with a precision rate of 100% for the datasets *Wine* and *FedEco* which can be explained by the presence of more random dependence structure within these synthetic data compared to the initial data. Indeed, if the data input to a generative model does not follow a realistic dependence structure and therefore follows purely random processes, data generation methodologies cannot produce observations close enough to fool the controller model. These results are however intended to be reassuring for the Extended Boltzmann Machine which seems to preserve the dependence structure of real data in the synthetic data that it generates, which opens a vast field of possibilities for the creation of scenarios according to more deep statistical properties.

Table 4. Evaluation of sequential stability from RBM and Gaussian Copula

Means of accuracy rate calculated from 100 simulations for each dataset				
Dataset	Gaussian Copula	RBM	δ RBM	δ Gaussian Copula
Iris	41.4 ± 1.8	47.9 ± 4.6	−8%	−26%
Wine	36.3 ± 0.9	100 ± 0	−0.5%	+89.7%
Cancer	98.3 ± 1.2	100 ± 0	+5.7%	+61.3%
FedEco	46.3 ± 5.6	100 ± 0	+10.2%	+63.9%
Student	41.6 ± 7.2	59 ± 3.2	−7.5%	−41%
Heart	43.2 ± 1.3	49.5 ± 3.1	0%	−22.4%
AutoMPG	45.7 ± 0.8	50 ± 2.8	+2.7%	−6%
StatLog	96.7 ± 0	52.4 ± 2	−0.9%	+11%
Ecoli	60 ± 2	51.5 ± 3.3	−29%	−48.5%
Breast	43.6 ± 1.2	48.2 ± 3.5	0%	+2.7%

5 Conclusion

To conclude, the first results obtained with the Extended Boltzmann Machine, are promising and open the potential use of RBMs to process non binary data. Indeed, the modifications to the RBM improve its generalization capacity in a multivariate context, while preserving certain dependency structures present in the initial data. However, it is necessary to remain cautious despite its performance, as the Extended Boltzmann Machine has some limits that require further analysis. Indeed, assumptions made in development, such as the order of qualitative variables during their encoding, the initialization of weights and biases according to a uniform distribution, or the use of a Gaussian generator sometimes inadequate with the empirical characteristics of the data, need to be more fully studied to demonstrate the stability of the results produced. In addition, the Extended Boltzmann Machine relies on several hyperparameters, including the number of neurons in the hidden layer and the number of Gibbs samples made for each observation, which require more in-depth analyses that should be conducted in a controlled experimental setting, then on real data, in order to better understand the behavior of the algorithm and demonstrate the relevance of the values we used. In addition, limitations related to computation times have also been noted, especially when used on large databases, which constitutes a barrier to its application. We therefore plan to continue our research by carrying out new studies in order to validate the technical and statistical assumptions to propose a modified version of RBMs, capable of deal with complex issues in various professional sectors.

References

1. Bidaurrazaga, A., Pérez, A., Santana, R.: Structural restricted Boltzmann machine for image denoising and classification (2023)
2. Borisov, V., Broelemann, K., Kasneci, E., Kasneci, G.: DeepTLF: robust deep neural networks for heterogeneous tabular data. Int. J. Data Sci. Anal. **16**(1), 85–100 (2023)
3. Chow, C., Liu, C.: Approximating discrete probability distributions with dependence trees. IEEE Trans. Inf. Theory **14**(3), 462–467 (1968)
4. Dankar, F.K., Ibrahim, M.K., Ismail, L.: A multi-dimensional evaluation of synthetic data generators. IEEE Access **10**, 11147–11158 (2022)
5. Fischer, A., Igel, C.: An introduction to restricted Boltzmann machines. In: Alvarez, L., Mejail, M., Gomez, L., Jacobo, J. (eds.) CIARP 2012. LNCS, vol. 7441, pp. 14–36. Springer, Heidelberg (2012). https://doi.org/10.1007/978-3-642-33275-3_2
6. Fischer, A., Igel, C.: Training restricted Boltzmann machines: an introduction. Pattern Recogn. **47**(1), 25–39 (2014)
7. Hinton, G.E.: Training products of experts by minimizing contrastive divergence. Neural Comput. **14**(8), 1771–1800 (2002)
8. Hinton, G.E., Sejnowski, T.J.: Optimal perceptual inference. In: The IEEE Conference on Computer Vision and Pattern Recognition, vol. 448, pp. 448–453. Citeseer (1983)
9. Knight, W.R.: A computer method for calculating Kendall's tau with ungrouped data. J. Am. Stat. Assoc. **61**(314), 436–439 (1966)
10. Kondratyev, A., Schwarz, C.: The market generator. Available at SSRN 3384948 (2019)
11. Kullback, S., Leibler, R.: On information and sufficiency. Ann. Math. Stat. **22**(1), 79–86 (1951)
12. Lailler, A., Cohen, S.: Improving market data generation with restricted Boltzmann machines. Available at SSRN 4020037 (2022)
13. Lezmi, E., Roche, J., Roncalli, T., Xu, J.: Improving the robustness of trading strategy backtesting with Boltzmann machines and generative adversarial networks. arXiv preprint arXiv:2007.04838 (2020)

14. Lichman, M.: UCI Machine Learning Repository. University of California, Irvine, School of Information and Computer Sciences (2013). UCI Machine Learning Repository. http://archive.ics.uci.edu/ml. http://archive.ics.uci.edu/ml
15. Lilliefors, H.W.: On the Kolmogorov-Smirnov test for normality with mean and variance unknown. J. Am. Stat. Assoc. **62**(318), 399–402 (1967)
16. Lin, W.C., Tsai, C.F.: Missing value imputation: a review and analysis of the literature (2006–2017). Artif. Intell. Rev. **53**, 1487–1509 (2020)
17. Miao, J., Wang, J., Zhang, D., Miao, Q.: Improved generative adversarial network for rotating component fault diagnosis in scenarios with extremely limited data. IEEE Trans. Instrum. Meas. **71**, 1–13 (2021)
18. Montanez, A., et al.: SDV: an open source library for synthetic data generation. Ph.D. thesis, Massachusetts Institute of Technology (2018)
19. Patki, N., Wedge, R., Veeramachaneni, K.: The synthetic data vault. In: IEEE International Conference on Data Science and Advanced Analytics (DSAA), pp. 399–410 (2016). https://doi.org/10.1109/DSAA.2016.49
20. Sanyal, A., Kusner, M., Gascon, A., Kanade, V.: TAPAS: tricks to accelerate (encrypted) prediction as a service. In: The 35th International Conference on Machine Learning, pp. 4490–4499. PMLR (2018)
21. Sobieszczanski-Sobieski, J.: Overcoming the Bellman's curse of dimensionality in large optimization problems. In: Pan American Congress of Applied Mechanics. No. NAS 1.15: 102662 (1990)
22. Xu, X., Yoneda, M.: Multitask air-quality prediction based on LSTM-autoencoder model. IEEE Trans. Cybern. **51**(5), 2577–2586 (2019)

Compact Solution Representation in Qualitative Constraint-Based Reasoning

Yakoub Salhi(✉) and Michael Sioutis(✉)

LIRMM UMR 5506, Université de Montpellier & CNRS, Montpellier, France
{yakoub.salhi,michael.Sioutis}@lirmm.fr

Abstract. In the framework of Qualitative Spatio-Temporal Reasoning (QSTR), we can consider constraints like x {*is above* $\vee$ *is under*} y, and combinations thereof, to represent and reason about spatial or temporal information in an intuitive, human-like way. QSTR becomes particularly important in view of possible lack, uncertainty, and/or imperfection of metric data, as treating such quantitative information qualitatively would provide more leeway to perform sound reasoning. Adding to the usefulness of QSTR, in this short paper, we introduce the notion of *multi-scenario* for representing solutions of networks of qualitative spatio-temporal constraints in a compact manner, as a means to assessing and enhancing the explainability and robustness of AI systems that involve spatio-temporal information. Further, we prove certain theoretical properties pertaining to this novel notion, and even introduce some robustness measures relating to our notion of multi-scenario.

Keywords: Qualitative Constraints · Spatial and Temporal Reasoning · Solution Representation · Explainability · Robustness

1 Introduction

Representing and reasoning about spatial or temporal information in an intuitive, human-like manner, is the area of study of Qualitative Spatio-Temporal Reasoning (QSTR), a prolific symbolic AI framework that spans various fields, such as knowledge representation and reasoning, logic, cognition, and mathematics [5,11]. As illustration, within the QSTR framework, we can consider a relation like x {*is above* $\vee$ *is under*} y, which does not necessitate providing any quantitative information, yet reads rather natural. Such intuitive relations, and combinations thereof, can be modeled as a qualitative constraint network (QCN), a simplified example of which is provided in Fig. 1. Here, we introduce a concept for representing qualitative solutions, i.e., *scenarios*, of QCNs in a compact manner, as a means to assessing and enhancing the explainability and robustness of AI systems that involve QSTR.

S. Destercke et al. (Eds.): SUM 2024, LNAI 15350, pp. 421–428, 2025.
https://doi.org/10.1007/978-3-031-76235-2_31

Context and Motivation

We position our study here in the context of *implicants*, as they have been introduced and used in propositional logic [13,14]; specifically, an implicant can be viewed as a partial model of a propositional formula, every possible full extension of which becomes a (complete) model of the formula. Further, an implicant is said to be *prime* if none of its proper subsets is an implicant. Prime implicants are important in various domains, for example, in knowledge compilation [2,4,19] and diagnosis [9,18], and most recently in explainable AI via the compilation of ML classifiers into Boolean circuits [3,6–8,20]. In addition to explainability, prime implicants can convey information about the robustness of an AI system, in the sense that they serve as sufficient knowledge for drawing a conclusion, and their size relates to the robustness (tolerance to perturbation) of that conclusion [22,23]. Motivated by the explainability and robustness aspects of prime implicants, we aim to identify a similar concept in QSTR. For example, we can consider ML classifiers that are compiled into QCNs, much like as in the ongoing research involving Boolean circuits that we referenced earlier. In this case, the set of scenarios of a QCN corresponds to the set of positive decisions, and a QCN-based prime implicant notion could be used to explain and assess the robustness of these decisions.

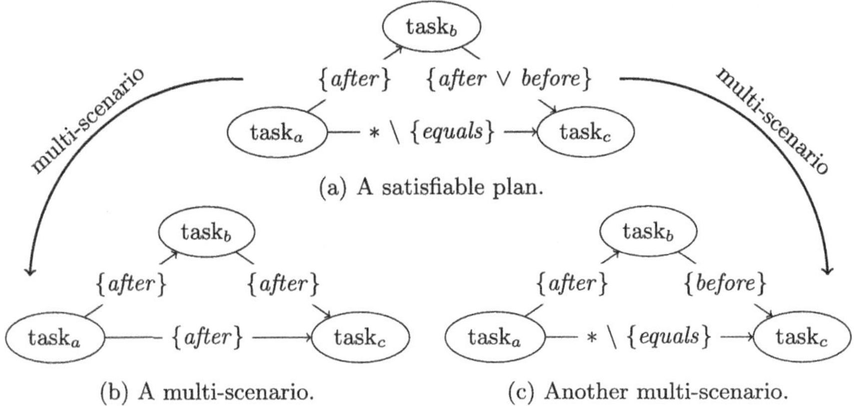

Fig. 1. Extracting multi-scenarios from a simplified qualitative constraint network (QCN) represening a temporal plan, where $*$ denotes any possible relation; in a multi-scenario *any* combination of base relations is a valid configuration, and hence each multi-scenario is weakly globally consistent (see Definition 3), but not vice versa, e.g., consider the original QCN.

Related Work and Contribution

So far, [17] presents the sole QCN-based prime implicant notion for use with qualitative spatio-temporal information, called *prime scenario*. Specifically, a prime scenario is defined as a minimal set of constraints in a given QCN that can only yield scenarios of that QCN. For example, let us consider Fig. 1, and the QCN in Fig. 1a there in particular. Then, the set {task$_a$ {*after*} task$_b$, task$_b$ {*after*} task$_c$} corresponds to a prime scenario of the QCN, as we can entail task$_a$ {*after*} task$_c$ from this set and recover a scenario of the QCN. However, the set {task$_a$ {*after*} task$_b$, task$_b$ {*before*} task$_c$} does not correspond to a prime scenario of the QCN, as it can lead to a configuration that is not permitted, viz., task$_a$ {*equals*} task$_c$; in this case, the reader can easily see that we need a prime scenario for every scenario of the QCN. Further, to recover scenarios of the original QCN some reasoning has to be performed, which can be a hindrance to time-critical AI systems. Different to the concept of prime scenario, we introduce the notion of *multi-scenario*, a generalized scenario where every possible combination of base relations leads to a (classical) scenario of a given QCN. Interestingly, a multi-scenario of a QCN forms a (minimal and) weakly globally consistent subQCN of the former [10] (see also Definition 3), with the added benefit of producing scenarios in a completely straightforward manner. As illustration, let us revisit Fig. 1 and observe the QCN and its multi-scenarios there. Indeed, it can be seen that we can cover the entire scenario space of the original QCN with two multi-scenarios. Relating to our motivation discussion earlier, a multi-scenario can provide a sufficient explanation for a series of ML decisions compiled into a QCN, e.g., decisions pertaining to generating temporal plans as in the figure, by capturing one of the largest sets of such plans in terms of set inclusion. Thus, we think that multi-scenarios can be considered as the suitable counterparts to prime implicants.

2 Preliminaries

A binary qualitative spatial or temporal constraint language is based on a finite set B of *jointly exhaustive and pairwise disjoint* relations, called *base relations* [11] and defined over an infinite domain D (e.g., $\mathbb{R}$).[1] The set B can be used to represent the definite knowledge between any two entities of D (e.g., for $\mathsf{D} = \mathbb{R}$, we could have $\mathsf{B} = \{$*less than*, *equals*, *greater than*$\}$, with the usual semantics). The set B contains the identity relation Id, and is closed under the *converse* operation ($^{-1}$). Indefinite knowledge can be specified by a union of possible base relations, and is represented by the set containing them. Hence, 2^{B} represents the total set of relations. The set 2^{B} is equipped with the usual set-theoretic operations of union and intersection, the converse operation, and the *weak composition* operation denoted by the symbol $\diamond$ [11]. For all $r \in 2^{\mathsf{B}}$, $r^{-1} = \bigcup\{b^{-1} \mid b \in r\}$. The weak composition ($\diamond$) of two base relations $b, b' \in \mathsf{B}$ is defined as the smallest (i.e., most restrictive) relation $r \in 2^{\mathsf{B}}$ that includes $b \circ b'$, or, formally, $b \diamond b' = \{b'' \in \mathsf{B} \mid b'' \cap (b \circ b') \neq \emptyset\}$, where $b \circ b' = \{(x, y) \in \mathsf{D} \times \mathsf{D} \mid \exists z \in \mathsf{D}$ such that $(x, z) \in b \wedge (z, y) \in b'\}$ is the (true) composition of b and b'. Finally, for all $r, r' \in 2^{\mathsf{B}}$, $r \diamond r' = \bigcup\{b \diamond b' \mid b \in r,\ b' \in r'\}$.

[1] Formally, we have that $\bigcup\{b \in \mathsf{B}\} = \mathsf{D} \times \mathsf{D}$ (jointly exhaustive), and $\forall b, b' \in \mathsf{B}$ such that $b \neq b'$, we have that $b \cap b' = \emptyset$ (pairwise disjoint).

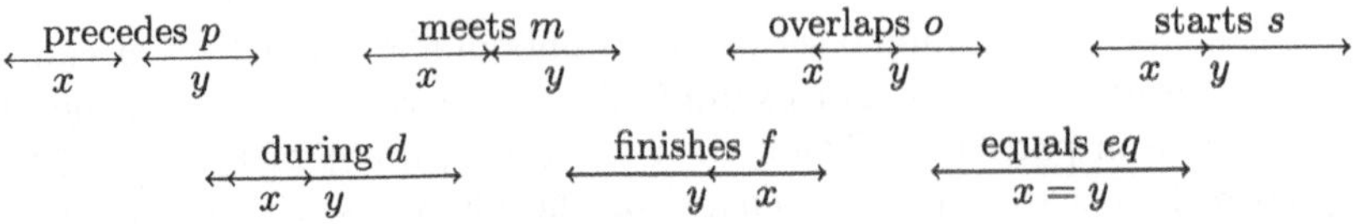

Fig. 2. A representation of the 13 base relations b of IA, each one relating two potential intervals x and y as in $x\ b\ y$; the converse of b, i.e., b^{-1}, can be denoted by bi and is omitted in the figure.

As illustration, consider the well-known qualitative temporal constraint language of Interval Algebra (IA), introduced by Allen [1]. IA considers time intervals on the real line, and the set of base relations $\mathsf{B} = \{eq\ (= \mathsf{Id}),\ p,\ pi,\ m,\ mi,\ o,\ oi,\ s,\ si,\ d,\ di,\ f,\ fi\}$ to encode knowledge about temporal relations between such intervals, as described in Fig. 2. As another example, the Region Connection Calculus [15] considers spatial regions and topological relations between them.

To represent and reason about qualitative spatio-temporal information, we can use (among other ways) a *qualitative constraint network (QCN)*, a simplified example of which is shown in Fig. 1; we recall the following definition:

Definition 1. *A Qualitative Constraint Network (QCN) is a tuple* (V, C) *where:*

- $V = \{v_1, \ldots, v_n\}$ *is a finite set of variables over some infinite domain* D *(e.g., time points or 2D regions);*
- *and* C *is a mapping* $V \times V \rightarrow 2^{\mathsf{B}}$ *s.t.,* $\forall v \in V$, $C(v, v) = \{\mathsf{Id}\}$, *and,* $\forall v, v' \in V$, $C(v, v') = (C(v', v))^{-1}$.

A QCN $\mathcal{N} = (V, C)$ is *trivially inconsistent* iff $\exists v, v' \in V$ s.t. $C(v, v') = \emptyset$.

A *solution* of a QCN $\mathcal{N} = (V, C)$ is a mapping $f : V \rightarrow \mathsf{D}$ s.t. $\forall v, v' \in V$, $\exists b \in C(v, v')$ s.t. $(f(v),\ f(v')) \in b$; $\mathcal{N}$ is said to be *consistent* iff it admits a solution.

A *subQCN* (refinement) $\mathcal{N}'$ of $\mathcal{N}$, denoted by $\mathcal{N}' \subseteq \mathcal{N}$, is a QCN (V, C') such that, $\forall u, v \in V$, $C'(u, v) \subseteq C(u, v)$.

A QCN $\mathcal{N} = (V, C)$ is *atomic* iff $\forall v, v' \in V$, $C(v, v') = \{b\}$ with $b \in \mathsf{B}$.

A *scenario* of a QCN $\mathcal{N} = (V, C)$ is a consistent atomic subQCN $\mathcal{S}$ of $\mathcal{N}$, and a *partial scenario* of $\mathcal{N}$ on $V' \subseteq V$ is a scenario restricted to constraints involving only V'.

We recall the definition of $\diamond$-consistency [11,16], which entails consistency for all variable triples in a QCN.

Definition 2. *Given a QCN* $\mathcal{N} = (V, C)$, $\mathcal{N}$ *is* $\diamond$-consistent, *iff,* $\forall \{v_i, v_j, v_k\} \in V$, $C(v_i, v_j) \subseteq C(v_i, v_k) \diamond C(v_k, v_j)$.

Finally, the following property from [10] will be useful in the sequel.

Definition 3. *A QCN* $\mathcal{N} = (V, C)$ *is* weakly globally consistent *iff, for any* $V' \subset V$, *every partial scenario of* $\mathcal{N}$ *on* V' *can be extended to a partial scenario of* $\mathcal{N}$ *on* $V' \cup \{v\} \subseteq V$, *for any* $v \in V \backslash V'$.

3 Multi-scenarios

In this section, we introduce the novel notion of this work, namely, that of *multi-scenario* of a QCN. In short, a multi-scenario in itself is a QCN for which every possible combination of base relations leads to a scenario of it. Thus, we argue that multi-scenarios can be considered as the suitable counterparts to prime implicants in propositional logic [13,14]. It may be tempting to consider a *direct* application of prime implicant computation approaches to our context, but there are fundamental dinstictions between the two frameworks that make such a feat impossible. First of all, QSTR deals with binary relations between variables, whilst propositional logic with truth values of variables. Yet, we can address this issue by associating a propositional variable with each base relation and each pair of variables of a QCN; in fact, this is typical in a SAT encoding [12,21]. However, we would be still missing the crucial semantics of QSTR, as, for example, for three qualitative constraints x $\{after\}$ y, y $\{after\}$ z, and x $\{before \vee after\}$ z, a prime implicant for the respective SAT encoding $p_{xy}^{after} \wedge p_{yz}^{after} \wedge (p_{xz}^{before} \vee p_{xz}^{after})$ could be the model $\langle p_{xy}^{after} = \text{TRUE}, p_{yz}^{after} = \text{TRUE}, p_{xz}^{before} = \text{TRUE} \rangle$; this is obviously right from the perspective of propositional logic, but at the same time obviously wrong from the perspective of QSTR.

Let us now formally define the notion of multi-scenario of a QCN as follows.

Definition 4 (Multi-Scenario). *A* multi-scenario *of a QCN $\mathcal{N}$ is a subQCN $\mathcal{M}$ of $\mathcal{N}$ s.t. (P) every atomic subQCN of $\mathcal{M}$ is a scenario of $\mathcal{N}$; $\mathcal{M}$ is said to be* maximal *if there is no subQCN $\mathcal{M}'$ of $\mathcal{N}$ that satisfies Property (P) and $\mathcal{M} \subsetneq \mathcal{M}'$.*

We say that a multi-scenario $\mathcal{M}$ of a QCN $\mathcal{N}$ *covers* a scenario $\mathcal{S}$ of $\mathcal{N}$ if $\mathcal{S}$ is a scenario of $\mathcal{M}$.

Example 1 As illustration, a QCN and a set of multi-scenarios that *cover* all of its scenarios (i.e., qualitative solutions) are provided in Fig. 1.

The following proposition is a direct consequence of the definition of multi-scenario:

Proposition 1 (Anti-Monotonicity). *If $\mathcal{M}$ is a multi-scenario of a QCN $\mathcal{N}$, then every not trivially inconsistent subQCN of $\mathcal{M}$ is a multi-scenario.*

An example that illustrates the concept of anti-monotonicity follows.

Example 2 Let us focus on the multi-scenario depicted in Fig. 1c; we can verify that we can obtain another multi-scenario by removing any other base relation(s) from the constraint between task_a and task_c (as long as we keep the subQCN not trivially inconsistent of course).

Further, we can obtain the following complexity result:

Proposition 2. *The problem of deciding whether a subQCN of a given QCN is a multi-scenario of the latter is tractable.*

Proof. Let $\mathcal{N} = (V, C)$ be a QCN and $\mathcal{M} = (V, C')$ a subQCN of $\mathcal{N}$. We know that $\diamond$-consistency is sufficient to check the consistency of an atomic QCN. Furthermore, a subQCN is a multi-scenario of a given QCN if *all* their atomic sub-QCNs are consistent. Consequently, $\mathcal{M}$ is a multi-scenario iff for every $i, j, k \in V$, every $b \in C'(i,k)$, and every $b' \in C'(k,j)$, $C'(i,j) \subseteq b \diamond b'$ holds. This property can be checked using a poly-time procedure, cf. [16].

An example that illustrates the proof of the aforementioned result follows.

Example 3. Let us revisit the multi-scenario depicted in Fig. 1c; we can verify, for instance, that the constraint between task_a and task_c is included in the weak composition of every possible combination of base relations of the remaining two constraints, i.e., $* \setminus \{equals\} \subseteq after \diamond before$ (here, the remaining constraints are defined by singleton relations, so this is also the only possible combination of base relations). Verifying this property for every constraint in a given QCN can be done in cubic time in the number of variables of the QCN, and one can leverage the procedure formalized in [16] for this purpose.

Next, we can obtain the following result due to Propositions 1 and 2:

Proposition 3. *The problem of deciding whether a multi-scenario of a given QCN is maximal is tractable.*

Finally, by the definition of multi-scenario we obtain the following result, which shows how this notion relates to weak global consistency:

Lemma 1. *Every multi-scenario is weakly globally consistent and, hence, also $\diamond$-consistent.*

4 Robustness Measures

In this section, we introduce some robustness measures pertaining to our notion of multi-scenario. In what follows, $\mathsf{MMS}(\mathcal{N})$ denotes the set of maximal multi-scenarios (MMSs) of a QCN $\mathcal{N}$, $\mathsf{Scenarios}(\mathcal{N})$ the set of scenarios of $\mathcal{N}$, $|\mathcal{N}|$ the size of $\mathcal{N}$, i.e., its number of base relations, and $\mathcal{N}_\top$ the *universal QCN*, where every constraint is universal (i.e., $= \mathsf{B}$).

- $R_1(\mathcal{N}) = \max\{|\mathcal{M}| \mid \mathcal{M} \in \mathsf{MMS}(\mathcal{N})\}$
- $R_2(\mathcal{N}) = \max\{|\mathsf{Scenarios}(\mathcal{M})| \mid \mathcal{M} \in \mathsf{MMS}(\mathcal{N})\}$

Intuitively, our measures relate a reduction of conflicts among base relations of a QCN to an increase of its robustness. We can then prove the following result that establishes some well-behaveness properties for our measures:

Proposition 4. *The following properties are satisfied for R_i, where $i \in \{1, 2\}$:*

1. *for any inconsistent QCN $\mathcal{N}$, $R_i(\mathcal{N}) = 0$;*
2. *for any (consistent) QCN $\mathcal{N}$, $R_i(\mathcal{N}) \leq R_i(\mathcal{N}_\top)$;*
3. *for all two QCNs $\mathcal{N}$ and $\mathcal{N}'$ with $\mathsf{Scenarios}(\mathcal{N}) = \mathsf{Scenarios}(\mathcal{N}')$, $R_i(\mathcal{N}) = R_i(\mathcal{N}')$;*
4. *for all two QCNs $\mathcal{N}$ and $\mathcal{N}'$ with $\mathsf{Scenarios}(\mathcal{N}) \subseteq \mathsf{Scenarios}(\mathcal{N}')$, $R_i(\mathcal{N}) \leq R_i(\mathcal{N}')$.*

5 Conclusion and Future Work

In this short paper, we proposed and studied the notion of *multi-scenario* for representing solutions of networks of qualitative constraints in a compact manner, as a way to assessing and enhancing the explainability and robustness of AI systems that involve spatio-temporal information.

Future work consists of implementing and evaluating methods for identifying multi-scenarios in networks of qualitative constraints, developing a target compilation language based on the notion of multi-scenario, and controlling certain permissible combinations between base relations of a multi-scenario.

Acknowledgement. The work was partially funded by the Agence Nationale de la Recherche (ANR) for the "Hybrid AI" project that is tied to the chair of Dr. Sioutis, and the I-SITE program of excellence of Université de Montpellier that complements the ANR funding.

Disclosure of Interests. The authors have no competing interests to declare that are relevant to the content of this article.

References

1. Allen, J.F.: Maintaining knowledge about temporal intervals. Commun. ACM **26**, 832–843 (1983)
2. Cadoli, M., Donini, F.M.: A survey on knowledge compilation. AI Commun. **10**, 137–150 (1997)
3. Darwiche, A., Hirth, A.: On the (complete) reasons behind decisions. J. Log. Lang. Inf. **32**, 63–88 (2023)
4. Darwiche, A., Marquis, P.: A knowledge compilation map. J. Artif. Intell. Res. **17**, 229–264 (2002)
5. Dylla, F., et al.: A survey of qualitative spatial and temporal calculi: algebraic and computational properties. ACM Comput. Surv. **50**, 7:1–7:39 (2017)
6. Ignatiev, A., Narodytska, N., Marques-Silva, J.: Abduction-based explanations for machine learning models. In: AAAI (2019)
7. Ignatiev, A., Narodytska, N., Marques-Silva, J.: On relating explanations and adversarial examples. In: NeurIPS (2019)
8. Izza, Y., Marques-Silva, J.: On explaining random forests with SAT. In: IJCAI (2021)
9. de Kleer, J., Mackworth, A.K., Reiter, R.: Characterizing diagnoses and systems. Artif. Intell. **56**, 197–222 (1992)
10. Li, S., Long, Z., Liu, W., Duckham, M., Both, A.: On redundant topological constraints. Artif. Intell. **225**, 51–76 (2015)
11. Ligozat, G.: Qualitative Spatial and Temporal Reasoning. Wiley, ISTE (2013)
12. Pham, D.N., Thornton, J., Sattar, A.: Modelling and solving temporal reasoning as propositional satisfiability. Artif. Intell. **172**, 1752–1782 (2008)
13. Quine, W.V.: The problem of simplifying truth functions. Am. Math. Mon. **59**, 521–531 (1952)

14. Quine, W.V.: A way to simplify truth functions. Am. Math. Mon. **62**, 627–631 (1955)
15. Randell, D.A., Cui, Z., Cohn, A.: A spatial logic based on regions and connection. In: KR (1992)
16. Renz, J., Ligozat, G.: Weak composition for qualitative spatial and temporal reasoning. In: CP (2005)
17. Salhi, Y., Sioutis, M.: Prime scenarios in qualitative spatial and temporal reasoning. In: TIME (2023)
18. Schlobach, S., Huang, Z., Cornet, R., van Harmelen, F.: Debugging incoherent terminologies. J. Autom. Reason. **39**, 317–349 (2007)
19. Schrag, R.: Compilation for critically constrained knowledge bases. In: AAAI (1996)
20. Shih, A., Choi, A., Darwiche, A.: A symbolic approach to explaining Bayesian network classifiers. In: IJCAI (2018)
21. Westphal, M., Wölfl, S.: Qualitative CSP, finite CSP, and SAT: comparing methods for qualitative constraint-based reasoning. In: IJCAI (2009)
22. Willot, H., Destercke, S., Belahcène, K.: Explaining robust classification through prime implicants. In: SUM (2022)
23. Willot, H., Destercke, S., Belahcène, K.: Prime implicants as a versatile tool to explain robust classification. In: ISIPTA. In: Proceedings of Machine Learning Research, vol. 215, pp. 461–471 (2023)

Credibility-Limited Revision for Epistemic Spaces

Kai Sauerwald(✉)

FernUniversität in Hagen, Artificial Intelligence Group, 58084 Hagen, Germany
kai.sauerwald@fernuni-hagen.de

Abstract. We consider credibility-limited revision in the framework of belief change for epistemic spaces, permitting inconsistent belief sets and inconsistent beliefs. In this unrestricted setting, the class of credibility-limited revision operators does not include any AGM revision operators. We extend the class of credibility-limited revision operators in a way that all AGM revision operators are included while keeping the original spirit of credibility-limited revision. Extended credibility-limited revision operators are defined axiomatically. A semantic characterization of extended credibility-limited revision operators that employ total preorders on possible worlds is presented.

Keywords: Epistemic Space · Epistemic State · Credibility-Limited Revision · Non-Prioritized Revision · AGM Revision · Extended · Inconsistency

1 Introduction

Much research in belief change theory is on the change of logical theories [8]. A well-known and widely accepted approach for the revision of logical theories is revision by Alchourron, Gärdenfors and Makinson [1] (AGM), which realizes the famous principle of minimal change. Another belief change operation in this setting is credibility-limited revision by Hansson, Fermé, Cantwell and Falappa [13]. This class of operations implements the idea that an (AGM) revision should performed only when the newly arriving information is credible and if the information is not credible, the agent's beliefs are not altered. Intuitively, credibility-limited revision is a generalization of AGM revision; when one considers all potential information as credible, one would expect that a credibility-limited revision *is* an AGM revision.

Apart from the classical setting of theory change, belief change is considered in the more general setting of belief change over epistemic states by Darwiche and Pearl [3,17,20]. In this setting, which has wide applications in iterated belief change [8], one does not only consider the beliefs of an agent but also considers extra logical information that guides the belief change process as part of the representation. To deal with this expressive setting, both above-mentioned kinds

S. Destercke et al. (Eds.): SUM 2024, LNAI 15350, pp. 429–444, 2025.
https://doi.org/10.1007/978-3-031-76235-2_32

of belief changes have been adapted to this setting, i.e., AGM revision by Darwiche and Pearl [3] and credibility limited revision by Booth, Fermé, Konieczny and Pino Pérez [2]. A recent clarification of the Darwiche and Pearl framework is the framework of *belief change for epistemic spaces* [20]. Agents' epistemic states are bound to a specific type of representation, and an epistemic space is an abstraction that describes the whole room of all possible epistemic states of an individual agent. Belief change operators for an epistemic space reside within these representational bounds. We consider what is called here the unrestricted framework of belief change for epistemic spaces, which means that inconsistent beliefs are permitted, these are often neglected but not always [9].

This paper starts with the observation that when using the unrestricted framework of belief changes for epistemic spaces, the given notion of credibility-limited revision *does not* behave very well in the unrestricted case; all AGM revision operators are excluded, and inconsistent belief sets cannot be handled. We deal with this observation by providing the following results, which are also the main contributions[1]:

- [Extended Credibility-Limited Revision] We define *extended credibility-limited revision*, which builds upon credibility-limited revision by Booth et al. [2]. For this, we consider the axiomatic description of credibility-limited revision by Booth et al. and identify one postulate that makes these exclude AGM revision operators and incompatible with inconsistent beliefs. For defining extended credibility-limited revision, we add two postulates to the original postulates by Booth et al. for credibility-limited revision (and remove the postulate which makes them incompatible with AGM revision). The additional postulates ensure that operators are excluded which do not match the intuition of credibility-limited revision.
- [Semantic Characterization] A semantical characterization of extended credibility-limited revision. This characterization is given in terms of functions that assign total preorders to epistemic states, i.e., in the same style as the Darwiche-Pearl representation theorem for revision [3], respectively as in the semantic characterization of credibility-limited revision by Booth et al. [2].
- [Genuineness] We show that extended credibility-limited revisions are a genuine extension of credibility-limited revisions by Booth et al. [2] that include all AGM revision operators.

The paper contains the proofs for all propositions and theorems given here[2]. The next section gives the background on propositional logic and order theory. In Sect. 3, we present epistemic spaces, as well as AGM revision operators for epistemic states [3] and credibility-limited revision operators by Booth et al. [2]. We observe in Sect. 4 that credibility-limited revision for epistemic spaces does not include AGM revisions for epistemic spaces. Section 5 introduces extended

[1] Some of these results are already part of the dissertation thesis by the author [17].

[2] We include only a sketch proof for Theorem 13. The full proof can be found in the arxiv version of this paper https://arxiv.org/abs/2409.07119.

credibility-limited revision and we consider a semantic characterization of this class of operators. An example of extended credibility-limited revision is given in Sect. 6, and we consider some properties of extended credibility-limited revision operators. The last section, Sect. 7, summarises the results presented here.

Before starting with the main content of the paper, we consider some remarks. This paper is mainly developed from a technical perspective, and after the introduction we do *not* focus on discussing applications and implications of the results given here and delegate such a discussion to different paper. From a theoretical perspective, we should be interested in considering belief changes on *arbitrary* epistemic spaces and arbitrary inputs, as we do in this paper. A rationale is that this allows us to study belief change independent of specific representations of epistemic states, respectively, in a way that the results apply to *all* possible representations, including those with inconsistent belief sets. Doing so has the advantage that the theory applies to application scenarios that have not been anticipated. One application of this is employing belief change operators as descriptional theories, which is, in my opinion, a prerequisite for using belief change theory in, e.g., approaches like cognitive logics [16]. In that sense, the purpose of this paper goes beyond just generalizing credibility-limited revision; it exemplifies how to generalise belief change operators to arbitrary inputs and representations.

2 Background

Let Σ be a non-empty finite propositional signature whose elements are called atoms. With $\mathcal{L}$ we denote the set of all propositional formulas over Σ defined as usually using Boolean connectives. We assume that the tautology $\top$ and the falsum $\bot$ are elements of $\mathcal{L}$. The set of all Σ-interpretations is denoted by Ω and we write interpretations as strings of atoms from Σ where an bar over an atom indicates that this atom is mapped to `false` and otherwise to `true`. For instance, the interpretation $\omega = a\bar{b}c$ maps a to `true` and b to `false` and c to `true`. The models relation $\models$ between interpretations and formulas is defined as usually and with $[\![\alpha]\!] = \{\omega \in \Omega \mid \omega \models \alpha\}$ we denote the set of all models of α. We say a formula α logically entails a formula β, written $\alpha \models \beta$, if $[\![\alpha]\!] \subseteq [\![\beta]\!]$ holds. These notions are lifted to sets of formulas $X \subseteq \mathcal{L}$ as usually, i.e., $[\![X]\!] = \bigcap_{\alpha\in X} [\![\alpha]\!]$ and $X \models \alpha$ if $[\![X]\!] \subseteq [\![\alpha]\!]$. We say that $X \subseteq \mathcal{L}$ is deductively closed if $X = \mathrm{Cn}(X)$, whereby $\mathrm{Cn}(X) = \{\alpha \in \mathcal{L} \mid X \models \alpha\}$ is the closure under logical entailment. With $\mathcal{L}^{\mathrm{Bel}}$ we denote the set of all deductively closed sets. For $\alpha \in \mathcal{L}$ we define $X + \alpha = \mathrm{Cn}(X \cup \{\alpha\})$. Moreover, for $M \subseteq \Omega$ we define $\mathrm{Th}(M) = \{\alpha \in \mathcal{L} \mid M \subseteq [\![\alpha]\!]\}$. A formula $\alpha \in \mathcal{L}$, respectively a set $X \subseteq \mathcal{L}$, is called consistent if $[\![\alpha]\!] \neq \emptyset$, respectively $[\![X]\!] \neq \emptyset$. A total preorder $\preceq$ on subset $M \subseteq \Omega$ is a relation $\preceq \subseteq M \times M$ such that $\preceq$ is total, i.e., for all $\omega_1, \omega_2 \in M$ holds $\omega_1 \preceq \omega_2$ or $\omega_2 \preceq \omega_1$, and transitive, i.e., for all $\omega_1, \omega_2, \omega_3 \in M$ holds that $\omega_1 \preceq \omega_2$ and $\omega_2 \preceq \omega_3$ imply $\omega_1 \preceq \omega_3$. Note that totality implies that $\preceq$ is reflexive, i.e., $\omega \preceq \omega$ holds for all $\omega \in M$. A total preorder $\ll$ on $M \subseteq \Omega$ is called a *linear order*, if $\ll$ is antisymmetric, i.e., for all $\omega_1, \omega_2 \in M$ holds that

$\omega_1 \ll \omega_2$ and $\omega_2 \ll \omega_1$ imply $\omega_1 = \omega_2$. The set of minimal elements of $X \subseteq \Omega$ with respect to $\preceq$ is $\min(X, \preceq) = \{\omega \in X \mid \omega \preceq \omega' \text{ for all } \omega' \in X\}$ and $\simeq$ denotes the equivalent part of $\preceq$.

3 Background on Belief Change for Epistemic Spaces

In this work, we model agents by the means of logic. Deductive closed sets of formulas, which we denote from now as *belief set*, represent deductive capabilities. The interpretations represent worlds that the agent is capable to imagine. The following notion describes the space of epistemic possibilities of an agent's mind in a general way.

Definition 1 ([20]; **adapted).** A tuple $\mathbb{E} = \langle \mathcal{E}, \mathrm{Bel} \rangle$ is called an *epistemic space* if $\mathcal{E}$ is a non-empty set and $\mathrm{Bel} : \mathcal{E} \to \mathcal{L}^{\mathrm{Bel}}$.

We call the elements of $\mathcal{E}$ *epistemic states* and use $[\![\Psi]\!]$ as shorthand for $[\![\mathrm{Bel}(\Psi)]\!]$. Within this framework belief change operators are transitions from one epistemic state to another when new beliefs are received, i.e., belief change operators for an epistemic space $\mathbb{E}$ are global objects, functions on all epistemic states in the mathematical sense.

Definition 2. *A* belief change operator for an epistemic space $\mathbb{E} = \langle \mathcal{E}, \mathrm{Bel} \rangle$ *is a function* $\circ : \mathcal{E} \times \mathcal{L} \to \mathcal{E}$.

The framework of belief change for epistemic spaces can be instantiated to often-considered settings of belief change. When $\mathcal{E}$ is the set of all belief sets over $\mathcal{L}$ and $\mathrm{Bel}(\Psi) = \Psi$, one obtains the classical setting of theory change [1], respectively the setting considered by Katsuno and Mendelzon [15]. In iterated belief change, typical instantiations for $\mathbb{E}$ are ranking functions by Spohn [21] or total preorders [3]. The notion of an epistemic space by Schwind et al. [20] slightly differs from the notion here insofar that here, we *do permit* inconsistent beliefs (cf. Definition 1). For that reason we denote the framework considered here as *unrestricted*. We can (nearly) obtain the restricted setting by considering only consistent formulas and demanding that an epistemic space $\mathbb{E} = \langle \mathcal{E}, \mathrm{Bel} \rangle$ satisfies the following condition:

$$\text{If } \Psi \in \mathcal{E}, \text{ then } \mathrm{Bel}(\Psi) \neq \mathrm{Cn}(\bot) \qquad \text{(global consistent)}$$

Clearly, to study types of belief changes, one restricts the space of all belief change operators for an epistemic spaces to specific classes of operators. In the following, we consider such classes of operators.

AGM Revision. Revision is the process of incorporating new beliefs into an agent's belief set while maintaining consistency, whenever this is possible. We use an adaptation of the AGM postulates for revision [1] for the framework of epistemic spaces [3], which is inspired by the approach of Katsuno and Mendelzon [15]. A belief change operator $*$ for an epistemic space $\mathbb{E} = \langle \mathcal{E}, \mathrm{Bel} \rangle$ is called an *(AGM) revision operator for* $\mathbb{E}$ if the following postulates are satisfied [3]:

(R1) $\alpha \in \text{Bel}(\Psi * \alpha)$
(R2) $\text{Bel}(\Psi * \alpha) = \text{Bel}(\Psi) + \alpha$ if $\text{Bel}(\Psi) + \alpha$ is consistent
(R3) If α is consistent, then $\text{Bel}(\Psi * \alpha)$ is consistent
(R4) If $\alpha \equiv \beta$, then $\text{Bel}(\Psi * \alpha) = \text{Bel}(\Psi * \beta)$
(R5) $\text{Bel}(\Psi * (\alpha \wedge \beta)) \subseteq \text{Bel}(\Psi * \alpha) + \beta$
(R6) If $\text{Bel}(\Psi * \alpha) + \beta$ is consistent, then $\text{Bel}(\Psi * \alpha) + \beta \subseteq \text{Bel}(\Psi * (\alpha \wedge \beta))$

AGM revision is well-known for realizing the principle of minimal change on the prior beliefs when revising. Note that AGM revision in the setting epistemic spaces is expressible, as the model is Turing complete [18]. However, in some epistemic spaces no AGM revision operator exist at all [19].

Credibility-Limited Revision. Credibility-limited revision was introduced by Hansson et al. [13] and restricts the process of revision to credible beliefs. To deal with epistemic states, *credibility-limited revision* was adapted by Booth et al. [2]. A belief change operator $\circledast$ for an epistemic space $\mathbb{E} = \langle \mathcal{E}, \text{Bel} \rangle$ is called an *credibility-limited revision operator for* $\mathbb{E}$ if the following postulates are satisfied [2]:

(CL1) $\alpha \in \text{Bel}(\Psi \circledast \alpha)$ or $\text{Bel}(\Psi \circledast \alpha) = \text{Bel}(\Psi)$
(CL2) $\text{Bel}(\Psi \circledast \alpha) = \text{Bel}(\Psi) + \alpha$ if $\text{Bel}(\Psi) + \alpha$ is consistent
(CL3) $\text{Bel}(\Psi \circledast \alpha)$ is consistent
(CL4) If $\alpha \equiv \beta$, then $\text{Bel}(\Psi \circledast \alpha) = \text{Bel}(\Psi \circledast \beta)$
(CL5) If $\alpha \in \text{Bel}(\Psi \circledast \alpha)$ and $\alpha \models \beta$, then $\beta \in \text{Bel}(\Psi \circledast \beta)$
(CL6) $$\text{Bel}(\Psi \circledast (\alpha \vee \beta)) = \begin{cases} \text{Bel}(\Psi \circledast \alpha) \text{ or} \\ \text{Bel}(\Psi \circledast \beta) \text{ or} \\ \text{Bel}(\Psi \circledast \alpha) \cap \text{Bel}(\Psi \circledast \beta) \end{cases}$$

The postulate (CL1) is known as *relative success* and denotes that either the agent keeps its prior beliefs (falling back to prior beliefs) or the belief change is successful in achieving the success condition of revision (the beliefs get accepted for revision). Through (CL2), known as *vacuity*, new beliefs are just added when they are not in conflict with $\text{Bel}(\Psi)$. The postulate (CL3), also known as *strong consistency* [13], ensures consistency, and by (CL4) the operator has to implement independence of syntax. Postulate (CL5) guarantees that when the revision by a belief α is successful, then it is also successful for every more general belief β. The trichotomy postulate (CL6) guarantees decomposability of revision of disjunctive beliefs.

4 Observations on AGM Revision and Credibility-Limited Revision in the Unrestricted Framework

The approach for credibility-limited revision for epistemic spaces, as given by Booth et al. (cf. Sect. 3), is made with the restriction to consider only consistent beliefs. In the unrestricted framework of epistemic spaces, we also permit inconsistent beliefs, and next, we observe now that in these cases, no credibility-limited revision exists at all.

Proposition 3. *Let $\mathbb{E} = \langle \mathcal{E}, \mathrm{Bel} \rangle$ be an epistemic space and let $\circledast$ be a belief change operator for $\mathbb{E}$. If $\mathbb{E}$ is not globally consistent, then $\circledast$ is not a credibility-limited revision operator.*

Proof. If $\mathbb{E}$ is not globally consistent, then there is some epistemic state $\Psi_\bot \in \mathcal{E}$ with $[\![\Psi_\bot]\!] = \emptyset$, i.e., $\mathrm{Bel}(\Psi_\bot) = \mathrm{Cn}(\bot)$. Suppose now that $\circledast$ is a credibility-limited revision operator. Because of that $\circledast$ satisfies (CL1) and (CL3). From (CL3), we obtain that $[\![\Psi_\bot \circledast \bot]\!] \neq \emptyset$ holds. This is a contradiction, because due to (CL1), we also have that $[\![\Psi_\bot \circledast \bot]\!] = \emptyset$ holds. □

When consider belief changes in the unrestricted framework of epistemic spaces, we observe that AGM revision operators are not credibility-limited revision operators.

Proposition 4. *Let $\mathbb{E} = \langle \mathcal{E}, \mathrm{Bel} \rangle$ be an epistemic space. Every AGM revision operator for $\mathbb{E}$ is not a credibility-limited revision operator for $\mathbb{E}$.*

Proof. For each AGM revision operator $*$ for epistemic spaces holds $[\![\Psi * \bot]\!] = \emptyset$ due to (R1) (as in the setting of theory change). Because of that, $*$ violates (CL3), as (CL3) demands that $[\![\Psi * \bot]\!] \neq \emptyset$ holds. Consequently, $*$ is not a credibility-limited revision operator. □

To describe Proposition 4 from the viewpoint of classes of operators, we define the respective classes of operators. With $\mathsf{AGMRev}(\mathbb{E})$ we denote the class of all AGM revision operators for $\mathbb{E}$, i.e., $\mathsf{AGMRev}(\mathbb{E}) = \{ * : \mathcal{E} \times \mathcal{L} \to \mathcal{E} \mid * \text{ satisfies (R1)-(R6)} \}$, and with $\mathsf{CLRev}(\mathbb{E})$ we denote the class of all credibility-limited revision operators for $\mathbb{E}$, i.e., $\mathsf{CLRev}(\mathbb{E}) = \{ \circledast : \mathcal{E} \times \mathcal{L} \to \mathcal{E} \mid \circledast \text{ satisfies (CL1)-(CL6)} \}$. Proposition 4 yields the following results.

Corollary 5. *For each epistemic space $\mathbb{E}$ holds:*

$$\mathsf{AGMRev}(\mathbb{E}) \cap \mathsf{CLRev}(\mathbb{E}) = \emptyset$$

5 Extended Credibility-Limited Revision

In the following, we extend credibility-limited revision [2] so that AGM revision operators are not excluded in the unrestricted framework of epistemic spaces and that operators exist, even when inconsistent beliefs are permitted. At first, we will observe that just dropping (CL3) on the postulate side will include belief change operators with undesired behaviour. We introduce two postulates that exclude operators with undesired behaviour, which are meant to replace (CL3). By employing these postulates we define extended credibility-limited revision. This sections ends with a semantic characterization of extended credibility-limited revision.

5.1 Credibility-Limited Revision Without (CL3)

In Sect. 4, we showed that AGM revision operators are not credibility-limited revision operators in the unrestricted stetting of belief change for epistemic spaces and when inconsistent beliefs are permitted, no credibility-limited revision operator exists. The cause for this is the postulate (CL3) of credibility-limited revision, e.g., AGM revision operators are incompatible with the postulate (CL3). However, excluding (CL3), respectively by just taking (CL1), (CL2), and (CL4)–(CL6), we would observe drastic consequences, because we would permit operators that would yield randomly inconsistent states for certain inputs. The following example contains a fairly simple operator which has such a behaviour.

Example 6. Let $\Sigma = \{a\}$ and let $\mathbb{E}_{\perp,a} = \langle \mathcal{E}, \mathrm{Bel} \rangle$ be the epistemic space given by:

$$\mathcal{E} = \{\Psi_\perp, \Psi_a\} \qquad [\![\Psi_\perp]\!] = \emptyset \qquad [\![\Psi_a]\!] = \{ab\}\ .$$

Note that the function Bel is implicitly defined via $\mathrm{Bel}(\Psi) = \mathrm{Th}([\![\Psi]\!])$. We define a belief change operator $\circledast$ for $\mathbb{E}_{\perp,a}$ as follows:

$$\Psi \circledast \alpha = \begin{cases} \Psi_a & \text{if } [\![\alpha]\!] = \{a\} \\ \Psi_\perp & \text{if } [\![\alpha]\!] = \{\overline{a}\} \text{ or } [\![\alpha]\!] = \emptyset \\ \Psi & \text{if } [\![\alpha]\!] = \{a, \overline{a}\} \end{cases}$$

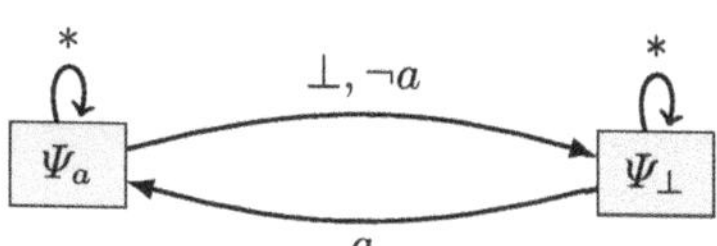

Fig. 1. Graphical representation of the operator $\circledast$ given in Example 6.

Figure 1 illustrates $\circledast$ graphically. We make two observations regarding $\circledast$:

Observation I. There are situations where $\circledast$ yields an inconsistent belief set for a consistent formula (on a consistent belief set), e.g., we have $[\![\Psi_a \circledast \neg a]\!] = \emptyset$.

Observation II. There are situations where $\circledast$ yields a consistent belief set for a consistent formula α (on an inconsistent belief set) and yields an inconsistent belief set for some consequences of α, e.g., we have $[\![\Psi_\perp \circledast a]\!] = \{a\}$ and $[\![\Psi_\perp \circledast \top]\!] = \emptyset$.

Indeed, we obtained the intended behaviour.

Proposition 7. *The operator* $\circledast$ *from Example 6 satisfies* (CL1)–(CL6) *except for* (CL3).

Proof. Violation of (CL3) is given by Example 6. From the definition of $\circledast$ we obtain that $\circledast$ satisfies (CL1), (CL2), and (CL4). We show satisfaction of (CL5) and (CL6):

(CL5) Note that we have $\alpha \in \mathrm{Bel}(\Psi)$ for each $\Psi \in \mathcal{E}$ and for each $\alpha \in \mathcal{L}$. Consequently, $\circledast$ satisfies (CL5).

(CL6) Let $\gamma = \alpha \vee \beta$. For $\alpha \equiv \beta$ we obtain $[\![\Psi \circledast \gamma]\!] = [\![\Psi \circledast \alpha]\!] = [\![\Psi \circledast \beta]\!]$ from (CL4). In the following we assume $[\![\alpha]\!] \neq [\![\beta]\!]$. Observe that this implies $[\![\gamma]\!] \neq \emptyset$. Next, we consider two subcases for $\Psi \in \mathcal{E}$:

$\Psi = \Psi_{\perp}$. Observe that we have $[\![\Psi \circledast \varphi]\!] = [\![\Psi]\!] = \emptyset$ for all φ with $[\![\varphi]\!] \neq \{a\}$. Consequently, if $a \notin [\![\gamma]\!]$, then we obtain $[\![\Psi \circledast \gamma]\!] = [\![\Psi]\!] = [\![\Psi \circledast \alpha]\!] = [\![\Psi \circledast \beta]\!] = \emptyset$. If $[\![\gamma]\!] = \{a\}$, then we have $[\![\Psi \circledast \gamma]\!] = \{a\}$ and we obtain from $[\![\alpha]\!] \neq [\![\beta]\!]$ that either $[\![\alpha]\!] = \{a\}$ or $[\![\beta]\!] = \{a\}$. Thus, we obtain either $[\![\Psi \circledast \gamma]\!] = [\![\Psi \circledast \alpha]\!]$ or $[\![\Psi \circledast \gamma]\!] = [\![\Psi \circledast \beta]\!]$ by (CL4). We consider the remaining case of $\{a\} \subsetneq [\![\gamma]\!]$. Then we have $[\![\Psi \circledast \gamma]\!] = \emptyset$. From $[\![\alpha]\!] \neq [\![\beta]\!]$ we obtain that $[\![\alpha]\!] \neq \{a\}$ or $[\![\beta]\!] \neq \{a\}$ holds. Thus, we obtain either $[\![\Psi \circledast \gamma]\!] = [\![\Psi \circledast \alpha]\!]$ or $[\![\Psi \circledast \gamma]\!] = [\![\Psi \circledast \beta]\!]$.

$\Psi = \Psi_a$. Observe that we have $[\![\Psi \circledast \varphi]\!] = [\![\Psi]\!] = \{a\}$ for all φ with $[\![\varphi]\!] \not\subseteq \{\overline{a}\}$. Consequently, if $\overline{a} \notin [\![\gamma]\!]$, then we obtain $[\![\Psi \circledast \gamma]\!] = [\![\Psi]\!] = [\![\Psi \circledast \alpha]\!] = [\![\Psi \circledast \beta]\!] = \emptyset$. If $[\![\gamma]\!] = \{\overline{a}\}$, then we have $[\![\Psi \circledast \gamma]\!] = \emptyset$ and we obtain from $[\![\alpha]\!] \neq [\![\beta]\!]$ that either $[\![\alpha]\!] = \{\overline{a}\}$ or $[\![\beta]\!] = \{\overline{a}\}$. Thus, we obtain either $[\![\Psi \circledast \gamma]\!] = [\![\Psi \circledast \alpha]\!]$ or $[\![\Psi \circledast \gamma]\!] = [\![\Psi \circledast \beta]\!]$ by (CL4). We consider the remaining case of $\{\overline{a}\} \subsetneq [\![\gamma]\!]$. Then we have $[\![\Psi \circledast \gamma]\!] = \{a\}$. From $[\![\alpha]\!] \neq [\![\beta]\!]$ we obtain that $[\![\alpha]\!] \neq \{\overline{a}\}$ or $[\![\beta]\!] \neq \{\overline{a}\}$ holds. Thus, we obtain either $[\![\Psi \circledast \gamma]\!] = [\![\Psi \circledast \alpha]\!]$ or $[\![\Psi \circledast \gamma]\!] = [\![\Psi \circledast \beta]\!]$.

In summary, $\circledast$ satisfies (CL1)–(CL6) except for (CL3). □

5.2 Defining Extended Credibility-Limited Revision

For extended credibility-limited revision we replace (CL3) by postulates that prevent the behaviour given in Observation I and Observation II in Example 6. The first postulate is

(CL3wcp) If $\mathrm{Bel}(\Psi \circledast \alpha)$ is inconsistent, then $\mathrm{Bel}(\Psi)$ or α is inconsistent.

which is already known in its contrapositive formulation,

(WCP) If $\mathrm{Bel}(\Psi)$ and α are consistent, then $\mathrm{Bel}(\Psi \circledast \alpha)$ is consistent.

as *weak consistency preservation* [13,14]. The postulate (CL3wcp) states that the inconsistency of the result of a change on Ψ by α is rooted in inconsistency of either $\mathrm{Bel}(\Psi)$ or α. Moreover, we will assume satisfaction of the following postulate:

(CL3u) If $\mathrm{Bel}(\Psi \circledast \alpha)$ is consistent and $\alpha \models \beta$, then $\mathrm{Bel}(\Psi \circledast \beta)$ is consistent.

The postulate (CL3u) states that the consistency of a change on Ψ by α is inherited "upward" to all changes on Ψ by consequences of α. Regarding our observations in Example 6: the postulate (CL3wcp) prevents situations like in Observation I, and the postulate (CL3u) rules out situations mentioned in Observation II of Example 6. Considering (CL3), (CL3u), and (CL3wcp) yields directly the interrelation of these postulates.

Proposition 8. *Let* $\mathbb{E} = \langle \mathcal{E}, \mathrm{Bel} \rangle$ *be an epistemic space and* $\circ$ *be a belief change operator for* $\mathbb{E}$. *If* $\circ$ *satisfies (CL3), then* $\circ$ *satisfies (CL3u) and (CL3wcp).*

Proof. Suppose that $\circ$ satisfies (CL3). Then, the antecedent of (CL3wcp) is never fulfilled, and hence, (CL3wcp) is always satisfied by $\circ$. For (CL3u), observe that the consequent of (CL3u) is always fulfilled by $\circ$. Consequently, (CL3u) is always satisfied by $\circ$. □

Given these postulates, we define extended credibility-limited revision operators for epistemic spaces in the following as operators that satisfy (CL1), (CL2), (CL3wcp), (CL3u) and (CL4)–(CL6). For the sake of clarity, we give this set of postulates its own naming.

Definition 9 (Extended Credibility-Limited Revision). *Let* $\mathbb{E} = \langle \mathcal{E}, \mathrm{Bel} \rangle$ *be an epistemic space. A belief change operator* $\circledast$ *for* $\mathbb{E}$ *is an* extended credibility-limited revision operator for $\mathbb{E}$ *if* $\circledast$ *satisfies:*

(ECL1) $\alpha \in \mathrm{Bel}(\Psi \circledast \alpha)$ *or* $\mathrm{Bel}(\Psi \circledast \alpha) = \mathrm{Bel}(\Psi)$
(ECL2) $\mathrm{Bel}(\Psi \circledast \alpha) = \mathrm{Bel}(\Psi) + \alpha$ *if* $\mathrm{Bel}(\Psi) + \alpha$ *is consistent*
(ECL3) *If* $\mathrm{Bel}(\Psi \circledast \alpha)$ *is inconsistent, then* $\mathrm{Bel}(\Psi)$ *or* α *is inconsistent*
(ECL4) *If* $\mathrm{Bel}(\Psi \circledast \alpha)$ *is consistent and* $\alpha \models \beta$, *then* $\mathrm{Bel}(\Psi \circledast \beta)$ *is consistent*
(ECL5) *If* $\alpha \equiv \beta$, *then* $\mathrm{Bel}(\Psi \circledast \alpha) = \mathrm{Bel}(\Psi \circledast \beta)$
(ECL6) *If* $\alpha \in \mathrm{Bel}(\Psi \circledast \alpha)$ *and* $\alpha \models \beta$, *then* $\beta \in \mathrm{Bel}(\Psi \circledast \beta)$

(ECL7) $\mathrm{Bel}(\Psi \circledast (\alpha \vee \beta)) = \begin{cases} \mathrm{Bel}(\Psi \circledast \alpha) \text{ or} \\ \mathrm{Bel}(\Psi \circledast \beta) \text{ or} \\ \mathrm{Bel}(\Psi \circledast \alpha) \cap \mathrm{Bel}(\Psi \circledast \beta) \end{cases}$

5.3 Semantic Characterization

Next, we characterize extended credibility-limited revision operators semantically. Booth et al. [2] proposed to use faithful assignments to capture the class of credibility-limited revision operators. In the following, we present an extended version of their assignments, which are meant to capture extended credibility-limited revision operators.

Definition 10 ((Extended) Credibility-Limited Assignment). *Let* $\mathbb{E} = \langle \mathcal{E}, \mathrm{Bel} \rangle$ *be an epistemic space. A function* $\Psi \mapsto (\preceq_\Psi, C_\Psi, b_\Psi)$ *is called an (extended)* credibility-limited assignment for $\mathbb{E}$ *if* $C_\Psi \subseteq \Omega$ *is a set of interpretations with* $[\![\Psi]\!] \subseteq C_\Psi$, *and* $\preceq_\Psi$ *is a total preorder over* C_Ψ, *and* $b_\Psi \in \{\top, \bot\}$ *for all* $\Psi \in \mathcal{E}$ *such that the following holds:*

($\mathrm{CLA}_\bot$) *If* $b_\Psi = \bot$, *then* $C_\Psi = \Omega$.

(Extended) credibility-limited assignments carry two kinds of information. First, C_Ψ describes semantically all consistent beliefs denoted as credible and b_Ψ represents whether an inconsistent formula is considered as credible or not. Note that b_Ψ is an extension to the assignments considered by Booth et al. [2]. Second, the total preorder $\preceq_\Psi$ serves the same purpose as in Katsuno-Mendelzon characterzation of revision [15]; representing the preferences of the agent. Note that $\preceq_\Psi$ might be a relation over a strict subset of Ω.

Definition 11. *Let* $\mathbb{E} = \langle \mathcal{E}, \text{Bel} \rangle$ *be an epistemic space. A credibility-limited assignment* $\Psi \mapsto (\preceq_\Psi, C_\Psi, b_\Psi)$ *for* $\mathbb{E}$ *is called* faithful *if the following holds:*

(CLFA1) *If* $\omega_1 \in [\![\Psi]\!]$ *and* $\omega_2 \in [\![\Psi]\!]$, *then* $\omega_1 \simeq_\Psi \omega_2$
(CLFA2) *If* $\omega_1 \in [\![\Psi]\!]$ *and* $\omega_2 \notin [\![\Psi]\!]$, *then* $\omega_1 <_\Psi \omega_2$

We connect credibility-limited assignments with belief change operators by the following notion of compatibility [4].

Definition 12. *A credibility-limited assignment* $\Psi \mapsto (\preceq_\Psi, C_\Psi, b_\Psi)$ *is called* (credibility-limited) revision-compatible *with a belief change operator* $\circledast$ *if the following holds:*

$$[\![\Psi \circledast \alpha]\!] = \begin{cases} \min([\![\alpha]\!], \preceq_\Psi) & \textit{if } [\![\alpha]\!] \cap C_\Psi \neq \emptyset \\ \emptyset & \textit{if } [\![\alpha]\!] = \emptyset \textit{ and } b_\Psi = \bot \\ [\![\Psi]\!] & \textit{otherwise} \end{cases} \qquad \text{(revision-compatible)}$$

Given the notion of revision-compatibility, we will now show that faithful credibility-limited assignments fully capture extended credible-limited revision operators for epistemic states.

Theorem 13. *Let* $\mathbb{E} = \langle \mathcal{E}, \text{Bel} \rangle$ *be an epistemic space and let* $\circledast$ *be a belief change operator for* $\mathbb{E}$. *Then* $\circledast$ *is an extended credibility-limited revision operator for* $\mathbb{E}$ *if and only if there is a faithful credibility-limited assignment* $\Psi \mapsto (\preceq_\Psi, C_\Psi, b_\Psi)$ *that is revision-compatible with* $\circledast$.

Proof (sketch). Overall, the proofs follows a similar structure as the proof for the semantic characterization of (non-extended) credibility-limited revision by Booth et al. [2]. Their proof is conceptually extended by dealing with inconsistency and adapted to deal with the two different postulates (CL3u) and (CL3wcp). For the $\Rightarrow$-direction, one has to give a construction of an faithful credibility-limited assignment $\Psi \mapsto (\preceq_\Psi, C_\Psi, b_\Psi)$ that is revision-compatible with $\circledast$. The construction used in the full proof works as follows[3]. We set C_Ψ as follows

$$C_\Psi = \{\omega \mid [\![\varphi_\omega]\!] = [\![\Psi \circledast \varphi_\omega]\!]\}, \qquad \text{(see [2, Remark 1])}$$

for each $\Psi \in \mathcal{E}$, where φ_ω denotes a formula with $[\![\varphi_\omega]\!] = \{\omega\}$. If $[\![\Psi]\!] \neq \emptyset$ and $\bot \in \text{Bel}(\Psi \circledast \bot)$, then we set $b_\Psi = \bot$; otherwise we set $b_\Psi = \top$. For each $\Psi \in \mathcal{E}$ let $\preceq_\Psi \subseteq C_\Psi \times C_\Psi$ be the relation such that

$$\omega_1 \preceq_\Psi \omega_2 \text{ if and only if } \omega_1 \in [\![\Psi \circledast \varphi_{\omega_1,\omega_2}]\!]$$

holds, where $\varphi_{\omega_1,\omega_2}$ denotes a formula with $[\![\varphi_{\omega_1,\omega_2}]\!] = \{\omega_1, \omega_2\}$. □

[3] The full proof can be fund the in arxiv version https://arxiv.org/abs/2409.07119.

6 Example and Properties

In the following, we consider an example for an extended credibility-limited revision operator and demonstrate the semantic characterization by Theorem 13.

Example 14. Let $\Sigma = \{a, b\}$ and let $\mathbb{E}_{\text{ex}} = \langle \mathcal{E}, \text{Bel} \rangle$ be the epistemic space where $\mathcal{E} = \{\Psi_\bot, \Psi_{\{ab\}}, \Psi_{\{\overline{a}b\}}, \Psi_{\{a\overline{b}\}}, \Psi_{\{\overline{a}\overline{b}\}}, \Psi_{\{\overline{a}b,a\overline{b}\}}\}$ is a set of epistemic states with:

$$\llbracket \Psi_\bot \rrbracket = \emptyset \qquad \llbracket \Psi_{\{\overline{a}b,a\overline{b}\}} \rrbracket = \{\overline{a}b, a\overline{b}\}$$
$$\llbracket \Psi_{\{ab\}} \rrbracket = \{ab\} \qquad \llbracket \Psi_{\{a\overline{b}\}} \rrbracket = \{a\overline{b}\}$$
$$\llbracket \Psi_{\{\overline{a}b\}} \rrbracket = \{\overline{a}b\} \qquad \llbracket \Psi_{\{\overline{a}\overline{b}\}} \rrbracket = \{\overline{a}\overline{b}\}$$

In the following, we obtain an extended credibility-limited revision operator $\circledast$ for $\mathbb{E}_{\text{ex}}$ by specifying a faithful credibility-limited assignment that is revision-compatible with $\circ$. We use the following linear order $\ll$ on Ω:

$$ab \ll \overline{a}b \ll a\overline{b} \ll \overline{a}\overline{b}$$

We specify $\Psi \mapsto (\preceq_\Psi, C_\Psi, b_\Psi)$ stepwise. We start by providing C_Ψ for each $\Psi \in \mathcal{E}$, which encodes semantically the set of those formulas that are considered as credible:

$$C_{\Psi_\bot} = \emptyset \qquad C_{\Psi_{\{\overline{a}b,a\overline{b}\}}} = \{\overline{a}b, a\overline{b}, \overline{a}\overline{b}\}$$
$$C_{\Psi_{\{ab\}}} = \Omega \qquad C_{\Psi_{\{a\overline{b}\}}} = \{ab, a\overline{b}\}$$
$$C_{\Psi_{\{\overline{a}b\}}} = \{\overline{a}b\} \qquad C_{\Psi_{\{\overline{a}\overline{b}\}}} = \{\overline{a}b, a\overline{b}, \overline{a}\overline{a}\}$$

We set $b_\Psi = \top$ for each $\Psi \in \mathcal{E} \backslash \{\Psi_{\{ab\}}\}$, and set $b_{\Psi_{\{ab\}}} = \bot$. Meaning For each $\Psi \in \mathcal{E} \backslash \{\Psi_{\{\overline{a}\overline{b}\}}\}$ we set $\preceq_\Psi \subseteq (C_\Psi \times C_\Psi)$:

$$\preceq_\Psi = ((\ll \cap (C_\Psi \times C_\Psi)) \backslash (C_\Psi \times \llbracket \Psi \rrbracket)) \cup (\llbracket \Psi \rrbracket \times C_\Psi),$$

i.e., $\preceq_\Psi$ is the total preorder on C_Ψ such that $\min(C_\Psi, \preceq_\Psi) = \llbracket \Psi \rrbracket$ and the remaining elements in $C_\Psi \backslash \llbracket \Psi \rrbracket$ are ordered according to $\ll$. For $\Psi_{\{\overline{a}\overline{b}\}}$, we specify $\preceq_{\Psi_{\{\overline{a}\overline{b}\}}} \subseteq (C_{\Psi_{\{\overline{a}\overline{b}\}}} \times C_{\Psi_{\{\overline{a}\overline{b}\}}})$ as follows:

$$\overline{a}\overline{b} \preceq_{\Psi_{\{\overline{a}\overline{b}\}}} \overline{a}b \qquad \overline{a}b \preceq_{\Psi_{\{\overline{a}\overline{b}\}}} \overline{a}b \qquad a\overline{b} \preceq_{\Psi_{\{\overline{a}\overline{b}\}}} \overline{a}b$$
$$\overline{a}\overline{b} \preceq_{\Psi_{\{\overline{a}\overline{b}\}}} a\overline{b} \qquad \overline{a}b \preceq_{\Psi_{\{\overline{a}\overline{b}\}}} a\overline{b} \qquad a\overline{b} \preceq_{\Psi_{\{\overline{a}\overline{b}\}}} a\overline{b}$$
$$\overline{a}\overline{b} \preceq_{\Psi_{\{\overline{a}\overline{b}\}}} \overline{a}\overline{b}$$

Because (CLA$_\bot$), (CLFA1), and (CLFA2) are satisfied, $\Psi \mapsto (\preceq_\Psi, C_\Psi, b_\Psi)$ is a faithful credibility-limited assignment. A belief change operator $\circledast$ for $\mathbb{E}_{\text{ex}}$ that is revision-compatible with $\Psi \mapsto (\preceq_\Psi, C_\Psi, b_\Psi)$ is then:

$$\Psi \circledast \alpha = \begin{cases} \Psi_{\llbracket \Psi \rrbracket \cap \llbracket \alpha \rrbracket} & \text{if } \llbracket \Psi \rrbracket \cap \llbracket \alpha \rrbracket \neq \emptyset \\ \Psi_\bot & \text{if } \llbracket \Psi \rrbracket = \emptyset \text{ and } \Psi = \Psi_{\{ab\}} \\ \Psi_{\{\overline{a}b,a\overline{b}\}} & \text{if } \{\overline{a}b, a\overline{b}\} \subseteq \llbracket \alpha \rrbracket \text{ and } \Psi = \Psi_{\{\overline{a}\overline{b}\}} \\ \Psi_{\min(\llbracket \alpha \rrbracket, \ll)} & \text{otherwise} \end{cases}$$

By Theorem 13, we obtain that $\circledast$ is an extended credibility-limited revision operator. A graphical representation of this operator is given in Fig. 2.

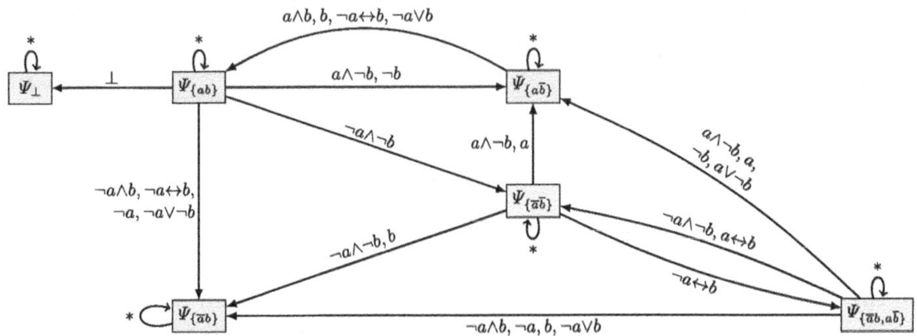

Fig. 2. Graphical representation of the extended credibility-limited revision operator $\circledast$ given in Example 14.

Note that $\circledast$ in Example 14 has properties that AGM revision operators do not have. The beliefs accepted for revision are not the full language $\mathcal{L}$. The selection of beliefs accepted for revision is done individually for each epistemic state. Inconsistent beliefs are only accepted for revision in selected epistemic states. Moreover, $\circledast$ in Example 14 demonstrates that in contrast to the credibility-limited revision operators considered by Booth et al. [2] (cf. Section 3), extended credibility-limited revision operators, as defined in Definition 9, are able to deal with inconsistent input and with inconsistent epistemic states, and therefore make use of the full unrestricted framework of belief change for epistemic spaces, as introduced in Sect. 3.

The following proposition points out that our generalization approach is successful in the sense that every AGM revision operator for epistemic states is indeed an extended credibility-limited revision operator in the sense of Definition 9.

Proposition 15. *Let $\mathbb{E} = \langle \mathcal{E}, \mathrm{Bel} \rangle$ be an epistemic space and let $*$ be a belief change operator for $\mathbb{E}$. The operator $*$ is an AGM revision operator for $\mathbb{E}$ if and only if $*$ is an extended credibility-limited revision operator for $\mathbb{E}$ which is revision-compatible with some faithful credibility-limited assignment $\Psi \mapsto (\preceq_\Psi, C_\Psi, b_\Psi)$ where $C_\Psi = \Omega$ and $b_\Psi = \bot$ holds for each $\Psi \in \mathcal{E}$.*

Proof. (idea) We consider both directions of the claim independently.

$\Rightarrow$ Suppose that $*$ is an AGM revision operator for $\mathbb{E}$. We use the credibility-limited assignment $\Psi \mapsto (\preceq_\Psi, C_\Psi, b_\Psi)$ given by:

$$C_\Psi = \Omega \qquad b_\Psi = \bot \qquad \omega_1 \preceq_\Psi \omega_2 \text{ if } \omega_1 \in [\![\Psi * \varphi_{\omega_1,\omega_2}]\!]$$

The proof by Darwiche and Pearl [3, Thm. 9] yields that $\Psi \mapsto (\preceq_\Psi, C_\Psi, b_\Psi)$ is faithful and revision-compatible with $\circledast$.

$\Leftarrow$ Suppose that $*$ is an extended credibility-limited revision operator for $\mathbb{E}$ and $\Psi \mapsto (\preceq_\Psi, C_\Psi, b_\Psi)$ is as given above. We obtain that $*$ satisfies (R2) and (R4), because (R2) coincides with (ECL1) and (R4) coincides with (ECL5). Because $C_\Psi = \Omega$ and $b_\Psi = \bot$ holds, by considering the revision-compatibility one sees easily that (R1) and (R3) are satisfied by $*$. To see that (R5) and (R6) are satisfies by $*$, use that (ECL7) is equivalent to (R5) and (R6) whenever (R1)–(R4) are satisfied [11]. $\square$

Next, we show that extended credibility-limited revision really extends credibility-limited revision for epistemic states as advertised in Sect. 1. Therefore, we use Theorem 13 to characterize operators that satisfy (CL1)–(CL6), including (CL3), when there is no epistemic state with an inconsistent belief set (see global consistency, defined on p. 4). Note that this is close to the setting originally considered by Booth et al. [2].

Proposition 16. *Let $\mathbb{E} = \langle \mathcal{E}, \mathrm{Bel} \rangle$ be a global consistent epistemic space and let $\circledast$ be a belief change operator for $\mathbb{E}$. The operator $\circledast$ satisfies (CL1)–(CL6) if and only if there is a faithful credibility-limited assignment $\Psi \mapsto (\preceq_\Psi, C_\Psi, b_\Psi)$ that is revision-compatible with $\circledast$ such that $b_\Psi = \top$ for each $\Psi \in \mathcal{E}$.*

Proof. We consider both directions independently.

$\Rightarrow$ If $\circledast$ satisfies (CL1)–(CL6), then (ECL1)–(ECL7) are satisfied (as (ECL3) and (ECL4) are implied by (CL3)). By Theorem 13, there exists some faithful credibility-limited assignment $\Psi \mapsto (\preceq_\Psi, C_\Psi, b_\Psi)$ that is revision-compatible with $\circledast$. Let $\Psi \in \mathcal{E}$ be an epistemic state. From (CL3) we obtain that $[\![\Psi \circledast \alpha]\!] \neq \emptyset$ for each $\alpha \in \mathcal{L}$. Consequently, we have $b_\Psi = \top$, as otherwise we would obtain $[\![\Psi \circledast \bot]\!] = \emptyset$ by revision-compatibility.

$\Leftarrow$ Suppose there is a credibility-limited assignment $\Psi \mapsto (\preceq_\Psi, C_\Psi, b_\Psi)$ that is revision-compatible with $\circledast$ such that $b_\Psi = \top$ for each $\Psi \in \mathcal{E}$. By Theorem 13, we obtain that $\circledast$ satisfies (CL1), (CL2), and (CL4)–(CL6). For satisfaction of (CL3) observe that by revision-compatibility we obtain $[\![\Psi \circledast \bot]\!] \neq \emptyset$ due to $b_\Psi = \top$ for each $\Psi \in \mathcal{E}$. For all consistent formulas α we have $[\![\Psi \circledast \alpha]\!] \neq \emptyset$ due to the global consistency of $\mathcal{E}$. $\square$

With $\mathsf{ECLRev}(\mathbb{E})$ we denote the set off all extended credibility-limited revision operators for $\mathbb{E}$, i.e., $\mathsf{ECLRev}(\mathbb{E}) = \{ \circledast : \mathcal{E} \times \mathcal{L} \to \mathcal{E} \mid \circledast \text{ satisfies (ECL1)-(ECL7)} \}$. The next proposition summarizes the interrelation between the class of extended credibility-limited revisions operators, the class of credibility-limited revision operators and the class of AGM revision operators in the framework of epistemic spaces.

Proposition 17. *For each epistemic space $\mathbb{E}$ holds:*

$$\mathsf{AGMRev}(\mathbb{E}) \subseteq \mathsf{ECLRev}(\mathbb{E})$$
$$\mathsf{CLRev}(\mathbb{E}) \subseteq \mathsf{ECLRev}(\mathbb{E})$$

Proof. The statement $\mathsf{AGMRev}(\mathbb{E}) \subseteq \mathsf{ECLRev}(\mathbb{E})$ is a direct consequence of Proposition 15. From Proposition 16, we obtain $\mathsf{CLRev}(\mathbb{E}) \subseteq \mathsf{ECLRev}(\mathbb{E})$ whenever $\mathbb{E}$ is a global consistent epistemic spaces. In all cases where $\mathbb{E}$ is not a global consistent epistemic spaces, we obtain $\mathsf{CLRev}(\mathbb{E}) = \emptyset$ from Proposition 3. □

7 Conclusion

In this paper, we considered belief changes in the unrestricted framework of epistemic spaces, which means inconsistent beliefs are permitted. Credibility-limited revision as defined by Booth et al. [2] does not extend well to this unrestricted setting, as AGM revision operators are not included and no operators exist when an epistemic state is present that has inconsistent beliefs. Extended credibility-limited revision operators are introduced, and we show that this class of operators deals with the before-mentioned problems. All AGM revision operators are also extended credibility-limited revision operators and extended credibility-limited revision operators do exists for epistemic spaces with inconsistent epistemic states. Furthermore, a semantic characterization of extended credibility-limited revision is presented. The approach here might serve as a prototype of how to deal with inconsistent beliefs in the framework of epistemic spaces. Especially, when considered other approach to belief change, e.g., like update [5,14] and other kinds of non-prioritized belief change [6,7,12], in the framework of epistemic spaces.

Finally, I like to remark that, independently, Grimaldi, Martinez and Rodriguez [10], made a similar approach to extending credibility-limited revision, that also uses (WCP), but does not use (CL3u) to deal with inconsistent belief sets. A comparison of both approaches could be insightful.

Acknowledgments. I thank the anonymous reviewers for their valuable hints and comments that helped me to improve this paper. The research reported here was partially supported by the Deutsche Forschungsgemeinschaft (DFG, grant 465447331, project “Explainable Belief Merging”, EBM).

References

1. Alchourrón, C.E., Gärdenfors, P., Makinson, D.: On the logic of theory change: partial meet contraction and revision functions. J. Symbolic Logic **50**(2), 510–530 (1985). https://doi.org/10.2307/2274239
2. Booth, R., Fermé, E.L., Konieczny, S., Pino Pérez, R.: Credibility-limited revision operators in propositional logic. In: Brewka, G., Eiter, T., McIlraith, S.A. (eds.) Proceedings of the 30th International Conference on Principles of Knowledge Representation and Reasoning (KR 2012), pp. 116–125. AAAI Press (2012)
3. Darwiche, A., Pearl, J.: On the logic of iterated belief revision. Artif. Intell. **89**, 1–29 (1997). https://doi.org/10.1016/S0004-3702(96)00038-0

4. Falakh, F.M., Rudolph, S., Sauerwald, K.: Semantic characterizations of AGM revision for Tarskian logics. In: Governatori, G., Turhan, A. (eds.) RuleML+RR 2022. LNCS, vol. 13752, pp. 95–110. Springer, Cham (2022). https://doi.org/10.1007/978-3-031-21541-4_7
5. Fermé, E., Gonçalves, S.: On the logic of theory change iteration of KM-update. Int. J. Approximate Reason. **162**, 109005 (2023). https://doi.org/10.1016/J.IJAR.2023.109005
6. Fermé, E.L., Hansson, S.O.: Selective revision. Stud. Logica. **63**(3), 331–342 (1999). https://doi.org/10.1023/A:1005294718935
7. Fermé, E.L., Hansson, S.O.: Shielded contraction. In: Williams, M.A., Rott, H. (eds.) Frontiers in Belief Revision, pp. 85–107. Springer, Dordrecht (2001). https://doi.org/10.1007/978-94-015-9817-0_4
8. Fermé, E.L., Hansson, S.O.: Belief Change - Introduction and Overview Springer Briefs in Intelligent Systems. Springer, Cham (2018). https://doi.org/10.1007/978-3-319-60535-7
9. Fermé, E.L., Wassermann, R.: On the logic of theory change: iteration of expansion. J. Braz. Comput. Soc. **24**(1), 8:1–8:9 (2018). https://doi.org/10.1186/s13173-018-0072-4
10. Grimaldi, D.A., Martinez, M.V., Rodríguez, R.O.: Moderated revision. Int. J. Approximate Reasoning **166**, 109126 (2024). https://doi.org/10.1016/J.IJAR.2024.109126
11. Gärdenfors, P.: Knowledge in Flux: Modeling the Dynamics of Epistemic States. MIT Press, Cambridge (1988)
12. Hansson, S.O.: A survey of non-prioritized belief revision. Erkenntnis **50**(2), 413–427 (1999). https://doi.org/10.1023/A:1005534223776
13. Hansson, S.O., Fermé, E.L., Cantwell, J., Falappa, M.A.: Credibility limited revision. J. Symbolic Logic **66**(4), 1581–1596 (2001). https://doi.org/10.2307/2694963
14. Katsuno, H., Mendelzon, A.O.: On the difference between updating a knowledge base and revising it. In: Allen, J.F., Fikes, R., Sandewall, E. (eds.) Proceedings of the 2nd International Conference on Principles of Knowledge Representation and Reasoning (KR 1991), pp. 387–394. Morgan Kaufmann (1991)
15. Katsuno, H., Mendelzon, A.O.: Propositional knowledge base revision and minimal change. Artif. Intell. **52**(3), 263–294 (1992). https://doi.org/10.1016/0004-3702(91)90069-V
16. Ragni, M., Kern-Isberner, G., Beierle, C., Sauerwald, K.: Cognitive logics - features, formalisms, and challenges. In: Giacomo, G.D., Catalá, A., Dilkina, B., Milano, M., Barro, S., Bugarín, A., Lang, J. (eds.) Proceedings of the 24nd European Conference on Artificial Intelligence (ECAI 2020). Frontiers in Artificial Intelligence and Applications, vol. 325, pp. 2931–2932. IOS Press (2020). https://doi.org/10.3233/FAIA200459
17. Sauerwald, K.: Semantics of Belief Change Operators for Intelligent Agents: Iteration, Postulates, and Realizability, Dissertations in Artificial Intelligence, vol. 352. IOS Press (2022). https://doi.org/10.3233/DAI352
18. Sauerwald, K., Beierle, C.: Iterated belief change, computationally. In: Kern-Isberner, G., Lakemeyer, G., Meyer, T. (eds.) Proceedings of the 19th International Conference on Principles of Knowledge Representation and Reasoning (KR 2022), pp. 410–414 (2022). https://doi.org/10.24963/kr.2022/42
19. Sauerwald, K., Thimm, M.: The realizability of revision and contraction operators in epistemic spaces. In: Proceedings of the 21st International Conference on Principles of Knowledge Representation and Reasoning (KR 2024), pp. 665–670 (2024). https://doi.org/10.24963/kr.2024/62

20. Schwind, N., Konieczny, S., Pino Pérez, R.: On the representation of darwiche and pearl's epistemic states for iterated belief revision. In: Kern-Isberner, G., Lakemeyer, G., Meyer, T. (eds.) Proceedings of the 19th International Conference on Principles of Knowledge Representation and Reasoning (KR 2022), pp. 320–330 (2022). https://doi.org/10.24963/kr.2022/32
21. Spohn, W.: Ordinal conditional functions: a dynamic theory of epistemic states. In: Harper, W., Skyrms, B. (eds.) Causation in Decision, Belief Change, and Statistics, II, pp. 105–134. Kluwer Academic Publishers (1988)

Towards a Dialogue Game-Based Semantics for Extended Abstract Argumentation Frameworks Based on Indecision-Blocking

Yamil Osvaldo Omar Soto[1], Cristhian Ariel David Deagustini[2,3], Maria Vanina Martinez[3(✉)], and Gerardo Ignacio Simari[1,4]

[1] Department of Computer Science and Engineering, Universidad Nacional del Sur (UNS) & Institute for Computer Science and Engineering (ICIC UNS-CONICET), San Andres 800, Bahia Blanca, Buenos Aires, Argentina
{yamil.soto,gis}@cs.uns.edu.ar

[2] Area de Agentes y Sistemas Inteligentes, Fac. de Cs. de la Adm., Universidad Nacional de Entre Rios (FCAD UNER), Monseñor Tavella 1424, Concordia, Entre Rios, Argentina
ariel.deagustini@uner.edu.ar

[3] Artificial Intelligence Research Institute (IIIA-CSIC), Carrer Can Planas s/n, Campus UAB, Bellaterra, Barcelona, Catalonia, Spain
vmartinez@iiia.csic.es

[4] Arizona State University, 699 S Mill Ave, Tempe, AZ 85281, USA

Abstract. Dialogue game-based semantics for abstract argumentation are relevant for several reasons. From a theoretical point of view, they provide a different perspective regarding extension-based or labeling-based approaches to study the theoretical properties of the argument evaluation process. From a more practical perspective, they allow us to examine whether or not an argument belongs to a given extension (or labeling) without computing an entire (set of) extension(s) or labeling(s), and guide the development of efficient algorithms. This last point is significant in the context of the development of argumentation-based knowledge representation and reasoning tools for real-world applications. In this paper, we expand the dialogue game-based semantics available for extended abstract argumentation frameworks, a generalization of abstract argumentation frameworks where two kinds of defeat are considered, proper and blocking, and the sub-argument relation is taken into account. The novel dialogue game-based semantics we propose is inspired in a specific interpretation of cycles that considers them as an indecision, a situation in which we do not have enough information to decide the status of the arguments involved.

Keywords: Argumentation semantics · Dialogue game-based semantics · Argumentation lines · Extended Abstract Argumentation Frameworks

S. Destercke et al. (Eds.): SUM 2024, LNAI 15350, pp. 445–460, 2025.
https://doi.org/10.1007/978-3-031-76235-2_33

1 Introduction and Related Work

Dialogue game-based semantics for abstract argumentation [25,26] represents the dialectical nature of argumentation as two-party zero-sum games where the first move corresponds to the proponent, who moves an initial argument that they wish to put to the test. Then, the opponent, the proponent's counterpart, and the proponent take turns in moving arguments attacking their counterpart's last move.

Dialogue game-based semantics for abstract argumentation are relevant for several reasons. From a theoretical point of view, they provide a different perspective regarding extension-based or labeling-based approaches to study the properties of the argument evaluation process. From a practical perspective, they allow us to examine whether or not an argument belongs to a given extension (or labeling) without computing an entire (set of) extension(s) or labeling(s), and guide the development of algorithms. Several techniques can be applied to dialogue games, allowing the development of efficient algorithms for the evaluation of arguments [1,8,9,20,31]. The latter point is significant in the context of the development of argumentation-based knowledge representation and reasoning tools for real-world applications [5,6,10,11].

In this paper, we propose a novel dialogue game-based semantics for *extended abstract argumentation frameworks* (EAFs) [22–24], a generalization of abstract argumentation frameworks (AFs) [15] in which defeat relations are defined through a notion of attack and a preference relation for arguments. This extension allows to distinguish more refined situations of defeats – *e.g.*, proper and blocking defeats – as well as to take into account the sub- and super-argument relations. Our motivation to introduce novel semantics is the lack of dialogue game-based semantics satisfying some specific properties regarding the treatment of cycles and their impact on other arguments. Next, we present the intuitions behind the behavior we seek for a dialogue game-based semantics.

Arguments in cycles might affect the acceptance status of each other, a situation that does not hold in an acyclic chain of arguments. Additionally, if we consider only those arguments involved in a cycle, there is a lack of information to decide among them. For these reasons, we will consider cycles as *potential indecisions* (see [3] for other possible interpretations of cycles). To evaluate the status of an argument, particularly those included in a cycle, all related arguments with those in the cycle must be considered; that is, arguments from outside the cycle defeating arguments in it, arguments defeating the defeaters, and so on. However, it may be the case that even considering all related arguments we do not have enough information to favor some arguments in the cycle over others. In this last case, we say the cycle represents an *effective indecision*, and we call it a *deadlock*. Since, according to our interpretation, every cycle represents a (potential or effective) indecision, all must be treated in a balanced manner (for some well-known problems regarding the unequal treatment of cycles see [2,4,28]).

If considering arguments related to a cycle we have enough information to resolve the indecision represented by the cycle (*i.e.*, we can decide the status of its arguments), then the argument evaluation process can proceed straightforwardly. However, if a cycle persists (*i.e.*, it is a deadlock) an important question arises: how should the status of an argument defeated by arguments involved in a deadlock be affected? Note that there exist two kinds of arguments defeated by arguments in a deadlock: those *internal* to the deadlock, *i.e.*, involved in it, and those *external* to the deadlock. For admissibility-based semantics, arguments that are internal and external to a deadlock will be *skeptically rejected*; that is, they might at most belong to some extension but not to all of them. However, we consider that indecision must not be used to decide the status of an argument; consequently, we agree with rejecting arguments that are internal to a deadlock, but we consider that defeats over external arguments should be ineffective – *i.e.*, they should not affect the status of the defeated arguments. We call this intuitive principle *indecision-blocking*. Similar ideas were first introduced for argumentation frameworks in recent work [12–14].

Furthermore, beyond our abstract and theoretical motivation, there are concrete and practical reasons to support indecision-blocking. For example, the *in dubio pro reo* penal law principle, which states that in the presence of contradictory evidence, we must judge in favor of the accused, shares similar intuitions[1]. Let us consider the following *floating argument* [21] example to clarify the relation between these two ideas.

Example 1. Let $F = (Ar, Att)$ be an argumentation framework, where $Ar = \{\mathscr{A}, \mathscr{B}, \mathscr{C}\}$ and $Att = \{(\mathscr{B}, \mathscr{A}), (\mathscr{C}, \mathscr{A}), (\mathscr{B}, \mathscr{C}), (\mathscr{C}, \mathscr{B})\}$.

Let us instantiate F in a legal context, such that $\mathscr{A}$ is a presumptive argument about the innocence of some individual, and that $\mathscr{B}$ and $\mathscr{C}$ are two contradictory pieces of evidence against $\mathscr{A}$. If we consider the in dubio pro reo principle in Example 1, arguments $\mathscr{B}$ and $\mathscr{C}$ must be rejected, and $\mathscr{A}$ accepted. This coincides with our intended result for a dialogue game-based semantics embedding the indecision-blocking principle.

Motivated by the above and, to the best of our knowledge, a lack of a dialogue game-based semantics embedding the indecision-blocking principle, we develop a novel, skeptical, balanced treatment of cycles, and indecision-blocking-inspired dialogue game-based semantics that we call *ib-acceptability*. Additionally, we compare ib-acceptability with two well-known dialogue game-based semantics: the grounded semantics and the default semantics of the *DeLP* formalism [18] (for a comparison between these semantics, see [17]), which we refer to as *gr-acceptability* and *df-acceptability*, respectively. We use the term *acceptability* for these semantics since they are presented through distinct definitions of the concept of *acceptable*[2] *argumentation line* (see Sect. 3).

[1] We do not claim that indecision-blocking accurately represents the in dubio pro reo principle in abstract argumentation, but simply that the latter is a motivation for the former.

[2] The concept of *acceptability* refers to a property defined over argumentation lines that should not be confused with acceptability as defined by Dung in [15], which we call *defense*.

We focus only on df-acceptability and gr-acceptability for several reasons. There exist similar definitions to df-acceptability (*e.g.*, [7,19]); however, df-acceptability is the most commonly used in [18] and its extensions. Both df-acceptability and gr-acceptability are *skeptical semantics*, which are of our special interest. There exist other dialogue game-based semantics that can be easily adapted to argumentation line terms, such as the argumentation games for the *pairwise cogency semantics* [4] or for the *credulous preferred semantics* [26]. Nevertheless, the above-mentioned semantics are credulous ones. Lastly, there also exist other dialogue game-based semantics that we do not cover here as they are more challenging to adapt into argumentation line terms, being meta-games rather than dialogue games (*e.g.*, [4]), or because they involve the exchange of entire extensions rather than individual arguments (*e.g.*, [26]).

We study the behavior of the dialogue game-based semantics through several examples; though we recognize the limitations of example-based comparisons (see, for instance [30,35]), due to the lack of a framework to compare dialogue game-based semantics for abstract argumentation, we limit our analysis to this approach.

Our main contributions are: (i) we propose a modified version of EAFs inspired by some insight from structured argumentation; (ii) we introduce a novel dialogue game-based semantics via an alternative version of the notion of acceptable argumentation line; (iii) we analyze, based on specific cases, the behavior of different dialogue game-based semantics by setting a specific kind of argumentative reasoning embedding the indecision-blocking principle.

This paper is organized as follows. In Sect. 2 we propose a modified version of EAFs. In Sect. 3 we recap the bundle set-based approach to semantics. We discuss different definitions of acceptability focusing on ib-acceptability and analyze them through specific examples in Sect. 4. Finally, conclusions and future lines of research are delineated in Sect. 5.

2 Extended Abstract Argumentation Framework

In this section, we briefly introduce an alternative version of extended abstract argumentation frameworks (EAFs). We modify the notions of conflict relation, conflict inheritance, core conflict (here called *disagreement sub-argument*), preference relation among arguments, and defeat. Our decisions are inspired by several properties of Defeasible Logic Programming (DeLP) [18], a well-studied formalism that provides interesting insights due to its structured nature. For this reason, this work can be considered a proposal that goes from the particular and concrete characteristics of DeLP to the general and abstract aspects of the EAFs. For a detailed presentation of EAFs as originally introduced, see [22–24]. EAFs are generalizations of argumentation frameworks formed by four elements: a set of arguments, and three relations between those arguments.

Definition 1. *An* extended abstract argumentation framework *(EAF) is a tuple of the form* $(Ar, \sqsubseteq, Att, Pref)$, *where* Ar *is a finite set of arguments,* $\sqsubseteq$ *is the sub-*

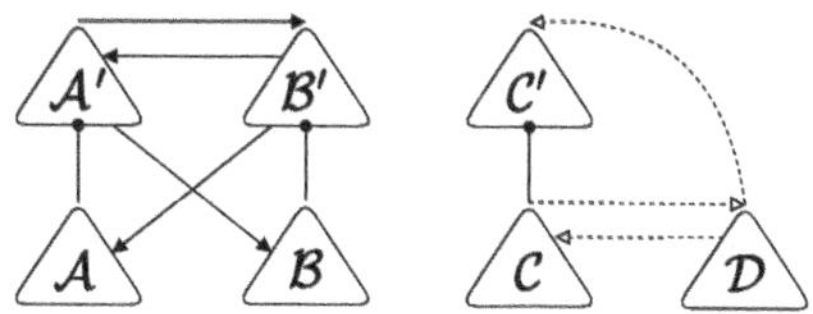

Fig. 1. Illustrating different kinds of defeats.

argument (super-argument) relation, Att is an irreflexive binary relation between arguments, and Pref is a preference relation among arguments.

Symbol $\sqsubseteq$ denotes the sub-argument (super-argument) relation: $\mathscr{A} \sqsubseteq \mathscr{B}$ means that argument $\mathscr{A}$ (resp $\mathscr{B}$) is a sub-argument (resp., super-argument) of argument $\mathscr{B}$ (resp., $\mathscr{A}$). Any argument $\mathscr{A}$ is considered to be a sub-argument and a super-argument of itself. $\mathscr{A} \sqsubset \mathscr{B}$ denotes that $\mathscr{A}$ (resp., $\mathscr{B}$) is a proper sub-argument (resp., super-argument) of $\mathscr{B}$ (resp., $\mathscr{A}$). Regarding the attack relation, it propagates to super-arguments; that is, if $(\mathscr{A}, \mathscr{B}) \in Att$, then $(\mathscr{A}, \mathscr{B}') \in Att$, for any super-argument $\mathscr{B}'$ of $\mathscr{B}$.

Definition 2. *Let $F = (Ar, \sqsubseteq, Att, Pref)$ be an EAF and let $(\mathscr{A}, \mathscr{B}) \in Att$. The* disagreement sub-argument *of $(\mathscr{A}, \mathscr{B})$ is an argument $\mathscr{B}' \sqsubseteq \mathscr{B}$ such that $(\mathscr{A}, \mathscr{B}') \in Att$ and there exists no argument $\mathscr{B}'' \sqsubset \mathscr{B}'$ such that $(\mathscr{A}, \mathscr{B}'') \in Att$.*

The relation *Pref* is introduced in the framework to be used to evaluate arguments, modeling a preference criterion based on a measure of strength of arguments.

Definition 3. *Let $F = (Ar, \sqsubseteq, Att, Pref)$ be an EAF. The* preference relation *among arguments Pref is a binary relation over $Ar \times Ar$. Let $\mathscr{A}, \mathscr{B}, \mathscr{B}' \in Ar$ such that $(\mathscr{A}, \mathscr{B}) \in Att$ and $\mathscr{B}'$ is the disagreement sub-argument of $(\mathscr{A}, \mathscr{B})$. Then, we have:*

1. *If $(\mathscr{A}, \mathscr{B}') \in Pref$ and $(\mathscr{B}', \mathscr{A}) \notin Pref$ then $\mathscr{A}$ is preferred to $\mathscr{B}$, denoted $\mathscr{A} \succ \mathscr{B}$.*
2. *If $(\mathscr{A}, \mathscr{B}') \in Pref$ and $(\mathscr{B}', \mathscr{A}) \in Pref$ then $\mathscr{A}$ and $\mathscr{B}$ are equally preferred arguments, denoted $\mathscr{A} \equiv \mathscr{B}$.*
3. *If $(\mathscr{A}, \mathscr{B}') \notin Pref$ and $(\mathscr{B}, \mathscr{A}') \notin Pref$ then $\mathscr{A}$ and $\mathscr{B}$ are incomparable arguments, denoted $\mathscr{A} \bowtie \mathscr{B}$.*

The preference relation allows us to establish defeat relations between arguments. In condition 1, a concrete preference is made ($\mathscr{A} \succ \mathscr{B}$), then it is said that $\mathscr{A}$ is a *proper defeater* of $\mathscr{B}$, denoted $\mathscr{A} \hookrightarrow \mathscr{B}$. In conditions 2 and 3, two arguments have the same relative preference ($\mathscr{A} \equiv \mathscr{B}$) or are incomparable ($\mathscr{A} \bowtie \mathscr{B}$). In those cases, argument $\mathscr{A}$ is a *blocking defeater* [27,33] of $\mathscr{B}$, denoted $\mathscr{A} \leadsto \mathscr{B}$. Note that the above is result of $\mathscr{A} \leadsto \mathscr{B}'$ ($\mathscr{B}' \leadsto \mathscr{A}$). However, we do not know if $\mathscr{B} \leadsto \mathscr{A}$ is the case because we do not know if $(\mathscr{B}, \mathscr{A}) \in Att$ and, if such attack relation exists in F, we neither know what the preference relation is between $\mathscr{B}$

and the disagreement sub-argument of $(\mathscr{B}, \mathscr{A})$. An argument $\mathscr{A}$ is said to be a *defeater* of an argument $\mathscr{B}$ if $\mathscr{A} \hookrightarrow \mathscr{B}$ or $\mathscr{A} \rightsquigarrow \mathscr{B}$. If $\mathscr{A}$ is a defeater of $\mathscr{B}$ and $\mathscr{B}$ is a defeater of $\mathscr{A}$, then both are *mutual defeaters*. If $\mathscr{A}$ is a defeater of $\mathscr{B}$ and $\mathscr{B}$ is not a defeater of $\mathscr{A}$, then $\mathscr{A}$ is a *strict defeater* of $\mathscr{B}$.

Example 2 illustrates different kinds of defeats. We represent EAFs as graphs, in which, to keep them as simple as possible, we only show proper sub-argument relationships, and we show the defeat relations instead of the attack and preference relations from which they are established. Solid, dotted, and ending-dot lines represent proper defeat, blocking defeat, and proper sub-argument relationships.

Example 2. Let $F = (Ar, \sqsubseteq, Att, Pref)$ be an EAF where $Ar = \{\mathscr{A}, \mathscr{A}', \mathscr{B}, \mathscr{B}', \mathscr{C}, \mathscr{C}', \mathscr{D}\}$; $\mathscr{A} \sqsubset \mathscr{A}'$, $\mathscr{B} \sqsubset \mathscr{B}'$, $\mathscr{C} \sqsubset \mathscr{C}'$; $\mathscr{A}' \hookrightarrow \mathscr{B}$, $\mathscr{B}' \hookrightarrow \mathscr{A}$, $\mathscr{A}' \hookrightarrow \mathscr{B}'$, $\mathscr{B}' \hookrightarrow \mathscr{A}'$, $\mathscr{C} \rightsquigarrow \mathscr{D}$, $\mathscr{D} \rightsquigarrow \mathscr{C}$, and $\mathscr{D} \rightsquigarrow \mathscr{C}'$. Figure 1 illustrates this example.

$\mathscr{A}'$ (resp., $\mathscr{B}'$) is a proper and strict defeater of $\mathscr{B}$ (resp., $\mathscr{A}$). As the defeat relation is established considering sub-arguments, from the above we have that $\mathscr{A}'$ and $\mathscr{B}'$ are proper and mutual defeaters. $\mathscr{C}$ and $\mathscr{D}$ are blocking and mutual defeaters. $\mathscr{D}$ is a blocking and strict defeater of $\mathscr{C}'$.

Note that in EAFs the defeat relation plays the role that the attack relation plays in AFs. In this manner, we can adapt the definitions of *defense* and *admissibility* as follows. Given an EAF $F = (Ar, \sqsubseteq, Att, Pref)$, a set of arguments $Args \subseteq Ar$ *defends* an argument $\mathscr{A} \in Ar$ if and only if for all $\mathscr{B} \in Ar$ such that $\mathscr{B}$ defeats $\mathscr{A}$, there exists $\mathscr{C} \in Args$ such that $\mathscr{C}$ defeats $\mathscr{B}$. Let $D(Args) = \{\mathscr{A} \in Ar \mid Args \text{ defends } \mathscr{A}\}$ be the set of all arguments defended by $Args$. A set of arguments $Args \subseteq Ar$ is *conflict-free* if and only if for all $\mathscr{A}, \mathscr{B} \in Args$ it holds that $(\mathscr{A}, \mathscr{B}) \notin Att$. A set of arguments $Args \subseteq Ar$ is *admissible* if and only if $Args$ is conflict-free and $Args \subseteq D(Args)$.

3 Argumentation Lines and Bundle Set-Based Semantics

We now introduce the machinery necessary to adapt dialogue game-based semantics to EAFs following the bundle set-based approach introduced in [34].

Definition 4. *Let $F = (Ar, \sqsubseteq, Att, Pref)$ and let $\mathscr{A} \in Ar$. An* argumentation line, *or simply* line, *for or rooted in $\mathscr{A}$ is a sequence of arguments of the form $\lambda = [\mathscr{A}_1, \ldots, \mathscr{A}_n]$ where $\mathscr{A} = \mathscr{A}_1$ and each $\mathscr{A}_i$ is defeated by $\mathscr{A}_{i+1}$.*

We adopt the following notation convention: capital letters $\mathscr{A}, \mathscr{B}, \mathscr{C}, \ldots$ denote different arguments, while $\mathscr{A}_i$ denotes an argument in the i-th position of some line λ. The following presents df-acceptability, the default definition of acceptability [18]. Given an argumentation line λ, the *supporting* (resp., *interfering*) set of λ, denoted $S(\lambda)$ (resp., $I(\lambda)$), is the set of all arguments in λ in an odd (resp., even) position.

Definition 5. *An argumentation line λ is* df-acceptable *if and only if:*

(Fin) λ is finite.
(Con) The supporting (resp., interfering) set is conflict-free.
(NrS) No argument $\mathscr{A}_j$ in λ is a sub-argument of an argument $\mathscr{A}_i$ in λ, where $i < j$.
(MPaB) For all i, such the argument $\mathscr{A}_i$ is a blocking defeater for $\mathscr{A}_{i-1}$, if $\mathscr{A}_{i+1}$ exists, then $\mathscr{A}_{i+1}$ is a proper defeater for $\mathscr{A}_i$.

*The abbreviations above stand for finitude (*Fin*), concordance (*Con*), non-reintroduction of sub-arguments (*NrS*), and move a proper defeater after a blocking defeater (*MPaB*).*

Definition 6. *Let $\lambda = [\mathscr{A}_1, \ldots, \mathscr{A}_n]$ be an argumentation line. A* segment *of λ is a sequence $\lambda' = [\mathscr{A}_i, \ldots, \mathscr{A}_j]$ with $1 \leqslant i \leqslant j \leqslant n$. A segment $\lambda' = [\mathscr{A}_i, \ldots, \mathscr{A}_j]$ of λ is* initial *if $i = 1$, and* proper *if it is initial and $j < n$.*

An argumentation line λ is *exhaustive* if there exists no line λ' such that λ is a proper segment of λ'. Otherwise, λ is *partial.* A *bundle set* is a set of argumentation lines such that no line in the set is a proper segment of another line in the set.

Definition 7. *Let F be an EAF. A set $\Lambda(\mathscr{A})$ of lines rooted in argument $\mathscr{A}$ is a* bundle set *for $\mathscr{A}$ if and only if there are no $\lambda, \lambda' \in \Lambda(\mathscr{A})$ such that λ is a proper segment of λ'.*

We denote $\Lambda(\mathscr{A})$ simply as Λ when there is no need to mention $\mathscr{A}$. There exist several kinds of bundle sets—the following definition formalizes two of them.

Definition 8. *Let $\Lambda(\mathscr{A})$ be a bundle set from an EAF F. $\Lambda(\mathscr{A})$ is* exhaustive *if and only if it is the set of all exhaustive lines rooted in $\mathscr{A}$,* i.e., *there does not exist λ'' rooted in $\mathscr{A}$ such that: (i) λ'' is not an initial segment of any $\lambda \in \Lambda$, and (ii) λ' is a proper segment of λ'', where λ' is an initial segment of some $\lambda \in \Lambda(\mathscr{A})$. Otherwise, $\Lambda(\mathscr{A})$ is* partial.

Example 3. Let $F = (Ar, \sqsubseteq, Att, Pref)$ be an EAF where $Ar = \{\mathscr{A}, \mathscr{B}, \mathscr{C}, \mathscr{D}, \mathscr{E}, \mathscr{F}, \mathscr{G}, \mathscr{H}\}$ and $\mathscr{B} \hookrightarrow \mathscr{A}, \mathscr{C} \hookrightarrow \mathscr{A}, \mathscr{D} \hookrightarrow \mathscr{B}, \mathscr{E} \hookrightarrow \mathscr{B}, \mathscr{F} \hookrightarrow \mathscr{C}, \mathscr{G} \hookrightarrow \mathscr{C}, \mathscr{H} \hookrightarrow \mathscr{F}$. The set of all exhaustive lines rooted in $\mathscr{A}$ is: $\lambda_0 = [\mathscr{A}, \mathscr{B}, \mathscr{D}]$, $\lambda_1 = [\mathscr{A}, \mathscr{B}, \mathscr{E}]$, $\lambda_2 = [\mathscr{A}, \mathscr{C}, \mathscr{F}, \mathscr{H}]$ and $\lambda_3 = [\mathscr{A}, \mathscr{C}, \mathscr{G}]$. The bundle set $\Lambda = \{\lambda_0, \lambda_1, \lambda_2, \lambda_3\}$ is exhaustive; intuitively, this is because it is the set of all exhaustive lines rooted in $\mathscr{A}$. The bundle set $\Lambda' = \{\lambda_1, \lambda_3\}$ is partial because neither λ_0 nor λ_2 are in Λ'. The bundle set $\Lambda'' = \{\lambda_0, \lambda_1, \lambda_2', \lambda_3\}$, where $\lambda_2' = [\mathscr{A}, \mathscr{C}, \mathscr{F}]$, is partial because λ_2' is partial (it is a proper segment of λ_2).

The exhaustive bundle set for any given argument is unique. If we only consider df-acceptable lines in Definition 8, then an *exhaustive and df-acceptable* bundle set is the set of all df-acceptable lines rooted in some argument $\mathscr{A}$ such that no line can be further extended while satisfying df-acceptability. In Sect. 4, in addition to exhaustive and df-acceptable bundle sets, we will refer to *exhaustive and gr-acceptable* and to *exhaustive and ib-acceptable* bundle sets, which are essentially the same as the former ones but changing the definition of acceptability satisfied by the lines.

Definition 9. *Let Λ be a bundle set from a EAF F. The* marking *of Λ is obtained by marking each $\mathscr{A}_i$ in all $\lambda \in \Lambda$ as follows:*

1. *Let $\lambda = [\mathscr{A}_1, \ldots, \mathscr{A}_n] \in \Lambda$, then $mark(\mathscr{A}_n, \lambda) = U$.*
2. *Let $\lambda = [\mathscr{A}_1, \ldots, \mathscr{A}_i, \ldots, \mathscr{A}_n] \in \Lambda$, where $1 \leqslant i < n$. Let $E = \{\lambda' \in \Lambda \,|\, [\mathscr{A}_1, \ldots, \mathscr{A}_i]$ is a proper segment of $\lambda'\}$. We have $mark(\mathscr{A}_i, \lambda) = U$ if and only if for every line $\lambda' \in E$ it holds that the defeater $\mathscr{A}_{i+1}$ of $\mathscr{A}_i$ is such that $mark(\mathscr{A}_{i+1}, \lambda') = D$. Otherwise, $mark(\mathscr{A}_i, \lambda) = D$.*

In the notation "$mark(\mathscr{A}_i, \lambda)$", λ is explicitly mentioned since $\mathscr{A}_i$ may appear in different lines, positions, and with different markings. The argument in which the bundle set Λ is rooted is denoted with $root(\Lambda)$, and its marking is denoted $mark(root(\Lambda))$.

Definition 10. *Let $F = (Ar, \sqsubseteq, Att, Pref)$ be an EAF. An argument $\mathscr{A} \in Ar$ is* accepted *according to df-acceptability in F if and only if there exists an exhaustive and df-acceptable bundle set $\Lambda(\mathscr{A})$ such that $mark(root(\Lambda(\mathscr{A}))) = U$.*

Since the EAFs presented here are inspired by DeLP and adopt the bundle set-based approach to semantics first introduced for that formalism, the results we present here can eventually be adapted for DeLP – this is the topic of future work.

4 Alternative Notions of Acceptability for Argumentation Lines

Below, we propose an alternative definition of acceptable argumentation line, called *gr-acceptable*, which adapts the grounded game [17,26,29]. Essentially, in this version of the grounded game, proponent and opponent arguments are equated with arguments in the supporting and interfering sets. Additionally, the notions of proper and blocking defeat are replaced with those of strict defeat and defeat.

Definition 11. *An argumentation line λ is* gr-acceptable *if and only if:*

- *(Fin) λ is finite.*
- *(NrSS) For all $\mathscr{A}_i, \mathscr{A}_j \in S(\lambda)$ such that $i \neq j$, it follows that $\mathscr{A}_i \neq \mathscr{A}_j$.*
- *(SdS) For all $\mathscr{A}_i \in I(\lambda)$, if $\mathscr{A}_{i+1} \in S(\lambda)$ exists, then $\mathscr{A}_{i+1}$ is a strict defeater of $\mathscr{A}_i$.*

The above abbreviations stand for finitude (Fin), non-repetition of arguments in the supporting set (NrSS), and strict defeats by the supporting set (SdS).

Definition 12. *Let $F = (Ar, \sqsubseteq, Att, Pref)$ be an EAF. An argument $\mathscr{A} \in Ar$ is* accepted *according to gr-acceptability in F if and only if there exists an exhaustive and gr-acceptable bundle set $\Lambda(\mathscr{A})$ such that $mark(root(\Lambda(\mathscr{A}))) = U$.*

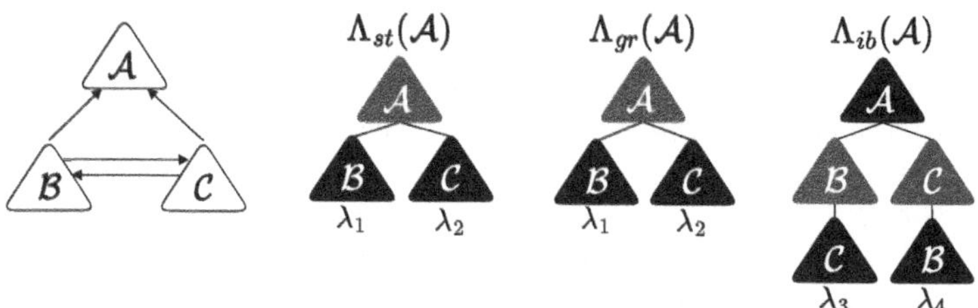

Fig. 2. Illustrating Example 4. On the right side, $\Lambda_{st}(\mathscr{A})$, $\Lambda_{gr}(\mathscr{A})$ and $\Lambda_{ib}(\mathscr{A})$ are the exhaustive and acceptable (w.r.t., Definitions 5, 11, and 14, resp.) bundle sets for $\mathscr{A}$. U-marked (resp., D-marked) arguments are colored green (resp., red). (Color figure online)

Example 4 adapts Example 1 to EAFs. We do not say anything about sub-argument relations as no proper sub-arguments are considered. Example 2 (left) illustrates Example 4, and Fig. 2 (right) shows the exhaustive bundle sets for argument $\mathscr{A}$ according to the different definitions of acceptability explored in this paper. Bundle sets are depicted as trees, where, given a bundle set Λ, $root(\Lambda)$ is the label of the root node of a tree $\mathscr{T}$, and for each node N of $\mathscr{T}$ labeled with an argument $\mathscr{B}$, N has a child N' labeled with $\mathscr{C}$ for each argument $\mathscr{C}$ defeating $\mathscr{B}$ and sharing the same line, *i.e.*, there exists $\lambda \in \Lambda$ such that $\lambda = [\ldots, \mathscr{B}, \mathscr{C}, \ldots]$. For a formal definition, please see [32].

Example 4. Let $F = (Ar, \sqsubseteq, Att, Pref)$ be an EAF where $Ar = \{\mathscr{A}, \mathscr{B}, \mathscr{C}\}$ and the defeat relations are $\mathscr{B} \hookrightarrow \mathscr{A}$, $\mathscr{C} \hookrightarrow \mathscr{A}$, $\mathscr{B} \hookrightarrow \mathscr{C}$, and $\mathscr{C} \hookrightarrow \mathscr{B}$. This is illustrated in Fig. 2.

Let us analyze the case of argument $\mathscr{A}$ in Example 4. Let $\lambda_1 = [\mathscr{A}, \mathscr{B}]$, $\lambda_2 = [\mathscr{A}, \mathscr{C}]$. $\Lambda_{st}(\mathscr{A}) = \Lambda_{gr}(\mathscr{A}) = \{\lambda_1, \lambda_2\}$, *i.e.*, the exhaustive and df-acceptable and the exhaustive and gr-acceptable bundle sets for $\mathscr{A}$ are the same (this does not hold in general as we will see later). Although $\Lambda_{st}(\mathscr{A})$ and $\Lambda_{gr}(\mathscr{A})$ share the same structure and both reject $\mathscr{A}$, *i.e.*, $mark(root(\Lambda_{st}(\mathscr{A}))) = mark(root(\Lambda_{gr}(\mathscr{A}))) = D$, they do so for different reasons. In $\Lambda_{st}(\mathscr{A})$, the *Con* principle prevents $\mathscr{C}$ (resp., $\mathscr{B}$) from being introduced after $\mathscr{B}$ (resp., $\mathscr{C}$) in λ_1 (resp., λ_2). In $\Lambda_{gr}(\mathscr{A})$, the *SdS* principle prevents $\mathscr{C}$ (resp., $\mathscr{B}$) from being introduced after $\mathscr{B}$ (resp., $\mathscr{C}$) in λ_1 (resp., λ_2).

According to a dialogue game-based semantics embedding the indecision-blocking principle, the results from Example 4 are unsatisfactory. To address this issue, we propose an alternative definition of acceptability: *indecision-blocking acceptability* (*ib-acceptability*, for short). Before that, we need to formalize the notion of *cycle* for EAFs considering the sub-argument relation.

Definition 13. *An argumentation line* $\lambda = [\mathscr{A}_1, \ldots, \mathscr{A}_n]$ *from am EAF* F *is a* cycle *if and only if* $\mathscr{A}_n$ *is a super-argument for the disagreement sub-argument of* $(\mathscr{A}_2, \mathscr{A}_1)$.

Definition 14. *An argumentation line* λ *is* ib-acceptable *if and only if:*

(Fin) *λ is finite.*

(NrDA) *No argument $\mathscr{A}_j$ in λ is a super-argument for the disagreement sub-argument of arguments $(\mathscr{A}_{i+1}, \mathscr{A}_i)$, where $\mathscr{A}_i$ and $\mathscr{A}_{i+1}$ are in λ, and $i, i{+}1 < j$.*

(MdM) *For all i such that argument $\mathscr{A}_i$ is a mutual defeater of $\mathscr{A}_{i-1}$, if $\mathscr{A}_{i+1}$ exists, then $\mathscr{A}_{i+1}$ is a proper and strict defeater for $\mathscr{A}_i$.*

(WCon) *No argument $\mathscr{A}_j \in S(\lambda)$ (resp., $\mathscr{A}_j \in I(\lambda)$) is defeated by some argument $\mathscr{A}_i \in S(\lambda)$ (resp., $\mathscr{A}_i \in I(\lambda)$), where $i < j$.*

*The abbreviations stand for finitude (*Fin*), non-reintroduction of defeated arguments (*NrDA*), mutual defeaters management (*MdM*), and weak concordance (*WCon*).*

Let us explore the principles in Definition 14. Note that, in an EAF like the one described in Example 4, according to gr-acceptability, the defeat of $\mathscr{B}$ (resp., $\mathscr{C}$) over $\mathscr{A}$ is effective (*i.e.*, it produce the rejection of $\mathscr{A}$) although $\mathscr{B}$ (resp., $\mathscr{C}$) is rejected according to gr-acceptability. Then, to block indecision, ib-acceptability must treat defeats coming from cycles differently than gr-acceptability. In this sense, our intended behavior is similar (although not the same) to the treatment carried out by df-acceptability on defeats by blocking-defeated arguments. To see this, consider the following example:

Example 5. Let $F = (Ar, \sqsubseteq, Att, Pref)$ be an EAF where $Ar = \{\mathscr{A}, \mathscr{B}, \mathscr{C}\}$ and the defeat relations are $\mathscr{B} \hookrightarrow \mathscr{A}$, $\mathscr{B} \hookrightarrow \mathscr{C}$, and $\mathscr{C} \hookrightarrow \mathscr{B}$; this is a simplified version of Example 4. Then, we have the exhaustive and df-acceptable and the exhaustive and gr-acceptable bundle sets $\Lambda_{st}(\mathscr{A}) = \{\lambda\}$ and $\Lambda_{gr}(\mathscr{A}) = \{\lambda'\}$, where $\lambda = [\mathscr{A}, \mathscr{B}, \mathscr{C}]$, $\lambda' = [\mathscr{A}, \mathscr{B}]$, and $mark(root(\Lambda_{st}(\mathscr{A}))) = U$ and $mark(root(\Lambda_{gr}(\mathscr{A}))) = D$. In this example, argument $\mathscr{A}$ is accepted in F according to df-acceptability, *i.e.*, the defeat from $\mathscr{B}$ over $\mathscr{A}$ does not affect the status of the last one, contrary to what happens according to gr-acceptability.

For the above, we take df-acceptability as the basis for ib-acceptability. However, as we already noticed in Example 4, df-acceptability does not behave as we intended. In Example 4, this is due to the *Con* principle; however, dispensing with this principle is not an option due to the fallacious behavior that it may produce (see [18,33] for more details) – nevertheless, we can weaken it to obtain our intended behavior without obtaining fallacious results. Towards this end, we add the *WCon* principle, which allows introducing conflicting arguments in the supporting (resp., interfering) set only if the argument to be introduced is not defeated by a previous argument also in the supporting (resp., interfering) set. Note that if the introduced argument defeats an argument previously introduced in the same set, this defeat will be considered in another line. Then, its effect on the root argument will be considered, and we thus avoid fallacious situations.

Furthermore, we replace the *NrS* principle with the *NrDA* principle that, intuitively, guarantees that arguments based on previously defeated arguments are not introduced in the line, and at the same time allows the introduction of non-defeated sub-arguments of already introduced arguments.

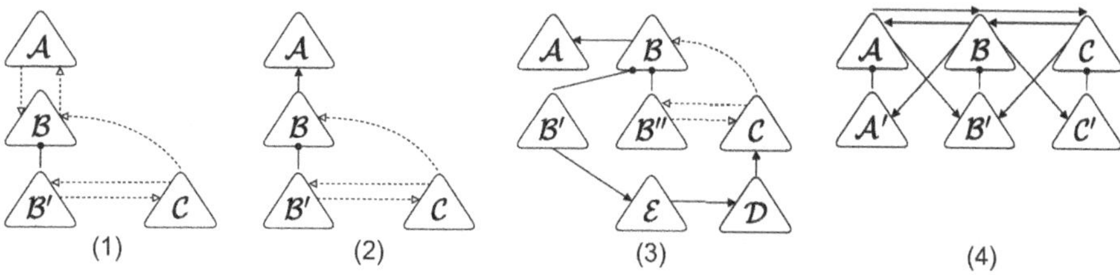

Fig. 3. Illustrating Examples 6 (graphs 1 and 2), 7 (graph 3), and 8 (graph 4). Solid, dotted, and ending-dot lines represent the proper defeat, blocking defeat, and sub-argument relations, resp.

We adhere to the treatment of df-acceptability of blocking defeaters (*MPaB* principle); however, we observe the following: although it can be argued that mutually proper defeaters are different from mutually blocking defeaters, both are cycles according to Definition 13. We thus consider it appropriate to treat both in the same manner. In this way, the *MdM* principle of ib-acceptability extends the management of blocking defeaters in df-acceptability to cycles of length two. We adopt this criterion because df-acceptability allows a sequence of mutual and proper defeaters, although they are a sequence of cycles, and as we want the defeat coming from a cycle not to be effective, we consider that treating proper and mutual defeaters in the same manner as blocking defeaters is a better option for our intended reasoning.

Definition 15. *Let $F = (Ar, \sqsubseteq, Att, Pref)$ be an EAF. An argument $\mathscr{A} \in Ar$ is* accepted *according to ib-acceptability in F if and only if there exists an exhaustive and ib-acceptable bundle set $\Lambda(\mathscr{A})$ such that $mark(root(\Lambda(\mathscr{A}))) = U$.*

Consider again Example 4, and let $\lambda_3 = [\mathscr{A}, \mathscr{B}, \mathscr{C}]$ and $\lambda_4 = [\mathscr{A}, \mathscr{C}, \mathscr{B}]$. The exhaustive and ib-acceptable bundle set for $\mathscr{A}$ is $\Lambda_{ib}(\mathscr{A}) = \{\lambda_3, \lambda_4\}$ and we have that $mark(root(\Lambda_{ib}(\mathscr{A}))) = U$. This is our intended outcome for $\mathscr{A}$. Figure 2 (right) illustrates $\Lambda_{ib}(\mathscr{A})$. Below we explain the difference in the result w.r.t. the other versions of acceptability. Unlike df-acceptability, ib-acceptability allows the introduction of $\mathscr{C}$ after $\mathscr{B} \in \lambda_1$ (resp., $\mathscr{B}$ after $\mathscr{C} \in \lambda_2$) since the *Con* principle is replaced with the *WCon* principle. As opposed to gr-acceptability, ib-acceptability does not require the *SdS*-principle, *i.e.*, that an argument in the supporting set be a strict defeater of its predecessor, then allowing to introduce $\mathscr{C}$ after $\mathscr{B} \in \lambda_1$ (resp., $\mathscr{B}$ after $\mathscr{C} \in \lambda_2$).

It is well known that the grounded semantics, and consequently, gr-acceptability, is an admissible semantics. It is easy to note from argument $\mathscr{A}$ in Example 5 that neither df-acceptability nor ib-acceptability are admissible semantics.

4.1 Examining ib-acceptability via Specially Crafted Scenarios

In the following, we delve into the analysis of ib-acceptability through some specially designed EAFs, which are adaptations of some examples first presented in [17,33].

Example 6. Let $F = (Ar, \sqsubseteq, Att, Pref)$ be an EAF where $Ar = \{\mathscr{A}, \mathscr{B}, \mathscr{B}', \mathscr{C}\}$, $\mathscr{B}' \sqsubset \mathscr{B}$, and $\mathscr{A} \rightsquigarrow \mathscr{B}$, $\mathscr{B} \rightsquigarrow \mathscr{A}$, $\mathscr{B}' \rightsquigarrow \mathscr{C}$, $\mathscr{C} \rightsquigarrow \mathscr{B}'$, and $\mathscr{C} \rightsquigarrow \mathscr{B}$. Figure 3 (1) illustrates this example. Let $\lambda = [\mathscr{A}, \mathscr{B}]$, $\lambda' = [\mathscr{A}, \mathscr{B}, \mathscr{C}]$, and $\lambda'' = [\mathscr{A}, \mathscr{B}, \mathscr{C}, \mathscr{B}']$. The exhaustive and df-acceptable bundle set for $\mathscr{A}$ is $\Lambda_{st}(\mathscr{A}) = \{\lambda\}$, and $mark(root(\Lambda_{st}(\mathscr{A}))) = D$. Let us assume that now we have $\mathscr{B} \hookrightarrow \mathscr{A}$ rather than $\mathscr{A} \rightsquigarrow \mathscr{B}$ and $\mathscr{B} \rightsquigarrow \mathscr{A}$. This is illustrated in Fig. 3 (2). Now we have $\Lambda'_{st}(\mathscr{A}) = \{\lambda'\}$ and $mark(root(\Lambda'_{st}(\mathscr{A}))) = U$.

It was pointed out in [17] that the above is a counterintuitive situation because the strengthening of $\mathscr{B}$'s defeat on $\mathscr{A}$ turns $\mathscr{A}$ from rejected into accepted. Also, it was said that adopting gr-acceptability may be a good alternative because in both the above cases, we have the exhaustive and gr-acceptable bundle set $\Lambda_{gr}(\mathscr{A}) = \{\lambda''\}$ and $mark(root(\Lambda_{gr}(\mathscr{A}))) = D$. We consider that the above is debatable, at least according to the kind of reasoning we aim to represent, due to the propagation of indecision of gr-acceptability and the fact that it allows us to reintroduce already defeated arguments. To see this in more detail, note that in Fig. 3 (1), $\mathscr{A}$ and $\mathscr{B}$, and $\mathscr{B}'$ and $\mathscr{C}$, are mutual defeaters. According to what was said above about the treatment we want for mutual defeaters, in constructing an ib-acceptable line for argument $\mathscr{A}$, we consider correct to not introduce $\mathscr{C}$ after $[\mathscr{A}, \mathscr{B}]$ because although $\mathscr{C}$ is a strict defeater of $\mathscr{B}$, it is not a proper one. In the case shown in Fig. 3 (2), we can introduce $\mathscr{C}$ after $[\mathscr{A}, \mathscr{B}]$; however, we do not consider correct to introduce $\mathscr{B}'$ after $\mathscr{C}$ because it is the disagreement sub-argument between $\mathscr{C}$ and $\mathscr{B}$ in λ', *i.e.*, $\mathscr{B}'$ is an already defeated argument. The exhaustive and ib-acceptable bundle set for $\mathscr{A}$ in scenarios (1) and (2) coincides with $\Lambda_{st}(\mathscr{A})$ and $\Lambda'_{st}(\mathscr{A})$, respectively.

Example 7. Let $F = (Ar, \sqsubseteq, Att, Pref)$ be an EAF where $Ar = \{\mathscr{A}, \mathscr{A}', \mathscr{B}, \mathscr{B}', \mathscr{C}, \mathscr{C}'\}$; $\mathscr{A}' \sqsubset \mathscr{A}$, $\mathscr{B}' \sqsubset \mathscr{B}$, $\mathscr{C}' \sqsubset \mathscr{C}$; $\mathscr{A} \hookrightarrow \mathscr{B}'$, $\mathscr{A} \hookrightarrow \mathscr{B}$, $\mathscr{B} \hookrightarrow \mathscr{A}'$, $\mathscr{B} \hookrightarrow \mathscr{A}$, $\mathscr{B} \hookrightarrow \mathscr{C}'$, $\mathscr{B} \hookrightarrow \mathscr{C}$, $\mathscr{C} \hookrightarrow \mathscr{B}'$, $\mathscr{C} \hookrightarrow \mathscr{B}$. Figure 3 (4) illustrates this example.

In Example 7, the exhaustive and df-acceptable bundle set for $\mathscr{A}$ is $\Lambda_{st}(\mathscr{A}) = \{\lambda\}$, where $\lambda = [\mathscr{A}, \mathscr{B}, \mathscr{C}]$, and $mark(root(\Lambda_{st}(\mathscr{A}))) = U$. The exhaustive and ib-acceptable bundle set for $\mathscr{A}$ is $\Lambda_{ib}(\mathscr{A}) = \{\lambda'\}$, where $\lambda' = [\mathscr{A}, \mathscr{B}]$, and $mark(root(\Lambda_{ib}(\mathscr{A}))) = D$. We consider the latter to be a better behavior w.r.t. the indecision-blocking principle given that the cycle between $\mathscr{B}$ and $\mathscr{C}$ does not affect $\mathscr{A}$. In this case $\Lambda_{ib}(\mathscr{A})$ coincides with the exhaustive and gr-acceptable bundle set $\Lambda_{gr}(\mathscr{A})$.

Example 8. Let $F = (Ar, \sqsubseteq, Att, Pref)$ be an EAF where $Ar = \{\mathscr{A}, \mathscr{B}, \mathscr{B}', \mathscr{B}'', \mathscr{C}, \mathscr{D}, \mathscr{E}\}$; $\mathscr{B}' \sqsubset \mathscr{B}$, $\mathscr{B}'' \sqsubset \mathscr{B}$; $\mathscr{B} \hookrightarrow \mathscr{A}$, $\mathscr{C} \rightsquigarrow \mathscr{B}''$, $\mathscr{B}'' \rightsquigarrow \mathscr{C}$, $\mathscr{C} \rightsquigarrow \mathscr{B}$, $\mathscr{D} \hookrightarrow \mathscr{C}$, $\mathscr{E} \hookrightarrow \mathscr{D}$, $\mathscr{B}' \hookrightarrow \mathscr{E}$. Figure 3 (3) illustrates this example.

Example 8 is used in [17] to point out issues surrounding the *NrS* principle of df-acceptability. Here we argue how the *NrDA* principle of ib-acceptability improves the *NrS* principle. Intuitively, *NrDA* allows the introduction of arguments as long as they are not based on other defeated arguments, that is, super-arguments of disagreement sub-arguments of arguments previously considered in the line. We adopt this principle mainly for two reasons: (i) a super-argument of a non-disagreement sub-argument is an argument that has not been defeated; and (ii) we consider reintroducing a super-argument of an already defeated argument to be fallacious because by having a defeated argument as a sub-argument, the super-argument is also defeated.

Let us consider the case of argument $\mathscr{A}$ in Example 8. Let $\lambda = [\mathscr{A}, \mathscr{B}, \mathscr{C}, \mathscr{D}, \mathscr{E}]$, and $\lambda' = [\mathscr{A}, \mathscr{B}, \mathscr{C}, \mathscr{D}, \mathscr{E}, \mathscr{B}']$. The exhaustive and df-acceptable (resp., ib-acceptable) bundle set for $\mathscr{A}$ is $\Lambda_{st}(\mathscr{A}) = \{\lambda\}$ (resp., $\Lambda_{ib}(\mathscr{A}) = \{\lambda'\}$) with $mark(root(\Lambda_{st}(\mathscr{A}))) = U$ (resp., $mark(root(\Lambda_{ib}(\mathscr{A}))) = D$). According to df-acceptability, we cannot append $\mathscr{B}'$ after $\mathscr{E}$ because $\mathscr{B}'$ is a sub-argument of $\mathscr{B}$. However, note that $\mathscr{B}'$ is not a disagreement sub-argument. The result obtained using ib-acceptability is a more complete alternative because it considers the possibility of introducing a sub-argument of an already defeated argument as long as the sub-argument has not been defeated. The exhaustive and gr-acceptable bundle set $\Lambda_{gr}(\mathscr{A})$ coincides with $\Lambda_{ib}(\mathscr{A})$.

Through the previous examples we have shown how ib-acceptability successfully captures the indecision-blocking principle in various scenarios. However, to fully integrate the indecision-blocking principle into dialogue game-based semantics we must solve some problems related to cycle nesting. To see this, let us consider the following example. Let $F = (Ar, \sqsubseteq, Att, Pref)$ be an EAF where $Ar = \{\mathscr{A}.\mathscr{B}, \mathscr{C}, \mathscr{D}\}$ and $\mathscr{A} \hookrightarrow \mathscr{B}$, $\mathscr{B} \hookrightarrow \mathscr{A}$, $\mathscr{B} \hookrightarrow \mathscr{C}$, $\mathscr{C} \hookrightarrow \mathscr{D}$, $\mathscr{D} \hookrightarrow \mathscr{B}$. For the case of $\mathscr{C}$, we have the exhaustive and ib-acceptable bundle set $\Lambda_{ib}(\mathscr{C}) = \{\lambda\}$ where $\lambda = [\mathscr{C}, \mathscr{B}, \mathscr{A}]$ and $mark(root(\Lambda_{ib}(\mathscr{C}))) = U$. However, our intended result for a semantics fully integrating the indecision-blocking principle is $mark(root(\Lambda_{ib}(\mathscr{C}))) = D$ due to the fact that $\mathscr{C}$ is involved in a cycle. Something similar happens with argument $\mathscr{A}$, and our intended results hold for $\mathscr{B}$ and $\mathscr{D}$. This leaves open the definition of a dialogue game-based semantics fully embedding the indecision-blocking principle.

5 Conclusions and Future Work

We have proposed, inspired by structured argumentation, a modified version of EAFs, a generalization of AFs where the sub-argument relation is considered and the defeat relation is established through the attack relation and athe preference relation.

Through the notion of acceptability for argumentation lines and following a bundle set-based approach, we have revisited two dialogue game-based semantics, the default semantics of DeLP and the grounded semantics, which we call df-acceptability and gr-acceptability, respectively. Additionally, we proposed a novel definition of acceptability, ib-acceptability, motivated by the lack of a dialogue game-based semantics embedding the indecision-blocking principle. Finally, we carried out an example-based analysis of the behavior of the above-mentioned definitions of acceptability, concluding that ib-acceptability produces our intended outcomes in several scenarios of our interest. However, we also noted that indecision-blocking is not fully integrated into ib-acceptability, leaving the door open to future definitions of acceptability that fully incorporate the indecision-blocking principle.

This work is part of an ongoing line of research on dialogue game-based semantics that involves developing a principle-based framework to study and compare several definitions of acceptability proposed in the literature. Moreover, we seek to introduce novel ones that embed other reasonable argumentative principles and, consequently, are better suited to the requirement of specific argumentative reasoning scenarios. Another objective is to explore how ib-acceptability relates to other non-admissible semantics [4,14,16], due to the similarities between the principles embedded in these semantics and ib-acceptability. Finally, we are also interested in exploring how semantics are affected by incorporating more structured components in the framework.

Acknowledgements. We thank the anonymous reviewers, which helped improve the final version of this paper. This work was funded in Argentina in part by the following institutions: Universidad Nacional del Sur (UNS) under grant PGI 24/N057, Universidad Nacional de Entre Rios (PDTS-UNER 7066), CONICET under grant PIP 11220210100577CO, and Agencia Nacional de Promocion Cientifica y Tecnologica under grant PICT-2018-0475 (PRH-2014-0007). The authors also acknowledge support by the Spanish project PID2022-139835NB-C21 funded by MCIN/AEI/10.13039/501100011033, PIE 2023-5AT010, and iTrust (PCI2022-135010-2) CHIST-ERA under grant 2022/04/Y/ST6/00001.

References

1. Alfano, G., Greco, S., Parisi, F., Simari, G.I., Simari, G.R.: An incremental approach to structured argumentation over dynamic knowledge bases. In: Thielscher, M., Toni, F., Wolter, F. (eds.) Proceedings of KR, pp. 78–87. AAAI Press (2018)
2. Baroni, P., Caminada, M., Giacomin, M.: An introduction to argumentation semantics. Knowl. Eng. Rev. **26**(4), 365–410 (2011)
3. Bench-Capon, T.J.M.: Dilemmas and paradoxes: cycles in argumentation frameworks. J. Log. Comput. **26**(4), 1055–1064 (2016)
4. Bodanza, G.A., Tohmé, F.A., Simari, G.R.: Beyond admissibility: accepting cycles in argumentation with game protocols for cogency criteria. J. Log. Comput. **26**(4), 1235–1255 (2016)
5. Briguez, C.E., Budán, M.C., Deagustini, C.A.D., Maguitman, A.G., Capobianco, M., Simari, G.R.: Towards an argument-based music recommender system. In: Verheij, B., Szeider, S., Woltran, S. (eds.) Proceedings of COMMA 2012. Front. Artif. Intell. Appl., vol. 245, pp. 83–90. IOS Press (2012)
6. Briguez, C.E., Budán, M.C., Deagustini, C.A.D., Maguitman, A.G., Capobianco, M., Simari, G.R.: Argument-based mixed recommenders and their application to movie suggestion. Expert Syst. Appl. **41**(14), 6467–6482 (2014)
7. Buron Brarda, M.: Toma de decisiones individuales y colectivas en sistemas multiagente: un enfoque argumentativo con explicaciones interactivas (2023)
8. Capobianco, M., Simari, G.R.: A proposal for making argumentation computationally capable of handling large repositories of uncertain data. In: Godo, L., Pugliese, A. (eds.) SUM 2009. LNCS (LNAI), vol. 5785, pp. 95–110. Springer, Heidelberg (2009). https://doi.org/10.1007/978-3-642-04388-8_9

9. Cohen, A., Gottifredi, S., García, A.J.: A heuristic pruning technique for dialectical trees on argumentation-based query-answering systems. In: Cuzzocrea, A., Greco, S., Larsen, H.L., Saccà, D., Andreasen, T., Christiansen, H. (eds.) FQAS 2019. LNCS (LNAI), vol. 11529, pp. 101–113. Springer, Cham (2019). https://doi.org/10.1007/978-3-030-27629-4_12
10. Deagustini, C.A.D., Dalibón, S.E.F., Gottifredi, S., Falappa, M.A., Chesñevar, C.I., Simari, G.R.: Relational databases as a massive information source for defeasible argumentation. Knowl. Based Syst. **51**, 93–109 (2013)
11. Deagustini, C.A.D., Dalibón, S.E.F., Gottifredi, S., Falappa, M.A., Chesñevar, C.I., Simari, G.R.: Defeasible argumentation over relational databases. Argument Comput. **8**(1), 35–59 (2017)
12. Dondio, P.: A proposal to embed the in dubio pro reo principle into abstract argumentation semantics based on topological ordering and undecidedness propagation. In: Dondio, P., Longo, L. (eds.) Proceedings of AI^3. CEUR Workshop Proceedings, vol. 2296, pp. 42–56 (2018)
13. Dondio, P., Longo, L.: Beyond reasonable doubt: a proposal for undecidedness blocking in abstract argumentation. Intelligenza Artificiale **13**(2), 123–135 (2019)
14. Dondio, P., Longo, L.: Weakly complete semantics based on undecidedness blocking. CoRR **abs/2103.10701** (2021)
15. Dung, P.M.: On the acceptability of arguments and its fundamental role in nonmonotonic reasoning, logic programming and n-person games. Artif. Intell. **77**(2), 321–358 (1995)
16. Dvorák, W., Rienstra, T., van der Torre, L., Woltran, S.: Non-admissibility in abstract argumentation. In: Proceedings of COMMA. Front. Artif. Intell. Appl., vol. 353, pp. 128–139. IOS Press (2022)
17. García, A.J., Prakken, H., Simari, G.R.: A comparative study of some central notions of ASPIC+ and delp. Theory Pract. Log. Program. **20**(3), 358–390 (2020)
18. García, A.J., Simari, G.R.: Defeasible logic programming: an argumentative approach. Theory Pract. Log. Program. **4**(1–2), 95–138 (2004)
19. García, A.J., Simari, G.R.: Defeasible logic programming: Delp-servers, contextual queries, and explanations for answers. Argument Comput. **5**(1), 63–88 (2014)
20. Gottifredi, S., Rotstein, N.D., García, A.J., Simari, G.R.: Using argument strength for building dialectical bonsai. Ann. Math. Artif. Intell. **69**(1), 103–129 (2013)
21. Makinson, D., Schlechta, K.: Floating conclusions and zombie paths: two deep difficulties in the "directly skeptical" approach to defeasible inheritance nets. Artif. Intell. **48**(2), 199–209 (1991)
22. Martínez, D.C., García, A.J., Simari, G.R.: An abstract argumentation framework with proper and blocking defeaters. In: X CACIC (2004)
23. Martínez, D.C., García, A.J., Simari, G.R.: On acceptability in abstract argumentation frameworks with an extended defeat relation. In: Dunne, P.E., Bench-Capon, T.J.M. (eds.) Proceedings of COMMA. FAIA, vol. 144, pp. 273–278. IOS Press (2006)
24. Martínez, D.C., García, A.J., Simari, G.R.: Progressive defeat paths in abstract argumentation frameworks. In: Lamontagne, L., Marchand, M. (eds.) AI 2006. LNCS (LNAI), vol. 4013, pp. 242–253. Springer, Heidelberg (2006). https://doi.org/10.1007/11766247_21
25. McBurney, P., Parsons, S.: Dialogue games for agent argumentation. In: Simari, G.R., Rahwan, I. (eds.) Argumentation in Artificial Intelligence, pp. 261–280. Springer, Boston (2009). https://doi.org/10.1007/978-0-387-98197-0_13

26. Modgil, S., Caminada, M.: Proof theories and algorithms for abstract argumentation frameworks. In: Simari, G.R., Rahwan, I. (eds.) Argumentation in Artificial Intelligence, pp. 105–129. Springer, Boston (2009). https://doi.org/10.1007/978-0-387-98197-0_6
27. Pollock, J.L.: Defeasible reasoning. Cogn. Sci. **11**(4), 481–518 (1987)
28. Pollock, J.L.: Defeasible reasoning with variable degrees of justification. Artif. Intell. **133**(1–2), 233–282 (2001)
29. Prakken, H.: Dialectical proof theory for defeasible argumentation with defeasible priorities (preliminary report). In: Meyer, J.-J.C., Schobbens, P.-Y. (eds.) Formal Models of Agents. LNCS (LNAI), vol. 1760, pp. 202–215. Springer, Heidelberg (1999). https://doi.org/10.1007/3-540-46581-2_14
30. Prakken, H.: Intuitions and the modelling of defeasible reasoning: some case studies. In: Benferhat, S., Giunchiglia, E. (eds.) Proceedings NMR, pp. 91–102 (2002)
31. Rotstein, N.D., Gottifredi, S., García, A.J., Simari, G.R.: A heuristics-based pruning technique for argumentation trees. In: Benferhat, S., Grant, J. (eds.) SUM 2011. LNCS (LNAI), vol. 6929, pp. 177–190. Springer, Heidelberg (2011). https://doi.org/10.1007/978-3-642-23963-2_15
32. Rotstein, N.D., Moguillansky, M.O., Simari, G.R.: Dialectical abstract argumentation: a characterization of the marking criterion. In: Boutilier, C. (ed.) Proceedings of IJCAI, pp. 898–903 (2009)
33. Simari, G.R., Chesnevar, C.I., García, A.J.: The role of dialectics in defeasible argumentation. In: XIV International Conference of the Chilenean Computer Science Society (1994)
34. Soto, Y.O., Deagustini, C.A.D., Martinez, M.V., Simari, G.I.: A mathematical conceptualization of bundle sets in defeasible logic programming. In: Proceedings of ACM/SIGAPP, pp. 779–786. SAC 2024, ACM (2024)
35. Vreeswijk, G.: Interpolation of benchmark problems in defeasible reasoning. In: Glas, M.D., Pawlak, Z. (eds.) Proc. WOCFAI, pp. 453–468. Angkor, 6, rue Maître-Albert, 75005 Paris, France (1995)

Transfer Learning Approach for High-Imbalance and Multi-class Classification of Fluorescence Images

Vincenzo Taormina(✉), Domenico Tegolo, and Cesare Valenti

Dipartimento di Matematica e Informatica, Università degli Studi di Palermo, Palermo, Italy
{vincenzo.taormina,domenico.tegolo,cesare.valenti}@unipa.it

Abstract. Recent advances in deep learning have often surpassed human performance in image classification. Among the most renowned cases, just think of the ImageNet Large Scale Visual Recognition Challenge competition. However, challenges persist in complex fields such as medical imaging. An example is the Human Protein Atlas which maps all human proteins in more than 171,000 images that makes a computation challenge due to high class imbalance. To address these challenges from a green perspective, we propose a transfer learning approach using Convolutional Neural Networks (CNNs) pre-trained on the ImageNet dataset. We use CNN layers as feature extractors, feeding the extracted features into a Support Vector Machine with a linear kernel. Our method combines both image-level and cell-level perspectives. Furthermore, at the cell level, we segment nuclei and extract the surrounding nuclear membrane area. The combination of the two perspectives shows promising classification performance with limited computational effort.

Keywords: Transfer Learning · Fluorescence Images · Multi-Class Classification · Convolutional Neural Network · Support Vector Machine

1 Introduction

In recent years, image classification with deep learning techniques has frequently outperformed human performance in various specialized domains. However, its performance can still be low in complex fields characterized by highly imbalanced datasets and multi-class classification with mixed-pattern labels. This is the case of the Human Protein Atlas (HPA) dataset [1]. The importance of automatically identifying mixed and rare protein patterns in microscope images is underscored by the competition hosted on Kaggle [2]. The competition's goal was to develop models with high performance that approach expert-level annotations, fast in prediction even on minimal hardware resources. Classifying these fluorescence images is complex due to high class imbalance, especially with rare classes, and the presence of numerous patterns that can also appear as multiple patterns in the same sample. Additionally, the diverse expression of proteins across different cell types, with three distinct regions (nucleus, nuclear membrane, and cytoplasm),

S. Destercke et al. (Eds.): SUM 2024, LNAI 15350, pp. 461–469, 2025.
https://doi.org/10.1007/978-3-031-76235-2_34

adds complexity to handling this information. These characteristics, coupled with the dataset's large size, make it complex and computationally challenging to process. Additionally, there is a desire to achieve performance comparable to domain experts while considering computational and energy efficiency from a green perspective.

The authors who proposed the HPA database and contest [1] summarize the numerous models presented by the various participants. The analysis carried out on the numerous contributions highlights the importance of assembling different pipelines of smaller models, trained both using the whole image and segmented cells. Aggarwal et al. in [3] are among the authors who have utilized the transfer learning approach from pre-trained networks. They retrained three well-known networks (VGG16, ResNet152, and DenseNet169) using a fine-tuning procedure.

In order to tackle the complex multi-class and high-class imbalance HPA dataset with an environmentally-friendly approach, we suggest using transfer learning with well-known pre-trained Convolutional Neural Networks (CNNs) from the ImageNet dataset [4]. Instead of fine-tuning, we utilize the network layers as feature extractors. The extracted features are then fed into a Support Vector Machine (SVM) with a linear kernel for classification. The main contributions of the literature on HPA image classification operate from two perspectives: at the image and cell levels. Therefore, We've combined these approaches to achieve the final classification. Additionally, we've segmented the nucleus and the surrounding nuclear membrane area to capture detailed cellular structures. Results indicate that our approach, achieves promising performance with a limited computational effort.

The remainder of this paper is organized as follows. In the next section, we provide a detailed description of the materials and methodology. This includes a discussion of the dataset, the pre-trained CNNs used, and the image preprocessing and segmentation processes. We conclude Sect. 2 with the methodology for feature extraction and classification with SVM. Section 3 presents the experimental results and discusses the findings including the impact of using image, nucleus and nuclear membrane levels perspectives. Finally, Sect. 4 concludes the paper and discusses future research.

2 Description of the Materials and Methods

2.1 Human Protein Atlas Dataset

The HPA dataset [1] coming from an open-access Swedish project, aims to create a freely available map of all human proteins in cells, tissues, and organs. It is a very large annotated collection of fluorescence microscopy images of human cells, comprising more than 42000 samples of cells captured using confocal microscopy. It is characterized by highly imbalanced multi-class data with 28 patterns that can appear in mixed forms. For each sample there are four images [1], the antibody-stained protein of interest (green channel) and three reference channels to outline the cell: microtubules (red channel), nucleus (blue channel) and endoplasmic reticulum (yellow channel). Figure 1 shows the four channels from a sample. The dataset comprises 31072 samples as public-training and 11702 as private-test. Overall, there are more than 171,000 images in grayscale format (8 bits per pixel) available in the original acquisition resolution (2048×2048 or 3072×3072) and in scaled version (512×512).

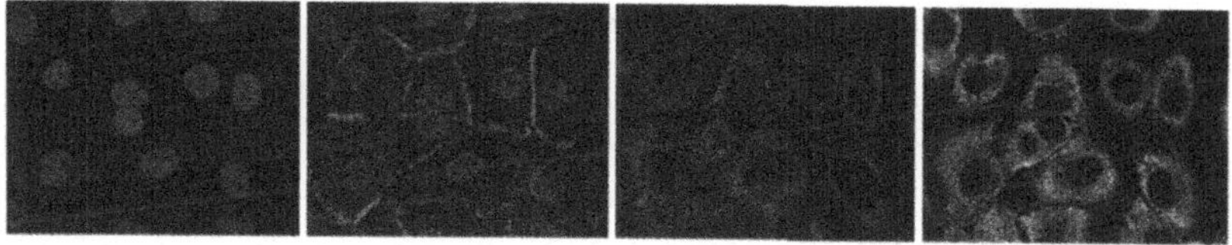

Fig. 1. The four channels composing a sample in the HPA dataset: the reference channels to outline the cell nucleus (blue), the antibody-stained protein of interest (green), microtubules (red), and endoplasmic reticulum (yellow). (Color figure online)

2.2 Pre-trained Convolutional Neural Networks

In this work we exploited the potential of well-known CNN architectures pre-trained on the very large ImageNet dataset. These networks have demonstrated their high performance in image classification, as shown in the ImageNet Large Scale Visual Recognition Challenge (ILSVRC) [5]. The main reason behind this success is that discriminative features for image classification can be intrinsically learned through the hierarchy of convolutional kernels within CNNs. An additional important characteristic is that they can also be fine-tuned on other databases [6]. This is significant from a green perspective because it allows for the reuse of already trained models, reducing the need for extensive computational resources and energy consumption required for training models from scratch. The property of transfer learning can be applied to CNNs either through fine-tuning, where the entire architecture is retrained on new data, or by freezing some of the layers and retraining only the weights of the last layers. Additionally, the layers of a pre-trained CNN can be used as feature extractors, capturing coarse details in the initial layers and finer details in the deeper layers. This transfer learning strategy, capable of extracting rich and generic features, allows these features to be utilized in a traditional classifier like a SVM, enabling tuning of the classifier's parameters even with limited data. Furthermore, by leveraging pre-trained models as feature extractors, we minimize the computational overhead and energy consumption, contributing to a more sustainable approach to deep learning. For this study, we used several pre-trained CNN architectures to extract features from the HPA images. The specific architectures utilized are listed in Table 1.

Table 1. Pre-trained CNN architectures and their respective number of layers.

Pre-trained CNNs	# Layers	Pre-trained CNNs	# Layers
AlexNet [7]	25	GoogleNet [12]	144
VGG-16 [8]	41	ResNet-50 [11]	177
VGG-19 [8]	47	Inception-v3 [13]	315
MobileNet-v2 [9]	53	ResNet-101 [11]	347
SequeezeNet [10]	68	DenseNet-201 [14]	708
ResNet-18 [11]	68	InceptionResNet-v2 [15]	824

2.3 Image Preprocessing and Segmentation

To apply transfer learning based on feature extraction using well-known CNN layers, the HPA images must be appropriately preprocessed before being fed into the CNNs. Specifically, each CNN requires a 3-channel image input with 8 bits per pixel for each channel. The required input sizes are 224 × 224 × 3 (MobileNet-v2, ResNet-18, ResNet-50, ResNet-101, DenseNet-201, GoogleNet, VGG-16, VGG-19), 227 × 227 × 3 (AlexNet, SqueezeNet), or 299 × 299 × 3 (Inception-v3, InceptionResNet-v2), depending on the CNN architecture.

At the image level, we resized with bicubic interpolation to the input CNNs image size, the scaled version images (512 × 512 pixels) to match the input size required by the CNNs. Since each sample contains four channels, in this work only the green channel was used that is the antibody-stained protein of interest, and we simply replicated the green channel three times.

At the cell level, segmentation is necessary. As suggested in [2], participants in the competition could use well-established methods such as the pretrained HPACellSegmentator [16] or well-knowen Cellpose [17]. The Cellpose Library provides several pretrained models for cell segmentation in microscopy images, which can be applied directly for accurate and fast results. We used the Cellpose Library via the Medical Imaging Toolbox in MATLAB R2024a [18]. We applied the pretrained Cellpose model "nuclei" instead of "cyto" or "cyto2" because "nuclei" is optimized for microscopy images where the edges of nuclei are well-highlighted, while the latter two work well in microscopy images with cytoplasmic cells. In our work, we utilized the blue channel of the scaled version images (512 × 512), which highlights only the cell nuclei. This is the reference channel to outline the cell nucleus [1], so we use this channel to apply the segmentation process and obtain a binary mask.

In addition to set the "nuclei" model and the average nucleus diameter to 55 pixels, we used default parameters and qualitatively appreciated the excellent segmentation results and execution speed. The following Fig. 2 shows two examples of segmentation starting from the blue channel and displaying the segmented nuclei highlighted in different colors. The choice to segment images in scaled version (512 × 512) compared to the original full-size images (2048 × 2048 or 3072 × 3072) was due to the execution speed (approximately 1 s versus 60 s in the same framework).

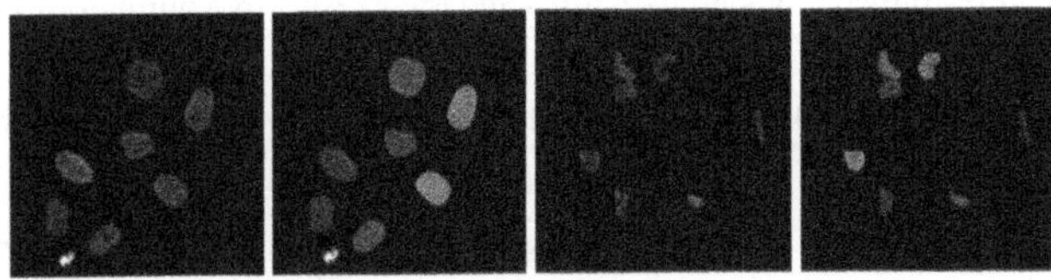

Fig. 2. Two examples of segmentation with pretrained Cellpose Library on the blue channel, showing the segmented nuclei highlighted in different colors. (Color figure online)

However, since the details of the cells appear pixelated in the scaled images, we apply the segmentation mask found in the scaled images to the original full-size images after resizing the mask to the respective full-size dimensions.

Our research focused on segmenting only one cell per sample to reduce computational load. We excluded segmented nuclei at the edges of the images and removed cells smaller than a minimum threshold of 100 pixels based on their area. We chose the cell with the median area from the remaining segmented cells. We then cropped the respective bounding box to extract the nucleus and set the background around the nucleus to zero. The average size of the bounding box side is 226 pixels, with a maximum of 392 pixels and a minimum of 96 pixels.

Given the particular interest in the nuclear membrane area, we also extracted a ring that incorporates this membrane. This process applies a morphological dilation operation to the nucleus mask and then subtracts the eroded version of the original mask. For dilation and erosion, we used a disk-shaped mask of 5×5.

After obtaining the segmentation mask (using the blue channel), the segmentation of the nuclei and the ring that incorporates the nuclear membrane is performed on the green channel, which represents the antibody-stained protein of interest. Figure 3 highlights an example of nucleus and nuclear membrane ring extraction. Moreover, at the cell level, after segmentation, nuclei and ring segmentations are resized to match the dimensions required as input for the CNNs. No additional preprocessing are applied to the images beyond resizing for CNNs input.

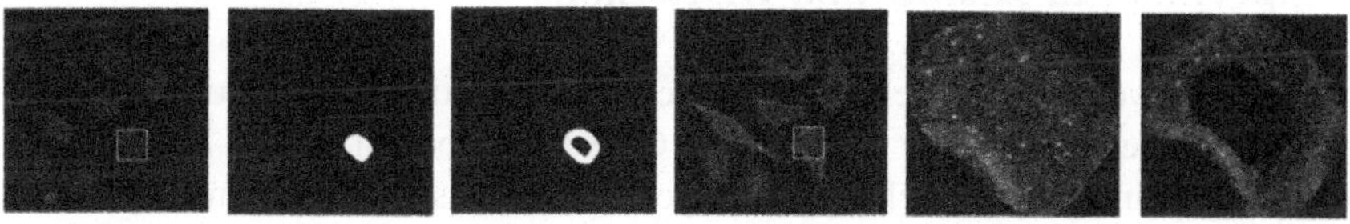

Fig. 3. Illustration of the segmentation process. (From left to right) Blue channel scaled image used for segmentation. Binary mask of a segmented nucleus. Mask of the segmented ring. Original full-size image of the green channel. Segmented nucleus. Segmented nuclear membrane.

2.4 Data Splitting, Features Extraction and Classification

This work focused on multi-class classification by discarding multi-label cases from the public training set, resulting in slightly over 15,000 samples from the original 31,000. This reduced set included 24 classes, as the "Lysosomes," "Microtubule Ends" and "Mitotic Spindle Microtubule" classes had no samples, and "Rods and Rings" had only one. We used 1,008 samples for training and the remaining 14,117 samples for testing. To address the highly imbalanced data, we aimed to balance the training set where possible, leaving the excess cases in the test set. Cases for the training set were randomly selected, with a maximum threshold of 45 samples for sufficiently numerous classes. For classes with fewer than 50 samples, we divided them 80% for training and 20% for testing. This approach balanced the training set while keeping the test set imbalanced. The strategy used for the training-validation-test chain was 5-fold cross-validation, where the 1,008 samples were divided into five blocks. Iteratively, one block was used for validation while the remaining blocks were used for training. The 14,117 samples were exclusively used for testing.

Feature extraction is performed on the green channel for both the image level and the nucleus and surrounding nuclear membrane. For each CNN, we selected the following final layers: "fc8" for AlexNet and VGG architectures; "fc_1000" for DenseNet and the three ResNet architectures; "predictions" for InceptionResNet-v2 and Inception-v3; "pool10" for SqueezeNet; "loss3-classifier" for GoogleNet and "logits" for MobileNet-v2. These layers enable the extraction of 1,000 features each, as the networks are trained to recognize 1,000 different objects as specified in the ImageNet competition [5]. Therefore, when we use these final layers as feature extractors, we utilize these 1,000 values as feature vectors.

The classification is performed by using an SVM with a linear kernel. The primary parameter to optimize in cross-validation is the penalty parameter "C," with the F1 score chosen as the figure of merit. One Against One (OAO) and One Against All (OAA) are well-known strategies for implementing multi-class classification. We chose the OAO strategy, which has demonstrated superior results in our experiments. As an ensemble strategy, we simply concatenated the feature vectors before training the SVM. This allowed us to test various combinations, both among the twelve CNNs and between the image level and the nucleus and nuclear membrane levels.

3 Experimental Results and Discussions

In this section, we present the classification results obtained using our proposed method. We evaluated the performance of individual CNNs contribution as well as various concatenation combinations. The classification performance is summarized in Table 2, which lists the F1-scores, the Accuracy (ACC) and the Balanced Accuracy (BAC) for the main cases analyzed. Concatenation methods generally outperformed individual CNNs, demonstrating the effectiveness of combining features from multiple sources. Specifically, the ensemble that combined features from both the image level and the nucleus and nuclear membrane levels achieved the highest F1-score and exceeded 70% in ACC and BAC. To perform these experiments, we utilized an HP Elitebook notebook with 16 GB of RAM, an Intel Core i7 8th Gen processor, and Windows 10 as the operating system. We extracted the features and performed the classification using MATLAB R2024a [18]. This setup demonstrates that our approach does not require particularly powerful hardware resources. The computation time for training varies based on the number of concatenated features. On average, training the SVMs in OAO mode takes about 330.42 s for 1,000 feature vectors. When concatenating 12,000 features, the average training time increases to 1,380.19 s. The maximum training time, when concatenating 36,000 features, is just over 1 h. It should be noted, as shown in other studies [3][6], that the computation times for fine-tuning a well-known CNN are significantly higher, and careful tuning of hyperparameters is required to avoid overfitting. For a rough comparison with fine-tuning a known CNN, we trained the VGG16 model by freezing the weights of approximately the first half of the layers in the architecture. The computation time for this process was about 6 h. We emphasize that our goal is to use a minimal computational resources by leveraging pre-trained CNNs, for a green approach. It's important to note that these results are based on a small training dataset of approximately 1,000 examples used to train a multiclass classifier with 24 classes. Additionally, only one cell per sample was considered. The significant improvement in performance observed with the ensemble of image, nucleus, and nuclear membrane levels highlights the efficacy of integrating multi-level features.

Table 2. Summary of classification results.

Method	F1-score	ACC	BAC
Image Level Best Layer	0,37	46,21	45,75
Image Level Concatenate Layers	0,48	59,16	59,20
Nucleus Level Concatenate Layers	0,31	42,85	41,03
Membrane Level Concatenate Layers	0,27	37,92	36,11
Nucleus & Membrane Concatenate Layers	0,34	45,45	44,97
Image & Nucleus & Membrane Concatenate Layers	**0,59**	**70,92**	**71,77**

4 Conclusions and Future Work

In this work, our primary aim was to provide a computationally efficient solution to the complex multi class and high-class imbalance HPA Dataset. We chose a green approach by reusing pre-trained CNNs as feature extractors to minimize both computational effort and the use of training data, knowing that training models from scratch or fine-tuning pre-trained CNNs would significantly increase resource consumption and data requirements. Our transfer learning approach has shown promising results in efficiently classifying the complex HPA dataset. By leveraging transfer learning with pre-trained CNNs and combining image-level with cell-level features, we achieved significant improvements in classification performance. The ensemble of features from both the nucleus and nuclear membrane levels proved particularly effective. Notably, our method operates efficiently on modest hardware resources.

Moving forward, our research will progress to enhance model robustness and performance. Key areas of focus include gradually expanding the training dataset to refine our classifiers and exploring additional optimization strategies, such as multi-label classification with a stacking approach. Since in multi-label images cells can exhibit different characteristics [1], we plan to implement a clustering-based strategy [19] and investigate other factor on traits all the cells segmented. We aim to investigate the impact of utilizing different combinations of four channels, beyond just the green channel, to assess their influence on the classification performance. This analysis can help to better understand the role of multi-channel data in enhancing the accuracy of classification models. Furthermore, we will assess the calibration properties [20] of the classifiers using metrics like Expected Calibration Error. We plan to extend our evaluations to larger datasets and diverse biological contexts, ensuring broad applicability and robust performance across various domains.

Acknowledgments. C. Valenti and V. Taormina are supported by the research fund of the University of Palermo: FFR 2024. Cesare Valenti is member of the "Gruppo Nazionale Calcolo Scientifico - Istituto Nazionale di Alta Matematica (GNCS-INdAM)". The research leading to these results was also supported by the European Union–NextGenerationEU through the Italian Ministry of University and Research under PNRR–M4C2–I1.3 Project PE_00000019 "HEAL ITALIA" to Domenico Tegolo, CUP B73C22001250006.

References

1. Ouyang, W., et al.: Analysis of the human protein atlas image classification competition. Nat. Methods **16**(12), 1254–1261 (2019)
2. Human Protein Atlas Image Classification, available online https://www.kaggle.com/competitions/human-protein-atlas-image-classification. Accessed 25 June 2024
3. Aggarwal, S., et al.: An artificial intelligence-based stacked ensemble approach for prediction of protein subcellular localization in confocal microscopy images. Sustainability **15**(2), 1695 (2023)
4. Deng, J., Dong, W., Socher, R., Li, L. J., Li, K., Fei-Fei, L.: Imagenet: a large-scale hierarchical image database. In: 2009 IEEE Conference on Computer Vision and Pattern Recognition, pp. 248–255. IEEE (2009)
5. Russakovsky, O., et al.: Imagenet large scale visual recognition challenge. Int. J. Comput. Vis. **115**, 211–252 (2015)
6. Taormina, V., Cascio, D., Abbene, L., Raso, G.: Performance of fine-tuning convolutional neural networks for HEP-2 image classification. Appl. Sci. **10**(19), 6940 (2020)
7. Krizhevsky, A., Sutskever, I., Hinton, G.E.: ImageNet classification with deep convolutional neural networks. Adv. Neural. Inf. Process. Syst. **2012**(25), 1097–1105 (2012)
8. Simonyan, K., Zisserman, A.: Very Deep convolutional networks for large-scale image recognition. ArXiv 2014, arXiv:1409.1556 (2014)
9. Sandler, M., Howard, A., Zhu, M., Zhmoginov, A., Chen, L. C.: Mobilenetv2: inverted residuals and linear bottlenecks. In Proceedings of the IEEE Conference on Computer Vision and Pattern Recognition, pp. 4510–4520 (2018)
10. Iandola, F.N., et al.: Squeezenet: Alexnet-level accuracy with 50× fewer parameters and <0.5 mb model size. arXiv 2016, arXiv:1602.07360 (2016)
11. He, K., Zhang, X., Ren, S., Sun, J.: Deep residual learning for image recognition. In: Proceedings of the IEEE Computer Society Conference on Computer Vision and Pattern Recognition, Las Vegas, NA, USA, 26 June–1 July 2016, pp. 770–778 (2016)
12. Szegedy, C., et al.: Going deeper with convolutions. In Proceedings of the IEEE Conference on Computer Vision and Pattern Recognition, Boston, MA, USA, 7–12 June 2015 (2015)
13. Szegedy, C., Vanhoucke, V., Ioffe, S., Shlens, J., Wojna, Z.: Rethinking the inception architecture for computer vision. In Proceedings of the IEEE Conference on Computer Vision and paTtern Recognition, pp. 2818–2826 (2016)
14. Huang, G., Liu, Z., Van der Maaten, L., Weinberger, K.Q.: Densely connected convolutional networks. In Proceedings of the IEEE Conference on Pattern Recognition and Computer Vision 2017, Honolulu, HI, USA, 21–26 July 2017, pp. 4700–4708 (2017)
15. Szegedy, C., Ioffe, S., Vanhoucke, V., Alemi, A.: Inception-v4, inception-resnet and the impact of residual connections on learning. In: Proceedings of the AAAI Conference on Artificial Intelligence, vol. 31, no. 1 (2017)
16. HPACellSegmentator. https://github.com/CellProfiling/HPA-Cell-Segmentation
17. Stringer, C., Wang, T., Michaelos, M., Pachitariu, M.: Cellpose: a generalist algorithm for cellular segmentation. Nat. Methods **18**, 100–106 (2021)

18. The MathWorks Inc. MATLAB version: R2024a (2024). https://www.mathworks.com
19. Vivona, L., et al.: Unsupervised clustering method for pattern recognition in IIF images. In: International. Image Processing, Applications and Systems, pp. 1–6, IEEE (2016)
20. Posocco, N., Bonnefoy, A.: Estimating expected calibration errors. In: Artificial Neural Networks and Machine Learning–ICANN 2021, p. 139–150 (2021)

Stability of Extensions in Incomplete Argumentation Frameworks

Anshu Xiong[1,2(✉)], Huan Zhang[1,2], and Songmao Zhang[1]

[1] Key Laboratory of MADIS, Academy of Mathematics and Systems Science, Chinese Academy of Sciences, 100190 Beijing, China
zhanghuan@amss.ac.cn, smzhang@math.ac.cn

[2] University of Chinese Academy of Sciences, 100049 Beijing, China
xionganshu21@mails.ucas.ac.cn

Abstract. Existing works on stability of incomplete argumentation frameworks (IAFs) discuss the status of an argument or a set of arguments under a given semantics during the changing of an IAF towards complete AFs. We argue that the stability of an IAF itself is worth studying, i.e., checking whether all extensions under a semantics are the same in every completion of the IAF. When an IAF becomes stable in this sense, there is no need to investigate the uncertain arguments or attacks within the IAF as in the end all its complete AFs will share the same extensions. We further present a relaxed notion called weak stability so that the same extensions are required solely within the shared arguments of every two completions rather than all arguments in the IAF. In this paper we study six common semantics of AF and show precise complexity results of checking the (weak) stability of an IAF, which turns to be more complex than checking a set of arguments whereas less complex than checking an argument. Meanwhile, we give a SAT encoding for the stability problems with $coNP$-c complexity.

Keywords: Abstract argumentation · Incomplete knowledge · Stability · Computational complexity

1 Introduction

The *Abstract Argumentation Framework (AF)* proposed by Dung in 1995 [13] represents the attack relations among a group of arguments where justified sets of arguments (called extensions) can be computed under various types of semantics. Uncertainty is introduced into AF either quantitatively as in [17–19] or qualitatively as in [6,9,11]. *Incomplete Argumentation Framework (IAF)* [6] is one of the most prominent proposals for modeling qualitative uncertainty. An IAF contains not only certain arguments and attacks as classical AFs but also uncertain arguments and attacks whose existence is unknown. A central notion called *completion* represents an AF obtained when all the uncertain elements of an IAF have been determined existent or not. By completions, the *verification*

S. Destercke et al. (Eds.): SUM 2024, LNAI 15350, pp. 470–485, 2025.
https://doi.org/10.1007/978-3-031-76235-2_35

problem of *extension* in AF can be adapted to *i-extension* [6] and *i*-extension* [15] in IAF which refer to a set of arguments being (part of) an extension in at least one or all completions of an IAF.

Initially studied in *structured argumentation frameworks* [26], the issue of *stability* for argumentation with uncertainty generally refers to that some problem of interest will have the same answer in the future despite the uncertainty in the current framework. The verification problems of IAFs discussed in [6,15] can be viewed as studying the stability of a set of arguments. Concretely, a set of arguments is stable w.r.t. an IAF I if it is either an extension or not in every completion of I. Subsequently, the stability of an argument in an IAF is studied in [5,23] and later a finer-grained version in [24], where an argument is stable w.r.t. an IAF I under a *justification status* i (ranging over $\{sceptical, credulous\} \times \{in, out, undec\}$ for a single argument) only when the justification status of the argument remains i in all completions of I. Meanwhile, *functionality* studied in [3] presents a more strictly stable status for a single argument, which specifies if an argument can be firmly decided in any completion.

Existing works on stability of IAF discuss the status of a given argument or a set of arguments. We argue that the stability of an IAF itself is worth studying, i.e., whether all the completions of a given IAF generate the same extensions under a type of semantics. Theoretically, this problem is essentially different from the stability of a single argument or a set of arguments as it is concerned with the stability of the whole sets of arguments. Further, considering the strictness of stability of extensions of an IAF, we present a relaxed notion named *weak stability* so that the same extensions are required solely within the shared arguments of every two completions rather than all arguments in the IAF. Practically, consider an interactive system for domain experts to construct an AF from a preliminary IAF. At each step they decide the existence of one uncertain entity, either an attack or an argument. If at certain point, we can tell that the current IAF has become stable, then there is no need for the experts to do any input as long as the purpose of the experts is to obtain extensions. In addition, when the IAF becomes weakly stable though not stable yet, the experts just need to focus on deciding the uncertain arguments and ignore the uncertain attacks, since any completions with the same arguments of a weakly stable IAF will share the same extensions no matter how the uncertain attacks turn. Such argumentative settings can benefit from the notion of (weak) stability of extensions in IAF.

The contribution of this paper consists of a study on stability and weak stability of extensions in IAF, where we give precise complexity results of checking these two kinds of stability under six common semantics[1] consisting of conflict-free, admissible, stable [2], complete, grounded, and preferred semantics. Particu-

[1] Following [6] we regard conflict-freeness and admissibility as common semantics, whereas originally in the Dung's paper [13] they are basic notions to define semantics.

[2] Note that the notion of stable semantics is irrelevant to the stability problem of IAF studied in the literature and this paper.

larly, for conflict-free and grounded semantics, we find tractable methods which can be conducted in polynomial time for checking both kinds of stability. For other types of semantics, we prove the hardness of their stability problems by reduction from relevant known hard problems of AFs. Further for admissible, stable and complete semantics, we provide a SAT encoding, which can be used to solve their stability problems by directly calling a SAT solver.

2 Preliminaries

Argumentation Framework (AF). An *abstract argumentation framework* [13] is a directed graph $F = \langle \mathcal{A}, \mathcal{R} \rangle$ where $\mathcal{A}$ represents a set of considered arguments and $\mathcal{R} \subseteq \mathcal{A} \times \mathcal{A}$ the set of attacks between arguments. We say that a *attacks* b if $(a, b) \in \mathcal{R}$, a *attacks* $S \subseteq \mathcal{A}$ if there exists an argument $b \in S$ s.t. $(a, b) \in \mathcal{R}$, and the meaning of S *attacks* a is analogous. Given a set of arguments $S \subseteq \mathcal{A}$, let $S_F^+ = \{a \in \mathcal{A} \mid S \text{ attacks } a\}$, $S_F^- = \{a \in \mathcal{A} \mid a \text{ attacks } S\}$, $F_{\downarrow S} = \langle S, \mathcal{R} \cap (S \times S) \rangle$ representing the *subgraph* of F on S, and further given a set of attacks $\mathcal{R}'$, let $\mathcal{R}'|_S = \mathcal{R}' \cap (S \times S)$. We say that S *defends* a if all the attackers of a are attacked by S and use the so-called *characteristic function* $\Gamma_F(S)$ [13] to denote all the arguments that S defends, i.e. $\{a \in \mathcal{A} \mid S \text{ defends } a \text{ in } F\}$.

Semantics of AF. A *semantics* σ is a function whose input is an AF $F = \langle \mathcal{A}, \mathcal{R} \rangle$ and output $\sigma(F)$ is a set of subsets of $\mathcal{A}$, so-called *σ-extensions* and any element of $\sigma(F)$ is called an *σ-extension* of F. In this paper we consider *conflict-free, admissible, stable, complete, preferred and grounded* semantics (abbr. `cf`, `ad`, `st`, `co`, `pr`, `gr`) originally proposed in [13].

Definition 1. *Given an AF $F = \langle \mathcal{A}, \mathcal{R} \rangle$ and $S \subseteq \mathcal{A}$,*

1. $S \in \mathtt{cf}(F)$ *iff* $\forall a, b \in S, (a, b) \notin \mathcal{R}$,
2. $S \in \mathtt{ad}(F)$ *iff* $S \in \mathtt{cf}(F)$ *and* $S \subseteq \Gamma_F(S)$,
3. $S \in \mathtt{st}(F)$ *iff* $S \in \mathtt{cf}(F)$ *and* $S_F^+ = \mathcal{A} \backslash S$,
4. $S \in \mathtt{co}(F)$ *iff* $S \in \mathtt{ad}(F)$ *and* $\forall a \in \Gamma_F(S), a \in S$,
5. $S \in \mathtt{pr}(F)$ *iff* S *is* $\subseteq$*-maximal in* $\mathtt{ad}(F)$, *and*
6. $S \in \mathtt{gr}(F)$ *iff* $S = \bigcup_{i=0}^{\infty} \Gamma_F^i(\emptyset)$, *where* $\Gamma_F^0(\emptyset) = \emptyset$ *and* $\Gamma_F^{i+1}(\emptyset) = \Gamma_F(\Gamma_F^i(\emptyset))$.

Now we recall the IAF that expands AF with qualitative uncertainty.

Definition 2 (IAF). *[6] An incomplete argumentation framework (IAF) is a quadruple $\langle \mathcal{A}, \mathcal{A}^?, \mathcal{R}, \mathcal{R}^? \rangle$ where $\mathcal{A}, \mathcal{A}^?$ are disjoint sets of arguments and $R, R^?$ are disjoint subsets of $(\mathcal{A} \cup \mathcal{A}^?) \times (\mathcal{A} \cup \mathcal{A}^?)$. $\mathcal{A}$ (resp., $\mathcal{R}$) is the set of arguments (resp., the set of attacks) that are known to definitely exist, while $\mathcal{A}^?$ (resp., $\mathcal{R}^?$) contains all possible additional arguments (resp., all possible additional attacks) not (yet) known to exist.*

Note that Definition 2 does not impose any constraints on the connections between certain and uncertain elements, e.g., uncertain arguments can issue certain attacks.

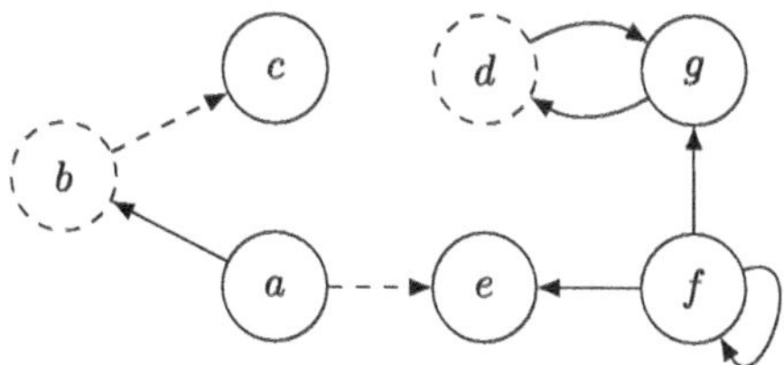

Fig. 1. An example of IAF I_0 (uncertain arguments and attacks are depicted using dashed circles and lines, respectively)

The notion of *completion* [6] represents the resolution of uncertainty in IAFs. A completion of an IAF is an AF that contains all certain information of IAF and each uncertain element in IAF is decided as existent or not. Precisely, the upper bound of the number of possible completions of an IAF is $2^{|\mathcal{A}^?|+|\mathcal{R}^?|}$.

Definition 3 (Completion). *[6] Given an IAF $I = \langle \mathcal{A}, \mathcal{A}^?, \mathcal{R}, \mathcal{R}^? \rangle$, a completion of I is an abstract argumentation framework $\langle \mathcal{A}', \mathcal{R}' \rangle$ with $\mathcal{A} \subseteq \mathcal{A}' \subseteq \mathcal{A} \cup \mathcal{A}^?$ and $\mathcal{R}|_{\mathcal{A}'} \subseteq \mathcal{R}' \subseteq (\mathcal{R} \cup \mathcal{R}^?)|_{\mathcal{A}'}$.*

In [15], the i^*-*extensions* are proposed as specifications of i-*extensions* in [6], where the problems of stability of a set of arguments in IAFs are solved as the verification problems about i^*-extensions and the corresponding complexity results are given.

Definition 4 (i^*-extension). *[15] Given an IAF $I = \langle \mathcal{A}, \mathcal{A}^?, \mathcal{R}, \mathcal{R}^? \rangle$ and a semantics σ, a set of arguments S is a possible (resp.,necessary) i^*-extension of I under σ iff, for at least one (resp., for every) completion C of I, S is a σ-extension of C. Let* IPosVer$^\sigma(I, S)$ *(resp.,* INecVer$^\sigma(I, S)$*) be the problem of checking whether S is a possible (resp.,necessary) i^*-extension of I under σ. We say that S is stable w.r.t. I under σ iff* INecVer$^\sigma(I, S)$ *= true or* IPosVer$^\sigma(I, S)$ *= false, i.e. S is necessary or impossible to be σ-extension in every completion of I.*

Given an IAF $I = \langle \mathcal{A}, \mathcal{A}^?, \mathcal{R}, \mathcal{R}^? \rangle$ and a set of arguments $S \subseteq \mathcal{A} \cup \mathcal{A}^?$, similarly to notations for AF, we give the following notations:

- $S_I^+ = \{a \in \mathcal{A} \cup \mathcal{A}^? \mid \exists b \in S,\ (b, a) \in \mathcal{R}\}$;
- $S_I^- = \{a \in \mathcal{A} \cup \mathcal{A}^? \mid \exists b \in S,\ (a, b) \in \mathcal{R}\}$; and
- $S_I^u = \{a \in \mathcal{A} \cup \mathcal{A}^? \mid \forall b \in S,\ (b, a) \notin \mathcal{R} \cup \mathcal{R}^?\}$.

Example 1. Consider the IAF I_0 given in Fig. 1. Let $S = \{a, c\}$, and we have $S_{I_0}^+ = S_{I_0}^- = \{b\}$ and $S_{I_0}^u = \{d, f, g\}$. One can see that S is a necessary i^*-extension under gr semantics, thus S is stable w.r.t. I_0 under gr. On the other hand, $S \cup \{d\}$ is not stable w.r.t. I_0 under co semantics, based on the fact that $S \cup \{d\}$ is a possible i^*-extension but not a necessary i^*-extension under co since the uncertainty of d leads to its nonexistence in some completions of I_0.

3 Stability of Extensions in IAFs

Now we propose the notion of *stability* of IAFs, or precisely the stability of all extensions under a semantics in IAFs. Intuitively, we consider an IAF to be stable under semantics σ once all of its completions possess the same σ-extensions.

Definition 5 (Stability of IAF). *Given an IAF $I = \langle \mathcal{A}, \mathcal{A}^?, \mathcal{R}, \mathcal{R}^? \rangle$ and a semantics σ, I is stable under σ iff for any two completions C_1 and C_2 of I, $\sigma(C_1) = \sigma(C_2)$.*

In order to check the stability of an IAF on extensions, we can alternatively check the stability of all of its sets of arguments. Once each set of arguments is stable, those turning to be an extension of some completion are also necessarily extensions of all completions, which means that the extensions of each completion are the same. And whenever an unstable set of arguments is detected, there exist at least two completions with different extensions, hence the IAF is not stable. Formally, we have:

Proposition 1. *Given an IAF $I = \langle \mathcal{A}, \mathcal{A}^?, \mathcal{R}, \mathcal{R}^? \rangle$ and semantics σ, I is stable under σ iff $\forall S \subseteq \mathcal{A} \cup \mathcal{A}^?$, S is stable w.r.t. I under σ.*

Attentive readers may find that the condition for the stability of an IAF to hold in Definition 5 is strict. Specifically, i) it requires that all completions must share the same extensions despite that they may have different arguments, and ii) it implies the impossibility for a set containing uncertain arguments to be an extension, since such set must be unstable and leads the unstability of extensions of the IAF. Furthermore, it means that every uncertain argument is not acceptable in completions, as shown by Proposition 2.

Proposition 2. *Given a stable IAF $I = \langle \mathcal{A}, \mathcal{A}^?, \mathcal{R}, \mathcal{R}^? \rangle$ under semantics σ, for any set of arguments $S \subseteq \mathcal{A} \cup \mathcal{A}^?$ s.t. $S \cap \mathcal{A}^? \neq \emptyset$, S is not a σ-extension in any completion of I.*

To relax the strictness, we propose a weak version of stability of IAF named *weak stability* such that i) for any two completions of the IAF, only extensions composed of the shared arguments of two completions are required to be the same, and ii) sets containing uncertain arguments can become extensions in completions.

Definition 6 (Weak Stability). *Given an IAF $I = \langle \mathcal{A}, \mathcal{A}^?, \mathcal{R}, \mathcal{R}^? \rangle$ and semantics σ, I is weakly stable under σ iff for any two completions $C_1 = \langle \mathcal{A}_1, \mathcal{R}_1 \rangle$ and $C_2 = \langle \mathcal{A}_2, \mathcal{R}_2 \rangle$ of I, $\{S \in \sigma(C_1) \mid S \subseteq \mathcal{A}_{12}\} = \{S \in \sigma(C_2) \mid S \subseteq \mathcal{A}_{12}\}$ where $\mathcal{A}_{12} = \mathcal{A}_1 \cap \mathcal{A}_2$.*

Weak stability requires that every two completions of the given IAF possess the same extensions solely under the range of their shared arguments rather than all arguments in the original IAF. Extensions falling out of the shared arguments of two completions are doomed different and allowed to occur. For this reason, weak stability solely requires that each set of arguments has the same verification result in those completions where the arguments in this set are all existent, formally shown by the following proposition.

Proposition 3. *Given an IAF* $I = \langle \mathcal{A}, \mathcal{A}^?, \mathcal{R}, \mathcal{R}^? \rangle$ *and semantics* σ, *I is weakly stable under* σ *iff* $\forall S \subseteq \mathcal{A} \cup \mathcal{A}^?$, *S is stable w.r.t.* I_S *under* σ *where* $I_S = \langle \mathcal{A} \cup S, \mathcal{A}^? \backslash S, \mathcal{R}, \mathcal{R}^? \rangle$.

Example 2. The IAF I_0 in Fig. 1 is not stable under co since $\{a, c, d\}$ is not stable w.r.t. I_0 under co. However, I_0 is weakly stable under complete semantics. For instance, completions containing d have co-extensions $\{a, c\}$ and $\{a, c, d\}$, whereas completions not containing d have co-extension $\{a, c\}$, satisfying the condition of weak stability. Note that uncertain argument d becomes acceptable in those completions that have the extension $\{a, c, d\}$. Going back to the interactive setting mentioned in Sect. 1, if experts want to obtain the co-extensions of an ending AF from I_0, the uncertain attacks (a, e) and (b, c) do not need to be decided since I is weakly stable. After experts decide the existence of d, the IAF will become stable and the left uncertain entities $(a, e), (b, c)$ and b can be ignored.

One can see that weak stability is discussed locally within every two completions whereas stability takes a global view requiring all completions to have the same extensions. Obviously, stable IAFs are special cases of weakly stable IAFs and when considering IAFs without uncertain arguments these two kinds of stability coincide.

4 Computational Issues

4.1 Complexity Results

In this section, we investigate complexity of the decision problems called σ-STABILITY (resp., σ-WEAK-STABILITY) asking whether a given IAF is stable (resp., weakly stable) under a type of semantics σ. We will show that σ-STABILITY and σ-WEAK-STABILITY have the same complexity under all six common semantics discussed in this paper.

Recall Propositions 1 and 3 that have provided a method at an exponential level for the problems by enumerating and checking the stability of all sets of arguments in an IAF. As a matter of fact, for $\sigma \in \{\texttt{cf}, \texttt{gr}\}$, there are tractable, equivalent conditions which can be checked in polynomial time. We will first discuss these two semantics, and then for $\sigma \in \{\texttt{ad}, \texttt{st}, \texttt{co}, \texttt{pr}\}$, we will give the precise upper and lower bounds of complexity of both kinds of stability.

Proposition 4 shows that for cf-STABILITY, we need to check the conflict relations between every two certain arguments and make sure that each uncertain argument is self-attacked, otherwise the set composed of a single uncertain argument without self-attack turns not stable in completions under cf. Differently for cf-WEAK-STABILITY, we solely need to check the conflict relations between every two arguments.

Proposition 4. *Given an IAF* $I = \langle \mathcal{A}, \mathcal{A}^?, \mathcal{R}, \mathcal{R}^? \rangle$,

1. *I is stable under* cf *iff (i) for any two arguments* $a, b \in \mathcal{A}$, $(a, b) \in \mathcal{R}^?$ *implies* $(b, a) \in \mathcal{R}$, *and (ii) for any argument* $a \in \mathcal{A}^?$, $(a, a) \in \mathcal{R}$; *and*

2. *I is weakly stable under* $\texttt{cf}$ *iff for any two arguments* $a, b \in \mathcal{A} \cup \mathcal{A}^?$, $(a, b) \in \mathcal{R}^?$ *implies* $(b, a) \in \mathcal{R}$.

Since the number of all possible argument pairs of an IAF is bounded by a square level, checking two kinds of stability under $\texttt{cf}$ is of polynomial complexity.

Corollary 1. *Both* $\texttt{cf}$-STABILITY *and* $\texttt{cf}$-WEAK-STABILITY *are in P.*

Now we discuss the $\texttt{gr}$ semantics. Since any AF has a unique $\texttt{gr}$-extension, we slightly abuse the notation and use $\texttt{gr}(F)$ to denote the $\texttt{gr}$-extension of an AF F from now on. Note that any weakly stable IAF is also stable under $\texttt{gr}$ because the $\texttt{gr}$-extension of completions without uncertain arguments must be the $\texttt{gr}$-extension of all completions. Hence the complexity of $\texttt{gr}$-WEAK-STABILITY is equivalent to $\texttt{gr}$-STABILITY and we only discuss the latter in the following.

Note that the $\texttt{gr}$-STABILITY problem inherently differs from INECVER$^{\texttt{gr}}$ although both are concerned with the unique $\texttt{gr}$-extension. Given an IAF I, INECVER$^{\texttt{gr}}$ asks whether a given set of arguments is necessary i^*-extension of I under $\texttt{gr}$, whereas $\texttt{gr}$-STABILITY asks whether there exists a set of arguments S of I such that INECVER$^{\texttt{gr}}(I, S) = true$ holds. Such a set is unknown and in the worst case all sets of arguments need to be enumerated in order to answer the $\texttt{gr}$-STABILITY problem.

We will show that there exists only one candidate set for check as it is the uniquely necessary i^*-extension under $\texttt{gr}$. Before constructing this set, we introduce two useful notions, *strong defense* from [22] and *strong characteristic function* from [21] on which our construction is based. These two notions for IAFs are adapted from the notions of defense and characteristic function in AFs.

Definition 7 (Strong Defense). *[22] Given an IAF $I = \langle \mathcal{A}, \mathcal{A}^?, \mathcal{R}, \mathcal{R}^? \rangle$, a set of arguments $S \subseteq \mathcal{A} \cup \mathcal{A}^?$ and an argument $a \in \mathcal{A} \cup \mathcal{A}^?$, S strongly defends a iff $\forall b \in \mathcal{A} \cup \mathcal{A}^?$ s.t. $(b, a) \in \mathcal{R} \cup \mathcal{R}^?, b \in (S \cap \mathcal{A})^+_I$ holds.*

We use the notion of strong defense to specify strong characteristic function. Differently from [21], we limit the range of the function from $\mathcal{A} \cup \mathcal{A}^?$ to $\mathcal{A}$ so as to generate the candidate set mentioned above.

Definition 8 (Strong Characteristic Function). *Given an IAF $I = \langle \mathcal{A}, \mathcal{A}^?, \mathcal{R}, \mathcal{R}^? \rangle$ and a set of arguments $S \subseteq \mathcal{A} \cup \mathcal{A}^?$, the strong characteristic function on S w.r.t. I is defined as $\overline{\Gamma}_I(S) = \{a \in \mathcal{A} \mid S$ strongly defends a in $I\}$.*

The work in [22] used strong defense to define new extension-based semantics on IAFs whereas failed to explore the connection between strong defense in an IAF and defense in the completions of the IAF. We identify such a connection in the following lemma, i.e., a certain argument which is strongly defended by a set of certain arguments in an IAF is necessary to be defended by this set in any completion of the IAF and vice versa.

Lemma 1. *Given an IAF $I = \langle \mathcal{A}, \mathcal{A}^?, \mathcal{R}, \mathcal{R}^? \rangle$ and a set of arguments $S \subseteq \mathcal{A}$, for each $a \in \mathcal{A}$, $a \in \overline{\Gamma}_I(S)$ iff for any completion C of I, $a \in \Gamma_C(S)$.*

Lemma 1 presents an equivalent condition for a set of certain arguments S necessarily defending an argument a in an IAF I. Complementarily, we give a version of possibly defending, which states that a is possibly defended by S in some completions only when each certain argument that is impossible to be attacked by S must not attack a in I.

Lemma 2. *Given an IAF* $I = \langle \mathcal{A}, \mathcal{A}^?, \mathcal{R}, \mathcal{R}^? \rangle$ *and a set of arguments* $S \subseteq \mathcal{A}$, *for each* $a \in \mathcal{A}$, $\forall b \in \mathcal{A} \cap S_I^u, (b, a) \notin \mathcal{R}$ *iff there is a completion* C *of* I, $a \in \Gamma_C(S)$.

Just as the monotonicity of characteristic function in AFs, one can easily find that the strong characteristic function is monotonic according to Definition 8.

Lemma 3. *Given an IAF* $I = \langle \mathcal{A}, \mathcal{A}^?, \mathcal{R}, \mathcal{R}^? \rangle$, $\forall S, S' \subseteq \mathcal{A} \cup \mathcal{A}^?$, $S \subseteq S'$ *implies* $\overline{\Gamma}_I(S) \subseteq \overline{\Gamma}_I(S')$.

In [13], gr-extension of an AF is generated in an iterative way based on Γ, and we use the same way to construct the unique candidate set E based on $\overline{\Gamma}$ in the subsequent theorem. We will show that an IAF I is stable under gr only when such non-self-attacked arguments as neither included in E nor attacked by E in I are attacked by a certain argument that is impossible to be attacked by E, i.e. those arguments are impossible to be defended by E according to Lemma 2. In fact, once E satisfies the above condition, E turns to be the unique necessary i^*-extension under gr, thus we can conclude that the stability of I is decided by whether E is necessary i^*-extension of I under gr.

Theorem 1. *Given an IAF* $I = \langle \mathcal{A}, \mathcal{A}^?, \mathcal{R}, \mathcal{R}^? \rangle$, *$I$ is stable under* gr *iff* $\forall a \in (\mathcal{A} \cup \mathcal{A}^?) \backslash (E \cup E_I^+)$ *that* $(a, a) \notin \mathcal{R}$, $\exists b \in \mathcal{A} \cap E_I^u$ *s.t.* $(b, a) \in \mathcal{R}$ *where* $E = \bigcup_{i=0}^{\infty} \overline{\Gamma}_I^i(\emptyset)$.

Proof. It is easy to see that $E \subseteq \mathcal{A}$ by the construction of E, which means that all arguments of E are existent in every completion of I. Before proving the theorem, we show two auxiliary conclusions: i) for any completion C of I, $E \subseteq \texttt{gr}(C)$ and ii) $\overline{\Gamma}_I(E) \subseteq E$. Firstly we prove i). We use E_i to denote $\overline{\Gamma}_I^i(\emptyset)$ and show $\forall i \geq 0, E_i \subseteq \texttt{gr}(C)$ by induction over i. In case of $i = 0$, we immediately obtain $E_0 = \emptyset \subseteq \texttt{gr}(C)$. Now assume that $\forall 0 \leq i \leq k, E_i \subseteq \texttt{gr}(C)$, then we have to show $E_{k+1} \subseteq \texttt{gr}(C)$. Since $E_{k+1} = \overline{\Gamma}_I(E_k)$ and $E_k \subseteq \mathcal{A}$, $E_{k+1} \subseteq \Gamma_C(E_k)$ holds by Lemma 1. According to assumption that $E_k \subseteq \texttt{gr}(C)$ and by the monotonicity of Γ, we have $E_{k+1} \subseteq \Gamma_C(\texttt{gr}(C))$, thus $E_{k+1} \subseteq \texttt{gr}(C)$ since gr-extension is also co-extension. Now we have $\forall i \geq 0, E_i \subseteq \texttt{gr}(C)$, thus $E = \bigcup_{i=0}^{\infty} E_i \subseteq \texttt{gr}(C)$ holds. Now we prove ii). According to the monotonicity of $\overline{\Gamma}$, it can be shown that for any integer $i \geq 0$, $E_i \subseteq E_{i+1}$. Since the number of arguments is limited, there must exist an integer n such that $E = E_n$ and $E_n = E_{n+1}$ by Lemma 3. Thus $\overline{\Gamma}_I(E) = \overline{\Gamma}_I(E_n) = E_{n+1} \subseteq E$. Now we prove Theorem 1. For brevity, let $S = (\mathcal{A} \cup \mathcal{A}^?) \backslash (E \cup E_I^+)$ in the following.
($\Leftarrow$) We show that for any completion C of I, $gr(C) = E$. For argument $a \notin E \cup E_C^+$, obviously $a \notin E_I^+$, then $a \in S$ holds. If $(a, a) \in \mathcal{R}$, then $a \notin \Gamma_C(E)$ and if not, by assumption and Lemma 2 we also have $a \notin \Gamma_C(E)$. Hence $a \notin \Gamma_C(\texttt{gr}(C))$

since $E \subseteq \mathtt{gr}(C)$ by i) and Γ is monotonic. Thus there is no argument not included in E but defended by E in C, yielding $E = \mathtt{gr}(C)$ by the construction process of gr-extension.
($\Rightarrow$) Firstly we prove that iii) $\forall b \in S, \exists c \in S$ *s.t.* $(c,b) \in \mathcal{R} \cup \mathcal{R}^?$, i.e. every argument of S will be possibly attacked by some arguments in S. Assume there exists an argument $b \in S$ s.t. $\forall c \in S, (c,b) \notin \mathcal{R} \cup \mathcal{R}^?$, then b is defended by E in the completion $C_1 = \langle \mathcal{A} \cup \{b\}, \mathcal{R}|_{\mathcal{A}\cup\{b\}}\rangle$ of I. Since $E \subseteq \mathtt{gr}(C_1)$, $b \in \mathtt{gr}(C_1)$ holds. If $b \notin \mathcal{A}$, then there exists a completion C_2 of I where b is nonexistent, thus $b \notin \mathtt{gr}(C_2)$. On the other hand, if $b \in \mathcal{A}$, then $b \notin E^u_I$ otherwise $b \in \overline{\Gamma}_I(E)$ and $b \in E$ by ii) yielding $b \notin S$, a contradiction occurs. Thus there is an argument $e \in E$ s.t. $(e,b) \in \mathcal{R} \cup \mathcal{R}^?$, then let $C_2 = \langle \mathcal{A}, \mathcal{R}|_{\mathcal{A}} \cup (e,b)\rangle$ where b is attacked by $E \subseteq \mathtt{gr}(C_2)$, hence $b \notin \mathtt{gr}(C_2)$. Thus there always exist two completions C_1 and C_2 of I that $b \in \mathtt{gr}(C_1)$ and $b \notin \mathtt{gr}(C_2)$, contradicting that I is stable under gr. Now we show the right direction of Theorem 1. Assume there exists an argument $a \in S$ without self-attack such that $\forall b \in \mathcal{A} \cap E^u_I, (b,a) \notin \mathcal{R}$, which is equivalent to that $\forall b \in \{a\}^-_I \cap \mathcal{A}, \exists e \in E$ *s.t* $(e,b) \in \mathcal{R} \cup \mathcal{R}^?$. Let $C_1 = \langle \mathcal{A} \cup \{a\}, \mathcal{R}|_{\mathcal{A}\cup\{a\}} \cup \mathcal{R}_a\rangle$ where $\mathcal{R}_a = \{(e,b) \mid e \in E, b \in \{a\}^-_I \cap \mathcal{A}$ and $(e,b) \in \mathcal{R}^?\}$. Observe that the certain arguments attacking a in I are attacked by E in C_1 and the uncertain arguments will be nonexistent, thus $a \in \Gamma_{C_1}(E)$, yielding $a \in \mathtt{gr}(C_1)$ since $E \subseteq \mathtt{gr}(C_1)$. Next consider the completion $C_2 = \langle \mathcal{A} \cup \mathcal{A}^?, \mathcal{R} \cup \mathcal{R}'\rangle$ of I where $\mathcal{R}' = \{(b,c) \in \mathcal{R}^? \mid b \in S$ or $c \in S\}$. By iii) we have that every argument of S is attacked by S itself in C_2. Note that S is not attacked by E in I, hence any argument of S is not attacked by E in C_2. Thus E does not defend any argument of S in C_2. And since $E \subseteq \mathtt{gr}(C_2)$, $E = \mathtt{gr}(C_2)$ holds by the construction process of gr-extension, yielding $a \notin E$. Consequently, there exist two completions C_1 and C_2 of I that $\mathtt{gr}(C_1) \neq \mathtt{gr}(C_2)$, contradicting that I is stable under gr.

In the above proof we have shown that there must exist an integer n so that $E_n = E_{n+1} = E$, thus the construction of E will stop at the $n+1$ step. And checking the properties needed for E is in polynomial time, so we can conclude that gr-STABILITY is in P.

Corollary 2. *Both* gr-STABILITY *and* gr-WEAK-STABILITY *are in* P.

Example 3. In the IAF I_0 of Fig. 1, $\overline{\Gamma}^1_{I_0}(\emptyset) = \{a\}$ and $\overline{\Gamma}^2_{I_0}(\emptyset) = \overline{\Gamma}^3_{I_0}(\emptyset) = \{a,c\}$, thus E constructed by Theorem 1 is $\{a,c\}$. The non-self-attacked arguments neither included in E nor attacked by E are d,e,g, and $S^u_{I_0} = \{d,f,g\}$. Arguments e,f,g are all attacked by f and d is attacked by g, thus E is necessary to be gr-extension and I_0 is stable under gr.

Now we give the complexity results of stability under $\sigma \in \{\mathtt{ad},\mathtt{st},\mathtt{co},\mathtt{pr}\}$.

Upper bounds. For $\sigma \in \{\mathtt{ad},\mathtt{st},\mathtt{co}\}$, both IPOSVER$^\sigma$ and INECVER$^\sigma$ are in P [15], thus the complementary problem of σ-STABILITY is in NP according to Proposition 1, yielding that σ-STABILITY is in $coNP$. Similarly by Proposition 3, we can conclude that σ-WEAK-STABILITY is also in $coNP$.

For pr-STABILITY, we also consider its complementary problem. An IAF $I = \langle \mathcal{A}, \mathcal{A}^?, \mathcal{R}, \mathcal{R}^?\rangle$ is not stable under pr semantics iff $\exists S_1, S_2 \in \mathcal{A} \cup \mathcal{A}^?$, $\exists C_1, C_2$ that are completions of I, such a condition as $S_1 \notin \mathtt{ad}(C_1)$ or $S_1 \subset S_2 \in \mathtt{ad}(C_1)$, $S_1 \in \mathtt{ad}(C_2)$, and $\forall S' \in \mathtt{ad}(C_2), S_1 \not\subset S'$ holds. Observe that this condition takes the form of two existential quantifiers (which can be collapsed to one) followed by an universal quantifier. Further, as the verification under ad on AF can be solved in polynomial time, the complementary problem of pr-STABILITY is in Σ_2^p, thus pr-STABILITY is in Π_2^p. Analogously, one can give an equivalent condition in the above form to check the complementary problem of pr-WEAK-STABILITY by Proposition 3, thus pr-WEAK-STABILITY is also in Π_2^p.

Next we show the lower bounds for the two kinds of stability under $\sigma \in \{\mathtt{ad}, \mathtt{st}, \mathtt{co}, \mathtt{pr}\}$ by reducing from some known hard problems in AFs.

Lower Bounds. For $\sigma \in \{\mathtt{ad}, \mathtt{st}, \mathtt{co}\}$, we show *coNP*-hardness of σ-STABILITY by reducing from the complementary problem of the existence of nonempty σ-extension problem denoted as $Exist_\sigma^{\neg\emptyset}$ which is NP-hard [14]. And we prove Π_2^p-hardness of pr-STABILITY by reducing from the skeptical acceptance problem under pr denoted as $Skept_{\mathtt{pr}}$ which is Π_2^p-hard [14].

Before presenting the following Theorems 2 and 3, we need to introduce a conclusion about AFs under $\sigma \in \{\mathtt{ad}, \mathtt{st}, \mathtt{co}, \mathtt{pr}\}$ stating that when considering the verification problem of a set of arguments S under σ in an AF F, the effects from any arguments attacked by S can be ignored, and thus we can alternatively consider the verification of S in the subgraph of F without those arguments.

Lemma 4. *Given an AF $F = \langle \mathcal{A}, \mathcal{R} \rangle$ and a set of arguments $S \subseteq \mathcal{A}$, under $\sigma \in \{\mathtt{ad}, \mathtt{st}, \mathtt{co}, \mathtt{pr}\}$, $S \in \sigma(F)$ iff $\exists S' \subseteq S_F^+$, $S \in \sigma(F_{\downarrow \mathcal{A} \setminus S'})$ holds.*

Given an instance F of $Exist_\sigma^{\neg\emptyset}$, we construct an instance I without uncertain arguments of σ-STABILITY by adding a new certain argument w to attack all arguments of F and making the attacks from F to w to be uncertain. We claim that for $\sigma \in \{\mathtt{ad}, \mathtt{st}, \mathtt{co}\}$ there is no nonempty σ-extension of F only when I is stable under σ, shown by Theorem 2. Such a reduction can also be applied to σ-WEAK-STABILITY, as given an instance I without uncertain arguments, I is weakly stable iff I is stable.

Theorem 2. *Given an AF $F = \langle \mathcal{A}, \mathcal{R} \rangle$, let IAF $I = \langle \mathcal{A} \cup \{w\}, \emptyset, \mathcal{R} \cup \mathcal{R}_w, \mathcal{R}^? \rangle$ where $w \notin \mathcal{A}$ is a new argument, $\mathcal{R}_w = \{(w, a) \mid a \in \mathcal{A}\}$ and $\mathcal{R}^? = \{(a, w) \mid a \in \mathcal{A}\}$. Then for $\sigma \in \{\mathtt{ad}, \mathtt{st}, \mathtt{co}\}$, $Exist_\sigma^{\neg\emptyset}(F) = false$ iff I is stable under σ.*

Proof. By the construction of IAF I, we have that for any completion C of I, $C_{\downarrow \mathcal{A}} = F$ holds since all attack relations between arguments in $\mathcal{A}$ of I are certain and they are the same as in F. Meanwhile, $\{w\}$ is a st-extension of C as w attacks all arguments in I except itself.
($\Rightarrow$) We show that for any completion C of I, $\{w\}$ is the unique nonempty σ-extension of C and we just need to show the uniqueness. Assume there exists a nonempty set of arguments $S \neq \{w\}$ such that $S \in \sigma(C)$, then $S \subseteq \mathcal{A}$ since w has conflicts with all other arguments in C. And since $S \in \mathtt{ad}(C)$ and w attacks S in C, $\{w\} \subseteq S_C^+$ holds. Thus $S \in \sigma(C_{\downarrow \mathcal{A}}) = \sigma(F)$ by Lemma 4, contradicting $Exist_\sigma^{\neg\emptyset}(F) = false$. ($\Leftarrow$) Assume $Exist_\sigma^{\neg\emptyset}(F) = true$, i.e. there

exists a nonempty σ-extension $S \subseteq \mathcal{A}$ in F. Consider the two completions C_1 and C_2 of I so that $C_1 = \langle \mathcal{A} \cup \{w\}, \mathcal{R} \cup \mathcal{R}_w \cup \mathcal{R}_S \rangle$ where $\mathcal{R}_S = \{(a, w) \mid a \in S\}$ and $C_2 = \langle \mathcal{A} \cup \{w\}, \mathcal{R} \cup \mathcal{R}_w \rangle$. Since S attacks $\{w\}$ in C_1 and $S \in \sigma(F) = \sigma(C_{1\downarrow\mathcal{A}})$, we have $S \in \sigma(C_1)$ by Lemma 4. Then observe that S is attacked by w but does not attack back in C_2, thus $S \notin \mathtt{ad}(C_2)$, yielding $S \notin \sigma(C_2)$ under $\sigma \in \{\mathtt{st}, \mathtt{co}\}$. Hence $\sigma(C_1) \neq \sigma(C_2)$, contradicting that I is not stable under σ.

Corollary 3. *For $\sigma \in \{\mathtt{ad}, \mathtt{st}, \mathtt{co}\}$, both σ-STABILITY and σ-WEAK-STABILITY are coNP-c.*

Given an instance (F, c) of $Skept_{pr}$ where F is an AF and c is an argument in F, to construct an instance I of $\mathtt{pr}$-STABILITY from F, we just need to add an attack from c to w on the basis of the construction in Theorem 2. We claim that $Skept_{pr}(F, c) = true$ only when I is stable under $\mathtt{pr}$. Note that there are none uncertain arguments in I, thus such a reduction can also be applied to $\mathtt{pr}$-WEAK-STABILITY.

Theorem 3. *Given an AF $F = \langle \mathcal{A}, \mathcal{R} \rangle$ and an argument $c \in \mathcal{A}$, let IAF $I = \langle \mathcal{A} \cup \{w\}, \emptyset, \mathcal{R} \cup \mathcal{R}_w \cup \{(c, w)\}, \mathcal{R}^? \rangle$ where $w \notin \mathcal{A}$ is a new argument, $\mathcal{R}_w = \{(w, a) \mid a \in \mathcal{A}\}$, and $\mathcal{R}^? = \{(a, w) \mid a \in \mathcal{A}\} \backslash \{(c, w)\}$. Then $Skept_{pr}(F, c) = true$ iff I is stable under* $\mathtt{pr}$.

Proof. Due to the proof of Theorem 2, for any completion C of I we have $C_{\downarrow\mathcal{A}} = F$ and $\{w\}$ is a $\mathtt{st}$-extension of C.
($\Rightarrow$) Note that by the definition of $Skept_{pr}$ in [14], when an AF F has no nonempty $\mathtt{pr}$-extension, for any argument a of F, $Skept_{pr}(F, a) = true$. When $\mathtt{pr}(F) = \{\emptyset\}$, for any completion C of I, $\{w\}$ is the unique $\mathtt{pr}$-extension for which the proof is analogous to the right direction in Theorem 2. Then when $\mathtt{pr}(F) \neq \{\emptyset\}$, we show that for any completion C of I, $\mathtt{pr}(C) = \mathtt{pr}(F) \cup \{\{w\}\}$: ($\supseteq$) For any set of arguments $S \in \mathtt{pr}(F)$, c attacks w in I and $c \in S$ by assumption, hence $\{w\} \subseteq S^+_C$. Since $S \in \mathtt{pr}(F) = \mathtt{pr}(C_{\downarrow\mathcal{A}})$, $S \in \mathtt{pr}(C)$ by Lemma 4. And $\{w\} \in \mathtt{st}(C) \subseteq \mathtt{pr}(C)$, hence $\mathtt{pr}(F) \cup \{\{w\}\} \subseteq \mathtt{pr}(C)$. ($\subseteq$) For any set of arguments $S \in \mathtt{pr}(C)$, if $S \nsubseteq \mathcal{A}$, then $S = \{w\}$ otherwise contradicting that S is conflict-free. On the other hand, if $S \subseteq \mathcal{A}$, then $\{w\} \subseteq S^+_C$ since w attacks S in C and $S \in \mathtt{ad}(C)$. Then $S \in \mathtt{pr}(C_{\downarrow\mathcal{A}}) = \mathtt{pr}(F)$ follows by Lemma 4, thus $\mathtt{pr}(C) \subseteq \mathtt{pr}(F) \cup \{\{w\}\}$ holds. Consequently, any completions of I always have the same $\mathtt{pr}$-extensions, thus I is stable under $\mathtt{pr}$.
($\Leftarrow$) Assume that there exists a set of arguments $S \subseteq \mathcal{A}$ such that $S \in \mathtt{pr}(F)$ and $c \notin S$. Consider the two completions C_1 and C_2 of I so that $C_1 = \langle \mathcal{A} \cup \{w\}, \mathcal{R} \cup \mathcal{R}_w \cup \mathcal{R}_S \rangle$ where $\mathcal{R}_S = \{(a, w) \mid a \in S\}$ and $C_2 = \langle \mathcal{A} \cup \{w\}, \mathcal{R} \cup \mathcal{R}_w \rangle$. Since S attacks $\{w\}$ in C_1 and $S \in \mathtt{pr}(F) = \mathtt{pr}(C_{1\downarrow\mathcal{A}})$, we have $S \in \mathtt{pr}(C_1)$ by Lemma 4. Then observe that S is attacked by w but does not attack back it in C_2, thus $S \notin \mathtt{ad}(C_2)$, yielding $S \notin \mathtt{pr}(C_2)$. Consequently, there exist two completions C_1 and C_2 of I such that $\mathtt{pr}(C_1) \neq \mathtt{pr}(C_2)$, contradicting that I is stable under $\mathtt{pr}$.

Corollary 4. *Both* $\mathtt{pr}$-STABILITY *and* $\mathtt{pr}$-WEAK-STABILITY *are Π_2^p-c.*

4.2 SAT Encoding

In this subsection we provide a SAT encoding similarly as [5] for deciding the (weak) stability of IAFs under $\sigma \in \{\texttt{ad}, \texttt{st}, \texttt{co}\}$ by a single call to a SAT solver. The problems under these three semantics are all *coNP*-c. Given an IAF $I = \langle \mathcal{A}, \mathcal{A}^?, \mathcal{R}, \mathcal{R}^? \rangle$, for all $a \in \mathcal{A} \cup \mathcal{A}^?$ and $(a,b) \in \mathcal{R} \cup \mathcal{R}^?$, we initially use variables y_a^1 and $r_{a,b}^1$ to encode a completion $C_1 = \langle \mathcal{A}_1, \mathcal{R}_1 \rangle$ of I such that $y_a^1 = true$ iff $a \in \mathcal{A}_1$, and $r_{a,b}^1 = true$ iff $(a,b) \in \mathcal{R}_1$. Similarly, we use y_a^2 and $r_{a,b}^2$ to encode another completion C_2 of I, and all the following formulas for C_1 have a version for C_2 obtained by substituting each superscript 1 with 2. In addition, we use variables x_a, z_a^1 for all $a \in \mathcal{A} \cup \mathcal{A}^?$ to represent a set of arguments S such that $x_a = true$ iff $a \in S$, and $z_a^1 = true$ iff $a \in S_{C_1}^+$ respectively. For $\sigma \in \{\texttt{cf}, \texttt{ad}, \texttt{st}, \texttt{co}\}$, the formula $\varphi_\sigma^1(I)$ encodes that $\varphi_\sigma^1(I) = true$ iff $S \in \sigma(C_1)$:

$$\varphi_{\texttt{cf}}^1(I) = \bigwedge_{(a,b) \in \mathcal{R} \cup \mathcal{R}^?} \left(r_{a,b}^1 \rightarrow (\neg x_a \vee \neg x_b) \right),$$

$$\varphi_{\texttt{ad}}^1(I) = \varphi_{\texttt{cf}}^1(I) \wedge \bigwedge_{a \in \mathcal{A} \cup \mathcal{A}^?} \bigwedge_{(b,a) \in \mathcal{R} \cup \mathcal{R}^?} \left((x_a \wedge r_{b,a}^1 \rightarrow z_b^1) \right),$$

$$\varphi_{\texttt{st}}^1(I) = \varphi_{\texttt{cf}}^1(I) \wedge \bigwedge_{a \in \mathcal{A} \cup \mathcal{A}^?} \left((y_a^1 \wedge \neg x_a) \rightarrow z_a^1 \right),$$

$$\varphi_{\texttt{co}}^1(I) = \varphi_{\texttt{ad}}^1(I) \wedge \bigwedge_{a \in \mathcal{A} \cup \mathcal{A}^?} \left(y_a^1 \wedge \Big(\bigwedge_{(b,a) \in \mathcal{R} \cup \mathcal{R}^?} r_{b,a}^1 \rightarrow z_b^1 \Big) \rightarrow x_a \right).$$

The formula $\varphi_?^1(I)$ encodes the condition of C_1 to be a completion of I while $\varphi_z^1(I)$ encodes the condition for the validity of z_a^1 for each $a \in \mathcal{A} \cup \mathcal{A}^?$. In addition, we need the formula $\varphi_{xy}^1(I)$ to encode whether $S \subseteq \mathcal{A}_1$:

$$\varphi_?^1(I) = \bigwedge_{a \in \mathcal{A}} y_a^1 \wedge \bigwedge_{(a,b) \in \mathcal{R}} r_{a,b}^1 \wedge \bigwedge_{(a,b) \in \mathcal{R}^?} \left(r_{a,b}^1 \rightarrow (y_a^1 \wedge y_b^1) \right),$$

$$\varphi_z^1(I) = \bigwedge_{a \in \mathcal{A} \cup \mathcal{A}^?} \left(z_a^1 \leftrightarrow \bigvee_{(b,a) \in \mathcal{R} \cup \mathcal{R}^?} (x_b \wedge r_{b,a}^1) \right),$$

$$\varphi_{xy}^1(I) = \bigwedge_{a \in \mathcal{A} \cup \mathcal{A}^?} x_a \rightarrow y_a^1.$$

Finally according to Propositions 1 and 3, we can integrate the above formulas to summarize the encoding of σ-STABILITY and σ-WEAK-STABILITY:

Proposition 5. *Given an IAF $I = \langle \mathcal{A}, \mathcal{A}^?, \mathcal{R}, \mathcal{R}^? \rangle$ and semantics $\sigma \in \{\texttt{ad}, \texttt{st}, \texttt{co}\}$, let $\varphi_C(I) = \varphi_?^1(I) \wedge \varphi_?^2(I) \wedge \varphi_z^1(I) \wedge \varphi_z^2(I)$, and we have:*

1. *the formula $\varphi_C(I) \wedge \varphi_\sigma^1(I) \wedge \varphi_{xy}^1(I) \wedge (\neg\varphi_\sigma^2(I) \vee \neg\varphi_{xy}^2(I))$ is unsatisfiable iff I is stable under σ; and*

2. *the formula* $\varphi_C(I) \wedge \varphi_\sigma^1(I) \wedge \varphi_{xy}^1(I) \wedge \neg\varphi_\sigma^2(I) \wedge \varphi_{xy}^2(I)$ *is unsatisfiable iff* I *is weakly stable under* σ.

For stability problem under pr, there is no way to solve it by a single call to a SAT solver since its complexity is up to Π_2^p-c. As an alternative solution, an algorithm based on CEGAR in [8] can be designed by iteratively finding an unstable set of arguments under pr, viewed as a counterexample. Due to space limits, we will not discuss the implementation in detail here.

5 Discussion and Conclusions

Table 1 lists all the complexity results of the stability problems of IAFs on a single set, a single argument and all extensions under six semantics. According to the theorems and propositions obtained in Sect. 4, we summarize our results in the rightmost column where checking stability and weak stability of IAFs on extensions under each discussed semantics hold the same complexity. Through those results, we conclude that checking the (weak) stability of an IAF turns to be more complex than checking a set of arguments whereas less complex than checking an argument. Note that checking the stability of a single argument is of the hardest complexity as its enumeration involves all completions and their extensions.

Under most semantics we discussed, the complexity results of (weak) stability problems are higher than P since the stability of extensions is decided by the stability of all sets of arguments [6,15]. Interestingly for grounded semantics, through finding that there is an unique candidate set needed for deciding its stability and thus enumerating all sets of arguments can be avoided, we prove that the complexity of gr(-WEAK)-STABILITY is in P. Meanwhile, we actually obtain an alternative method for the necessary i^*-extension verification problem under gr: given an IAF I and a set of certain arguments S, S is a necessary i^*-extension of I under gr iff S becomes the E in Theorem 1 and satisfies the properties required there. Differently from [6,15], our method directly checks structure information of an IAF rather than construct specific completions.

Obviously, the stability of an argument [3,5,23,24] cannot imply the stability of extensions on IAFs. On the other hand, when an IAF is stable, the *in* stable arguments [24] are definite which are the acceptable arguments of any completion because the extensions of each completion are the same. However, the stability of IAF does not imply the *out* or *undec* stability [24] of arguments since the attacks from extensions cannot be determined although the extensions are determined. For the same reason, the arguments satisfying *determinism, totality, and functionality* [3] in the IAF cannot be determined neither, except the skeptically accepted arguments in each completion.

As a generalization of IAF, *Probabilistic AF (PrAF)* [19] gives uncertain arguments and attacks a value representing the probability of their existence. The work of [1,12,16] addressed the problem of computing the probability of extensions and acceptability. One can see that a prAF can be transformed into an IAF where the uncertain elements are those with existence probability < 1 in the prAF. When such an IAF is detected to be stable, the probability of a set of arguments becoming an extension and/or an argument becoming acceptable can be easily decided to be either 0 or 1.

The problem about extensions of a changing framework was also discussed in the dynamics research of AF. The preservation problem [7,25] asks which change can preserve an original extension, the enforcement problem [4,10] asks how an AF can be modified to enable a non-extension to become an extension, and the recomputation problem [2,20] studies how to efficiently compute extensions of a changing AF by divide-and-conquer and incremental ways. A dynamic AF under modifications within a given argumentation universe can be viewed as an IAF. When such an IAF is stable, the preservation requirement becomes fulfilled for each set of arguments whereas the enforcement problem turns unsolvable. Moreover, there is no need to recompute extensions when the AF changes since any modification of its structure will not affect its extensions.

To sum up, we address the stability and weak stability of all extensions in IAFs and study the computational issues of the corresponding decision problems under six common semantics. Such stability problems are distinct from previously presented ones, although the purposes are all to avoid unnecessary investigation of uncertain elements. In the future we intend to study these problems under other semantics besides Dung's and examine the interactive argumentation setting mentioned in the first part of the paper. Moreover, the (weak) stability under different semantics could be related, e.g., the stability of an IAF under `ad` entails its stability under `pr` since preferred extensions are the inclusion-maximal admissible sets of arguments, whereas such an entailment does not hold for weak stability. We leave a systematic study of these relations to our future work.

Table 1. Complexity of various types of stability in IAFs.

σ	a set of arguments [6,15]		an argument [5,23,24]		extensions
	impossible	necessary	credulous-in	sceptical-in	
`cf`	P	P	P	trivial	P
`gr`	P	P	coNP-c	coNP-c	P
`ad`	P	P	Π_2^p-c	trivial	coNP-c
`st`	P	P	Π_2^p-c	coNP-c	coNP-c
`co`	P	P	Π_2^p-c	coNP-c	coNP-c
`pr`	Π_2^p-c	coNP-c	Π_2^p-c	Π_2^p-c	Π_2^p-c

References

1. Alfano, G., Calautti, M., Greco, S., Parisi, F., Trubitsyna, I.: Explainable acceptance in probabilistic and incomplete abstract argumentation frameworks. Artif. Intell. **323**, 103967 (2023)

2. Alfano, G., Greco, S., Parisi, F.: Incremental computation in dynamic argumentation frameworks. IEEE Intell. Syst. **36**(6), 80–86 (2021)
3. Alfano, G., Greco, S., Parisi, F., Trubitsyna, I.: Incomplete argumentation frameworks: properties and complexity. In: Proceedings of the AAAI Conference on Artificial Intelligence, vol. 36, pp. 5451–5460 (2022)
4. Baumann, R., Brewka, G.: Expanding argumentation frameworks: enforcing and monotonicity results. COMMA **10**, 75–86 (2010)
5. Baumeister, D., Järvisalo, M., Neugebauer, D., Niskanen, A., Rothe, J.: Acceptance in incomplete argumentation frameworks. Artif. Intell. **295**, 103470 (2021)
6. Baumeister, D., Neugebauer, D., Rothe, J., Schadrack, H.: Verification in incomplete argumentation frameworks. Artif. Intell. **264**, 1–26 (2018)
7. Boella, G., Kaci, S., van der Torre, L.: Dynamics in argumentation with single extensions: abstraction principles and the grounded extension. In: Sossai, C., Chemello, G. (eds.) ECSQARU 2009. LNCS (LNAI), vol. 5590, pp. 107–118. Springer, Heidelberg (2009). https://doi.org/10.1007/978-3-642-02906-6_11
8. Clarke, E., Grumberg, O., Jha, S., Lu, Y., Veith, H.: Counterexample-guided abstraction refinement for symbolic model checking. J. ACM (JACM) **50**(5), 752–794 (2003)
9. Coste-Marquis, S., Devred, C., Konieczny, S., Lagasquie-Schiex, M.C., Marquis, P.: On the merging of Dung's argumentation systems. Artif. Intell. **171**(10–15), 730–753 (2007)
10. Coste-Marquis, S., Konieczny, S., Mailly, J.G., Marquis, P.: Extension enforcement in abstract argumentation as an optimization problem. In: Twenty-Fourth International Joint Conference on Artificial Intelligence (2015)
11. Dimopoulos, Y., Mailly, J.G., Moraitis, P.: Control argumentation frameworks. In: Proceedings of the AAAI Conference on Artificial Intelligence, vol. 32 (2018)
12. Dondio, P.: Toward a computational analysis of probabilistic argumentation frameworks. Cybern. Syst. **45**(3), 254–278 (2014)
13. Dung, P.M.: On the acceptability of arguments and its fundamental role in nonmonotonic reasoning, logic programming and n-person games. Artif. Intell. **77**(2), 321–357 (1995)
14. Dvorák, W., Dunne, P.E.: Computational problems in formal argumentation and their complexity. Handb. Formal Argumentation **1**, 631–687 (2018)
15. Fazzinga, B., Flesca, S., Furfaro, F.: Revisiting the notion of extension over incomplete abstract argumentation frameworks. In: IJCAI, pp. 1712–1718 (2020)
16. Fazzinga, B., Flesca, S., Parisi, F.: On the complexity of probabilistic abstract argumentation frameworks. ACM Trans. Comput. Logic (TOCL) **16**(3), 1–39 (2015)
17. Hunter, A.: Probabilistic qualification of attack in abstract argumentation. Int. J. Approximate Reason. **55**(2), 607–638 (2014)
18. Hunter, A., Thimm, M.: Probabilistic reasoning with abstract argumentation frameworks. J. Artif. Intell. Res. **59**, 565–611 (2017)
19. Li, H., Oren, N., Norman, T.J.: Probabilistic argumentation frameworks. In: Modgil, S., Oren, N., Toni, F. (eds.) TAFA 2011. LNCS (LNAI), vol. 7132, pp. 1–16. Springer, Heidelberg (2012). https://doi.org/10.1007/978-3-642-29184-5_1
20. Liao, B., Jin, L., Koons, R.C.: Dynamics of argumentation systems: a division-based method. Artif. Intell. **175**(11), 1790–1814 (2011)
21. Mailly, J.G.: Extension-based semantics for incomplete argumentation frameworks: grounded semantics and principles. In: Bouraoui, Z., Vesic, S. (eds.) ECSQARU 2023. LNCS, vol. 14294, pp. 84–94. Springer, Cham (2023). https://doi.org/10.1007/978-3-031-45608-4_7

22. Mailly, J.G.: Extension-based semantics for incomplete argumentation frameworks: properties, complexity and algorithms. J. Log. Comput. **33**(2), 406–435 (2023)
23. Mailly, J.G., Rossit, J.: Stability in abstract argumentation. arXiv preprint arXiv:2012.12588 (2020)
24. Odekerken, D., Bor, A., Bex, F.: Stability and relevance in incomplete argumentation frameworks. In: Computational Models of Argument: Proceedings of COMMA 2022 , vol. 353, p. 272 (2022)
25. Rienstra, T., Sakama, C., van der Torre, L., Liao, B.: A principle-based robustness analysis of admissibility-based argumentation semantics. Argument Comput. **11**(3), 305–339 (2020)
26. Testerink, B., Odekerken, D., Bex, F.: A method for efficient argument-based inquiry. In: Cuzzocrea, A., Greco, S., Larsen, H.L., Saccà, D., Andreasen, T., Christiansen, H. (eds.) FQAS 2019. LNCS (LNAI), vol. 11529, pp. 114–125. Springer, Cham (2019). https://doi.org/10.1007/978-3-030-27629-4_13

Author Index

S. Destercke et al. (Eds.): SUM 2024, LNAI 15350, pp. 487–488, 2025.
https://doi.org/10.1007/978-3-031-76235-2

SPRINGER NATURE

GPSR Compliance

The European Union's (EU) General Product Safety Regulation (GPSR) is a set of rules that requires consumer products to be safe and our obligations to ensure this.

If you have any concerns about our products, you can contact us on ProductSafety@springernature.com

In case Publisher is established outside the EU, the EU authorized representative is:

Springer Nature Customer Service Center GmbH
Europaplatz 3
69115 Heidelberg, Germany

The manufacturer's authorised representative in the EU is Springer Nature Customer Service Centre GmbH, Europaplatz 3, 69115 Heidelberg, Germany. If you have any concerns regarding our products, please contact ProductSafety@springernature.com

Printed and bound by CPI Group (UK) Ltd, Croydon, CR0 4YY
24/04/2026
02096365-0016